AF746810

SEDIMENTATION, TECTONICS
AND EUSTASY

# Sedimentation, Tectonics and Eustasy

## SEA-LEVEL CHANGES AT ACTIVE MARGINS

Edited by David I.M. Macdonald

SPECIAL PUBLICATION NUMBER 12 OF THE
INTERNATIONAL ASSOCIATION OF SEDIMENTOLOGISTS
PUBLISHED BY BLACKWELL SCIENTIFIC PUBLICATIONS
OXFORD LONDON EDINBURGH BOSTON
MELBOURNE PARIS BERLIN VIENNA

Editorial Offices:
Osney Mead, Oxford OX2 0EL
25 John Street, London WC1N 2BL
23 Ainslie Place, Edinburgh EH3 6AJ
3 Cambridge Center, Cambridge
Massachusetts 02142, USA
54 University Street, Carlton
Victoria 3053, Australia

Other Editorial Offices:
Arnette SA
2, rue Casimir-Delavigne
75006 Paris
France

Blackwell Wissenschaft
Meinekestrasse 4
D-1000 Berlin 15
Germany

Blackwell MZV
Feldgasse 13
A-1238 Wien
Austria



First published 1991

Set by Setrite Typesetters, Hong Kong
Printed and bound in Great Britain by
The Alden Press, Oxford



DISTRIBUTORS

Marston Book Services Ltd
PO Box 87
Oxford OX2 0DT
(*Orders*: Tel: 0865 791155
Fax: 0865 791927
Telex: 837515)

USA
Blackwell Scientific Publications, Inc.
3 Cambridge Center
Cambridge, MA 02142
(*Orders*: Tel: 800 759–6102)

Canada
Oxford University Press
70 Wynford Drive
Don Mills
Ontario M3C 1J9
(*Orders*: Tel: 416 441–2941)

Australia
Blackwell Scientific Publications
(Australia) Pty Ltd
54 University Street
Carlton, Victoria 3053
(*Orders*: Tel: 03 347–0300)

British Library
Cataloguing in Publication Data

Sedimentation, tectonics and eustasy.
1. Sedimentation 2. Tectonics
I. Macdonald, David II. Series
551.304

ISBN 0–632–03017–8

Library of Congress
Cataloging-in-Publication Data

Sedimentation, tectonics, and eustasy/edited by David Macdonald.
p. cm. — (Special publication number 12 of the International Association of Sedimentologists)
Includes bibliographical references and index.
ISBN 0–632–03017–8
1. Sedimentation and deposition. 2. Geology, Stratigraphic. 3. Geology, Structural. 4. Plate tectonics. 5. Sea-level.
I. Macdonald, David (David I.M.) II. Series: Special publication . . . of the International Association of Sedimentologists; no. 12.
QE571.S4185 1991
551.3′03 — dc20

## Contents

# Preface

The papers in this Special Publication arise from a discussion meeting, 'Sea-level changes at active plate margins: processes and products', sponsored by the International Association of Sedimentologists (IAS) and the Geological Society of London, and organized by the British Sedimentological Research Group (BSRG). The meeting was held in London in May 1989. Forty-seven papers and eleven posters were presented, of which 25 are published here. Only one paper in this volume (by Y. Saito) was not presented at the meeting.

The idea for the conference came from a paper by Anthony Hallam and co-workers, published in the *Geological Magazine* (**123**, 425–435) in 1986, where they suggested that the effects of global sea-level change could be recognized in the Cretaceous of the Andean back-arc basin. This prompted me to wonder whether such global effects were common in tectonically active basins, given the background of recent seismic and sequence-stratigraphic studies which had emphasized the role of eustatic sea-level fluctuations in controlling sedimentation patterns. Most research was on the relatively simple interaction of sea level with lithosphere stretching and thermal subsidence in ensialic sedimentary basins and at passive continental margins. However, the basin inputs to the sequence-stratigraphic accommodation model (relative sea-level change, tectonic subsidence or uplift, and sediment added) were common to all basins and could be applied to active plate margins.

Originally, three major themes were planned for the conference:

**1** Underlying mechanisms and rates of relative sea-level change at active plate margins.

**2** The interaction of eustatic and tectonic processes at modern margins and the resulting sedimentary products.

**3** Recognition of the products in the sedimentary record, and possible criteria for distinguishing global eustatic from local tectonic effects.

It soon became apparent, however, that tectonically active basins on passive margins and in foreland settings were just as problematical, and the scope of the conference was widened to include some of these papers. In recognition of this wider scope, the title of this volume (*Sedimentation, Tectonics and Eustasy*) reflects the three major processes whose interaction produces the sedimentation patterns we study.

If such a large, diverse conference as this could have been said to have had a definite conclusion, it was that the products of sea-level fluctuations can be recognized in most tectonically active basins; in some cases at least, the origin of these was clearly independent of basin tectonism. There was a general feeling that these are preferentially found during periods of relative quiescence in the tectonic history of the basin, and that active tectonism during sedimentation swamped the eustatic signal.

The major disagreements concerned high-frequency sea-level changes, particularly their dating and mechanism of formation. The paper by Fortuin and de Smet suggests that adjacent parts of the same arc can record very different histories of vertical movement, calling doubt on the possibility of synchronous sequence boundaries being formed. A number of papers (e.g. Pirrie *et al.*) identified the common problem of coincidence in time of a sea-level fall with a fluctuation in the global cycle chart. This creates a dilemma as to the origin of the fluctuation. To put it crudely: how many coincidences constitute an adequate test of the 'Cycle Chart'? As the paper by Carter and others rightly reminds us, the so-called 'Vail' or 'Haq *et al.*' curves are models, subject to refinement and testing. Detailed, *published*, biostratigraphy is needed worldwide otherwise these curves will become self-fulfilling prophecies. There were also major disagreements as to the origin of high-frequency cycles, with no one prepared to commit themselves. Bilal Haq makes the, albeit tentative, suggestion that these cycles may be glacio-eustatic, at least in the Cretaceous and Tertiary. Perhaps the major input into this debate in the future will come from palaeoclimatologists.

A conference and publication of this size depends on the help of many people, to whom I owe a great debt of gratitude. As the subject embraces the whole realm of sedimentary geology, it was recognized from the start that contributions from geophysicists, structural geologists and palaeontologists would be vital, and I am grateful to the IAS Bureau for their wholehearted support in stepping outside the boundaries of pure sedimentology. I would like to thank Phil Allan, Poppe de Boer and Maurice Tucker in particular. Amerada Hess Ltd, the BSRG, the IAS, and The Royal Society all provided generous financial support, which enabled us to invite a wide variety of keynote speakers. The organization of the meeting was lightened by the great assistance I received from the British Antarctic Survey, and I am grateful to its Director, David Drewry, and Chief Geologist, Mike Thomson, for their support. Many of my colleagues here provided support and encouragement and I am grateful to them all, particularly Gill Clarke, Gill McDonnell and Hanni Willan, who have performed many of the tedious tasks, and kept smiling. The 69 referees were prompt and conscientious: I just wish they could have agreed more often! My greatest debt is to Tina and Eleanor, for their love and support.

David I.M. Macdonald
*British Antarctic Survey*
*Cambridge*

# Models and Tests of Sea-Level Change

*Spec. Publs. int. Ass. Sediment.* (1991) **12**, 3–39

# Sequence stratigraphy, sea-level change, and significance for the deep sea

BILAL U. HAQ

*Division of Ocean Sciences, National Science Foundation, Washington, DC 20550, USA*

## ABSTRACT

Sequence-stratigraphic models conceptualize the deposition of genetically related sediment packages along continental margins as a response to various phases of the cycle of relative change of sea level. Regional tectonics (subsidence/uplift), eustasy, and rate of sediment supply interact to produce the measure of relative change of sea level. During a complete cycle of sea-level change, characteristic stratal patterns resulting within siliciclastic systems comprise: a basal, basinward prograding wedge of lowstand (or shelf-margin) systems tract; an intermediate, retrogradational landward back-stepping, transgressive systems tract; and an upper, aggrading to prograding, highstand systems tract. In siliciclastic systems most of the terrigenous material is delivered to the basin during the lowstand. Stratal patterns in carbonate systems are comparable to those in the siliciclastic systems; much of the carbonate is transferred to the basin during lowstand progradation. However, carbonate can also be exported off-bank during late highstand, when productivity on the bank top is high while accommodation is reduced. Both lowstand and late highstand progradation contribute to lateral growth of the platform. In hybrid siliciclastic–carbonate systems, lowstands may be dominated by clastic deposition, while carbonate accumulation may occur both during lowstand and highstand times.

The history of the development of the basic conceptual framework of sequence stratigraphy, and the efforts that led to the documentation of sea-level variations deciphered from sequence analysis of subsurface and outcrop data, is outlined. The resulting history of sea-level change and its relationship to the sedimentary patterns of the deep sea are also discussed. A climatic feedback model of an alternating erosive/corrosive response to sea-level fall and rise is offered. Erosion on the sea floor is dominant during low sea-stands, and canyon incision on the margins during major falls of sea level. Carbonate dissolution on the sea floor is enhanced during times of maximum flooding of the shelves and during early highstand when carbonate is conserved on the inner shelves and nearshore. During the late highstand phase, following reduced accommodation on the banks, carbonate is again exported to the basins, abating the corrosive cycle. Distinction between early and late highstand phases, and between late highstand and lowstand progradation is, therefore, critical to understanding the response of the carbonate system and deep-sea sedimentary patterns to sea-level change along continental margins.

## INTRODUCTION

Continental margins demarcate the broad zone where continental and oceanic crust make contact. Throughout their geological history these margins see extensive tectonic and eustatic activity, persisting as major depocentres of terrigenous sediments brought down by streams, and as sites of authigenic carbonate precipitation. The sedimentary record along the margins is thus a product of complex interaction between sea-level change, regional tectonics (subsidence/uplift), climate and sediment supply. Because of the commonly steep topography at the edge of the continents, the margins may be characterized by drift, downslope movement, and mass transport of sediments, which further complicate the nature of their sedimentary record.

The stratigraphic models of response of the shelf-slope sediments to changes in relative sea level are an outgrowth of the need to assemble a 'process-based' framework within which the complex sedimentary record of the margins can be evaluated to aid exploration geology. The early efforts at shaping these concepts have, therefore, come largely from the industrial research laboratories. These attempts have resulted in a conceptual framework within which downdip and lateral changes in facies can be explained in terms of the processes that spawn them, and the rock record can be organized into time-constrained genetic packages. The appeal of the sequence-stratigraphic concepts for explorationists also lies in the predictive nature of the models, and this is especially useful in frontier areas where accurate pre-drill estimates of geological parameters can often make the difference between success and failure. Sequence stratigraphy is an integrated approach that draws from all branches of stratigraphy, and thus has important implications for all its subfields. It holds the potential to revolutionize the way we look at the sedimentary record.

Sequence-stratigraphic models envision depositional sequences as composites of genetically related packages of sediments deposited during the various phases of a cycle of sea-level change. Although these concepts owe their origin to seismic stratigraphy (Vail *et al.*, 1977), the new concepts are equally applicable to the subsurface (seismic and well-log), as well as outcrop data. The documentation of sea-level changes in sedimentary sections around the world, particularly in outcrop sections, has led to a new generation of sea-level curves for the Mesozoic and Cenozoic (Haq *et al.*, 1987,1988a), which represent a significant improvement in resolution of events and their timing over the earlier curves based on seismic data alone (Vail *et al.*, 1977). These curves are, however, working paradigms based on global stratigraphic data of varying quality, and refinement in some parts of the chronoeustatic framework is to be expected as better stratigraphic and chronologic data are gathered.

The purpose of this paper is to outline the development of the basic concepts of sequence stratigraphy and their implications, and the basis of the new sea-level curves. Although the documentation of sea-level changes is largely based on sections along passive margins and interior basins (because of the relative tectonic simplicity of these areas), the concepts of sequence stratigraphy are equally applicable to active margins. Tectonic events act to either amplify or suppress the eustatic signature within the sedimentary record. In tectonically active areas, special attention may therefore be necessary to discriminate between tectonic and eustatic signals.

## Historical overview

For nearly two centuries the classical subdivisions of the stratigraphic record that form the basis of geological 'stages' (and higher stratigraphic categories) in Europe and elsewhere were identified as unconformity-bounded entities that represented cycles of sedimentation, comprising transgressive and regressive phases. The recognition of transgressive and regressive facies throughout Europe and their apparent synchroniety persuaded Suess (1906) to propose that sea-level rises and falls may be global (eustatic) in origin.

In the first half of the twentieth century, stratigraphers in many parts of the world continued to describe epochal subdivisions, or stages (often of local or regional extent only), as facies-based sedimentary episodes, representing *de facto* cycles of sea-level change. The formal concept of the depositional sequence had to wait until 1950 when Sloss applied the term to unconformity-bounded cycles of sedimentation of the North American cratonic interior. Sloss' (1950,1963,1988) sequences, however, were perceived as sedimentary records of major tectonic cycles punctuating the Palaeozoic of the North American craton. Four sequences (Sauk, Tippecanoe, Kaskaskia, and Absaroka) encompassed the entire Palaeozoic. In his 1963 paper, Sloss characterized sequences as lithostratigraphic units of higher category than groups or supergroups, traceable over major areas of the continent.

The new depositional sequence is not only an order of a magnitude smaller than Sloss', but also differs in its definition. It was defined by Mitchum *et al.* (1977) as a stratigraphic unit composed of a relatively conformable succession of genetically related strata bounded by unconformities or their correlative conformities. This concept evolved from an attempt to delineate seismic-stratigraphic criteria for the identification of sedimentary sequences. On seismic profiles regional unconformities could be identified through physical relationship of strata, and patterns of onlap, toplap, truncation, and basinward shift of coastal onlap could be used as objective criteria to demarcate sequence boundaries. Rises and falls of sea level, related to eustatic changes, were seen as the driving mechanism behind the formation of characteristic geometric patterns comprising sequences along continental margins. The depositional sequence was perceived as a basic stratigraphic unit, and the associated surfaces as time boundaries with chronostratigraphic significance

(Mitchum *et al.*, 1977).

The analysis of depositional sequences based on geometric relationship of strata, in particular the changes in patterns of coastal onlap on seismic lines from different regions, provided Vail *et al.* (1977) with the basic data to construct sea-level curves for the Phanerozoic. The resolution of events on seismic-stratigraphically based sea-level curves was, however, limited by the resolution of the seismic data. On seismic profiles only features that are several tens of metres and larger in thickness are resolvable. Another limitation of the sea-level curves based on seismic data alone was the paucity of first-order age dating of seismic reflectors. Although palaeontological ties were available from drill-holes in some areas, by necessity the ages of seismic horizons often had to be extrapolated from distant areas. In addition to these limitations, many of the seismic data that formed the basis of these curves remained locked up in the industrial coffers as proprietary information.

Although seismic-stratigraphic techniques and the concept of the seismic depositional sequence caught on in both industry and academia, the above limitations generated considerable criticism of the sea-level curves. The curves were questioned for the lack of adequate corrections for regional subsidence and possible overestimation of the magnitude of sea-level changes (Pitman, 1978; Watts & Steckler, 1979), for the paucity of documentation in support of global synchroneity of sea-level change events, and for the timing of certain events and their significance for the deep-sea basins (Mörner, 1981; Hallam 1984; Parkinson & Summerhayes, 1985; Miall, 1986). Pitman (1978), in particular, pointed out that the rate of change of sea level (which is the result of interplay between subsidence and eustasy) is more important than the eustatic sea-level change itself in determining the nature of sequences along the margins. He presented a simple model that could be used for predicting stratification patterns, given the rates of regional subsidence, sediment supply, and sea-level change.

The growth of the new sequence-stratigraphic concepts was partially a reaction to these criticisms and partially owing to the need to reduce reliance on proprietary seismic and well log data. This led to the development of alternative criteria for the identification of sea-level changes in easily accessible sections, where the lessons learned from seismic stratigraphy could be applied to outcrops in the public domain. The outcrop sections also offered more easily retrievable and better biochronostratigraphic information than subsurface sections.

The early theoretical models of the formation of siliciclastic sequences and their bounding surfaces were first developed in the mid 1970s and later presented in a report to Exxon Production Research Company (EPR) by Jervey in the early 1980s (released and published in Jervey, 1988). Jervey employed seismic-stratigraphic experience and modelled the interaction of subsidence, sea-level change, and sediment influx for an Atlantic-type margin in the context of a smoothly fluctuating sea level along a sinusoidal curve. From these simulations he concluded that the seismic sequence boundaries were located in time near the inflection points of the sea-level curve, where the rate of change is maximum. The internal facies distribution, as well as the geometry of depositional surfaces, are related to the changes in the accommodation potential (space available for sediment accumulation between the sea floor and base level) which is a function of subsidence and sea-level rise or fall. Jervey also observed that when the rate of sea-level fall is highest, sand-prone, turbiditic lowstand deltaic facies are most likely to be deposited. Conversely, when the sea level rises rapidly, clayey facies accumulate in low-energy coastal plain environments, and authigenic minerals such as glauconite and phosphorite, as well as organic-rich shales, may be deposited offshore owing to lack of clastic influx.

During the early 1980s a large group of research professionals at EPR were engaged in the development of various aspects of sequence stratigraphy to enhance its use as a predictive tool in exploration geology. The near-academic environment of this industrial research facility and the availability of abundant resources, combined with the presence of a large pool of practising seismic stratigraphers led by Peter Vail, made it an ideal place for the advancement of sequence-stratigraphic concepts, their testing in various settings, and the documentation of sequences in many parts of the world.

The theoretical models of Jervey were further refined and detailed by Posamentier *et al.* (1988). The response of the siliciclastic depositional system to changing sea level and accommodation potential along a shelf–slope profile, as well as updip along the fluvial profile, was considered in greater detail. The concept of 'systems tracts' of Brown & Fisher (1977) as the linkage of contemporaneous depositional systems was redefined in the context of building blocks of sequences as genetically related sediment packages.

The application of these concepts, in particular the recognition of relative changes of sea level on well-logs and outcrops by EPR stratigraphers, made it possible to independently identify sequences in subsurface and surface sections and correlate them to synthetic seismograms. The concept of stacking patterns of parasequences (higher frequency flooding events) within systems tracts was advanced by van Wagoner (1985) and van Wagoner *et al.* (1988). The identification of relative changes of sea level by characteristic facies stacking patterns associated with different systems tracts forms the basis of the application of sequence-stratigraphic concepts to well-log and outcrop data.

The application of sequence stratigraphy to carbonate depositional systems was undertaken by Sarg (1988), based on models developed in Guadalupian sections of the Permian Basin of New Mexico (Sarg & Lehman, 1986).

Both the clastic and carbonate accumulation models were tested by a large number of early practitioners of sequence stratigraphy at EPR and applied in exploration efforts in many parts of the world, including offshore Australia, Indonesia, Paris Basin, USA Gulf Coast and Western Interior, and Latin America, among others. These applications and test cases led to continued refinement of the models, and the realization that for each particular locale the models need to be adapted to account for local subsidence and sediment supply histories and basin-floor topography.

The documentation of sea-level cycles represented by type 1 and type 2 sequences (resulting from major and minor sea-level falls, respectively) was a large group effort spearheaded by J. Hardenbol and P. Vail. However, during the course of assembly of global sequence-stratigraphic data, the need for a chronostratigraphic system capable of tying these observations together was seen as a prerequisite. The wide diversity of approaches and varying rigour in the treatment of radiometric and biostratigraphic data in existing regional chronostratigraphic schemes made correlation of widely separated strata difficult. This effort to develop an internally consistent chronostratigraphy went hand in hand with global documentation of Mesozoic and Cenozoic sequences (Haq *et al.*, 1987,1988a).

Parallel to these activities, the application of the new concepts in resolving biostratigraphic issues was also being explored. Condensed sections were first recognized in seismic data (Vail *et al.*, 1984). The study of biofacies changes within various systems tracts accompanying sea-level rises and falls led to the recognition of the importance of the condensed sections and their associated downlap surfaces. Of special importance was their use in understanding and predicting the accumulation and preservation of organic matter, and the concentration of authigenic minerals, as well as pelagic fossils most useful in chronostratigraphy (Baum & Vail, 1988; Loutit *et al.*, 1988).

The downturn in the petroleum economy in the mid 1980s resulted in a significant scaling down of these research efforts at EPR and other industrial laboratories. However, the release of much of the conceptual information in papers published in 1987–1988 has resulted in wide exposure of the concepts of sequence stratigraphy. The major effort for further advancement of these concepts has now shifted to the academia, helped by the exodus of research scientists from industry.

## KEY CONCEPTUAL MODELS

This section describes the basic models of sequence stratigraphy. These depositional concepts need to be tailored for each location, taking into consideration the regional rate of subsidence (or emergence), rate of sediment supply, and topography of the basin floor (typical shelf–slope setting versus ramp setting, and isolated platforms). Observations along continental margins in sections with typical shelf–slope profiles suggest that the response also differs when the system is siliciclastics-dominated, versus a system with mixed clastic–carbonate deposition. On isolated carbonate platforms, with their relatively pure carbonate accumulation, the basic models are valid, but the role of the high- and lowstand times in exporting sediments off the platform is somewhat different than in the siliciclastic system.

In view of the rapidly proliferating terminology of sequence stratigraphy it may be useful to review briefly the definitions of the terms now widely, but sometimes improperly, used in stratigraphic literature. These definitions are taken from Mitchum (1977), Mitchum *et al.* (1977), Brown & Fisher (1977), Haq *et al.* (1987,1988a), van Wagoner *et al.* (1988), Jervey (1988), Loutit *et al.*, (1988), Posamentier *et al.* (1988), Sarg (1988), and Vail *et al.* (1991) among others (see also the Appendix for common sequence-stratigraphic terms in French, German, and Spanish translations).

*Sequence stratigraphy* is that branch of stratigraphy

which facilitates local, regional, and interregional correlations by subdividing the rock record into a succession of *depositional sequences* composed of genetically related strata. The depositional packages of sediments comprising *sequences* are bounded by surfaces of erosion or non-deposition or their lateral non-hiatal continuities (*conformities*). Sequences are the basic units of sequence stratigraphy and are interpreted to have been deposited during a complete cycle of relative sea-level change. A complete *cycle of sea-level change* is bracketed between two sea-level falls, extending from a point just prior to one inflection point of maximum rate of sea-level fall to the subsequent rise to a point preceeding the inflection point of the next fall.

Each phase of sea-level cycle, i.e. lowstand, transgression, highstand, and regression, produces a distinct package of genetically related sediments known as a depositional *systems tract*, which comprises linkage of contemporaneous depositional systems. Systems tracts are characterized by distinctive collections of lithofacies. The systems tracts of different phases can be distinguished by the types of bounding surfaces, stratal geometry, position within the sequence, and the stacking configuration of parasequences. *Parasequences* are the basic building blocks of systems tracts. They represent higher frequency flooding events that can occur in all systems tracts and comprise relatively conformable successions of genetically related beds bounded by marine flooding surfaces or their lateral continuities. A different set of systems tracts will result if the sequence is a type 1 or type 2.

The type of sequence is determined by the rate of relative fall of sea level. When the rate of eustatic fall exceeds the rate of regional subsidence at the *offlap break* (break in the slope of the previous highstand) a relative sea-level fall will result. (Offlap break is a term suggested by Vail *et al.* (1991) to replace terms used previously, such as *depositional-shoreline break* or *shelf break*, to avoid confusing the break in the slope of the previous highstand with the physiographic shelf-slope break. On wide shelves the offlap break is coincident with the maximum storm wave base, but on narrow shelves it may coincide with the physiographic shelf edge.) The exposed and incised proximal shelf will result in a *type 1 sequence boundary* (the sequence overlying this boundary will be a *type 1 sequence*). When the rate of eustatic fall does not exceed the rate of subsidence, the outer portion of the shelf seaward of the offlap break will not be exposed, and a *type 2 sequence boundary* will result (the sequence overlying a type 2 boundary is a *type 2 sequence*) (Fig. 1).

A type 1 sequence is typically composed of *lowstand, transgressive*, and *highstand systems tracts*. In type 2 sequences the lowstand systems tract is replaced by the *shelf margin systems tract* (Fig. 1). The distinction between lowstand and shelf margin systems tracts will become clear after the explanation of the depositional events during a complete cycle of sea-level change given in the following section.

The combined influence of regional tectonics (subsidence/uplift) and eustasy determines the accommodation potential for the sediments and the distribution of facies within the systems tracts. *Accommodation* is the space available for the accumulation of sediments between the sea floor and sea level. In the following sections the depositional models for siliciclastic and carbonate systems will be discussed separately.

## SILICICLASTIC SYSTEM

Posamentier *et al.* (1988) have discussed sea-level control of clastic deposition on the continental shelf in considerable detail. Here the dynamics of clastic deposition will be reviewed only briefly, along with some variations to models by changing the parameters that control the sedimentary patterns. To appreciate the formation of a sequence it is instructive to follow the evolution of a complete cycle of sea-level change, starting with a sea-level fall from a relative highstand position, followed by a rise and a subsequent fall.

### Siliciclastic depositional cycle

During the highstand of sea level the accommodation potential for the sediment accumulation on the shelf decreases progressively as the space between base level and sea floor is filled, or when less space is being created per unit time owing to decreasing rate of relative sea-level rise. At this time the aggradationally stacked parasequences of the rising sea level during transgression and early highstand give way to increasingly progradational parasequences of the later highstand time (Fig. 1). When the accommodation potential is diminished nearshore, the shoreline starts regressing and the sediments prograde further out into the basin. This means that the top of the highstand systems tract will be associated with widespread fluvial deposition above the shore-

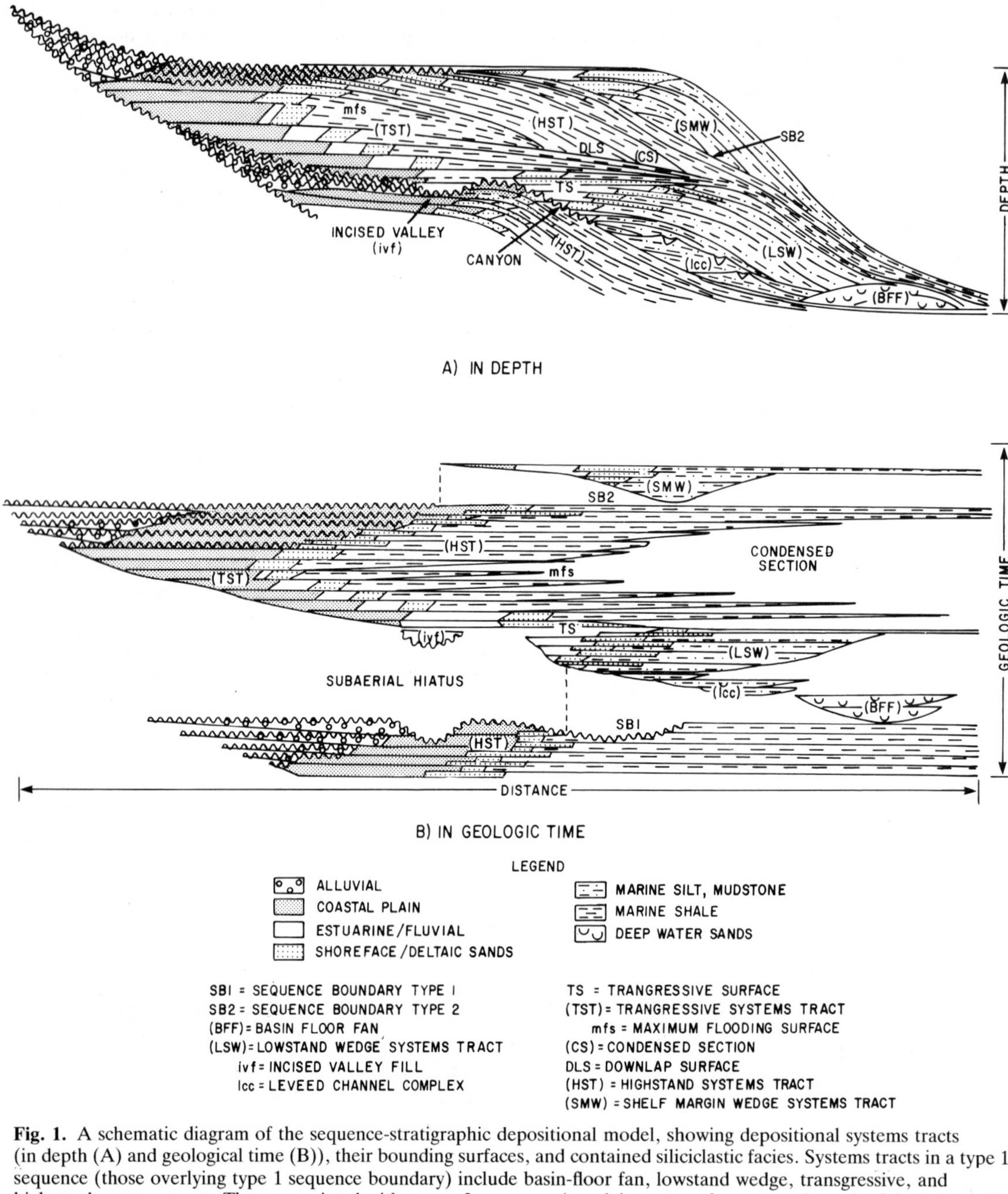


**Fig. 1.** A schematic diagram of the sequence-stratigraphic depositional model, showing depositional systems tracts (in depth (A) and geological time (B)), their bounding surfaces, and contained siliciclastic facies. Systems tracts in a type 1 sequence (those overlying type 1 sequence boundary) include basin-floor fan, lowstand wedge, transgressive, and highstand systems tracts. Those associated with a type 2 sequence (overlying a type 2 sequence boundary) include shelf-margin wedge, transgressive, and highstand systems tracts.

line (Posamentier & Vail, 1988). Once sea level falls from this position, the deposition of a new sequence begins.

### *Sea-level fall and lowstand systems tract in siliciclastics*

The lower boundary of a sequence is marked by the

start of a relative fall of sea level. When the rate of eustatic fall outstrips the rate of subsidence and the sea level falls below the offlap break, the shelf above the break is exposed, resulting in a type 1 sequence boundary. However, both subaerial and submarine erosion can produce truncation at the sequence boundary. If the sea-level fall is extensive and rapid and falls below the physiographic shelf edge, the entire shelf may be exposed subaerially. The resulting sequence boundary may be depicted by an abrupt basinward shift of coastal onlap. The lowstand systems tract begins to be deposited once the relative sea level starts to fall (Posamentier *et al.*, 1988; see Figs 1, 2).

The subaerial exposure of the shelf above the offlap break initiates erosion and fluvial incision as the stream equilibrium profiles are lowered (Posamentier & Vail, 1988), leading to development of a prominent unconformity. Directly above the unconformity the basal lithology may display erosional valleys filled with braided stream deposits.

Withdrawal of the sea below the offlap break initially leads to the formation a *lowstand fan complex* (*basin-floor* and *slope fans*) (Figs 1, 2). As the sea level continues to fall, the shelf incision may initiate canyon cutting into the shelf above the offlap break and on the slope below it. The canyon acts as a conduit for sediments, which are point-sourced directly on to the slope or the basin floor to form *slope* or *basin-floor fans*. The basin-floor fans are deposited directly on to the basin floor and may appear as sandy mounds or lobes. They may be accompanied by active canyon erosion. The slope fans, on the other hand, are accompanied by debris flows and slumps and are deposited above the basin floor. A distinct discontinuity separates the older basin-floor fan from the younger slope fan. Slope fans are formed once the relative sea level begins to rise and the canyons start backfilling with aggradational leveed channels and attached lobes (Vail *et al.*, 1991).

Submarine *canyons* are common features along present-day continental margins. Canyon cutting into the shelf and slope is more likely to be initiated during major sea-level falls. For example, seven of the largest submarine canyons in the world have been documented to have been cut into the continental margins of the Bering Sea during late Cenozoic glacial lowstand times (Carlson & Karl, 1988). The heavy sediment load during falling sea level induces turbidity currents with greater erosive power to cut and deepen canyons. During the lowstand phase the canyons can channel turbidity currents and debris flows and their dense sediment load

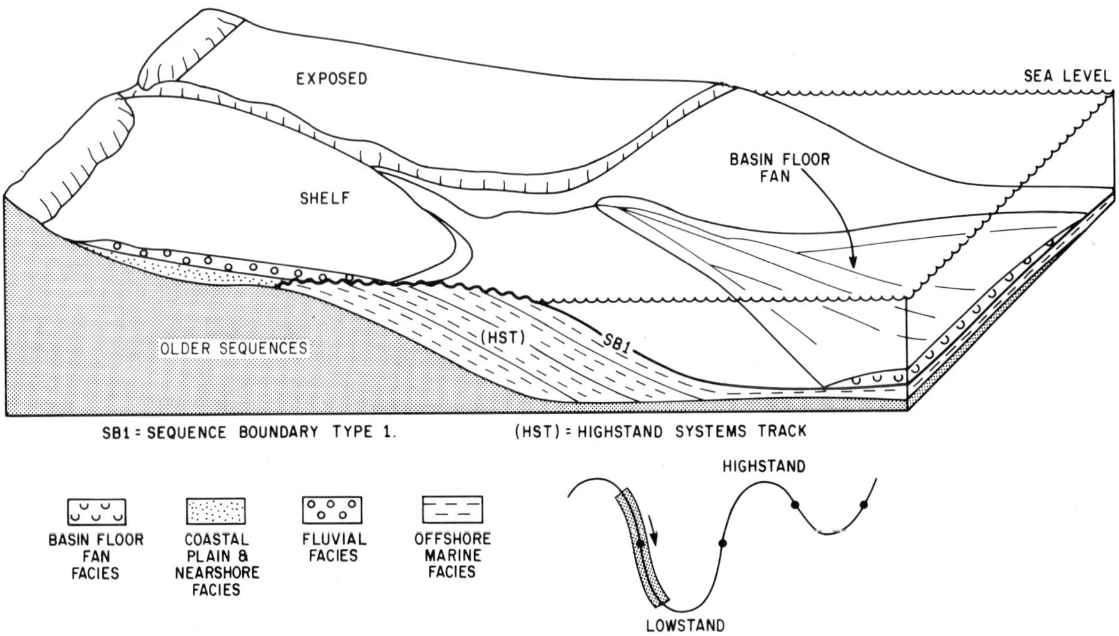


**Fig. 2.** Siliciclastic system. Schematic block diagram showing a major sea-level fall and the formation of type 1 sequence boundary, overlain by a basin-floor fan during the early phase of lowstand time (modified after Posamentier *et al.*, 1988).

to the basin floor. Canyons may also be initiated through localized slope failure of unconsolidated sediments and subsequent headward erosion until the depression breaches the shelf break, gaining access to the shelf sediments, which may further cut the depression, initiating a canyon (Farre *et al.*, 1983). Mountain (1987) has documented several late Cretaceous–Paleogene slope failure events on the New Jersey margin that were initiated following collapse of underlying carbonates, causing regional hiatuses. These scars, however, did not develop into major canyons because they failed to gain access to a major source of clastic detritus (Mountain, 1987). The modern Mississippi canyon is inferred to have formed following retrogressive slope failure and mass slumping of sediments (Coleman *et al.*, 1983), most likely during glacial lowstand time.

Once initiated, the canyons may continue to be active through the subsequent sea-level rise, especially during the early stages of lowstand progradation. Lowstand progradation may fill the heads of the canyons and end canyon formation. If not filled, the sediment-channelling role of the canyons is considerably curtailed past lowstand owing to significantly reduced sediment influx. The study of present-day submarine canyons, when the sea level is relatively high, reveals that the canyons remain modestly active through bio-erosion of the canyon walls and updip gullies (Sanford *et al.*, 1990). Sediment in the canyons also may be resuspended through the action of internal tides focused along canyon axes. These turbid plumes may move basinward and be redeposited (Gardner, 1989).

The basin-floor fans are dominated by sand-prone sediments, which may consist of non-consolidated or winnowed sand lobes, massive sand mounds redeposited by bottom currents (Mutti, 1985; Mutti & Normark, 1987).

The slope fans are characterized by complex turbiditic and slump facies, including thin-bedded turbidites, hemipelagic shales, and massive sandy channel fills in the upper part, to laminated levee overbank turbidites, attached channel lobes with turbiditic sands, and fine-grained turbidites in the lower part (Vail *et al.*, 1991).

The *lowstand wedge* member of the *lowstand prograding complex* (Figs 1, 3) is a wedge-shaped body of sediments composed of progradationally to aggradationally stacked parasequences characterized by shallowing upward lowstand deltas and incised-valley fill (van Wagoner *et al.*, 1988). The lowstand wedge is a regressive unit dominated by coastal-belt

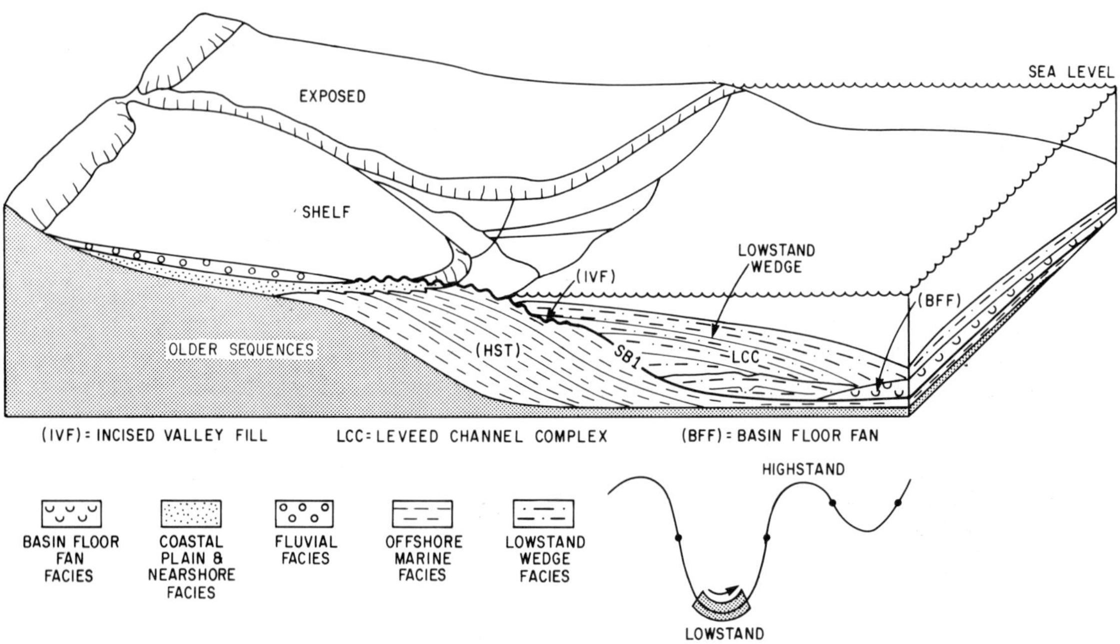


**Fig. 3.** Siliciclastic system. Later phase of the lowstand and the formation of incised valley system, leveed channel complex (within the slope fan), and the lowstand wedge systems tract (modified after Posamentier *et al.*, 1988).

sand deposits that onlap landward and prograde basinward, downlapping on the basin-floor fan. The toes of the prograding units may interfinger with shingled turbidites, which can occur attached or detached from the clinoforms. Slumps may also characterize the early stages of lowstand wedge formation in areas of high sedimentation rates (Vail *et al.*, 1991).

Relative sea-level rise causes cessation of stream incision and existing incised valleys may begin to be filled with coarser braided-stream sediments. The incised valleys may also be filled later during the accumulation of the transgressive systems tract. The later part of the lowstand system tract is characterized by deltaic deposits that prograde over the incised valley fill and early part of the lowstand wedge. The later stages may also contain evidence of slumps that may continue into the early part of the transgressive systems tract (Vail *et al.*, 1991).

The lowstand systems tract is topped by a transgressive surface (Fig. 1). During the lowstand phase the shelf space may increase by the accretion of the lowstand wedge deposits beyond the offlap break of the previous highstand. Aggradation increases upwards while rate of regression of the shelf decreases until the rate of new shelf space added outpaces the sediment supply and a transgression of the shoreline occurs (Posamentier & Vail, 1988). The deposition of the lowstand systems tract concludes at the first major transgression and the change from fore-stepping progradational facies to back-stepping retrogradational facies.

### *Sea-level fall and shelf margin wedge systems tract in siliciclastics*

If the rate of eustatic sea-level fall is less than the rate of subsidence at the offlap break and no relative sea-level fall is apparent, or if the sea-level fall is slow and the sea does not retreat beyond the offlap break, a type 2 sequence boundary results. This boundary is marked by an unconformity due to subaerial exposure shoreward of the offlap break, but lacks both erosion associated with stream rejuvenation (van Wagoner *et al.*, 1988) and the development of a basin-floor fan. The *shelf margin wedge* is the lowest, regressive, depositional unit of the type 2 sequence, which overlies a type 2 sequence boundary (see Figs 1, 6). It is typified by progradational to aggradational parasequences deposited on the outer part of the shelf. The lower boundary of the shelf margin wedge can be rec-

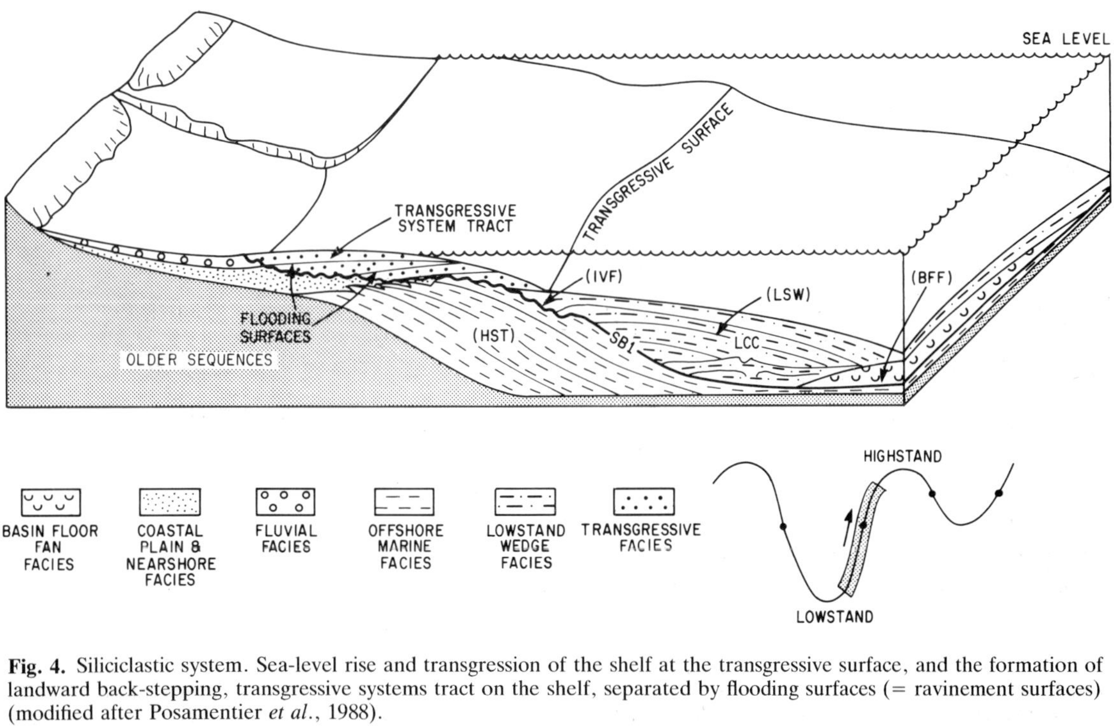


**Fig. 4.** Siliciclastic system. Sea-level rise and transgression of the shelf at the transgressive surface, and the formation of landward back-stepping, transgressive systems tract on the shelf, separated by flooding surfaces (= ravinement surfaces) (modified after Posamentier *et al.*, 1988).

ognized by the sudden basinward shift of coastal onlap. This boundary is an erosional unconformity only on the proximal part of the shelf, characterized by coastal plain and/or paralic/deltaic sediments overlying fluvial deposits. Basinward on the shelf the unconformity becomes a conformity and the boundary is more subtle. A change from rapidly progradational parasequences of the underlying highstand systems tract to the overlying progradational to aggradational parasequences marks the boundary (Posamentier & Vail, 1988). The distal part of the shelf-margin wedge is similar to the lowstand wedge, and topped in both cases by the transgressive surface and overlain by retrogradational transgressive deposits.

*Sea-level fall on ramp setting in siliciclastics*

When the depositional margin is ramp-like, as is often the case in basins in continental interior (epeiric) seas, the lowstand systems tract is relatively thin and consists of lower and upper units separated by incised valley fill. Stream rejuvenation and sediment bypass of the coastal plain during relative sea-level fall characterize the lower progradational unit, which may be truncated at the top. Lower shoreface or offshore facies are deposited within this unit. Stream incision may be common, because the sea level falls relatively rapidly on a ramp, leading to cutting of incised valleys into the lower prograding unit, which may be filled with coarser grained sediments, ranging from sands to conglomerates, during deposition of the upper unit (Vail *et al.*, 1991). The lower unit is analogous to the basin-floor fan in slope settings.

The deposition of the upper unit is accompanied by filling of incised valleys and shoreline progradation. This creates a lowstand systems tract comprised of incised valley fill updip and two sets of progradationally stacked parasequences downdip with an erosional surface in between (van Wagoner *et al.*, 1988).

*Sea-level rise and transgressive systems tract in siliciclastics*

As the eustatic sea-level rise outpaces the regional subsidence, the first transgression above the lowstand complex is marked by the *transgressive surface* (Figs 1, 4). As the transgression proceeds on to the shelf each successive flooding event may be marked by individual transgressive surfaces of erosion, or *ravinement surfaces* (Weimer, 1984). Collectively,

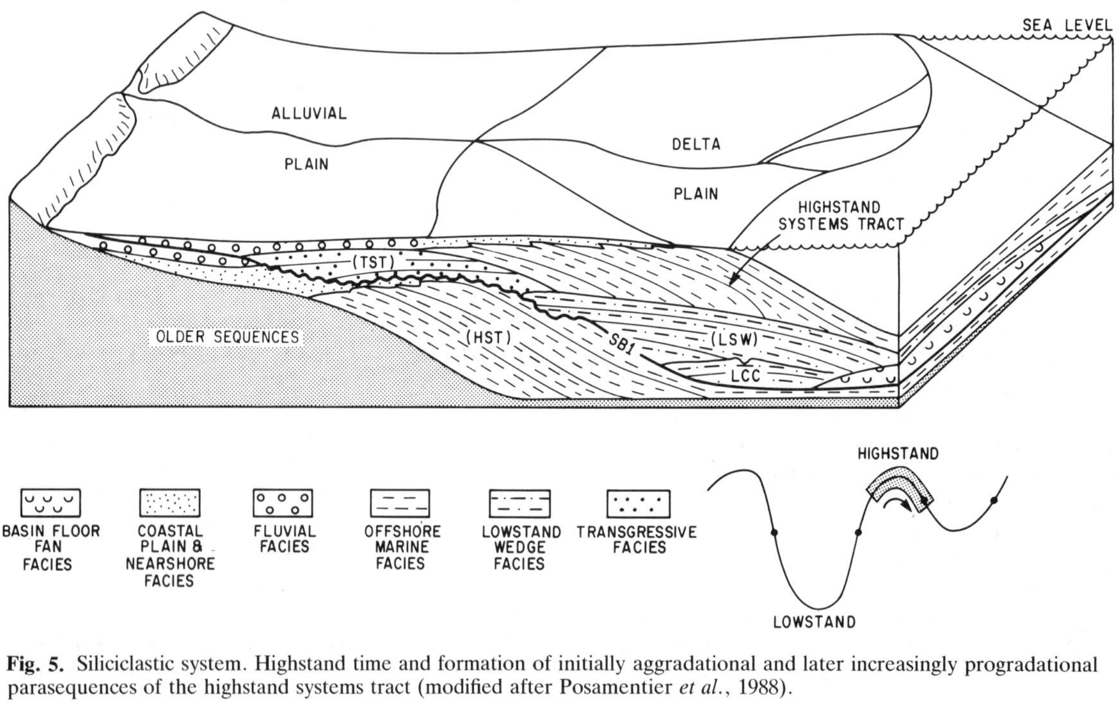


**Fig. 5.** Siliciclastic system. Highstand time and formation of initially aggradational and later increasingly progradational parasequences of the highstand systems tract (modified after Posamentier *et al.*, 1988).

these back-stepping surfaces of erosion can be confused with the transgressive surface, which is the first surface of major flooding on the shelf after a lowstand and may also be an erosional surface. Sediment starvation basinward may lead to burrowing at the transgressive surface, and shoreward the basal lithology above it may display lag gravel, rip-up clasts of underlying lithologies, and fossil shell debris (Baum & Vail, 1988).

Some authors have argued (e.g. Galloway, 1989) that the transgressive surface forms a better boundary between two depositional sequences, because marine flooding is recorded widely on the shelf and adjacent slope representing a depositional hiatus. Using this concept for the sequence boundaries would, however, introduce major time gaps (represented by the much larger non-depositional and/or erosional hiatuses that follow sea-level fall) in the middle of the depositional sequences. The duration of the hiatuses represented by transgressive surfaces are relatively minor when compared with the duration represented by the widespread unconformities at sequence boundaries as defined here. On the proximal shelf the transgressive surface can also be confused easily with the time-transgressive ravinement surfaces of erosion, which step up in time with each back-stepping parasequence that depict different flooding events during transgression.

The transgressive surface is overlain by landward back-stepping (retrogradational), transgressive facies that constitute the *transgressive systems tract* (Figs 1, 4), the middle parcel of sediments in both type 1 and type 2 sequences. The retrogradational parasequences represent consecutive flooding events of the shelf. Parasequences in this systems tract onlap landward on to the sequence boundary; basinward they downlap on to each other to form the back stepping pattern. As the transgression proceeds, terrigenous sediments are sequestered increasingly in more landward position, and hemipelagic sediments begin to blanket the shelf at the trailing edge forming parasequences that thicken landward and thin basinward. Each parasequence is associated with a ravinement surface at its base. The parasequences of the proximal portion are typically composed of marsh or lake sediments. Occurrence of coals is most common in the transgressive systems tract (Vail *et al.*, 1991). Transgressive facies are more widespread in type 2 sequences than in type 1 sequences, because in a type 1 sequence the shelf is flooded only when the sea level is rising relatively rapidly, shifting the depocentres landward more abruptly than in a type 2 sequence. Deposition above a type 1 sequence boundary results in the deficiency of transgressive sediments (Posamentier & Vail, 1988).

Transgressive facies continue to be deposited progressively further updip (landward) until the time of the maximum flooding of the shelf, marked by a surface, the *maximum flooding surface* (Fig. 1), that manifests itself as the *downlap surface* on seismic profiles. The transgressive systems tract thins basinward and fines upwards, forming a *condensed section* at the maximum flooding surface owing to sediment starvation as the depocentres move landward (Fig. 1). When the maximum flooding of the shelf is complete and the high sea-stand stabilizes, the highstand facies begin to accumulate. As the accommodation potential nearshore decreases, the shoreline once again begins to regress basinward. The prograding parasequences of the highstand systems tract are deposited over the transgressive deposits on the upper shelf and over lowstand deposits on lower shelf and slope, separated from them by the downlap surface and its associated condensed section.

### *Condensed section*

Loutit *et al.* (1988) have discussed in detail the significance of the condensed section to continental margin stratigraphy. Whilst the shelf is starved of sediments during the time of maximum flooding as the land-derived sediments are sequestered nearshore, a physically *condensed section* is created on the outer shelf and slope (see Fig. 14). During this time pelagic or hemipelagic sediments may be deposited over broad areas of the shelf. Within a depositional sequence the condensed section occurs partly at the top of the transgressive systems tract and partly within the highstand systems tract. In the latter it is represented by the toes of the prograding clinoforms where the highstand deposits downlap on to the transgressive deposits of the outer shelf and the toes of the back stepping transgressive systems tract deposits (the *downlap surface*). The relative duration of the sediment starvation within the condensed section increases basinward, until beyond the area of terrigenous influence, the biogenic deposition in the deeper basins can be thought of as a series of stacked condensed sections (Haq *et al.*, 1988a).

The condensed section is defined as thin marine stratigraphic units consisting of pelagic to hemipelagic sediments characterized by very slow sedimentation rates (Loutit *et al.*, 1988). Marine condensed sections are created by sediment starvation, and are therefore commonly characterized by apparent hiatuses, thin zones of burrowed, and somewhat lithified, beds (omission surfaces), or hardgrounds (Baum & Vail, 1988; Loutit *et al.*, 1988). Sediment starvation also may concentrate authigenic minerals (phosphorite, glauconite), pelagic fossils (diverse planktonic and benthic assemblages), organic matter, and trace elements. The concentration of minerals containing radiogenic elements (U, Th) in the condensed section produces strong positive anomalies on the gamma-ray well log readings. On a shelf to basin depositional profile a condensed section may range from a fossil-rich zone on the proximal shelf to a biostratigraphic gap in the basin. In the deep basins the condensed section (time of maximum flooding of the shelf) may be represented by a dissolution surface (see section on implications for deep-sea sedimentation).

*Highstand systems tract in siliciclastics*

During high sea-stand, after the maximum flooding of the shelf, the deposition of the highstand systems tract is marked initially by aggradationally stacked parasequences (Fig. 5). As the accommodation space nearshore is filled, the shoreline begins to regress and the dominantly aggradational pattern changes to a progradational stacking pattern in the remainder of the highstand systems tract. This systems tract is bounded at its base by the downlap surface and the condensed section, and its top is a sequence boundary, representing a sea-level fall and the beginning of a new cycle of sedimentation (Fig. 5). The highstand deposits are typically a vertical succession of upward-coarsening (or upward shoaling) parasequences with increased terrigenous input upwards. Late highstand progradation is accompanied by subaerial deposition of fluvial sediments on the proximal shelf (discussed in detail by Posamentier & Vail, 1988). The highstand progradational complex is composed of thin-bedded coastal and delta plain sediments. Coarser sediments are trapped in the fluvial complex, and thus the highstand deltaic sediments may be finer grained than their lowstand counter-parts (Vail *et al.*, 1991).

The highstand systems tract is commonly the most widespread genetically-related sediment package on the shelf, often prograding out and draping the slope as well. Major sea-level falls (type 1 sequence boundaries) and concomitant shelfal erosion often wipe out a part or complete record of the previous sedi-

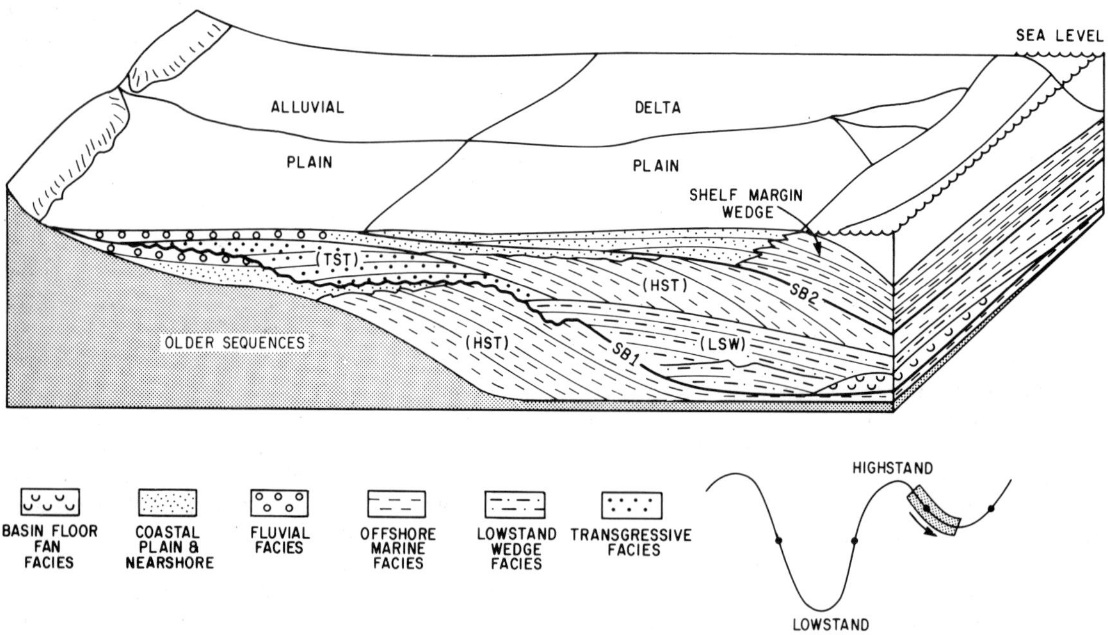


**Fig. 6.** Siliciclastic system. Minor sea-level fall and the formation of type 2 sequence boundary, overlain by the offlap break-draping shelf-margin wedge systems tract (the lowest unit of a type 2 sequence) (modified after Posamentier *et al.*, 1988).

mentary cycle(s). The record of sea-level rise, on the other hand, may be preserved intact, or may at least be represented by the maximum landward marine incursion, which may be the only available evidence of sea-level change. This creates the apparent impression in the sedimentary record that the sea-level rises are gradual, whereas sea-level falls are abrupt. An example of this pattern of preferred transgressive systems tract preservation is provided in the Neogene North Carolina sequences (Riggs *et al.*, 1990; Snyder *et al.*, 1990).

Erosion and extensive hiatuses can also occur during the condensed section and highstand phases, caused by the landward migration of longshore ocean currents such as the Gulf Stream. If the currents are strong enough and impinge the shelf, they can induce extended erosion of the shelf. This may have been the case in the Miocene when the palaeo-Gulf Stream strengthened and migrated over the mid-Atlantic shelf during the highstand times, causing regional hiatuses and erosion of the transgressive systems tract (Popenoe, 1985).

The depositional cycle (sequence) is completed when the relative sea level falls once again.

## Variations to the conceptual models

The conceptual models of the deposition sequences in the preceding pages make some necessary assumptions for the sake of simplicity. These include the assumptions of a relatively constant regional subsidence with an overall increase in subsidence rate toward the basin (i.e. zero at the hinge point and maximum near the continent–ocean crust boundary), a uniform rate of sediment supply, and a sinusoidal pattern of sea-level change. The resultant shape of the individual systems tract packages could vary significantly if any of these factors or the physiographic outline of the depositional margin were to change.

Although the subsidence rate can safely be assumed to remain constant over limited time intervals, any abrupt variation in the rate will terminate the ongoing cycle of sedimentation with similar abruptness, and start a new cycle. An acceleration in the rate of subsidence may stimulate an apparent sea-level rise, whereas a sudden slowing down of subsidence will mimic a sea-level fall. An extreme case would be the uplift and subaerial exposure of a margin, removing the region completely from the marine sedimentation cycle and causing widespread subaerial erosion. Thus, tectonic events will tend to amplify or subdue the stratigraphic expression of sequence cycles. Whether tectonics work in tandem with, or against, eustasy will determine the resultant stratigraphic record (see Vail *et al.* (1991) for a discussion of the tectonic and eustatic signatures).

Experience from various sedimentary basins indicates, however, that eustatic sea-level fluctuations occur at a higher frequency than changes in regional subsidence. The signals due to these two contributory factors should therefore be easily discriminated. The observation that similar systems-tract patterns occur at the same time in widely varied sedimentary basins also suggests a commonality of control by a mechanism such as eustasy (Posamentier & Vail, 1988).

When the rate of supply of terrigenous sediments fluctuates, it mainly influences the basinward limit of deposition (Posamentier & Vail, 1988). The seaward extent of progradation of lowstand and highstand systems tracts will vary with the rate of sediment supply. Theoretically, excess sediment at the time of landward transgression, if available, could overwhelm the nearshore accommodation rapidly, shortening the duration of the transgressive systems tract, and promoting early regression of the shoreline.

Although the trend of the curve of the cycle of relative sea-level change is assumed to be sinusoidal, each cycle may vary considerably depending on local conditions. The shape of the curve, which reflects the direction and rate of sea-level variation and local tectonic history, will influence the resultant shape of the systems tracts within the sequences. Embry (1988) has discussed several possible shapes of the curve based on variations in rates and tectonics.

In addition, the conceptual models described above are more clearly applicable along the margins near the point source of sediments. Away from streams and rivers discharging their loads into the basins, the systems tracts will be expressed differently. Fans and lowstand systems tract thin out away from the point-sourced areas (Fig. 7). In areas remote from terrigenous influences, only incised valley fill may represent lowstand deposits. Transgressive deposits may also be thinner away from this point source; they may be marked by a thin veneer of transgressive facies at the top of incised valley fills. In areas with little terrigenous influx, only the condensed section may be prominent (Fig. 7).

Similarly, in siliciclastic systems, highstand deposition also diminishes away from the point source of sediments. The highstand sediments prograde a relatively short distance. An example of this is

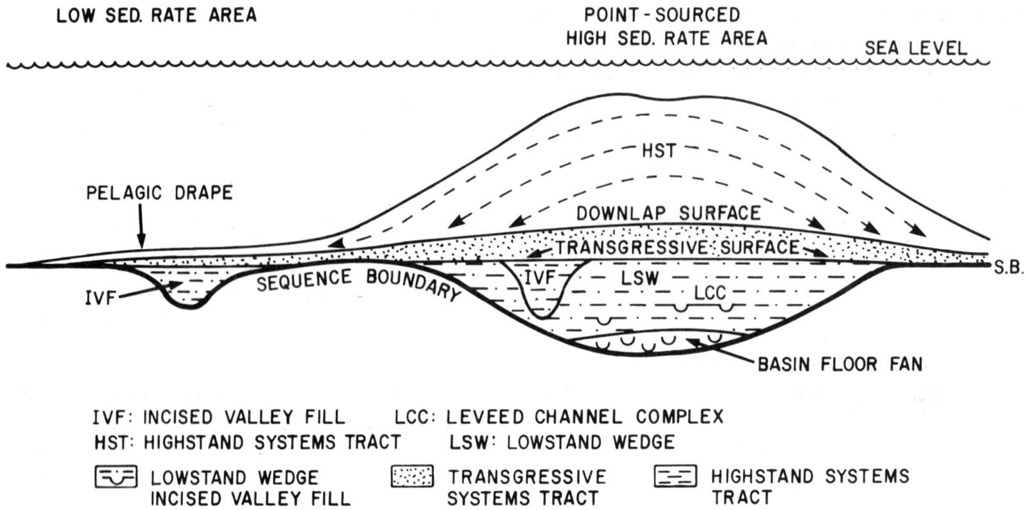


**Fig. 7.** An across-dip (along strike) schematic view of the depositional systems tracts, contrasting the area of high sedimentation rate (near a point source of sediments) with the area of low sedimentation rate (which shows terrigenous influence only during lowstand time). In low sedimentation rate areas the sequence boundary, transgressive surface, and downlap surface may merge into each other laterally (modified after Loutit *et al.*, 1988).

illustrated by the present-day Mississippi Delta, where away from the deltaic influence the condensed section is still exposed on the sea floor, and in some areas is coincident with the sequence boundary (the exposure surface) below it (Loutit *et al.*, 1988; Fig. 7).

In practice, away from point-sourced areas of little or no terrigenous influence, the definition of the third and higher order cycles may become increasingly obscure on seismic data. In these areas the second-order (major) cycles (*sensu* Haq *et al.*, 1988a) are the only recognizable sequences on seismic lines (see e.g. Greenlee & Moore, 1988). Conversely, in areas of high clastic input, such as near the mouths of major rivers, fourth and higher order sea-level perturbations can produce complete sequences ('simple sequences' of Vail *et al.*, 1991) that emulate systems-tract development of the third-order cycles of areas with relatively lower sedimentation rates, as in the Quaternary of the Gulf of Mexico (see e.g. Beard *et al.*, 1982).

Although minor sea-level events (*sensu* Haq *et al.*, 1988a) in areas of moderate sedimentation rates commonly may not be recognizable on seismic lines, they are readily identified in outcrops and well logs. Basinward shift of coastal onlap is difficult to demonstrate in minor sequences; however, the sequence boundary, even in relatively thin sections, can be identified in outcrops and well logs by a number of criteria. These include, changes in the stacking patterns of parasequences, truncation below the boundary, onlap on to the boundary, and basinward shift in facies. The basinward shift in facies is typified by abrupt change from marine to non-marine, or from deeper marine to shallower marine sediments. When these criteria can be demonstrated regionally, a sea-level fall is indicated, ruling out facies change due to autocyclicity (Haq *et al.*, 1988b).

## CARBONATE SYSTEM

Sarg (1988) has presented a detailed discussion of the application of sequence-stratigraphic concepts to carbonate systems. Here these concepts, as applied to the shallow marine carbonates of the continental margins, are reviewed briefly in the light of what has been learned from the siliciclastic-dominated systems. The deep marine carbonates (oozes, chalks), which are also influenced by sea-level changes, albeit indirectly, are discussed separately (see the section on implications for deep-sea sedimentation).

The shallow-marine carbonate system differs from the siliciclastic system in that carbonate sediments are largely formed organically over the depocentres rather than being transported. This means that in addition to regional subsidence and sea-level change, marine climate and hydrography also play relatively important roles in controlling carbonate deposition. Temperature is an important factor in the distribution of both carbonates and evaporites, but carbonate facies are also sensitive to changes in water depth, chemistry, and circulation.

Carbonates occur largely in tropical and temperate areas as aggrading to prograding deposits along the margins of the basins, or on isolated platforms (Wilson, 1975). The basin-margin deposits can occur as banks or broad ramps. Sarg (1988) distinguishes the two basin-rim carbonate depositional settings (Fig. 8) as regional platforms or ramps with slopes of less than 5° (here referred to simply as *ramps*), and regional banks with 5–35° slopes (herein referred to as *banks*). Regional ramps are built along existing gentle slopes with no distinct offlap break (such as in the interior basins) and show aggradational to progradational growth patterns. Regional banks develop on relatively steeper existing slopes, which may possess distinct offlap breaks. Bank sediments characteristically prograde basinward (Fig. 8). Isolated platforms occur on elevated blocks and can

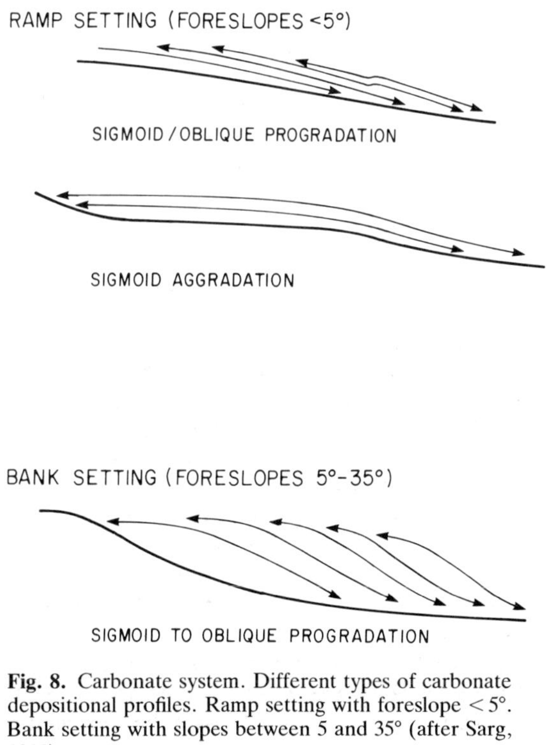


**Fig. 8.** Carbonate system. Different types of carbonate depositional profiles. Ramp setting with foreslope $< 5°$. Bank setting with slopes between 5 and 35° (after Sarg, 1988).

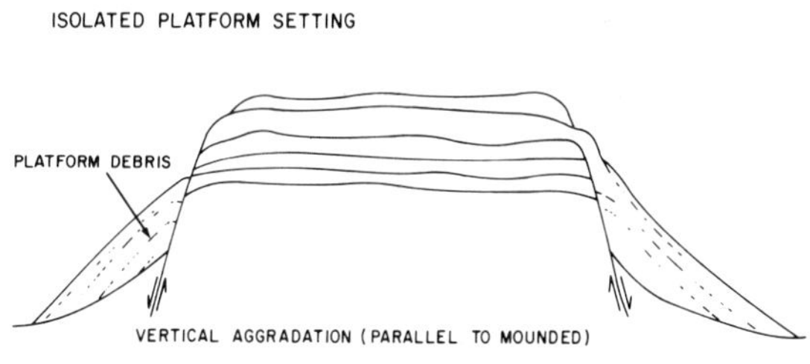


**Fig. 9.** Isolated carbonate platform setting showing vertical aggradation on platform top and off-bank debris shed during the late highstand (modified after Sarg, 1988).

cover a large area (e.g. the Great Bahama Platform). These elevated features have steep margins and relatively flat tops. Shallow-marine carbonate deposition occurs over much of their areal extent, attempting to accrete to prevailing sea level (Fig. 9).

Carbonate deposition takes place in facies belts. Several such facies belts can be identified along marginal banks (Wilson, 1975), ranging from supra- to intertidal facies, shallow-marine facies, shelfal facies, bank-margin facies, foreslope facies, and basinal facies (Fig. 10). The supra- to intertidal facies are typically represented by minor upward-shoaling parasequences. Shallow-marine facies are also upward-shoaling, but include rich faunal and floral elements and evidence of bioturbation. In more restricted environments, evaporitic lagoons may form in these settings, and reefs may build up locally in shallow shelfal areas. The bank margin facies can be complex lithologically, and their parasequences may be capped by regional exposure surfaces, and may contain submarine hardgrounds.

The foreslope facies occur only when a distinct break in grade is present, such as in bank margin settings. These facies consist of bedded lime-mudstone with megaslumps and re-cemented debris, and may be largely composed of material shed from adjacent banks. Carbonate turbidites may also be present. This precludes the occurrence of well-developed parasequences in foreslope facies. Basinal facies may consist of interbedded carbonate and siliciclastic deposits. The toes of the basinward dipping foreslope clinoforms thin towards the basin (Sarg, 1988).

As mentioned earlier, climate can be an important determinant of the type of peritidal carbonate facies that will accumulate. In arid climates evaporites develop along with shelfal carbonates if the basin is restricted. Aridity/humidity also determines the extent of secondary diagenesis during the time of

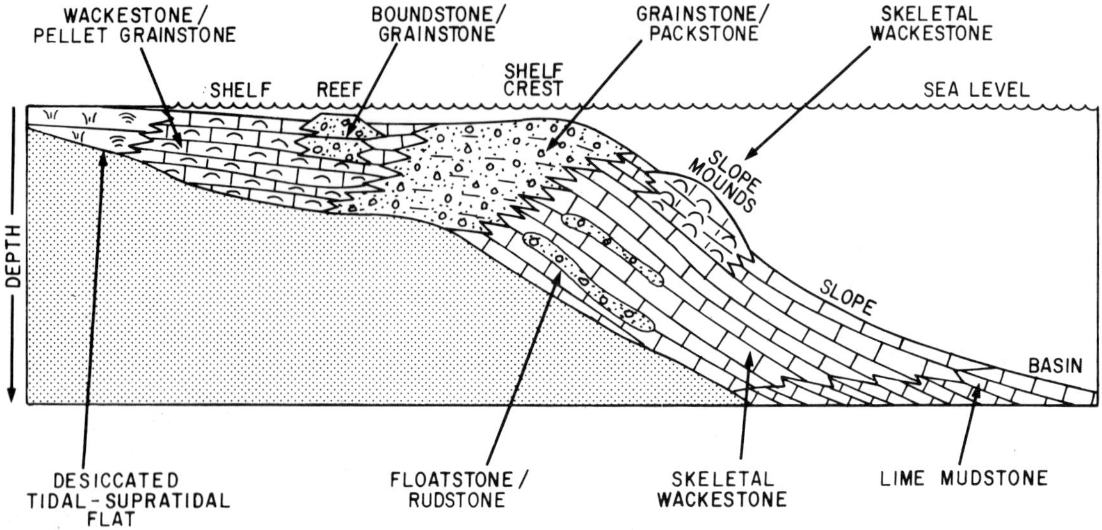


**Fig. 10.** Carbonate system. Carbonate facies belts along a continental margin bank setting (after Sarg, 1988).

low seastands and exposure of the shelf (Sarg, 1988).

### Carbonate depositional cycle

Sea-level change exercises a primary control over carbonate productivity. Some carbonate may accumulate by non-skeletal fixation (e.g. ooids, peloidal mud), but it is largely produced by organisms within the photic zone, and thus limited to the upper 100 m or so of the water column. The production is maximum in the upper 10 m, however, declining rapidly below that level (Schlager, 1981). Thus the rate of sea-level change (rapid versus slow rise) is influential in determining the ability of the carbonate system to keep pace with the changing sea level. For example, a rate of rise of 10 m/Ka or less would allow many modern coral reefs to keep pace with the rising sea level, since most coral reefs can grow at those rates, some growing as fast as 15 m/Ka (Grigg & Epp, 1987). In the long term, carbonate accumulation is controlled not only by the overlying productivity, but also by the local and regional hydrographic conditions, including temperature, salinity, oxygen content, and nutrients, as well as the accommodation potential.

#### *Carbonate lowstand systems tract*

In a carbonate-dominated system a sea-level fall below the offlap break will generate a type 1 sequence boundary (Fig. 11B). The exposure of the bank margin encourages erosion, leading to piling of carbonate megabreccias and debris/sand at the foot of the slope as basin-floor fans. These massive debris flows most likely occur when the bank margins collapse at the beginning of the sea-level fall. However, bioclastic basin-floor fans are relatively rare in carbonate systems. Slope fans also occur in carbonate systems and show a stratal geometry similar to siliciclastic slope fans, consisting of leveed channels and attached lobes (Vail *et al.*, 1991). Deep erosional incisions of the shelf during lowstand time may initiate canyons, much like those in a siliciclastics-dominated system.

Concomitant with slope-front erosion during the generation of a type 1 sequence boundary, the onshore lens of fresh water migrates basinward and initiates extensive regional diagenesis of the carbonate rocks deposited during a previous highstand of sea level. Following a major sea-level drop (75–100 m or more) for an extended period of time and given enough rainfall, the meteoric lens may reach deep into the subsurface and underlying sequences (Sarg, 1988). Karst development, dissolution, and reprecipitation of calcite cement becomes widespread in such a system. These major events are relatively rare (major and some medium type 1 sequence boundaries of Haq *et al.* (1987)). Major karst surfaces are also caused by tectonic enhancement of sequence boundaries.

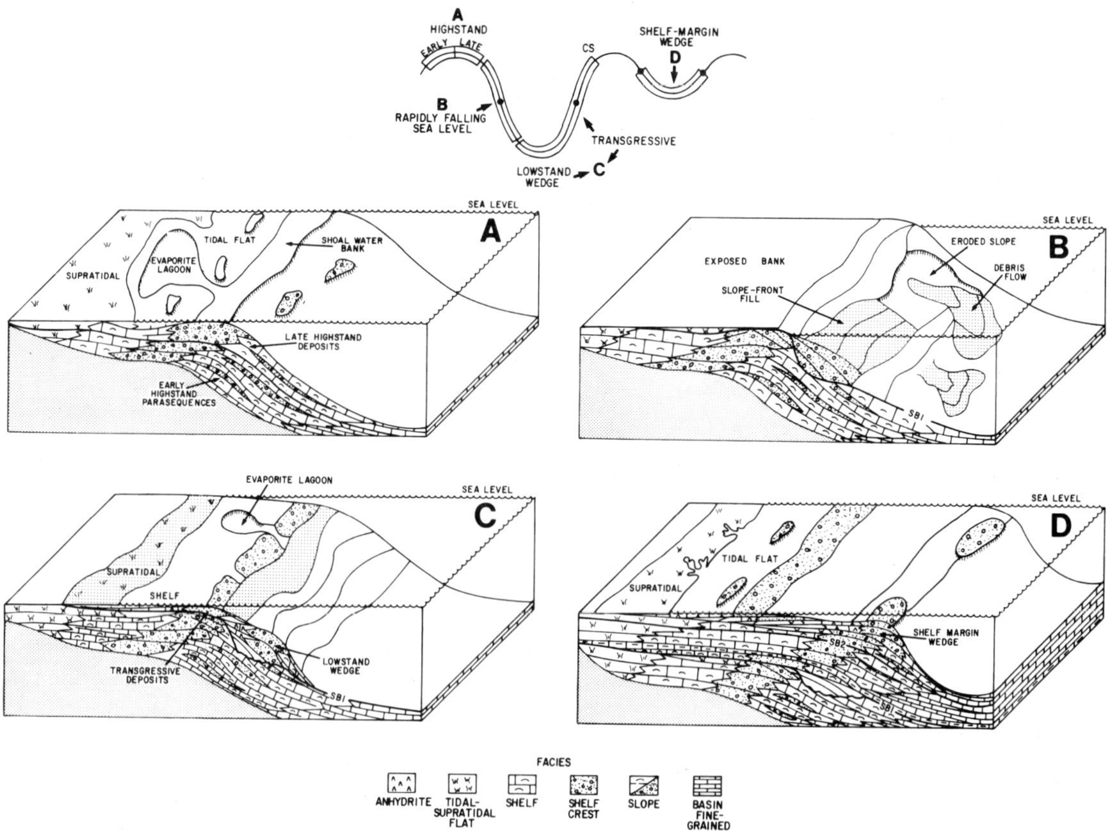


**Fig. 11.** Carbonate system. The Sarg model of carbonate facies deposition in a bank setting. (A) Early highstand (aggradational to progradational) and late highstand (progradational) parasequences of the highstand systems tract. (B) Major sea-level fall, exposure and erosion of the shelf, and resulting slope-front debris flows. (C) Lowstand time followed by relative sea-level rise and the deposition of lowstand wedge, and retrogradational, transgressive systems tracts. (D) Minor sea-level fall and deposition of shelf-margin wedge, followed by relative sea-level rise (modified after Sarg, 1988).

More common are the smaller scale type 1 sequence boundaries (some medium and all minor type 1 sequence boundaries of Haq *et al.* (1987)) when sea-level lowerings are relatively moderate and for shorter durations. In such cases the meteoric lens remains in the shallower part of the shelf and does not become well established, causing less widespread diagenesis.

As in the siliciclastic system, when the sea level does not fall beyond the offlap break and remains above or at the bank margin level, a type 2 sequence boundary is generated (Fig. 11D). According to Sarg (1988), in such cases the primary meteoric influence is restricted to the inner shelf or bank and similar to that at the smaller scale type 1 boundary. Shelfal and slope-front erosion is also of minor importance at type 2 sequence boundaries.

Once the sea-level fall reverses itself, carbonate lowstand deposits begin to accumulate (Fig. 11C), and a prograding complex builds the bank basinward. The lowstand wedge onlaps landward near the offlap break. When the sea level begins to rise after a type 2 sequence boundary, the *bank margin wedge* (analogous to shelf-margin wedge of siliciclastic system) begins to form and onlaps landward of the offlap break. Shingled bioclastic turbidites or debris flows may occur at the toes of the prograding lowstand clinoforms. Slumps may also be present in the basinal part of the lowstand complex (Vail *et al.*, 1991).

*Carbonate transgressive systems tract*

At the conclusion of the lowstand time, when sea level begins to rise more rapidly, the shoreline transgresses rapidly landward and at least the proximal lowstand deposits are succeeded laterally in the landward direction by the retrogradational parasequences of the transgressive systems tract (Fig. 11C). The drowning of the shelf and the landward movement of the depocentres that accompany the transgression starve the bank of sediments and result in the development of the low sedimentation-rate condensed section. In carbonate systems the condensed section is often a shaley micrite package with submarine hardgrounds. In well-oxygenated conditions over the shelf, a slower sea-level rise allows carbonate production to keep pace with the increasing accommodation space and skeletal grainstone-rich 'keep-up' parasequences are deposited. In conditions that are adverse to carbonate productivity, or if the sea level rises too rapidly, 'catch-up' type parasequences may be deposited that cannot keep pace with the rate of sea-level rise (see below for further description of keep-up and catch-up systems).

*Carbonate highstand systems tract*

Although the major part of carbonate progradation in ramp and bank settings occurs during the lowstand, the late highstand phase is also characterized by progradation. Once the sea level has risen to its apex and has stabilized, highstand deposits begin to accumulate. On the shelf the highstand parasequences are aggradational during the rising phase of the sea level, but become increasingly progradational during the late highstand. When the rate of sea-level rise slows, the shoreline begins to regress and parasequences prograde incrementally further basinward (Fig. 11A).

Two end members of shallow marine highstand carbonate systems have been recognized, i.e. the 'keep-up' and the 'catch-up' systems (Kendall & Schlager, 1981; Sarg, 1988). Keep-up carbonates are deposited rapidly in areas of high productivity where the system can keep up with the rising sea level. They are generally mud-poor and grain-rich with only small amounts of early cementation (Sarg, 1988). Catch-up carbonates are deposited when carbonate productivity is low or the rate of relative sea-level rise is rapid so that carbonate deposition cannot keep pace with it. Carbonates therefore show slow accumulation rates, which may result in widespread early submarine cementation owing to the potentially long duration in which the cement can precipitate. They are relatively more mud-prone than keep-up carbonates. Once the accommodation space is reduced in the late part of the highstand phase, the catch-up carbonates may change into keep-up type (see Sarg (1988) for examples of keep-up and catch-up carbonates). If relative sea-level rise is rapid, keep-up carbonates may give up and drown.

The cycle of carbonate deposition ends with the next sea-level lowering and the formation of a new sequence boundary.

## Isolated platform carbonates

The free-standing carbonate platforms that are remote from terrigenous influence respond somewhat differently to sea-level fluctuations than both the carbonate banks and ramps, as well as the siliciclastic-dominated systems. During highstand, when the platform is flooded and carbonate production is high, the excess carbonate mud and sand

ISOLATED PLATFORM (KEEP-UP GROWTH AT RIMS, CATCH-UP IN INTERIOR)

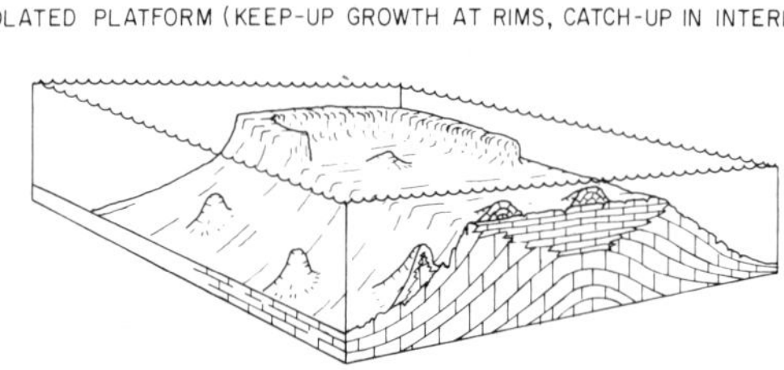

ISOLATED PLATFORM (KEEP-UP TYPE GROWTH)

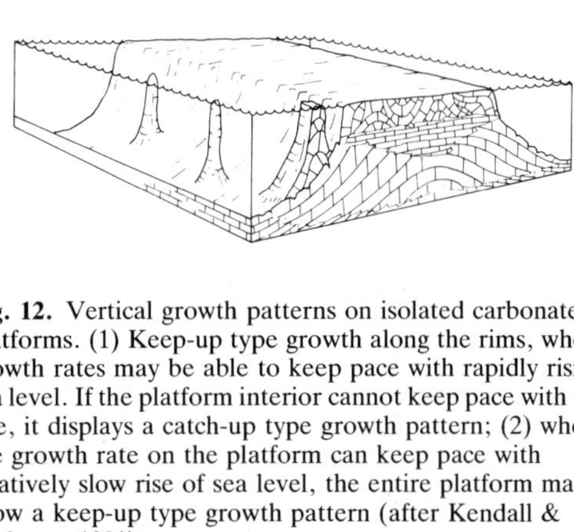

**Fig. 12.** Vertical growth patterns on isolated carbonate platforms. (1) Keep-up type growth along the rims, where growth rates may be able to keep pace with rapidly rising sea level. If the platform interior cannot keep pace with the rise, it displays a catch-up type growth pattern; (2) when the growth rate on the platform can keep pace with relatively slow rise of sea level, the entire platform may show a keep-up type growth pattern (after Kendall & Schlager, 1981).

may be shed off the platform, if the accommodation on the platform is low, resulting in increased carbonate accumulation rates both on and off the platform. Conversely, during lowstands carbonate production is much reduced on the platform. If the platform is subaerially exposed during the lowstand, karst development, chemical erosion, and freshwater diagenesis become dominant (Kendall & Schlager, 1981; Droxler & Schlager, 1985). Major carbonate deposition during the lowstand takes place off-platform as lowstand systems tract.

The highstand shedding of carbonate to the basin in response to reduced accommodation on the platform top is analogous to late highstand basinward progradation in the siliciclastic system. The platform (and shelfal) highstand-shedding and progradation is also more likely to occur during the late highstand phase. There need be little or no off-platform (or off-bank) shedding during the transgressive phase and early highstand when accommodation on platform top or shelf bank is high for the accumulation of carbonate. Only when this space is reduced, either following high productivity and rapid accumulation or during the late highstand, will the carbonate begin to be transferred to the deeper basin through off-flank shedding or progradation.

Drowned carbonate platform tops can display either the keep-up type of sediment accumulation or they may give up if rate of relative sea-level rise is too rapid (Fig. 12). The latter type of parasequences are characterized by shallow-water carbonates, overlain by hardgrounds and deeper marine sediments. When the rims of the platform can grow faster than the rest of the platform, they may show keep-up type of parasequences, while the platform centre may first give up and then try to catch up with it (Fig. 12). Conversely, when the rate of sea-level rise is slower and/or carbonate production high, the keep-up type of accumulation of shallow-water carbonates occurs over much of the platform top (Kendall & Schlager, 1981). Lagoonal facies of the inner platform may be able to keep up with the platform margin facies in such a case.

The well-studied Great Bahama Bank provides some illuminating examples of modern and ancient off-platform sediment accumulation. Wilber *et al.* (1991) have documented wedge-shaped aragonite mud packages on the present-day leeward side of western Great Bahama Bank as the highstand-shed material derived from the platform interior. These are modern analogues to the ancient prograded bank margin sequences documented by Eberli & Ginsberg (1987,1989), who demonstrated that the modern Great Bahama Bank has formed from the coalescence of three smaller platforms. Both lowstand and late highstand progradation were responsible for lateral platform growth. The intermediate seaways were first filled by the debris shed from the platforms, but once the slopes of the platform margins were reduced considerably, progradation began, and each successive cycle produced discrete lowstand and highstand progradational packages (Fig. 13). The number of these progradational events is almost the same as the number of sea-level fluctuations of the eustatic curves of Haq *et al.* (1987). In addition, the trends in the onlap (i.e. the extent of landward shift versus basinward restriction) in the global curve and the bank sequences were identical (Eberli & Ginsburg, 1987).

Aggrading sequences developed on the Bahama Bank when the sea level rose much above the platform top and the productivity was high (Fig. 13).

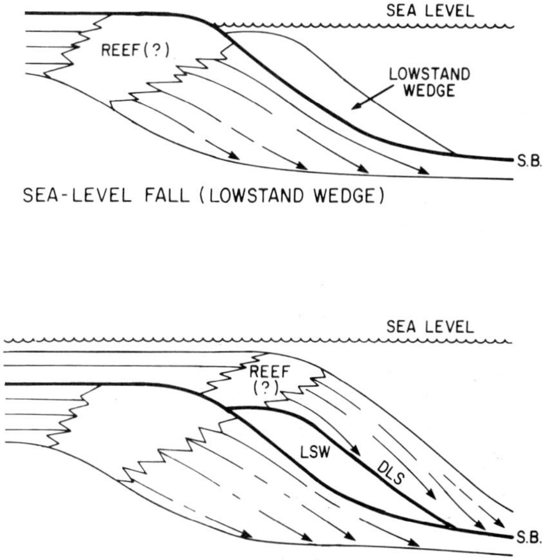


**Fig. 13.** Lateral growth patterns of isolated carbonate platforms. A fall of sea level leads to lateral growth by the basinward development of a lowstand wedge. After the sea-level rise and resumption of production on the platform top, the platform may first grow vertically (during early highstand when accommodation is high), and later begin to build outward by progradation (in later highstand). The progradation of sediments may progressively cover the reef growth on the platform edge (after Eberli & Ginsburg, 1989).

However, during events when the sea level barely flooded the platform, or during late highstands and lowstands, the lack or elimination of accommodation on top of the platform led to progradation as excess sediment was transported off-platform, resulting in outbuilding of platform margins (Eberli & Ginsberg, 1989).

Examples of prograding margins and platform growth are also provided by the Triassic carbonates of the Dolomites (Bosellini, 1984), by the Albian carbonates of La Gandara Platform of northern Spain (Garcia-Mondejar & Fernandez-Mendiola, 1989), the Aptian platform carbonates of the Maestrat Basin in eastern Spain (Salas, 1989), and the Paleogene carbonate platforms of the south Pyrenees (van den Hurk *et al.*, 1989).

## MIXED SILICICLASTIC–CARBONATE SYSTEM

In settings where a terrigenous source of sediments is near a shallow-water carbonate bank or ramp, the two systems may function together. Such mixed systems are quite common in the sedimentary record and respond to the changes in relative sea level by transporting siliciclastics to the basin largely during lowstands, but depositing carbonates during both the lowstands and highstands.

In such settings if the sea level falls below the offlap break or beyond the distally steepened edge of the ramp (type 1 sequence boundary), the exposure of the shelf or upper ramp curtails carbonate production in these areas. The siliciclastics, however, continue to be conveyed to the basin floor through incised valleys, much like in the pure siliciclastic system, to form basin-floor or slope fans. If the sea-level lowering is more modest and does not drop below the offlap break (type 2 sequence boundary) the lowstand deposits more typically consist of mixed carbonates (shed from upper shelf) and siliciclastics, with megabreccia slumps (Dolan, 1989).

During sea-level highstand, on the other hand, the rise of sea above the shelf allows resumption of carbonate production and the deposition of predominantly carbonate parasequences during the early highstand when the accommodation on the shelf is plentiful. During later highstand more mixed carbonate/shale parasequences are deposited as the accommodation decreases and the shoreline recedes basinward.

A sudden uplift in the hinterland (following a tectonic event), or upstream coalescence of rivers that might result in increased terrigenous supply, can overwhelm the shallow-water carbonate system. The excessive detrital material will favour basinward building of fans and siliciclastic progradational wedges.

In modern settings near steep-sided platform margins the shallow-water carbonate sediments may be shed as mounded debris flows during the highstand time, imitating lowstand fans.

Examples of mixed carbonate–siliciclastic depositional systems are provided by the well-studied Permian basins of Texas (Sarg, 1988), and the Permo-Triassic sequences of the Salt Range in Pakistan (Haq *et al.*, 1988a), where lowstand deposits tend to be largely terrigenous, and highstand deposits shallow marine carbonates. Other examples of hybrid systems can be found in Zullo & Harris (1987), Dolan (1989), Arias *et al.* (1989), Pujalte *et al.* (1989), and in the paper by Luterbacher *et al.*

## RECOGNITION OF SEQUENCES IN OUTCROP SECTIONS

In an ideal situation, when both subsurface and outcrop information is available, sequence-stratigraphic analysis should be carried out first on well logs and seismic data to provide a regional overview of the area. Once a regional depositional framework has been established, it can be combined with outcrop studies in the area to pin down facies relationships within the regionally identified genetic packages. More often, however, all three types of data are not available for all areas. Individual outcrop sections commonly do not provide a clear representation of the genetic relationships between vertically stacked facies. In the absence of subsurface data, detailed study of several sections in the area can lead to a broader regional understanding and more reliable systems-tract definition. Many of the global stratigraphic data on which the new cycle charts (Haq *et al.*, 1987) are based come from such outcrop studies.

In outcrops, the various system tracts (lowstand wedge, transgressive, highstand, and shelf margin wedge) can be distinguished between three prominent surfaces (Fig. 14). The most readily identifiable of these is the *transgressive surface*, which marks the first major flooding and the rapid sea-level advance over the shelf. A transgressive surface is quite conspicuous in the field because it is often

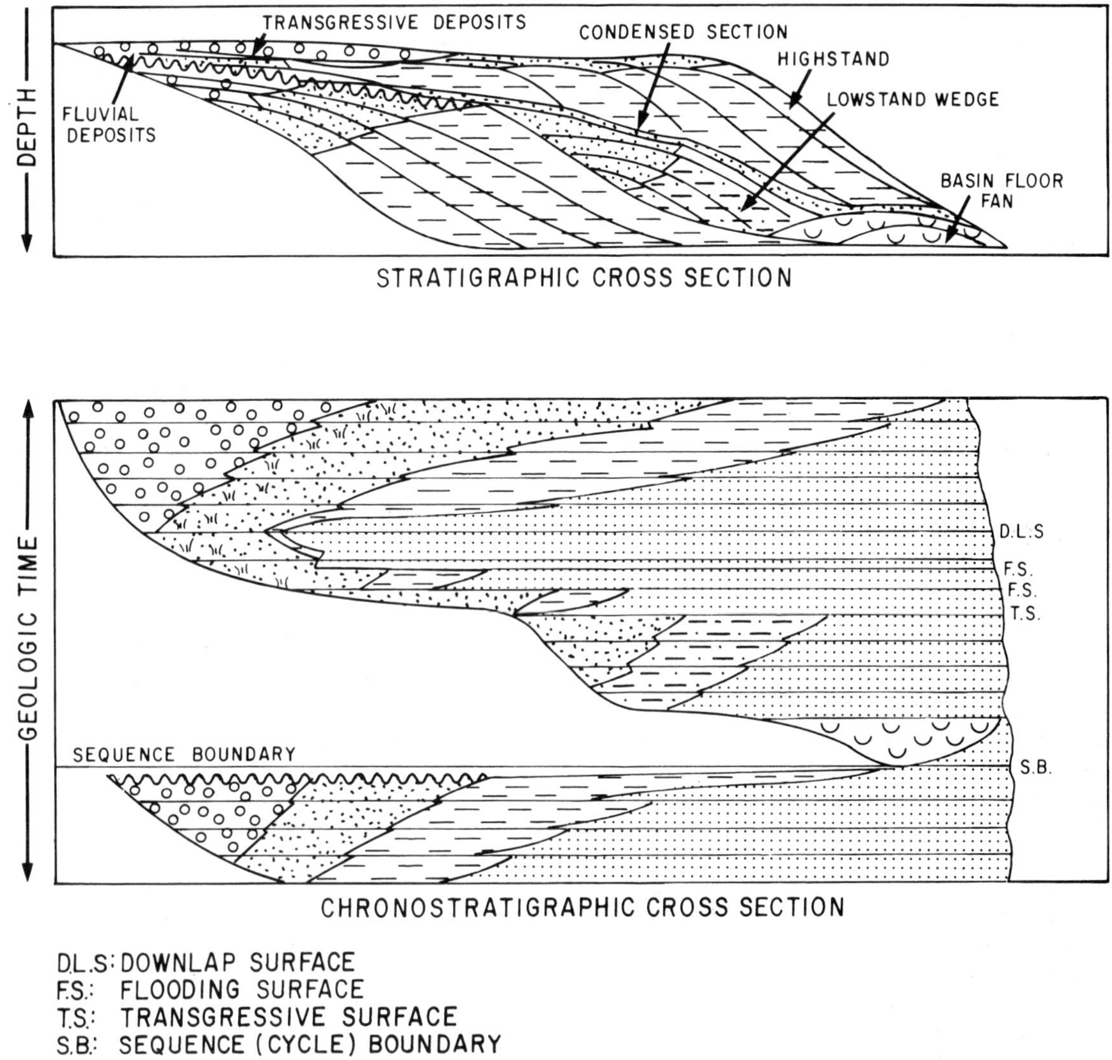


**Fig. 14.** Stratigraphic (depth) and chronostratigraphic (time) cross-sections through the record of a complete type 1 sequence, showing the position of the three prominent depositional surfaces that help identify systems tracts in outcrop sections, namely, the transgressive surface, which is followed by numerous pulses of increasingly shoreward flooding marked by flooding (or ravinement) surfaces, the downlap surface (also the surface of maximum flooding of the shelf), and the sequence boundary. The depth section also shows the extent of the condensed section at the downlap surface (after Loutit *et al.*, 1988).

associated with marked lithologic changes, e.g. from lower shoreface or non-marine below to nearshore marine above (or on the shelf, from sandy to more shaley). The lowstand sediments below the transgressive surface are characterized by the sediments of the most regressive phase of the sequence (Haq *et al.*, 1988a). When lowstand deposits are lacking (as on the most proximal part of the shelf, see Fig. 15A), the transgressive surface is coincident with the underlying unconformable portion of the sequence boundary. In the vicinity of the incised valleys, however, the transgressive surface may be separated from the sequence boundary by the thick intervening sediment fill of incised valleys. The

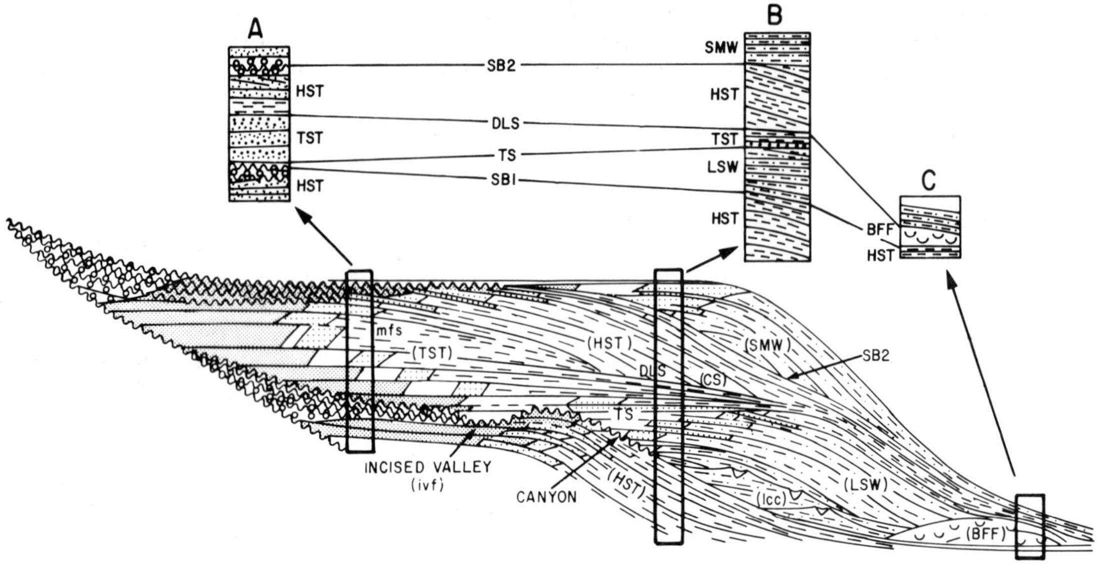


**Fig. 15.** Stratigraphic cross-section illustrating the variations in vertically stacked facies as determined by the position of the outcrop section along the shore to basin profile. (A) Proximal palaeoshelf position. Sequence boundary and transgressive surface are coincident and marked by unconformity, overlain by transgressive deposits and aggradational early highstand facies. (B) Position of the section just off the palaeoshelf edge. Sequence boundary and transgressive surface are now separated by the proximal part of the lowstand wedge facies, and overlain by the distal transgressive deposits, a well-defined condensed section, and the late highstand distal clinoforms. (C) Section near the distal part of the lowstand wedge. At this position the sequence boundary is a relatively conformable surface, overlain by fan deposits and thinned-out lowstand facies, draped by pelagic sediments representing distal highstand facies.

proximal coincidence of the two surfaces, and the conspicuous lithologic changes at the transgressive surface over the remainder of the shelf, nevertheless, make it likely that these surfaces can be confused with each other in outcrop sections.

The second easily recognizable surface in outcrops is the *maximum flooding surface* on the shelf. This surface manifests itself as the *downlap surface* on the distal part of the shelf, and is particularly well displayed in seismic profiles (Fig. 14). It is associated with the period of sediment starvation (*condensed section*) on the middle and outer shelf and slope, when the terrigenous material is sequestered nearshore in response to maximum landward incursion of the sea. The condensed section lies partly within the transgressive and partly within the highstand systems tracts. It is typically marked by concentrations of pelagic sediment components, rich glauconitic and phosphoritic layers, organic-rich 'black' layers, and marine hardgrounds caused by lithification which may show extensive burrowing. Palaeontological gaps (rather than true unconformities) are most often associated with the condensed section. These features may also lead to confusion between the downlap surface and the sequence boundary.

The third prominent surface is the *sequence boundary*, expressed as the basinward (downward) shift of coastal onlap, or by truncation on seismic profiles. In outcrop sections the sequence boundary may be an obvious unconformity or its correlative conformity, depending on the position of the section along the shelf-to-basin profile and the rate of sea-level fall. If the position of the section is on the proximal part of the shelf, the lack of lowstand deposits (see Fig. 15A) means that the unconformable part of the sequence boundary coincides with the overlying transgressive surface. Basinward, the sequence boundary becomes conformable and may be characterized by a change from interbedded progradational deposits of the late highstand below, to more massive aggradational deposits above. An occurrence of sandy facies basinward of limestone facies, or an abrupt appearance of sands in the distal, basinal location, also should signal the proximity of a sequence boundary.

In any given outcrop, the vertical succession of facies also will depend on the position of the section along the palaeoshelf to slope profile. As illustrated in Fig. 15A, a section on the proximal shelf may comprise retrogradational–transgressive deposits directly overlying the coincident transgressive surface and sequence boundary, topped by a relatively thin veneer of nearshore highstand facies. An outcrop just beyond the palaeoshelf edge (Fig. 15B), on the other hand, could record all of the representative facies and surfaces associated with the cycle of sea-level change, i.e. sequence boundary marked by an unconformity, overlain by proximal part of the lowstand wedge deposits (including the possible record of incised-valley fill) marked at its top by the transgressive surface and overlain by the distal part of the transgressive facies, a condensed section marked by a downlap surface, and the distal part of the highstand systems tract. This sequence in turn may be terminated at the top by a type 1 or 2 sequence boundary, overlain by another lowstand or shelf-margin wedge facies.

An outcrop near the distal limit of the palaeoslope (Fig. 15C) may only preserve a record of the sequence boundary depicted by a relatively conformable surface, overlain by the basin-floor and slope fan facies that may be covered by a veneer of basin sediments representing distal toes of the highstand and shelf-margin wedge clinoforms.

These conspicuous variations in the vertical facies relationship within relatively short distance of each other underscore the need for studying several across-dip sections over the region before interpreting sequence stratigraphy of the area.

## DOCUMENTATION OF SEA-LEVEL FLUCTUATIONS

The new generation of sea-level curves (Haq *et al.*, 1987,1988a) were based largely on the study of sequences in outcrop sections that augmented the earlier results from subsurface data. The documentation was obtained from outcrops on the continental margins and interior basins in many parts of the world. The bulk of the data, however, came from western Europe and the trans-Tethys region, the USA Gulf and Atlantic coasts and the Western Interior Seaway, New Zealand and Australia, Pakistan, and Arctic Islands of Bjørnoya and Svalbard. Since the publication of the cycle charts, sequence-stratigraphic data have also begun to accumulate from other regions of the world (see e.g. Embry, 1988; see the paper by Legaretta & Uliana). Detailed lists of sections that contributed to the documentation can be found in Haq *et al.* (1988a, Appendices C–F).

### Stage stratotypes

An important aspect of this documentation has been the integration of sequence-stratigraphic data from stratotype sections of the northwestern European stages into the new cycle charts. European stages form the basis for the internationally accepted chronostratigraphy. Thus, special effort was made to incorporate these data, following study of bio- and lithofacies and physical stratigraphic relationships of the Mesozoic and Cenozoic stage stratotype sections, where available, to effect first-order correlations with standard chronostratigraphy (see Haq *et al.*, 1988a, Appendices C–F).

### Time scales

The incorporation of sequence-stratigraphic data from stage stratotypes and other well-known reference sections was also in part an attempt to establish a reliable chronostratigraphy with as wide a global validity as possible. During the course of assembling global sequence-stratigraphic data it became apparent that a wide diversity of approaches and rigour in the treatment of chronostratigraphic data had been used in different existing time scales. To effect meaningful global correlations, the need for an internally consistent time scale and magneto-biochronology became clear prerequisites. A chronology was needed that would integrate geochronologic, magneto-, bio-, and sequence-stratigraphic information, and reconcile known Earth historic facts without making unnecessary assumptions for broad temporal extrapolations.

Radiometric and bio- and magneto-stratigraphic data, which provide the building blocks on which the global chronostratigraphic framework is based, are replete with numerous sources of errors. For example, radiometric data are severely constrained by several limitations inherent in the samples and the various available radiometric techniques. Plutonic rocks, which perhaps provide the most reliable dates owing to their greater chemical stability, are the most difficult to place stratigraphically. Volcanic extrusives (e.g. lavas) may suffer from the presence of excessive radiogenic $^{40}$Ar, among other problems. Bentonites are products of alteration and often acquire a bias due to contamination. Glauconites, which are perhaps best constrained stratigraphically, are biased by syngenetic and post-depositional diagenetic sources of errors. Many of these uncertainties can be alleviated by using careful criteria in the selection of samples and techniques (see Odin (1982) for details).

There is also ongoing controversy between geochronologists about the use of high-temperature (plutonic and volcanic rocks and bentonites) versus low-temperature (glauconite) radiometric dates, and the way they are used for time calibration. High-temperature dates are often preferred over glauconite dates, because the latter are considered to be more prone to alteration. Glauconites are affected by inherent geochemical problems that bias the dates obtained from them towards somewhat younger apparent ages. However, such problems also exist with most high-temperature minerals that bias dates towards both older or younger apparent ages (discussed in Odin (1982) and Haq *et al.* (1988a)). Nevertheless, reliable dates can be obtained when glauconite samples that have high potassium content are carefully chosen to select grains in which the influence of alteration can be mineralogically discounted (Keppens & Pasteels, 1982). A recent comparison of available glauconite and non-glauconite dates for the last 125 Ma (Craig *et al.*, 1989) has clearly shown that differences between them are small and that glauconites can yield reliable dates for this time interval.

A prime reason for using glauconite dates for which reliability can be ascertained, is that glauconites with greater potassium concentration are most likely to occur in the condensed section (and the underlying transgressive facies). These facies also provide the best biostratigraphic information, leading to direct, first-order, age-dating for the calibration of biozones in the continental margin sections (Loutit *et al.*, 1988).

Some geochronologists still prefer to use a few selected radiometric dates as 'tie-points', employed as anchors, for interpolating the intervening intervals according to some assumption of constancy of rates (constant seafloor spreading, sedimentation, or evolutionary rates). Such an approach, though an easy way out, ignores possible excursions from the linear trends that are likely to be inherent in the data. In our view, a qualified use of a large set of analytically acceptable dates with known stratigraphic limits can provide important constraints for the time scale construction. This will identify trends in the data that deviate from linearity, and therefore better able to approximate reality, than the use of isolated anchors used to nail-down long segments of the stratigraphic column (Haq *et al.*, 1988b).

Similarly, biostratigraphic data can often inherit bias due to the varying level of rigour in data collection, divergent taxonomic concepts, sampling problems, and due to undetected biogeographic variations and discontinuities in the stratigraphic record. Palaeomagnetic data are also prone to interpretive inaccuracies. Since these data are dependent on biostratigraphic–radiometric calibration, the bias in the latter can be a source of interpretive error in the palaeomagnetic data. Sea floor magnetic anomaly data often can be complicated by the presence of fracture zones, anomaly skewness, and by ridge jumps and sudden changes in spreading centres. Asymmetrical spreading along opposite flanks of a spreading centre also may lead to interpretive errors. In addition, sampling and stratigraphic problems peculiar to magnetic polarity data can introduce significant bias (see Haq *et al.* (1988a) for a complete discussion of the common sources of errors).

In view of the array of sources of uncertainties that can bias the resultant schemes of geological calibration, it appears more reasonable that an approach based on multiple correlative systems, and one that does not overlook a large body of empirical and analytical data, or make unnecessary assumptions, has the best chance of reducing inconsistencies introduced due to any singular technique, and producing more realistic time scales. The effort to establish a robust chronostratigraphy with these attributes has been discussed in detail in Haq *et al.* (1988a).

Refined stratigraphic resolution is of crucial importance in attempts to tie sequence-stratigraphic data from different parts of the world for constructing eustatic sea-level curves. The sequence documentation (Haq *et al.*, 1988a) consists of rigorous pattern matching of sequences and systems tracts in different basins, including litho- and biofacies analyses, which can help weed out local events and ensure the retention of consistent and widely distributed events. The biostratigraphic zonal resolution on which such correlations are commonly based is at best of the order of half to one million years, but is often less accurate. The frequency of third-order events on the cycle charts sometimes may exceed the available biostratigraphic resolution. In such cases, our methodology does not attempt to correlate singular events in widely separated areas. Instead, a series of sea-level events, with their characteristic systems-tract patterns, and constrained between well-established stratigraphic datums, are correlated with similar series of events in other areas that show the same characteristic patterns of systems tract development, as well as number of sequences, and constrained between similar datums. Biostratigraphic data alone may not help correlate all events from one area to the other, but the reproducibility of the number of sequences and characteristic parasequence stacking patterns within systems tracts lend considerable confidence to such correlations (Haq *et al.*, 1988b).

Nevertheless, the sea-level curves need to tested rigorously in places other than our reference areas (Haq *et al.*, 1988a, Appendices C–F). As better, multiple, correlative methods, and perhaps, new and independent systems, are pursued, it will be inevitable that the eustatic model based on the above-mentioned criteria will be refined, especially where it concerns many of the minor and some medium-magnitude events. This process is well underway and several additional events have already been indentified in the Triassic and Cretaceous.

## Mesozoic–Cenozoic cycle charts

The left-hand half of Fig. 16 depicts the history of sea-level change during the Cenozoic. Triassic, Jurassic, and Cretaceous cycle charts can be found in Haq *et al.* (1987,1988a) and are not reproduced here. In the cycle charts the absolute time (in Ma) is tied to magnetochronostratigraphy, standard chronostratigraphic subdivisions (epochs/series, stages), biochronostratigraphy (biochronozones of fossil groups most useful for each period), sequence-chronostratigraphy (cycle nomenclature, coastal onlap patterns, ages of sequence boundaries and downlap surfaces, and depiction of systems tracts), and eustatic sea-level curves, both short and long term (see Haq *et al.* (1988a) for a detailed description of various segments of the cycle charts).

The cycle terminology in the sequence-chronostratigraphy segment of the charts includes a single first-order sequence chronozone. The Phanerozoic is divided into two megasequence-set chronozones, and the interval from the latest Permian to Holocene comprises the upper Phanerozoic megasequence set. The Mesozoic and Cenozoic, however, encompass three megasequences (megacycles), i.e. the upper part of Absaroka, the Zuni, and the Tejas (proper-name terminology is modified from the American Indian tribal name sequences of Sloss (1963)).

The second-order sequence chronozones are the supersequences (supercycles). Seven supercycle sets and 27 supercycles are identified, all terminating in major sequence boundaries. The third-order chronozones include sequences (third-order cycles) that terminate in either major, medium, or minor sequence boundaries. A total of 121 sea-level change events are recognized in the upper Phanerozoic megasequence set. Of these, 19 events terminate in relatively major magnitude falls of sea level, 43 in medium magnitude falls, and 59 in minor sea-level falls. In areas of average to low sedimentation rates, only the major and some medium magnitude cycle boundaries can be recognized on seismic profiles. Often, only detailed well-log and outcrop analyses can resolve the minor and many medium magnitude third-order cycles. In areas of high sedimentation rates, however, even the smaller third-order and fourth-order sequences (which are composed of several tens of metres of sediments) can be resolved on seismic lines, such as in the Pleistocene of the Gulf of Mexico.

On the cycle charts, type 1 sequence boundaries (shown as 'saw-toothed' boundaries on the coastal onlap curves) where basin-floor fans have been documented are indicated with the letter 'F' in the systems tracts column. The unshaded triangle within each onlap event represents the condensed section (also distinguished as relatively major, medium, or minor, by the relative thickness of the line representing the downlap surface associated with the condensed section).

### *Magnitude of sea-level fluctuations*

The relative magnitudes of sea-level rises and falls as depicted on the cycle charts are the best approximation of these changes compared with the present-day global mean sea level. The long-term changes in relative magnitude are estimated using the method described in Hardenbol *et al.* (1981) with the Turonian high value adopted from Harrison (1986). The relative magnitude of the third-order (short term) sea-level changes are calculated from coastal aggradation of the lowstand systems tracts (corrected for subsidence and compaction) in seismic and sequence-stratigraphic data.

The method of determining sea-level changes from lowstand coastal aggradation, however, can yield variable results when applied in different areas, depending on differences in the regional setting and subsidence rates. For example, Greenlee & Moore (1988) estimated that Neogene sea-level rises on the Atlantic (New Jersey and Alabama) margin of the USA were much lower than those depicted on the eustatic curves (by as much as 100 m or more in the middle Miocene). Their estimates of sea-level falls were, however, close to those on the eustatic curves.

Hay *et al.*'s (1989) mass-balanced palaeogeographic reconstructions based on restoring Late Cenozoic sediment masses of the Gulf of Mexico to the source areas of western central North America and correcting for isostasy, compaction, subsidence and sea-level change, produced palaeoshorelines that do not agree with the shorelines indicated by marine deposits. Their study concluded that the amplitude of sea-level changes in the Gulf of Mexico were on average about half of those on the eustatic curves. This suggests that isostatic readjustment may play an important role in amplitude calculations, since all other corrections had been considered in the eustatic curves.

It is obvious that the estimates of sea-level highs and lows on the eustatic curves are much less rigorous than the other aspects of the data depicted on the cycle charts. This is due to the fact that all of the existing methods for determining amplitude changes are hampered by several limitations. The lack of a practical, fixed, reference point against which these variations can be measured, and the uncertainties involving the reconstruction of past hypsometric curves, mean that transgressions of the land by the sea at different times cannot be used effectively as a measure of the magnitude of sea-level change. Burton *et al.* (1987) and Kendall & Lerche (1988) have discussed the problems of estimating absolute sea-level change from various techniques. Numerical modelling efforts indicate that, at best, only a measure of the sum of subsidence and eustasy can be obtained (i.e. relative change of sea level) with the existing methods (Burton *et al.*, 1987).

Estimates of the amplitude of sea-level variations from stable isotopic data are impeded by the fact that for the Mesozoic and much of the Cenozoic, the oxygen isotopic changes are not dominantly related to changes in global ice volume. Even for the Pleistocene, the prediction of accurate sea-level amplitude changes from oxygen isotopic data suffers from the difficulty in separating the ice-volume due to sea-ice versus continental-ice and in knowing the isotopic fractionation in each. Added to these is the difficulty in accurately estimating the isostatic response. Beyond the late Neogene glaciations, there is also the problem of separating the contribution of temperature change to the isotopic signal from the ice-volume effect, although Matthews & Poore (1980; see also Prentice & Matthews, 1988) are convinced that the record of $\delta^{18}O$ variations of tropical planktonic foraminifera reflect ice-volume changes for at least the last 40 Ma. They arrived at this conclusion, however, by assuming a relatively constant tropical sea-surface temperature (around 28°C) since that time, which has not been proven independently.

Shackleton & Opdyke (1973) estimated that a change of 0.1/ml in $\delta^{18}O$ was representative of about 10 m of sea-level change. A similar figure (0.11/ml) was arrived at by Fairbanks & Matthews (1978) for the Pleistocene. Using this estimate would drop the overall sea level in the Cenozoic by some 364 m (Williams, 1988). Estimates from the eustatic curves are, however, of the order of about a 225–250 m overall drop in the Cenozoic (Haq *et al.*, 1987, Fig. 2). Even the lower figure has been considered as too high by some investigators based on independent evidence (e.g. Hay *et al.*, 1989). This suggests that a significant component of the $\delta^{18}O$ signal must be due to temperature and/or other (still unknown?) factors. This may explain why the oxygen isotopic curves show little detailed relationship prior to the Pleistocene and no similarity to the eustatic curves prior to late Neogene.

From the above discussion it is apparent that the estimates of the changes in the magnitude of global sea level may have to be modified as more accurate and independent methods of determining the amplitude become available.

### *Tithonian change in the nature of the sea-level curve*

One of the most prominent features of the Mesozoic–Cenozoic sea-level curves (see Haq *et al.* (1988a) for a complete Mesozoic–Cenozoic cycle chart) is the observation that in the Early Triassic through to Late Jurassic (Scythian through to early Kimmeridgian) interval almost all of the sea-level fluctuations (both rises and falls) are gradual and occur over relatively longer spans of time. The third-order cycles in this interval average about one every 3 Ma. With the exception of four cycle boundaries in the Triassic and three in the Early–Middle Jurassic, which are type 1, all other's are type 2 sequence boundaries, i.e. events in which sea level did not fall beyond the depositional offlap break. Lack of documented fans is also a prominent feature of this interval.

In the late Kimmeridgian (Tithonian) there is a profound change in the style of sea-level change. The rises and falls become more pronounced and the average duration decreases from 3 Ma/cycle to about 1.6 Ma/cycle. Forty of the 87 cycles in the Tithonian to Holocene interval show characteristic type 1 sequence boundaries, 70% of which have been documented to be associated with the deposition of off-shore fans. Canyon-cutting events on the shelf, which were relatively unknown previously, have also become more familiar occurrences in the stratigraphic record since the Tithonian.

The record of Late Jurassic to Holocene sea-level fluctuations is also characterized by cycles of progressively landward back-stepping aggradational sequences followed by increasingly regressing progradational sequences. These cycles are expressed at the amplitude and time-scale of second-order 'supercycles'. The supercycles develop the more typical alternating retrogradational–progradational style after the Tithonian.

It is obvious that the nature of sea-level variations, related to the mechanism by which sea-level rises and falls, changed dramatically in the Late Jurassic. From the character of late Neogene changes, it can be inferred that the eustatic events that are driven by the waxing and waning of glacial ice produce more abrupt rises and falls and more frequent type 1 sequence boundaries. Could the change from relatively gradual to more pronounced variations and more numerous type 1 sequences (although still less than 47%) since late Kimmeridgian be related to the development of enough land ice to effect this transformation? The known geological record does not bear out this contention.

## SEA-LEVEL CHANGE AND IMPLICATIONS FOR DEEP-SEA SEDIMENTATION

Sea-level oscillations directly impact the continental margins where the landward or basinward movement of depocentres creates the characteristic back-stepping and fore-stepping stratal patterns discussed in the previous sections. Only in cases of extreme highstands when the seas invade inland areas, or major sea-level falls when terrigenous material may be delivered to the deep basins, does the sea-level change directly influence areas other than the margins. At other times (during average type 1 or type 2 sea-level cycles), how are the deep-sea sedimentation patterns influenced by these fluctuations?

The changes in sea level have the circuitous capacity to indirectly influence deep-sea sedimentation through: (1) sequestering or releasing siliciclastics and carbonates along the margins to facilitate changes in the ocean nutrient levels and chemistry, (2) causing climatic changes and intensifying or weakening vertical and horizontal thermal gradients of the ocean, and (3) through the changes in the thermal structure, causing changes in both surface and deep-water circulation that control erosive action along the margins and in the deep sea.

As stated earlier, in terms of sequence stratigraphy, the deep-sea record can be thought of as vertically stacked condensed sections, characterized by pelagic sediments. Only when a deep-sea basin is in the vicinity of a terrigenous sediment source can it receive siliciclastics in significant volume, which will be preferentially delivered to the basin during the lowstand times (although slope failure may be able to contribute allochthonous sediments to the foot of the slope during highstand as well). An example is provided by the Deep Sea Drilling Project (DSDP) Site 219 near the southern tip of India. Late Paleocene siliciclastic turbidites of lowstand origin, overlain by carbonates shed from the adjacent shallow Maldive banks possibly during the highstands, have been recorded at this site (Dolan, 1989). In the absence of a terrigenous source, the comparatively uniform pelagic record of the deep basins is punctuated only by stratigraphic gaps of various durations. Many of these gaps (non-depositional or erosional hiatuses) have been documented to be of ocean-wide extent (Barron & Keller, 1982; Keller & Barron, 1987; Keller *et al.*, 1987).

In order to discern the deep-sea response to sea-level changes along the ocean margins, the available information from various sources on widespread deep-sea and continental-margin seismic reflectors (caused either by dissolution of carbonate or by erosion), the occurrence of major canyon-cutting episodes along the margins, the times of known hiatuses (inferred from gaps indicated by missing fossil biozones in the deep-sea record), the microplankton latitudinal migrationary events (interpreted as response to warming and cooling of surface waters), and composite benthic $\delta^{18}O$ curves (palaeoclimatic trends) have been plotted against the Cenozoic coastal onlap and sea-level curves in Fig. 16.

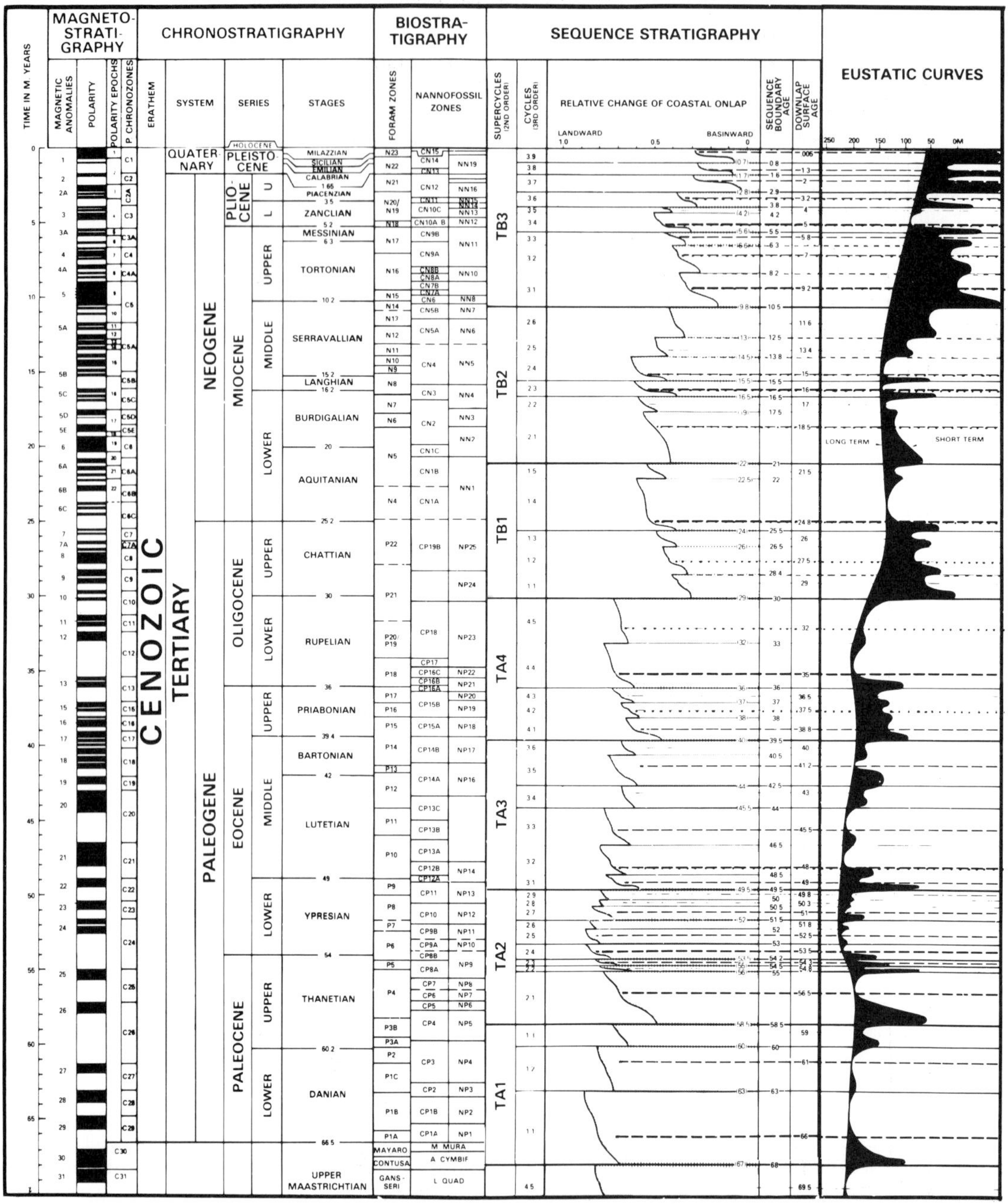


**Fig. 16.** Comparison of Cenozoic sea-level fluctuations with major deep-sea seismic reflectors and major canyon-cutting events, times of known widespread deep-sea hiatuses (separated as predominantly erosional or dissolution-caused), microplankton latitudinal migrationary events (interpreted as a response to cooling or warming of surface waters), and benthic oxygen isotopic curves from the low-latitude Pacific and mid- to low-latitude Atlantic (to show major palaeoclimatic trends). The data are drawn from various sources (see text for details).

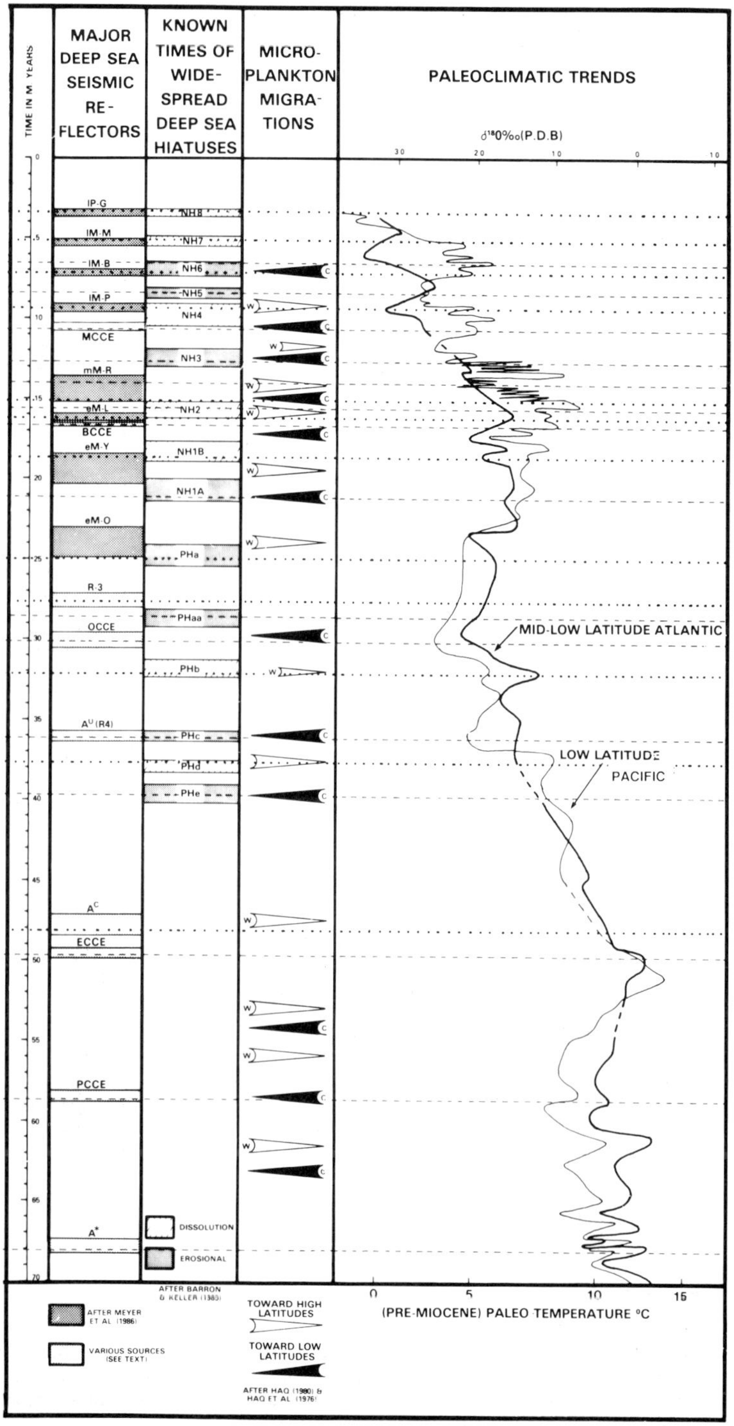
TIME IN M YEARS
MAJOR DEEP SEA SEISMIC RE-FLECTORS
KNOWN TIMES OF WIDE-SPREAD DEEP SEA HIATUSES
MICRO-PLANKTON MIGRA-TIONS
PALEOCLIMATIC TRENDS
IP-G
IM-M
IM-B
IM-P
MCCE
mM-R
eM-L
BCCE
eM-Y
eM-O
R-3
OCCE
A^U (R4)
A^C
ECCE
PCCE
A*
NH8
NH7
NH6
NH5
NH4
NH3
NH2
NH1B
NH1A
PHa
PHaa
PHb
PHc
PHd
PHe
DISSOLUTION
EROSIONAL
MID-LOW LATITUDE ATLANTIC
LOW LATITUDE PACIFIC
(PRE-MIOCENE) PALEO TEMPERATURE °C
AFTER MEYER ET AL (1986)
VARIOUS SOURCES (SEE TEXT)
AFTER BARRON & KELLER (1980)
TOWARD HIGH LATITUDES
TOWARD LOW LATITUDES
AFTER HAQ (1980) & HAQ ET AL (1976)

The column displaying major deep-sea seismic reflectors also includes four major canyon-cutting events. The Neogene reflectors (marked eM-O to lP-G) have been identified by Mayer *et al.* (1986) as being of regional extent in the equatorial Pacific. These authors show that the reflectors can be linked to carbonate dissolution or diagenesis caused by global reorganizations of surface- and bottom-water circulation and changes in ocean chemistry. Other regionally identifiable seismic horizons in the latest Cretaceous and Paleogene (horizons A*, $A^c$, $A^u$ (R4), and R3) that have been documented from the USA east coast are also plotted (see e.g. Tucholke & Mountain, 1979; Vail *et al.*, 1980; Tucholke & Laine, 1982; Miller *et al.*, 1985; Mountain & Tucholke, 1985). The major canyon-cutting events (CCE) on the continental margins that originate following major sea-level falls include the events in mid-Paleocene (PCCE) corresponding to the major sequence boundary at 58.5 Ma, in the early Eocene (ECCE) at the major sequence boundary at 49.5 Ma, in the mid-Oligocene (OCCE) at the major 30 Ma sequence boundary, and in the late middle Miocene (MCCE) corresponding to the major boundary at 10.5 Ma.

The column to the right of the seismic reflectors shows the ocean-wide Paleogene (PHa-PHe) and Neogene (NH1a-NH8) deep-sea hiatuses documented by Barron & Keller (1982), Keller & Barron (1987), and Keller *et al.* (1987). Hiatuses (gaps in sedimentation) can be caused by active removal of sediments by erosive bottom currents, by non-deposition (or depositional starvation, such as on the outer shelf and slope during transgressions), or by dissolution and diagenesis of predominantly calcareous sediments. The gaps caused by dissolution can be pervasive, but are usually of shorter duration than those caused by sediment removal. The longer duration erosional hiatuses, on the other hand, are capable of wiping out the record of several previous events. For this reason erosional events are also difficult to date with any precision — their exact time of initiation remains unknown.

In Fig. 16 the hiatuses have been distinguished as being in two main categories, dominantly erosional and those caused by dissolution of carbonate. The oldest Paleogene hiatus (PHe) documented by Keller *et al.* (1987) from Atlantic, Pacific, and Indian Oceans also has an equivalent in the Gulf of Mexico (Angstadt *et al.*, 1983).

The next column lists the calcareous microplankton migrationary events (interpreted as a response to cooling (C) or warming (W) of surface waters) that have been identified in the Atlantic Ocean (Haq *et al.*, 1977; Haq, 1980). During the cooling events high-latitude assemblages invade the low latitudes, while during warming episodes warm and temperate assemblages migrate to higher latitude areas.

The final column summarizes the benthic foraminiferal $\delta^{18}O$ curves from the low-latitude Pacific (after Douglas & Savin, 1973,1975; Woodruff *et al.*, 1981), and low to mid-latitude Atlantic (after Boersma & Shackleton, 1977,1978; Boersma *et al.*, 1979) in order to display the broad palaeoclimatic trends (see also Miller *et al.* (1987) for a synthesis of Cenozoic oxygen isotopic data). The overall rapid decline in palaeotemperature near the Eocene–Oligocene boundary is now generally ascribed to the development of a major ice cap on Antarctica (Shackleton & Kennett, 1975). The pre-Oligocene world, on the other hand, is considered to be largely ice-free.

The attempted correlations in Fig. 16 reveal some important general trends. All supercycle (second-order cycle) boundaries (at 68, 58.5, 49.5, 39.5, 30, 21, and 10.5 Ma) are either associated with major erosional seismic reflectors (unconformities) or canyon-cutting events along the margins, or both. Most of these events also occur, where such documentation is available, near climatic cooling events as indicated by microplankton migrationary patterns from high latitudes towards temperate areas. Oxygen isotopic trends confirm these observations.

When they are compared with the sea-level oscillations, the equatorial Pacific carbonate dissolution/diagenesis-linked seismic reflectors (Mayer *et al.*, 1986) are all associated with times of transgression and the formation of condensed sections. The deep-sea hiatuses, on the other hand, seem to be centred around sequence boundaries (sea-level falls) if the hiatuses are identified as gaps caused by bottom-water erosion, or conversely can be linked to periods of transgression and maximum flooding of the shelves when the hiatuses are clearly due to dissolution. Erosional hiatuses are commonly also associated with climatic cooling events, while corrosional hiatuses occur during climatic warming.

It is apparent from the available evidence that dissolution of carbonate in the deep sea is favoured during the sea-level rise and a part of the ensuing early sea-level highstands. Conversely, erosion, both along the margins and in the deeper parts of the basins, in enhanced during episodes of sea-level fall and lowstands. Major falls (supercycle boundaries) are also associated with widespread canyon incision on the continental shelves.

Several working hypotheses have been proposed for the observed relationships between sea-level high- and lowstands and the deep sea. Berger (1970) and Berger & Winterer (1974) offered the basin-shelf carbonate fractionation model to account for increased dissolution and hiatuses during high sea-stands. Milliman (1974) compared the carbonate deposition during the interglacial (highstand) and glacial (lowstand) times and suggested that shallow-water carbonate accumulation during interglacials leads to a deficiency of carbonate in ocean waters, raising the calcite compensation depth (CCD) (and by implication increased deep-sea dissolution). During the glacials, when shallow-water areas are subaerially exposed, the transfer of carbonate to the oceans will lower the CCD (and reduce deep-sea dissolution).

Vail *et al.* (1980) maintained that unconformities, whether on the shelves, the slope, or the ocean floor, preferentially develop during lowstands, when shifts in depositional patterns related to lowered sea-level take place. Haq *et al.* (1987) explained the observed patterns of erosion on the sea floor during lowstands and dissolution during highstands partially through a climatic feedback mechanism. They suggested that highstands not only lead to sequestering of clastics and carbonates on the inner shelves, shoaling of the CCD, and increased dissolution in the deep basins, but the elevated sea levels also cause climatic equitability, weakening of latitudinal and vertical thermal gradients, and reduced bottom-water activity, leading to decreased potential for erosional hiatuses. During sea-level falls, on the other hand, terrigenous as well as carbonate detritus bypass the shelves into the basins, resulting in increased carbonate in the oceans and depressed CCD. However, the climatic inequitability and intensified thermal gradients during lowstands ensure increased bottom-water activity and attendent erosion on the sea floor.

Keller *et al.* (1987) found a correlation between widespread, but relatively short-lived, dissolution hiatuses in the Paleogene with times of global cooling and periods of transgression and highstands. This is, however, somewhat contradictory to the scenario suggested by Fig. 16, where dissolution-caused reflectors and hiatuses are not only centred around times of maximum flooding of the shelves, but also most commonly occur during periods of global warming, rather than cooling.

Open ocean biota, particularly plankton, seem to show an empirical relationship to sea-level changes. Haq (1973) presented quantitative data from one group of calcareous plankton which demonstrated that plankton diversity increased during the highstands and the spread of epicontinental seas, while widespread orogenic events and falls of sea level were followed by reductions in the number of taxa. He argued that the smaller temperature contrast and overall equitable climates during the highstands and the expansion of ecospace favour growth and speciation. Large-scale sea-level falls, on the other hand, result in greater temperature contrast, extreme climate and loss of ecospace, conditions that are less favourable for the growth and speciation of marine plankton.

Fischer & Arthur (1977) also saw an empirical relationship between times of high seastands, higher diversities amongst pelagic organisms, and increased community complexity. Like Haq (1973), they also ascribed these 'polytaxic' phases to more equitable climates, reduced oceanic convection, and greater niche availability. The opposite relationship was seen for periods of lowstands when sharper temperature gradients and increased current activity lead to lower diversity 'oligotaxic' assemblage make-up. They postulated that productivity would increase in the open ocean during lowstands, and on the epicontinental seas during highstands.

A more inclusive model of the influence of sea-level oscillations on deep-sea sedimentary patterns is given below — the discussion draws upon previous propositions and takes into consideration the observations offered in the foregoing sections.

### The erosive–corrosive cycle

Erosion along the continental margins and on the basin floor occurs predominantly during the fall of sea level and the ensuing lowstand. Erosion is enhanced during low seastands in two ways: first, the steepened stream gradients and the increased sediment load of the incised valley system during the lowstands lead to greater erosive potential and increased frequency of turbidity (and contour) currents. Major sea-level falls may be associated with canyon-cutting on the margins or reactivation and further deepening of existing canyons. Clastics and carbonates bypass the shelves to the basins through the incised valleys. When the shelf is exposed subaerially (following type 1 sequence boundary), erosion of the shelf top adds to the sediment load being transferred to the basin. Thus, potentially large amounts of sediment can be removed from the shelf/slope and redistributed during lowstands. Slope erosion due to sediment creep, slope failure, and mass movement of sediments downslope also contributes to the redistribution process.

Second, erosion in the deep sea is also enhanced through climatic deterioration and the intensification of thermal gradients. Increased albedo during lowstands favours extreme climates, and this, in turn, leads to enhanced thermal contrast between land and sea, between high and low latitudes, and between surface and bottom water. Extreme climates also increase weathering and erosion on land. The coincidence of lowstands and cooling climates (glacials since late Neogene) would also enhance the production of cold erosive bottom waters in polar regions leading to widespread erosional hiatuses.

The higher influx of carbonate in the deep sea during lowstands would depress the CCD and favour decreased dissolution (the deeper facies, below the depressed CCD, continue to show signs of dissolution). The influx of terrigenous material into the basins translates as increased nutrient supply favouring increased productivity over the open ocean, which can help to further suppress the CCD.

This pattern will change abruptly when the eustatic sea level begins to rise rapidly and the shelves are flooded. This cuts off the terrigenous sediment supply, reduces the carbonate influx into the basins, and reverses the climate feedback loop at the same time.

The 'erosive' part of the 'erosive–corrosive' cycle explains the observation that erosional hiatuses are centred around major type 1 sequence boundaries and canyon incision in the shelf is initiated during major falls of sea level, i.e. second-order cycle (supercycle) boundaries.

Following the first major transgression and flooding of the shelf, the sediment-supply spigot is turned off as the depocentres move landward. The sequestering of siliciclastics and carbonates on the proximal shelves depletes both the nutrient and carbonate budgets of the ocean. The locus of productivity shifts from the open ocean to the shelves where carbonate and organic matter accumulate in large quantities. The decreased carbonate in the open ocean elevates the CCD, inducing widespread carbonate dissolution on the sea floor.

The open-ocean carbonate deficiency phase, however, may last only through the period of transgression and early highstand of sea level. In the late highstand phase, once the accommodation on shelves or carbonate platforms is reduced through aggradation, the sediments begin to prograde further out over the distal part of the shelf, or shed-off the banks and platforms, over lowstand wedges of the previous cycle, into deeper basins. In late highstand the carbonate flux to the deep sea increases, and the locus of productivity moves offshore as the shoreline begins to regress. Thus, the early to late highstand is the transition period from corrosive mode to relative high rates of sediment accumulation.

The sea-level rise and decreased albedo lead to weakening of vertical and latitudinal thermal gradients during the highstand. The reduced thermal contrast and coincidence of highstands with periods of global warming lower both the potential for bottom-water generation in the polar regions and the intensity of bottom currents. The lack of sediment-laden influx along the margins and the cut-back in bottom current activity favour reduced erosion along the margins and the deep sea.

During sea-level rise and maximum flooding of the shelves the oceanic oxygen-minimum zone may rise enough to abut the shelf in some areas. The increased productivity of the shelves, combined with the greater preservational potential for organic matter in the expanded oxygen-minimum zone, makes the transgressive systems tract, the condensed section, and the early highstand systems tract as the most likely units for the occurrence of hydrocarbon source rocks.

The above scenario explains the general coincidence of dissolution-related reflectors and hiatuses in the deep sea with the periods of transgression and early highstand. For the reasons argued above, late highstands are perhaps the best of times for deep-sea sedimentation adjacent to continental margins. The record of these intervals hold the highest potential for being preserved relatively intact when both the erosive and corrosive forces are considerably weakened.

The erosive–corrosive model can conceivably operate at the third-order, as well as higher order, cycle levels, and should be testable, particularly for the last glacio-eustatic event where the sequence-stratigraphic data as well as chronostratigraphic resolution are adequate to differentiate between the various phases of the sea-level cycle.

## ACKNOWLEDGEMENTS

The sequence-stratigraphic models and sea-level curves presented in this paper were developed on passive continental margins and interior basins. The concepts are, however, equally applicable to active margins when tectonic and eustatic signatures in the sedimentary record can be discriminated: thus, the invitation to contribute this paper to the symposium on *Sea-level Change at Active Margins: Processes and Products*. The invitation by the convener, David Macdonald, is gratefully acknowledged, as is the generous donation from Amerada Hess Ltd to cover my travel costs.

The developments in sequence stratigraphy were brought about by the effort of many EPR scientists over a period of two decades. The synthesis presented here is a summary of this effort. I am grateful to many colleagues for continued discussions on the subject, and 'show-and-tell' examples of sequences in the field over several years. This paper benefited from reviews by Robert Ginsburg, John Ladd, John Milliman, Gregory Mountain, and Peter Vail, though none of them necessarily agree with all of its contents. I also acknowledge the help of Miguel Uliana and Christian Betzler in putting together the appendix of common terms of sequence stratigraphy.

## REFERENCES

Angstadt, D.M., Austin, J.A., Jr. & Buffler, R.T. (1983) Deep-sea erosional unconformity in the southeastern Gulf of Mexico. *Geology*, **11**, 215–218.

Arias, C., Masse, J.P. & Vilas, L. (1989) Secuencias deposicionales en el Aptiense-Albiense P.P. del Prebetico. El Ejemplo de las Sierras del Carche y Sopalmo (Prov. De Murcia). *XII Congr. Espanol Sediment.*, Bilbao, Simposios, pp. 33–42.

Baum, G.R. & Vail, P.R. (1988) Sequence stratigraphic concepts applied to Paleogene outcrops, Gulf and Atlantic Basins. In: *Sea-level Changes: an Integrated Approach* (Eds Wilgus, C.K., Hastings, B.S., Kendall, C.G.St.C., Posamentier, H., Ross, C.A. & van Wagoner, J.). Soc. Econ. Paleontol. Mineral., Spec. Publ. 42, 309–328.

Barron, J.A. & Keller, G. (1982) Widespread Miocene deep-sea hiatuses: coincidence with periods of global cooling. *Geology* **10**, 577–581.

Beard, J.H., Sangree, J.B. & Smith, L.A. (1982) Quaternary chronology, paleoclimate, depositional sequences, and eustatic cycles. *Bull., Am. Assoc. Petrol. Geol.* **66**, 158–169.

Berger, W. (1970) Biogenous deep-sea sediments: fractionation by deep sea circulation. *Geol. Soc. Am. Bull.* **81**, 1385–1402.

Berger, W. & Winterer, E.L. (1974) Plate stratigraphy and the fluctuating carbonate line. *Spec. Publ. int. Sediment Ass.* **1**, 11–48.

Boersma, A. & Shackleton, N.J. (1977) Tertiary oxygen and carbon isotope stratigraphy, Site 357 (mid-latitude South Atlantic). *Initial Reports of the Deep Sea Drilling Project* Vol. 39, pp. 911–924. US Government Printing Office, Washington, DC.

Boersma, A. & Shackleton, N.J. (1978) Oxygen and carbon isotopic record through the Oligocene, DSDP Site 366, Equatorial Atlantic. *Initial Reports of the Deep Sea Drilling Project*, Vol. 41, pp. 957–962. US Government Printing Office, Washington, DC.

Boersma, A., Hall, M.A. & Given, Q. (1979) Carbon and oxygen isotope variations at DSDP Site 384 (North Atlantic) and some Paleocene paleotemperature and carbon isotope variations in the Atlantic Ocean. *Initial Reports of the Deep Sea Drilling Project*, Vol. 43, pp. 695–717. US Government Printing Office, Washington, DC.

Bosellini, A. (1984) Progradation geometries of carbonate platforms: examples from the Triassic of the Dolomites, northern Italy. *Sedimentology* **31**, 1–24.

Brown, L.F. & Fisher, W.L. (1977) Seismic-stratigraphic interpretation of depositional systems. In: *Seismic Stratigraphy – Applications to Hydrocarbon Research* (Ed. Payton, C.E.). Am. Assoc. Petrol. Geol. **26**, 213–248.

Burton, R., Kendall, C.G.St.C. & Lerche, I. (1987) Out of our depth: on the impossibility of fathoming eustasy from the stratigraphic record. *Earth Sci. Rev.* **24**, 237–277.

Carlson, P.R. & Karl, H.A. (1988) Development of large submarine canyons in the Bering Sea, indicated by morphologic, seismic, and sedimentologic characteristics. *Geol. Soc. Am. Bull.* **100**, 1594–1615.

Coleman, J.M., Prior, D.B. & Lindsay, J.F. (1983) Deltaic influences on shelf edge instability processes. *Soc. Econ. Paleontol. Mineral., Spec. Publ.* **33**, 121–137.

Craig, L.E., Smith, A.G. & Armstrong, R.L. (1989) Calibration of the geologic time scale: Cenozoic and late Cretaceous glauconite and nonglauconite dates compared. *Geology* **17**, 830–832.

Dolan, J.F. (1989) Eustatic and tectonic controls on deposition of hybrid siliciclastic/carbonate basinal cycles: discussion with examples. *Bull., Am. Assoc. Petrol. Geol.* **73**(10), 1233–1246.

Douglas, R.G. & Savin, S.M. (1973) Oxygen and carbon isotope analyses of Cretaceous and Tertiary foraminifera from the Central North Pacific. *Initial Reports of the Deep Sea Drilling Project*, Vol. 17, pp. 590–605. US Government Printing Office, Washington, D.C.

Douglas, R.G. & Savin, S.M. (1975) Oxygen and carbon isotope analyses of Tertiary and Cretaceous microfossils from the Shatsky Rise and other sites in North Pacific Ocean. *Initial Reports of the Deep Sea Drilling Project*, Vol. 32, pp. 509–520. US Government Printing Office, Washington, DC.

Droxler, A.W. & Schlager, W. (1985) Glacial versus interglacial sedimentation rates and turbidite frequency in the Bahamas. *Geology* **13**, 799–802.

Embry, A.F. (1988) Triassic sea-level changes: evidence

from the Canadian Arctic Archipelago. *Soc. Econ. Paleontol. Mineral., Spec. Publ.* **42**, 249–259.

Eberli, G.P. & Ginsburg, R.N. (1987) Segmentation and coalescence of Cenozoic carbonate platforms, northwestern Great Bahama Bank. *Geology* **12**, 75–79.

Eberli, G.P. & Ginsberg, R.N. (1989) Cenozoic progradation of northwestern Great Bahama Bank, a record of lateral platform growth and sea-level fluctuations. *Soc. Econ. Paleontol. Mineral., Spec. Publ.* **44**, 339–352.

Fairbanks, R.G. & Matthews, R.K. (1978) The marine oxygen isotopic record in Pleistocene coral, Barbados, West Indies. *Quaternary Res.* **10**, 181–196.

Farre, J.A., McGregor, B.A., Ryan, W.B.F. & Robb, J.M. (1983) Breaching the shelfbreak: passage from youthful to mature phase in submarine canyon evolution. *Soc. Econ. Paleontol. Mineral., Spec. Publ.* **33**, 25–39.

Fischer, A.G. & Arthur, M.A. (1977) Secular variations in the pelagic realm. *Soc. Econ. Paleontol. Mineral., Spec. Publ.* **25**, 19–50.

Galloway, W.E. (1989) Genetic stratigraphic sequences in basin analysis. I: Architecture and genesis of flooding-surface bounded depositional units. *Bull., Am. Assoc. Petrol. Geol.* **73**(2), 125–142.

Garcia-Mondejar, J. & Fernandez-Mendiola, P.A. (1989) Secuencias y asociaciones de sistemas (Systems Tracts) en la serie carbonatada Albiense entre La Gandara y Las Machorras (Cantabria y Burgos). *XII Congr. Espanol Sediment.*, Bilbao, Simposios, pp. 55–64.

Gardner, W.D. (1989) Baltimore Canyon as a modern conduit of sediments to the deep sea. *Deep-Sea Res.* **36**(3), 323–358.

Greenlee, S.M. & Moore, T.C. (1988) Recognition and interpretation of depositional sequences and calculation of sea-level changes from stratigraphic data — offshore New Jersey and Alabama Tertiary. In: *Sea-level Changes: an Integrated Approach* (Eds Wilgus, C.K., Hastings, B.S., Kendall, C.G.St.C., Posamentier, H., Ross, C.A. & van Wagoner, J.) Soc. Econ. Paleontol. Mineral., Spec. Publ. 42, 329–355.

Grigg, R.W. & Epp, D. (1989) Critical depth for the survival of coral islands: effects on the Hawaiian Archipelago. *Science* **243**, 638–641.

Hallam, A. (1984) Pre-Quaternary sea-level changes. *Ann. Rev. Earth Planet. Sci.* **12**, 205–243.

Haq, B.U. (1973) Transgressions, climatic change and the diversity of calcareous nannoplankton. *Mar. Geol.* **15**, M25–M30.

Haq, B.U. (1980) Biogeographic history of Miocene calcareous nannoplankton and paleoceanography of the Atlantic Ocean. *Micropaleontology* **26**, 414–443.

Haq, B.U., Premoli-Silva, I. & Lohmann, G.P. (1977) Calcareous plankton paleobiogeographic evidence for major climatic fluctuations in the Early Cenozoic Atlantic Ocean. *J. Geophys. Res.* **82**, 3861–3876.

Haq, B.U., Hardenbol, J. & Vail, P.R. (1987) Chronology of fluctuating sea levels since the Triassic. *Science* **235**, 1156–1167.

Haq, B.U., Hardenbol, J. & Vail, P.R. (1988a) Mesozoic and Cenozoic chronostratigraphy and eustatic cycles. In: *Sea-level Changes: an Integrated Approach* (Eds Wilgus, C.K., Hastings, B.S., Kendall, C.G.St.C., Posamentier, H., Ross, C.A. & van Wagoner, J.) Soc. Econ. Paleontol. Mineral., Spec. Publ. 42, 71–108.

Haq, B.U., Hardenbol, J., Vail, P.R. & van Wagoner, J.C. (1988b) Sea level history — technical comments and response. *Science* **241**, 596–602.

Hardenbol, J., Vail, P.R. & Ferrer, J. (1981) Interpreting paleoenvironments, subsidence history and sea-level changes of passive margins from seismic and biostratigraphy. *Oceanologica Acta Suppl.* **3**, 33–44.

Harrison, C.G.A. (1986) Long-term eustasy and epeirogeny in continents. *Studies in Geophysics*, pp. 111–131. US National Academy of Sciences, Washington, DC.

Hay, W.W., Shaw, C.A. & Wold, C.N. (1989) Mass-balanced paleogeographic reconstructions. *Geol. Rundsch.* **78**, 207–242.

Jervey, M.T. (1988) Quantitative geological modeling of siliciclastic rock sequences and their seismic expression. In: *Sea-level Changes: an Integrated Approach* (Eds Wilgus, C.K., Hastings, B.S., Kendall, C.G.St.C., Posamentier, H., Ross, C.A. & van Wagoner, J.) Soc. Econ. Paleontol. Mineral., Spec. Publ. 42, 47–70.

Keller, G. & Barron, J.A. (1983) Paleoceanographic implications of Miocene hiatuses. *Geol. Soc. Am. Bull.* **94**, 590–613.

Keller, G. & Baron, J.A. (1987) Palaeodepth distribution of Neogene deep sea hiatuses. *Paleoceanography* **2**, 697–713.

Keller, G., Herbert, T., Dorsey, R., D'Hondt, S., Johnsson, M. & Chi, W.R. (1987) Global distribution of Paleogene hiatuses. *Geology* **15**, 199–203.

Kendall, C.G.St.C. & Lerche, I. (1988) The rise and fall of eustasy. In: *Sea-level Changes: an Integrated Approach* (Eds Wilgus, C.K., Hastings, B.S., Kendall, C.G.St.C., Posamentier, H., Ross, C.A. & van Wagoner, J.) Soc. Econ. Paleontol. Mineral., Spec. Publ. 42, 3–17.

Kendall, C.G.St.C. & Schlager, W. (1981) Carbonate and relative changes in sea level. *Mar. Geol.* **44**, 181–212.

Keppens, E. & Pasteels, P. (1982) A comparison of rubidium–strontium and potassium–argon apparent ages on glauconites. In: *Numerical Dating in Stratigraphy* (Ed. Odin, G.S.) pp. 225–239. Wiley, New York.

Loutit, T.S., Hardenbol, J., Vail, P.R. & Baum, G.R. (1988) Condensed sections: the key to age dating and correlation of continental margin sequences. In: *Sea-level Changes: an Integrated Approach* (Eds Wilgus, C.K., Hastings, B.S., Kendall, C.G.St.C., Posamentier, H., Ross, C.A. & van Wagoner, J.) Soc. Econ. Paleontol. Mineral., Spec. Publ. 42, 183–215.

Matthews, R.K. & Poore, R.Z. (1980) Tertiary $\delta^{18}O$ record and glacio-eustatic sea-level fluctuations. *Geology* **8**, 501–504.

Mayer, L.A., Shipley, T.H. & Winterer, E.L. (1986) Equatorial Pacific seismic reflectors as indicators of global oceanographic events. *Science* **233**, 761–764.

Miall, A.D. (1986) Eustatic sea level changes interpreted from seismic stratigraphy: a critique of the methodology with particular reference to the North Sea Jurassic record. *Bull., Am. Assoc. Petrol. Geol.* **70**(2), 131–137.

Miller, K.G., Fairbanks, R.G. & Mountain, G.S. (1987) Tertiary oxygen isotope synthesis, sea level history, and continental margin erosion. *Paleoceanography* **2**, 1–19.

MILLER, K.G., MOUNTAIN, G.S. & TUCHOLKE, B.E. (1985) Oligocene glacio-eustasy and erosion on the margins of the North Atlantic. *Geology* **13**, 10–13.

MILLIMAN, J.D. (1974) *Marine Carbonates*. Springer-Verlag, Heidelberg, pp. 242–249.

MITCHUM, R.M., JR. (1977) Glossary of terms used in seismic stratigraphy. In: *Seismic Stratigraphy – Applications to Hydrocarbon Research* (Ed. Payton, C.E.) Mem., Am. Assoc. Petrol. Geol. 26, 205–212.

MITCHUM, R.M., JR., VAIL, P.R. & THOMPSON, S. (1977) The depositional sequence as a basic unit for stratigraphic analysis. In: *Seismic Stratigraphy – Applications to Hydrocarbon Research* (Ed. Payton, C.E.) Mem., Am. Assoc. Petrol. Geol. 26, 53–62.

MÖRNER, N.A. (1981) Revolution in Cretaceous sea level analysis. *Geology* **9**, 344–346.

MOUNTAIN, G.S. (1987) Cenozoic margin construction and destruction offshore New Jersey. *Cushman Found. Foraminiferal Res., Spec. Publ.* **24**, 57–83.

MOUNTAIN, G.S. & TUCHOLKE, B.E. (1985) Mesozoic and Cenozoic geology of the U.S. Atlantic continental slope and rise. In: *Geologic Evolution of the United States Atlantic Margin* (Ed. Poag, W.C.) pp. 293–341. Van Nostrand Reinhold, New York.

MUTTI, E. (1985) Turbidite systems and their relation to depositional sequences. In: *Provenance of Arenites* (Ed. Zuffa, G.G.) pp. 65–93. NATO–ASI series, Riedel, Dordrecht.

MUTTI, E. & NORMARK, W.R. (1987) Comparing examples of modern and ancient turbidites systems: problems and concepts. In: *Deep-water Clastic Deposits and Case Histories*. (Eds Leggett, J.K. & Zuffa, G.G.) pp. 1–35. Graham and Trottman, London.

ODIN, G.S. (Ed.) (1982) *Numerical Dating in Stratigraphy*. Wiley, New York, 1040 pp.

PARKINSON, N. & SUMMERHAYES, C.P. (1985) Synchronous global sequence boundaries. *Bull., Am. Assoc. Petrol. Geol.* **69**, 685–687.

PITMAN, W.C., III (1978) Relationship between eustasy and stratigraphic sequences of passive margins. *Geol. Soc. Am. Bull.* **89**, 1389–1403.

POPENOE, P. (1985) Cenozoic depositional and structural history of North Carolina margin from seismic stratigraphic analysis. In: *Geological Evolution of the United States Atlantic Margin* (Ed. Poag, C.W.) pp. 125–187. Van Nostrand Reinhold, New York.

POSAMENTIER, H.W. & VAIL, P.R. (1988) Eustatic controls on clastic deposition. II – sequence and systems tract models. In: *Sea-level Changes: an Integrated Approach* (Eds Wilgus, C.K., Hastings, B.S., Kendall, C.G.St.C., Posamentier, H., Ross, C.A. & van Wagoner, J.) Soc. Econ. Paleontol. Mineral., Spec. Publ. 42, 125–154.

POSAMENTIER, H.W., JERVEY, M.T. & VAIL, P.R. (1988) Eustatic controls on clastic deposition. I – conceptual framework. In: *Sea-level Changes: an Integrated Approach* (Eds Wilgus, C.K., Hastings, B.S., Kendall, C.G.St.C., Posamentier, H., Ross, C.A. & van Wagoner, J.) Soc. Econ. Paleontol. Mineral., Spec. Publ. 42, 109–124.

PRENTICE, M.L. & MATTHEWS, R.K. (1988) Cenozoic ice-volume history: development of a composite oxygen isotope record. *Geology* **16**, 963–966.

PUJALTE, V., ROBLES, S., ORUE-ETXEBARRIA, X., ZAPATA, M. & GARCIA-PORTERO, J. (1989) Influencia del eustatismo y tectonica en la genesis de secuencias y macrosecuencias deposicionales del Maastrichtiense Superior-Eoceneo Inferior de la cuenca Vasca. *XII Congr. Espanol Sediment.*, Bilbao, Simposios, pp. 147–156.

RIGGS, S.R., SNYDER, S.W., SCOTT, W. & HINE, A.C. (1990) Stratigraphic framework for cyclical deposition of Miocene sediments in the Carolina phosphogenic province. In: *Phosphate Deposits of the World*, vol. 3 (Eds Burnett, W.C. & Riggs, S.R.) pp. 381–395. Cambridge University Press, Cambridge.

SALAS, R. (1989) El modelo de plataforma carbonatada Aptiense Superior del Tipo Mola de Xert en la cuenca del Maestrat (Cordillera Iberica Oriental). *XII Congr. Espanol Sediment.*, Bilbao, Simposios, pp. 75–84.

SANFORD, M.W., KUEHL, S.A. & NITTROUER, C.A. (1990) Modern sedimentary processes in the Wilmington Canyon area, U.S. East Coast. *Mar. Geol.* **92**, 205–226.

SARG, J.F. (1988) Carbonate sequence stratigraphy. In: *Sea-level Changes: an Integrated Approach* (Eds Wilgus, C.K., Hastings, B.S., Kendall, C.G.St.C., Posamentier, H., Ross, C.A. & van Wagoner, J.) Soc. Econ. Paleontol. Mineral., Spec. Publ. 42, 155–182.

SARG, J.F. & LEHMANN, P.J. (1986) Lower-middle Guadalupian facies and stratigraphy of San Andreas/ Grayburg formations, Permian Basin, Guadalupe Mountains, New Mexico and Texas. In: *Field Guide Book, Permian Basin Section*. Society of Economic Paleontologists and Mineralogists, Tulsa, Publ. 86 (26).

SCHLAGER, W. (1981) The paradox of drowned reefs and carbonate platforms. *Geol. Soc. Am. Bull.* **92**, 197–211.

SHACKLETON, N.J. & KENNETT, J.P. (1975) Paleotemperature history of the Cenozoic and the initiation of Antarctic glaciation: oxygen and carbon isotopic analyses in DSDP Sites 277, 279, and 281. *Initial Reports of the Deep Sea Drilling Project*, Vol. 29, pp. 743–755. US Government Printing Office, Washington, DC.

SHACKLETON, N.J. & OPDYKE, N.D. (1973) Oxygen isotopic and paleomagnetic stratigraphy of Equatorial Pacific core V28–238: oxygen isotope temperature and ice volume on a $10^5$ year scale. *Quaternary Res.* **3**, 39–55.

SLOSS, L.L. (1950) Paleozoic stratigraphy in the Montana area. *Bull., Am. Assoc. Petrol. Geol.* **34**, 423–451.

SLOSS, L.L. (1963) Sequences in cratonic interior of North America. *Geol. Soc. Am. Bull.* **74**, 93–114.

SLOSS, L.L. (1988) Forty years of sequence stratigraphy. *Geol. Soc. Am. Bull.* **100**, 1161–1665.

SNYDER, S.W., HINE, A.C. & RIGGS, S.R. (1990) The seismic stratigraphic record of shifting Gulf Stream flow paths in response to Miocene glacioeustasy: implications for phosphogenesis along the North Carolina continental margin. In: *Phosphate Deposits of the World*, vol. 3, (Eds Burnett, W.C. & Riggs, S.R.) pp. 396–423. Cambridge University Press, Cambridge.

SUESS, E. (1906) *The Face of the Earth*, vol. 2. Clarendon Press, Oxford, 556 pp.

TUCHOLKE, B.E. & MOUNTAIN, G.S. (1979) Seismic stratigraphy, lithostratigraphy and paleosedimentation patterns in the western North Atlantic. *Am. Geophys. Union, Maurice Ewing Ser.* **3**, 58–86.

TUCHOLKE, B.E. & LAINE, E.P. (1982) Neogene and Quaternary development of lower continental rise off the central U.S. East Coast. In: *Studies in Continental Margin Geology* (Eds Watkins, J.S. & Drake, C.L). Mem., Am. Assoc. Petrol. Geol. 34, 295–305.

Vail, P.R., Mitchum, R.M, Jr., Todd, R.G., Widmier, J.M., Thompson, S., Sangree, J.B., Bubb, J.N. & Hatlelid, W.G. (1977) Seismic stratigraphy and global changes of sea level, parts 1–6. In: *Seismic Stratigraphy – Applications to Hydrocarbon Research* (Ed. Payton, C.E.) Mem., Am. Assoc. Petrol. Geol. 26, 49–133.

Vail, P.R., Mitchum, R.M., Shipley, T.H. & Buffler, R.T. (1980) Unconformities of the North Atlantic. *Philos. Trans. R. Soc. London, Ser. A***294**, 137–155.

Vail, P.R., Hardenbol, J. & Todd, R.G. (1984) Jurassic unconformities, chronostratigraphy and sea level changes from seismic stratigraphy and biostratigraphy. In: *Carbonate Depositional Environments* (Eds Scholle, P.A., Bebout, D.G. & Moore, C.H.) Mem., Am. Assoc. Petrol. Geol. **33**, 129–144.

Vail, P.R., Audemard, F., Bowman, S.A., Eisner, P.N. & Perez-Cruz, G. (1991) The stratigraphic signatures of tectonics, eustasy and sedimentation. In: *Cycles and Events in Stratigraphy* (Eds Seilacher, D. & Eisner, G.). Springer-Verlag (in press).

van den Hurk, A.M., Betzler, C., Eichenseer, H. & Luterbacher, H.-P. (1989) Origen y evolucion de las plataformas y cuencas carbonatadas del Paleogeno Surpirenaico (factores de control). *XII Congr. Espanol Sediment.*, Bilbao, Simposios, pp. 97–105.

van Wagoner, J.C. (1985) Reservoir facies distribution as controlled by sea-level change. *Abstract and Poster Session, Midyear Meeting*, Golden, CO, pp. 91–92. Society of Economic Paleontologists and Mineralogists, Tulsa.

van Wagoner, J.C., Posamentier, H.W., Mitchum, R.M., Jr., Vail, P.R., Sarg, J.F., Loutit, T.S. & Hardenbol, J. (1988) An overview of the fundamentals of sequence stratigraphy and key definitions. In: *Sea-level Changes: an Integrated Approach* (Eds Wilgus, C.K., Hastings, B.S., Kendall, C.G.St.C., Posamentier, H., Ross, C.A. & van Wagoner, J.) Soc. Econ. Paleontol. Mineral., Spec. Publ. 42, 39–46.

Watts, A.B. & Steckler, M.S. (1979) Subsidence and eustasy at the continental margins of eastern North America. *Am. Geophys. Union, Maurice Ewing Ser.* **3**, 218–239.

Weimer, R.J. (1984) Relation of unconformities, tectonics and sea-level changes. Cretaceous of Western Interior, U.S.A. In: *Seismic Stratigraphy – Applications to Hydrocarbon Exploration* (Ed. Payton, C.E.) Mem., Am. Assoc. Petrol. Geol. 26, 7–35.

Wilber, R.J., Milliman, J.D. & Halley, R.B. (1990) Accumulation of Holocene banktop sediment on the western slope of Great Bahama Bank: modern progradation of a carbonate megabank. *Geology* **18**, 970–974.

Williams, D.F. (1988) Evidence for and against sea-level changes from stable isotopic record of the Cenozoic. In: *Sea-level Changes: an Integrated Approach* (Eds Wilgus, C.K., Hastings, B.S., Kendall, C.G.St.C., Posamentier, H., Ross, C.A. & van Wagoner, J.) Soc. Econ. Paleontol. Mineral., Spec. Publ. 42, 31–38.

Wilson, J.L. (1975) *Carbonate Facies in Geological History*. Springer-Verlag, New York, 471 pp.

Woodruff, F., Savin, S.M. & Douglas, R.G. (1981) Miocene stable isotopic record: a detailed deep Pacific Ocean study and its paleoclimatic implications. *Science* **212**, 665–668.

Zullo, V.A. & Harris, W.B. (1987) Sequence stratigraphy, biostratigraphy, and correlation of Eocene through Miocene strata of North Carolina. *Cushman Found. Foraminiferal Res., Spec. Publ.* **24**, 197–214.

**Appendix 1.** Common sequence-stratigraphy terms

| English | French | German | Spanish |
|---|---|---|---|
| Accommodation space | Espace disponible | Verfügbarer Platz | Espacio para acomodación |
| Basin-floor fan | Cone sous-marin | Beckenboden Fächer | Abanico de fondo de cuenca |
| Coastal onlap | Biseau d'aggradation cotière | Küstenaggradationsüberlappung | Bisel de agradacion costera (traslapo costero) |
| Condensed section | Intervalle condensé | Kondensierter Abschnitt | Intervalo condensado |
| Conformity | Concordance | Konkordanz | Concordancia (conformidad) |
| Depositional sequence | Séquence de depot | Ablagerungssequenz | Secuencia deposicional |
| Downlap | Biseau de progradation | Progradationsüberlappung | Bisel de progradación |
| Downlap surface | Surface basale de progradation | Basale Progradationsüberlappungsfläche | Superficie basal de progradación |
| Downward shift | Déplacement vers le bassin | Beckenwärtige Verlagerung | Migracion centripeta |
| Eustatic sea level | Niveau eustatique | Eustatischer Meeresspiegel | Nivel marino eustático |
| Flooding surface | Surface d'inondation | Flutungsfläche | Superficie de inundación |
| Highstand systems tract | Cortège de haut niveau | Hochstand System Bündel | Cortejo sedimentario de nivel alto |
| Lowstand wedge | Prisme de bas niveau | Tiefstand Keil | Cuña de nivel bajo |
| Lowstand wedge systems tract | Cortège de bas niveau | Tiefstand Keil System Bündel | Cortejo de cuña de nivel bajo |
| Onlap | Biseau d'aggradation | Aggradationsüberlappung | Bisel de agradación (traslapo) |
| Parasequence | Paraséquence | Parasequenz | Parasecuencia |
| Sequence-stratigraphy | Stratigraphie séquentielle | Sequenz Stratigraphie | Estratigrafia secuencial |
| Shelf margin wedge | Prisme de bordure de plate-forme | Schelfrandenkeil | Cuña de borde de plataforma |
| Shelf margin wedge systems tract | Cortège de bordure de plate-forme | Schelfrandenkeil System Bündel | Cortejo sedimentario de borde de plataforma |
| Systems tract | Cortège sédimentaire | System Bündel | Cortejo sedimentario |
| Transgressive surface | Surface de transgression | Transgressive Fläche | Superficio de transgresión |
| Transgressive systems tract | Cortège transgressif | Transgressives System Bündel | Cortejo sedimentario de transgresión |
| Unconformity | Discordance | Diskordanz | Discordancia |

*Spec. Publs. int. Ass. Sediment.* (1991) **12**, 41–65

# Application of global sea-level and sequence-stratigraphic models in Southern Hemisphere Neogene strata from New Zealand

R.M. CARTER, S.T. ABBOTT, C.S. FULTHORPE, D.W. HAYWICK
*and* R.A. HENDERSON

*Department of Geology, James Cook University of North Queensland, Townsville, Queensland 4811, Australia*

## ABSTRACT

Current research on sedimentary facies development related to eustasy focuses on two models. The *global sea-level model* (GSM) comprises a best-estimate curve of Phanerozoic eustasy. Up to seven orders of change can be distinguished on the GSM, encompassing magnitudes of variation in sea level up to several hundred metres on wavelengths between tens of thousands (seventh-order cycles) and tens of millions of years (first-order cycles). The *sequence-stratigraphic model* (SSM) summarizes the stratigraphic architecture of sediments deposited during a single major third-order sea-level cycle, recognizing three major facies assemblages grouped into lowstand, transgressive, and highstand systems tracts. Passive margins adjacent to plate boundaries form ideal sites at which to test these sea-level models. Gentle uplift there exposes inner basin margins and may preserve the shoreline of the highstand systems tract on coastal terraces, as well as excellent sections through the transgressive and highstand systems tracts in coastal cliffs or ranges, and lowstand systems tracts near the shelf-edge offshore.

Analysis of seismic records of Neogene platform–clinoform shelf deposits from the Canterbury Basin indicates a lack of correspondence between sequence development in the southwest Pacific and the GSM. Maximum marine transgression occurred during the mid-Oligocene, when the GSM indicates the largest lowstand of the Phanerozoic. Nine seismic sequences occur between 9 and 13 Ma, a time when the GSM indicates only 2.5 third-order global sea-level cycles.

In contrast, tests of the SSM using data from the modern New Zealand shelf and from the Wanganui and Hawke's Bay Basins indicate a close match between the SSM and cyclothemic sediments deposited during the 40 000 and 100 000 year long, c. 80–130 m magnitude, fifth and sixth-order cycles of the Plio-Pleistocene. The cyclothems studied were deposited near the basin margins, and comprise an erosion surface (sequence boundary) overlain by shallow marine gravel–sand–mud with transported fossils (transgressive systems tract), an offshore facies shell-bed with *in situ* fossils (maximum flooding surface), and sometimes an upwards coarsening (shallowing) siltstone–sandstone section (highstand systems tract). In Hawke's Bay, the lower cyclothems contain fluvial facies of the lowstand systems tract preserved just above the sequence boundary. There is also a general correspondence between the SSM and the sedimentary architecture of the 80 Ma long, local relative sea-level cycle that controlled the development of the Canterbury Basin passive margin.

We conclude that, as tested in the Southern Hemisphere: (1) the mid-Oligocene and late Miocene parts of the GSM are at best imprecise, and at worst in error; (2) the SSM is a powerful and accurate descriptive model of sedimentary facies development that occurs under sea-level cycles as different in scale as 40 000 years (sixth order) and 80 Ma (first order); and (3) more precise tests of both the GSM and SSM require close transects of continuously cored Ocean Drilling Project sites across well-developed seismic sequence systems.

## INTRODUCTION

A world-wide phase of reinterpretation of sedimentary basins commenced with the publication of the American Association of Petroleum Geologists Memoir 26 on seismic stratigraphy (Vail *et al.*, 1977). Twelve years later, this reinterpretation, based upon a postulated global sea-level curve, continues apace, with many authors reporting apparent matches between the global curve and their data (Poag & Ward, 1987; Greenlee & Moore, 1988; Rahmanian *et al.*, 1990). However, other researchers have questioned both the theoretical basis for deriving a global sea-level curve (e.g. Parkinson & Summerhayes, 1985; Miall, 1986; Burton *et al.*, 1987; Galloway, 1989a; Christie-Blick *et al.*, 1990), or the accuracy of the published Exxon curve as applied to particular regions (Hubbard *et al.*, 1985a,b; Carter, 1988a; Hubbard, 1988).

Lack of publication of the seismic and well data used to derive the global curve is a source of particular concern, and has made it impossible for non-Exxon researchers to check the derivation of the sea-level curve. It is therefore particularly important that the Exxon curve be assessed against public domain data on a global scale. The ready availability of high quality open-file offshore seismic data, excellent onland outcrop of Cenozoic strata, a detailed biostratigraphic framework, and a location distant from the continental margins that provided the base-data for the Exxon sea-level curve, make the New Zealand continental margin an appropriate location for such tests in the Southern Hemisphere. We therefore summarize in this paper the results of New Zealand tests of both the 'global' sea-level curve and of the theory of sequence analysis with which it is associated.

## MODELS OF GLOBAL SEA-LEVEL CHANGE AND SEQUENCE ARCHITECTURE

Two different conceptual models underlie the application of sequence stratigraphy by Vail and his co-workers: one model relates to presumed sea-level behaviour through time; the other model relates to the stratigraphic record produced during a single sea-level cycle. Though the two models are inter-related they are logically distinct, and we believe that is important to test them separately.

### The global sea-level model

The first model (Fig. 1) corresponds to the global sea-level model (GSM; Table 1) initially published by Vail *et al.* (1977), and later modified by Haq *et al.* (1987). This 'global' curve was derived primarily from the analysis of seismic profiles in terms of the unconformity bounded units termed sequences. In essence, the GSM curve comprises a prediction (or retrodiction) of the behaviour of eustatic sea level through time derived from a globally averaged coastal onlap chart.

**Table 1.** Commonly used abbreviations for sea-level and systems tract terminology

| | |
|---|---|
| *Conceptual models* | |
| GSM | Global sea-level model |
| SSM | Sequence-stratigraphic model |
| *Systems tracts* | |
| LST | Lowstand systems tract |
| TST | Transgressive systems tract |
| HST | Highstand systems tract |
| MFS | Maximum flooding surface |
| *Clinoform features* | |
| SB | Shelf break |
| DSB | Depositional shoreline break |
| CBP | Clinoform break-point |

The GSM displays systematic patterns of sea-level variation at different wavelengths (= timespans). Thus Vail *et al.* (1977) recognized first, second, and third-order cycles with timespans ranging between many tens of millions of years for first-order cycles to a few million years for third-order cycles (Fig. 1; Table 2).

Fifth and lower order sea-level cycles have not so far been depicted as part of the Exxon global curve. They are, however, well documented in Plio-Pleistocene oceanic piston-cores and Ocean Drilling Project (ODP) hydraulic piston cores, especially those analysed in terms of the standard oxygen isotope stages (e.g. Shackleton & Opdyke, 1973, 1976; Nelson *et al.*, 1985; Williams *et al.*, 1988). Compelling evidence exists that the late Neogene glacio-eustatic fifth-order sea-level cycles typically exceeded 100 m in amplitude (e.g. Cronin, 1983; table 3 in Carter & Johnson, 1986).

Fourth-order cycles were also absent from the original GSM (Vail *et al.*, 1977). However, fourth-order cycles were later depicted over the early Eocene part of the Haq *et al.* (1987) version, and have recently been claimed by van Wagoner & Mitchum (1989) to be widespread through the stratigraphic column. Other studies have identified a stratigraphic signature appropriate to fourth-order cycles in Eocene and mid-Miocene onland outcrops (Kidwell, 1984; Plint, 1988) and in mid-late Miocene offshore seismic profiles (Fulthorpe & Carter, 1989). In contrast to fifth and third-order cycles, which

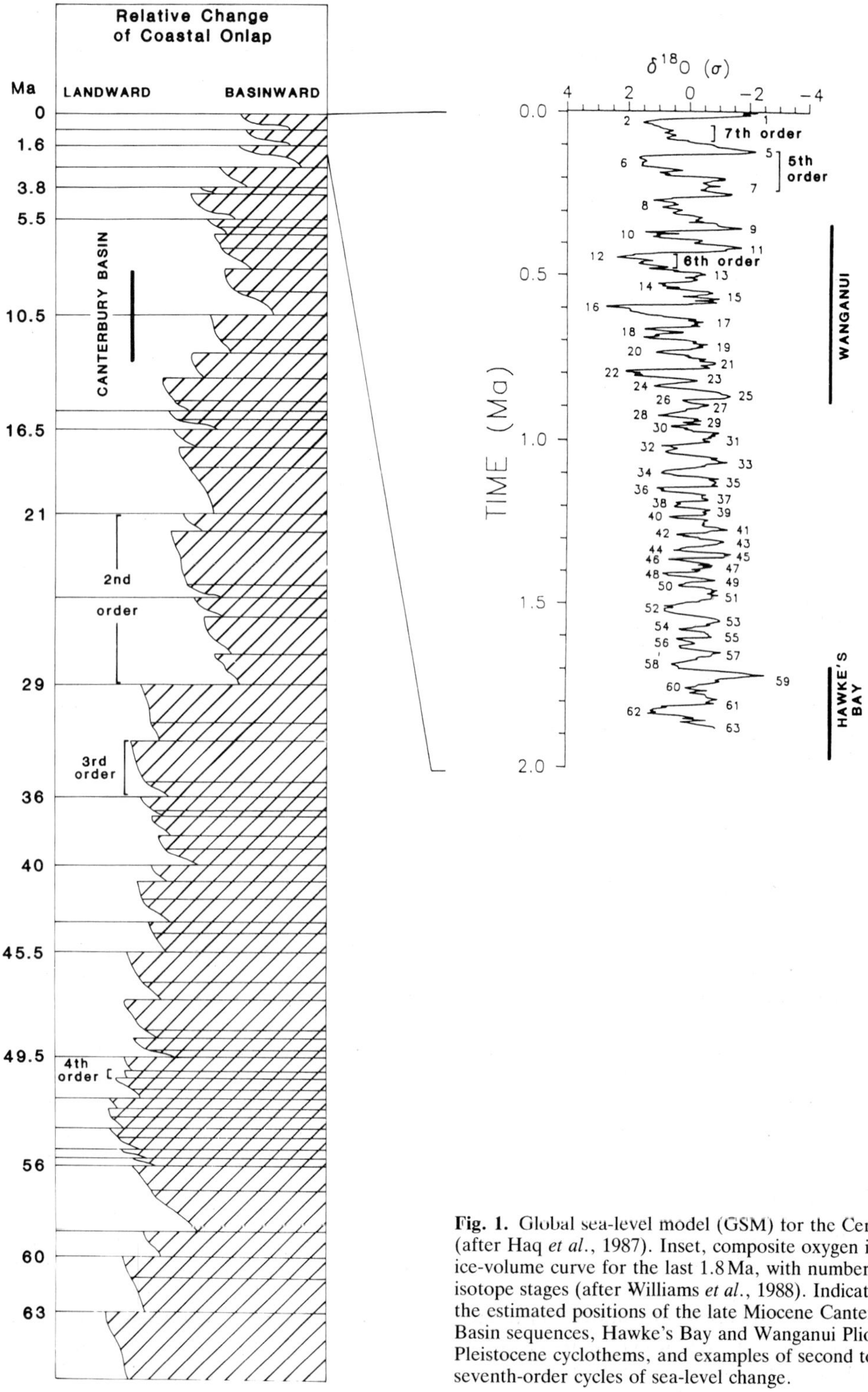


**Fig. 1.** Global sea-level model (GSM) for the Cenozoic (after Haq *et al.*, 1987). Inset, composite oxygen isotope ice-volume curve for the last 1.8 Ma, with numbered isotope stages (after Williams *et al.*, 1988). Indicated also, the estimated positions of the late Miocene Canterbury Basin sequences, Hawke's Bay and Wanganui Plio-Pleistocene cyclothems, and examples of second to seventh-order cycles of sea-level change.

**Table 2.** The fundamental orders of cycles of sea-level fluctuation

| Order | Timespan | Dominant cause | Described by |
|---|---|---|---|
| 1st | 100 Ma | Ocean-basin volume changes | Pitman, 1978; Kominz, 1984 |
| 2nd | 5–100 Ma | Thermo-tectonic subsidence | Vail *et al.*, 1977; Haq *et al.*, 1987 |
| 3rd | 1–5 Ma | Eustasy of unknown type | Vail *et al.*, 1977; Haq *et al.*, 1987 |
| 4th | 0.3–0.6 Ma | Eustasy of unknown type | Kidwell, 1984; Plint, 1988; Fulthorpe & Carter, 1989; van Wagoner & Mitchum, 1989 |
| 5th | c. 100 ka | Glacio-eustasy (eccentricity) | Emiliani, 1955; Shackleton & Opdyke, 1973,1976 |
| 6th | c. 40 ka | Glacio-eustasy (tilt) | Williams, 1988 |
| 7th | c. 20 ka | Glacio-eustasy (precession) | Hays *et al.*, 1976 |
| Infra-7th | < 20 ka | Eustasy–sediment supply | Boyd *et al.*, 1988; Anderson & Thomas, 1989 |

have respective known and claimed amplitudes of > 100 m each, fourth-order cycles seem to be of much smaller amplitude — at most a few tens of metres, and perhaps as little as a few metres. It is therefore apparent that the effects of fourth-order cycles will generally only be conspicuous in outcrop in shallow water, basin-margin facies, such as those described by Kidwell (1984) and Plint (1988).

These and other studies indicate that sea-level cyclicity occurs at all scales down to timespans as short as a few thousand years. It is perhaps significant that the higher frequency cycles correspond at least in part to Milankovitch frequencies, including those for obliquity (fifth order), tilt (sixth order), and precessesion (seventh order) (e.g. Hays *et al.*, 1976; Ruddiman *et al.*, 1989).

**The sequence-stratigraphic model**

The second model (see Fig. 1 in paper by B.U. Haq) summarizes the idealized stratigraphic architecture of sediment deposited during a single sea-level cycle, i.e. comprises a sequence-stratigraphic model (SSM). An alternative model proposed by Galloway (1989a; after Frazier, 1974) has similar architecture to the Exxon SSM, but locates the sequence boundaries along the maximum flooding surface corresponding to successive sea-level highs, rather than along subaerial unconformities corresponding to successive sea-level lows. The difference is conceptual rather than practical, and we have accordingly preferred to apply the more widely used Exxon SSM herein.

Though developed from an analysis of Exxon third-order cycles, it has become apparent that the Exxon SSM is applicable to sediments deposited during any sea-level cycle between the first and infra-seventh orders inclusive.

**What is the fundamental sequence level?**

The development of the theory of sequence analysis has proceeded from higher to lower level orders of sea-level change. Thus the first Exxon sea-level chart (Vail *et al.*, 1977) dealt with third-order cycles, the first major revision (Haq *et al.*, 1987) incorporated some fourth-order cycles, and the most recent statement (van Wagoner & Mitchum, 1989) indicates that fourth-order cycles are the basic building block for the GSM. Furthermore, we show later in this paper that Plio-Pleistocene fifth-order cycles correspond closely to the Exxon SSM, and there is abundant evidence from studies of post-glacial sediments that the same facies patterns are characteristic of sea-level cycles as short as a few thousand years, i.e. infra-seventh order (e.g. Boyd *et al.*, 1988; Anderson & Thomas, 1989).

At each stage in the development of the SSM, the term parasequence has been used to refer to minor sedimentary rhythms within a sequence. For example, van Wagoner *et al.* (1988) defined a parasequence as 'a relatively conformable succession of genetically related beds or bedsets bounded by marine-flooding surfaces and their correlative surfaces', and commented that 'parasequences and parasequence sets are the fundamental building blocks of sequences'. Thus initially parasequences were subdivisions of third-order cycles (Vail *et al.*, 1977), whereas in the latest statement they are subdivisions of fourth-order cycles (van Wagoner & Mitchum, 1989).

This history suggests:

**1** that both sequences and parasequences are expedient concepts rather than rigorously defined stratigraphic entities, since their scale can apparently be changed at will;

**2** that sequence analysis has developed from coarser

to finer time periods, raising the question 'What is the fundamental level at which sequence analysis should be applied?'

It is not self-evident that the SSM at successive orders can be adequately represented as the sum of a set of topologically similar SSMs that comprise parasequences of the next higher order. For such a pattern to hold would require that each increasingly lower order sequence be controlled by an ever larger sea-level cycle. Since it is likely that in going from fifth to fourth-order cycles there is actually a *decrease* in the magnitude of the inferred controlling sea-level cycle, it is apparent that the general model is inadequate. There is also the additional difficulty that in areas of high sediment supply the thickness of sixth or fifth-order sequences may greatly exceed 'typical' third-order sequence thickness. For instance, combined transgressive systems tract (TST) and highstand systems tract (HST) delta complexes on the post-glacial Mississippi shelf are 10–50 m thick and extend 200 km downdip and more than 400 km along strike, and Holocene sediment beneath the modern delta alone is more than 175 m thick (Boyd *et al.*, 1988; Coleman & Roberts, 1988).

We conclude that it is unlikely that sequences at all orders correspond to a 'Russian doll' stacked set, whereby the SSM applies at any level and the sequence at that level is viewed as forming from a large number of finer sequences (parasequences) of the next higher order. None the less, some orders of sequence do indeed embrace lower orders, e.g. the major first-order thermo-tectonic cycle that incorporates the complete sedimentary history of the Canterbury Basin (described below), which includes examples of second, third, fourth, and probably fifth-order sequences.

One possible way of dealing with the scale problem would be to accept that the fundamental 'building block' sequence for derivation of the GSM is of a different order during different geological epochs. Thus the building block might be viewed as fifth order in the late Pleistocene, fourth order in the Miocene and Eocene, and third order in the Cretaceous, consequent upon the rates and magnitudes of eustatic sea-level variation at those times. Alternatively, these differences may be apparent only, and result from decreasing stratigraphic resolution with greater age.

The Exxon SSM may in some such fashion be able to accommodate the problems of scale. None the less, there remains the potentially irreconcilable problem of establishing a truly global sea-level curve (GSM), because of the difficulty of extracting an independent eustatic signature from local sea-level curves, which are also controlled in part by sediment supply and subsidence (cf. Watts, 1982; Galloway, 1989a).

## TESTING THE GLOBAL SEA-LEVEL MODEL

Tests of the Exxon GSM may take one of two forms:

**1** compilation of independent stratigraphic data from a particular region, such as the vertical occurrence of unconformities in an offshore well, and comparison of the pattern with that of the model curve

**2** derivation of an independent coastal onlap curve for a particular region, using sequence analysis on seismic sections, and its comparison with the model curve.

We have used the second technique in an analysis of high-quality, open-file seismic data from the Canterbury Basin, eastern South Island.

### Canterbury Basin

The Canterbury Basin represents a passive margin sequence that accumulated on the eastern margin of the New Zealand Plateau after spreading commenced on the mid-Pacific Rise in the Late Cretaceous (Figs 2, 3). The basin developed by thermal subsidence of the shoulder of the adjacent Bounty Trough–Great South Basin, which together comprise a major re-entrant rift basin in the eastern New Zealand margin (Sporli, 1980; Carter, 1988b).

#### *First–second-order sedimentary cycles*

The sediments preserved in the Canterbury Basin fall into four megasequences, recording phases of Cretaceous rift-valley fill, Cretaceous–early Oligocene marine transgression, a mid-Oligocene condensed sequence, and Miocene–Recent progradation (Fig. 3). A similar history characterizes the nearby rift axis, the Great South Basin. Carter (1988a,b) has discussed this eastern New Zealand sedimentary sequence in detail, and subdivided the four megasequences into eight regional sedimentary phases (Carter, 1988a) (Table 3). He concluded that sedimentation occurred under conditions of steadily rising relative sea level (due to thermal subsidence), and that the eight sedimentary phases resulted 'from

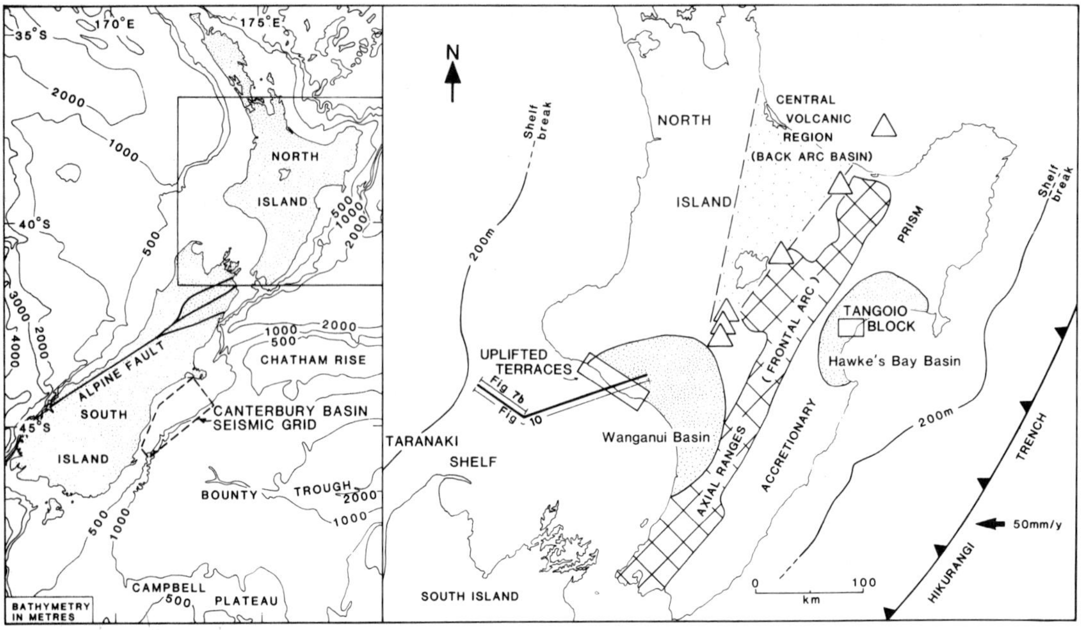


**Fig. 2.** Location map for the Canterbury Basin (Cretaceous–Recent passive margin) and the Quaternary Wanganui and Hawke's Bay Basins, located on either side of the rapidly uplifting central ranges of North Island.

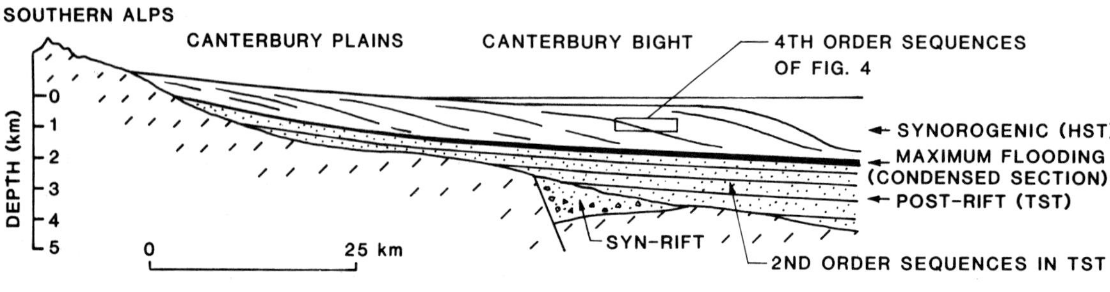


**Fig. 3.** Summary post-rift stratigraphy of the Canterbury Basin passive margin, showing position of the Paleogene second-order seismic sequences, the late Miocene fourth-order seismic sequences, and interpreted as a major transgressive–regressive cycle (first-order sequence) deposited in response to the balance between long-term thermo-tectonic subsidence and sediment supply.

the interplay of local tectonics, sediment supply, and oceanographic regime' (Carter, 1988a, p. 102). Further, in the case of the mid-late Oligocene condensed strata, which accumulated at the peak of regional Cenozoic transgression, there is a spectacular mismatch with the predicted 29 Ma lowstand of the GSM (cf. Carter, 1985). In contrast, however, Vail *et al.* (1989) and Loutit *et al.* (1988, p. 192) have interpreted the same sequence, and particularly its condensed Oligocene parts, as consistent with the Exxon GSM and as 'a spectacular example of the effects of subsidence and sea-level movements on stratal geometry'.

There is no doubt that sedimentary rhythms at the scale of first- and major second-order cycles occur on the eastern New Zealand passive margin. However, the above summary shows how difficult it is to assess objectively whether such rhythms were controlled by the hypothetical GSM or by local factors. Nevertheless, one result of the interplay between

**Table 3.** Major stratigraphic and sedimentary phases for the eastern South Island passive margin, as subdivisions of a major thermo-tectonic cycle (the Creteceo-Cenozoic Kaikoura Synthem) (amended after Carter, 1988a)

| Phase | Group | Environment | GSM model |
|---|---|---|---|
| Syn-orogenic | | | |
| Phase 8c | Offshore Otakou equivalent | Epicontinental platform | Highstand ST |
| Phase 8b | Otakou | Shelf sediment prism | Highstand ST |
| Phase 8a | Maori Bottom | Valley fill–coastal plain | Highstand ST |
| Condensed | | | |
| Phase 7 | Kekenodon | Epicontinental platform | Maximum flooding |
| Post-rift | | | |
| Phase 6 | Onekakara (offshore) | Offshore biopelagic | Transgressive ST |
| Phase 5 | | Mid-outer shelf | Transgressive ST |
| Phase 4 | | Shoreline–inner shelf | Transgressive ST |
| Phase 3 | Onekakara (paralic) | Coastal plain | Transgressive ST |
| Syn-rift | | | |
| Phase 2 | Matakea (coal measure) | Rift fill | Non-marine |
| Phase 1 | Matakea (breccia–conglomerate) | Rift fill | Non-marine |

thermo-tectonism and sediment supply on this margin is that the basin fill comprises a single first-order cycle, which bears close similarity to the Vail SSM for third-order sequences (see Fig. 1 in paper by B.U. Haq; Table 3). Some other passive margin sections show a similar analogy with the whole SSM, for instance the USA New Jersey margin, which exhibits a first-order sea-level cycle similar to that of the Canterbury Basin (e.g. Greenlee & Moore, 1988). Other well-studied margins exhibit a more complex sedimentary architecture, only some parts of which are analogous to major subdivisions of the SSM. For example, the early Tertiary parts of the Gulf of Mexico margin have an architecture in which sediment supply has generally dominated thermo-tectonism, leading to a dominantly 'HST' architecture (Galloway, 1989b), whereas younger (Coleman & Roberts, 1988) and older (DSDP Leg 96) parts of the Gulf sequence exhibit more complex sedimentary patterns (A.H. Bouma, personal communication).

Our conclusion is, therefore, that at the scale of first- and major second-order cycles the development of the eastern New Zealand sequence has been controlled dominantly by local thermo-tectonics (cf. Watts, 1982), sediment supply, and oceanographic regime. The effects of global sea-level changes, where present, are confined to modulation of the detailed stratigraphy.

### *Second–fourth-order seismic sequences*

The marine megasequences of the eastern South Island margin contain smaller seismic units (Figs 3, 4). Within the post-rift transgressive megasequence, these units are similar in scale to second-order cycles of the Exxon GSM (Fulthorpe & Carter, unpublished data). Within the synorogenic megasequence, the seismic units are finer scale still, and apparently equate with fourth-order cycles.

A detailed study has recently been undertaken of the seismic sequences within the mid-late Miocene part of the Canterbury Basin (Fulthorpe & Carter, 1989). Ten dip profiles from the central part of the basin were interpreted in detail. Eleven onlap surfaces were distinguished, each characterized by truncation of underlying reflectors and/or onlap of overlying reflectors. The onlap surfaces thus distinguished were interpreted as unconformities bounding depositional sequences of the type described by Vail (1987), and chronostratigraphic diagrams and local onlap curves were then derived.

Comparison of the resulting local onlap curves with each other, and with the Exxon GSM, showed (Fig. 4):

**1** significant differences between individual chronostratigraphic charts derived from parallel seismic lines located 12 km apart

**2** the apparent presence of nine cycles on the basin-average chart over a period where the GSM shows only 2.5 third order cycles.

Fulthorpe & Carter (1989) concluded from these data that the sequences studied were of fourth-order type. If they were sea-level controlled, then the late Miocene part of the Exxon GSM is in error, or at least lacks important detail. Alternatively, some of the sequences might have been deposited in response

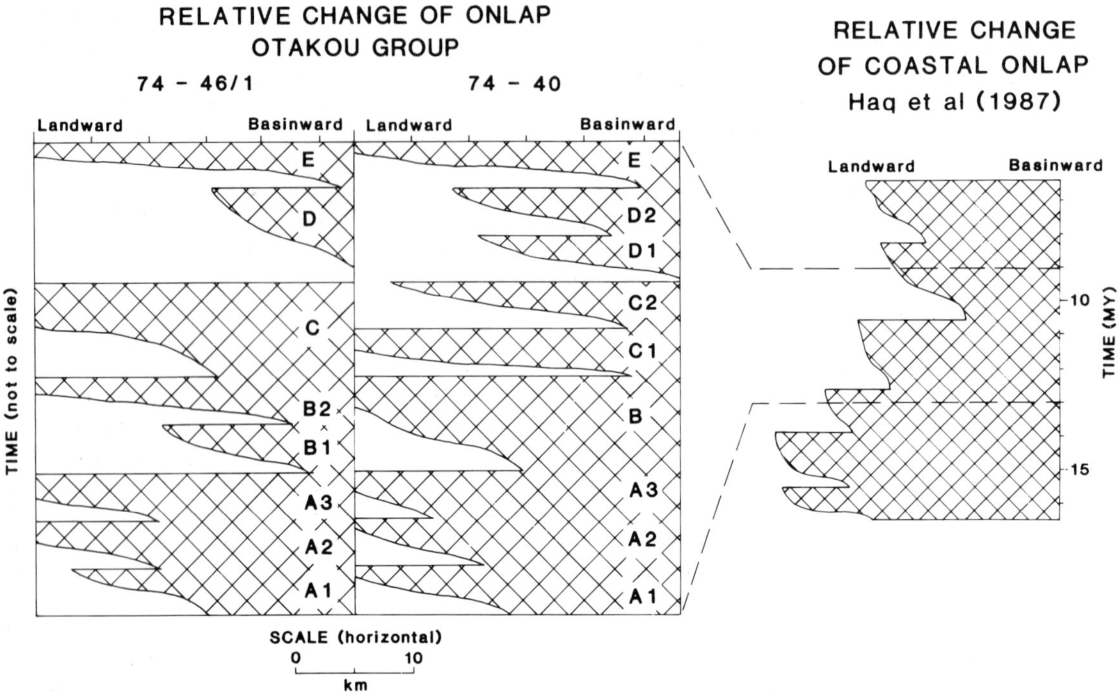


**Fig. 4.** Comparison of relative change of onlap from two closely spaced parallel seismic lines in the Canterbury Basin and approximate correlation with the mid-late Miocene part of the global sea-level model of Haq *et al.* (1987).

to basinal factors, such as variations in sediment supply. However, no criteria are available to distinguish sequences controlled by eustatism from those controlled by other factors.

Sequences of the type and scale ascribed by Vail *et al.* (1989) and van Wagoner & Mitchum (1989) to third and fourth-order global eustatism indubitably occur on the eastern South Island passive margin. However, as for first–second-order cycles, it is again difficult to assess the relationship between these sequences and the Exxon GSM. None the less, in so far as we are able to test it, we conclude that the mid-late Miocene part of the GSM is either incorrect or inadequate as a representation of global sea-level signature, since it does not match the stratigraphic signature preserved on this well-studied Southern Hemisphere passive margin.

## TESTING THE SEQUENCE-STRATIGRAPHIC MODEL

The sequence-stratigraphic model, as presently formulated, predicts the distribution of sedimentary facies that should occur during a single third-order cycle of the GSM curve. The model is therefore derived by inference from a coastal onlap curve, which is itself inferential. It follows that an independent test of the SSM cannot be made against data that have been used in derivation of the SSM. Rather, a proper test of the SSM must examine the facies associated with a sea-level cycle for which reality can be demonstrated from evidence extrinsic to the SSM. We conclude that sediments deposited during the Plio-Pleistocene glacio-eustatic changes, including those that characterize the modern (post-glacial) shelf, offer the best opportunity to test the SSM.

We therefore present below a comparison between the sedimentary architecture of modern shelf sediments and Plio-Pleistocene cyclothems from New Zealand, and the Exxon SSM.

### Features of the sequence-stratigraphic model

The SSM predicts that sediments associated with a single sea-level cycle occur within three major facies packets, termed systems tracts, which occur on opposing sides of the contemporary shelf edge (see Fig. 1 in paper by B.U. Haq: the application of the term 'shelf edge' in sequence stratigraphy has been controversial. Its usage is summarized in the Appendix). The cycle for such a 'type 1' sequence is defined as starting at a relative lowstand with sea level located below the shelf edge; the accompanying lowstand systems tract (LST) includes submarine fan, slope, and lowstand shoreline sediments. Subsequently, during sea-level rise, the shoreline crosses the shelf break and migrates landwards across the low-gradient shelf, accompanied by deposition of shoreline–shelf facies of the transgressive systems tract (TST). Finally, at the termination of sea-level rise, the shoreline–shelf facies of the highstand systems tract (HST) prograde seawards across the maximum flooding surface (MFS), which itself comprises a condensed sequence or paraconformity.

Posamentier & Vail (1988), after Vail & Todd (1981), also recognize a 'type 2' sequence, which is controlled by a smaller sea-level fall that does not cross the shelf edge, and term the corresponding lowstand deposits the shelf-margin systems tract. Though it is undoubtedly easier to recognize a sequence boundary that corresponds to sea-level withdrawal over the shelf edge, we do not accept the utility of recognizing type 1 and type 2 sequences because of the arbitrary nature of their distinction. For instance, the shelf break is located at widely varying depths on different margins depending upon the interplay of local subsidence and sediment supply (e.g. Mougenot *et al.*, 1983). Also, the established pattern of late Neogene glacio-eustatism demonstrates the existence of a complete spectrum of sea-level changes, from magnitudes of only a few metres to c. 150 m or more (cf. Fig. 1). Deposition of a stratigraphic sequence at the appropriate scale will accompany any of these sea-level cycles, and no useful distinction can be made between 'entirely onshelf' (type 2) and 'onshelf–offshelf' (type 1) sequence types.

## The post-glacial shelf

The modern shorelines of the world are located at an interglacial highstand position. The period of time during which the shoreline has been located at its present position varies depending upon local glacio- and hydro-isostatic effects (e.g. Clark & Lingle, 1979). However, away from the zones of major glacio-isostasy the sea level reached its modern position about 6500 years ago. At such locations, which include eastern New Zealand and eastern Australia, the 'Holocene' highstand systems tract has been developing from near the present shoreline for several thousand years. Furthermore, there is now good evidence for some shelves that the pre-6500 years BP post-glacial transgression was episodic (Frazier, 1974; Sanders & Kumar, 1975; Stubblefield *et al.*, 1984; Carter *et al.*, 1986), which has important implications for the detailed distribution of TST facies.

On the inner shelf off both eastern New Zealand and northeastern Australia, a shoreface-connected prism of terrigenous sediment up to 30 m thick has been prograding seawards for the last 6500 years and represents the contemporary HST (Fig. 5). Seawards of the HST, the sea floor is generally underlain by a < 1-m thick veneer of relict, palimpsest, and modern sediment above the Pleistocene, i.e. the MFS is represented at the sea floor by paraconformity and/or condensed sedimentation; locally, TST deposits (including 'drowned' shorelines) may occur between the underlying sequence boundary and the MFS condensed section. Finally, the shelf edge and upper slope are characterized by shoreline features and sediments of the LST deposited during the peak of the last glaciation (cf. in New Zealand, Herzer, 1981; Carter *et al.*, 1985; Carter & Carter, 1986; in Australia, Johnson & Searle, 1986; Carter & Johnson, 1986).

The pattern of shelf–upper-slope sedimentology on the Australian and New Zealand shelf (Fig. 5) appears at first sight to be dramatically different from the Exxon SSM (see Fig. 1 in paper by B.U. Haq). In fact the difference is superficial, being caused by the extended width of the shelf and the (geologically) short duration of the Holocene highstand. The two patterns are topologically equivalent, and — given a longer timespan and a narrower shelf — the modern pattern would match the SSM closely, differing only in the relative thinness of the TST. A similarly close match between the SSM and the sediments preserved on a modern shelf–slope transect has been reported for the Gulf of Mexico by Boyd *et al.* (1988).

## Plio-Pleistocene cyclothems in North Island, New Zealand

The North Island of New Zealand is situated at the junction between a transform and a subducting plate boundary and is flanked by two large late Neogene sedimentary basins, each with several kilometres of

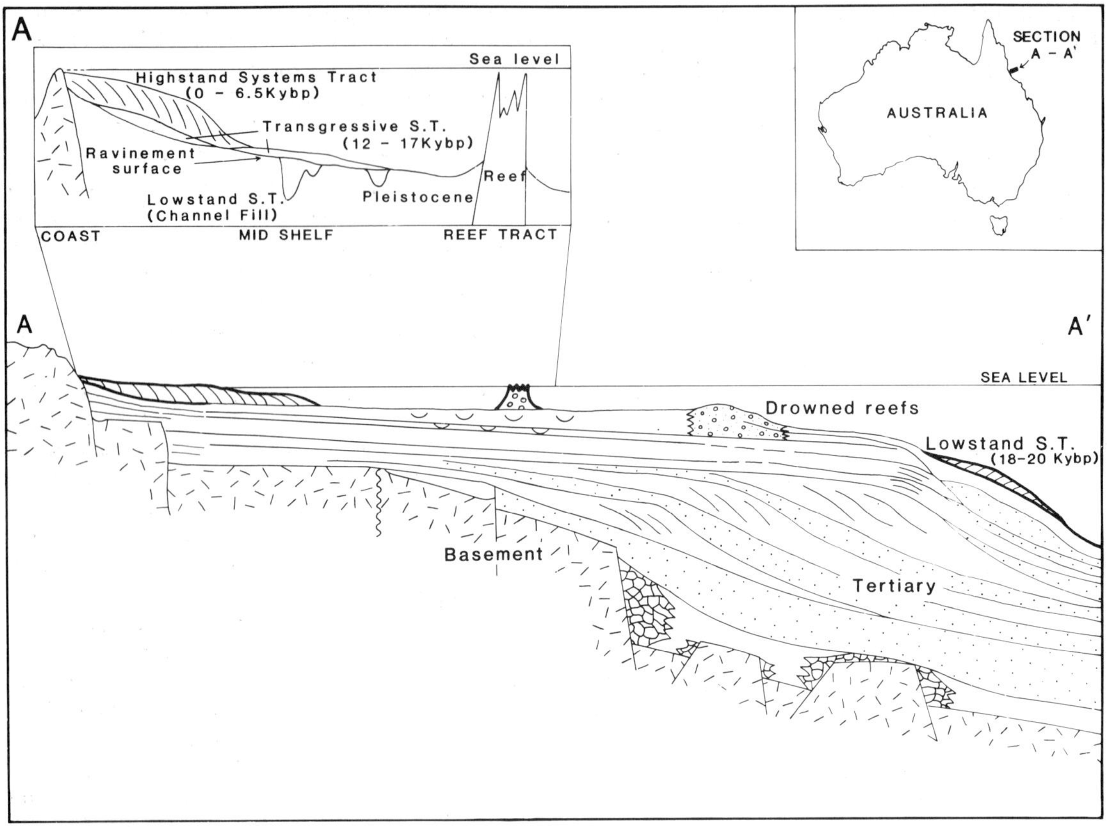


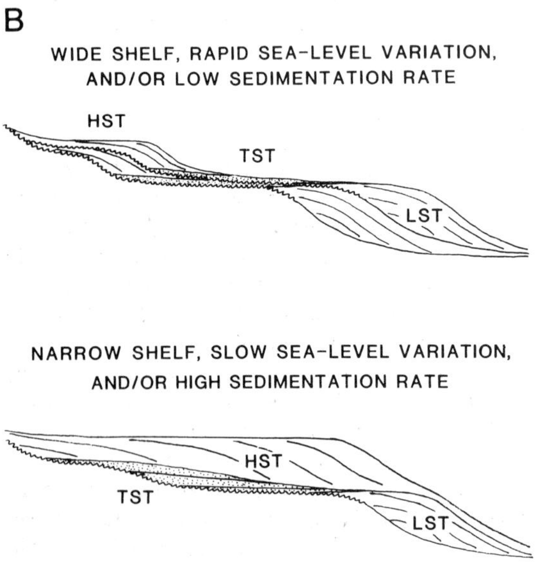


**Fig. 5.** (A) Interpretation of a tropical Southern Hemisphere shelf transect (Great Barrier Reef margin, northeastern Australia) in terms of the sequence-stratigraphic model of Vail (1987). (B) Schematic interpretation of (A) under differing conditions of physiography, sediment supply, and sea-level change (for references, see accompanying text).

basin fill, the Hawke's Bay and Wanganui Basins (Fig. 2). The Hawke's Bay Basin complex represents the upper fore-arc basin within the Neogene accretionary prism associated with Neogene subduction along the Hikurangi plate margin (Lewis, 1980). Driving subsidence is concentrated in simple synclinal sags located between rising, fault bounded, basement cored, sea-floor anticlinal ridges (Lewis, 1971,1973), resulting in individual sub-basins 20–50 km wide. In contrast, the Wanganui Basin is located in a back-arc setting, but only c. 200 km above the descending Pacific plate. A sequence analysis of the complete Cenozoic stratigraphy in the vicinity has been presented by Shell Oil Company (1987). According to Stern & Davey (1989), the Wanganui Basin comprises an early flexural, compressional phase in the evolution of a fully fledged back-arc basin, with driving subsidence provided from the vertical component of shear across a locked plate interface.

Despite their different origins, the common location of the Wanganui and Hawke's Bay Basins near a plate boundary is fortuitous with respect to studies of their Plio-Pleistocene fill. Uplift at the coastline in both basins approximates 0.5 m/ka (Pillans, 1986), resulting in excellent exposure of gently seaward-dipping Plio-Pleistocene strata in coastal cliffs and hills.

### *Wanganui Basin*

Fundamental studies by earlier workers, particularly Fleming (1953), Pillans (1983), and Beu & Edwards (1984), have established the cyclothemic nature of the Plio-Pleistocene strata of the Wanganui Basin and their general correlation with the standard oxygen isotope stages (e.g. Shackleton & Opdyke, 1976). Though the sequence goes back to 2 Ma and beyond, we describe below particularly that part between c. 0.95 and 0.35 Ma.

Individual cyclothems are between 5 and 22 m thick, and comprise an erosional surface overlain by a shallow water, gravel–sand–mud lithofacies (base-of-cycle association) followed by a deeper water mid-cycle shell-bed and a top-of-cycle siltstone lithofacies (Abbott, unpublished data). Ten such unconformity bounded cyclothems are recognized (Fig. 6). The base-of-cycle association is dominated by shoreface and shallow innermost-shelf marine lithologies, including cross-bedded shell gravel with mud drapes, flaser-bedded sandstone, and lenticular to wavy bedded sandstone/siltstone; only the shell gravel is fossiliferous, comprising abraded shells derived from underlying beds and/or transported in from contemporary shallow marine environments. Shoreface environments are represented by well-sorted parallel-laminated or cross-bedded sandstone with characteristic intertidal and shallow subtidal fossil species. The mid-cycle shell-beds comprise both biogenic and mixed sedimentological–biogenic types (terminology after Kidwell *et al.*, 1986) rich with *in situ*, soft-bottom, offshore fossil assemblages. Top-of-cycle siltstones include streaky laminated siltstone (largely barren of fossils) and masssive fossiliferous and sandy siltstone.

We interpret the bounding unconformity at the base of each cyclothem as marking the lower sequence boundary, and the mid-cyclothem silty shell-bed as representing a condensed deposit associated with the maximum flooding surface (Fig. 6b). Thus the gravel–sand lithofacies assemblage represents an onlapping TST, and the siltstone assemblage a downlapping HST.

The Wanganui Basin is apparently unique in the world in preserving not only the inner mid-shelf part of each cyclothem (sequence), but also the coastal-plain–shoreline part of the HST on uplifted inland terraces (Fig. 7a; Pillans, 1983), and the outer-shelf–upper-slope LSTs, which are seen on seismic profiles through the modern shelf edge (the 'giant foresets' formation of Pilaar & Wakefield, 1978; cf. Beggs & Thrasher, 1988) (Fig. 7b). Based on this information, and on the modern shelf slope of c. 1:700, we estimate that the Wanganui coastal outcrop cyclothems represent deposition c. 5–20 km from the contemporary shoreline, and MFS water depths of c. 15–50 m.

### *Hawke's Bay Basin*

Similar cyclothemic sediments that straddle the Plio-Pleistocene boundary are well exposed north of Napier in central Hawke's Bay (Fig. 2). Five 30–150-m thick cyclothems have been mapped in detail, and correlate broadly with isotope stages Me to Md of Gardner (1982) or stages 64–59 of Williams *et al.* (1988) (Beu & Edwards, 1984).

Though superficially similar, the Hawke's Bay cyclothems differ from their Wanganui counterparts in important ways. In particular, they are significantly thicker overall; contain non-marine facies representing river channels graded to the LST in the lower part of the sequence; and include coarser shoreface facies gradationally above each HST silt-

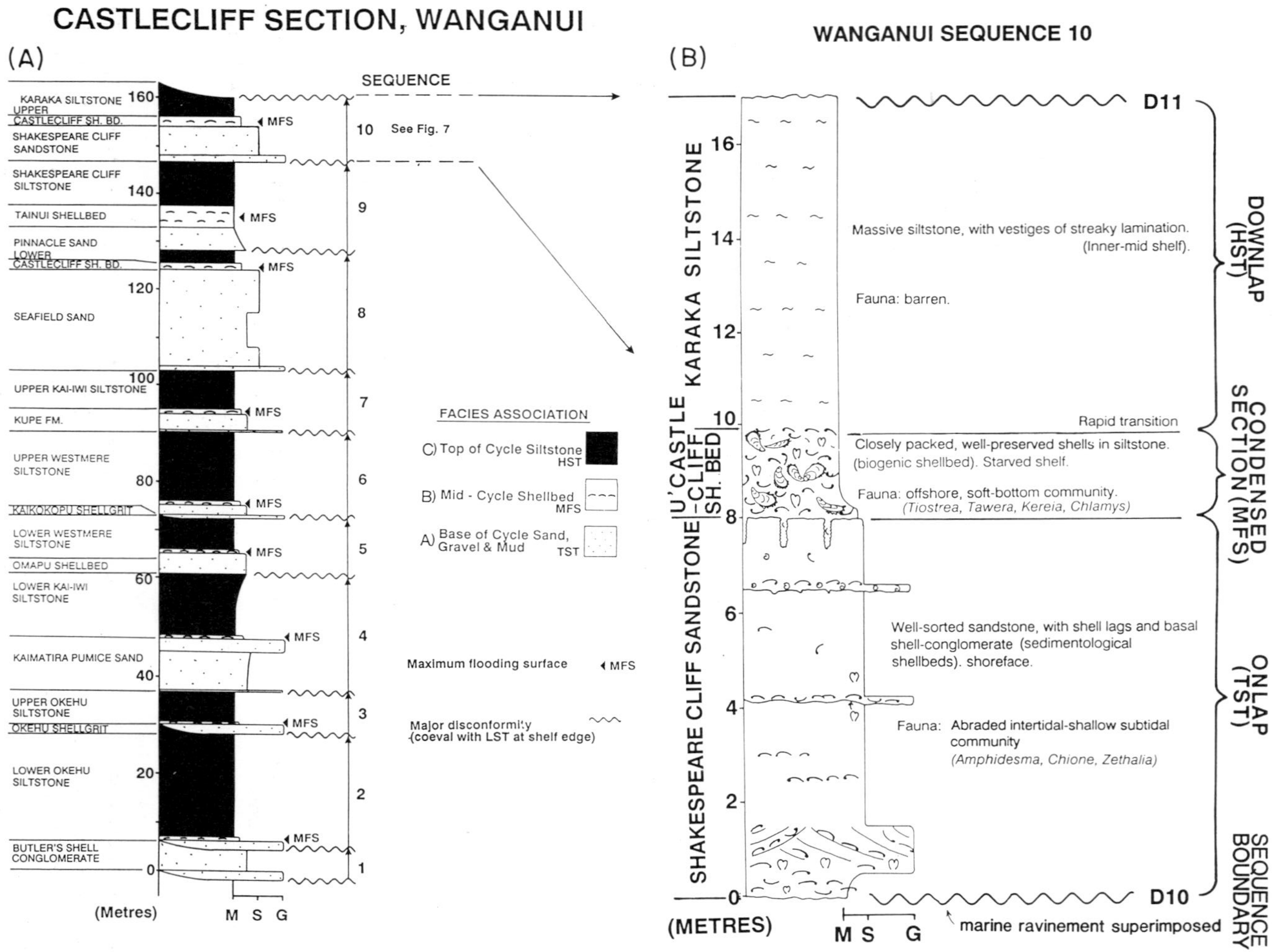


**Fig. 6.** (A) Summary column of mid-Pleistocene cyclothems in the Wanganui coast section. (B) Detail of cyclothem 10, with interpretation in terms of the sequence-stratigraphic model. Note that sequences 1–6 were deposited during 40 000 year (sixth-order) sea-level cycles, whereas sequences 7–10 were deposited during 100 000 year (fifth-order) cycles (cf. Fig. 1).

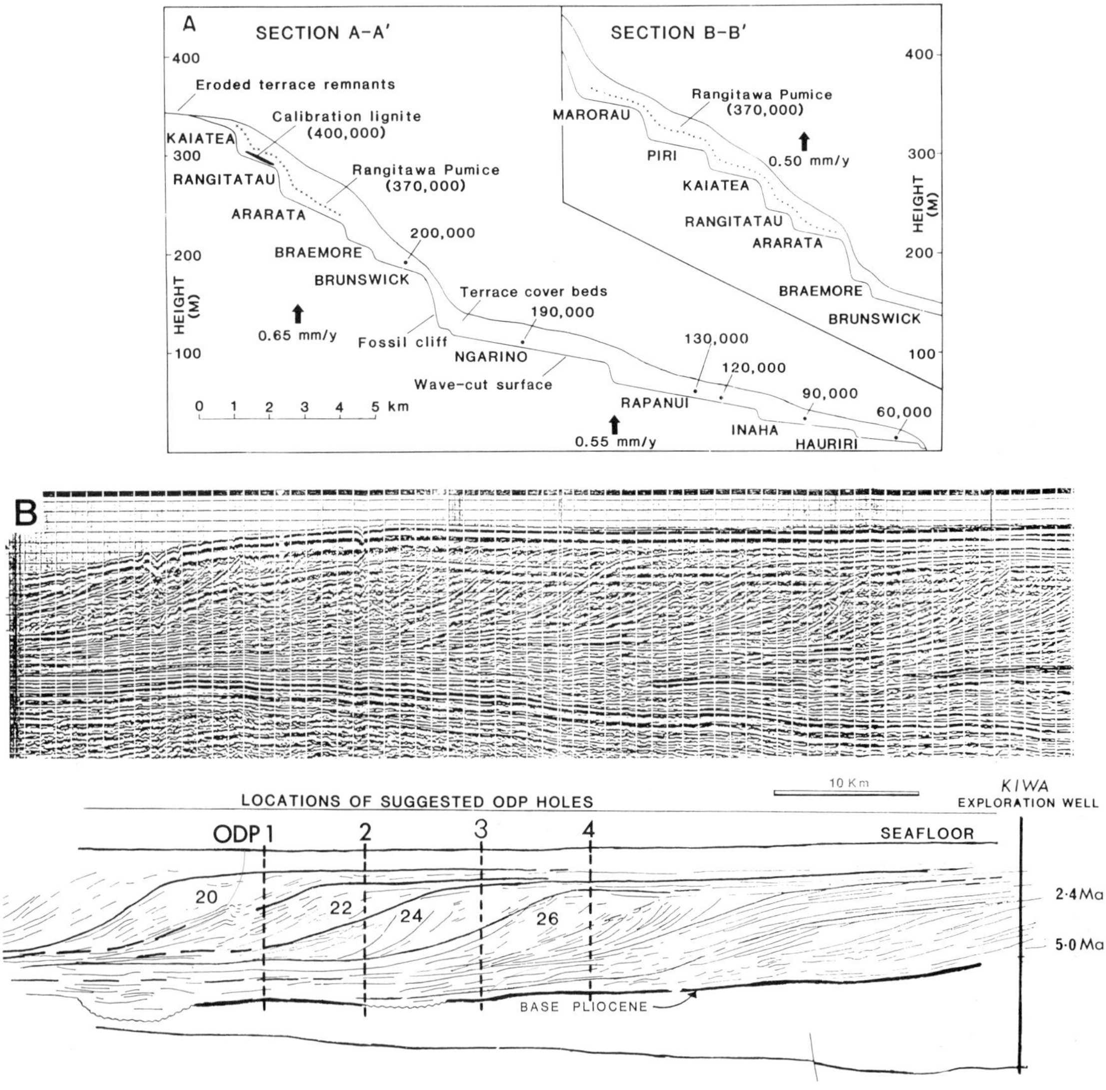


**Fig. 7.** (A) Configuration of the coastal terrace system representing highstand systems tracts for isotope stages 5 (Rapanui Terrace) to 19 (Marorau Terrace), with indicated amino-acid racemization and fission-track dates (after Pillans, 1983); (B) configuration of the Pleistocene lowstand systems tracts located at the Taranaki–Wanganui shelf edge (New Zealand Geological Survey open-file seismic line 72–169, interpretation after Beggs & Thrasher, 1988).

stone unit (Figs 8, 9) (Haywick *et al.*, 1991). Thus where the gravel–sand parts of the Wanganui cyclothems predominantly belong to the TST and were deposited under a rising sea level, the coarser grained facies in Hawke's Bay represent mostly HSTs and were probably deposited under a static–falling sea level. These characteristics are, in general, consistent with the Hawke's Bay cyclothems representing a nearer shore location than those exposed at the Wanganui coast.

Analysis of the foraminiferal faunas from the siltstones of the HST confirms the presumed upward-shallowing nature of each siltstone (Fig. 9) (Haywick & Henderson, unpublished data). The microfaunal assemblages present can be equated broadly to those present on the modern New Zealand shelf, with

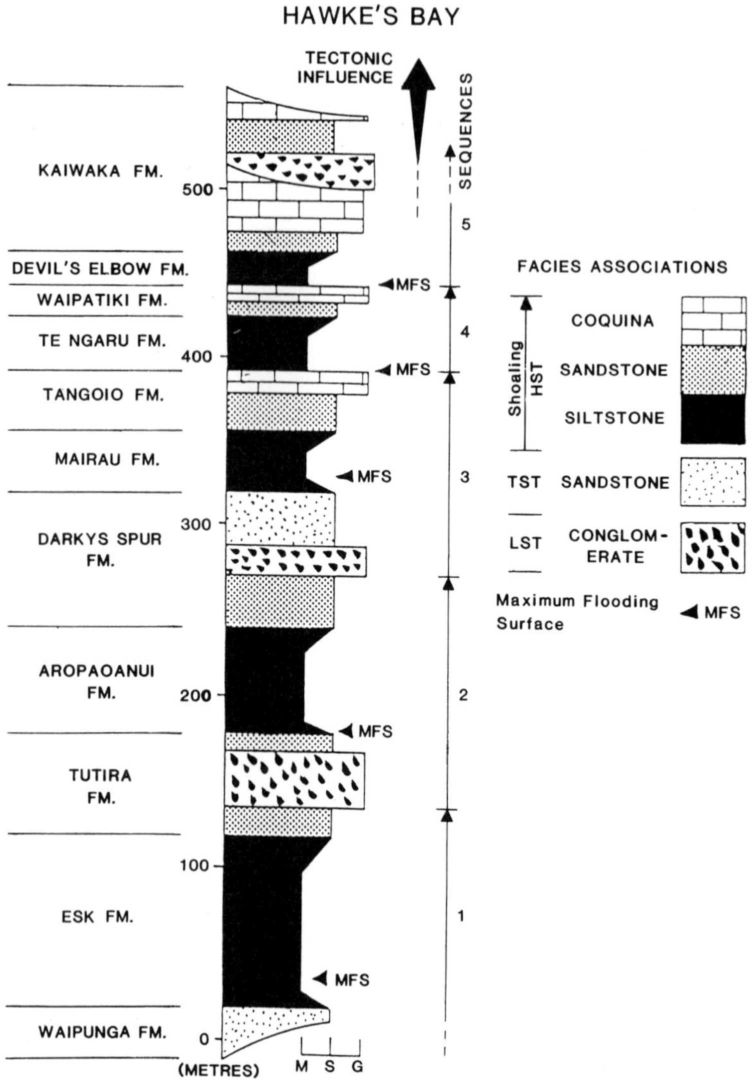


**Fig. 8.** Summary column of Plio-Pleistocene cyclothems from the Tangoio area, Hawke's Bay Basin (after Haywick *et al.*, 1991). Five sequences are identifiable, though the top of the highest (S5) is masked by the effects of regional uplift and emergence. Note that each sequence was deposited during a 40 000 year (sixth-order) sea-level cycle (cf. Fig. 1).

rotaliid-porcellaneous forms (*Ammonia*, *Elphidium*, *Cibicides*) dominating inner shelf stations, and a *Uvigerina–Siphotextularia–Sigmoilopsis* assemblage characterizing the more offshore shelf. Interestingly, the vertical succession of assemblages from some siltstones indicates a maximum depositional depth (= MFS) located a few metres above their base (Fig. 9) rather than at the basal contact with the underlying coarser grained LSTs/TSTs. These data show that the MFS is not located at the point of obvious change in the section, and may indicate also that further pulses of shoreline advance occurred after the initial TST/HST drowning.

## Correspondence between Plio-Pleistocene cyclothems and the SSM

It is apparent that the stratigraphic architecture of the fifth and sixth-order Plio-Pleistocene sequences (cyclo-thems) of the Wanganui and Hawke's Bay Basins corresponds closely to the Exxon sequence-stratigraphic model (Fig. 10). Arising from this correspondence, we note the following points of general significance.

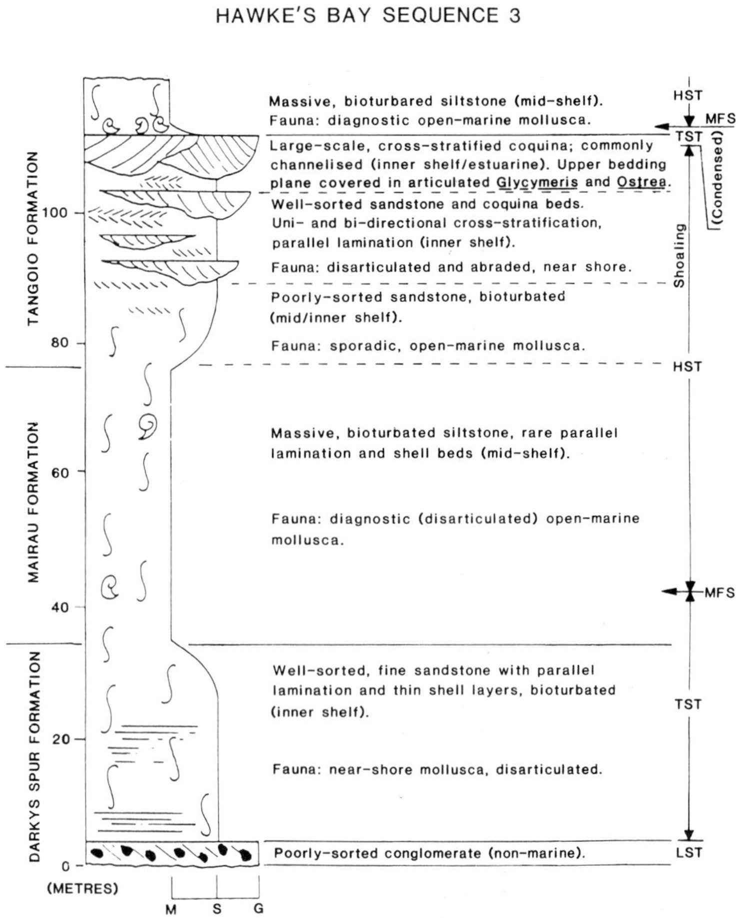


**Fig. 9.** Detail of Hawke's Bay sequence 3, with interpretation in terms of the sequence-stratigraphic model.

## *Symmetry of sea-level cycles*

An important difference of symmetry exists between Plio-Pleistocene fifth-order cycles and third-order Exxon coastal-onlap cycles. Whereas the third-order Exxon cycles comprise longer periods of rise separated by near instantaneous falls, the fifth-order cycles comprise longer periods of fall punctuated by very rapid rises — the glacial 'terminations' of Broeker & van Donk (1970) (cf. Fig. 1). This difference presently lacks explanation, but it demonstrates that the Exxon third and fourth-order cycles cannot be glacio-eustatic unless there were pre-Pliocene glacial cycles of very different periodicity and symmetry from those of the Plio-Pleistocene.

## *Robustness of the SSM*

The SSM grew out of an analysis of third-order sequences, and its theory has been developed in refined detail based upon a sawtooth coastal-onlap curve controlled by an idealized, symmetrical sea-level cycle (Posamentier & Vail, 1988; Posamentier *et al.*, 1988). That the model fits so well with the observed stratigraphy developed under fifth-order sea-level cycles of known asymmetry, as tested in the Wanganui and Hawke's Bay Basins, indicates its robustness. The precise shape of the eustatic curve is apparently not a strong control on facies development at the systems tract level, however important it may be in modulating intra-systems tract sedimentary detail.

## *Effect of sea-floor slope*

The regional sea-floor slope exercises a strong control on the cross-strike stratigraphy of a cyclothem. A low sea-floor slope (say, $\leq 1$ m/km) will lead to a reduced or absent TST, a 'layer cake' configuration

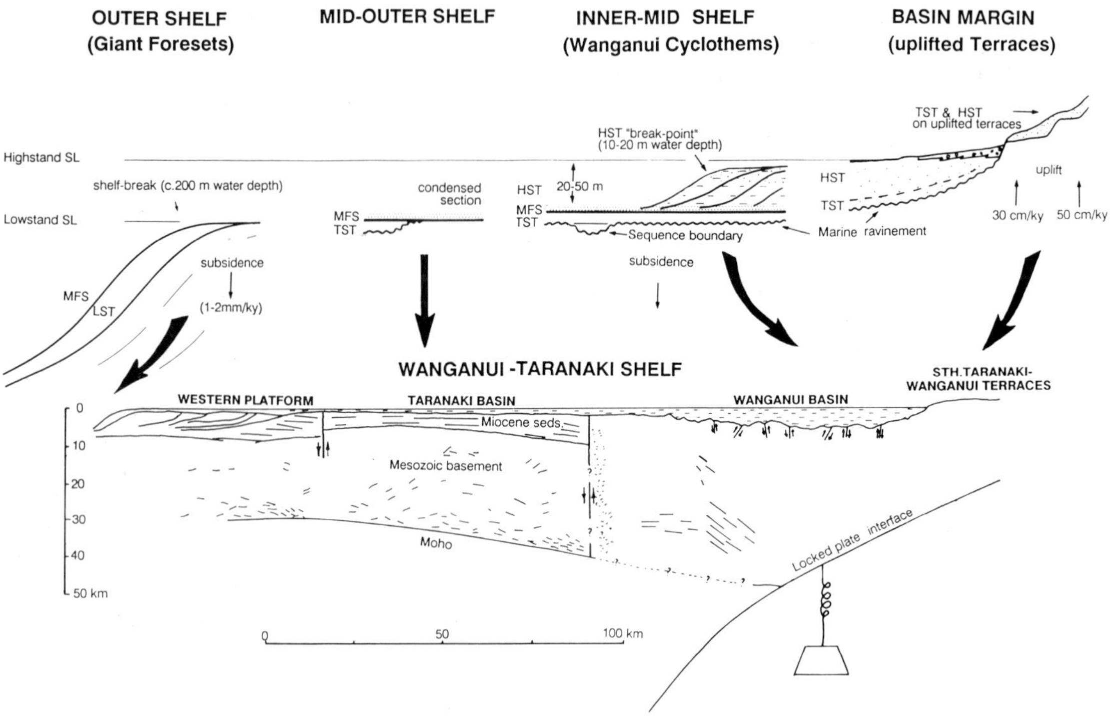


**Fig. 10.** Summary cross-section of the Wanganui Basin and adjacent shelf, showing a close correspondence with the sequence-stratigraphic model (tectonic interpretation after Stern & Davey, 1989).

of TST/HST sediments, and a small depth contrast between nearby cross-strike sections. A high sea-floor slope (say 5 m/km), conversely, may lead to a thick TST and to a high depth contrast between nearby sections.

### *Effect of location across the basin shelf*

Within a cyclothem, and for a given sea-floor slope, the detailed stratigraphy varies markedly across the depositional strike and away from the contemporary highstand shoreline (Fig. 10). Sections representative of highstand nearshore locations are characterized by packets of coarse grained, shallow-marine sediment at the top of their HST (e.g. Hawke's Bay cyclothems). Sections located on the middle shelf are frequently thinner and may contain a well-developed MFS shell-bed at the base of a conspicuous deeper water siltstone (e.g. Wanganui coastal cyclothems). Though not yet described from New Zealand, a more offshore section still, on the outer shelf, should be characterized by a relatively thin sequence overall, with a relatively thick MFS shell-bed and with or without an underlying TST.

### *Controls on the development of the TST*

Earlier sedimentary models have represented the TST in oversimplified fashion as either a continuous 'transgressive sand' blanket (e.g. Swift *et al.*, 1972) or as continuously on-stepping parasequences (e.g. Vail, 1987). In fact, the development of the TST is dependent particularly upon the nature of the antecedent surface, the rate of supply of sediment, and the nature and rate of sea-level rise. Where fluvial channels graded to the LST shoreline are established, the advancing sea-level causes aggradation at the mouth, leading to the development of estuarine and mouth-bar facies above and adjacent to the LST channel fill. As shown for the Great Barrier Reef shelf by Johnson *et al.* (1982), this may lead to the anomaly of raised sea-floor relief above the former channels. Other mechanisms that encourage deposition of a thick TST include a low rate of sea-level rise, or, particularly, pauses or minor falls during an overall rise, as are known to have occurred during the last post-glacial melting. Conversely, extremely rapid sea-level rise across a non-incised shelf will lead to the circumstance where only a shell-lag pavement separates the sequence boundary from any overlying MFS or HST strata. Thus, in apparent contradistinction to the terminology (transgressive ST), each paracycle in a thick TST might actually be deposited during an intra-sequence sea-level stasis or minor fall.

### *Diagnostic faunal assemblages*

Facies associated with the TST and HST are characterized not only by their sedimentology but also by distinctive fossil assemblages (Table 4). The TST and nearshore HST contain either worn and transported shell-beds, often with remanié forms (association A1), or *in situ* faunas that are diagnostic of intertidal or shallow subtidal conditions (association A2). In contrast, the MFS condensed shell-bed comprises an offshore fauna (associations B1–B4), the soft-bottom elements of which also occur scattered *in situ* through the overlying HST silts (association A3).

## INTERACTION OF SEQUENCE DEVELOPMENT AND TECTONICS

### Tectonics as a sedimentary control

The location of the North Island Plio-Pleistocene basins near an active plate boundary (Fig. 2) raises the possibility that their cyclothemic nature has been controlled by periodic pulses of tectonic activity. Though tectonic uplift has indeed played an important role in uplifting and displaying the cyclothems (e.g. Lewis, 1971; Pillans, 1983), the following evidence indicates that their fundamental control was glacio-eustatism.

**1** The cyclothems studied occur within extensive homoclinal sequences that are internally undeformed and of considerable areal extent. Thus Haywick *et al.* (1991) have mapped five cyclothems in Hawke's Bay over an undeformed area of 450 km$^2$, and Lewis (1971,1973) illustrated high-resolution seismic profiles from offshore that show gently dipping and unfaulted Plio-Pleistocene sediment is quite charac-

**Table 4.** Fossil shell associations in relation to systems tract position (Abbott & Carter, unpub. data)

| Shell association | Characteristic species | Matrix | Environment | Systems tract |
|---|---|---|---|---|
| Type A1 | Largely reworked shells, especially *Mactra tristis & Paphies subtriangulata; Pseudechinus* | Coarse sand and gravel | Tidal bars and shoals | TST or HST |
| Type A2 | *Austrovenus, P. australe, Macomona, Mactra tristis* | Silty sand | Estuarine flats | TST or HST |
| | *Dosinia* spp., *Mactra* spp., *Spisula, Fellaster* | Fine–med. sand | Sandy, ocean beach | TST or HST |
| | *Divaricella, Serratina, Myadora, Antisolarium* | Very fine sand | Subtidal ocean shoreface | TST or HST |
| | *Maoricolpus, Purpurocardia, Tawera* | Granular sand | Tidal channels; scoured | TST or HST |
| Type A3 | *Stiracolpus, Poirieria, Kereia, Poirieria, Pelicaria, Nemocardium, Chlamys celator, Atrina* | Silt | Inner-middle shelf (soft bottom) | Mainly HST |
| | *Atrina, Tiostrea, Chlamys* | Sandy silt | Inner-middle shelf (clumps, hard bottom) | Mainly HST |
| Type B1 | *Tiostrea, Dosina, Zenatia* | Muddy coarse sand | Starved innermost shelf | Mid-cycle |
| Type B2 | *Tiostrea, Dosina, Caryocorbula, Kereia, Zenatia, Myadora kaiiwiensis* | Sandy silt | Starved innermost shelf | Mid-cycle |
| Type B3 | Similar to B4, but parallel shells of *Purpurocardia, Tawera, Austrofusus, Pecten, Dosina,* Brachiopods | Sandy silt | Starved inner shelf | Mid-cycle |
| Type B4 | *Tiostrea, Chlamys gemmulata, Barbatia, Glaphyrina, Aeneator,* cellariform bryozoans | Silt | Starved inner-middle shelf | Mid-cycle |

teristic of the 20–50-km-wide synforms that occur between actively uplifting basement horsts. In the Wanganui Basin, Anderton (1981, fig. 5) has shown that there has been little deformation of the Wanganui–Taranaki shelf since the time of his reflector A, estimated to be equivalent to the top of the local Nukumaruan Stage (c. 1.4 Ma). Sediments younger than this (which includes those of the Wanganui coastal section) essentially underlie the shelf with horizontal dips (cf. Goldberg & Carter, 1979). Exceptions occur at the shoreline, where successive cyclothems are tilted gently seawards because of basement uplift further east, and at the shelf edge, where successive LSTs have a gentle seawards depositional dip.

**2** An interlocking web of evidence, including biostratigraphy (Beu & Edwards, 1984), fission-track dating (Seward, 1976,1979), palaeomagnetism (Turner & Kamp, 1989), oxygen isotope data (Stevens & Vella, 1981), amino-acid racemization (Pillans, 1983), and sedimentation rates (Abbott, unpublished data), indicates a close correlation between the cyclothemic record and the established isotope curve that proxies for Plio-Pleistocene ocean volume (e.g. Williams *et al.*, 1988).

**3** This correlation implies an average length for cyclothems of either 40 000 or 100 000 years, with probable rates of intercyclothem post-glaciation sea-level rise of 8–30 m/ka (cf. Cronin, 1983; Carter *et al.*, 1986; Snyder *et al.*, 1989), and concomitant shoreline shifts at rates up to 30 km/ka. Against this pattern, the return rate of large earthquakes, and their probable displacements, are generally too small to have exercised major control. For instance, in the plate boundary zone of the North Island main ranges the maximum uplift rate is 3.9 m/ka, associated with a return period of about 1300 years for earthquakes causing displacement of $\geqslant 2.5$ m (Moore, 1987). Uplift rates in the flanking sedimentary basins are correspondingly less, generally 0.1–0.5 m/ka (e.g. Pillans, 1986).

**4** For a maximum length of 50 000 years for the sea-level changes that precede or follow a glaciation (and which are marked by the sequence boundaries), a maximum local tectonic uplift of 5–25 m is indicated over the same timespan. This estimate (in itself inflated — since it is unlikely that all displacement at a particular locality is concentrated along a single fault) is small compared with the likely 80–130 m magnitude of the glacio-eustatic sea-level oscillations.

We conclude that though tectonic activity had local effects on the distribution of sedimentary facies within a cyclothem, as for instance the thicker (but still cyclothemic) section south of the Upokongaro Fault at the Wanganui coast (Beu *et al.*, 1986), it is unlikely to have been the major control on the development of the cyclothems themselves. None the less, hinterland tectonics and volcanic activity have been powerful controls on the location of the axis of tilting (= shoreline) and the rate of terrigenous sediment supply to the Wanganui Basin, thereby influencing both the thickness of cyclothemic units and their overall stratigraphic trend.

### Long-term tectonostratigraphic trends

In both Wanganui and Hawke's Bay, the vertical sequence of cyclothems encompasses a long-term stratigraphic signature, i.e. successive cyclothems show progressive changes in character, particularly in the ratio of TST:HST strata (e.g. Figs 6a, 8).

Both basins have been located throughout the Plio-Pleistocene in a flanking position with respect to the main axis of plate boundary uplift, which runs north-northeast through the central North Island (Fig. 2). During their history, therefore, the inner edge of each basin has been subjected to a progressive seawards tilting, in response to the uplift gradient away from the plate boundary (e.g. Pillans, 1986), resulting in the exposure of the gently dipping Plio-Pleistocene cyclothems at the coast (cf. Fig. 10).

Progressive tilting of this kind imposed on a system with a static sea level will tend to result over time in the location of the shoreline near the axis of tilting (Lewis, 1974, p. 58). If sea level is fluctuating markedly, however, as during the Plio-Pleistocene, then shorelines that represent successive highstands will be located progressively seawards by an amount that depends upon the local uplift gradient. The evidence for this has been clearly presented for the Wanganui Basin by Pillans (1983), who has shown that the cliff-line at the inner edge of successively younger highstand terraces has moved seawards at a rate of 1–5 km/cycle between isotope stages 17 and 1. The shift in shoreline that occurs between successive glacio-eustatic cycles is generally accompanied by a seawards shift of the major sedimentary facies of the TST and HST.

It follows that steady-state tilting of the margin of a basin affected by glacio-eustatism will produce progressive changes in the detailed character of successive cyclothems at a location within the basin, except in the special case where the location lies at the axis of tilting. If the axis lies to seaward, and tilting is upwards, then successive cyclothems will become thicker, more sand-dominated, and possess a thinner MFS condensed shell-bed — all features consistent with the seawards displacement of the shoreline through time. Alternatively, if tilting on the landward edge of the basin is downward, i.e. depression rather than uplift, then these trends will be reversed in response to the steady landwards retrogression of the shoreline. The effect on sedimentary facies will be less marked, in general, for locations seaward of the axis of tilting, where the main effect will be slight increases in water depth above that produced by the background subsidence rate for the margin (i.e. increased 'accommodation' *sensu* Vail (1987)).

It is obvious that differing stratigraphic patterns of cyclothem character can be produced by changes in the rate or direction of tilting, or by lateral movements in the position of the axis of tilting. The New Zealand evidence suggests, however, that dramatic changes of this type do not usually occur over short periods, i.e. over less than many hundred thousand years, or several fifth-order cycles. For instance, the long-term patterns shown by the Wanganui and Hawke's Bay sections can be interpreted in terms of simple and gradual shifts of the local axis of tilting (Figs 6a, 8).

## CONCLUSIONS

The studies summarized above lead us to the following conclusions.

### *The global sea-level model*

The limited accuracy of biostratigraphic age determination in well-cuttings makes it difficult, and perhaps impossible, to adequately test the accuracy of the Exxon global sea-level model in frontier offshore basins that contain only a limited number of wells. None the less, a test using excellent seismic data from New Zealand suggests that the mid-late Miocene part of the Exxon GSM does not represent a global signature, since the curve does not correspond to the well-developed sequence pattern preserved within the passive margin sediments of the Canterbury Basin.

### *The sequence-stratigraphic model*

The pattern of post-glacial sedimentation on Southern Hemisphere shelves, and patterns of Plio-Pleistocene cyclothemic sedimentation in the Wanganui and Hawke's Bay Basins, corresponds closely to the Exxon sequence-stratigraphic model. The model is also able to be applied to sedimentary packets deposited under sea-level cycles between first order (e.g. Canterbury Basin) and infra-seventh order (e.g. the modern shelf). The Exxon SSM is therefore a powerful general tool with which to analyse the history of any shallow-marine sedimentary sequence deposited under the influence of a sea-level cycle.

### *The pre-Pliocene absence of Cenozoic cyclothems*

We have shown that fifth and sixth-order sea-level cycles of glacio-eustatic origin impose a conspicuous cyclothemic character on their associated sedimentary sequences. A similar cyclothemic character has also been demonstrated for Plio-Pleistocene sediments in the Northern Hemisphere by Clifton *et al.* (1988), and has long been known also to characterize the late Palaeozoic glacial epoch (e.g. Weller, 1930; Boardman & Heckel, 1989). Despite the voluminous literature on many Cenozoic basins, we are not aware of any published study that documents similar cyclothems in Cenozoic strata older than Pliocene. For instance, all New Zealand shelf sedimentary basins exhibit a characteristic change to cyclothemic sedimentation in the late Neogene. This fact is a powerful argument against the pre-Pliocene third-order cycles of the Exxon GSM being controlled by glacio-eustatism of the Plio-Pleistocene type.

### **Concluding comments**

We believe that earlier investigators have underestimated the difficulty of making an objective test of either the global sea-level curve or of the sequence-stratigraphic model. Elements of circular reasoning are present in some earlier studies where authors have matched aspects of their stratigraphic data with the presumed global curve.

Our studies lead us to have considerable confidence in the correctness and power of the Exxon sequence-stratigraphic model as applied to sea-level controlled, cyclothemic sequences. Ironically, however, our demonstration of this is based upon analysis of fifth and sixth-order Plio-Pleistocene cycles that are not part of the current Exxon GSM curve. At the same time, we suspect that the Exxon 'Global' sea-level curve, in general, represents a patchwork through time of many different local relative sea-level curves. None the less, and despite our own analyses, we concede that there has probably not yet been an adequate independent test of the Exxon GSM. Such a test must await the drilling of a close transect of continuously cored ODP sites across well-developed sequence systems in both hemispheres.

## ACKNOWLEDGEMENTS

We acknowledge gratefully the advice and help of Alan Beu, Tony Edwards, Peter Kamp, Mac Beggs, and Brad Pillans. We thank also the residents of the Wanganui and Tangoio areas, and particularly Glenys Wilson, Audrey and Bill Cox, and the Arnold family, for their generous help with accommodation and sustenance during field work. The New Zealand Geological Survey and British Petroleum (N.Z.) Ltd kindly provided access to the seismic information used. This research has been financially supported by Special Research Grants from James Cook University, and by the Australian Research Council. The draft manuscript benefited greatly from critical readings by Drs A.H. Bouma, P.J. Brenchley, and D.I.M. Macdonald. The senior author also thanks the International Association of Sedimentologists for a travel grant, which enabled him to attend the Sea-level Meeting at the Geological Society of London in 1989.

## REFERENCES

Anderson, J.B. & Thomas, M.A. (1989) High-resolution seismic stratigraphy of inner continental shelf, Texas Gulf Coast: long term and short term cyclicity. *Abstracts, GCAGS/SEPM October 25–27 Meeting*, p. 1180. American Association of Petroleum Geologists, Tulsa, OK.

Anderton, P.W. (1981) Structure and evolution of the South Wanganui Basin, New Zealand. *N. Z. J. Geol. Geophys.* **24**, 39–63.

Beggs, J.M. & Thrasher, G.P. (1988) Plio/Pleistocene progradation of the Taranaki shelf margin and the character of sea level lowstand deposits near the shelf break. *Geological Society of New Zealand, Annual Conference*, Hamilton, Abstracts, p. 35.

Beu, A.G. & Edwards, A.R. (1984) New Zealand Pleistocene and Late Pliocene glacio-eustatic cycles. *Palaeogeogr. Palaeoclimatol. Palaeoecol.* **46**, 119–142.

Beu, A.G., Edwards, A.R. & Pillans, B.J. (1986) A review of New Zealand Pleistocene stratigraphy, with emphasis on the marine rocks. In: *Proceedings 1st International Colloquium on Quaternary Stratigraphy of Asia and Pacific Area*, Osaka (Eds Itihara, M. & Kamei, T.) pp. 250–269.

Boardman, D.R. & Heckel, P.H. (1989) Glacial-eustatic sea-level curve for early Late Pennsylvanian sequence in north-central Texas and biostratigraphic correlation with curve for midcontinent North America. *Geology* **17**, 802–805.

Boyd, R., Suter, J. & Penland, S. (1988) Implications of modern sedimentary environments for sequence stratigraphy. In: *Sequences, Stratigraphy, Sedimentology: Surface and Subsurface*, (Eds James, D.P. & Leckie, D.A.). Can. Soc. Petrol. Geol. Mem. 15, 33–36.

Broeker, W.S. & van Donk, J. (1970) Insolation changes, ice volumes and the $O^{18}$ record in deep-sea cores. *Rev. Geophys. Space Phys.* **8**, 169–188.

Burton, R., Kendall, C.G. St.C. & Lerche, I. (1987) Out of our depth: on the impossibility of fathoming eustasy from the stratigraphic record. *Earth Sci. Rev.* **24**, 237–277.

Carter, L. & Carter, R.M. (1986) Holocene evolution of the nearshore sand wedge, South Otago continental shelf, New Zealand. *N. Z. J. Geol. Geophys.* **29**, 413–424.

Carter, R.M. (1985) The mid-Oligocene Marshall Paraconformity, New Zealand: coincidence with global eustatic sea-level fall or rise? *J. Geol.* **93**, 359–371.

Carter, R.M. (1988a) Plate boundary tectonics, global sea-level changes and the development of the Eastern South Island Continental Margin, New Zealand, Southwest Pacific. *Mar. Petrol. Geol.* **5**, 90–107.

Carter, R.M. (1988b) Post-breakup stratigraphy of the Kaikoura Synthem (Cretaceous–Cenozoic), continental margin, southeastern New Zealand. *N. Z. J. Geol. Geophys.* **31**, 405–429.

Carter, R.M. & Johnson, D.P. (1986) Sea-level controls on the post-glacial development of the Great Barrier Reef, Queensland. *Mar. Geol.* **71**, 137–164.

Carter, R.M., Carter, L., Williams, J. & Landis, C.A. (1985) Modern and relict sedimentation on the Otago continental shelf. *N. Z. Oceanogr. Inst. Mem.* **93**, 43 pp.

Carter, R.M., Carter, L. & Johnson, D.P. (1986) Submergent shorelines in the SW Pacific: evidence for an episodic post-glacial transgression. *Sedimentology* **33**, 629–649.

Christie-Blick, N., Mountain, G.S. & Miller, K.G. (1990) Seismic stratigraphic record of sea-level change. In: *Studies in Geophysics*. US Academy of Science, Washington, DC (in press).

Clark, J.A. & Lingle, C.S. (1979) Predicted relative sea-level changes (18 000 years BP to present) caused by late glacial retreat of the Antarctic ice sheet. *Quaternary Res.* **11**, 279–298.

Clifton, H.E., Hunter, R.E. & Gardner, J.V. (1988) Analysis of eustatic, tectonic and sedimentologic influences on transgressive and regressive cycles in the Upper Cenozoic Merced Formation, San Francisco, California. In: *New Perspectives in Basin Analysis* (Eds Kleinspehn, K.L. & Paola, C.), pp. 109–128.

Coleman, J.M. & Roberts, H.H. (1988) Sedimentary development of the Louisiana continental shelf related to sea level cycles: Part I — sedimentary sequences. *Geo-Mar. Lett.* **8**, 63–108.

Cronin, T.M. (1983) Rapid sea level and climate change: evidence from continental and island margins. *Quaternary Sci. Rev.* **1**, 177–214.

Emiliani, C. (1955) Pleistocene temperatures. *J. Geol.* **63**, 538–578.

Fleming, C.A. (1953) The geology of the Wanganui Subdivision, Waverley and Wanganui sheet districts (N137 & N138). *N. Z. Geol. Surv. Bull.* **52**, 1–362.

Frazier, D.E. (1974) Depositional episodes: their relationship to the Quaternary stratigraphic framework in the northwestern portion of the Gulf Basin. *Univ. Tex., Bur. Econ. Geol. Circ.* **74–1**, 28 pp.

Fulthorpe, C.S. & Carter, R.M. (1989) Test of seismic sequence methodology on a southern hemisphere passive margin: the Canterbury Basin, New Zealand. *Mar. Petrol. Geol.* **6**, 348–349.

Galloway, W.E. (1989a) Genetic stratigraphic sequences in basin analysis. I: architecture and genesis of flooding-surface bounded depositional units. *Bull., Am. Assoc. Petrol. Geol.* **73**, 125–142.

Galloway, W.E. (1989b) Genetic stratigraphic sequences in basin analysis. II: application to northwest Gulf of Mexico Cenozoic basin. *Bull., Am. Assoc. Petrol. Geol.* **73**, 143–154.

Gardner, J.V. (1982) High-resolution carbonate and organic-carbon stratigraphies for the late Neogene and Quaternary from the western Caribbean and eastern equatorial Pacific. In: *Initial Reports of the Deep Sea Drilling Project*, vol. 68, (Eds Prell, W.L. & Gardner, J.V.) pp. 347–364. US Government Printing Office, Washington, DC.

Goldberg, L. & Carter, R.M. (1979) A Pleistocene molluscan fauna beneath the Maui-A gas platform, offshore Taranaki (Note). *N. Z. J. Geol. Geophys.* **22**, 407–409.

Greenlee, S.M. & Moore, T.C. (1988) Recognition and interpretation of depositional sequences and calculation of sea-level changes from stratigraphic data — offshore New Jersey and Alabama Tertiary. In: *Sea-level Changes: an Integrated Approach* (Eds Wilgus, C.K., Hastings, B.S., Kendall, C.G.St.C., Posamentier, H., Ross, C.A. & van Wagoner, J.). Soc. Econ. Paleontol. Mineral. Spec. Publ. 42, 329–353.

Haq, B.U., Hardenbol, J. & Vail, P.R. (1987) Chronology of fluctuating sea levels since the Triassic. *Science* **235**, 1156–1167.

Hays, J.D., Imbrie, J. & Shackleton, N.J. (1976) Variations in the earth's orbit: pacemaker of the ice ages. *Science* **194**, 1121–1132.

Haywick, D.W., Lowe, D.A., Henderson, R.A., Beu, A.G. & Carter, R.M. (1991) Plio-Pleistocene (Nukumaruan) lithostratigraphy of the Tangoio block, central Hawke's Bay, New Zealand. *N. Z. J. Geol. Geophys.* (in press).

Herzer, R.H. (1981) Late Quaternary stratigraphy and sedimentation of the Canterbury continental shelf, New Zealand. *N. Z. Oceanogr. Inst. Mem.* **89**, 1–71.

Hubbard, R.J. (1988) Age and significance of sequence boundaries on Jurassic and Early Cretaceous rifted continental margins. *Bull., Am. Assoc. Petrol. Geol.* **72**, 49–72.

Hubbard, R.J., Pape, J. & Roberts, D.G. (1985a) Depositional sequence mapping as a technique to establish tectonic and stratigraphic framework and evaluate hydrocarbon potential on a passive continental margin. In: *Seismic Stratigraphy, II: an Integrated Approach to Hydrocarbon Exploration* (Eds Berg, O.R. & Woolverton, D.G.). Mem., Am. Assoc. Petrol. Geol. 39, 79–91.

Hubbard, R.J., Pape, J. & Roberts, D.G. (1985b) Depositional sequence mapping to illustrate the evolution of a passive continental margin. In: *Seismic Stratigraphy, II: an Integrated Approach to Hydrocarbon Exploration* (Eds Berg, O.R. & Woolverton, D.G.). Mem., Am. Assoc. Petrol. Geol. 39, 93–115.

Johnson, D.P. & Searle, D.E. (1986) Post-glacial seismic stratigraphy, central Great Barrier Reef, Australia. *Sedimentology* **31**, 335–352.

Johnson, D.P., Searle, D.E. & Hopley, D. (1982) Positive relief over buried post-glacial channels, Great Barrier Reef Province, Australia. *Mar. Geol.* **46**, 149–159.

Kidwell, S.M. (1984) Outcrop features and origin of basin margin unconformities in the lower Chesapeake Group (Miocene), Atlantic coastal plain. In: *Interregional Unconformities and Hydrocarbon Accumulation* (Ed. Schlee, J.S.). Mem., Am. Assoc. Petrol. Geol. 36, 37–58.

Kidwell, S.M., Fursich, F.T. & Aigner, T. (1986) Conceptual framework for the analysis and classification of fossil concentrations. *Palaios 1 Res. Rep.* 228–238.

Kominz, M.A. (1984) Oceanic ridge volumes and sea-level change — an error analysis. In: *Interregional Unconformities and Hydrocarbon Accumulation* (Ed. Schlee, J.S.). Mem., Am. Assoc. Petrol. Geol. 36, 109–127.

Lewis, K.B. (1971) Growth rate of folds using tilted wave-planed surfaces: coast and continental shelf, Hawke's Bay, New Zealand. *R. Soc. N. Z. Bull.* **9**, 225–231.

Lewis, K.B. (1973) Erosion and deposition on a tilting continental shelf during Quaternary oscillations of sea level. *N. Z. J. Geol. Geophys.* **16**, 281–301.

Lewis, K.B. (1974) The continental terrace. *Earth Sci. Rev.* **10**, 37–71.

Lewis, K.B. (1980) Quaternary sedimentation on the Hikurangi oblique-subduction and transform margin, New Zealand. *Spec. Publs. Int. Ass. Sediment.* **4**, 171–189.

Loutit, T.S., Hardenbol, J., Vail, P.R. & Baum, G.R. (1988) Condensed sections: the key to age determination and correlation of continental margin sequences. In: *Sea-level Changes: an Integrated Approach* (Eds Wilgus, C.K., Hastings, B.S., Kendall, C.G.St.C., Posamentier, H., Ross, C.A. & van Wagoner, J.). Soc. Econ. Paleontol. Mineral., Spec. Publ. 42, 183–213.

Martin, A.K. & Flemming, B.W. (1986) The Holocene shelf sediment wedge off the south and east coast of South Africa. In: *Shelf Sands and Sandstones* (Eds Knight, R.J. & McLean, J.R.). Can. Soc. Petrol. Geol. Mem. 11, 27–44.

Miall, A.D. (1986) Eustatic sea level changes interpreted from seismic stratigraphy: a critique of the methodology with particular reference to the North Sea Jurassic record. *Bull., Am. Assoc. Petrol. Geol.* **70**, 131–137.

Moore, P.R. (1987) Age of the raised beach ridges at Turakirae Head, Wellington: a reassessment based on new radiocarbon dates. *J. R. Soc. N. Z.* **17**, 313–324.

Mougenot, D., Boillot, G. & Rehault, J.-P. (1983) Prograding shelf-break types on passive continental margin: some European examples. *Soc. Econ. Paleontol. Mineral., Spec. Publ.* **33**, 61–77.

Nelson, C.S., Hendy, C.H., Jarrett, G.R. & Cuthbertson, A.M. (1985) Near-synchroneity of New

Zealand alpine glaciations and Northern Hemisphere continental glaciations during the past 750 kyr. *Nature* **318**, 361–363.

Parkinson, N. & Summerhayes, C. (1985) Synchronous global sequence boundaries. *Bull., Am. Assoc. Petrol. Geol.* **69**, 685–687.

Pilaar, W.F.H. & Wakefield, L.L. (1978) Structural and stratigraphic evolution of the Taranaki Basin, offshore North Island, New Zealand. *Assoc. Petrol. Exploration. Aust. J.* **18**, 93–101.

Pillans, B. (1983) Upper Quaternary marine terrace geochronology and deformation, South Taranaki, New Zealand. *Geology* **11**, 292–297.

Pillans, B. (1986) A late Quaternary uplift map for North Island, New Zealand. *R. Soc. N. Z. Bull.* **24**, 409–417.

Pitman, W.C. (1978) Relationship between eustacy and stratigraphic sequences of passive margins. *Geol. Soc. Am. Bull.* **89**, 1389–1403.

Plint, A.G. (1988) Global eustacy and the Eocene sequence in the Hampshire Basin, England. *Basin Res.* **1**, 11–22.

Poag, C.W. & Ward, L.W. (1987) Cenozoic unconformities and depositional super sequences of North Atlantic continental margins. *Geology* **15**, 159–162.

Posamentier, H.W. & Vail, P.R. (1988) Eustatic controls on clastic deposition, II — sequence and systems tract models. In: *Sea-level Changes: an Integrated Approach* (Eds Wilgus, C.K., Hastings, B.S., Kendall, C.G.St.C., Posamentier, H., Ross, C.A., & van Wagoner, J.). Soc. Econ. Paleontol. Mineral., Spec. Publ. 42, 125–154.

Posamentier, H.W., Jervey, M.T. & Vail, P.R. (1988) Eustatic controls on clastic deposition, I — conceptual framework. In: *Sea-level Changes: an Integrated Approach* (Eds Wilgus, C.K., Hastings, B.S., Kendall, C.G.St.C., Posamentier, H., Ross, C.A. & van Wagoner, J.). Soc. Econ. Paleontol. Mineral., Spec. Publ. 42, 109–124.

Rahmanian, V.D., Moore, P.S., Mudge, W.J. & Spring, D.E. (1990) Sequence stratigraphy and the habitat of hydrocarbons, Gippsland Basin, Australia. In: *Classic Petroleum Provinces* (Ed. Brooks, J.). Geol. Soc. London, Spec. Publ. (in press).

Ruddiman, W.E., Raymo, M.E., Martinson, D.G., Clement, B.M. & Backman, J. (1989) Pleistocene evolution: Northern Hemisphere ice sheets and North Atlantic Ocean. *Paleoceanography* **4**, 353–412.

Sanders, K.E. & Kumar, N. (1975) Evidence of shoreface retreat and in-place 'drowning' during Holocene submergence of barriers, shelf off Fire Island, New York. *Geol. Soc. Am. Bull.* **86**, 65–76.

Seward, D. (1976) Tephrostratigraphy of the marine sediments in the Wanganui Basin, New Zealand. *N. Z. J. Geol. Geophys.* **19**, 9–20.

Seward, D. (1979) Comparison of zircon and fission-track ages from tephra horizons. *Geology* **7**, 479–482.

Shackleton, N.J. & Opdyke, N.D. (1973) Oxygen isotope and palaeo-magnetic stratigraphy of equatorial Pacific core V28–238: oxygen isotope temperatures and ice volumes in a $10^5$ and $10^6$ year scale. *Quaternary Res.* **3**, 39–55.

Shackleton, N.J. & Opdyke, N.D. (1976) Oxygen isotope and paleomagnetic stratigraphy of Pacific core V28–239: late Pliocene to latest Pleistocene. In: *Investigations of Late Quaternary Paleoceanography and Paleoclimatology* (Eds Cline, R.M. & Hays, J.D.). Geol. Soc. Am. Mem. 145, 449–464.

Shell Oil Company (1987) Seismic sequences as applied to basin-fill analysis: Taranaki Basin, New Zealand. In: *Atlas of Seismic Stratigraphy* (Ed. Bally, A.W.). Am. Assoc. Petrol. Geol. Stud. Geol. 27 (1), 53–71.

Snyder, S.W., Hine, A.C. & Riggs, S.R. (1989) Subsidence versus eustatic flooding rates: their relative roles in controlling preservation of coastal and shelf sequences. *Abstracts, International Geological Congress*, Washington, **3**, 145.

Sporli, K.B. (1980) New Zealand and oblique-slip margins: tectonic development up to and during the Cainozoic. In: *Sedimentation in Oblique-slip Mobile Zones* (Eds Ballance, P.F. & Reading, H.G.). Spec. Publs. Int. Ass. Sediment. 4, 147–170.

Stern, T.A. & Davey, F.J. (1989) Crustal structure and origins of basins formed behind the Hikurangi subduction zone, New Zealand. In: *Origin and Evolution of Sedimentary Basins and their Energy and Mineral Resources* (Ed. Price, R.A.). International Union of Geodesy & Geophysics and American Geophysical Union, Geophys. Monogr. 48, 73–85.

Stevens, K.F. & Vella, P. (1981) Palaeoclimatic interpretation of stable isotope ratios in molluscan fossils from middle Pleistocene strata, Wanganui, New Zealand. *Palaeogeogr. Palaeoclimatol. Palaeoecol.* **34**, 257–265.

Stubblefield, W.L., McGrail, D.W. & Kersey, D.G. (1984) Recognition of transgressive and post-transgressive sand ridges on the New Jersey continental shelf. *Soc. Econ. Paleontol. Mineral. Spec. Publ.* **3**, 1–23.

Swift, D.J.P., Kofoed, J.W., Salisbury, F.P. & Sears, P. (1972) Holocene evolution of the shelf surface, south and central Atlantic shelf of North America. In: *Shelf Sediment Transport: Process and Pattern* (Eds Swift, D.J.P. *et al.*) pp. 499–574. Dowden, Hutchinson & Ross, Stroudsburg, PA.

Turner, G.M. & Kamp, P.J.J. (1990) Paleomagnetic location of the Jaramille Subchron and Brunhes. Matuyama transition in the Castlecliffian stratotype section. *Earth Planet Sci. Lett.* **100**, 42–50.

Vail, P. (1987) Seismic stratigraphy interpretation using sequence stratigraphy, part 1: seismic stratigraphy interpretation procedure. In: *Atlas of Seismic Stratigraphy* (Ed. Bally, A.W.). Am. Assoc. Petrol. Geol. Stud. Geol. 27, 1–10.

Vail, P.R. & Todd, R.G. (1981) Northern North Sea Jurassic unconformities, chronostratigraphy and sea-level changes from seismic stratigraphy. In: *Petroleum Geology of the Continental Shelf, Northwest Europe* (Eds Cling, J.V. & Hobson, C.D.) pp. 216–235. Institute of Petroleum, London.

Vail, P.R., Mitchum, R.M. & Thompson, S. (1977) Seismic stratigraphy and global changes of sea level, part 4: global cycles of relative changes of sea level. In: *Seismic Stratigraphy — Applications to Hydrocarbon Exploration* (Ed. Payton, C.E.). Mem., Am. Assoc. Petrol. Geol. 26, 83–97.

Vail, P.R., Hardenbol, J. & Todd, R.G. (1984) Jurassic unconformities, chronostratigraphy, and sea-level changes from seismic stratigraphy and biostratigraphy. In: *Interregional Unconformities and Hydrocarbon Accumulation* (Ed. Schlee, J.S.). Mem., Am. Assoc.

Petrol. Geol. 36, 129–144.

Vail, P.R., Audemard, F., Bartek, L.R., Bowman, S., Coterill, K., Emmet, P.A., Liu, C., Perez, G., Ross, M.R. & Wu, S. (1989) The global stratigraphic signature of the Neogene. *Abstracts, Sea-level Changes at Active Plate Margins*, p. 12. Geological Society of London and International Association of Sedimentologists, Discussion Meeting.

van Wagoner, J.C. & Mitchum, R.M. (1989) High-frequency sequences and their stacking patterns. *Abstracts, International Geological Congress*, Washington, **3**, 284.

van Wagoner, J.C., Posamentier, H.W., Mitchum, R.M., Vail, P.R., Sarg, J.F., Loutit, T.S. & Hardenbol, J. (1988) An overview of the fundamentals of sequence stratigraphy and key definitions. In: *Sea-level Changes: an Integrated Approach* (Eds Wilgus, C.K., Hastings, B.S., Kendall, C.G.St.C., Posamentier, H., Ross, C.A. & van Wagoner, J.). Soc. Econ. Paleontol. Mineral., Spec. Publ. 42, 39–45.

Watts, A.B. (1982) Tectonic subsidence, flexure and global changes of sea level. *Nature* **297**, 470–474.

Weller, J.M. (1930) Cyclic sedimentation of the Pennsylvanian Period and its significance. *J. Geol.* **38**, 97–135.

Williams, D.F. (1988) Evidence for and against sea-level changes from the stable isotopic record of the Cenozoic. In: *Sea-level Changes: an Integrated Approach* (Eds Wilgus, C.K., Hastings, B.S., Kendall, C.G.St.C., Posamentier, H., Ross, C.A. & van Wagoner, J.). Soc. Econ. Paleontol. Mineral., Spec. Publ. 42, 30–36.

Williams, D.F., Thunell, R.C., Tappa, E., Domenico, R. & Raffi, I. (1988) Chronology of the Pleistocene oxygen isotope record: 0–1.88 m.y. b.p. *Palaeogeogr. Palaeoclimatol. Palaeoecol.* **64**, 221–240.

## APPENDIX: SHELF-BREAK TERMINOLOGY

In its original manifestation, recognition of a type 1 Exxon sequence boundary required documentation of a sea-level fall beneath the contemporary 'shelf edge', thereby producing a shelf-wide erosional unconformity (Vail *et al.*, 1977, 1984). In some cases, this was interpreted to mean the true edge of the continental shelf; in others, to represent the seaward break-point of the contemporary shore-connected coastal sediment prism. Confusion between these two concepts is not trivial since the shelf edge is often relict and located at depths >100 m (e.g. Mougenot *et al.*, 1983), whereas the coastal sediment prism is generally in quasi-equilibrium and has a seaward edge rarely deeper than c. 20 m.

In acknowledgment of this problem, Vail (1987) and Posamentier & Vail (1988) have recently substituted the term 'depositional shoreline break' (DSB) for use in Exxon-type sequence analysis, defining it as 'the concurrent physiographic break, landward of which the sea floor is at or near base level and seaward of which the sea floor is below base level. The break may occur at or seaward of the shoreline. Consequently, it refers to an active depositional physiographic break as opposed to a pre-existing or relict physiographic break at the edge of the continental shelf' (Posamentier & Vail, 1988, p. 125). Unfortunately this redefinition merely compounds the confusion, since (1) some modern shelf breaks are active rather than relict features (e.g. Lewis, 1973), and (2) seismic profiles contain numerous examples of both the depositional shoreline break and the shelf-edge break, without it often being apparent which is which.

An analysis of the modern shelf in terms of the sequence-stratigraphic model can be applied to clarify the confusion regarding the terms shelf break and DSB. Within the modern shelf system there are at least three separate points at which a seawards facing break in depositional slope can occur:

**1** at the seaward edge of the HST shore-connected sediment prism, the 10–20 m depth of which is controlled largely by local wave base;

**2** at the seaward edge of a HST cape-connected longshore bar or submergent spit, generally at depths of c. 5–40 m (e.g. Carter *et al.*, 1985, fig. 9; Martin & Fleming, 1986, fig. 10);

**3** at the shelf edge, the c. 50–200 m depth of which is controlled by the long-term interplay of sediment supply, subsidence and eustasy (e.g. Mougenot *et al.*, 1983).

In addition, during the fifth-order cycles of the Plio-Pleistocene, episodic transgression or regression may lead to the deposition of perched coastal sediment bodies — each with a DSB — at any position on the shelf (Fig. 11), resulting in the preservation of more than three clinoform break-points.

All these features may appear on seismic records as geometrically equivalent breaks in depositional slope (clinoform break), or as ramps, and either of the two shore-connected features may in some cases coincide with the third, i.e. with the shelf edge. Furthermore, there is no obvious way to identify such features unequivocally on a single profile, except where they occur in pairs along a single reflector. It follows that genetic terminology must be eschewed, and we suggest the descriptive terms 'clinoform break-point' (CBP) or 'ramp edge' be used, as appropriate.

A 'double clinoform' shelf depositional pattern is particularly characteristic of the early highstand

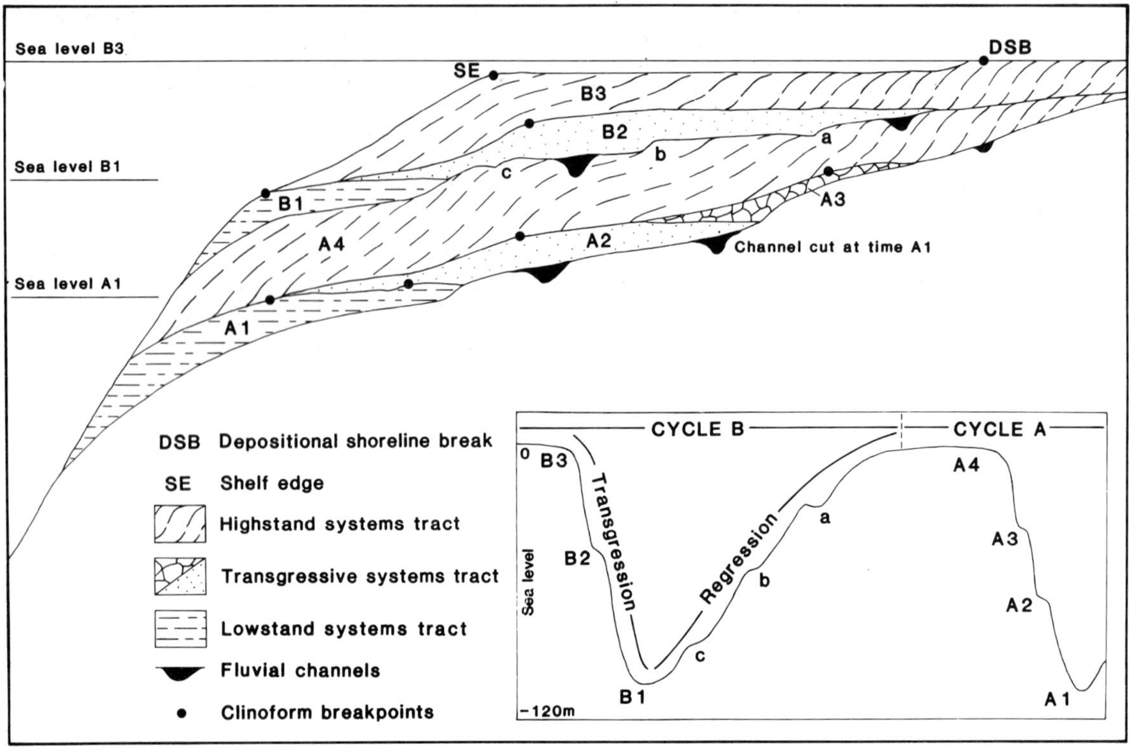


**Fig. 11.** Diagrams illustrating the nature of the depositional shoreline break (DSB) and shelf edge with respect to a sequence-stratigraphic interpretation of typical Plio-Pleistocene fifth-order sea-level cycles (amended after Carter *et al.*, 1986).

during periods of glacio-eustasy, as during the present interglacial (Fig. 11). In areas of high sediment supply, the DSB and the shelf break may both be visible on multifold seismic profiles (e.g. the Gulf of Mexico; Boyd *et al.*, 1988). More usually, the DSB is 10 m or less thick, making it undetectable except on high-resolution 3.5 kHz or similar profiles (thus discrepant interpretations are potentially possible between 3.5 kHz/outcrop studies and those based on multifold seismic profiles). During a glacial lowstand, the shore-connected prism break-point (now the LST) may come to coincide with the shelf-edge break-point, or even be located seawards of it on the upper slope (in which case a 'double clinoform' will again be apparent).

During non-glacial periods of Earth history, the shelf edge in general marks the bypass point for fine-grained sediment derived from a contemporary but distant shoreline, resulting in a 'single clinoform' depositional pattern on the shelf. However, a double clinoform pattern could also occur at such times if sea level/subsidence relations favoured a transgression of the shoreline at a rate faster than the contemporary rate of coastal progradation.

The simple Exxon SSM assumes the occurrence of a single clinoform break-point and that the magnitude of the causatory sea-level cycle is sufficient to locate the LST below the true shelf break. Clearly, this is not always the case. In cases of ambiguity it is therefore preferable to use the term 'clinoform break-point' rather than 'shelf break' (which is genetic) or 'roll-over' (which has a specific and different meaning in petroleum geology) in reference to these features.

## REFERENCES

Boyd, R., Suter, J. & Penland, S. (1988) Implications of modern sedimentary environments for sequence stratigraphy. In: *Sequences, Stratigraphy, Sedimentology: Surface and Subsurface* (Eds James, D.P. & Leckie, D.A.). Can. Soc. Petrol. Geol. Mem. 15, 33–36.

Carter, R.M., Carter, L., Williams, J. & Landis, C.A. (1985) Modern and relict sedimentation on the Otago continental shelf. *N. Z. Oceanogr. Inst. Mem.* **93**, 43 pp.

Lewis, K.B. (1973) Erosion and deposition on a tilting continental shelf during Quaternary oscillations of sea level. *N. Z. J. Geol. Geophys.* **16**, 281–301.

Mougenot, D., Boillot, G. & Rehault, J.-P. (1983) Prograding shelf-break types on passive continental margin: some European examples. *Soc. Econ. Paleontol. Mineral., Spec. Publ.* **33**, 61–77.

Posamentier, H.W. & Vail, P.R. (1988) Eustatic controls on clastic deposition, II — sequence and systems tract models. In: *Sea-level Changes: an Integrated Approach* (Eds Wilgus, C.K., Hastings, B.S., Kendall, C.G.St.C., Posamentier, H., Ross, C.A. & van Wagoner, J.). Soc. Econ. Paleontol. Mineral., Spec. Publ. 42, 125–154.

Vail, P. (1987) Seismic stratigraphy interpretation using sequence stratigraphy, part 1: seismic stratigraphy interpretation procedure. In: *Atlas of Seismic Stratigraphy* (Ed. Bally, A.W.). Am. Assoc. Petrol. Geol. Stud. Geol. 27, 1–10.

Vail, P.R., Mitchum, R.M. & Thompson, S. (1977) Seismic stratigraphy and global changes of sea level, part 4: global cycles of relative changes of sea level. *Mem., Am. Assoc. Petrol. Geol.* **26**, 83–97.

Vail, P.R., Hardenbol, J. & Todd, R.G. (1984) Jurassic unconformities, chronostratigraphy, and sea-level changes from seismic stratigraphy and biostratigraphy. In: *Interregional Unconformities and Hydrocarbon Accumulation* (Ed. Schlee, J.S.). Mem., Am. Assoc. Petrol. Geol. 36, 129–144.

# Rates, Magnitudes and Processes of Sea-Level Change

*Spec. Publs. int. Ass. Sediment.* (1991) **12**, 69–78

# High-level marine terraces in western and southern New Zealand: indicators of the tectonic tempo of an active continental margin

D.G. BISHOP

*New Zealand Geological Survey, Dept. of Scientific and Industrial Research, Private Bag, Dunedin, New Zealand*

## ABSTRACT

The techniques and results of recent studies of flights of high-level marine terraces, some extending to 1700 m above sea level, are described and discussed. A numerical correlation model relating terrace levels in New Zealand to the 'type' sequence in New Guinea gives results consistent with those derived from morphostratigraphic correlation with a partly dated New Zealand sequence. Both methods yield uplift rates of c. 3–8 mm/yr, in line with those determined by independent methods, and indicate an increased rate along the Alpine Fault, perhaps resulting from a change in relative motion between the Indo–Australian and Pacific plates, from about 135 000–140 000 years ago.

Greater attention to morphostratigraphy could enable a more satisfactory resolution of conflicting interpretations and allow more convincing correlations to be made.

## INTRODUCTION

The position of South Island, New Zealand, astride the Indo–Australian/Pacific plate boundary results in a high degree of neotectonic activity, including rapid modern and longer term uplift rates. The tectonic setting, coupled with its position in a large expanse of ocean in the westerly wind belt of the 'Roaring Forties', results in an environment that is particularly favourable to the formation and preservation of marine terraces, and their subsequent elevation to high altitudes.

This review discusses the results of recent work on sequences of very high-level (> 300 m) marine terraces. Numerous studies of low-level raised terraces have been made, particularly in the north and east of North Island, where fission-track dating and loess and tephrochronology have provided good age control (Pillans, 1986). The high-level terrace remnants, particularly of the west and south of South Island are more enigmatic. Absolute dating is seldom possible, and meaningful resolution of the sequences is difficult. Although disagreement exists in some areas, results consistent with independent data have been obtained by several workers and the potential of such analyses to unravel the late Quaternary tectonic history of active coastlines is considerable. Nevertheless, studies could be improved or have wider applicability if attention was given to recording relevant parameters in a uniform fashion.

### Tectonic setting

The on-land expression of the Indo–Australian/Pacific plate boundary in South Island is the Alpine Fault (Fig. 1). Oblique convergence of about 45 mm/yr (Walcott, 1979) across this feature has resulted in a component of shortening normal to the fault of about 20 mm/yr and in the uplift of the Southern Alps. The Alps are sharply bounded in the west by the dramatic faultline scarp of the Alpine Fault. Their uplift was underway at 7 Ma in the south and 5 Ma in the north, and appears to have accelerated in the last 1 Ma (Kamp *et al.*, 1989). The first influx of greywacke and schist gravels derived from the mountains appeared in west coast basins about 2.5 Ma ago. Total uplift is of the order of 25 km (Cooper, 1980). Uplift rates of between 5 and 20 mm/yr have been estimated from a variety of

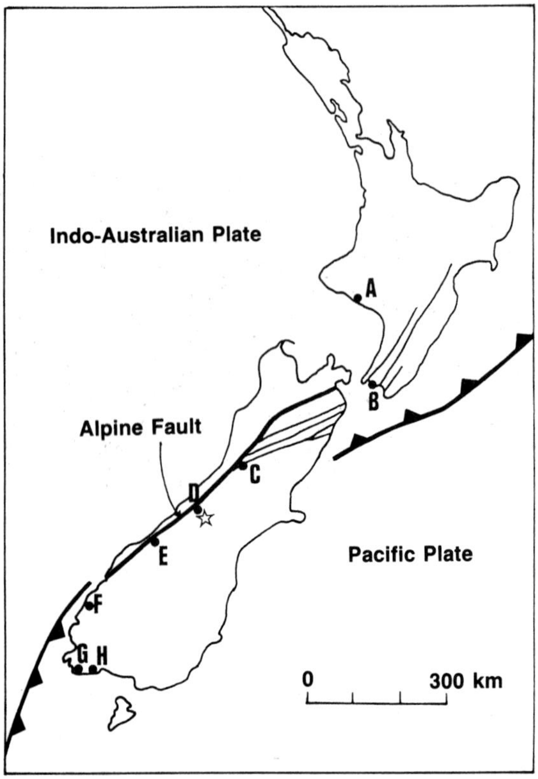


**Fig. 1.** Location map and tectonic setting of high-level marine terraces discussed. Area of Southern Alps with peaks over 3000 m is indicated by star. A, Taranaki; B, Cape Turakirae; C, Kaniere; D, Fox-Franz Josef; E, Haast; F, Caswell Sound; G, Fiordland: H, Waitutu.

sources (e.g. Suggate, 1963,1968; Cooper & Bishop, 1979; Wellman, 1979; Adams, 1980). Recent apatite and zircon fission-track studies indicated an average rate of 6.2 mm/yr for the last million years adjacent to the Alpine Fault in the central Southern Alps (Kamp *et al.*, 1989).

In the south, the discrete Fiordland massif, a crystalline complex of Palaeozoic and Mesozoic age (including Cretaceous granulites: Mattinson *et al.*, 1986), has also undergone major uplift, estimated at 10 km during the last 4–5 Ma (Oliver, 1980).

### Terrace nomenclature

The nomenclature for description of raised marine terraces is shown in Fig. 2. Of particular importance is the position of the strandline (the inland edge of the abrasion platform). Unfortunately this is perhaps the most difficult feature to locate accurately in the field, being obscured typically by cover beds, and/or rockfall, landslide, and slopewash deposits. Where cover beds have been stripped, the abrasion surface, cliffline, and strandline may be degraded. The errors introduced by substituting the position of the break of slope for terraces either mantled by cover beds, or stripped and degraded may be relatively insignificant in many cases, but nevertheless contribute to the uncertainty regarding the total uplift.

### Early work

Cooper & Bishop (1979) drew attention to a prominent flat bench or step (Fig. 3) cut in several adjacent schist ridges running out to the faultline scarp of the Alpine Fault at Okuru, near Haast. These steps are at an altitude of 1450 m above sea level (a.s.l.) and are up to 1.3 km wide. A less prominent step about 300 m wide is also developed at 960 m. Scattered, rounded, quartz granules with surface textures consistent with abrasion in a high-energy marine environment, and comparable to textures of pebbles on the modern beach, were recovered from several of the steps. It was concluded therefore that the steps were the remnants of strongly uplifted marine terraces. No indication of the age was available, but an estimate of 192 000 and 290 000 years for the lower and upper steps, respectively, was made on the assumption that a locally determined uplift rate of 5 mm/yr for the last 10 000 years was representative of the long-term rate (Cooper & Bishop, 1979). Alternatively, uplift rates of

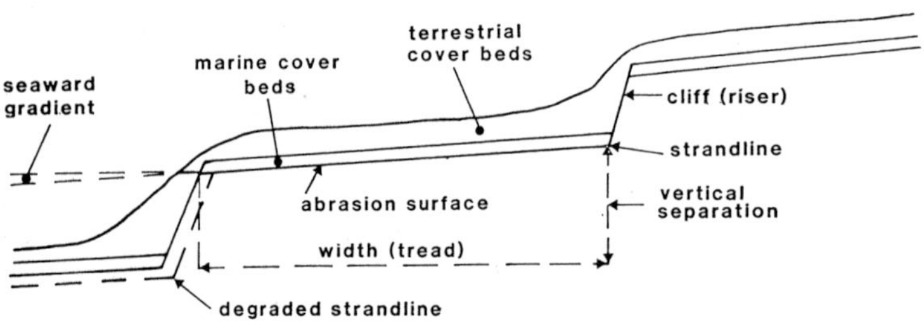


**Fig. 2.** Recommended nomenclature for the description of raised marine terraces.

**Fig. 3.** Ridges of the Southern Alps rising from the Alpine Fault (at base of slope) near Haast have prominent flat steps at 960 m (1) and 1450 m (2). Rounded quartz granules on them support their interpretation as elevated remnants of former marine terraces.

7.7 mm/yr or 11.6 mm/yr were calculated on the basis that either the 960 m or the 1450 m terrace corresponded to the 125 000 year terrace that is developed in many parts of the world.

## The global sequence

Over the next few years, the increasing availability of data from uplifted shore platforms and reef complexes led Bull (1984) to develop a technique for correlation of global sequences of marine terraces (Table 1), based primarily on the U–Th dated coral reef 'type' sequence of Huon Peninsula, New Guinea (Bloom *et al.*, 1974; Chappell, 1974,1983), the correlation of the reef sequences with the marine oxygen isotope stages (Shackleton & Mathews, 1977; Chappell & Shackleton, 1986), and the world-wide occurrence of marine terraces of broadly synchronous ages, notably about 80 000, 100 000, 120 000, 220 000, and 330 000 years (Bull, 1984; Ward, 1985). Bull argued that even though entire flights of terraces on some coastlines may lack any form of dateable deposit, consideration of the number of terraces and their altitudinal spacing gave each sequence a signature that allowed it to be uniquely correlated against 'the overall (global) population of dated terraces'. This hypothesis relies on the spacing of terraces within a flight being primarily determined by the altitude of formation of each terrace and on the intervals of time between successive periods of high sea level, neither of which has remained constant through time. A third, and critical, condition is that local uplift rates are uniform for long periods of

**Table 1.** The global terrace sequence, based on the sequence of marine terraces from Huon Peninsula, New Guinea. Simplified from Bull (1984)

| Oxygen stage number | Age (ka) | Altitude formed (m) |
|---|---|---|
| 1A | 6 | 0 |
| 2A | 30 | −42 |
| 3A | 40 | −37 |
| 3B | 46 | −37 |
| 3C | 57 | −29 |
| 3D | 64 | −26 |
| 4A | 76 | −46 |
| 5A | 83 | −13 |
| 5B | 94 | −20 |
| 5C | 103 | −10 |
| 5D | 120 | + 6 |
| 5E | 133 | + 5 |
| 6A | 176 | −21 |
| 6B | 202 | −17 |
| 7A | 214 | − 3 |
| 7B | 242 | −28 |
| 8A | 286 | −46 |
| 8B | 305 | −27 |
| 9A | 320 | + 4 |
| 9B | 336 | + 4 |

time, at least when averaged over periods of several tens of thousands of years.

Bull also noted that the number of terraces present also depends on the uplift rate, rates of 1–2 mm/yr being needed before a full sequence can be preserved. At lower uplift rates some terraces formed during relatively low sea-level peaks will be swamped by subsequent higher sea levels. Bull also argues that potential complications resulting from temporal variations or distortions of the marine geoid (Morner, 1976) or different responses of different parts of the crust to melting icesheets do not appear to have had a significant effect, at least in the timespan (c. 500 000 years) involved in the formation of late Quaternary marine terraces. Correlation with the global sequence also involved incorporating correction for the height of formation of the terrace relative to present-day sea level, a very significant adjustment for young terraces at low uplift rates. A 40 000 year terrace uplifted at 1 mm/yr for instance, would have a measured altitude of 3 m a.s.l., but a eustatic correction of 37 m, equal to the postulated height of the 40 000 year sea-level peak below present-day sea level. Fortunately the relative size of the correction decreases as uplift rates rise and terrace ages increase. Nevertheless there is a case for basing the correlation on those terraces formed close to, or above, present-day sea level (Table 1). These are also the major transgressions and as such are more likely to have left prominent terraces.

## NEW ZEALAND TERRACES

### West coast

Bull & Cooper (1986) applied the technique of trial and error sequential matching (Fig. 4) of the terrace flight in question to the New Guinea (global) sequence at three sites along a 250 km stretch of the Alpine Fault, at Haast, Fox-Franz Josef, and Kaniere. At each site they found subtle remnants of a flight of 18 terraces occurring as notched spurs, flat-crested ridges, and concordant summits. Rounded quartz pebbles were recovered from 16 of the 18 levels, but not from intermediate positions. 'Best fit' correlations with the New Guinea sequences in each case yielded an uplift rate that showed an abrupt increase from about 135 000 years ago, from 2.6 to 5.5 mm/yr at Kaniere, from 3.2 to 7.8 mm/yr at Fox-Franz Josef, and from 3.6 to 5.8 mm/yr at Haast (Fig. 5). The relative extent of the terraces was not described, nor was a profile given. The prominent 1450 m step earlier described from Okuru (Cooper & Bishop, 1979) was allocated an age of 320 000 years, and the 960 m bench became the 176 000 year level.

This technique, its data, and its interpretation were disputed by Ward (1988a). Ward's criticisms were, however, rebutted by Bull & Cooper (in Ward, 1988a), and while the author retains reservations as

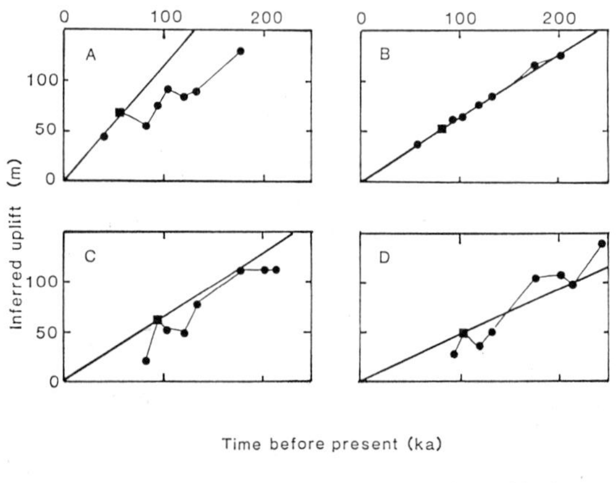


**Fig. 4.** Trial and error method for correlating with the global sequence, by allocating different ages to a terrace at 39 m. (A) allocated an age of 57 ka, (B) 83 ka, (C) 94 ka, and (D) 103 ka. Only one solution (B) gives a uniform uplift rate passing through the origin. Redrawn from Bull (1984, fig. 4).

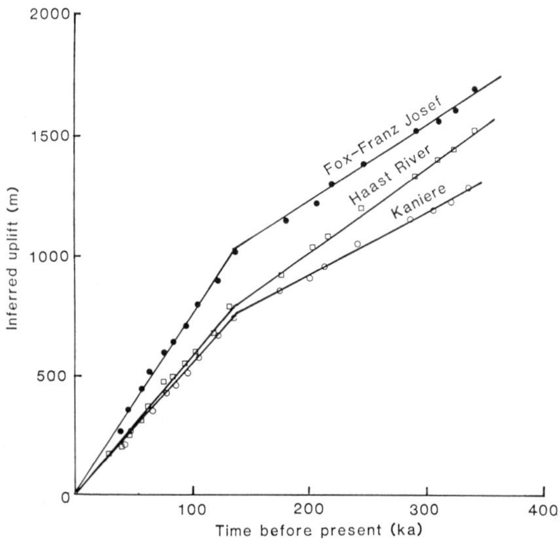


**Fig. 5.** Uplift rates inferred from flights of marine terrace remnants along the Alpine Fault. Redrawn from Bull & Cooper (1986, fig. 3).

to the origin of some of the steps, it must be admitted that their results are both reasonable and internally consistent.

### Taranaki

Meanwhile, Pillans (1983) had completed a major study of a flight of 12 undoubtedly marine terraces in south Taranaki, on the west coast of North Island. The Taranaki terraces rise to more than 300 m a.s.l. and extend up to 20 km inland from the coast, and are cut on relatively soft, shallow marine sediments of Plio-Pleistocene age. They can be traced laterally for more than 100 km.

The terraces are draped with up to 12 m of marine cover beds, including basal conglomerate, sand, and shell layers, in turn overlain by peat and a variety of littoral deposits including fluvial sediment, dune sand, loess, and tephra. Pillans was able to demonstrate deformation by broad doming including a seaward component of tilt, resulting in uplift rates increasing inland from 0.3 to 0.65 mm/yr. Numerous dates were obtained by radiocarbon, fission-track tephrochronology, and amino-acid epimerization of fossil wood. These dates allowed the ages of two terraces, at 120 000 and 400 000 years, to be estimated with confidence, and provided minimum ages for the remainder. With uplift rates determined, the ages of the remaining terraces were then calculated from a relation diagram using best-fit linear regressions (Fig. 6). The resulting ages, at 60 000, 80 000, 100 000, 120 000, 210 000, 310 000, 340 000, 400 000, 450 000, 520 000, 600 000, and 680 000 years are consistent both with the marine oxygen isotope stages and with the relevant part of the New Guinea sequence. The regression equations for the younger terraces also provide some independent confirmation of the height of formation of each terrace.

### Fiordland

A spectacular array of marine terraces is developed in Fiordland, in the far southwest of South Island (Bishop, 1985,1986) and figured in one of the first ever pictures of New Zealand, a sketch by Captain James Cook in 1770 (Begg & Begg, 1973). (A map of these terraces, printed in 1986, was submitted several years before the 1985 paper was written. The 1985 paper thus contains a later interpretation.)

The most prominent feature of the terrace sequence is the presence of two very extensive, wide, well-preserved, low-relief surfaces (H6 and H7) at 370 m and 500 m a.s.l. Together they form a bench some 10 km wide, complete with fossil stacks and islands. The higher terrace (H7) is backed by a distinct fossil cliffline (Fig. 7). Above these two major surfaces the remnants consist of small cut benches on mountain ridges; at lower altitudes the remnants are relatively narrow shelf-like surfaces. The terraces are cut on Ordovician sediments and hornfelses, Palaeozoic/Mesozoic granite, and a cover sequence of Cretaceous conglomerate and arkose.

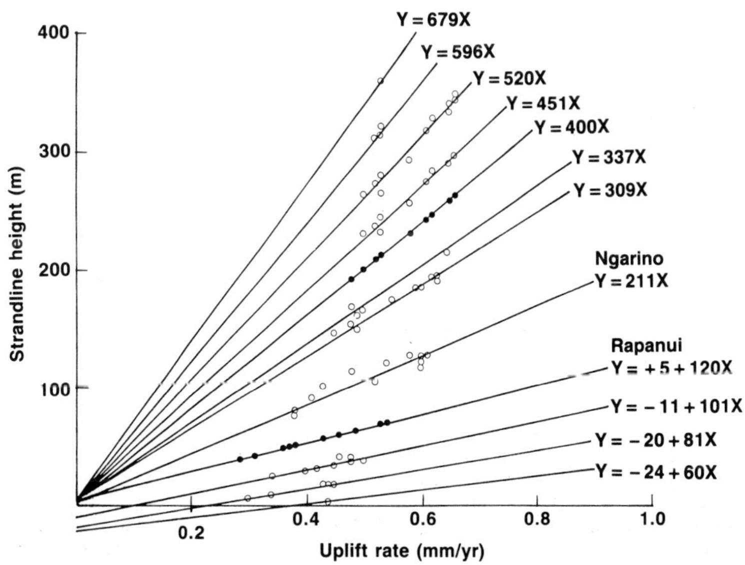


**Fig. 6.** Relation diagram for Taranaki terrace sequence, simplified from Pillans (1983, fig. 4).

**Fig. 7.** Raised marine terraces, southern Fiordland. The prominent shelf consists of terraces H6 and H7, which are backed by a fossil cliffline at an altitude of 500 m. Some higher remnants are arrowed.

Preservation is best on the older, harder rocks. An extensive shore platform with varying degrees of roughness is developed along much of the coastline, which is also characterized by stranded sea caves, notched headlands, and two-storied stacks. Except in coastal cliffs, the cover beds overlying the terraces are seldom exposed owing to the cover of dense forest, scrub, and swampy grassland. All deposits have been stripped from the higher remnants above H7. The apparent continuity for many kilometres to both the north and the east without change in altitude indicates that uplift has not been accompanied by tilting.

Three radiocarbon-dated wood samples (6820 ± 80, 3790 ± 70, 3620 ± 80 years BP) from or associated with the lowest raised surface, a very locally developed terrace 5–10 m a.s.l., indicate an uplift rate over the last 7000 years of about 1.5 mm/yr (Bishop, 1985).

The author's original interpretation of this sequence (Bishop, 1986) assumed that the 500 m terrace (H7) was equivalent to the 125 000 year global terrace and indicated an average uplift rate of c. 4 mm/yr and likely global ages for the prominent terraces of the rest of the sequence were calculated on this basis. Subsequently, however, Pillans' (1983) work was published and the striking morphostratigraphic similarity of the profiles of the terrace flights from the two areas was revealed (Fig. 8). Accordingly the author reinterpreted the sequence, with H6 being the correlative of the Rapanui (120 000 years) and H7 equivalent to the Ngarino terrace (210 000

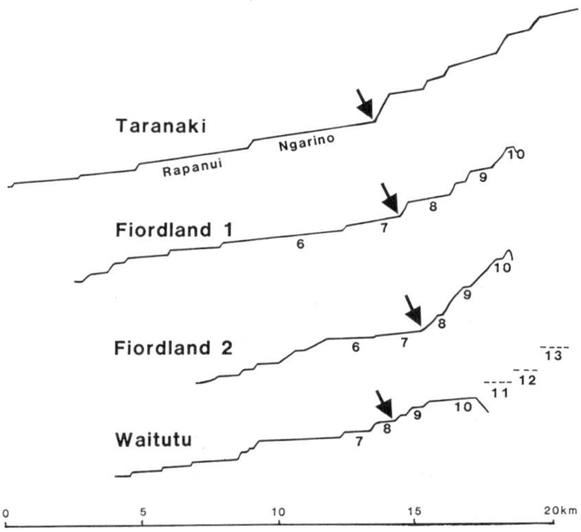


**Fig. 8.** Morphostratigraphic comparison of generalized terrace profiles from Taranaki, Fiordland, and Waitutu. The vertical exaggeration of the Fiordland and Waitutu profiles is 5, that of the Taranaki profile is 23, to compensate for the lower uplift rate in Taranaki (see Fig. 6). The arrowed strandline at the inland edge of a pair of prominent broad terraces separates each profile into a lower part with a low average gradient from an upper, steeper part, and is regarded as the key to correlation between sequences.

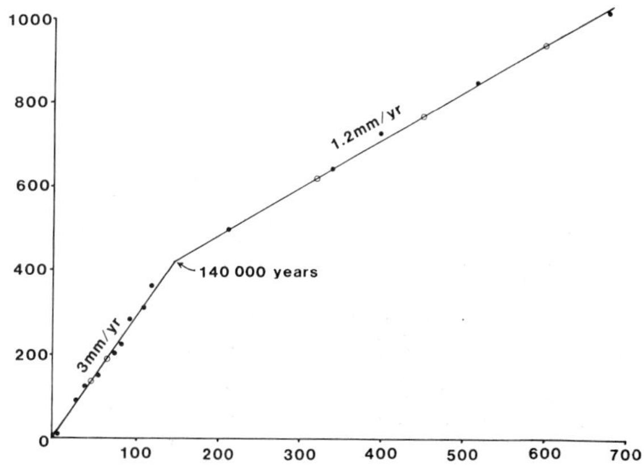


**Fig. 9.** Inferred uplift rate graph for Fiordland, assuming terraces H6 and H7 are the equivalents of the 120 000 and 210 000 terraces of the Taranaki sequence. Additional small terraces noted by Ward (1988b, table 3) are included. Open circles represent the absent 46 000 and 64 000 year terraces of the global sequence (Table 1), or terraces present in Taranaki but apparently not preserved in Fiordland.

years) of Taranaki and the rest of the sequence correlated more or less sequentially with that of New Guinea (Bishop, 1985). This indicated that the uplift rate fluctuated considerably (between 1.5 and 5.4 mm/yr) for the intervals between the formation of successive pairs of terraces, but averaged 3.2 mm/yr, and appeared to be higher from a time between the formation of H6 and H7, i.e. between 210 000 and 120 000 years ago, and the present.

For this review, the author has re-examined the data, and included the extra minor remnants noted by Ward (1988b) near the Andrew Burn. (Relatively minor or laterally impersistent features had previously been ignored.) Retaining the correlation of H6 and H7 and higher terraces with the Rapanui (120 000 years) and Ngarino (210 000 years) and older terraces of Taranaki, but using a Bull-type analysis for the terraces lower than H6, generates a set of ages in close agreement with the global sequence (Table 1). These indicate an increase in uplift rate since about 140 000 years ago (Fig. 9) very similar to that demonstrated by Bull & Cooper's (1986) results from the west coast. The Fiordland terraces thus appear to be consistent with both the Taranaki and west coast sequences, indicating both approaches (sequential correlation with either New Guinea or morphological correlation with Taranaki) yield compatible results.

### Waitutu district

In Bishop (1985), it was noted that the Fiordland terraces apparently extended some 25 km eastwards, across the Fiordland Boundary Fault and on to the soft, upper Tertiary siltstones of the Waitutu district, without any significant change in height. However, a subsequent analysis of the Waitutu district by Ward (1988b) concluded that the uplift rate there was only 1.1 mm/yr. Furthermore, Ward concluded that the Waitutu terraces, although not offset by the Hauroko Fault (Fiordland Boundary Fault of Bishop (1985,1986)), were significantly downwarped across a c. 5 km zone of poor terrace development west of the fault and in fact correlated with terraces at a lower altitude in Fiordland. The Fiordland uplift rate derived from this inferred relationship was only 0.6 mm/yr. Furthermore, there was no correspondence of terraces older than H5 or W9 between the two areas (Fig. 10).

This interpretation appears at variance with the geomorphic indications of the relatively rapid uplift

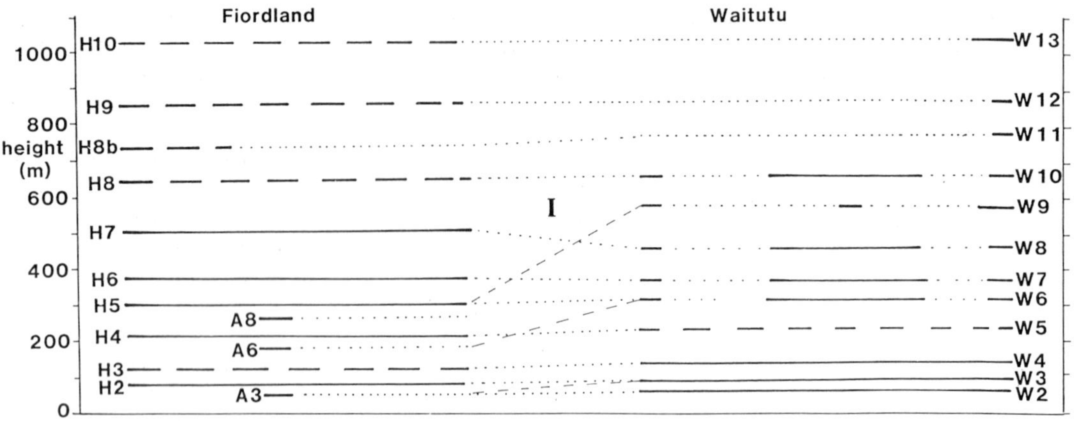


**Fig. 10.** Preferred correlation of Fiordland terraces H2–H10 (Bishop, 1985) and additional terraces A3, A6, and A8 from the Andrew Burn area (Ward, 1988b) with the Waitutu terraces W2–W13. The prominent surfaces are shown as solid lines, less prominent ones as dashes. The lateral extent is indicated diagrammatically by the length of the lines. The vertical error bar indicates the uncertainty (±30 m) in estimating the position of the strandlines. Correlations suggested by Ward are shown by light dashed lines.

of Fiordland with its steep, rugged coastline, deeply incised river valleys, and mountainous hinterland, compared with the lower, more subdued topography of the Waitutu area, even when allowing for the more resistant nature of the Fiordland rocks. The absence of Waitutu equivalents to the H6 and H7 terraces of Fiordland, which form such a prominent landscape feature, was also surprising.

Ward's correlation of the terraces of the two areas is debatable. Whereas the author mapped only the more prominent surfaces of Fiordland, Ward mapped the Waitutu district in much finer detail. Comparison of the overlapping portion of the two maps (Bishop, 1985, fig. 3; Ward, 1988b, fig. 2) shows that Fiordland terraces H2 and H3 include the Waitutu terraces W2, W3, and W4. Ward, however, also identified additional Fiordland terraces in the Andrew Burn area. Of these, A3 seems likely to be the equivalent of W2. The most probable correlation of the two sequences is shown in Fig. 10; errors, including individual bias, of up to ±30 m in estimating strandline altitudes from the 1:63 360 topographic maps of these heavily forested areas mean that the two sequences may be regarded as identical within the precision of the data. The absence of Waitutu equivalents of A6 and A8 (Fig. 10) may be because of their preservation on the harder rocks of Fiordland, but their destruction on the softer rocks of the Waitutu area.

Ward's (1988b) terrace ages are based on the relationship of the marine terrace sequence with a set of fluvioglacial terraces. The latter, however, with the exception of post-glacial deposits in the major valleys, are poorly developed, and confined

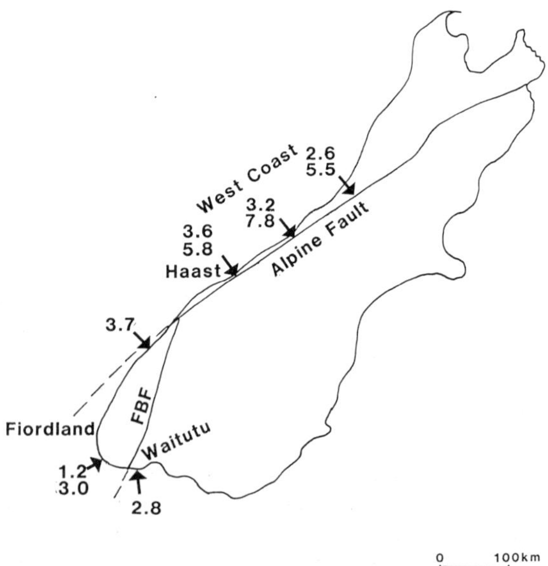


**Fig. 11.** Uplift rates in the west and south of South Island inferred from sequences of raised marine terraces. A double rate indicates an apparent change from the upper to the lower figure about 130 000–140 000 years ago. FBF, Fiordland Boundary Fault.

almost entirely to the relatively minor catchment of Crombie Stream; Group C terraces in fact are minute remnants shown only in a small subcatchment of Crombie Stream (Ward, 1988b, fig. 2). The nature of the terrace deposits is not described. Their assumed ages are based on discussions of the nature and extent of former glacial and drainage systems. Ward's generalized profile of the Waitutu marine terraces, however, is similar to both the Fiordland and Taranaki terrace profiles (Fig. 8), and indicates that Waitutu terraces W7 and W8 are the most likely correlatives of H6 and H7 in Fiordland, and the Rapanui and Ngarino terraces of Taranaki. On morphological criteria alone, W8 is a much more likely correlative of the Ngarino terrace than is the poorly developed W5, as was suggested by Ward.

In conclusion, therefore, the uplift rates of Fiordland and Waitutu are considered the same. Although the lower rate (1.1 mm/yr) favoured by Ward cannot be discounted, the author prefers the interpretation given in Fig. 9, on the strength of the morphostratigraphic correlation of both areas with Taranaki.

## CONCLUSIONS

Updated high-level marine terrace sequences in New Zealand can be interpreted provisionally either by direct correlation with the global terrace sequence, or by morphostratigraphic comparison with a partly dated local sequence. Both methods (alone or in combination) yield consistent results and indicate a change in uplift rate of sequences adjacent to the Alpine Fault about 135–140 thousand years ago (Fig. 11). A similar change reported from Vanuatu (Neef & Hendy, 1988) helps to confirm this result and suggests a change in convergence may have affected a significant part of the Indo–Australian/Pacific plate boundary about this time.

Careful attention to relative age data (degree of dissection, weathering, relationship to fluvioglacial deposits, degradation of former sea cliffs between terraces), to physical characteristics (width, lateral extent, seaward and lateral gradient, nature and thickness of cover beds), and to the overall characteristics of the sequence profile can help to confirm correlations. On coastlines of similar aspect, it is likely that certain terraces will tend to be more prominent than others. Identification of such surfaces can greatly assist in limiting correlation options with adjacent areas. More dating, however, is critical to resolve conflicting interpretations. Electron spin resonance or thermoluminescence dating of quartz grains in marine cover beds may be appropriate techniques where organic material is absent or unsuitable.

The uplift rates derived from these studies are consistent with those from other data (where available) or appropriate in terms of the inferred tectonic history. High-level marine terraces therefore offer a potentially detailed insight into both lateral and temporal aspects of the late Quaternary development of tectonically active coastlines, but more rigorous and uniform appraisals are needed, especially of features that can be quantitatively or semi-quantitatively assessed, if convincing conclusions are to be drawn.

A 'standard' high-level sequence on an active coastline elsewhere than in the western Pacific is urgently needed to supplement the New Guinea 'type' sequence, and in particular, to provide a set of dated terraces younger than 120 000 years. Darwin's observations (1846) of terrace flights extending to high altitudes in Chile suggest the west coast of South America could be a profitable study area.

## ACKNOWLEDGEMENTS

I am particularly grateful to Alan Cooper, Richard Norris, and Brad Pillans for helpful comments and discussion during the preparation of this chapter, and to the Geological Society of London and the International Association of Sedimentologists for the opportunity to present it in person at the discussion meeting "Sea-level changes at active plate margins" in May 1989.

## REFERENCES

Adams, J. (1980) Contemporary uplift and erosion of the Southern Alps, New Zealand. *Bull. Geol. Soc. Am.* **91**, 1–114.

Begg, A.C. & Begg, N.C. (1973) *Port Preservation*. Whitcoulls, Christchurch, 398 pp.

Bishop, D.G. (1985) Inferred uplift rates from raised marine surfaces, southern Fiordland, New Zealand. *N. Z. J. Geol. Geophys.* **28**, 243–251.

Bishop, D.G. (1986) Sheet B46 Puysegur. *Geological Map of New Zealand 1:50 000*. Department of Scientific and Industrial Research, New Zealand Geological Survey, Wellington.

Bloom, A.L., Broecker, W.S., Chappell, J., Matthews, R.K., & Mesolella, K.J. (1974) Quaternary sea level fluctuations on a tectonic coast: New $^{230}Th/^{234}U$ dates from the Huon Peninsula, New Guinea. *Quaternary Res.* **4**, 185–205.

Bull, W.B. (1984) Correlation of flights of global marine terraces. In: *Tectonic Geomorphology* (Eds Morisawa, M. & Hack, J.), Proceedings of the 15 Annual Geomorphology Symposium, State University of New York, Binghampton, pp. 129–152. Allen & Unwin, Boston.

Bull, W.B. & Cooper, A.F. (1986) Uplifted marine terraces along the Alpine Fault, New Zealand. *Science* **234**, 1225–1228.

Chappell, J. (1983) Revised sea-level record for the last 300 000 years from Papua New Guinea. *Search* **14**, 99–101.

Chappell, J. (1974) Geology of coral terraces, Huon Peninsula, New Guinea: a study of Quaternary tectonic movements and sea-level changes. *Bull. Geol. Soc. Am.* **85**, 553–570.

Chappell, J. & Shackleton, N.J. (1986) Oxygen isotopes and sea level. *Nature* **324**, 137–140.

Cooper, A.F. (1980) Retrograde alteration of chromium kyanite in metachert and amphibolite whiteschist from the Southern Alps, New Zealand, with implications for uplift on the Alpine Fault. *Contrib. Mineral. Petrol.* **75**, 153–164.

Cooper, A.F. & Bishop, D.G. (1979) Uplift rates and high-level marine platforms associated with the Alpine Fault at Okuru River, South Westland. *Bull. R. Soc. N.Z.* **18**, 35–46.

Darwin, C. (1846) *Geological Observations on South America*. Smith, Elder & Co, London, 279 pp.

Kamp, P.J.J., Green, P.F. & White, S.F. (1989). Fission track analysis reveals character of collisional tectonics in New Zealand. *Tectonics* **8**, 169–95.

Mattinson, J.M., Kimbrough, D.L. & Bradshaw, J.Y. (1986) Western Fiordland orthogneiss: Early Cretaceous arc magmatism and granulite facies metamorphism, New Zealand. *Contrib. Mineral. Petrol.* **92**, 383–392.

Morner, N.A. (1976) Eustasy and geoid changes. *J. Geol.* **84**, 123–151.

Neef, G. & Hendy, C. (1988) Late Pleistocene–Holocene acceleration of uplift rate in southwest Erromango Island, southern Vanuatu, South Pacific: relation to the growth of the Vanuatu mid-sedimentary basin. *J. Geol.* **96**, 481–494.

Oliver, G.J.H (1980) Geology of the granulite and amphibolite facies gneisses of Doubtful Sound, Fiordland, New Zealand. *N.Z. J. Geol. Geophys.* **23**, 27–41.

Pillans, B. (1983) Upper Quaternary marine terrace chronology and deformation, south Taranaki, New Zealand. *Geology* **11**, 292–297.

Pillans, B. (1986) A Late Quaternary uplift map for North Island, New Zealand. *Bull. R. Soc. N.Z.* **24**, 409–417.

Shackleton, N.J. & Mathews, R.Kü (1977) Oxygen isotope stratigraphy of late Pleistocene coral terraces in Barbados. *Nature* **268**, 618–620.

Suggate, R.P. (1963) The Alpine Fault. *Trans. R. Soc. N.Z.* **2**, 105–129.

Suggate, R.P. (1968) The Paringa Formation, Westland, New Zealand. *N.Z. J. Geol. Geophys.* **11**, 345–355.

Walcott, R.I. (1979) Plate motion and shear strain rates in the vicinity of the Southern Alps. *Bull. R. Soc. N.Z.* **18**, 5–12.

Ward; C.M. (1988a) New Zealand marine terraces: uplift rates (with response by W.B. Bull & A.F. Cooper). *Science* **240**, 803–805.

Ward, C.M. (1988b) Marine terraces of the Waitutu district and their relation to the late Cenozoic tectonics of the southern Fiordland region, New Zealand. *J. R. Soc. N.Z.* **18**, 1–28.

Ward, W.T. (1985) Correlation of east Australian Pleistocene shorelines with deep-sea core stages: a basis for a coastal chronology. *Bull. Geol. Soc. Am.* **96**, 1156–1166.

Wellman, H.W. (1979) An uplift map of New Zealand, and a model for uplift of the Southern Alps. *Bull. R. Soc. N.Z.* **18**, 13–20.

*Spec. Publs. int. Ass. Sediment.* (1991) **12**, 79–89

# Rates and magnitudes of late Cenozoic vertical movements in the Indonesian Banda Arc and the distinction of eustatic effects

A.R. FORTUIN *and* M.E.M. DE SMET*

*Geomarine Centre Amsterdam, Institute of Earth Sciences, Free University, P.O. Box 7161, 1007 MC Amsterdam, The Netherlands*

## ABSTRACT

Rates and magnitudes of late Cenozoic vertical movements in various sections of Banda Arc islands were calculated using the geohistory analysis technique. Short periods of uplift alternated with longer intervals of tectonic rest or subsidence. Sections with a high time-stratigraphic resolution show pulses of uplift with a duration between 0.1 and 1 Ma, and rates exceeding 5 m/ka. Such movements are of similar frequency to eustatic third-order sea-level changes, but an order of magnitude larger in vertical extent, and will therefore determine major unconformities (sequence boundaries) in tectonically unstable areas, such as continent–arc collision zones. Moreover, the spasmodic behaviour of tectonic movements, especially when shorter intervals of time are considered (on a scale of thousands of years), will cause local perturbations of the depositional sequences with regard to the eustatic cycle. Owing to this tectonic overprinting the eustatic signal will be difficult to disentangle from the tectonic signal, and the predictive value of seismic sequence analysis of sediments deposited along active margins is thus limited.

## INTRODUCTION

*I believe that where a deep-sea chart shows a complicated topography the adjoining coasts must show signs of upheaval* (G.A.F. Molengraaff, 1915a)

Eastern Indonesia owes its complex geology and topography to the particular setting at the junction of three converging and colliding major plates, the Indo–Australian Plate, the Pacific Plate, and the Eurasian Plate. Various completed and active collision belts can be distinguished, of which the Banda Arc (Fig. 1) represents the largest active belt. During the last few million years the northwards moving Australian continent and New Guinea entered the subduction complex bordering the Banda Sea (Hamilton, 1979; Bowin *et al.*, 1980), and this collision resulted in uplift and emergence of the outer Banda Arc islands.

The importance of young tectonic uplift was already noted at the beginning of this century (Verbeek, 1908; Molengraaff, 1912,1915a,b), when systematic geological investigation of Indonesia by the Dutch colonial government started. The widespread occurrence of raised coral reefs, or just their absence along faulted coasts, appealed to the imagination. Lack of dating possibilities, however, which also applied to adequate palaeontological dating of the underlying Neogene sediments, prevented more than a descriptive evaluation. An exclusion is the interesting, but clearly over-optimistic, guess of Molengraaff (1915b), who, knowing the presence of Quaternary reef terraces above 1000 m in Timor, suggested that differential uplift of 1500–2000 m might have occurred in Timor in 'may be the last 5000 years'. More recently, Tjia (1981) drew attention to the importance of spasmodic uplift in the Quaternary reef record of Indonesia by observing that long-term rates of uplift (in the order of tens of thousands of years) tend to be considerably lower than rates calculated for shorter time intervals.

During the joint Indonesian–Dutch oceano-

* Present address: University of London, Geol. Dept. Royal Holloway and Bedford New College, Egham Hill, Surrey, UK.

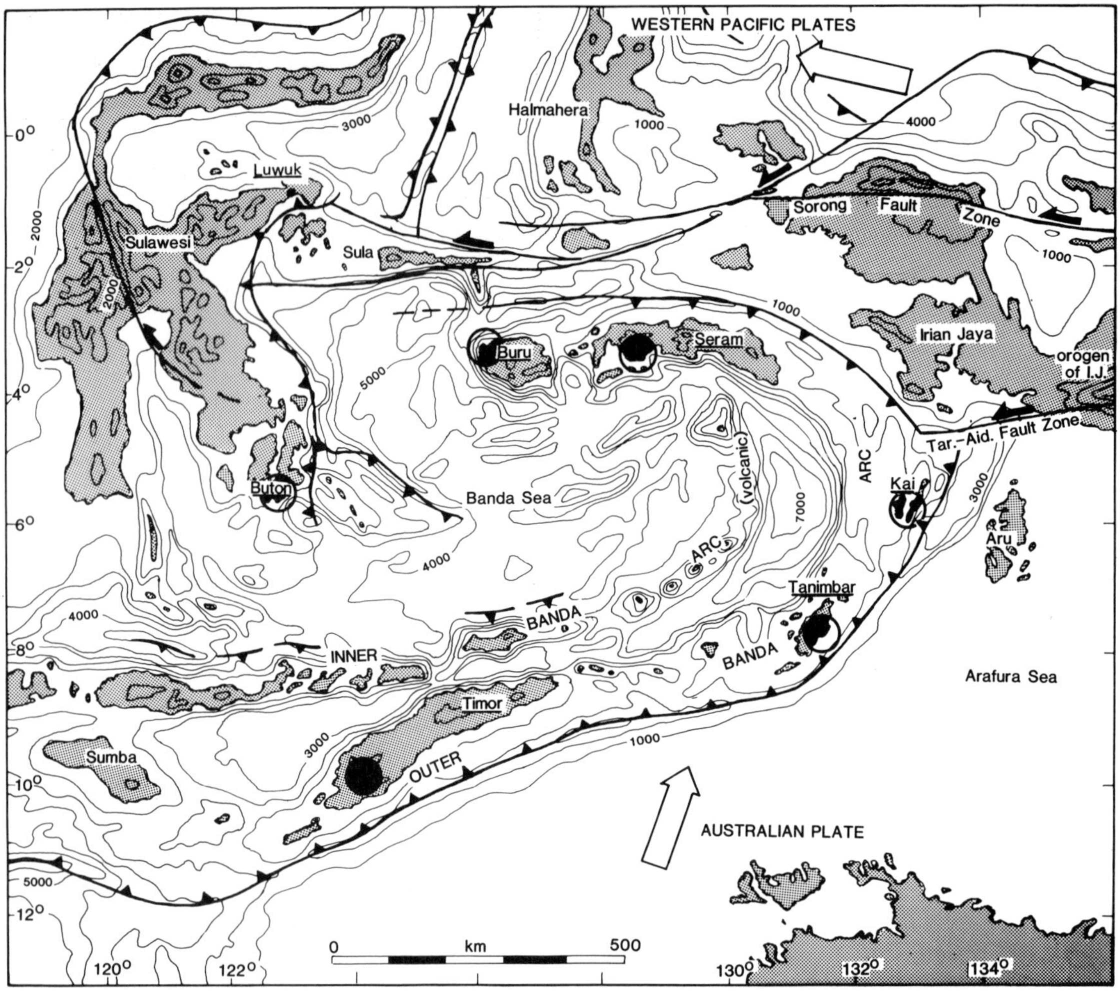


**Fig. 1.** General physiography of East Indonesia with indication of 1000 m topographic contours, illustrating major structural elements and relative plate motions. Study areas are encircled. The open arrows indicate motions of the Australian and Pacific plates with respect to the Banda Sea. Toothed lines indicate front of subduction zone.

graphic *Snellius-II* Expedition, various projects of seagoing and onshore geological research were carried out in and around the Banda Arc (results in van Hinte *et al.*, 1989). The *Snellius* GF 1–2 teams were primarily interested in rates and magnitudes of late Cenozoic vertical movements; they investigated key sections of uplifted basins presently exposed on the islands indicated (Fig. 1), and for each island the local tectonic and depositional setting was compared with the sequence of detected vertical motions. For details behind the data presented here, we refer to these island studies (Buton – Fortuin *et al.*, 1990; Buru – Fortuin *et al.*, 1988; Seram – de Smet *et al.*, 1989b; Kai – van Marle & de Smet, 1990; Tanimbar – de Smet *et al.*, 1989c; Timor – de Smet *et al.*, 1990). In addition, a geometrical model to reconstruct the last 10 Ma of motion of microplates within the Banda Arc has been presented (de Smet, 1989). The objective of this paper is to discuss some general results related to the topic of sea-level changes at active plate margins.

From island to island the setting and stratigraphic coverage of the late Cenozoic deposits investigated differs considerably and the stratigraphic overview (Fig. 2) reflects the varying depositional histories along the arc. Differences in the degree of exposure

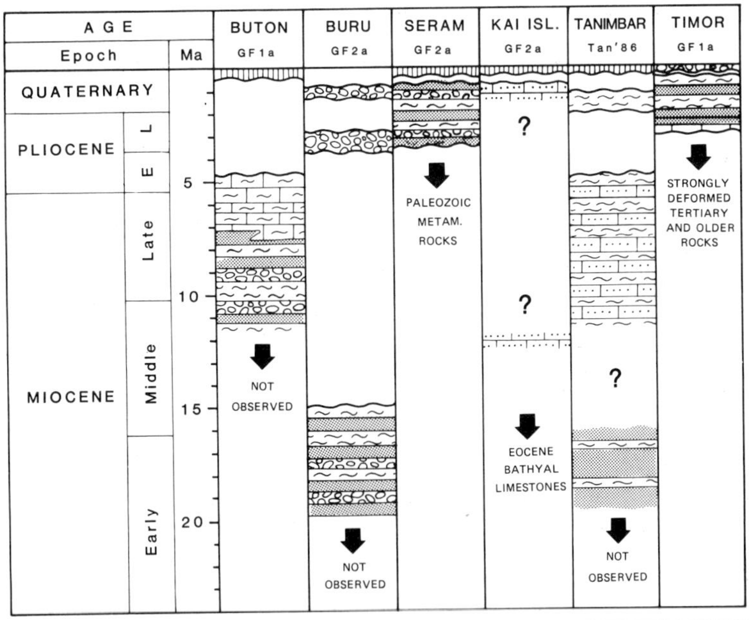


| | | BUTON | BURU | SERAM | KAI ISL. | TANIMBAR | TIMOR |
|---|---|---|---|---|---|---|---|
| STRATIGR. THICKNESS | ESTIMATED | 2000m | 1000m | 1500m | ? | 1000m | 600m |
| | OBSERVED | 350m | 250m | 1100m | 10m | 450m | 550m |
| NUMBER OF SAMPLES | | 99 | 51 | 179 | 9 | 120 | 230 |
| MICRO-FOSSIL ANALYSIS | Plank. Foram. | X | X | X | X | X | X |
| | Nannoplankton | X | | | X | | X |
| | Benth. Foram. | X | X | X | X | X | X |

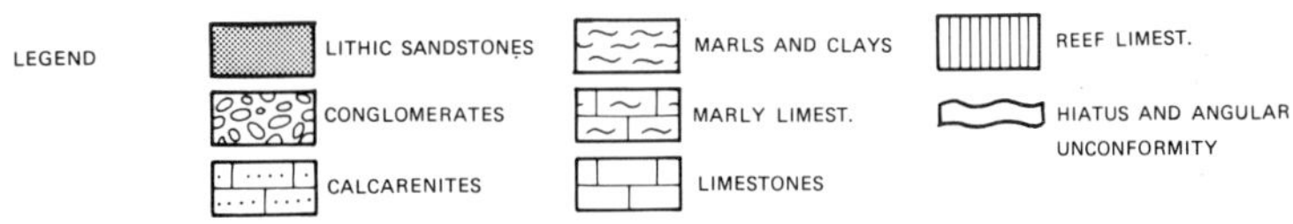


**Fig. 2.** Stratigraphic overview and summary of data collected during onshore campaigns GF 1a–2a of the Indonesian–Dutch *Snellius-II* Expedition and the Tanimbar Expedition (carried out under the auspices of the University of London Consortium for Geological Research in Southeast Asia). The uppermost half shows the overall stratigraphic development of the basins studied on the various islands. Blank intervals represent hiatuses. The number of samples collected in the field and thickness indications for observed and probably deposited sediments are a measure for the accuracy of the geohistory diagrams.

also play a role, as illustrated by strong deviations between the estimated total thickness of the stratigraphic interval studied on the various islands and the actual thickness that could be studied and sampled in the field. In Timor stratigraphic coverage is good, whereas on a heavily forested island, such as Buton, a composite stratigraphic column had to be made, based on biostratigraphic correlation of different sections. Kai Islands have the problem that Neogene sediments are hardly present or almost completely covered by elevated Quaternary reefs.

## RECONSTRUCTION OF THE VERTICAL MOVEMENTS

In order to obtain optimum data about the timing and magnitude of the vertical movements of the various island segments, the geohistory analysis technique (van Hinte, 1978,1983; Fig. 3) was applied. The best exposed or most accessible late Cenozoic sections on each island were selected and visited. These were logged, described, and systematically sampled at intervals of several metres. All samples collected in the field were studied for age and palaeowater-depth data. Microfaunal analysis of planktonic organisms (Foraminifera, coccoliths) provided essential age data. Detailed analysis of benthic foraminiferal associations and — this should be stressed — comparison with their present-day bathymetric distribution in the Banda basins (van Marle, 1988) enabled approximation of the depth of deposition at successive time intervals.

The resulting geohistory diagrams are not compensated for effects of compaction and isostasy due to sediment loading. This means that the indicated

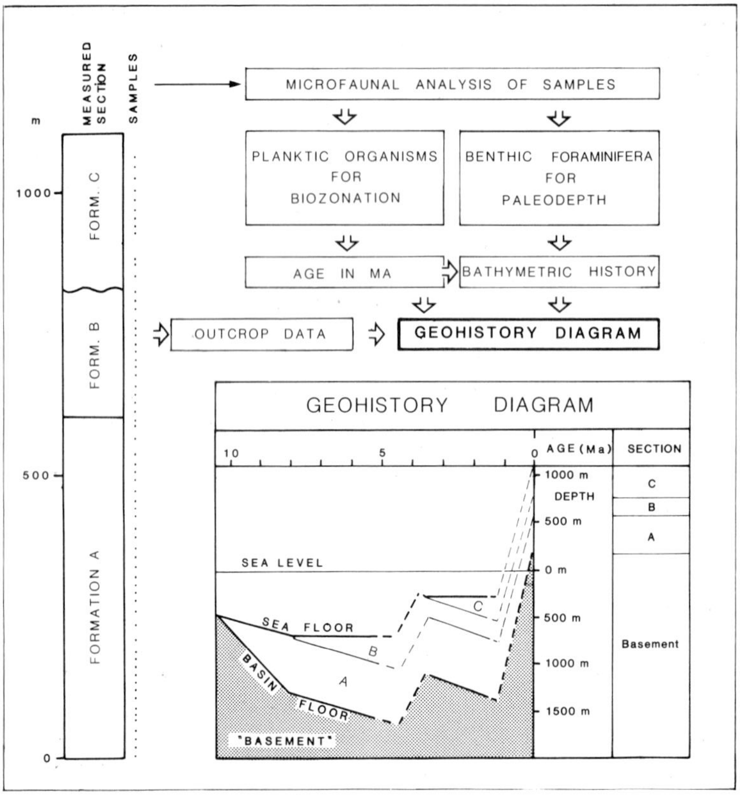


**Fig. 3.** Procedure of geohistory analysis applied to calculate vertical motions of the basin floor underlying the late Cenozoic sections studied. Sampled microfaunas are investigated for age and palaeowater-depth. By plotting these data in a time–depth framework, adding to each time reference point the amount of sediment added, it is possible to portray the vertical movements of the basin floor.

rates of subsidence may be somewhat higher than the actual rates. In our opinion such corrections would suggest an accuracy that is not in harmony with the simplifications involved. As discussed by de Smet *et al.* (1989a) factors such as the bathymetric range of the depth zones distinguished, the biostratigraphic resolution, and the quality of the measured and sampled stratal record influence the accuracy of the constructed geohistory diagrams. Irrespective of such shortcomings, however, the main tectonic changes will certainly be recorded. The rates of vertical movements calculated from these diagrams and presented in the next section should therefore be seen as approximations.

The influence of erosion on a section merits special attention. As soon as erosion takes place, sediment is removed and a hiatus formed. The resulting unconformity, on which we have to base our calculation of the average rate of uplift for the period involved, may cover a much longer period of non-deposition than the duration of the actual pulse of uplift. In other words, the onset of uplift is then no longer represented in the available record and the more sediment removed, the more moderate the calculated rate of uplift becomes. In fact the lowest rates of uplift calculated for the basin floors (Fig. 4), such as in Buru and Tanimbar, are due to considerable hiatuses in the sedimentary record, together with poorer dating possibilities.

## RESULTS

The assembled curves depicting the vertical motions of the basin floors studied in the various islands (Fig. 5) distinguish between motions calculated from measured sections and those for intervals of time represented by unconformities. For simplicity, the curves for palaeowater-depth and sediment accumulation, depicted in the various island studies, have been omitted. It is clear from Fig. 5 that the timing, rates, and magnitudes of vertical motions differed widely from place to place. The unifying trend, however, is the Pliocene–Quaternary uplift related to the emergence of the Banda Arc islands.

All uplift rates represent the vertical movements

| LATE CENOZOIC VERTICAL MOVEMENTS OUTER BANDA ARC ISLANDS | | |
|---|---|---|
| ISLAND | Highest rates calculated | |
| | Uplift | Subsidence |
| BUTON | 175 cm/ka | 100 cm/ka |
| BURU | 75 cm/ka | 20 cm/ka |
| SERAM | 250 cm/ka | 50 cm/ka |
| KAI ISL. | 300 cm/ka | ? |
| TANIMBAR ISL. | 50 cm/ka | 7.5 cm/ka |
| TIMOR | 1000 cm/ka | 20 cm/ka |

**Fig. 4.** Compilation of minimum rates of highest uplift or subsidence calculated for the various locations studied.

of local areas and are not valid for entire islands because the motions are differential on a regional scale. For example, in northwest Buru the basement has been tilted considerably. Differential uplift during the Pliocene and Quaternary was at least 2.5 km over a distance of 15 km. Differences in the rates of uplift given in Figs 4, 5 for this region are due to calculation of minimum uplift in the recorded section and maximum uplift calculated for more elevated parts.

On Seram and Timor the uplift recorded has the character of regional arching over distances of 10 km to more than 100 km. In southwest Seram (Fig. 6), compression resulted in the folding and uplift of the axis of a small Pliocene basin. The uplift of 250 cm/ka is based on the interpreted minimum uplift of the central fold axis. In the case of Timor, strong differential uplift is shown by very fast uplift of up to 1000 cm/ka calculated for the last 200 ka in central West Timor (Fig. 5) and uplift of only 3–50 cm/ka in coastal terraces up to 120 ka in age of west and northeast Timor (Chappell & Veeh, 1978; Tjokrosapoetro, 1978; Jouannic *et al.*, 1988).

A pronounced trend (Fig. 5) is that periods of subsidence have a longer duration than the periods of uplift. In combination with lower overall rates of subsidence than rates of uplift (Fig. 4) this results in a spasmodic characteristic of vertical motions.

Spasmodic uplift is best illustrated by the Timor example, where two relatively well-constrained phases of uplift of limited duration (of the order of 0.2 Ma) alternate with a nearly 2 Ma interval of slow subsidence during late Pliocene to early Pleistocene

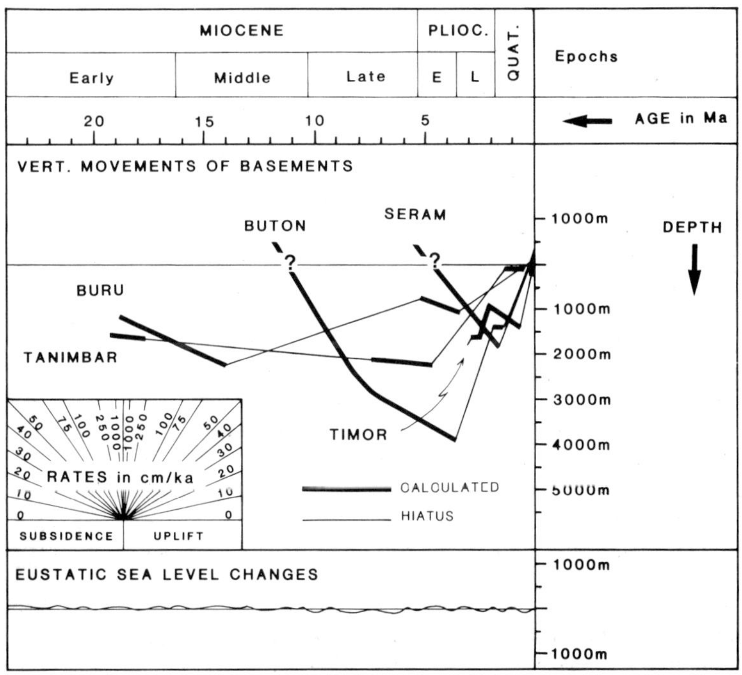


**Fig. 5.** Assembly of age–depth graphs showing the combined movements of the basin floors of the sections studied on the various Banda Arc islands. Thick lines are based on the sediments present, whereas the thin lines indicate the average vertical motion during the timespan of hiatuses in the record. Below, the curve for late Cenozoic eustatic sea-level changes (after Haq *et al.*, 1987) is indicated using the same scale. The inset enables direct reading of the rates of uplift or subsidence in centimetres per 1000 years (cm/ka).

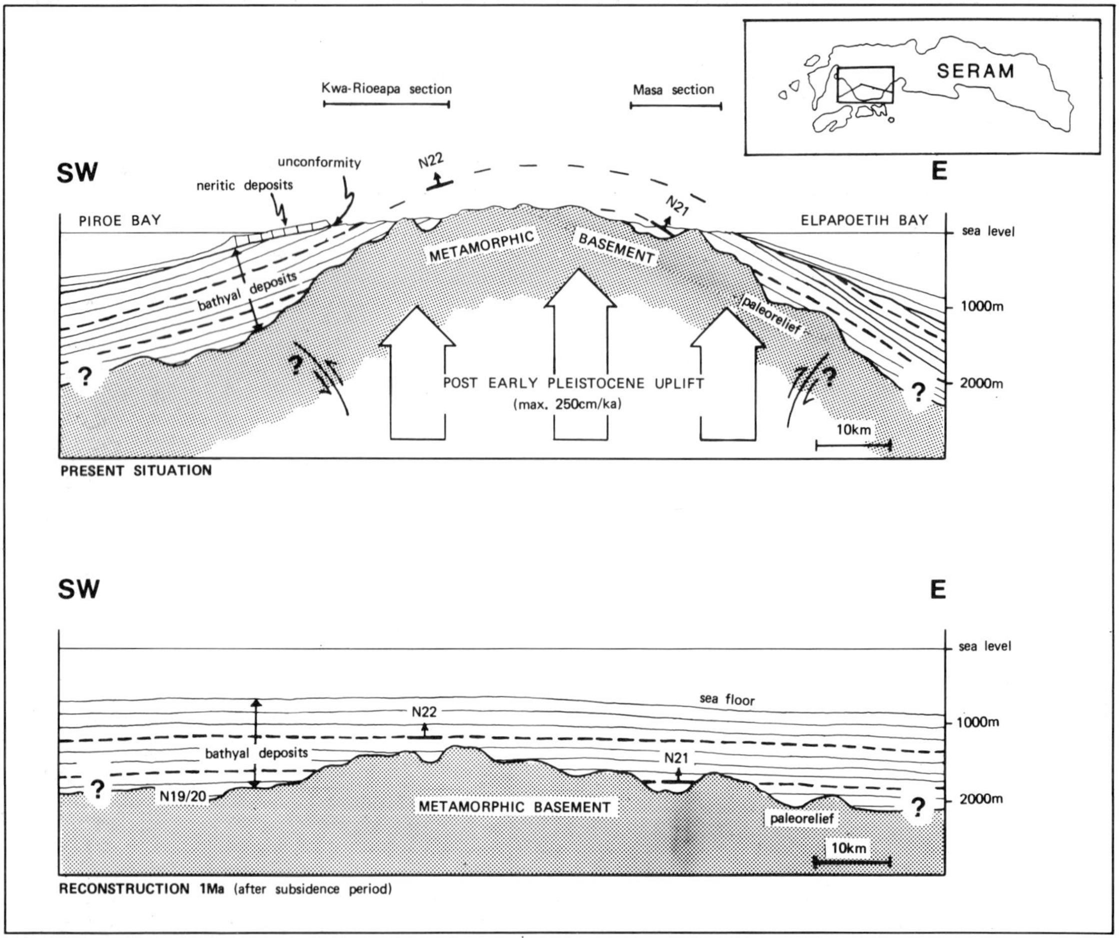


**Fig. 6.** Diagram illustrating the lateral variation of late Pleistocene antiform uplift movements in southwest Seram, as reconstructed from sections studied for geohistory analysis (after de Smet *et al.*, 1989b; inset shows location of the basin). Top: present situation. The deposits are dipping away from the most uplifted Palaeozoic basement rocks. Bottom: reconstruction at 1 Ma, when sediments accumulated at upper bathyal depths. The N numbers indicate the boundaries of late Pliocene to early Pleistocene planktonic foraminiferal biozones.

time. Here the shortest basement history, with good outcrop conditions, was studied. The more detailed geohistory diagram (Fig. 7) illustrates this better. It is based on a succession more than 500 m thick exposed along River Sabau in the Central Basin of central West Timor (de Smet *et al.*, 1990). These results differ considerably from calculation of uplift rates of Timor by Audley-Charles (1986), who based his calculation on average uplift since deposition of the oldest basinal deposits.

In the sections studied (Fig. 5), periods of uplift appear to be hardly covered by sediments, so that detailed information concerning the duration of uplift pulses has to remain scarce. Only in sections with a high time-stratigraphic resolution and small hiatuses, such as the Timor example, can the timing and rates be better constrained. The first phase of uplift in Timor is the only example we have of significant sedimentation during such an interval. Uplift of crustal parts of more than 1 km and within less than 1 Ma will, in general, result in vigorous erosion, especially in the case of emergence, so that such fast uplift pulses are hard to calculate from the stratal record.

Our results suggest that spasmodic behaviour of tectonic movements is an important feature in arc–

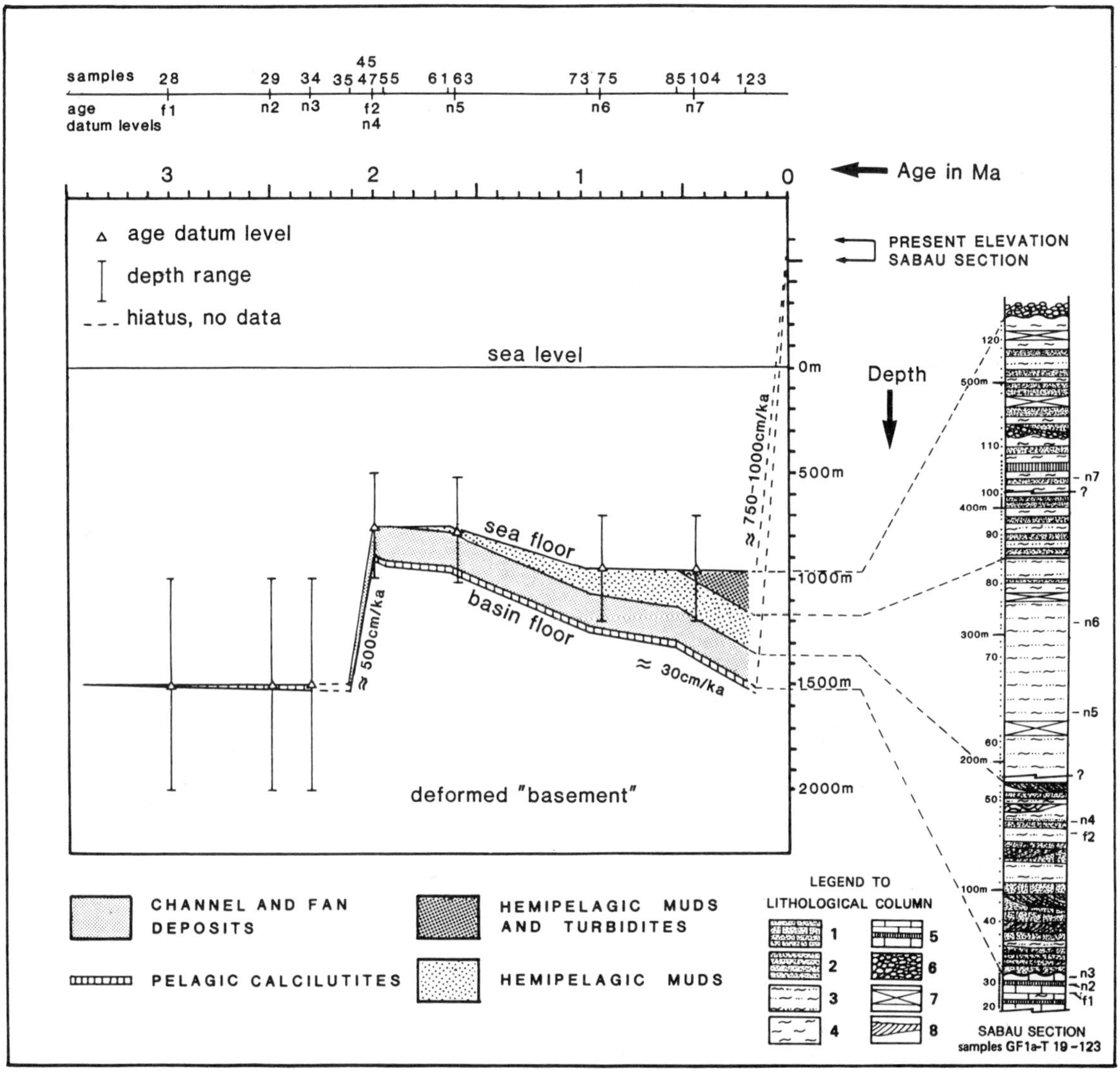


**Fig. 7.** Geohistory diagram of the Sabau River section, central West Timor, with indication of the position of age datum levels (**n** based on calcareous nannofossils and **f** on planktonic Foraminifera) and the possible bathymetric range of the benthic foraminiferal faunas at these points. The numbers on the legend to the lithological column indicate: 1, calcarenites; 2, sandy strata; 3, muddy deposits (mainly sandy marls); 4, marls, clayey marls; 5, pelagic calcilutite, with intercalated volcanic ash; 6, conglomeratic interval; 7, interval not exposed; 8, scour and fill structures. Figure modified after de Smet *et al.* (1990).

continent collision zones, which is easily removed from the stratal record owing to the erosional hiatuses resulting from such spasmodic uplift. That the impressive rate of uplift of central West Timor is not an isolated example is shown by similar rates for comparable time intervals from other collision belts, e.g. from the Himalayas (0.9 m/ka over 0.5 Ma, Zeitler *et al.*, 1982). Moreover, rapid thickening of a sedimentary pile owing to accretion to an advancing continental margin will, in general, result in rapid uplift.

For shorter intervals of time and on a local scale, spasmodic behaviour of tectonics is better known, for example amongst Quaternary geologists working on Holocene tectonics of convergent margins ('jerkiness', Bloom & Yonekura, 1985). Thommeret *et al.* (1981) convincingly showed that on a scale of 1 ka, upheaval of fault blocks may be followed by intervals of sinking. As discussed by Bloom & Yonekura (1985) and Berryman (1987), many workers on raised Quaternary terraces assume that the longer the timespan considered becomes (this is in the order of 10–100 ka), the more uniform the uplift seems to proceed. This, however, is not necessarily true and a *Snellius* team (Sumosusastro *et al.*, 1989) found evidence that a faulted flight of Quaternary terraces in the east arm of Sulawesi (Fig. 1) subsided intermittently (calculated on a 10 ka base) under conditions of overall uplift (up to 1.8 m/ka, calculated over the last 0.1–0.2 Ma for the fastest rising hinterland block). Therefore it seems realistic to assume that, at least near active plate margins, each order of magnitude of time is subjected to processes creating intricate patterns of uplift and subsidence. The tectonic reasons are open to discussion.

## TECTONICS AND EUSTATICS COMPARED

The concept of sequence stratigraphy (Vail *et al.*, 1977; Posamentier *et al.*, 1988; van Wagoner *et al.*, 1988) provides a powerful tool to better analyse time–rock relationships of the sedimentary record on passive margins. Sedimentary geometries, however, are the product of variations in rates of sediment supply, eustasy, and tectonic movements. As argued by Kendall & Lerche (1988), two of these processes must be specified in order to determine the third, which is hardly possible in geological practice. Therefore a variety of hypotheses exists to account for observed sedimentary geometries. On the other hand, extensive research on seismic facies development on passive margins by Vail and his co-workers, where tectonic subsidence is generally assumed to be constant with regard to eustatic fluctuations, has resulted in widely applied cycle charts (Haq *et al.*, 1987), presenting a chronology of inferred world-wide eustatic changes. However, the question arises, to what extent can these presumed global sea-level cycles be distinguished in the stratal record of unstable basins filled during orogenic processes.

The vertical tectonic changes recorded in the orogenetic outer Banda Arc are an order of magnitude larger than eustatic oscillations. At the bottom of Fig. 5 we have added, at the same scale, the curve of eustatic sea-level changes (Haq *et al.*, 1987). We did not incorporate these fluctuations in the geohistory diagrams above because oscillations of that size are beyond the limit of detection of palaeobathymetrical analysis of environments deeper than shelf. These fundamental differences in magnitude mean that in tectonically active areas the major unconformities, or sequence boundaries, will be determined rather by tectonics than by eustatics.

For the late Neogene time interval considered, the average duration of eustatic third-order cycles is about 0.8 Ma (P. Vail, personal communication), or about half of that for a period of relative rise or fall. Such tempos are of comparable duration to pulses of uplift registered in the Banda Arc, but an order of magnitude smaller. We therefore conclude that under such circumstances the eustatic signal will be difficult to disentangle from the tectonic signal. An exclusion has to be made for longer intervals of quiet subsidence.

An example illustrating the latter situation was observed in Buton, where orogenesis took place in the early–middle Miocene (Fortuin *et al.*, 1990) and turbidites become abruptly interbedded in late Miocene to early Pliocene pelagic marls and chalks. As this interval is of late Messinian age a relation with the well-known late Messinian sea-level drop is probable.

As outlined and modelled by Posamentier & Vail (1988), conditions of constant tectonic change (subsidence) with regard to the duration of global sea-level fluctuations result in a predictable succession of sequences bounded by unconformities, associated with eustatic fall inflection points. The foregoing, however, underlines that in tectonically active areas systems of different duration and intensity have their impact on the sedimentary record. This means that the resulting relative sea-level curve is dominated by tectonics and may show significant shifts in time of the major inflection points with regard to the cycle chart (Haq *et al.*, 1987), i.e. perturbations of the eustatic cycle may occur. We therefore conclude that sequence-stratigraphic analysis under such conditions loses much of its predictive value.

In the case of the well-documented Timor section (Fig. 7), deposition started during a relatively quiet period when pelagic calcilutites accumulated slowly at middle–lower bathyal depths (1000–2000 m). There was no indication at this stage of a hinterland being eroded. Around 2 Ma ago, sudden tectonic movements resulted in emergence of the northern parts of Timor and in the formation of the arc-parallel basin to which the studied section (Fig. 7) and others belong (de Smet *et al.*, 1990a). As a result of these movements, some submarine erosion took place, but deposition was soon resumed, though at much higher rates. Rapidly emplaced submarine channel and fan complexes along the base of slopes of this newly established outer arc basin were fed by debris from nearby carbonate shoals, together with fragments eroded from emerged allochthonous pre-Neogene units. When overall uplift waned, quiet conditions returned and hemipelagic marls and silts were deposited under conditions of gradual subsidence, as indicated by the benthic foraminiferal fauna. Finally, turbidites frequently entered the basin, until a second episode of fast uplift occurred, raising the studied section in about 0.2 Ma from some 1000 m depth to its present elevation of 500–600 m.

In this Timor section up to four sequence boundaries might be preserved according to the Haq *et al.* (1987) sea-level chart, notably at 3.0, 2.4, 1.6, and 0.8 Ma. We can indicate two distinct sequence boundaries, one around 2.1 Ma, when pelagic sedimentation was suddenly replaced by channel and fan deposits, and one around 0.2 Ma, which is the second uplift unconformity separating the marine sediments from overlying terrestrial conglomerates. The gradual influx of turbidites, which started in the section of Fig. 7 around 1.7 Ma, took place diachronously. Elsewhere in the basin, turbidites may occur laterally to the hemipelagic muds (Kenyon, 1974).

The 2.1 Ma level has been interpreted to be of tectonic origin (Kenyon, 1974; Audley-Charles, 1986). Around the 2.4 Ma eustatic low, bathyal pelagic chalks were still being deposited, far from clastic sources, so that this and the 3.0 Ma lowstand period probably cannot be reflected in the sedimentary record. After the first pulse of uplift waned, quiet deposition, dominantly of marls, followed and the newly emerged hinterland of proto-Timor cannot have been very far away. The timing around 2.0 Ma of this facies change might well indicate control by a eustatic rise. In that case one would also expect to find a sequence boundary around 1.6 Ma, due to the next eustatic fall, but no facies change was observed there.

Apart from possible age correlation errors these results seem an illustration of our conclusion that prediction of chronostratigraphic relations using the cycle chart (Haq *et al.*, 1987) in basinal settings in tectonically active areas is uncertain, at least before some information concerning the ages of the main (tectonic) unconformities is available.

As we did not study basin geometries we cannot give conclusive arguments as to whether the tectonic signal can be sufficiently eliminated from the sedimentary record, for example after careful correlation of numerous (seismic) sections of a basin. An argument in favour of such a possibility is that, although tectonics affect larger areas, their sedimentary expression varies from place to place and thus can be filtered out. We, however, believe that in those settings where crustal conditions are unstable and structures limited in space, such a filtering is hard to accomplish, mainly because of continuous lateral migration of zones of maximum uplift and subsidence. Waning uplift in one place, for instance, will most likely result in crustal relaxation and onset of gradual subsidence because of isostatic compensation. An upward fining sequence will then be deposited (of the transgressive systems tract type). Shortly afterwards, however, clastics may again invade this area, but now derived from a shifted locus of uplift. If such a change coincides with a eustatic lowstand the signals will reinforce each other, but it is equally possible that parts of the eustatic cycle become subdued, if not completely overprinted. Moreover, there is a probability that on a scale of 1–10 ka, tectonic oscillations in reverse of the overall trend will further complicate the recognition of eustatic trends.

## CONCLUSIONS

Most vertical movements in the Banda Arc are a matter of tilting processes and open folding on a scale of tens of kilometres. Vertical movements have a spasmodic character in time and space. Even with a net effect of uplift, longer periods of subsidence interrupt the periods of rapid uplift, which are relatively short, in the order of 0.1–1 Ma. With this duration the vertical movements, especially uplift, operate in the frequency band of eustatic cycles. Being an order of magnitude larger in vertical extent, tectonic changes and their timing rather than eustatics will, however, determine major unconformities (sequence boundaries) in these areas during the orogenic period.

The island sections investigated record mainly periods of subsidence, which will more easily reflect the eustatic signal. However, the complex interplay in time and space of tectonics and tectonic-related sediment supply will result in perturbations of the depositional sequences with regard to the eustatic cycle. This aspect seriously affects the predictive value of eustatic cycle charts along active margins. Fourth and fifth-order tectonic movements, that may be of reverse character to the overall trend, probably also disturb cycles.

## ACKNOWLEDGEMENTS

Research was carried out as part of the *Snellius-II* Expedition, organized by the Netherlands Council of Oceanic Research (SOZ) and by the Indonesian Institute of Science (LIPI); research funds were provided by the Dutch Organization for Scientific Research (NWO/WOTRO, project numbers 76–83 and 76–85). The authors gratefully acknowledge the support and contributions from many colleagues, both in Indonesia and the Netherlands. We especially thank J.E. van Hinte and S. Tjokrosapoetro for stimulating discussions, L.J. van Marle for providing the palaeobathymetrical data, S.R. Troelstra and M. Karmini for planktonic foraminiferal age determinations, and S. Hadiwasastra and K. Perch-Nielsen who provided the nannofossil ages. We are indebted to A.J. Barber for inviting one of us (M.D.S.) to participate in fieldwork on Tanimbar under the auspices of the University of London Consortium for Geological Research in Southeast Asia.

## REFERENCES

AUDLEY-CHARLES, M.G. (1986) Rates of Neogene and Quaternary tectonic movements in the southern Banda Arc based on micropaleontology. *J. Geol. Soc., London* **143**, 161–175.

BERRYMAN, K. (1987) Tectonic processes and their impact on the recording of relative sea-level changes. In: *Sea Surface Studies* (Ed. Devoy, R.J.N.) pp. 127–164. Croom Helm/Methuen, New York.

BLOOM, A.L. & YONEKURA, N. (1985) Coastal terraces generated by sea-level change and tectonic uplift. In: *Models in Geomorphology* (Ed. Woldenberg, M.J.) pp. 139–154. Allen & Unwin, Winchester, MA.

BOWIN, C.O., PURDY, C.R., JOHNSTON, G., SHOR, G., LAWVER, L., HARTONO, H.M.S. & JEZEK, P. (1980). Arc–continent collision in Banda Sea region. *Bull., Am. Ass. Petrol. Geol.* **64**, 868–915.

CHAPPELL, J. & VEEH, H.H. (1978) Late Quaternary tectonic movements and sea-level changes at Timor and Atauro island. *Geol. Soc. Am. Bull.* **89**, 356–368.

DE SMET, M.E.M. (1989) A geometrically consistent plate tectonic model for eastern Indonesia. In: *Proceedings of the Snellius-II Symposium, Theme I: Geology and Geophysics of the Banda Arc and Adjacent Areas* (Eds van Hinte, J.E., van Weering, Tj.C.E. & Fortuin, A.R.). Neth. J. Sea Res. 24, 173–183.

DE SMET, M.E.M., FORTUIN, A.R., TJOKROSAPOETRO, S. & VAN HINTE, J.E. (1989a) Late Cenozoic vertical movements of non-volcanic islands in the Banda Arc area. In: *Proceedings of the Snellius-II Symposium, Theme I: Geology and Geophysics of the Banda Arc and Adjacent Areas* (Eds van Hinte, J.E., van Weering, Tj.C.E. & Fortuin, A.R.). Neth. J. Sea Res. 24, 263–275.

DE SMET, M.E.M., SUMOSUSASTRO, P.A., SIREGAR, I., VAN MARLE, L.J., TROELSTRA, S.R. & FORTUIN, A.R. (1989b) Late Cenozoic geohistory of Seram, Indonesia. *Geol. Mijnbouw* **68**, 221–235.

DE SMET, M.E.M., VAN MARLE, L.J., CHARLTON, T., KAYE, S., FAENI, H. & TROELSTRA, S.R. (1989c) Late Cenozoic history of the island Yamclena, Tanimbar Archipelag., Indonesia. In: *Benthic foraminefera from the Banda Arc region, Indonesia, and their paleobathymetric significance for geologic interpretations of the Late Cenozoic sedimentary record.* (Ed. ván Marle., L.J.) Free University Press, Amsterdam, pp. 145–162.

DE SMET, M.E.M., FORTUIN, A.R., TROELSTRA, S.R., VAN MARLE, L.J., KARMINI, M., TJOKROSAPOETRO, S. & HADIWASASTRA, S. (1990) Detection of collision related vertical movements in the outer Banda Arc (Timor, Indonesia), using micropaleontological data. *J. SE Asian Earth Sci.* **4**, 337–356.

FORTUIN, A.R., DE SMET, M.E.M., SUMASUSASTRO, P.A., VAN MARLE, L.J. & TROELSTRA, S.R. (1988) Late Cenozoic geohistory of NW Buru, Indonesia and plate tectonic implications. *Geol. Mijnbouw* **67**, 91–105.

FORTUIN, A.R., DE SMET, M.E.M., HADIWASASTRA, S., VAN MARLE, L.J., TROELSTRA, S.R. & TJOKROSAPOETRO, S. (1990) Late Cenozoic sedimentary and tectonic history of South Buton, Indonesia. *J. SE Asian Earth Sci.*, **4**, 107–124.

HAMILTON, W. (1979) Tectonics of the Indonesian region. *U.S. Geol. Surv. Prof. Pap.* **1078**, 345 pp.

HAQ, B.U., HARDENBOL, J. & VAIL, P.R. (1987) Chronology of fluctuating sea levels since the Triassic. *Science* **235**, 1156–1167.

JOUANNIC, C., HOANG, C.T., HANTORO, W.S. & DELINOM, R.M. (1988) Uplift rate of coral reef terraces in the area of Kupang, West Timor: preliminary results. *Palaeogeogr. Palaeoclimatol. Palaeoecol.* **68**, 259–272.

KENDALL, C.G.ST.C & LERCHE, I. (1988) The rise and fall of eustasy. In: *Sea-level Changes: an Integrated Approach* (Eds Wilgus, C.K., Hastings, B.S., Kendall, C.G.St.C., Posamentier, H., Ross, C.A. & van Wagoner, J.) Soc. Econ. Paleontol. Mineral., Spec. Publ. 42, 3–17.

KENYON, C.S. (1974) *Stratigraphy and sedimentology of the Late Miocene and Quaternary deposits of Timor*. Unpublished PhD thesis, Imperial College, London, 291 pp.

MOLENGRAAFF, G.A.F. (1912) De jongste bodembewegingen op het eiland Timor en hunne beteekenis voor de geologische geschiedenis van den Oost Indischen Archipel. *Versl. Kon. Akad. Wet, Amsterdam, Wis-Natk. Afd., Verg. 29 Juni 1912*, pp. 121–132.

MOLENGRAAFF, G.A.F. (1915a) Folded mountain chains, overthrust sheets and block-faulted mountains in the East Indian Archipelago. *C.R. XIIth Int. Geol. Congr.*

*Toronto*, pp. 689–702.

MOLENGRAAFF, G.A.F. (1915b) Bodembewegingen en geologische tijdperken. *Verh. 15e Nat. Geneesk. Congr., Amsterdam*, pp. 1–8.

POSAMENTIER, H.W. & VAIL, P.R. (1988) Eustatic controls on clastic deposition. II – sequence and systems tract models. In: *Sea-level Changes: an Integrated Approach* (Eds Wilgus, C.K., Hastings, B.S., Kendall, C.G.St.C., Posamentier, H., Ross, C.A. & van Wagoner, J.). Soc. Econ. Paleontol. Mineral., Spec. Publ. 42, 125–154.

POSAMENTIER, H.W., JERVEY, M.T. & VAIL, P.R. (1988) Eustatic controls on clastic deposition, I – conceptual framework. In: *Sea-level Changes: an Integrated Approach* (Eds Wilgus, C.K., Hastings, B.S., Kendall, C.G.St.C., Posamentier, H., Ross, C.A. & van Wagoner J.). Soc. Econ. Paleontol. Mineral., Spec. Publ. 42, 109–124.

SUMOSUSASTRO, P.A., TJIA, H.D., FORTUIN, A.R. & PLICHT, J. VAN DER (1989) Quaternary reef record of differential uplift at Luwuk, Sulawesi east arm, Indonesia. In: *Proceedings of the Snellius-II Symposium, Theme I: Geology and Geophysics of the Banda Arc and Adjacent Areas* (Eds van Hinte, J.E., van Weering, Tj.C.E. & Fortuin, A.R.). Neth. J. Sea Res. 24 277–285.

THOMMERET, Y., LABOREL, J., MONTAGGIONI, L.F. & PIRAZZOLI, P.A. (1981) Late Holocene shoreline changes and seismotectonic displacements in western Crete (Greece). *Z. Geomorphol. N.F. Suppl.* **40**, 127–149.

TJIA, H.D. (1981) Examples of young tectonism in eastern Indonesia. In: *The Geology and Tectonics of Eastern Indonesia* (Eds Barber, A.J. & Wiryosujono S.). Geol. Res. and Dev. Centre, Spec. Publ. **2**, 89–104.

TJOKROSAPOETRO, S. (1978) Holocene tectonics on Timor island, Indonesia. *Bull. Geol. Surv. Indonesia* **4**, 49–63.

VAIL, P.R., MITCHUM, R.M., & THOMPSON, S. III (1977) Seismic stratigraphy and global changes of sea level, part 4: global cycles of relative changes of sea level. In: *Seismic Stratigraphy – Applications to Hydrocarbon Exploration* (Ed. Payton, C.E.). Mem., Am. Assoc. Petrol. Geol. 26, 83–97.

VAN HINTE, J.E. (1978) Geohistory analysis – application of micropaleontology in exploration geology. *Bull., Am. Ass. Petrol. Geol.* **62**, 201–222.

VAN HINTE, J.E. (1983) Synthetic seismic sections from biostratigraphy. In: *Studies in Continental Margin Geology* (Eds Watkins, J.S. & Drake, C.L.). Mem., Am. Ass. Petrol. Geol. 34, 675–685.

VAN HINTE, J.E., VAN WEERING, TJ.C.W. & FORTUIN, A.R. (Eds) (1989) *Proceedings of the Snellius-II Symposium, Theme I: Geology and Geophysics of the Banda Arc and Adjacent Areas*. Neth. J. Sea Res. **24** (2/3, 3/4).

VAN MARLE, L.J. (1988) Bathymetric distribution of benthic Foraminifera on the Australian–Irian Jaya continental margin, eastern Indonesia. *Mar. Micropaleontol.* **13**, 97–152.

VAN MARLE, L.J. & DE SMET, M.E.M. (1990) Notes on the Late Cenozoic history of the Kai Islands, eastern Indonesia. *Geol. Mijnbouw*. **69**, 93–103.

VAN WAGONER, J.C., POSAMENTIER, H.W., MITCHUM, R.M., VAIL, P.R., SARG, J.F., LOUTIT, T.S. & HARDENBOL, J. (1988) An overview of the fundamentals of sequence stratigraphy and key definitions. In: *Sea-level Changes: an Integrated Approach* (Eds Wilgus, C.K., Hastings, B.S., Kendall, C.G.St.C., Posamentier, H., Ross, C.A. & van Wagoner, J.). Soc. Econ. Paleontol. Mineral., Spec. Publ. **42**, 40–45.

VERBEEK, R.D.M. (1908) Rapport sur les Moluques. *Jaarb. Mijnw. in Ned. Oost Indië*, **XXXVII**, 844 pp.

ZEITLER, P.K., JOHNSON, N.M., NAESER, C.W. & TAHIRKHELI, R.A.K. (1982) Fission-track evidence for Quaternary uplift of the Nanga Parbat region, Pakistan. *Nature* **298**, 255–257.

*Spec. Publs. int. Ass. Sediment.* (1991) **12**, 91–103

# Depositional architecture of Quaternary fan-delta deposits of the Andean fore-arc: relative sea-level changes as a response to aseismic ridge subduction

S. FLINT†, P. TURNER‡ *and* E.J. JOLLEY‡,*

† *Department of Earth Sciences, University of Liverpool, P.O. Box 147, Liverpool, UK; and* ‡ *School of Earth Sciences, University of Birmingham, P.O. Box 363, Birmingham B15, UK*

## ABSTRACT

The structure, landform development, and sedimentary response of the coast of northern Chile over a distance of 600 km, between Antofagasta and Arica have been investigated to evaluate the influence of Nazca plate subduction dynamics on sea-level changes from Quaternary to Recent times. Sedimentological and geomorphological analysis of the coastal alluvial-fans–fan-deltas shows that the preserved sequences comprise poorly sorted, subangular gravels (subaerial alluvial fan), interbedded with complex, mixed conglomerate/sandstone units, which are characterized by greater textural maturity and copious marine shell debris. In the mixed sequences, conglomerate clasts are well rounded, moderately well-sorted and have a martix of the shell debris. The sandstones show trough bedding on a metre scale and abundant marine invertebrate burrows.

The present-day morphological organization of the coastal depositional systems comprises currently inactive alluvial fans, cut by canyons up to 10 m deep that act as sediment bypass conduits to sites of shoreline fan-delta progradation. Coupled with the organization of preserved subaerial/subaqueous sediments into crude cycles, these data suggest that relative sea-level changes within the Quaternary may have exerted a strong control on coastal fan development.

At Arica the coastal range is in net extension, characterized by extensional normal faulting and subsidence, in common with much of Chile. South of Arica uplift is recorded by marine terrace development and incision of alluvial fan surfaces; uplift reaches a maximum south of Iquique. The transition between regions in net subsidence and net uplift is marked by north-facing neotectonic normal fault scarps.

Variations in apparent uplift and subsidence are consistent with recently published oceanographic records on relative sea-level changes over a 30-year period. Our data suggest that these regionally variable patterns of coastal uplift along the north Chilean coast are controlled by the subduction of an aseismic ridge, which overprints eustatic sea-level fluctuations. Thus, subduction of oceanic plate heterogeneities may provide a mechanism for producing cyclicity in sedimentary sequences at a frequency equal to or higher than glacio-eustasy in fore-arc and possibly back-arc sedimentary basins. These sequences will be neither of global extent nor globally synchronous.

## INTRODUCTION

In the last decade the interdisciplinary science of basin analysis has produced advances in our understanding of the role of base-level changes in controlling the stratigraphy and architecture of basin-fill successions. Both global eustasy and regional tectonic activity are known to have major effects on basin-fill characteristics and facies architecture (Mitchell & Reading, 1986; Posamentier & Vail, 1988; Posamentier *et al.*, 1988). However, reliable distinction between the relative importance of tectonics and eustasy in controlling base level and thus architecture has not generally been made.

In this paper we investigate the structure, landform development, and sedimentary dynamics of the north Chilean Pacific margin to evaluate the influence of

* Present address: BP Exploration Company Ltd., St Vincent St., Glasgow, UK.

Nazca plate subduction processes on relative sea-level changes over Quaternary to Recent times.

## REGIONAL GEOLOGY

The coastline of northern Chile is dominated by the Coastal Cordillera, a longitudinally continuous mountain range, over 2000 m high in places (Fig. 1). This range comprises Jurassic volcanic rocks and minor intercalated sediments of the La Negra Formation (Garcia, 1967) and has been traditionally interpreted as marking the position of the Jurassic Andean volcanic arc (Coira *et al.*, 1982). The La Negra Formation is overlain unconformably, locally by Early Cretaceous basin fills (Flint *et al.*, 1986), but mainly by transgressive marine sandstones and coquina limestones of the Miocene–Pliocene La Portada Formation (Segerstrom, 1964; Ferraris & Dibiase, 1978). The La Portada Formation is in turn overlain by Quaternary alluvial fan and fan-delta deposits.

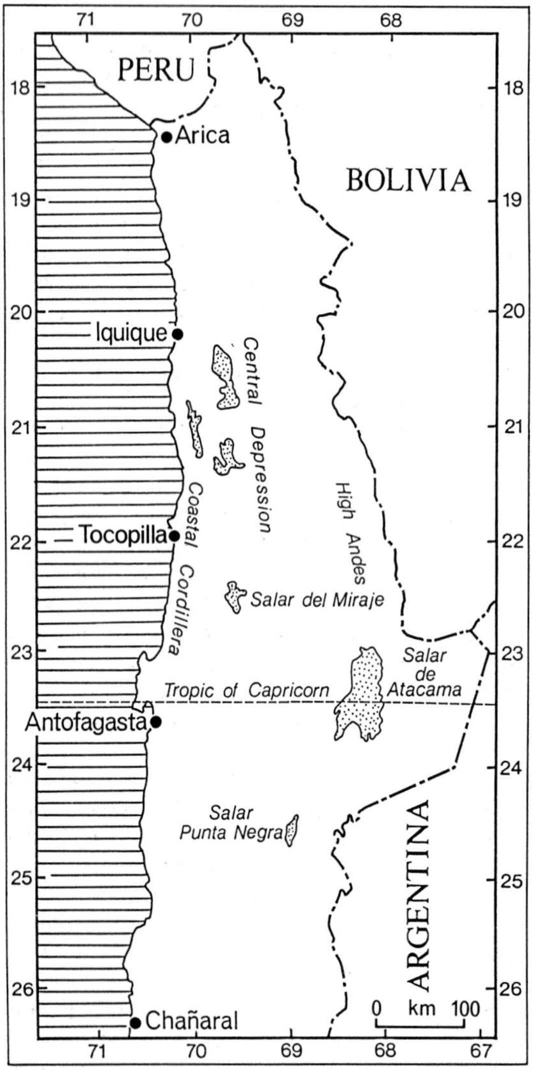


**Fig. 1.** The Andean margin of northern Chile showing main morphotectonic divisions of the fore-arc. Locations of Figs 2, 3, 7 are within this figure area.

Morphologically, the coastline consists of a series of bays, headlands, and distinctive terraces/wave-cut platforms. In this paper we describe the dynamic evolution of the Quaternary fan systems and integrate the field evidence with published marine geophysical and oceanographic data to elucidate the Quaternary–Recent relative sea-level history of the Chilean margin and possible driving mechanisms. A major feature in this region is the aridity of the climate. In the study area only one river, the Río Loa, flows through the Coastal Cordillera into the Pacific Ocean.

## SEDIMENTOLOGY OF THE COASTAL FAN/FAN-DELTA SYSTEMS

### Fan classification

The coastal alluvial fans of northern Chile were first documented by Charles Darwin (1876). We summarize here the first detailed sedimentological studies of these sequences (Jolley, unpublished data). The classification of the coastal fans is shown in Table 1. The fans can be divided conveniently into three main groups on the basis of catchment size.

**1** External fans, which are much larger and have access to external drainage from within and occasionally right through the Cordillera (e.g. the Río Loa fan; Fig. 2). Sixty per cent of the feeder canyons ($n = 40$) follow the trace of steep, E–W-trending normal faults that cut the Coastal Cordillera.

**2** Internal fans, which have a catchment area comprising the immediate watershed of the Coastal Cordillera.

**3** Side cones, which are small, steep scree cones with no true feeder canyon dominating the supply of sediment to the fan.

Based on the presence, numbers, size, and dis-

**Table 1.** Classification scheme for the coastal fans, with summary characteristics

| Catchment type | Percentage of total fans | Surface channel characteristics | Percentage of total |
|---|---|---|---|
| External fans | 18 | One major channel | 20 |
| | | A few major channels (< 10) | 33 |
| | | Abundant small channels | 47 |
| Internal fans | 35 | One major channel | 27 |
| | | A few major channels (> 10) | 50 |
| | | Abundant small channels | 23 |
| Side cones | 47 | Trenched upper fan | 32 |
| | | Gullied mid-fan | 68 |

tribution of the channels on the upper and middle fan surface, the external and internal fans can be subdivided into fan surfaces with one major channel, a small number (< 10) of large channels, or no dominant channels (but a distributary system of small channels). The side cones may exhibit trenching in the upper fan area or deep gullying in the middle/upper fan. The side cones typically have steep slopes, sometimes attaining a maximum angle of repose of 30°. A normal fan profile for a larger internal or external fan is an upper fan slope of 12°, some 7° across the middle fan, decreasing to 1–2° on the lower fan surface.

Out of 52 main locations studied between the mouth of the Río Loa southwards to the Mejillones Peninsula (Fig. 2), a total of 43 external and 79 internal fan systems were surveyed. However, many of the external systems are coalesced with subsidiary internal and side-cone fans.

## Coastal morphology and sedimentology

At Gatico, some 50 km south of Tocopilla (Figs 2, 3), a large fan serves as a type section for a complete Quaternary dispersal system and will be used to describe and interpret the major depositional environments.

### *Alluvial fans*

The alluvial fan at Gatico (location 35, Fig. 4A, B) is an external fan system, fed via a faulted, apical fan canyon. Side cones contribute to the total apron width of 6 km (Fig. 5A, B). The system has two terraces, one of 5 m and the second (younger) one of 1.2 m (Fig. 4). The upper terrace is incised by a series of channels that bypass material to secondary fans on the backshore.

Several major channels traverse the fan from apex to toe. Their position on the fan surface is controlled by fan lobe build-outs, which have deflected the channels on to their flanks or into lows between two merging lobes. The major channels are generally of low sinuosity, deepest at the fan apex (up to 10 m deep × 20 m wide), and have vertical margins with gravel levees up to 1.5 m high. These channels become less entrenched in the middle fan area (up to 4 m deep × 35 m wide), but are rejuvenated at the coast where incision over the coastal terraces can be up to 6 m. The sediments exposed in the channel banks are generally subangular gravel-pebble grade clasts of volcanic and intrusive rocks. The matrix is coarse sandstone and the sediments are poorly sorted, ungraded and with ill-defined *b*-axis imbrication.

Much of the fan area is inactive as it is bypassed via the deep channels. In places the surfaces are modified by complexes of small, sinuous channels or washouts (0.2–1.5 m deep). Many of the inter-channel areas have been reworked by aeolian processes, either actively deflating the fan surface or causing minor aeolian sand accumulations. An ochre-coloured soil characterizes the oldest inactive fan surfaces.

We believe that the fan sediments represent two episodes of development; an initial phase of fan progradation followed by the minor channelization and reworking of the fan surface during post-depositional, glacially induced pluvial periods.

### *Fan-deltas*

The Gatico fan-delta (Fig. 4A, B) comprises a series of lobes of clast-supported conglomerate located seaward of the channel mouths. A particularly well-developed system at Mantos del Pacifico (location 36, Fig. 3) some 3 km south of Gatico is fed from a canyon 18 m deep (Fig. 5D, E). Angular,

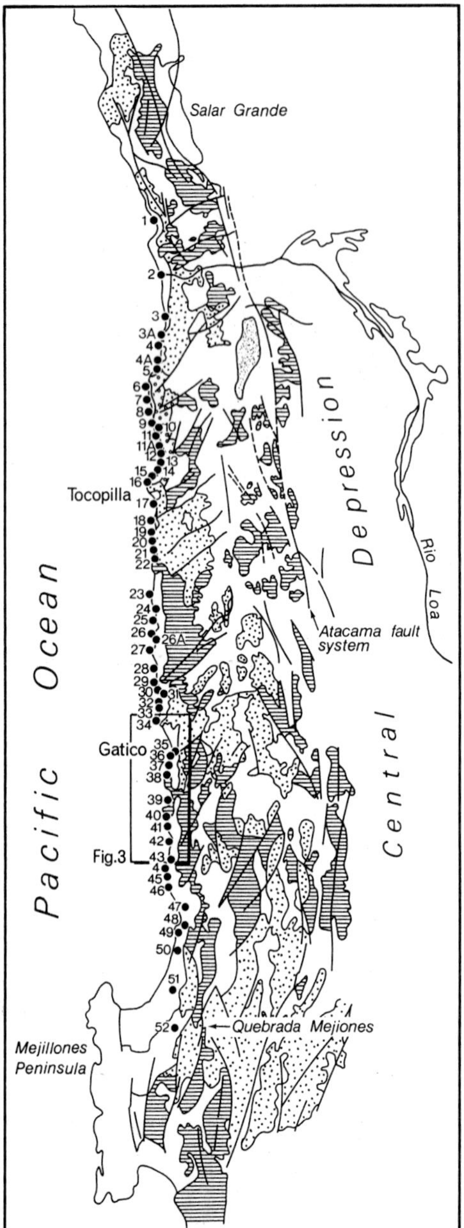


**Fig. 2.** Geological map of the study area (after Servicio Nacional de Geologia y Mineria, 1981), showing locations of the fan systems studied. Note the intense pattern of extensional normal faulting in the Coastal Cordillera.

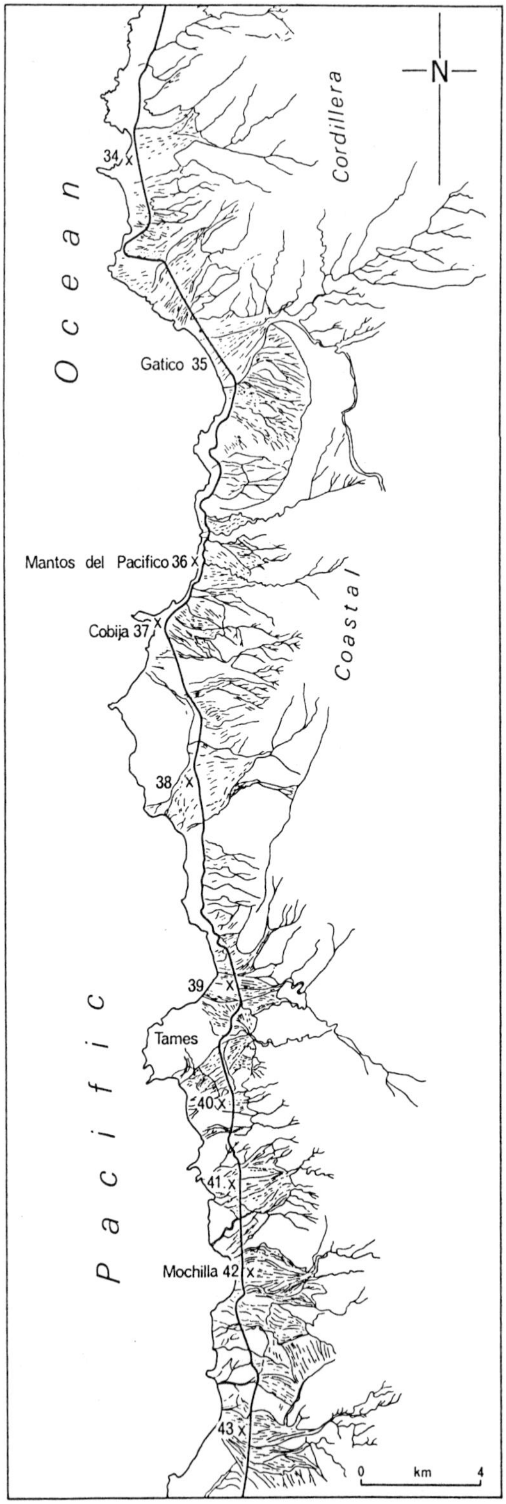


**Fig. 3.** Detailed map of the Gatico coastal segment showing the positions of specific locations referred to in the text. See Fig. 2 for position.

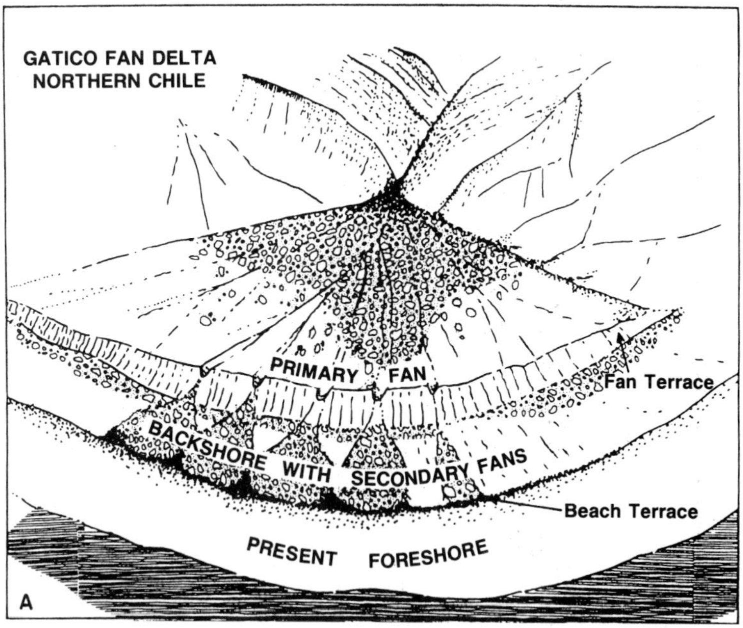


**Fig. 4.** (A) Line drawing of the Gatico coastal-fan–fan-delta system showing main depositional and erosional environments. (B) Summary log through the Gatico fan-delta, showing the main facies associations and interbedding of interpreted subaerial and subaqueous deposits in the exposed succession.

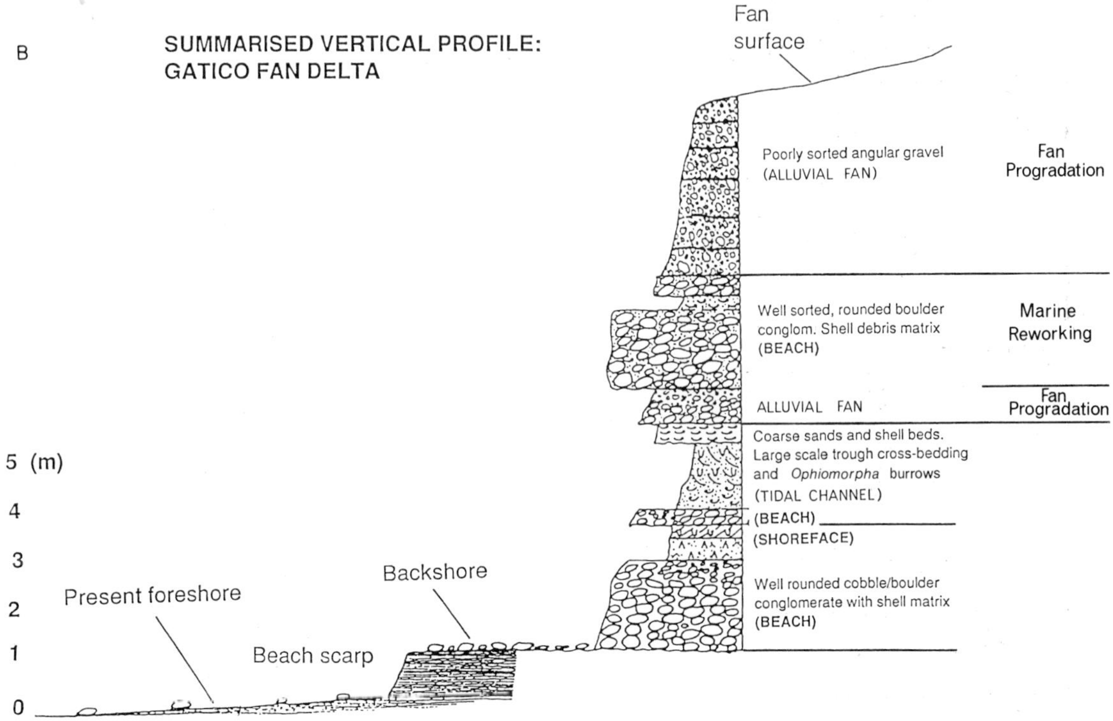

**Fig. 5.** (A) Morphology of the Chilean coast at Gatico (22.5°S), showing the Coastal Cordillera (Jurassic igneous sequences and sediments) with Late Tertiary to Recent coastal-fan–fan-delta systems shedding into the Pacific Ocean. View looking south. (B) View toward the Gatico fan apex (c. 2 km away). The fan surface comprises a dominantly inactive system that is bypassed via deep channels. (C) Fan surface, which is in balance between deflation and aeolian sand accumulation. Scattered boulders (up to 1 × 2 m) lie in shallow channels that were probably cut during a Quaternary pluvial period. (D) 'Lowstand' canyon incised into Late Tertiary/Quaternary alluvial fan sediments at Mancos del Pacifico (location 36, Figs 2, 3). This recent feature is c. 25 m deep. (E) Currently active fan-delta building an arrow-shaped platform into the Pacific Ocean. This system is fed by the canyon shown in Fig. 5D (just visible in immediate foreground). (F) Detail of fan-delta front (Fig. 5E). Angular, poorly sorted fan-delta gravels are prograding over well-rounded beach conglomerates (lower field of view). Small scarp is 0.5 m in height.

**Fig. 6.** (A) Marine terrace development at Hornitos (location 49, Fig. 2). Note that terraces are cut by later deep channels/canyons as the fans grade to new base-level. Telegraph pole in centre foreground is 4 m high. (B) Shallow-marine sandstone facies association in the uplifted Gatico fan-delta sequence. Large-scale trough and tabular cross-bedding with basal lags of volcanic cobbles are visible. Vertical face is c. 3 m high. (C) Uplifted uppermost Tertiary/Quaternary fan-delta deposits, south of Gatico. Cliff is c. 15 m high. Note younger canyon cutting the terrace in foreground left. (D) Detail of terrace material in Fig. 6A. Framework of well-rounded cobbles/pebbles of Jurassic volcanic rocks and Quaternary bivalve shells are contained in a matrix of shell debris.

poorly sorted breccias are prograding over the present-day beach, which consists of well-rounded, moderately well-sorted cobble/pebble conglomerates (Fig. 5E, F).

### *Shallow marine deposits*

The uplifted terraces (Fig. 6A) expose sandstone bodies up to 3 m thick, characterized by large-scale trough and planar tabular cross-bedding with common mud drapes to foresets and occasional pebble lags of volcanic rocks to bottomsets (Fig. 4B, 6B). The sandstones are medium grained, with up to 30% of invertebrate shell debris. Trace fossils include large *Ophiomorpha* burrows. The sequential position and internal characteristics of these sandstones are consistent with tidal channel deposits. Other sandstone bodies of up to 2 m thickness, which do not exhibit trough cross-bedding, may represent storm deposits.

The terraces also include horizons of well-rounded, moderately sorted pebble/cobble conglomerate (Fig. 6C) with a matrix of shell debris and occasional detrital, intact bivalve shells (Fig. 4B, 6D). These horizons are up to 2 m thick and identical in composition and gross geometry with the present-day beaches adjacent to fan systems. We thus interpret these conglomerates as ancestral reworked fan-delta/beach deposits.

The interbedding of subaerial alluvial fan gravels with the marine beach conglomerates and tidal/storm sandstones on a 3–7 m scale (Fig. 4B) indicates a crude cyclicity to the Quaternary depositional system. Such cyclicity could simply reflect the periodic switching of an alluvial fan lobe into the Gatico coastal area. However, it is likely that the Quaternary fans were similar, dominantly short-headed systems, to the present-day fans, which show little evidence of major lobe switching. Moreover, the present-day morphological organization of the coastal depositional systems comprises currently inactive alluvial fans, cut by canyons up to 10 m deep that act as sediment bypass conduits to sites of shoreline fan-delta progradation. Coupled with the organization of preserved subaerial/subaqueous sediments into crude cycles, these data suggest that relative sea-level changes within the Quaternary may have been a strong control on coastal fan development.

### Fan chronology and Quaternary glaciation

The presence of red beds throughout the Late Cretaceous–Recent geological record of northern Chile provides evidence for long-term desert environments in the Central Andes. However, there is abundant geomorphological and geological evidence for widespread glaciations at high altitudes and interglacial pluvial periods during the Pleistocene (Brüggen, 1950; Hollingworth & Guest, 1967; Caviedes & Paskoff, 1975; Clapperton, 1984). Quaternary pluvial periods are corroborated by the presence of lacustrine beach terraces over much of the Central Depression (Brüggen, 1950; Naranjo & Paskoff, 1982).

At least six Quaternary relative sea-level changes have been identified regionally in central and northern Chile, recorded by of a series of marine terraces (Herm, 1969; Paskoff, 1977). There is general agreement that the oldest terrace system correlatable along the whole northern–central Chile coast is Pliocene in age, based on gastropod biostratigraphy (Fuenzalida *et al.*, 1965; Herm, 1969). However, this method only allows differentiation of Pliocene from younger sequences as the Quaternary terraces contain fauna of modern species.

Dating of the coastal terraces between Gatico and Antofagasta was attempted by radiocarbon methods (Paskoff, 1973) and electron spin resonance (Radtke, 1985). These non-conclusive results suggest that a major period of terrace formation in this area may have taken place during a highstand associated with an interglacial period (c. 82 000 years BP).

The problems associated in dating these Andean terraces are discussed by DeVries (1988) who showed that the partly recrystallized nature of the shell material precludes amino-acid geochronology as used in other terrace systems (e.g. Dumas *et al.*, 1988). The age of the older terraces is beyond that of radiocarbon methods and DeVries (1988) adopted the relatively low-resolution gastropod biostratigraphy of Herm (1969).

### Terrace origin: eustasy versus local tectonics

The earliest interpretation of marine terrace formation, by Charles Darwin (1876), refers to rising of the land (in fact the Chilean coast) as the driving mechanism. However, marine terraces are a common component of the Quaternary stratal record over much of the world and clearly at least partly reflect a record of the glacial–interglacial fluctuations in global sea level. Recent, detailed work on western Pacific terrace systems in New Zealand (Ota *et al.*, 1988), Timor (Jouannic *et al.*, 1988), and Papua New Guinea (Crook, 1989) all testify to the eustatic origin of these phenomena. However, in all these areas the above studies have documented lateral height variations in terraces (relative to modern sea level) that clearly cannot be a function of global eustatic control.

Simple correlation along the Chilean coast shows that eustatic terrace cycles are overprinted by tectonic processes, as documented from the western Pacific. Three terraces are developed in the area between Coloso and Mejillones, near Antofagasta (Fig. 2). Observational data clearly show that the terraces are at different heights above present sea level, over distances of tens of kilometres. These height differences can be considerable; for example the oldest terrace in the area (330 000–345 000 years BP; Radtke, 1985) is currently 70 m above sea level at Coloso yet only 35 m above sea level at Mejillones, some 20 km to the northwest.

## DISCUSSION

The coastal fan/fan-delta systems of northern Chile preserve evidence of relative sea-level changes over uppermost Tertiary to Recent times. Reliable distinction between the importance of tectonics and eustacy in controlling relative sea level and thus basin-fill architecture has been the subject of much debate (Vail *et al.*, 1977; Miall, 1986; Posamentier & Vail, 1988). It has been argued that the *rate and*

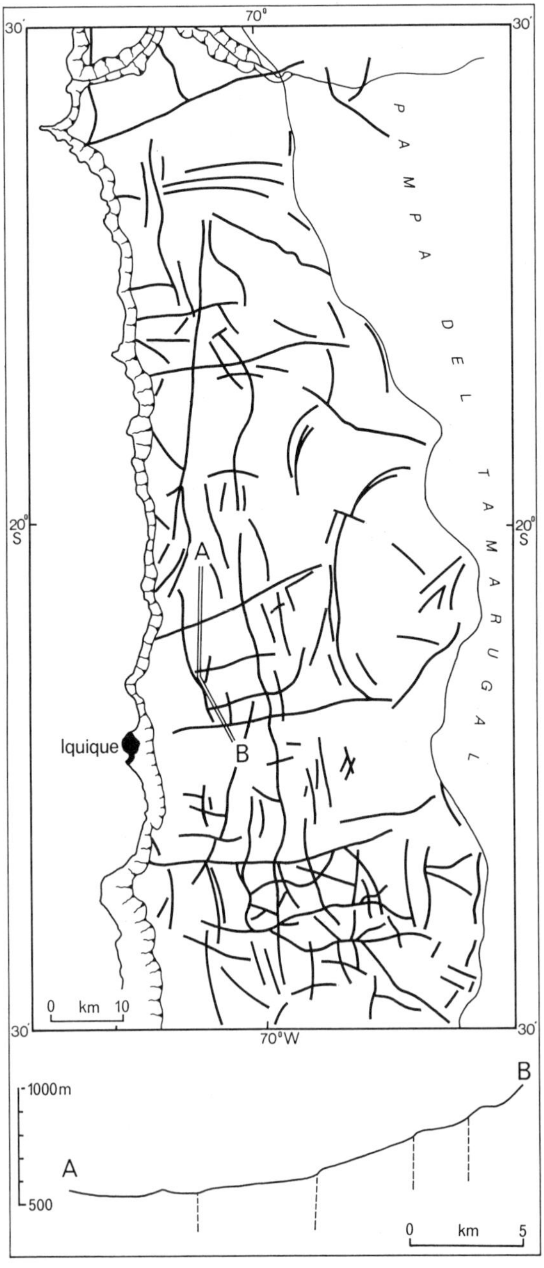


**Fig. 7.** Late Tertiary to Recent faulting pattern in the Iquique area. Note the E–W-trending normal faults that downthrow to the north, and the progressively increasing uplift to the south (after Mortimer & Sarić, 1972).

*frequency* of sea-level change allows differentiation between different allocyclic processes. Megacycles could thus represent either tectonic phenomena, such as thrust belt activity or changes in spreading rates/volumes, or major glaciations. Smaller scale, superimposed, higher frequency cyclicity is then argued to necessarily be of eustatic or autocyclic origin. Integration of sedimentological, stratigraphic, structural, and oceanographic data for the Chilean fan systems allows us to test this hypothesis at an active continental margin.

## Chilean coastal dynamics

At Arica (Fig. 1) the coastal range is in net extension, characterized by extensional normal faulting and subsidence, in common with much of the Chilean margin. Starting some 50 km south of Arica, the Andean fore-arc is characterized by a series of north-facing neotectonic normal fault scarps (Fig. 7); Mortimer, 1972; Mortimer & Sarić, 1972,1975). These E–W faults may represent segmentation of the fore-arc as they coincide with the transition between regions in net subsidence and net uplift.

Recently published tide-gauge records for the west coast of South America for the 30 year period between 1940 and 1970 (Aubrey *et al.*, 1988) provide an independent assessment of uplift at the Andean margin. The tide-gauge data (Fig. 8) show that most of the Andean margin from Colombia to central Chile is subsiding (negative values = land sinking relative to sea level). The exceptions to this trend are geographically restricted areas in Equador (Tumaco), Peru (Matarani), and Antofagasta, where land has risen relative to the Pacific ocean (i.e. a fall in relative sea level) by up to 3.5 mm over 30 years. Unfortunately the tide-gauge stations have a wide geographical spacing, but the data do support the above stratigraphic, morphological, and structural evidence for regionally variable patterns of coastal uplift. Furthermore, differential uplift reflected in similar deposits of the same age in northwest Peru has recently been documented (DeVries, 1988).

Plotting of the uplift pattern and tide-gauge readings on a map with Pacific bathymetric data reveals a good correlation between areas undergoing uplift and the intersection of aseismic ridges with the South American plate edge (Fig. 9A, B). Thus the uplift (measurable over a 30 year period) in specific areas of Equador and Peru is coincident with the Carnegie and Nazca ridges and the Antofagasta–Iquique sector coincides with the ongoing subduction of an unnamed ridge (Fig. 9A). Plate reconstructions indicate that aseismic ridge subduction has been an important component of the Tertiary Nazca plate

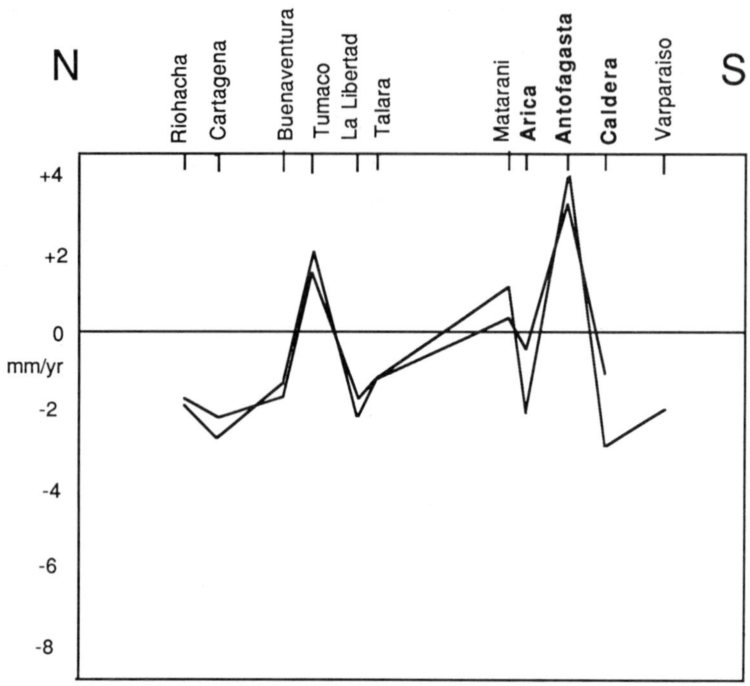


**Fig. 8.** Mean change of relative land level (= relative sea level) for the Andean margin from Riohacha (Atlantic coast of Colombia) to Valparaiso, Central Chile. Positive values indicate a rise in land level relative to the Pacific Ocean (i.e. a fall in relative sea level). The two lines represent results from regression analysis and eigenanalysis of the data. From Aubrey *et al.* (1988).

history (Cross & Pilger, 1982; Pilger, 1984) and several authors have suggested that subduction of such ridges may, on a large scale and over periods of several million years, affect arc volcanism and structural evolution (Nur & Ben-Avraham, 1981; Cross & Pilger, 1982).

The role of aseismic ridge subduction in producing *recent and ongoing* locally distributed uplift was first suggested by Aubrey *et al.* (1988). Our data further support the thesis that ridge irregularities or a series of seamounts/ridges may have been responsible for driving the high-frequency pattern of relative sea-level changes throughout the late Tertiary/Quaternary in northern Chile.

The subsidence and extensional tectonic regime in interridge areas, such as Arica (Figs 1, 9), is well documented (Katz, 1971), despite continuous subduction at the Andean margin since Jurassic times (Coira *et al.*, 1982). This fore-arc extension, typical for the 'cordilleran margin' type of Aubouin (1989), is marked in the study area by oblique NE–SW and N–S trending normal faults in the Coastal Cordillera (Fig. 2) and many neotectonic fault scarps in the Central Depression. This neotectonic style may be a result of subduction roll-back (Hartley *et al.*, 1988) or possibly a function of transtension resulting from the slightly oblique Nazca plate convergence angle with the Chilean margin.

### Sea-mount/ridge subduction in other areas

Extensive marine geophysical and geological studies of the Japan island-arc–subduction-zone system have recently yielded a wealth of data on subduction of ocean floor features, including seamounts and aseismic ridges. Several case studies have direct application to this discussion. Two actively subducting seamounts in the Nankai trough and at the Japan/Kuril trench junction can be demonstrated to be pushing fore-arc wedge material landward and upward, with deformation extending into the fore-arc basin fills (Yamazaki & Okamura, 1989). Similarly, in the Japan trench approximately half of the Daiichi Kashima seamount has been subducted over an estimated 150 000–250 000 years (Lallemand *et al.*, 1989).

Importantly, as seamount subduction continues, the earlier uplifted material undergoes gravity driven collapse or extension. Similar phenomena are documented from the Manilla trench system where

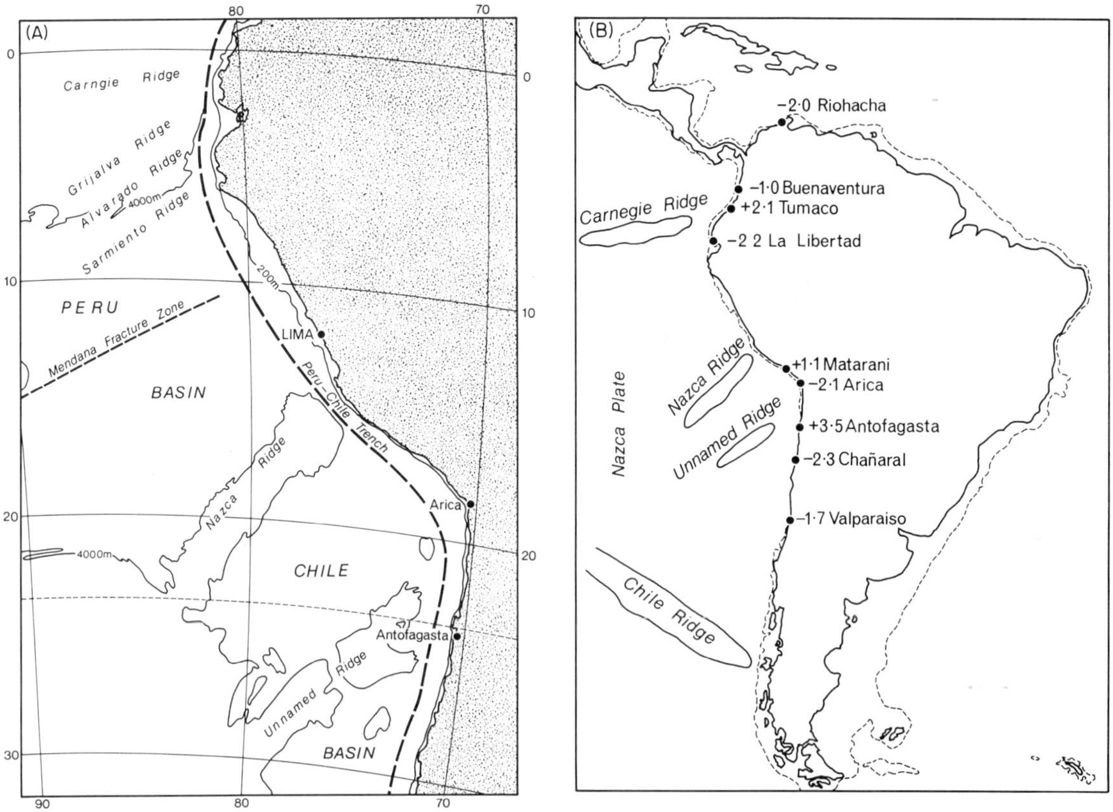


**Fig. 9.** (A) Map of the southeast Pacific Ocean/South American margin (modified after Kulm *et al.* (1981) and the American Association of Petroleum Geologists (1981)). Bathymetric data clearly delineate major named ridges intersecting the margin (e.g. Nazca Ridge) but also the unnamed ridge impinging on the Antofagasta–Iquique sector. (B) Data from Fig. 8 (Aubrey *et al.*, 1988) plotted in relation to the ridge systems show that the Andean margin is currently in extension/subsidence except for areas coincident with ridges, where active uplift is recorded from sea-level changes over a recent 30-year period.

continued oblique subduction of the South China Sea axial ridge is occurring. Subduction of 'normal' oceanic crust in areas to each side of the ridge has no obvious effect on the fore-arc while the subduction of the ridge has resulted in complex deformation and uplift over an area 40 km wide. The effect on the overriding plate edge has been to segment the original fore-arc basin into two sub-basins (Pautot & Rangin, 1989).

## Ridge subduction and fore-arc basin facies architecture

The fore-arc uplift–collapse process could result in a coupled lowering/raising of relative sea-level over relatively short periods (c. 100 000 years). Furthermore, as the above studies both show that negligible accretion/obduction occur during seamount chain/ridge subduction, the inherent morphological/structural irregularities in these structures could result in cycles of minor fore-arc uplift/subsidence at such 100 000–200 000 year frequencies. This tectonic mechanism could thus produce relative sea-level changes at a frequency equal to glacio-eustatic processes. These tectonically driven cycles will be restricted to arc-related sedimentary basins: they will not be of global extent nor globally synchronous. Further studies are required to develop criteria to distinguish the effects of ridge subduction from autocyclic or other tectonic phenomena.

## ACKNOWLEDGEMENTS

The work presented in this paper benefited from our discussions with Adrian Hartley, Graham Williams, Tim Buddin, and Phil Scanlan. Fieldwork in Chile was supported by the Royal Society and a NERC Research Fellowship to EJJ. Diagrams courtesy of Joe Lynch. The paper benefited from the comments of reviewers Mike Daly, Mike Bell, and editor David Macdonald.

## REFERENCES

American Association of Petroleum Geologists (1981) *Plate Tectonic Map of the Circum-Pacific Region, Southeast Quadrant*. AAPG, Tulsa, OK.

Aubouin, J. (1989) Some aspects of the tectonics of subduction zones. In: *Subduction Zones: the Kaiko Project* (Eds Cadet, J.P. & Uteda, S.). Tectonophysics 160, 1–21.

Aubrey, D.G., Emery, K.O. & Uchupi, E. (1988) Changing coastal levels of South America and the Caribbean region from tide-gauge records. *Tectonophysics* **154**, 269–284.

Bruggen, J. (1950) *Fundamentos de la Geologia de Chile*. Instituto de Geographico y Militar, Santiago, Chile, 347 pp.

Caviedes, C.N. & Paskoff, R. (1975) Quaternary glaciations in the Andes of north-central Chile. *J. Glaciol.* **14**, 155–170.

Clapperton, C.M. (1984) The glaciation of the Andes. *Quaternary Sci. Rev.* **2**, 83–155.

Coira, B., Davidson, J., Mpodozis, C. & Ramos, V. (1982) Tectonic and magmatic evolution of the Andes of northern Argentina and Chile. *Earth. Sci. Rev.* **18**, 303–332.

Crook, K.A.W. (1989) Quaternary uplift rates at a plate boundary, Lae urban area, Papua New Guinea. *Tectonophysics* **163**, 105–118.

Cross, T.A. & Pilger, R.H. Jr. (1982) Control of subduction geometry, location of magmatic arcs and tectonics of arcs and back-arc regions. *Geol. Soc. Am. Bull.* **93**, 545–562.

Darwin, C. (1876) *Geological Observations on the Volcanic Islands and Part of South America Visited During the Voyage of HMS Beagle*. Smith, Elder & Co., London, 647 pp.

de Vries, T.J. (1988) The geology of late Cenozoic marine terraces (tablazos) in northwest Peru. *J. South Am. Earth. Sci.* **1**, 121–136.

Dumas, B., Gueremey, P., Hearty, P.J., Lhenaff, R. & Raffy, J. (1988). Morphometric analysis and amino acid geochronology of uplifted shorelines in a tectonic region near Reggio Calabria, south Italy. *Palaeogeogr. Palaeoclimatol. Palaeoecol.* **68**, 273–289.

Ferraris, F. & Dibiase, F. (1978) Hoja Antofagasta. *Carta Geologíca de Chile, 1:250 000*. Servício Nacional de Geología y Mineria, Santiago, 42 pp.

Flint, S., Clemmey, H. & Turner, P. (1986) The Lower Cretaceous Way Group of northern Chile: an alluvial fan–fan-delta complex. *Sediment. Geol.* **46**, 1–22.

Fuenzalida, H., Cooke, R., Paskoff, R., Segerstrom, K. & Weischet, W. (1965) High stands of Quaternary sea-level along the Chilean coast. *Geol. Soc. Am. Spec. Pap.* **84**, 473–496.

Garcia, F. (1967) Geología del Norte Grande de Chile. *Proc. Symp. Geosin. Andino*. Soc. Geol. Chile, Publ. 3, 183 pp.

Hartley, A.J., Turner, P., Williams, G.D. & Flint, S. (1988) Palaeomagnetism of the Cordillera de la Costa, northern Chile: evidence for local forearc rotation. *Earth Planet. Sci. Lett.* **89**, 375–386.

Herm, D. (1969) Marines Pliozan und Pleistozan in Nord- und Mittet-Chile unter besonderer Berucksichtigung der Entwicklung der Mollusken-Faunen. *Zitteliana* **2**, 1–159.

Hollingworth, S.E. & Guest, J.E. (1967) Pleistocene glaciation in the Atacama desert, northern Chile. *J. Glaciol.* **6**, 749–751.

Jouannic, C., Chi-Trach Hoang, Wahyoe Soepri Hantoro & Delinom, R.M. (1988) Uplift rate of coral reef terraces in the area of Kupang, West Timor: preliminary results. *Palaeogeogr. Palaeoclimatol. Palaeoecol.* **68**, 259–272.

Katz, H.R. (1971) Continental margin in Chile — is tectonic style compressional or extensional? *Bull., Am. Assoc. Petrol. Geol.* **55**, 1753–1758.

Kulm, L.D., Prince, R.A., French, W., Johnson, S. & Masais, A. (1981) Crustal structure and tectonics of the central Peru continental margin and trench. In: *Nazca Plate: Crustal Formation and Andean Convergence* (Ed. Kulm, L.D.). Geol. Soc. Am. Mem. 154, 445–468.

Lallemand, S., Colotta, R. & von Huene, R. (1989) Subduction of the Daiichi Kashima seamount in the Japan trench. *Tectonophysics* **160**, 231–247.

Miall, A.D. (1986) Eustatic sea-level changes interpreted from seismic stratigraphy: a critique on the methodology with particular reference to the North Sea Jurassic record. *Bull., Am. Assoc. Petrol. Geol.* **70**, 131–137.

Mitchell, A.H.G. & Reading, H.G. (1986) Tectonics and sedimentation. In: *Sedimentary Environments and Facies* (Ed. Reading, H.G.) pp. 471–519. Blackwell Scientific Publications, Oxford.

Mortimer, C. (1972) The evolution of the continental margin of northern Chile. *Proc. 24th Int. Geol. Cong. Montreal*, Section 8, pp. 48–52.

Mortimer, C. & Sarie, N. (1972) Landform evolution in the coastal region of Tarapaca Province, Chile. *Rev. Geogr. Phys. Geol. Dynam.* **21**, 162–170.

Mortimer, C. & Sarie, N. (1975) Cenozoic studies in northernmost Chile. *Geol. Rundsch.* **64**, 395–420.

Naranjo, J.A. & Paskoff, R.P. (1982) Estratígrafia de las unidades sedimentárias Cenozoicas de la cuenca del Río Loa en la Pampa del Tamarugál, region de Antofagasta, Chile. *Rev. Geol. Chile.* **15**, 49–57.

Nur, A. & Ben-Avraham, Z. (1981) Volcanic gaps and the consumption of aseismic ridges in South America. *Geol. Soc. Am. Mem.* **154**, 729–740.

Ota, Y., Berryman, K.R., Hull, A.G., Miyauchi, T. & Iso, N. (1988) Age and height distribution of Holocene transgressive deposits in eastern North Island, New Zealand. *Palaeogeogr. Palaeoclimatol. Palaeoecol.* **68**, 135–151.

PASKOFF, R.P. (1973) Radiocarbon dating of marine shells taken from the north and central coast of Chile. *Proc. IX Int. INQUA Cong.*, Christchurch, New Zealand, pp. 281–282 (abstr.).

PASKOFF, R.P. (1977) Quaternary of Chile; the state of research. *Quaternary Res.* **8**, 8–21.

PAUTOT, G. & RANGIN, C. (1989) Subduction of the South China Sea axial ridge below Luzon (Philippines). *Earth Planet. Sci. Lett.* **92**, 57–69.

PILGER, R.H. JR. (1984) Cenozoic plate kinematics, subduction and magmatism: South American Andes. *J. Geol. Soc. London* **141**, 793–802.

POSAMENTIER, H.W. & VAIL, P.R. (1988) Eustatic controls on clastic deposition II — sequence and systems tract models. In: *Sea-level Changes: an Integrated Approach* (Eds Wilgus, C.H., Hastings, B.S., Kendall, C.G.St.C., Posamentier, H.W., Ross, C.A. & van Wagoner, J.C.). Soc. Econ. Paleontol. Mineral., Spec. Publ. 42, 109–124.

POSAMENTIER, H.W., JERVEY, M.T. & VAIL, P.R. (1988) Eustatic controls on clastic deposition, I — conceptual framework. In: *Sea-level Changes: an Integrated Approach* (Eds Wilgus, C.H., Hastings, B.S., Kendall, C.G.St.C., Posamentier, H.W., Ross, C.A. & van Wagoner, J.C.). Soc. Econ. Paleontol. Mineral., Spec. Publ. 42, 109–124.

RADTKE, U. (1985) Chronostratigraphie und neotektonik mariner terrassen in Nord- und MittelChile. *Proc. IV Congr. Geol. Chileno*, Antofagasta, Chile. **3**, 436–457.

SEGERSTROM, K. (1964) Quaternary geology of Chile: brief outline. *Geol. Soc. Am. Bull.* **75**, 157–170.

SERVICIO NACIONAL DE GEOLOGIA Y MINERIA (1980) *Mapa Geologíco de Chile, 1:1 000 000, Sheet 1*. Santiago, Chile.

VAIL, P.R., MITCHUM, R.M. JR., TODD, R.G., WIDMIER, J.M., THOMPSON, S. III., SANGREE, J.B., BUBB, J.N. & HATLELID, W.G. (1977) Seismic stratigraphy and global changes of sea-level (parts 1–6). In: *Seismic Stratigraphy and Applications to Hydrocarbon Exploration* (Ed. Payton, C.W.). Mem., Am. Assoc. Petrol. Geol. 26, 49–212.

YAMAZAKI, T. & OKAMURA, Y. (1989) Subducting seamounts and deformation of overriding forearc wedges around Japan. *Tectonophysics* **160**, 207–229.

*Spec. Publs. int. Ass. Sediment.* (1991) **12**, 105–119

# Volcano-tectonic control of offshore to tidal-flat regressive cycles from the Dunquin Group (Silurian) of southwest Ireland

R.J. SLOAN*,† *and* B.P.J. WILLIAMS‡

† *Department of Geology, The University, Bristol, BS8 1RJ, UK; and* ‡ *Department of Geology and Petroleum Geology, The University, Aberdeen AB9 1AS, UK*

## ABSTRACT

The Ferriter's Cove and Clogher Head Formations of the Dunquin Group (mid-Silurian) in Dingle, southwest Ireland, comprise a sequence of regressive cycles consisting of nearshore siliciclastic sediments and volcanic rocks. Each cycle records a transition from offshore, to shoreface, barrier, and back-barrier environments. The back-barrier area was a complex of intertidal flats and subtidal zones into which pyroclastic deposits and lavas were emplaced. There is no evidence of volcanism in the sub-barrier sequences. Each intercycle transgression was preceded by a volcanic episode. Decreasing water depth is attributed to sea-floor aggradation and gradual volcano-tectonic uplift prior to eruption, while post-volcanic magma chamber deflation triggered rapid subsidence and marine transgression. Evidence provided by two stratigraphically equivalent sections through the Ferriter's Cove and Clogher Head Formations suggests that the amount of subsidence was greater closer to the vent. Eustatic sea-level changes and variable rates of sediment supply were subordinate to volcano-tectonic control of relative sea level.

## INTRODUCTION

In nearshore sedimentary environments, relatively small variations in depositional depth may be expressed in the rock record by significant lithofacies variations. The Dunquin Group (mid-Silurian) of Dingle, southwest Ireland (Fig. 1) contains a sequence of stacked, offshore to tidal-flat regressive cycles marked by changes in grain size, sedimentary structures, biota, and degree of bioturbation, which reflect variations in palaeobathymetry and proximity to a prograding shoreline. Volcanic rocks are interstratified with these regressive cycles but occur only near the climax of each cycle. In this study we examine the links between the evidence of volcanism and the possible eustatic/depositional/tectonic controls on the regressive cycles.

The Dunquin Group (Holland, 1969) is late Wenlock to early Ludlow in age (Holland, 1988) and crops out in four inliers (Parkin, 1976) along the Dingle Peninsula, County Kerry, southwest Ireland and on a number of small islands (the Blasket Islands) off its western end, principally Beginish (R.J. Sloan, unpublished data) and Inishvickillane (Parkin, 1974). The group is best exposed in the Dunquin inlier (Fig. 2), a c. 1430 m-thick sequence of variably fossiliferous shallow- to non-marine siliciclastic sediments (Watkins, 1978) interbedded with pyroclastic rocks and lavas. The volcanic rocks range from basaltic to rhyolitic in composition, and form a non-bimodal subalkaline suite with geochemical characteristics (Sloan & Bennett, 1990) consistent with magma genesis at a destructive continental margin, although subduction of oceanic lithosphere may have ceased before eruption.

The Dingle Peninsula lies some 40 km south of the putative trace of the Iapetus Suture (Phillips *et al.*, 1975; McKerrow & Soper, 1989). Plate-tectonic reconstructions of the Wenlock of the British Isles vary in the timing of final closure of the Iapetus

* Present address: GAPS Geological Consultants, 13 Deodar Road, Putney, London SW15 2NP.

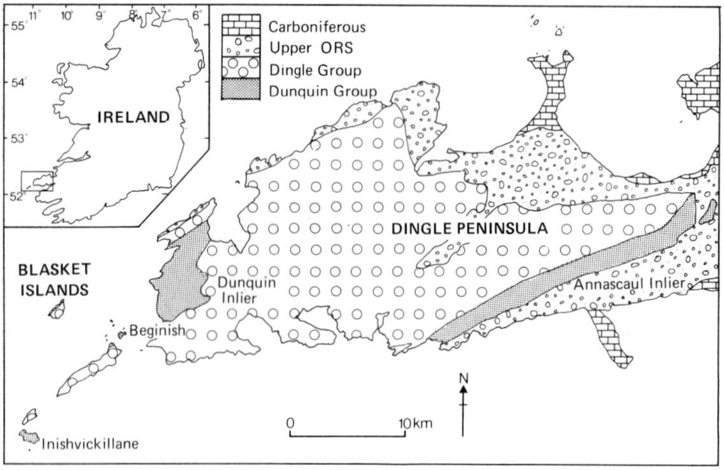


**Fig. 1.** Geological sketch map of the Dingle Peninsula, with inset showing location in Ireland.

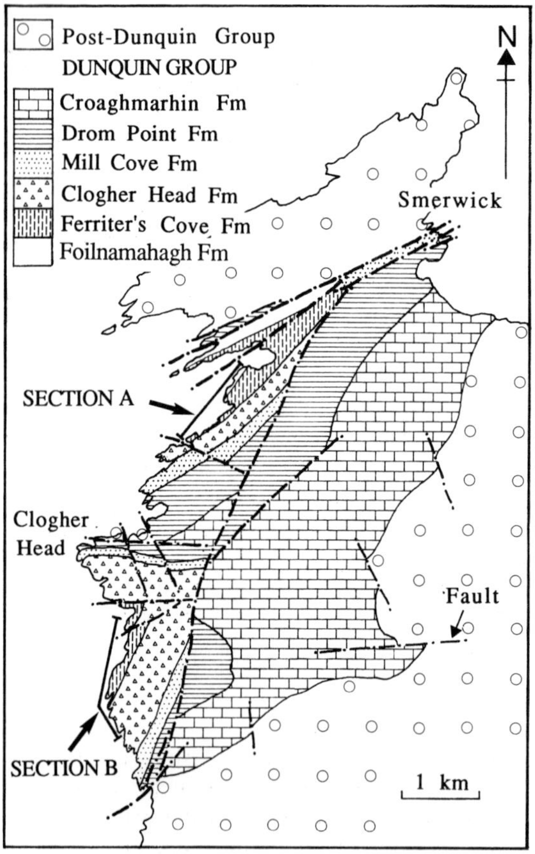


**Fig. 2.** Geological map of the Dunquin inlier, showing location of sections A and B (Fig. 8).

Ocean from Late Ordovician (e.g. Pickering *et al.*, 1989, and references therein) to Mid-Devonian (e.g. McKerrow & Soper, 1989). However, the apparent Llandovery to Wenlock overstep of the Iapetus Suture by turbidite systems in central Ireland (Murphy & Hutton, 1986) suggests that ocean lithospheric subduction had ceased by the Mid-Silurian. The Dunquin Group is considered to have been deposited in an epicontinental successor basin developed in the suture zone.

## STRATIGRAPHY AND PALAEOGEOGRAPHY

The Dunquin Group was subdivided by Holland (1969) into five formations: Ferriter's Cove Formation, Clogher Head Formation, Mill Cove Formation, Drom Point Formation, and Croaghmarhin Formation. Table 1 shows a revised stratigraphy of the group based upon detailed lithofacies logging and exposure mapping (Sloan, unpublished data). A new subdivision, the Foilnamahagh Formation, is herein defined as the lowest formation in the Dunquin Group, and consists predominantly of continental red beds and interpersed lavas and pyroclastic deposits. It replaces the basal 150 m of, and is distinct from, Holland's (1969) Ferriter's Cove Formation, which comprises a sequence 150–240 m thick of shallow-marine sediments and pyroclastic deposits. The Foilnamahagh Formation crops out only around Doon Point (Fig. 3) and also, owing to the effects of a major northeast-striking reverse fault that cuts through Ferriter's Cove, in a narrow strip on the south shore of this bay.

The Dingle region experienced both Caledonian

**Table 1.** Dunquin Group stratigraphy

| Holland (1969,1988) | This paper |
|---|---|
| Croaghmarhin Fm (400 m) | Croaghmarhin Fm (400 m) |
| Drom Point Fm (300 m) | Drom Point Fm (300 m) |
| Mill Cove Fm (100 m) | Mill Cove Fm (105 m) |
| Clogher Head Fm (200–600 m) | Clogher Head Fm (175–290 m) |
| Ferriter's Cove Fm (600 m) | Ferriter's Cove Fm (150–240 m) |
| | Foilnamahagh Fm (150 m) |

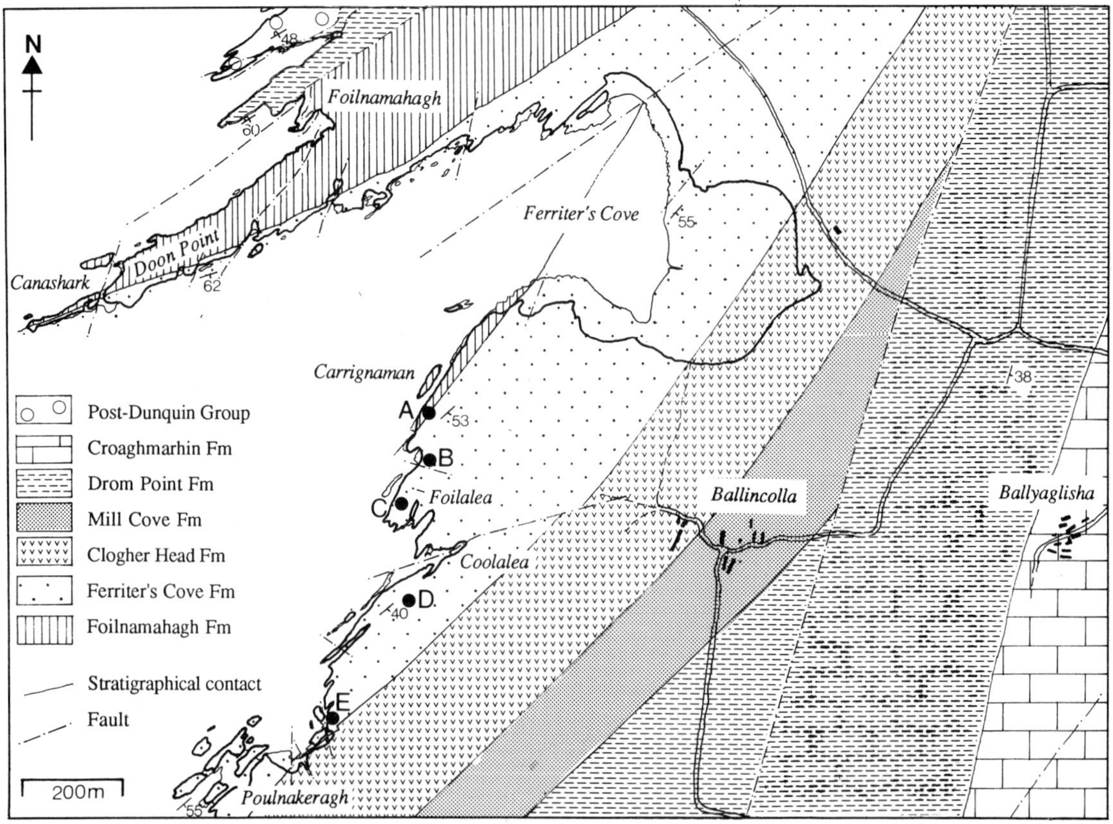


**Fig. 3.** Geological map of the northern end of the west coast section of Dunquin inlier, including section A. Letters A–E refer to stratigraphic positions shown in Fig. 8.

and Variscan deformation (Todd, 1989) with the effect that the Ferriter's Cove and Clogher Head Formations are exposed in three coastal sections on the limbs of an overturned fold couplet verging to the north-northwest. The northernmost section, between Foilnamahagh and Clogher Head, is herein termed section A (Figs 2, 3) and the southernmost, between Carrigcam and Cooshaun, is termed section B (Fig. 4). The central section, at Clogher Head itself, is overturned, deformed, and partly inaccessible; thus, correlation of this central section with sections A and B is uncertain and, consequently, it is not considered in any detail in this study.

According to Holland (1988), the Ferriter's Cove and Clogher Head Formations are biostratigraphically indistinguishable, and their contact should be placed at the base of 'a persistent thin rhyolitic rock' exposed in section A (Figs 2, 3). This unit (a welded

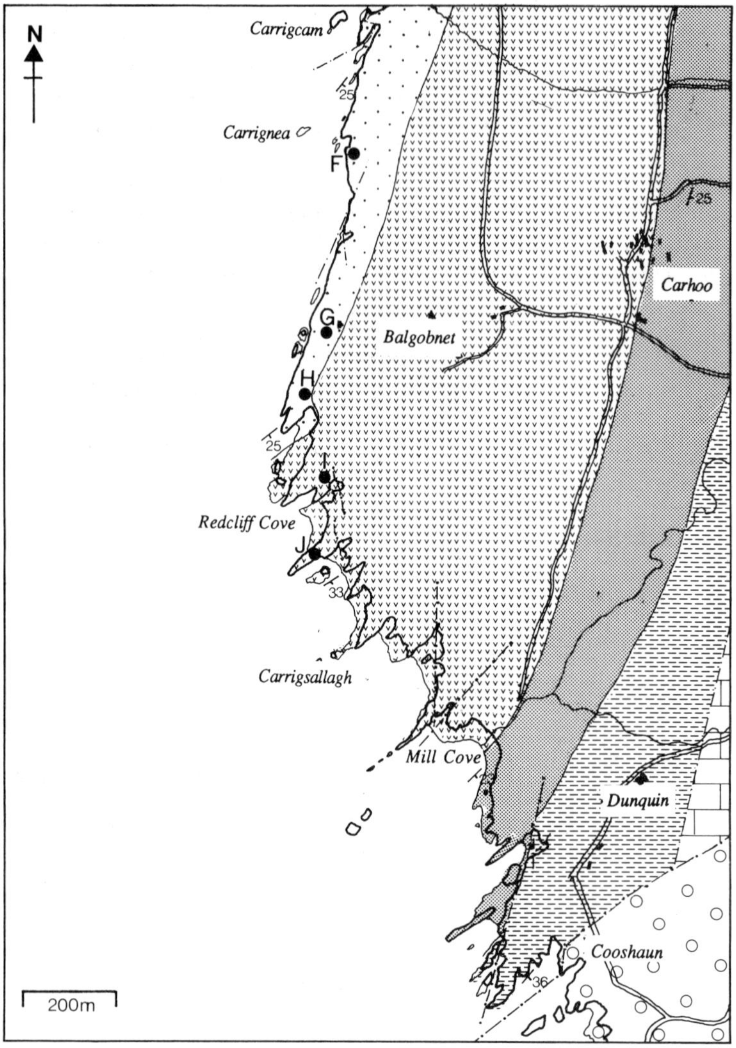


**Fig. 4.** Geological map of the southern end of the west coast of the Dunquin inlier, including section B. Letters F–J refer to stratigraphic positions shown in Fig. 8.

ignimbrite), however, makes a poor stratigraphic marker since it is laterally impersistent (traceable for only 800 m along strike), poorly exposed, and separates predominantly sedimentary sequences of remarkable similarity. It is recommended that the contact between the Ferriter's Cove and Clogher Head Formations should instead be placed at the base of a distinctive and unusual lahar deposit (*sensu* Fisher, 1984), exposed at two localities over 3.5 km apart (Figs 3, 4), at the base of a volcanic sequence over 25 m thick. This lahar deposit is 1–2 m thick, and contains abundant volcanic fragments (including pumice and broken crystals) and clasts of laminated siltstone up to 40 cm across, supported in a fine-grained matrix containing gastropods, possibly of the genus *Euomphalopterus* (J.S. Peel, personal communication, 1989). The effect of this correlation is that in section A, the basal 85 m, and in section B, the basal 135 m, of what Holland (1988) assigned to the Clogher Head Formation are here included in the Ferriter's Cove Formation.

Thick volcanic sequences on the islands of Beginish and Inishvickillane (400 m and over 900 m, respectively) were correlated by Sloan and Bennett (1990) with the volcanic rocks of the Dunquin inlier on the basis of close similarity in outcrop appearance, petrography, and geochemistry. Volcanic rocks of the Dunquin Group exposed in the Annascaul inlier

(Fig. 1), some 30 km west of the Dunquin inlier, are relatively thin (c. 80 m thick) and fine grained (Parkin, 1976). This southwesterly thickening in the volcanic successions suggests that the volcanic centre or centres may have lain to the west of Inishvickillane, although neither vent-related intrusions nor proximal volcanic facies are exposed. A palaeogeographic reconstruction (partly speculative) of the region in which the Dunquin Group was deposited is presented in Fig. 5. Volcanic highlands are shown flanked by alluvial fans shedding on to a broad coastal plain bordered by a shallow shelf. The regressive cycles described herein are exposed along the west coast of the Dunquin inlier and according to this palaeogeographic reconstruction were deposited some 15–20 km (not allowing for post-depositional shortening) from the vent or vents, on the flanks of a volcanic island (Bassett *et al.*, 1976) near to a fluctuating palaeoshoreline. Alluvial sediments, though uncommon, are present in the Foilnamahagh Formation and in the Mill Cove Formation (Sloan, unpublished data) and indicate that the barrier–lagoon system may have been fed by longshore drift from the mouths of rivers that tapped the proximal volcanic area. It is envisaged that this fluvial detritus was trapped in the barrier–lagoon system and did not feed directly on to the shelf. The sedimentary detritus throughout the Dunquin Group consists of reworked volcanic debris and subordinate bioclastic material. Regional palaeogeographic reconstructions (e.g. Zeigler, 1970; Phillips & Sevastopulo, 1986) show the

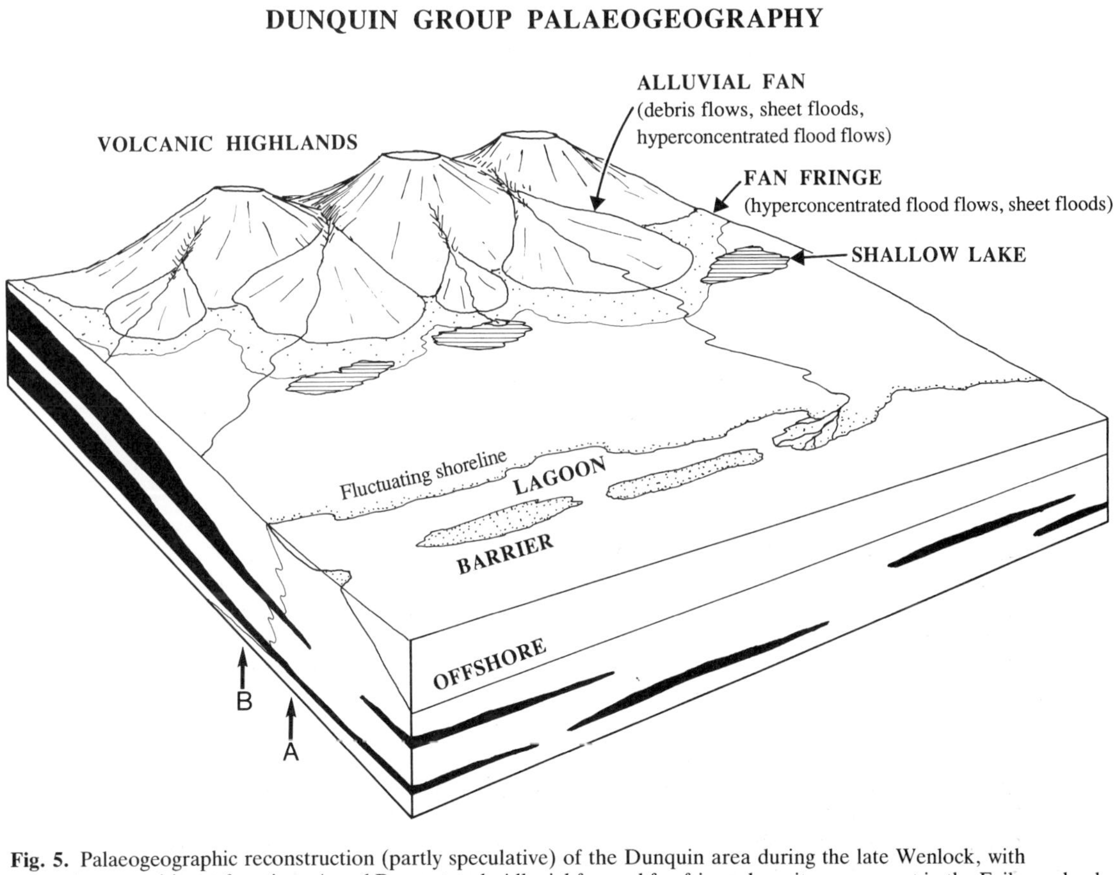


**Fig. 5.** Palaeogeographic reconstruction (partly speculative) of the Dunquin area during the late Wenlock, with approximate positions of sections A and B annotated. Alluvial fan and fan fringe deposits are present in the Foilnamahagh Formation and shallow lake and rare stream deposits in the Mill Cove Formation. The thick volcanic sequences in Inishvickillane and Beginish represent proximal volcanic facies. The vent–shoreline distance may have been 10–20 km.

Dunquin area as a volcanic island, with turbiditic greywacke sedimentation occurring in the surrounding epicontinental marine basin.

## REGRESSIVE CYCLES

The deposits representing the regressive cycles are generally 10–80 m thick and the base of each cycle is marked by an abrupt change from shallow water (lagoonal) or subaerial deposition to relatively deep water (off-shore or lower shoreface) sedimentation (Fig. 6). Each cycle may be divided into three parts: sub-barrier, barrier, and lagoonal deposits (Fig. 7). Six regressive cycles crop out in section A (Fig. 8), the upper four of which may be correlated with the four exposed in section B (Table 2).

### Sub-barrier sequences

The sub-barrier sequence is divided here into offshore, offshore–shoreface transition, and shoreface facies. Thorough bioturbation characterizes the offshore facies and although individual burrows are indistinct, abundant *Planolites* and *Teichichnus* burrows are evident locally. The sediments comprise muds and fine silts; material coarser than silt grade is rare. Bioclasts, mainly of brachiopods, crinoids, and corals, are dispersed in a matrix of mud, or less commonly concentrated in clast-supported layers of several millimetres up to 3 cm thick (see Fig. 9A). Upwards, there is a decrease in the intensity of

**Table 2.** Cycle thicknesses in sections A and B

| Cycle number | Total thickness (m) | | Sediment thickness (m) | | Volcanic thickness (m) | |
|---|---|---|---|---|---|---|
| | A | B | A | B | A | B |
| 6 | 151 | 218+ | 1 | 28 | 150 | 190+ |
| 5 | 79 | 88 | 55 | 27 | 24 | 61 |
| 4 | 10.3 | 56 | 10 | 14 | 0.3 | 42 |
| 3 | 27.4 | 47 | 27 | 33 | 0.4 | 14 |
| 2 | c. 70 | 20+ | ć.68 | n.e. | 2 | 20+ |
| 1 | 47 | n.e. | 37 | n.e. | 10 | n.e. |

n.e. = not exposed.

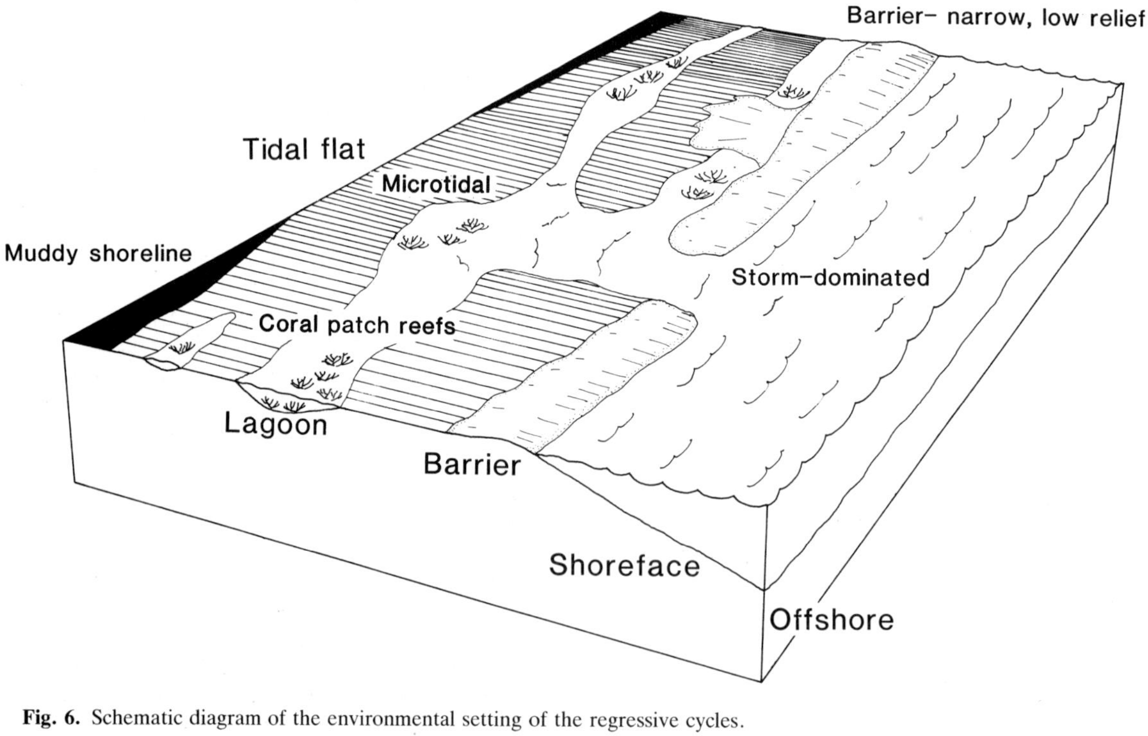


**Fig. 6.** Schematic diagram of the environmental setting of the regressive cycles.

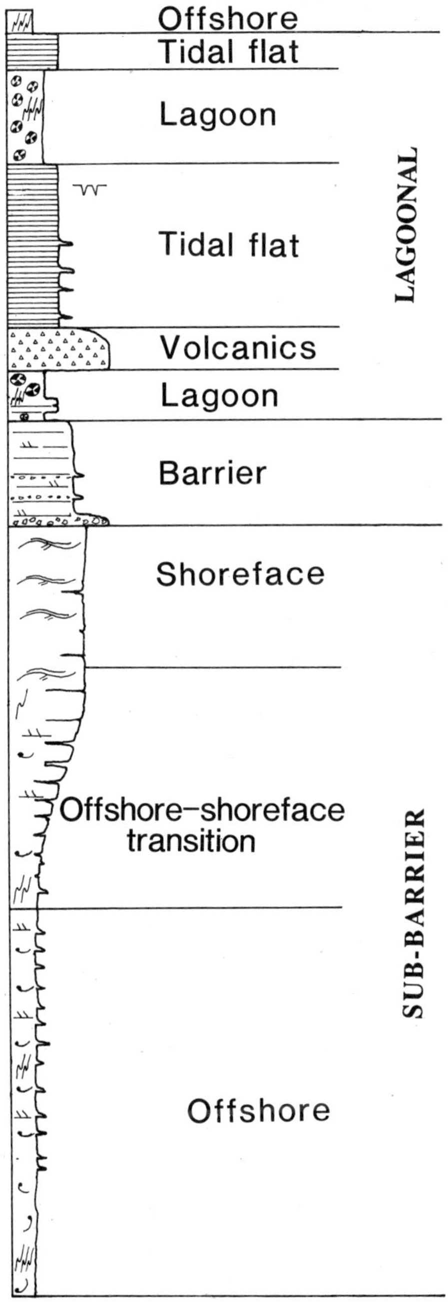


**Fig. 7.** Idealized log of a regressive cycle based on measured sections in section A. Cycles in section B differ in that offshore facies directly overlie the volcanic intervals without any intervening back-barrier deposits. The volcanic intervals vary from less than 1 m to over 190 m in thickness. Symbols as Fig. 8.

bioturbation, an increase in the frequency of shell layers and the appearance of graded and ripple-laminated silt bands 1–4 cm thick.

The base of the offshore–shoreface transition facies is marked by the first appearance of beds of very fine sand, 1–8 cm thick, which display planar and wave ripple lamination. The wavelength and amplitude of the ripples increase upwards and they grade up into hummocky and swaley cross-stratification (Harms *et al.*, 1975; Leckie & Walker, 1982; Fig. 9B). The facies is characterized by markedly less common shell layers than the offshore deposits, slight to moderate bioturbation, and a gradual upward decrease in the proportion of mud present.

The shoreface facies comprises hummocky and swaley cross-stratified (HCS and SCS), very fine to fine sandstone, the hummocks typically showing wavelengths of 60–100 cm and set heights of up to 8 cm. Locally, the HCS and SCS have wave-rippled tops and are overlain by bioturbated mud/silt. Upwards, the bioturbated and rippled tops of the units become scoured out by succeeding sands, with a resulting increase in HCS amalgamation. Shell layers are virtually absent.

Convincing evidence of storm deposition is provided by the presence of amalgamated HCS beds (Brenchley, 1985), allochthonous shell layers, and sharp-based, graded and rippled silt or sand layers. The sub-barrier sequences exposed in the Dunquin Group are similar to the transitions from modern shelf to coast deposits from the North Sea described by Aigner & Reineck (1982), in which variations in bioturbation, shell layer frequency, and changes in storm bed thickness, frequency, and amalgamation, were interpreted to indicate increasing proximity to the source of the storm sand (usually the shoreline) and/or palaeobathymetry. Seaward decrease in abundance of physical sedimentary structures and an increase in biogenic structures is fundamental to all shoreline sequences (Howard & Reineck, 1981) and is the result of increasing depth and its influence on the relative position of the wave base. The facies changes observed in the sub-barrier sequences are therefore interpreted to largely reflect increasing shoreline proximity and decreasing depth of deposition, possibly accelerated by volcano-tectonic uplift (see below). Evidence of volcanic activity in the sub-barrier sequences, e.g. tuffs (reworked or otherwise), clasts exceeding medium sand grade, or recognizable volcanic fragments, such as crystals or shards, are absent, indicating volcanic quiescence.

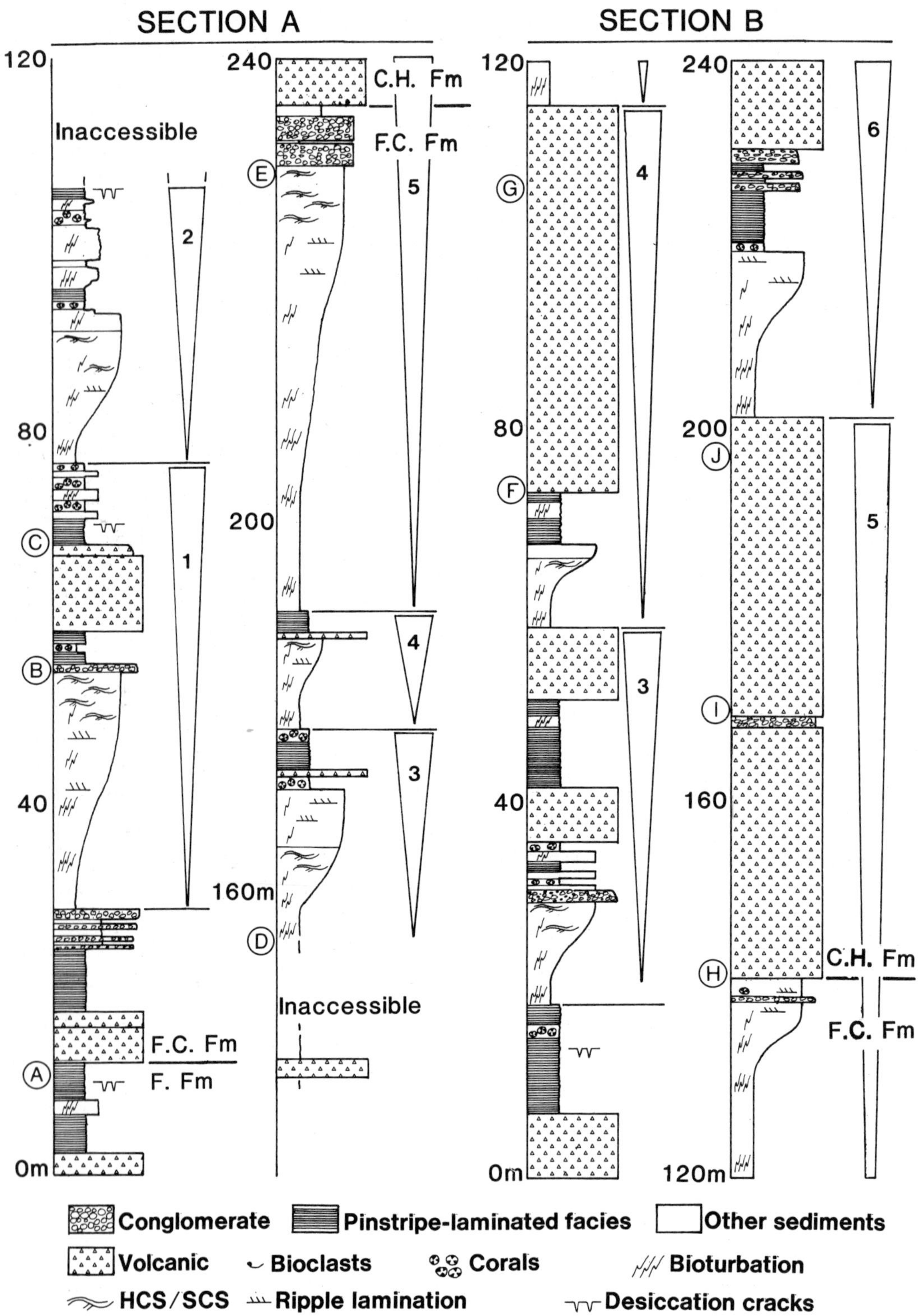
SECTION A
SECTION B
Inaccessible
C.H. Fm
F.C. Fm
F. Fm
120
80
40
0m
240
200
160m
120
80
40
0m
240
200
160
120m
A
B
C
D
E
F
G
H
I
J
1
2
3
4
5
6
Conglomerate
Pinstripe-laminated facies
Other sediments
Volcanic
Bioclasts
Corals
Bioturbation
HCS/SCS
Ripple lamination
Desiccation cracks

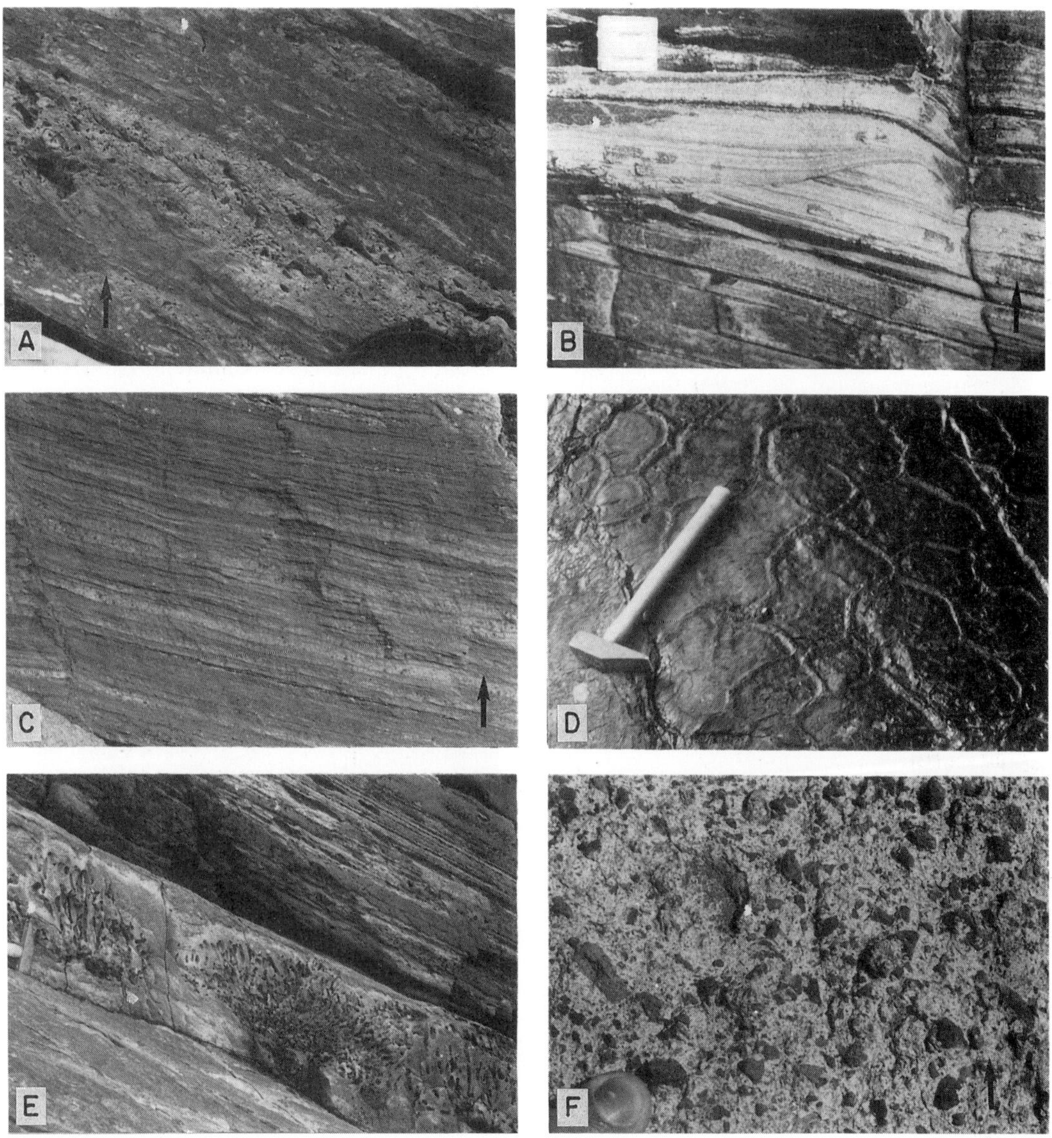


**Fig. 9.** (A) Offshore facies: weathered-out, storm-deposited shell layers containing brachiopod, coral, and crinoid bioclasts with graded coarse silt laminae in mud. Top of lens cap for scale. Arrow indicates way up. (B) Shoreface facies: swaley cross-bedded fine sand. (C) Pinstripe-laminated tidalites. Pale laminae of very fine sand alternate with mud laminae. Field of view 6 cm across. (D) Desiccation-cracked pinstripe-laminated tidal deposits. (E) Lagoonal facies: branching coral *Parastriatopora* in rare growth position. Individual colonies are up to 1 m across. (F) Edge-on view of a poorly welded pyroclastic flow deposit. Dark, angular clasts are lithics and more rounded, weathered fragments are pumice.

---

**Fig. 8.** *(opposite)* Two measured sections (sections A and B; Fig. 3 for locations). The cycles are numbered in order of deposition. Circled letters refer to localities marked on Figs 3, 4. F.Fm., Foilnamahagh Formation; F.C. Fm., Ferriter's Cove Formation; C.H. Fm., Clogher Head Formation. The climax of each regressive cycle is associated with a volcanic interval. Each cycle shows similar facies sequences: offshore, overlain by nearshore, overlain by lagoonal facies with volcanics.

## Barrier sequences

These are generally 1–4 m thick and comprise predominantly flat-laminated and rippled, very fine to medium sand, with occasional coarse sand- to granule-grade laminae. Also present are flat-bedded and planar cross-bedded, clast-supported conglomerates, up to 1 m thick, composed of well-rounded, moderately sorted, small to large pebbles. Sedimentary detritus in the barrier sequences consists of reworked volcanic debris and subordinate bioclastic material.

Fluvial sequences are present in the Foilnamahagh and Mill Cove Formations, suggesting that the barriers may have been fed by longshore drift from the mouths of rivers draining from catchment areas in the volcanic hinterland (Fig. 5). The preserved barrier sequences are thin with respect to the overlying lagoonal facies, suggesting that the barrier was narrow and of low relief. Accumulation of greater thicknesses may have been prevented by storm erosion of the barriers, the eroded barrier sediments providing a source for the sub-barrier storm sands (Elliott, 1986).

## Lagoonal sequences

The lagoonal (or back-barrier) sequences consist of intertidal-flat and subtidal lagoonal sediments and volcanic rocks, these three facies varying in stratigraphic order from cycle to cycle.

### *Tidal-flat facies*

These rocks comprise pinstripe-laminated units, 10–300 cm thick, made up of couplets of coarse silt or very fine sand, 0.5–2 mm thick, with a mud drape generally less than 1 mm thick (see Fig. 9C). Lamination is planar to undulating on a millimetre scale, ripple forms or cross-lamination are very rare and channel-fill sequences were not observed. At several localities polygonal desiccation cracks are developed (see Fig. 9D) and at one locality the laminites are reddened, burrowed, and contain incipiently developed calcrete (*sensu* Leeder, 1975).

The above features are interpreted to be the products of tidal-flat processes. The coarse silt/very fine sand of the pinstripe-laminated couplets was deposited by traction currents during ebb and flood tides with mud settling out of suspension during slack water (Klein, 1977; Kvale *et al.*, 1989). At times the tidal flat experienced periods of subaerial exposure, as indicated by desiccation-cracked tidal-flat muds and incipient pedogenesis. The absence of channels and scarcity of rippling suggest that tidal range was probably microtidal (less than 2 m).

### *Subtidal lagoonal facies*

This facies is closely associated with the intertidal-flat facies and typically comprises well-bioturbated mudrocks with thin laminae of very fine to medium sand. It is characterized by rich developments of the branching tabulate coral *Parastriatopora* (C.T. Scrutton, personal communication, 1989), which occur as fragments up to 20 cm across set in bioturbated muds and silts and forming lenses up to 20 m wide and 50 cm thick (see Fig. 9E). They are rarely seen in growth position.

*Parastriatopora* is generally the only fauna present (exceptions include one ostracod-bearing locality), suggesting that the corals may have grown in a stressful lagoonal environment experiencing fluctuating salinity and high suspended sediment content. Modern patch reefs are normally found in sheltered lagoons or gently shelving platforms where there is little intense surf action (Stoddart, 1969). For example, the fastest coral growth rates and the most extensive coral reefs in inshore embayments in Phuket, Thailand, recorded by Tudhope *et al.* (1988) were obtained from the muddiest, most turbid areas. The *Parastriatopora* colonies built up lagoonal patch reefs that were broken up during severe storms.

### *Volcanic deposits*

In marked contrast with the sub-barrier facies, the lagoonal sequences invariably include a volcanic interval. In section A, the volcanic intervals range in thickness from 0.3 to 150 m, and in section B from 14 to over 190 m (Table 1). Volcanic thicknesses are consistently greater in section B than in section A. For example, in cycle 4 the volcanic interval thickens from 0.3 m in section A to 42 m in section B (Fig. 8). The volcanic intervals are represented by single depositional units in cycles 2 and 4 of section A, but in all other cases by multi-unit sequences. The volcanic deposits include ignimbrites (*sensu* Sparks *et al.*, 1973), pyroclastic deposits, and a lava flow, but are dominated by ignimbrites, which typically show well-developed welding texures (Ragan & Sheridan, 1972; Peterson, 1979) and range in thickness from less than 1 m to c. 10 m (Fig. 9F). A cross-bedded surge deposit is also exposed in cycle 4, section B, locality G. Lithic and pumice clasts within these deposits are commonly small to medium pebble grade, rarely reaching over 30 cm in diameter. Several volcanic intervals were subaqueously emplaced, identified by the presence of very thinly bedded tuffs with sharp-based and loaded laminae in which lithics and crystals are normally graded and pumice reversely graded, and by porcellanous tuff bands showing ripple lamination interpreted to have formed by wave-reworking. The only lava flow in these sections is an andesite flow 35 m thick, which forms the topmost unit of cycle 5 in section B (Fig. 4, locality J). Geochemical evidence (Sloan & Bennett, 1990; Sloan, unpublished data) indicates that the pyroclastic rocks are predominantly subalkaline rhyolitic to dacitic in composition.

Volcanic intervals in section A are overlain by thin (less than 10 m) lagoonal sequences, which in turn are overlain by offshore facies of the succeeding cycle. For example, the volcanic interval in cycle 1 of section A (Fig. 3, locality C) comprises a fall unit 41 cm thick, overlain by an ignimbrite 870 cm thick, overlain by a second fall deposit, 26 cm thick; this is succeeded by sediments interpreted as tidal-flat deposits, including, at the base, a thin reworked volcanic pebbly sandstone. In section B, by contrast, the volcanic intervals are generally overlain abruptly by offshore facies without any intervening lagoonal sediments. For example, the topmost volcanic interval of the Ferriter's Cove Formation in section B (cycle 5) records, which at its base shows a transition from subaqueous to subaerial volcanic emplacement (evidence for subaerial emplacement in this instance includes the presence of a cross-bedded pyroclastic surge deposit, ballistically emplaced blocks, and an accumulation of densely welded ignimbrites 35 m thick). The upper surface of this predominantly subaerially emplaced volcanic sequence is directly overlain by typical offshore facies sediments (Fig. 4, locality G). The absence of lagoonal sediments may be attributed to their erosion in the shoreface zone following drowing of the volcanic land surface, or to the rapidity of submergence and of establishment of offshore conditions.

## DISCUSSION

Although the thickness of the cycles is variable, Fig. 8 shows that within each cycle there is a consistent vertical sequence of offshore, to shoreface, to barrier, to lagoonal facies with volcanic rocks. The absence of volcanic deposits or freshly reworked tuffs from the sub-barrier sequences indicates that these parts of the regressive cycles were deposited during periods of volcanic quiescence, contrasting with the abundant evidence of volcanic activity in the lagoonal deposits. Coarse, fluvially reworked volcanic detritus is considered to have been trapped in the barrier–lagoon systems (Figs 5, 6), and was not shed directly on to the marine shelf. Shoreface sands may have been derived through erosion of the barrier during storms. The depositional facies represented in the regressive cycles imply gradual shallowing, followed by more rapid transgressions that generally left no depositional record. The possible principal controls on relative sea level are threefold: (1) eustatic sea level variation, (2) the rate of sediment supply, and (3) tectonism.

**1** Eustatic sea-level variation: it is considered unlikely that the regressive cycles are the result of eustatic variation (cf. Bridges, 1976) as the late Wenlock is believed to have been a time of falling global sea level (McKerrow, 1979), while the association of the climax of each regressive cycle with a volcanic interval suggests that a more localized control on relative sea level was in operation.

**2** Sediment supply rate: many upward-coarsening, progradational, offshore to beach sequences, recording shoreline progradation have been documented (e.g. Galveston Island, Texas – Bernard *et al.*, 1962; San Blas, New Mexico – Curray & Moore, 1964). The Dunquin Group regressive cycles similarly record nearshore sea-floor aggradation and the seaward progradation of barrier–lagoon systems, with the consequent deposition of progressively shallower water facies. Lagoons, largely protected from storm erosion by barrier islands, became infilled and subaerially exposed. Such a scenario implies that rate of sediment supply was greater than the rate of subsidence. The transgressive episodes may have resulted from a change in the relative rates of subsidence and sediment supply: if the rate of sedimentation or rate of sediment supply were to fall to less than the rate of subsidence, a relative rise in sea level would have resulted. However, each transgression was preceded by volcanic eruptions that input a significant volume of fresh detritus into the basin. Given a constant rate of subsidence, this rapid increase in the rate of sediment supply would have led to a phase of regression rather than transgression. The absence of fresh volcanic debris from the sub-barrier sequences and the occurrence of volcanic intervals only near the climax of the regressive cycles cannot be coincidental and implies a causal link between regression and volcanism.

**3** Tectonism: it is considered most likely that the transgressions and regressions, rather than being controlled by eustatic variation or change in sediment supply rate, may have been controlled by periodic increases in the rate of subsidence, perhaps related to volcanism, i.e. volcano-tectonism. Volcano-tectonism is here taken to refer to all subsidence, uplift or lateral earth movements caused by magma intrusion and eruption. Volcano-tectonism is common to all volcanically active regions, but is particularly well documented from silicic and intermediate volcanic centres with large calderas, e.g. Valles caldera, California (Smith & Bailey, 1968), Long Valley caldera, California (Denlinger & Riley, 1984), Yellowstone caldera, Wyoming (Hildreth *et al.*, 1984). The classic resurgent cauldron model of Smith & Bailey (1968) envisages resurgence (or magma intrusion) into a subvolcanic cauldron causing doming and uplift (tumescence). The subsequent outpouring of tuffs and lavas triggers caldera collapse and subsidence (detumescence), which may affect an area tens of square kilometres in extent.

Two modern examples illustrate the wide lateral extent of the effects of volcano-tectonism: the 1914 eruption of the Sakara-jima volcano, Japan was followed by subsidence measured at 20 cm over 20 km away from the vent (Koto, 1916, cited in Hall, 1987); the Laguna de Bay volcanic centre of Luzon, Philippines (Wolf & Self, 1982), unlike the subaerial, continental setting of most calderas, is a nearshore feature situated in a very large (35 × 45 km) volcano-tectonic depression and eruption of ignimbrites from this centre has built an area of new land into the sea. In these examples, the effects of volcano-tectonism were felt the same order of distance from the volcanic centre as the interpreted distance involved in the Dunquin Group regressive cycles (15–20 km).

It is proposed that the regressive cycles of the Dunquin Group were deposited on the flanks of a volcanic island within a partially submerged volcano-tectonic depression (Fig. 10). During volcanically quiescent periods, offshore to nearshore storm and fair-weather sedimentation resulted in sea-floor aggradation, shoreline progradation, and the deposition of the bulk of the regressive cycles. Pre-volcanic magma intrusion (resurgence) accelerated sea-floor shoaling and the resultant regressive sedimentary packages are thinner, more condensed sequences than would have been deposited by simple sea-floor aggradation alone. Volcanism was followed by volcano-tectonic subsidence, rapid transgression, and the initiation of another regressive cycle.

The cycles in section A differ in several significant respects to those in section B (Table 2). The greater thickness of the volcanic intervals in section B may be attributed readily to the greater proximity of the former to the vent(s). Although the lithofacies present at the base of each cycle in both sections is virtually identical, the cycle thicknesses in section B are significantly greater than in section A. Therefore, the amount of subsidence may have been significantly greater in the area in which section B was deposited than in section A. This difference is in keeping with a model of volcano-tectonic subsidence in that the greatest subsidence would be expected closer to the vent(s). The volcanic intervals in section A are not overlain directly by offshore facies, but all contain intervening back-barrier sediments. Substorm wave-base water depths were not therefore established immediately upon cessation of volcanism, and post-eruption subsidence is interpreted to have been incremental rather than geologically instantaneous. Incremental, rather than abrupt, caldera collapse was postulated by Reedman *et al.* (1987) for the development of the Pitts Head Tuff (Ordovician) of North Wales.

It has not been possible to quantify accurately the relative sea-level changes owing to the effect on cycle thickness of unquantified variables such as uplift and subsidence during deposition of the cycles, eustatic sea-level variations, compaction, and deformation. However, assuming insignificant thickness variation due to the above factors, post-volcanic subsidence may have been of the order of 20–60 m — this figure is derived from the thickness of rocks between desiccation-cracked horizons and the base of the regressive cycles.

Although a pattern of gradual regression and more

**Fig. 10.** *(opposite)* Schematic diagram showing regressive phases as a result of sea-floor aggradation and volcano-tectonic uplift related to subvolcanic magma intrusion (phases 1 and 2) and transgression as a result of post-volcanism volcano-tectonic subsidence (phase 3). Subsidence was greater closer to the vent. A and B refer to the approximate locations of sections A and B.

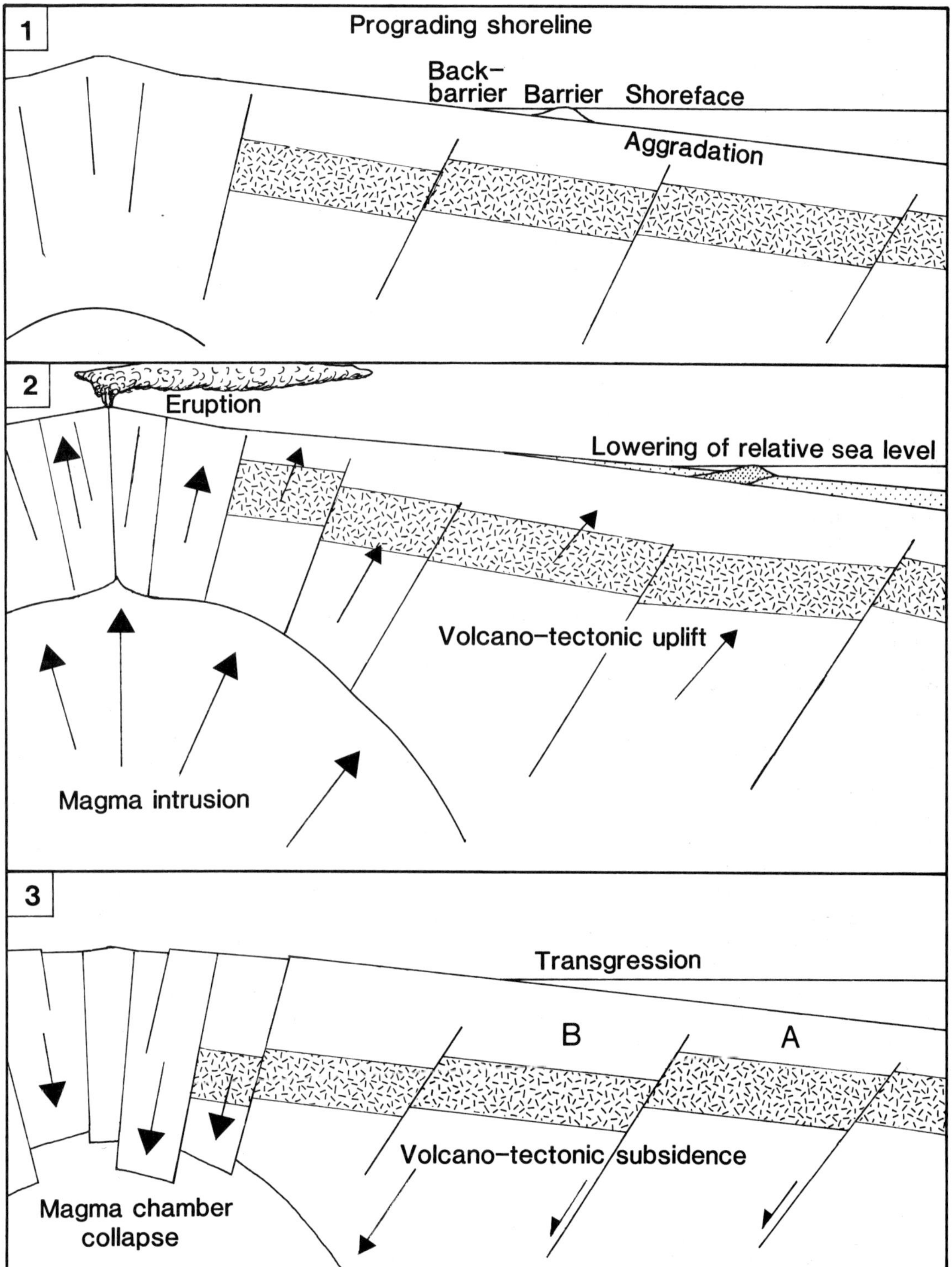
1
Prograding shoreline
Back-
barrier
Barrier
Shoreface
Aggradation
2
Eruption
Lowering of relative sea level
Volcano–tectonic uplift
Magma intrusion
3
Transgression
B
A
Volcano–tectonic subsidence
Magma chamber
collapse

rapid transgression is evident, owing to the low degree of biostratigraphic resolution and the lack of radiometric dates, the duration of these respective phases is unknown. An indication of the time taken for the deposition of the regressive cycles, may, however, be obtained by analogy with modern volcanic regions. Smith & Bailey (1968) argued that the duration of the pre-resurgent and resurgent periods of the Valles caldera (possibly equivalent to the regressive phases of the Dunquin Group) may have been less than 105 years, while the eruptive phases themselves were estimated at less than 10 years.

## CONCLUSIONS

**1** The Ferriter's Cove and Clogher Head Formations of the Dunquin Group comprise a sequence of regressive cycles, each of which records a transition from offshore, through barrier, to lagoonal environments. Facies variations in the sub-barrier sequences reflect decreasing palaeobathymetry and increasing proximity to a prograding shoreline. The barrier was narrow and of low relief, and the lagoonal area was a complex of intertidal flats and subtidal zones. The intercycle transgressions generally left no depositional record.

**2** Evidence of volcanism in the sub-barrier sequences is absent. Phases of volcanicity and transgressions were therefore episodic and penecontemporaneous.

**3** Regressions were the result of sea-floor aggradation and volcano-tectonic uplift due to magma emplacement. Intercycle transgressions were caused by post-volcanic magma chamber deflation and volcano-tectonic subsidence. The sequences in section B were deposited closer to the vent than those in section A, and experienced more sudden subsidence of greater magnitude than section A.

**4** The effects of the mid-Silurian eustatic sea-level fall and of the highly variable sediment supply rate (particularly the supply of fresh volcanic detritus) were subordinate to the effects of volcano-tectonism on relative sea-level.

## ACKNOWLEDGEMENTS

M.C. Bennett, S.P. Todd, T.J. Faulkner, M.J. Weedon, I.S. Stewart, and H. Mijnlieff are thanked for their help in developing the ideas presented here and in improving the script. We thank B.P. Kokelaar and J.R. Graham for their constructive criticisms of an earlier version of the manuscript. Much of the fieldwork by RS was carried out during tenure of a Department of Education for Northern Ireland Research Studentship at the University of Bristol.

## REFERENCES

Aigner, T. & Reineck, H.E. (1982) Proximality trends in modern storm sands from the Helgoland Bight (North Sea) and their implications for basin analysis. *Senckenbergiana Mar.* **14**, 183–215.

Bassett, M.K., Cocks, L.R.M. & Holland, C.H. (1976) The affinities of two endemic Silurian brachiopods from the Dingle Peninsula, Ireland. *Palaeontology* **19**, 615–625.

Bernard, H.A., LeBlanc, R.J. & Major, C.F. (1982) Recent and Pleistocene geology of southeast Texas. In: *Geology of the Gulf Coast and Central Texas, Guidebook of Excursions* (Eds Rainwater, E.H. & Zingula, R.P.) pp. 175–224. Houston Geol. Soc., Texas.

Brenchley, P.J. (1985) Storm influenced sandstone beds. *Modern Geol.* **9**, 369–396.

Bridges, P.H. (1976) Lower Silurian transgressive barrier islands, southwest Wales. *Sedimentology* **23**, 347–362.

Curray, J.R. & Moore, D.G. (1964) Holocene regressive littoral sand, Costa de Nayarit, Mexico. In: *Deltaic and Shallow Marine Deposits* (Ed. van Straaten, L.M.J.U.). Dev. Sedimentol. 1, 76–82.

Denlinger, R.P. & Riley, F. (1984) Deformation of Long Valley Caldera, Mono County, California, from 1975 to 1982. *J. Geophys. Res.* **89**, 8303–8314.

Elliott, T. (1986) Siliciclastic shorelines. In: *Sedimentary Environments and Facies* (Ed. Reading, H.G.) pp. 155–188. Blackwell Scientific Publications, Oxford.

Fisher, R.V. (1984) Submarine volcaniclastic rocks. In: *Marginal Basin Geology: Volcanic and Associated Sedimentary and Tectonic Processes in Modern and Ancient Marginal Basins* (Eds Kokelaar, B.P. & Howells, M.F.). Geol. Soc. London, Spec. Publ. 16, 5–27.

Hall, A. *Igneous Petrology*, Longman, Harlow, 573 pp.

Harms, J.C., Southard, J.B., Spearing, D.R. & Walker, R.G. (1975) Depositional environments as interpreted from primary sedimentary structures and stratification sequences. *Soc. Econ. Paleontol. Mineral., Short Course* **2**, 161 pp.

Hildreth, W., Christiansen, R.L. & O'Neill, J.R. (1984) Catastrophic isotopic modification of rhyolitic magma at times of caldera subsidence, Yellowstone Plateau Volcanic Field. *J. Geophys. Res.* **89**, 8339–8369.

Holland, C.H. (1969) Irish counterpart of Silurian of Newfoundland. In: *North Atlantic Geology and Continental Drift* (Ed. Kay, M.). Mem., Am. Assoc. Petrol. Geol. 12, 298–308.

Holland, C.H. (1988) The fossiliferous Silurian rocks of the Dunquin inlier, Dingle Peninsula, County Kerry, Ireland. *Trans. R. Soc. Edinburgh: Earth Sci.* **79**, 347–360.

Howard, J.D. & Reineck, H.E. (1981) Depositional facies

of high-energy beach-to-offshore sequence: comparison with low-energy sequence. *Bull., Am. Assoc. Petrol. Geol.*, **65**, 807–830.

KLEIN, G. deV. (1977) *Clastic Tidal Facies*. Continuing Education Publication Company, Champaign, IL, 149 pp.

KOTO, B. (1916) The great eruption of Sakura-jima in 1914. *J. Coll. Sci. Imp. Univ. Tokyo* **38**, 1–237.

KVALE, E.P., ARCHER, A.W. & JOHNSON, H.R. (1989) Daily, monthly, and yearly tidal cycles within laminated siltstones of the Mansfield Formation (Pennsylvanian) of Indiana. *Geology* **17**, 365–368.

LECKIE, D.A. & WALKER, R.G. (1982) Storm- and tide-dominated shorelines in Cretaceous Moosebar–Lower Gates interval-outcrop equivalents of deep basin gas trap in western Canada. *Bull. Am. Assoc. Petrol. Geol.* **66**, 138–157.

LEEDER, M.P. (1975) Pedogenic carbonate and flood sediment accretion rates: a quantitative model for alluvial and arid-zone lithofacies. *Geol. Mag.*, **112**, 257–270.

McKERROW, W.S. (1979) Ordovician and Silurian changes in sea level. *J. Geol. Soc. London* **136**, 137–146.

McKERROW, W.S. & SOPER, N.J. (1989) The Iapetus Suture in the British Isles. *Geol. Mag.* **126**, *1–8*.

MURPHY, F.C. & HUTTON, D.H.W. (1986) Is the Southern Uplands of Scotland really an accretionary prism? *Geology* **14**, 354–357.

PARKIN, J. (1974) Silurian rocks of Inishvickillane, Blasket Islands, Co. Kerry. *Sci. Proc. R. Dublin Soc. A* **5**, 277–291.

PARKIN, J. (1976) Silurian rocks of the Bull's Head, Annascaul and Derrymore Glen inliers, Co. Kerry. *Proc. R. Irish. Acad. Sect. B* **76**, 577–606.

PETERSON, D.W. (1979) Significance of the flattening of pumice fragments in ash-flow tuffs. *Geol. Soc. Am. Spec. Pap.* **180**, 195–204.

PHILLIPS, W.E.A. & SEVASTOPULO, G.P. (1986) The statigraphic and structural setting of Irish mineral deposits. In: *Geology and Genesis of Mineral Deposits in Ireland* (Eds Andrew, C.J., Crowe, R.W.A., Finlay, S., Pennell, W.M. & Pyne, J.F.) pp. 1–30. Irish Association for Economic Geology, Dublin.

PHILLIPS, W.E.A., STILLMAN, C.J. & MURPHY, T. (1975) A Caledonian plate tectonic model. *J. Geol. Soc. London* **132**, 579–609.

PICKERING, K.T., BASSETT, M.G. & SIVETER, D.J. (1989) Late Ordovician–early Silurian destruction of the Iapetus Ocean: Newfoundland, British and Scandinavia — a discussion. *Trans. R. Soc. Edinburgh: Earth Sci.* **79**, 361.

RAGAN, D.M. & SHERIDAN, M.F. (1972) Compaction of the Bishop Tuff, California. *Geol. Soc. Am. Bull.* **83**, 95–106.

REEDMAN, A.J., HOWELLS, M.F., ORTON, G. & CAMPBELL, S.D.G. (1987) The Pitts Head Tuff Formation: a subaerial to submarine welded ash-flow tuff of Ordovician age, North Wales. *Geol. Mag.* **124**, 427–439.

SLOAN, R.J. & BENNETT, M.C. (1990) Geochemical character of Silurian volcanism in SW Ireland. *J. Geol. Soc., London* **147**, 1051–60.

SMITH, R.L. & BAILEY, R.A. (1968) Resurgent cauldrons. *Geol. Soc. Am. Mem.* **116**, 613–662.

SPARKS, R.S.J., SELF, S. & WALKER, G.P.L. (1973) Products of ignimbrite eruptions. *Geology* **1**, 115–118.

STODDART, D.R. (1969) Ecology and morphology of Recent coral reefs. *Biol. Rev.* **44**, 433–498.

TODD, S.P. (1989) Caledonian tectonics and conglomerate sedimentation in Dingle, SW Ireland. Unpublished Ph.D. thesis, University of Bristol.

TUDHOPE, A.W., SCOFFIN, T.P. & BROWN, B.E. (1988) Effects of sedimentation on the growth of corals and coral reefs in Phuket, South Thailand. *Abstracts, British Sedimentologists Research Group 27th Annual Meeting*, British Antarctic Survey, Cambridge.

WATKINS, R. (1978) Silurian marine communities of Dingle, Ireland. *Palaeogeogr. Palaeoclimatol. Palaeoecol.* **23**, 79–118.

WOLFE, J.A. & SELF, S. (1982) Structural lineaments and Neogene volcanism in South-western Luzon. In: *The Tectonic and Geologic Evolution of the South-east Asia Seas and Islands*, Geophys. Monogr. Ser. **27**, 157–172.

ZEIGLER, A.M. (1970) Geosynclinal development of the British Isles during the Silurian period. *J. Geol.* **78**, 445–479.

# Neogene–Recent Margins

*Spec. Publs. int. Ass. Sediment.* (1991) **12**, 123–131

# Sea-level changes and sedimentary evolution during the Quaternary in the northwest Aegean continental margin, Greece

VASILIOS LYKOUSIS

*National Centre for Marine Research, Ag. Kosmas, 166 04 Athens, Greece*

## ABSTRACT

Continuous seismic profiling data have shown that five distinct unconformities can be traced below the shelf-break and slope environments of the northwest Aegean continental margin. These unconformities separate five oblique progradational delta sequences that have developed during lowstands of sea level within the five major Quaternary glacial stages (late Wurm, Riss, Mindel, Guntz, and Donau). The topset to foreset transition of these sequences, located at 120, 240, 430, 540, and c. 700 m below present sea-level, reflect continuous subsidence of the area, with a calculated rate of 0.93 ± 0.16 m/ka for the late Pleistocene.

A prograded sequence, occurring just below the Holocene transgressional surface, has been identified on 3.5 kHz profiles across the inner continental shelf. This suggests a minor sea-level lowstand (55–60 m below present) within the Riss–(late)Wurm interglacial period (isotopic stage 4.2).

Tectonic subsidence and sea-level changes during the Quaternary were the dominant factors affecting sedimentary evolution of the northwest Aegean continental margin, while tectonic uplift predominated in the onshore marginal areas.

## INTRODUCTION

The northwest Aegean continental margin and the associated Sporadhes Basin is the western part of the North Aegean trough (northeast Mediterranean), which constitutes the southern margin of the European plate (Le Pichon & Angelier, 1979). The region is thought to be a back-arc area relative to the Hellenic arc (Fig. 1) and tectonically is an asymetrical graben or half-graben (Brooks & Ferentinos, 1980) that has developed by continuous post-Miocene subsidence within an extensional tectonic regime (Le Pichon *et al.*, 1985). This extensional phase possibly began in late Miocene times (10 Ma ago) following the well-known mid-Miocene compressive phase in the greater area of the North Aegean trough (Lyberis, 1985), although Mercier (1979) proposed a subsequent compressional phase during the early–middle Pleistocene (3–0.9 Ma ago).

The structure of the northwest Aegean continental margin and Sporadhes Basin has been controlled mainly by NW–SE and ENE–WSW-trending major normal faults with a few kilometres of maximum vertical displacement (Brooks & Ferentinos, 1980; Lyberis, 1985). Active antithetic faults with throws up to a few tens of metres have been described in the continental slope of the area (Ferentinos *et al.*, 1981). The NW–SE major fault system separates the subsiding areas (mainly offshore) from the areas of intense uplift (mainly eastern and western surrounding land areas). Accordingly, during the Quaternary, the present NW–SE-elongated offshore area has subsided about 0.8–1.2 km (Faugeres & Robert, 1976; Faugeres, 1977; Lalechos & Savoyat, 1979) while the adjacent northwestern onshore areas (Mount Olympus) have been uplifted 0.5–1 km (Psilovikos, 1981), or 2 km according to Faugeres (1977). Considerable uplift during the Quaternary is demonstrated by the presence of extensive Pliocene deposits on the Kassandra and Khalkidiki Peninsulas to the northeast of the area under investigation.

The purpose of this paper is to study the relative

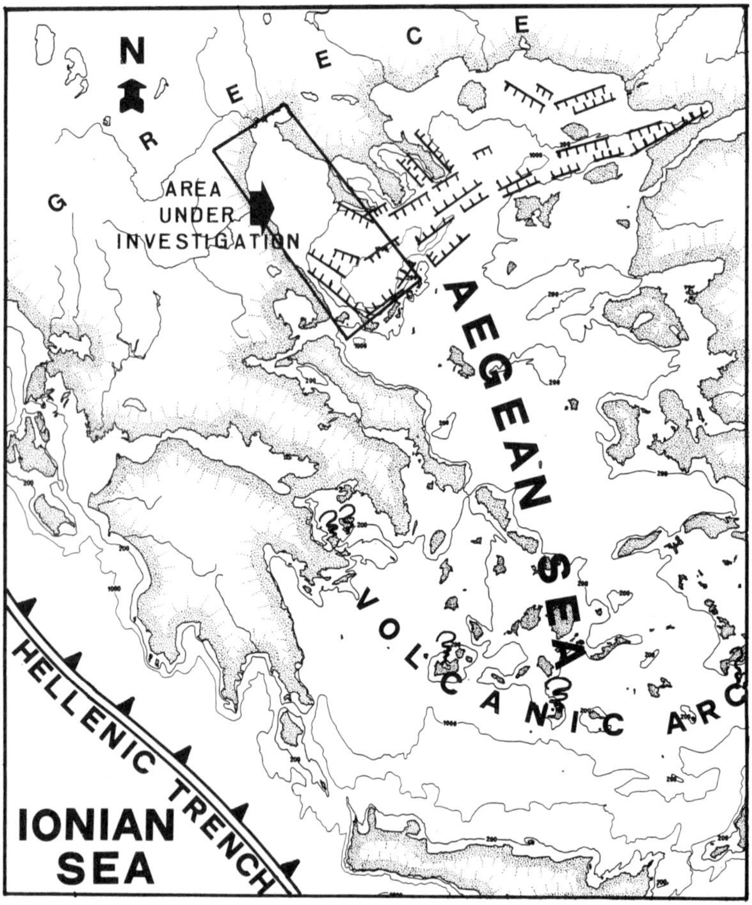


**Fig. 1.** Location of the northwest Aegean sea and geotectonic setting of the adjoining area (from Lyberis, 1985). Squiggle symbols on the volcanic arc indicate the active volcanic islands.

importance of tectonics and eustasy in the formation of Quaternary depositional sequences on the northwest Aegean continental margin. This is examined through high-resolution seismic data, stratigraphy, chronostratigraphic data from two offshore commercial boreholes, as well as onshore studies and observations.

## METHODS AND MATERIALS

The area under investigation has been surveyed offshore using high-resolution continuous seismic profiling systems (3.5 kHz, 9 kJ sparker), as well as onshore (surrounding land areas) by two field trips. Shallow seismic profiling data within the shelf, shelf break, and upper slope environments (c. 450 km) were obtained in 1986, operating an Ocean Research Equipment 3.5 kHz system from a trawler (Fig. 2). The selected sparker and 3.5 kHz profiles from the continental slope and basin (Sporadhes Basin) were collected during the 7/78 cruise of R.R.S. *Shackleton* in the northwest Aegean Sea. Position fixing on both cruises was carried out by radar.

## RESULTS

### Bottom morphology — bathymetry

The continental shelf occupies almost 40% of the study area (Fig. 2). To the northwest, the shelf is very smooth, strikes NW – SE, is fairly wide (90 km), and covers an area of about 4000 km$^2$. To the north, west, and south along the Kassandra and Magnisia Peninsulas and Sporadhes Islands, respectively, the shelf is narrow and the associated slope is steep owing to intense tectonic activity on the marginal faults.

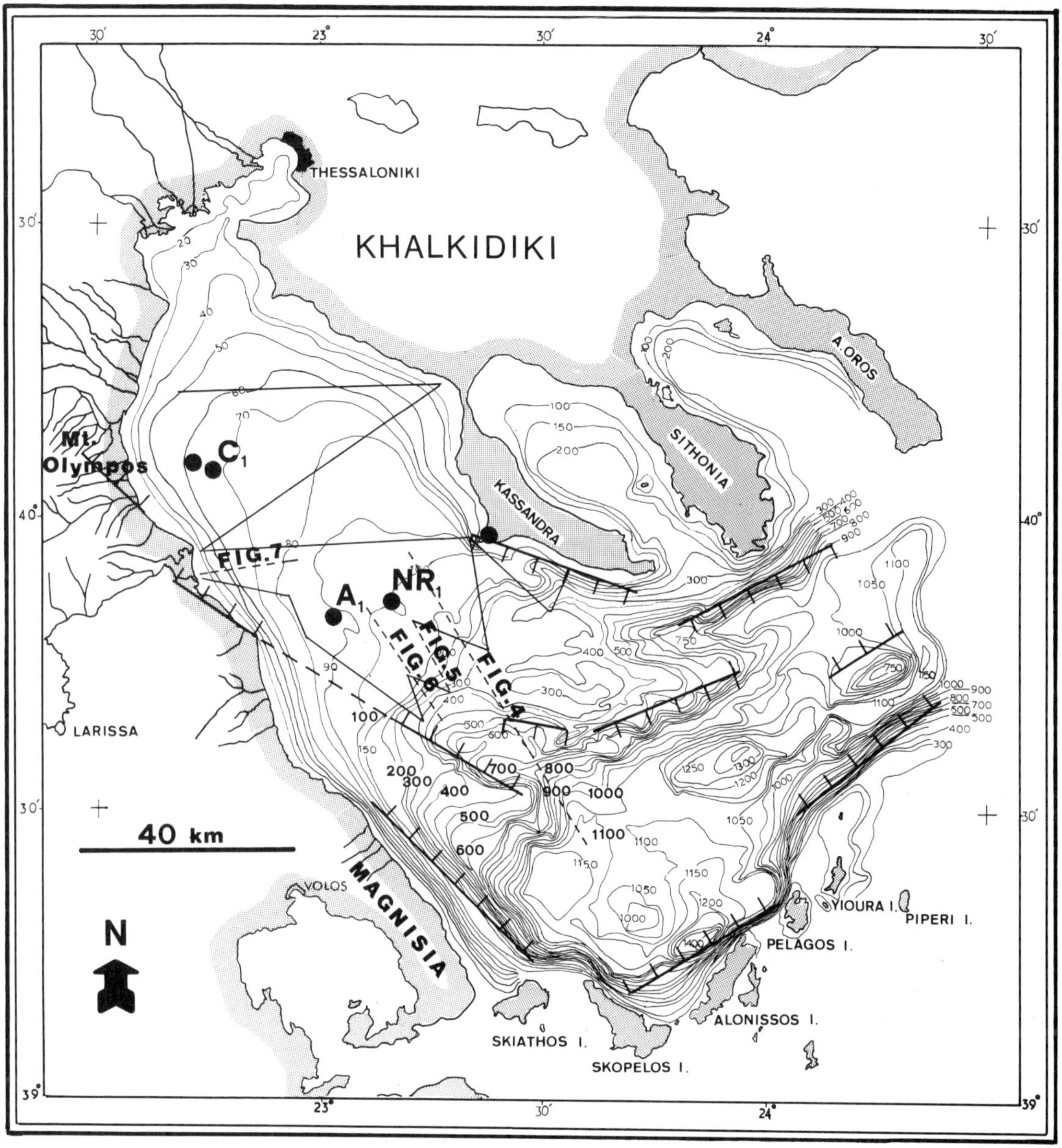


**Fig. 2.** Morphotectonic sketch-map of the area under investigation, showing seismic profiling tracks and commercial borehole sites. Seismic profiles of Figs 4–7 are also shown.

The slope is especially pronounced in the central region of the area under investigation. It is dissected by three submarine canyon and valley systems controlled by NW–SE-trending major faults (Fig. 2).

The Sporadhes Basin is a SSW–NNE and NW–SE fault-bounded marginal basin, exceeding 1000 m in water depth.

### Thickness of Quaternary sediments

The Quaternary sediments underlying the study area have been examined indirectly over the slope and basin (Sporadhes Basin) by seismic reflection and refraction methods and directly in the outer-shelf–shelf-break by four commercial boreholes (Texaco Co. and Petroleum Corporation of Greece).

Within the basin, refraction profiles (Needham *et al.*, 1973) have shown the presence of about 0.8 km of unconsolidated sediments (assumed $V_P$ = 2 km/s), of which at least the top 400 m are turbiditic sequences of Quaternary age (Brooks & Ferentinos, 1980). Beneath the continental slope the Quaternary sediments developed in a vertical succession of four prograded delta sequences with a total thickness of about 600 m (Brooks & Ferentinos, 1980).

The deep boreholes in the outer-shelf–shelf-break region revealed a thickness of about 0.8–1 km of Quaternary sediments (gravels, sands, marls, clays) overlying about 1.5 km of Pliocene sediments of similar composition (Faugeres & Robert, 1976) (Fig. 3).

## Quaternary sequence stratigraphy

Seismic stratigraphic analysis of the 3.5 kHz and sparker profiles (by the method of Mitchum *et al.*, 1977) from the slope and shelf of the northwest Aegean continental margin revealed a vertical succession of characteristic depositional sequences bounded by distinct unconformities.

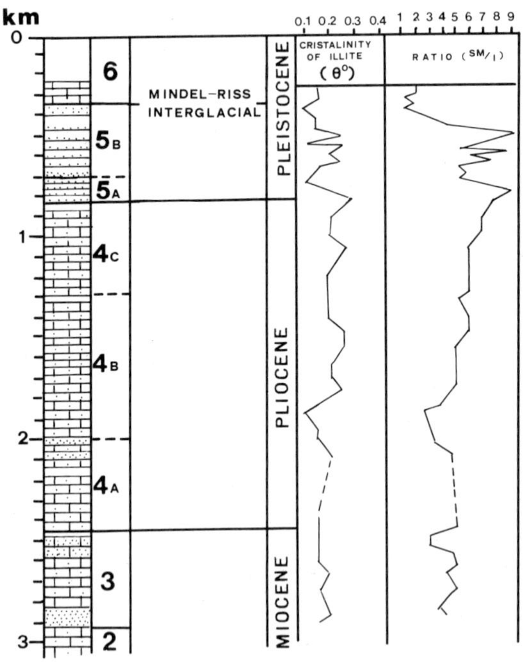


**Fig. 3.** Vertical succession of sediment sequences from Texaco Co. commercial drilling (borehole $A_1$) in the outer shelf (from Faugeres & Robert, 1976).

The seismic profiles (Figs 4, 5, 6) across the shelf-break and continental slope revealed the existence of five major unconformities ($U_1$–$U_5$), which define five successive oblique progradational sequences ($SP_1$–$SP_5$). These are interpreted as delta sequences that prograded basinward during lowstands of sea level (glacial stages), while the unconformities that separate the depositional sequences represent major transgressions during interglacial stages. Although these sequences have been affected by antithetic growth faults with throws up to 60 m (Brooks & Ferentinos, 1980), they can be traced clearly down to the Sporadhes Basin (Fig. 4). The maximum thickness of the vertical succession of the sequences (800–850 m) has been found beneath the outer continental shelf, while a few kilometres westwards (Fig. 2), borehole $A_1$ (Texaco Co.) revealed the Plio-Pleistocene boundary (Villafranchian) about 820 m below the sea floor (Faugeres & Robert, 1976). This suggests a Pleistocene age for the five depositional sequences ($SP_1$–$SP_5$). The lowest unconformity ($U_5$) may represent the Plio-Pleistocene boundary.

The topset to foreset transition of the shallowest delta sequence ($SP_1$), located beneath the shelf break in a water depth of 120 m (c. 160 ms), indicates the latest stage of delta evolution during the late glacial (Wurm) maximum (c. 18 000 years BP) and probably corresponds to the maximum fall of sea level (c. 100 m) in isotopic stage 2 (2.2) (Shackleton & Opdyke, 1976; Pisias *et al.*, 1984). Radiocarbon dates, together with the presence of shallow brackish benthic foraminifera in cores from the foreset beds from the similar delta sequences studied in the eastern Aegean Sea (Aksu & Piper, 1983), have confirmed the late Pleistocene age of these shelf-break delta sequences.

Approximate middle–late Pleistocene chronostratigraphic control on the seismic data can be obtained by correlation between borehole $A_1$ (Texaco Co.) and a seismic profile that passes within c. 6 km of it (Fig. 6). Unfortunately, strong sea-floor multiples on the shelf obscure deeper primary reflections, limiting the confidence with which such correlations can be made. However, in borehole $A_1$ the Mindel–Riss unconformity is located about 320 m below sea floor (b.s.f.) (Faugeres & Robert, 1976) and is interpreted to correlate with the seismic sequence boundary $U_2$. If it is accepted that the sequence boundaries $U_1$, $U_2$, and $U_5$ correlate with the Riss–(late) Wurm, Mindel–Riss, and Pliocene–Pleistocene transitions, and that the delta sequences

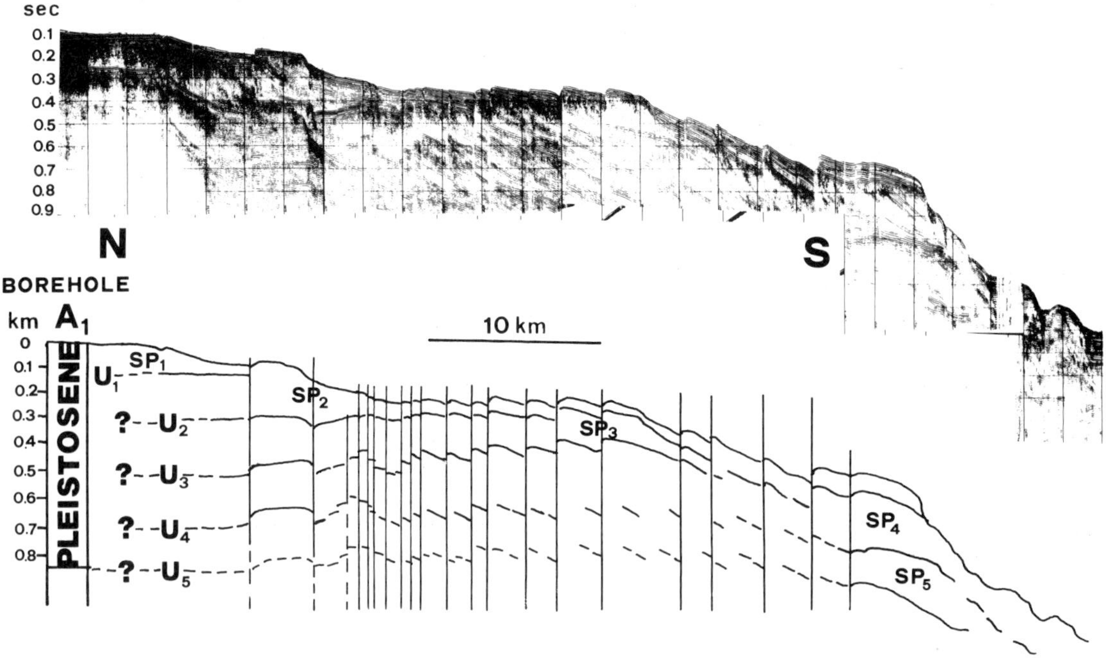


**Fig. 4.** Sparker profile from the outer shelf to Sporadhes Basin showing the vertical succession of the five Quaternary prograding sequences (modified from Brooks & Ferentinos, 1980).

$SP_1$ accumulated during the (late) Wurm glacial, then by analogy the $U_3$ and $U_4$ sequence boundaries probably correspond to the Guntz–Mindel and Donau–Guntz transitions.

The thickness of the deltaic sequences (up to 200 m) implies a significant contribution of tectonic subsidence in their evolution. Differential subsidence between the shelf and slope of the area is of rather minor importance (1–1.5 km and c. 1.5 km, respectively, during the Neogene) and is gradually increasing from the shelf to slope environments (Le Pichon *et al.*, 1985).

The lower edge of the topset to foreset transition of the delta sequence $SP_2$ (240 m below present sea

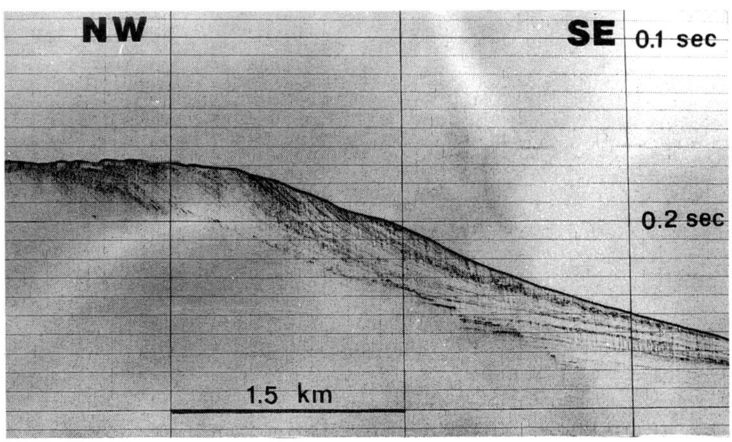


**Fig. 5.** 3.5 kHz seismic profile from the shelf-break showing the late glacial delta sequence.

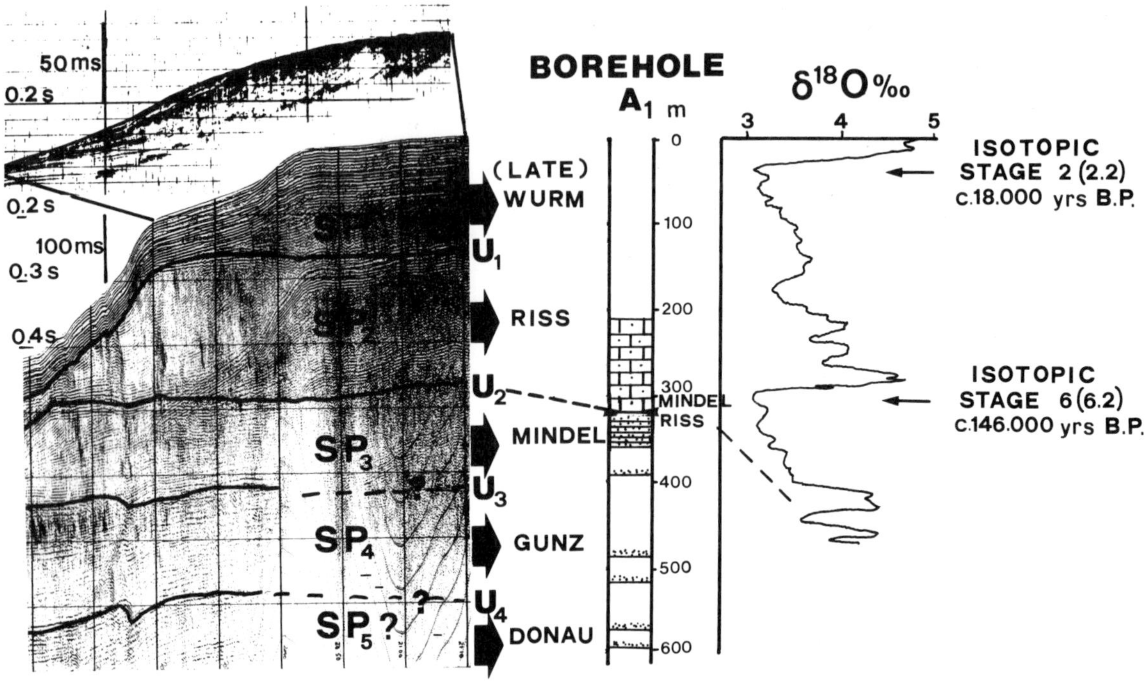


**Fig. 6.** Sparker profile from the outer-shelf–slope environment near borehole site $A_1$ showing the Pleistocene depositional sequences and their chronostratigraphic interpretation. ($\delta^{18}O$ record is from deep-sea cores, not this borehole).

level) is interpreted to represent the latest stage of delta progradation during the Riss glacial maximum (isotopic stage 6.2, c. 146 000 years BP). Utilizing the method of Aksu *et al.* (1987) we can estimate the relative subsidence from the Riss (c. 146 000 years BP) to (late) Wurm (c. 18 000 years BP) glacial maxima, by comparing the vertical displacement from topset to foreset transition of the two successive delta sequences, assuming similar lowstands of sea level, with an uncertainty of no more than 20 m. A subsidence rate of 0.93 ± 0.16 m/ka (120 ± 20 m subsidence/128 ka) can be calculated for the late Pleistocene, which is very similar to the subsidence rate (1.15 m/ka) estimated by Aksu *et al.* (1987) in the eastern Aegean Sea. Maximum growth fault offsets of 60–70 m of the sequence boundary $U_1$ (c. 130 000 years BP) suggest maximum vertical displacement of c. 0.45 m/ka due to gravitational instability of the Quaternary delta sequences.

The late Quaternary sedimentary evolution of the inner continental shelf in relation to sea-level changes has been deduced from the shallow depositional sequences revealed on 3.5 kHz profiles (Fig. 7). These 3.5 kHz profiles revealed four seismic sequences ($SF_1$–$SF_4$) and one regional unconformity ($U_0$). The lowest seismic sequence ($SF_4$) is a rather tangential oblique progradational seismic unit with toplap termination of the seismic reflectors at the upper sequence boundary ($U_0$). This unconformity ($U_0$) dips gently towards the shelf-break where it terminates as the upper boundary of the late Pleistocene (late Wurm) delta sequence. It is regarded as a late glacial regressional unconformity, developed by subaerial erosion during oxygen isotopic stage 3 (c. 60 000–27 000 years BP). Seismic sequence $SF_4$ is interpreted to represent delta progradation during a relative lowstand of sea level (c. 55–60 m) below present corresponding to oxygen isotopic stage 4.2 (c. 65 000 years BP). Subsequently, falling sea level during isotopic stage 3 caused the depocentre to migrate towards the outer shelf, and this inner shelf area was subjected to subaerial exposure and erosion producing unconformity $U_0$. The same unconformity may be coincident with the Holocene transgressional surface in this area.

Seismic sequence $SF_3$ consists of a few weak, discontinuous, subparallel reflectors or reflection-free areas (Fig. 7), suggesting fine-grained sediment

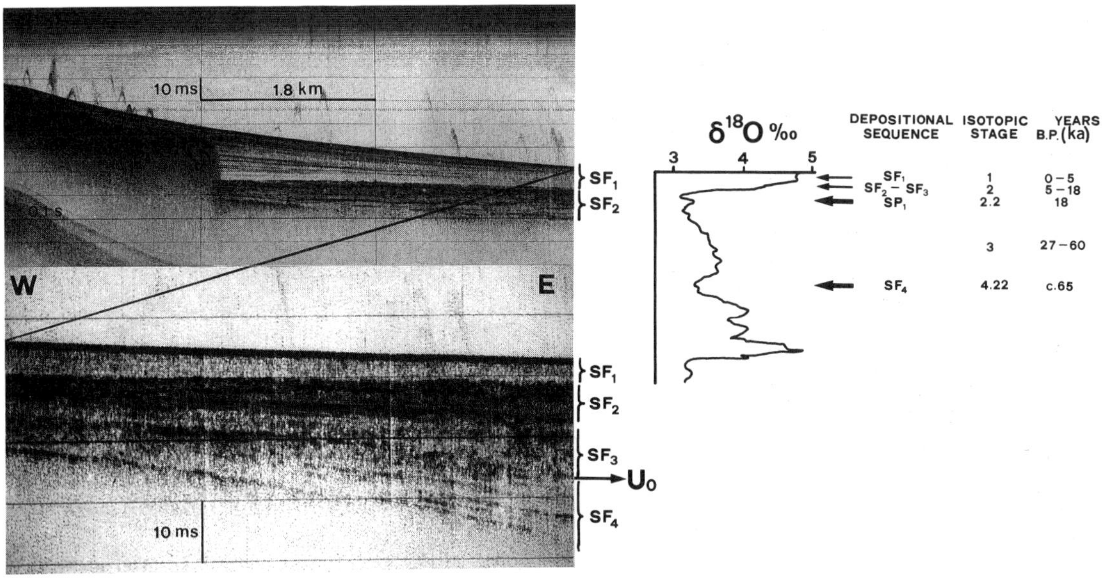


**Fig. 7.** 3.5 kHz seismic profile from the inner shelf showing the late Quaternary sediment sequences and their chronostratigraphic interpretation.

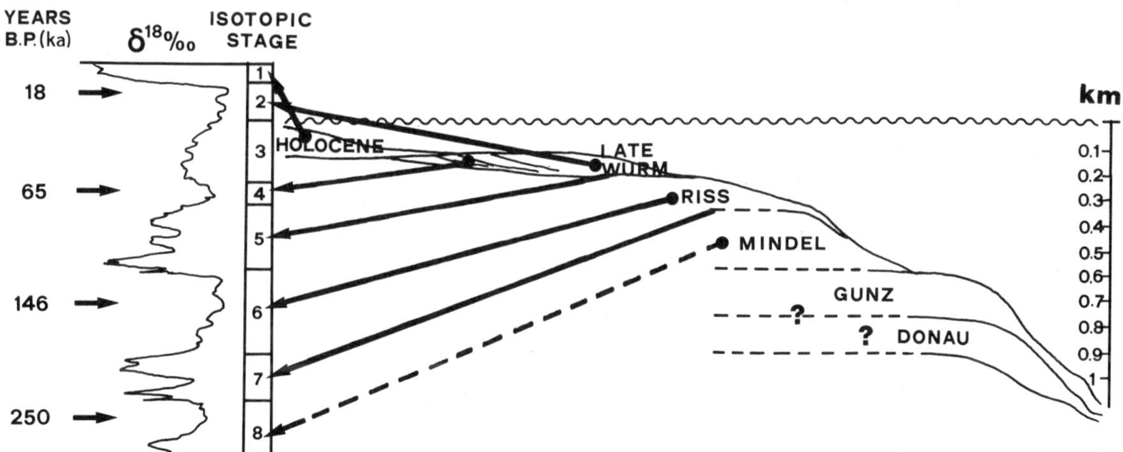


**Fig. 8.** Simplified illustration of Quaternary depositional sequences and their chronostratigraphic relationships.

deposition in a relatively quiet environment. In the Gulf of Mexico this type of sediment sequence just above the late glacial erosional surface has been interpreted as estuarine fill deposited during the early stages of the latest post-glacial (Holocene) transgression (Suter, 1986).

Seismic sequence $SF_2$ is characterized by parallel seismic reflection configurations with continuous, strong, and relatively low-amplitude reflectors (Fig. 7). This depositional sequence is interpreted to indicate deposition of relatively coarse-grained sediments (silty sand, muddy sand) in a shallow high-energy marine environment during the rapid Holocene transgression (isotopic stage 2).

The upper seismic sequence $SF_1$ displays a sigmoid–oblique progradation pattern (Holocene delta sequence) that changes gradually offshore to become reflection-free (bottomsets or distal shelf muds). This depositional sequence is interpreted to represent delta progradation during isotopic stage 1, since the sea surface rose to its present level (c. 6000 years BP).

The complete succession of the Quaternary depositional sequences and their chronostratigraphic interpretation is presented in Fig. 8.

### Onshore studies and observations

The onshore marginal zones of the northwest Aegean continental margin have been subjected to intense uplift. Recent uplift of 3–4 m is indicated by emerged strand lines along the Magnisia Peninsula, while Pleistocene uplift of a few hundred metres is demonstrated by the presence of extensive Pliocene deposits in Kassandra and Khalkidiki Peninsulas. Psilovikos (1981), studying the vertical succession of three generations of Quaternary alluvial fans on Mount Olympus, suggested uplift of 400–600 m during Villafranchian–middle Pleistocene and 150–350 m during the late Quaternary. The three generations of alluvial fans have been developed following intense uplift of the western marginal zones during the lower Villafranchian to early Pleistocene (first generation), lower–middle Pleistocene (second generation) and middle–late Pleistocene (third generation).

## CONCLUSIONS

**1** Both sea-level changes and tectonic activity have contributed to Quaternary sedimentary evolution in the northwest Aegean Sea. Sea-level fluctuations and tectonic subsidence are the predominant factors controlling the development of offshore sediment sequences, while tectonic uplift is the major factor in the onshore marginal zones.

**2** During the Quaternary, five major prograded delta sequences were constructed on the continental slope, reflecting five major sea-level lowstands during the Donau, Guntz, Mindel, Riss and (late) Wurm glacial maxima. Topset to foreset transition in these delta sequences located at c. 700, 540, 430, 240, and 120 m respectively, below the present sea level, suggest continuous subsidence, which has been estimated at 0.93 ± 0.16 m/ka for the late Pleistocene. Accordingly, the five sequence boundaries probably correspond to Riss–(late)Wurm, Mindel–Riss, Guntz–Mindel, Donau–Gunz, and Pliocene–Pleistocene transitions.

**3** A minor glacial stage within the Riss–(late)Wurm interglacial (isotopic stage 4.2) has been deduced from the presence of a delta sequence beneath the inner continental shelf for which the upper sequence boundary is the late Pleistocene regressional to Holocene transgressional unconformity. This example demonstrates that significant variations in the oxygen isotopic composition of foraminifera (isotopic stages) in deep-sea cores need not always correlate with major (maximum) lowstands and highstands of sea level detected on seismic reflection profiles. Limitations of the continuous seismic reflection and refraction methods (penetration–resolution) on the continental shelf could be responsible for the difficulty in defining the stratigraphic record of minor glacial stages and their relation to major glacial and interglacial stages during the Pleistocene. The limited extent and thickness of prograding sequences produced during minor glacial stages (due to their shorter duration and lower intensity), as well as subsequent erosion during major regressions, are probably additional factors restricting the detection of these stages.

## ACKNOWLEDGEMENTS

Professor M. Brooks and Dr R. Larter are gratefully acknowledged for their thorough reviews.

## REFERENCES

Aksu, A.E. & Piper, D.J.W. (1983) Progradation of the Late Quaternary Gediz delta, Turkey. *Mar. Geol.* **54**, 1–25.

Aksu, A.E., Piper, D.J.W. & Konuk, T. (1987) Late Quaternary tectonic and sedimentary history of outer Izmir and Candarli bays, Western Turkey. *Mar. Geol.* **76**, 89–104.

Brooks, M. & Ferentinos, G. (1980) Structure and evolution of the Sporadhes basin of the north Aegean trough, Northern Aegean sea. *Tectonophysics* **68**, 15–30.

Faugeres, L. (1977) Maissance et Development du Relief de l' Olympe (Grece). *Rev. Geogr. Phys. Geol. Dyn.* **2** (XIX), 7–26.

Faugeres, L. & Robert, C. (1976) Etude sedimentologique et mineralogique de deux forages du golfe Thermaique (Mer Egee). *Ann. Univ. Provence. Geol. Mediter.* **3**(4), 209–218.

Ferentinos, G., Brooks, M. & Collins, M.B. (1981)

Gravity-induced deformation on the north flank and floor of the Sporadhes basin of the North Aegean sea trough. *Mar. Geol.* **44**, 289–302.

LALECHOS, N. & SAVOYAT, E. (1979) La sedimentation Neogene dans le Fosse Nord Egeen. *Sixth Colloquium on the Geology of the Aegean Region*, 2, pp. 591–603. Institute Geological and Mineral Research, Athens.

LE PICHON, X. & ANGELIER, J. (1979) The Hellenic arc and trench system: a key to the neotectonic evolution of the Eastern Mediterranean area. *Tectonophysics* **60**, 1–42.

LE PICHON, X., LYBERIS, N. & ALVAREZ, F. (1985) Subsidence history of the North Aegean trough. In: *The Geological Evolution of the Eastern Mediterranean* (Eds Dixon, L. & Robertson, A.). Spec. Geol. Soc. London, Publ. 17, 727–741.

LYBERIS, N. (1985) Tectonic evolution of the North Aegean trough. In: *The Geological Evolution of the Eastern Mediterranean* (Eds Dixon, L. & Robertson, A.). Geol. Soc. London Spec. Publ. 17, 709–738.

MERCIER, J. (1979) Principal results of a neotectonic study of the Aegean arc and its location within the Eastern Mediterranean. *Sixth Colloquium on the Geology of the Aegean Region*, 3, pp. 1281–1291. Institute Geological and Mineral Research, Athens.

MITCHUM, R.M. JR., VAIL, P.R. & THOMPSON, S. III (1977) Seismic stratigraphy and global change of sea-level, Part 2: the depositional sequence as a basic unit for stratigraphic analysis. In: *Seismic Stratigraphy – Applications to Hydrocarbon Exploration* (Ed. Payton, C.E.). Mem., Am. Ass. Petrol. Geol. 26, 53–62.

NEEDHAM, D., LE PICHON, X., MELGUEN, M., PAUTOT, G., PERARD, V., AVEDIK, F. & CARRE, D. (1973) North Aegean trough: 1972 Jean Charcot cruise. *Bull. Geol. Soc. Greece* **10**, 152–153.

PISIAS, N.G., MARTINSON, D.G., MOORE, T.C., SHACKLETON, N.J., PRELL, W., HAYS, J. & BODEN, G. (1984) High resolution stratigraphic correlation on benthic oxygen isotopic records spanning the last 300 000 years. *Mar. Geol.*, **56**, 119–136.

PSILOVIKOS, A. (1981) *Geomorphological, morphogenetic, tectonic, sedimentological and climatic processes which led to the formation and evolution of composite alluvial fans of Olympus mountain, Greece.* Unpublished thesis, University of Thessaloniki, 158 pp.

SHACKLETON, N.J. & OPDYKE, N.D. (1976) Oxygen isotope and paleomagnetic stratigraphy of Pacific core V28–239. Late Pliocene to latest Pleistocene. In: *Investigation of the Late Quaternary Paleoceanography and Paleoclimatology* (Eds Cline, R.M. & Hays, J.D.). Geol. Soc. Am. Mem. 145, 449–464.

SUTER, J.R. (1986) Ancient fluvial systems and Holocene deposits, Southwestern Louisiana continental shelf. In: *Late Quaternary Facies and Structure, Northern Gulf of Mexico* (Ed. Berryhill, H.J.). Am. Assoc. Petrol. Geol., Stud. Geol. 23, 81–129.

*Spec. Publs. int. Ass. Sediment.* (1991) **12**, 133–150

# Sequence stratigraphy on the shelf and upper slope in response to the latest Pleistocene–Holocene sea-level changes off Sendai, northeast Japan

YOSHIKI SAITO

*Marine Geology Department, Geological Survey of Japan, 1-1-3 Higashi, Tsukuba, Ibaraki 305, Japan*

## ABSTRACT

The relation of sequence stratigraphy to modern sedimentary environments on a shelf and upper slope is dealt with in this paper. A depositional sequence consisting of three depositional units on the shelf off Sendai, northeast Japan, formed in response to latest Pleistocene–Holocene sea-level changes: unit 1, a coastal sequence and fluvial channel-fills deposited during the fall and lowstand of sea level in the latest Pleistocene; unit 2, a transgressive sand sheet and an onlap sequence deposited from 14 to 6 ka; and unit 3, a modern coastal sequence that developed at present-day sea level during the last 6 ka. The boundary between units 1 and 2 forms a transgressive surface. The surface is sharp on the shelf, and is defined by an erosional ravinement surface. To characterize the transgressive surface, the shelf and upper slope areas can be divided into four regions: inner, middle and outer shelf, and upper slope.

The transgressive sand sheet that widely overlies the transgressive surface is commonly 0.5–1.5 m thick, and consists of very poorly sorted pebbly and muddy sand, bearing mixed-habitat molluscan shell fragments, glauconite, and volcanic grains. This sand sheet is diachronous across the shelf.

At present, the condensed layer, or submarine hiatus, which is recognized above the sand sheet, is generated in response to a very low sedimentation rate in this offshore area. The main part of the condensed section is correlative with the transgressive sand in the outer shelf to upper slope areas. The presence of scattered pebbles or faunal mixing horizons in the condensed section shows the characteristics of the transgressive sand sheet.

## INTRODUCTION

The depositional sequence is a fundamental unit for interpretation of palaeosea-level changes and depositional regime from sedimentary rocks and seismic records (Vail *et al.*, 1977a,b, 1984; Haq *et al.*, 1987). Recently, depositional sequences in coastal and shelf areas formed in response to known late Pleistocene to Holocene sea-level changes have been studied in order to construct basic depositional models and to support seismic facies analysis (Nummedal *et al.*, 1987; Loutit *et al.*, 1988). In these studies, fundamental features used in analysis are the ravinement surface, an unconformity formed by shoreface erosion in coastal areas during a transgression (Bruun, 1962; Swift, 1968; Nummedal & Swift, 1987), and the condensed section, which is created in offshore areas under a very low sedimentation rate during a rapid rise in sea level (Vail *et al.*, 1984; Loutit *et al.*, 1988).

The basic relationships between sedimentary facies, seismic facies, depositional sequences, sequence boundaries, transgressive surface, and sea-level changes are now reasonably well defined (Kraft *et al.*, 1987; Wilgus *et al.*, 1988), and facies models that include sequence boundaries and transgressive surfaces created in response to sea-level changes have been published (e.g. Demarest & Kraft, 1987).

The depositional sequence that formed on the shelf off Sendai, northeast Japan, in response to latest Pleistocene to Holocene sea-level changes is dealt with in this paper. In particular, a transgressive surface and the suprajacent condensed section created in response to a rapid sea-level rise during the latest Pleistocene to Holocene are discussed.

The survey area, which was far from continental ice sheets during the last glacial, differs from areas in much of North America and northern Europe in its record of local Holocene sea-level changes induced by glacio- and hydro-isostacy (Walcott,

1972; Clark *et al.*, 1978). The highest sea level in the Holocene was recorded at about 6 ka in the Japanese Islands. Since then, local sea level has been falling with superimposed fluctuations of a few metres (Moriwaki, 1978). The rate of rise in sea level in the early Holocene was 1–2 cm/yr (Matsushima, 1987; Fujii & Nasu, 1988; Saito *et al.*, 1989b), about 10 times greater than that along the American Atlantic Shelf (Kraft *et al.*, 1987).

The survey area, the Sendai shelf, is relatively tectonically stable, despite the Japanese Islands active plate margin setting. In the subsiding areas around the Japanese Islands, stacked coastal sequences, each a few to several tens of metres thick and consisting of alternations of non-marine and marine deposits (Maenaka *et al.*, 1977; Kikuchi, 1977), accumulated together with associated shelf sequences (Okamura, 1989) in response to periodic sea-level changes with a period of about 100 ka and an amplitude of about 100 m during the last 1.0 Ma (Shackleton & Opdyke, 1973; Imbrie *et al.*, 1984). By contrast, the stratigraphy of the survey area reflects global sea-level changes since the middle Miocene (Yanagisawa *et al.*, 1989). Marine strata recognized in the coastal and shelf areas are the only present coastal sequence to have formed in response to the present high sea-level stand, whilst the ancient coastal sequence in the middle shelf formed during the lowest sea-level stand in the late Pleistocene.

## REGIONAL SETTING

### Shelf morphology

The survey area is located on the Pacific shelf of northeast Japan, about 200 km west of the Japan Trench Axis (Fig. 1). The shelf width is about 60 km. A well-defined shelf break occurs at a depth of about 145 m. Morphologically, the shelf area is

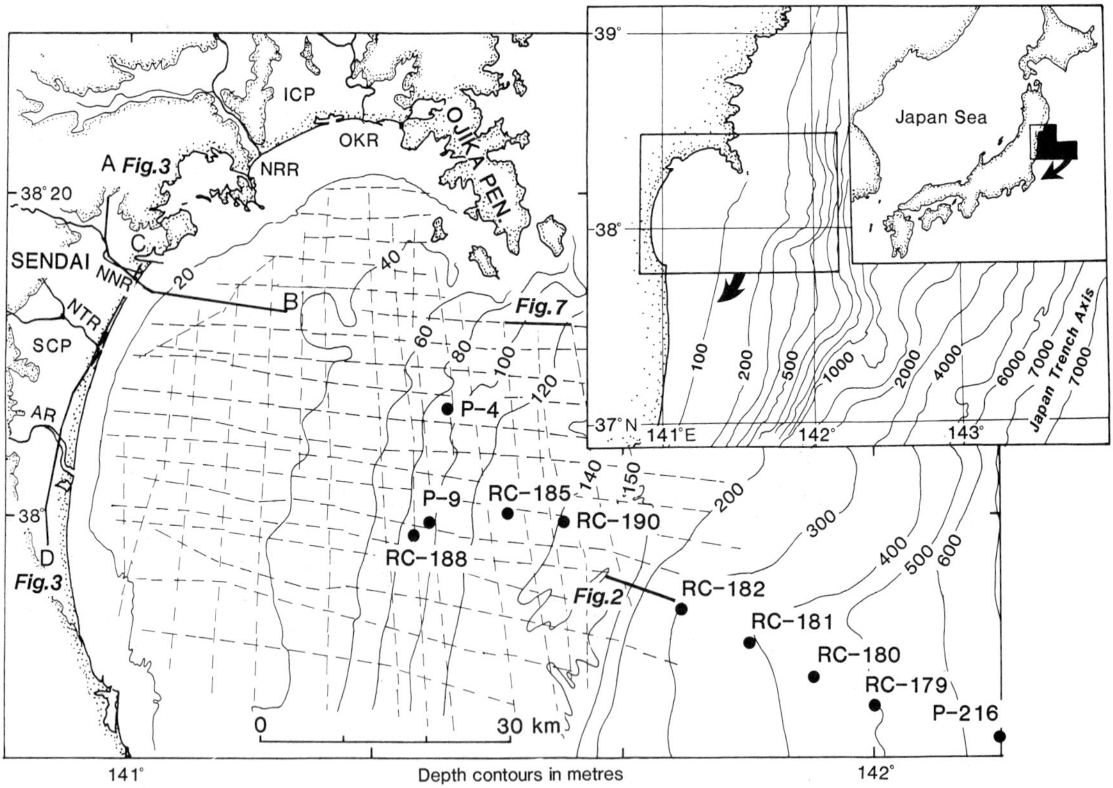


**Fig. 1.** Index map of the survey area, seismic survey lines (dashed lines), and the locations of cored samples. ICP, Ishinomaki coastal plain; SCP, Sendai coastal plain; OKR, Old Kitakami River; NRR, Naruse River; NNR, Nanakita River; NTR, Natori River; AR, Abukuma River.

divided into four parts (Saito, 1989a), shoreface (0–15 m), inner shelf platform (15 to c. 50 m), middle shelf slope (50–120 m), and outer shelf platform (120–145 m).

The shoreface consists of upper and lower segments. Longshore bars and troughs are recognized in the upper shoreface, in depths less than about 6 m. Some channels are found in the inner shelf, both transverse and normal to depth contours. In the middle shelf, breaks in slope parallel to depth contours are recognized. The convex-upward break at 97–107 m depth is widespread. Sand ridges on the outer shelf, with less than 10 m of relief and spacing of about 2 km, are regularly arranged with a NE–SW trend. These sand ridges are well developed in the northern part of the survey area (Saito, 1989b,d).

The internal structure of the middle to outer shelf is of the oblique-progradational type (Mitchum *et al.*, 1977). In addition, seaward-dipping clinoforms with a toplap geometry are recognized beneath the outer shelf and upper slope (Fig. 2; Saito *et al.*, 1989b). The upper slope is situated at the margin of a deep-water fore-arc basin (Yajima *et al.*, 1982).

#### Physical oceanographic conditions and surrounding coastal geology

The survey area is a storm-dominated, microtidal region. Mean tidal range is about 1 m. The mean significant wave during the last 15 years at Shin-Sendai Port is 0.79 m in wave height and 7.72 s in wave period (Sugawara *et al.*, 1986).

Holocene strand plains, the Ishinomaki and Sendai coastal plains, are well developed shoreward of the survey area (Fig. 1). These plains have largely formed since 6 ka, and have prograded seaward at a maximum rate of 1–1.2 km/ka (Matsumoto, 1984). Eight stratigraphic units, which overlie channels incised into older Tertiary strata, are recognized in boreholes on the plains (Fig. 3; Matsumoto, 1981). In ascending order, they are fluvial sandy gravel (BG), non-marine peaty clay (La), fluvial gravelly sand (Lb), non-marine mud (Ma), marine neritic mud (Mb), shoreface sand (Ua), foreshore and backshore sand (Ub), and floodplain mud (Uc). The boundary between the La and Lb units is an erosional surface formed during the late Pleistocene (Matsumoto, 1988; Saito, 1989d). The Mb–Ua–Ub sequence shows an upward-coarsening trend created by the progradation of coastal to neritic sediments since 6 ka (Matsumoto, 1984; Saito, 1987).

## METHODS

Field surveys were conducted by R/V *Hakurei-Maru* in 1981, salvage vessel No. 23 *Kaiko-Maru* in 1985, and R/V *Tokai-Nisei* in 1986. About 200 samples of surface sediments were collected with a Smith-McIntyre grab sampler, and approximately 50 cores were obtained. These samples have been subjected to grain size, chemical, palaeontological, sedimentary structure, textural, heavy mineral, and $^{14}C$ age analyses (Honza *et al.*, 1982; Aziz, 1986; Suzuki & Saito, 1988; Saito, 1989a,b; Saito *et al.*, 1989a,b).

Seismic surveys were carried out using an EG&G model 240 subtow-type uniboom, with an energy supply of 200 J, in 1985 and 1986, and a 3.5 kHz subbottom profiler in 1986. Seismic data lines are shown

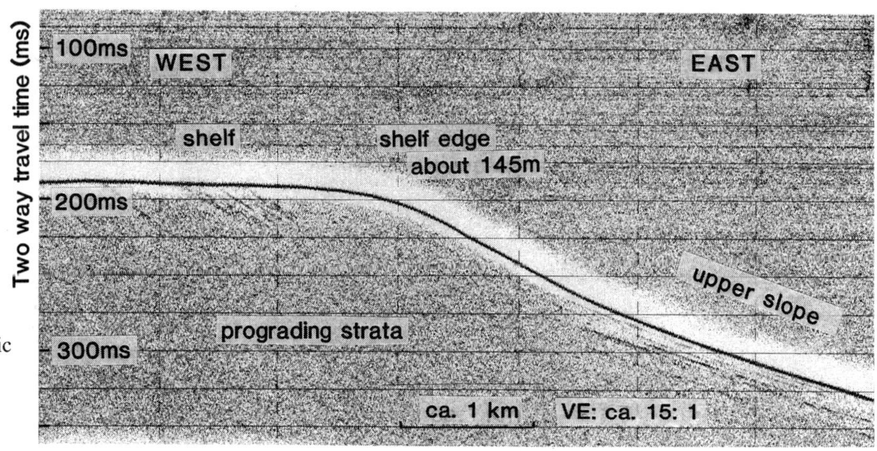


**Fig. 2.** A 3.5 kHz seismic record across the outer shelf and upper slope area. Location is shown in Fig. 1.

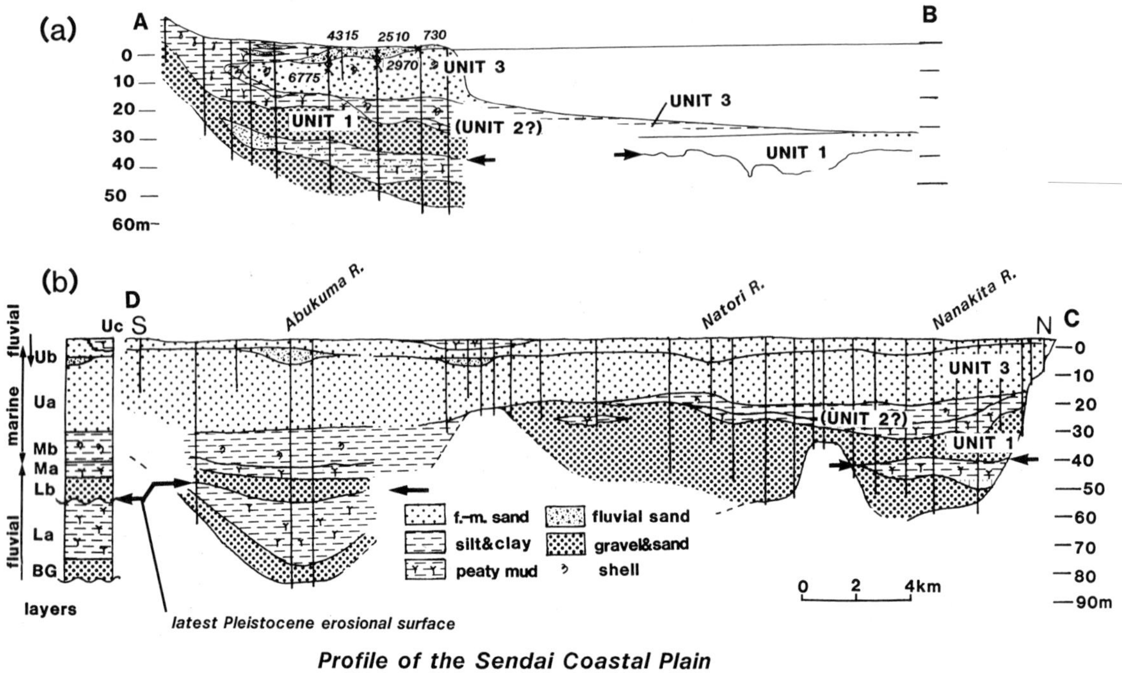


**Fig. 3.** (a) Geological profiles from Sendai coastal plain to the inner shelf (modified after Matsumoto, 1981 and Saito, 1989a). Numbers near crosses show $^{14}C$ dates (from Matsumoto, 1984). (b) Geological profile of the Sendai coastal plain (after Matsumoto, 1981 and Saito, 1989d). Locations are shown in Fig. 1.

in Fig. 1; their total length is approximately 1300 km. Ship speed was about 6 knots. The subsurface depths and thicknesses are based on a sound velocity of 1500 m/s (Saito, 1989d).

## RESULTS

The results of analyses of the surface sediments and the seismic surveys have already been reported by Aziz (1986), Suzuki & Saito (1988), Saito (1988,1989a,d), and Saito *et al.* (1989a,b). Here, outlines of stratigraphy, sediments, and depositional sequence are given. The relationship between these features and sea-level changes are discussed in a later section.

### Shelf sediments

Five types of surface sediment are recognized: modern shoreface sand, modern inner-shelf mud (neritic mud), modern middle-shelf mud, relict palimpsest mid- to outer-shelf sand, and modern residual inner-shelf gravelly sand (Fig. 4; Saito, 1988,1989a). The relict palimpsest sediments were deposited during the last transgression (the Jomon Transgression in Japan) from 14 to 6 ka. The modern residual sediment, which is derived both from the sediments deposited during the transgression and from fluvial sediment deposited during the fall and lowstand of sea level in the late Pleistocene, hardly contains any fine-grained particles less than about 100 μm and is estimated to be winnowed by present wave motion, which extends to 50–70 m during storms (Saito, 1988,1989a; Saito *et al.*, 1989b). The estimated storm wave-base at 50–70 m would divide the shelf into modern residual and relict palimpsest sediments (Saito, 1988).

### Stratigraphy

Three depositional units are discriminated in the uppermost sequence overlying the widely recognized erosional surface. Each unit is characterized by a combination of sedimentary facies, biofacies, internal structures, and distribution patterns. They are

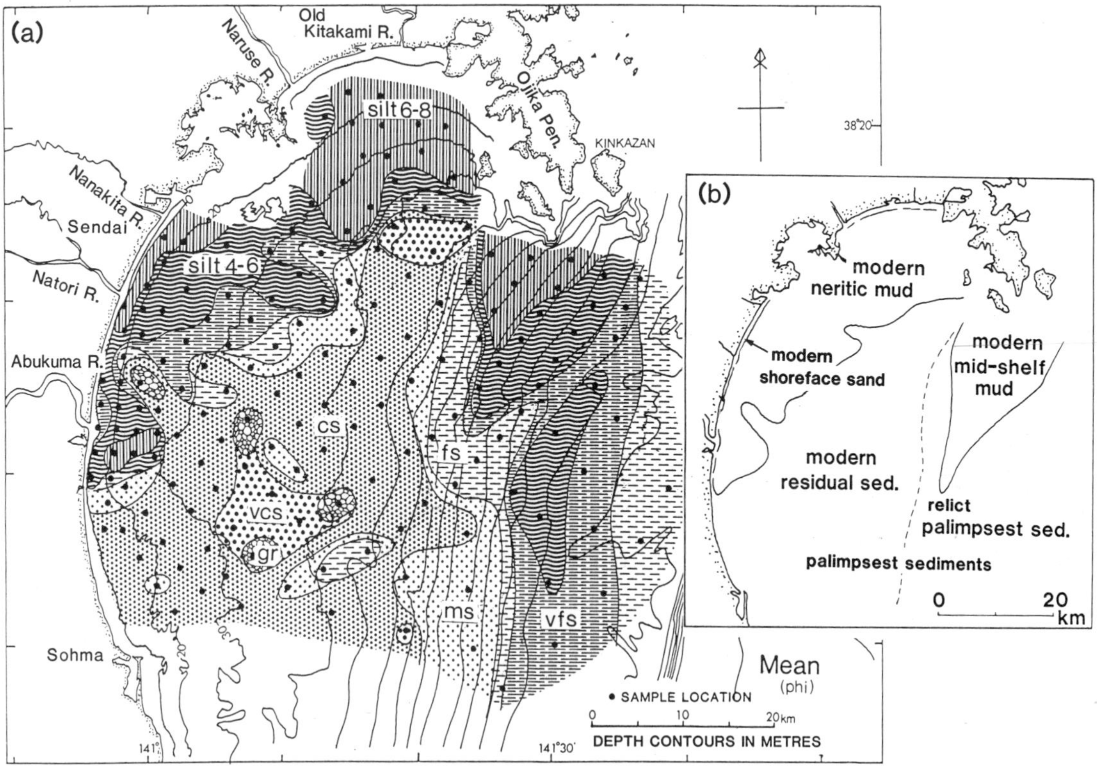


**Fig. 4.** Distribution of the surface sediments off Sendai (after Saito *et al.*, 1989b). (a) Mean diameter, (b) generalized distribution of surface sediments: gr, gravel; vcs, very coarse sand; cs, coarse sand; ms, medium sand; fs, fine sand; vfs, very fine sand; silt divided into 4–6 and 6–8 ϕ classes.

units 1, 2, and 3 in ascending order.

The base of unit 1 is correlated with the late Pleistocene erosional surface, and is continuous with the boundary between the La and Lb units of Matsumoto (1981) beneath the coastal plains (Fig. 3a). The erosional surface consists of buried channels and buried uplands in the inner shelf (Fig. 5). The buried channels can be traced to a depth of about 90 m. Some of these channels are completely buried, or are recognized as depressions in the inner shelf (see Figs 1, 11). Where water depth is greater than 90 m on the shelf, the base of unit 1 becomes a flat and smooth surface with a concave-upward morphology. This is characteristic of an abrasion platform (Fig. 5; Saito, 1989d).

Unit 1, which buries the erosional surface in the channels and covers the surface in the middle to outer shelf, consists of sediments deposited during the latest Pleistocene. Most of unit 1 in the inner shelf constitutes the lower part of the infills of buried channels, and is recognized as a dark-toned and chaotic facies on seismic records. In the middle shelf, in contrast, unit 1 is distributed parallel to the depth contours at depths of 90–110 m (Figs 6, 7). Unit 1 in the outer shelf is topographically considered to be represented by sand ridges that are regularly arranged, with NE–SW crests at an angle of approximately 35° to the contours at 80–100 m depth, and connected with unit 1 in the middle shelf (Fig. 6; Saito, 1989d).

Unit 2 is commonly recognized in cores from the shelf to upper slope (Fig. 8). This unit is 0.5–1.5 m thick in the cores, and is composed of very poorly sorted, muddy coarse sand, containing pebbles, mixed-habitat molluscan shells, glauconite, volcanic grains (Fig. 9; Saito *et al.*, 1989b) and mixed-habitat diatom (Saito *et al.*, 1989a) and ostracod assemblages (Itoh, 1989). In cores with well-preserved primary sedimentary structures, these sediments consist of thin alternations of mud and sand, several milli-

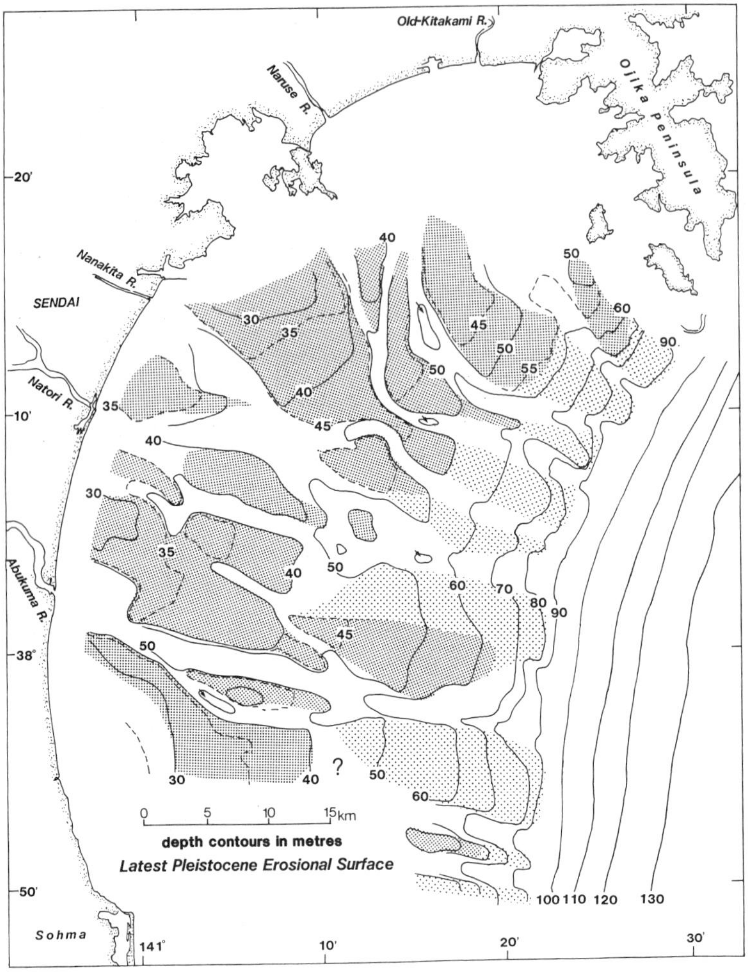


**Fig. 5.** Contour map of the depth below present sea level to the base of unit 1, the latest Pleistocene erosional surface, based on seismic reflection profiles. Heavy dotted areas indicate buried uplands; light dotted areas are inferred to be buried uplands. After Saito (1989d).

metres to a few centimetres thick; structures interpreted as scour-and-fill can be recognized at the base of sand layers (Fig. 9a, P-4 core). Sand laminae are commonly bioturbated, and therefore partly obscured. Pebbles, plant fragments, and molluscan shell fragments are scattered, but are more concentrated toward the base of this unit. On the other hand, unit 2 can only be recognized in some seismic records, as onlap seismic reflections in the area of high sedimentation rate near south of the Ojika Peninsula (Fig. 7). The area is located near and under the distribution of modern middle-shelf mud.

Unit 2 is correlated with the modern residual sediments in the inner shelf and with the relict palimpsest sediments in the middle to outer shelf. These sediments are continuous to the upper slope area (Fig. 8; Saito *et al.*, 1989b) and are relatively rich in glauconite and pumice grains (Fig. 10). On the middle to outer shelf these surface sediments have carbon:nitrogen ratios of 6 to 8. The values indicate that organic material in these sediments is predominantly planktonic in origin and that terrigenous input was very low (Saito *et al.*, 1989b).

Unit 3 comprises the modern coastal and modern middle-shelf sediments. The unit is about 20–30 m thick in the coastal area (Fig. 3) and is a few to several metres thick in the inner shelf region, except in channels, where it is about 10 m thick. In the middle shelf, unit 3 is distributed as a tongue shape south of the Ojika Peninsula, and the distribution coincides with that of middle-shelf mud (Fig. 4). In its northern part the unit is about 7 m thick, and thins southward (Saito, 1989a,d).

The ages of units 1–3 are estimated on the basis of stratigraphy, $^{14}C$, and tephra dates. The age of unit 1 spans the fall and lowstand of sea level in the

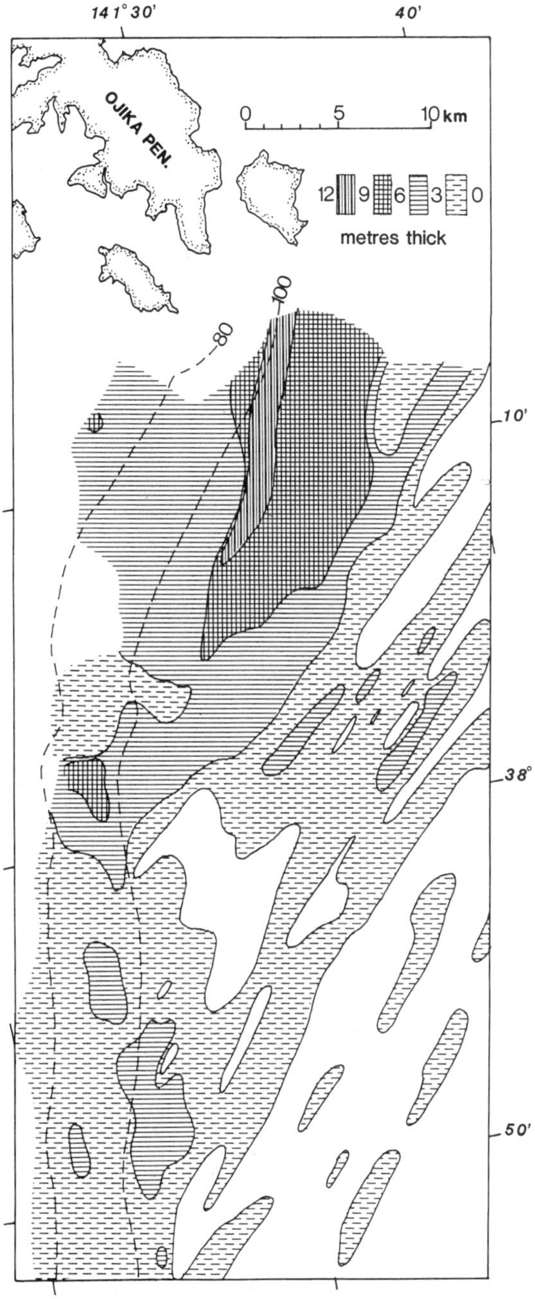


**Fig. 6.** Isopach map of the latest Pleistocene sediments, which are composed of mostly unit 1. Note the ancient shoreface-connected sand ridges.

latest Pleistocene; $^{14}C$ dates from peats in borehole samples from the coastal plains yield dates of 15.8, 26.3, 26.4, 32.1, 33.3, and >35 ka; these samples are considered to be taken from fluvial floodplain facies (Matsumoto, 1988). Middle-shelf samples of unit 1 from cores are dated at 13.4, 14.0, and 17.0 ka (Saito *et al.*, 1989a,b). These samples come from the ancient coastal plain on the middle shelf that formed during the lowest sea-level stand, for which the age is estimated to be about 20–14 ka (Shackleton *et al.*, 1983; Berger *et al.*, 1987). Unit 2 spans the period of rising sea level from the latest Pleistocene to the early Holocene, that is from 14 to 6 ka. This is based on $^{14}C$ dates of 5.6, 7.8, 7.8, and 8.1 ka from surface sediments in the inner shelf (Hattori, 1967), and 10.2, 10.1, 10.4, and 12.4 ka from cores in the middle to outer shelf (Hattori, 1967; Saito *et al.*, 1989b). Unit 3 was formed during the present stand in sea level since about 6 ka. Many $^{14}C$ dates of molluscan shells and peaty sediments in the Mb, Ua, Ub, and Uc units are younger than 6 ka (Matsumoto, 1984), and a thin tephra layer at the boundary between units 1 and 2 in core P-4 from the middle shelf can be correlated with the Chuseri Pumice erupted from Towada caldera, northern Honshu, at about 5.5 ka (Saito *et al.*, 1989a).

### Relict topography and depositional sequences

Relict topography formed during the low sea level in the late Pleistocene is found on the shelf (Fig. 11; Saito, 1989d). During the low sea-level stand, the present inner shelf area was subaerial, and composed of dissected channels, lowlands, and uplands (Fig. 5). The palaeo-strand plain developed the present middle shelf. The shelf was narrow, with a width of about 20 km, and the shelf edge was about 50 m deep. Shoreface-connected sand ridges were developed on the shelf, and extended to the shelf edge (Saito, 1989d).

The seaward edge of the palaeo-strand plain is recognized as a surficial break in slope on the present middle shelf (Fig. 11). Its depth ranges from 97 to 107 m. This break is located on units 1 and 2. The buried and partly surficial break on unit 1 only is not clear in comparison with the surficial one, and its depth ranges from about 105 to 125 m. This depth is 10–30 m deeper than the lowest sea level of 95 m (Saito *et al.*, 1989a).

## DISCUSSION

Relationships between sea-level changes, sequences, the transgressive surface, and the sand sheet are

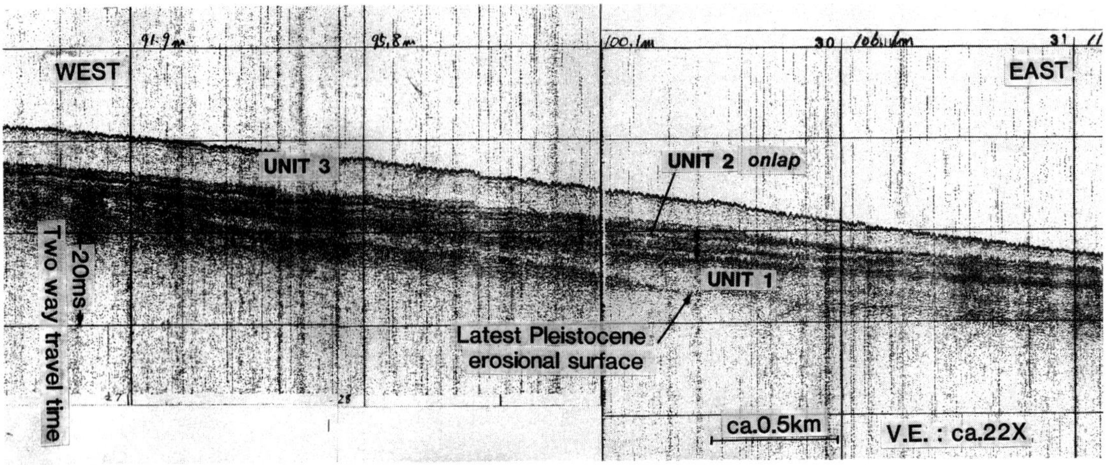


**Fig. 7.** Uniboom seismic record in the middle-shelf area. Location is shown in Fig. 1. Modified after Saito (1989d).

dealt with in this section. Figure 12 shows a sea-level curve for the survey area for the last 20 ka (Saito *et al.*, 1989b) and sea levels estimated from studies of uplifted marine terraces (Bloom & Yonekura, 1985; Kikuchi, 1988) and oxygen isotope records of deep-sea sediments (Shackleton *et al.*, 1983).

**Topographic changes and ravinement surface**

Shoreface erosion caused by sea-level rise is the most important process during transgression (Bruun, 1962), and the erosional surface created by this process is called a ravinement surface (Swift, 1968). The ultimate erosion depth of this process is reported to be about 10 m for modern coasts. For example, Belknap & Kraft (1985) found shoreface erosion of as much as 11–12 m on the Atlantic continental shelf off Delaware, and Penland & Suter (1983) obtained values of 7–10 m on the Louisiana inner continental shelf. Leatherman (1987) found more than 9 m of erosion on the Maryland coastal area, and Hands (1983) measured 7 m of erosion on the coastal area of Lake Michigan. On the other hand, Sunamura (1987) showed that the maximum erosion depth on the Pacific side of the Japanese Islands is about 40 m. The possible variation in shoreface erosion from 10 to 30 m into the ancient coastal sequence of unit 1 in the survey area is consistent with these data.

**Transgressive sand sheet**

Transgressive sand sheets deposited on shelves during a transgression have been widely recognized (e.g. Swift *et al.*, 1971; Maldonado *et al.*, 1983; Lee *et al.*, 1988; Saito *et al.*, 1989b).

The transgressive sand sheet off Sendai is 0.5–1.5 m thick, and is composed of very poorly sorted muddy and pebbly sand, and of partly bioturbated thin sand layers several millimetres to a few centimetres thick. This sediment contains mixed-habitat molluscan shell fragments, plant fragments, volcanic grains, and glauconite (Saito *et al.*, 1989b) as predicted for condensed sections by Vail *et al.* (1984) and Loutit *et al.* (1988). The mean depositional rate of the sand sheet off Sendai is 6–20 cm/ka, but actual rates during deposition of the sheet gradually decreased with the landward shift of the depositional site. At present, these sediments are exposed as the relict palimpsest sediments on the middle to outer shelf and the upper slope, and have been partly covered by the modern middle-shelf and basin muds. The relict palimpsest sediments correspond to an area of non-deposition, or of very low depositional rate at perhaps less than 1 cm/ka during the last 6 ka. These modern environments show a modern submarine hiatus created on the transgressive sand sheet.

A similar sand sheet on the storm-dominated shelf off Akita on the Japan Sea coast of northeast Japan

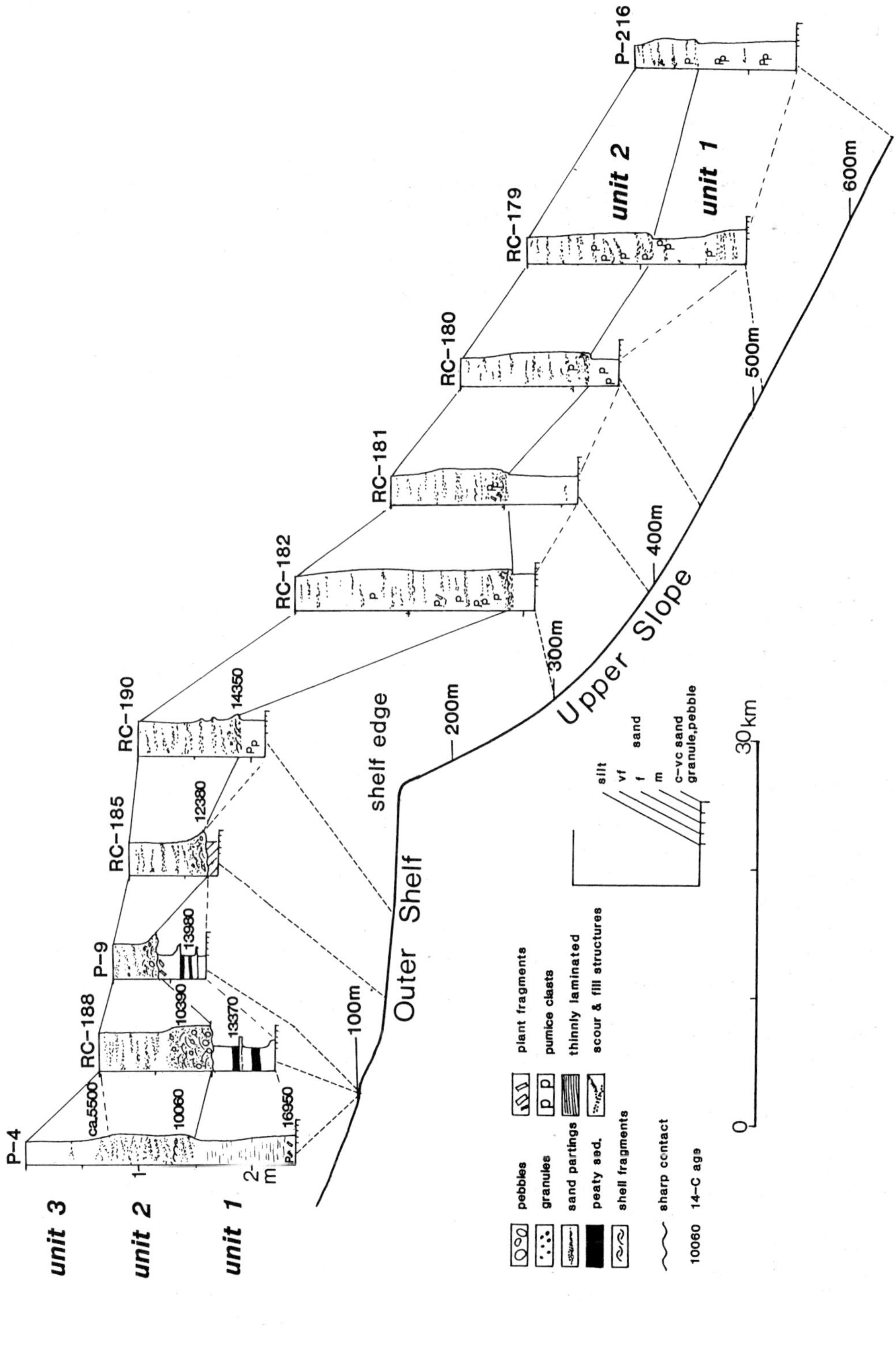


**Fig. 8.** Core logs arranged from the middle shelf to the upper slope, after Saito *et al.* (1989b). Locations are shown in Fig. 1.

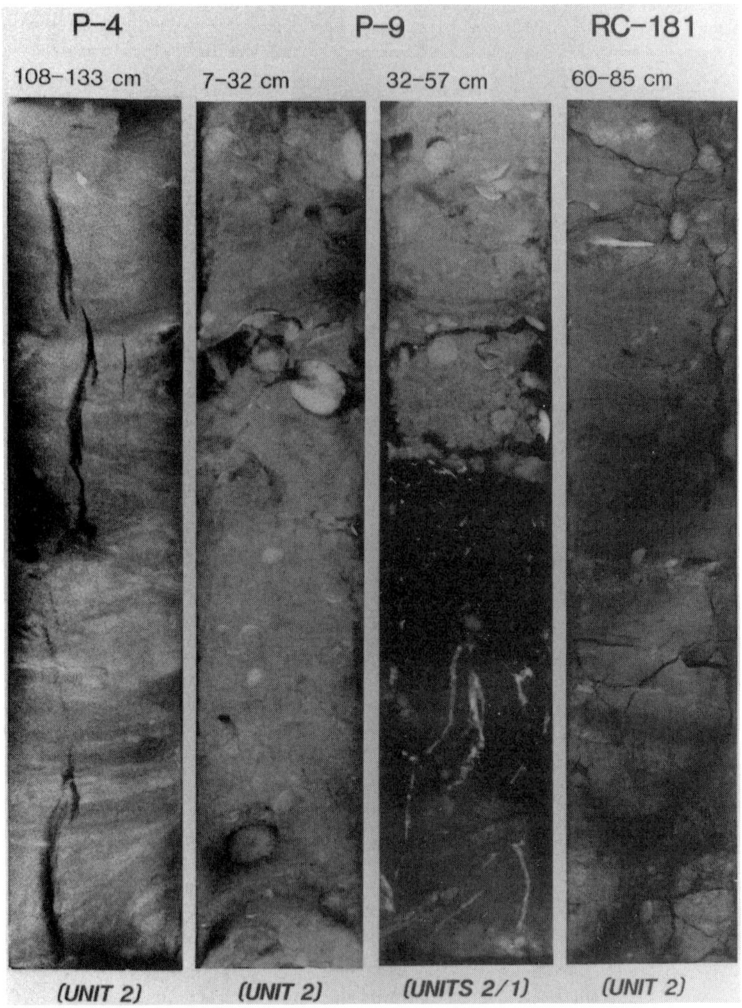


**Fig. 9.** (a) Soft X-radiographs (negative) of the transgressive sand sheet (unit 2) of the cores P-4, P-9, and RC-181. Each slab is 5 cm wide and 25 cm long. After Saito *et al.* (1989b).

is mostly comprised of glauconite pellets (Saito, 1989c). Both the Sendai and Akita transgressive sand sheets are believed to correspond to the condensed sections described by Vail *et al.* (1984) that form in response to rapid sea-level rise.

**Transgressive surface**

A transgressive surface is usually represented by a ravinement surface in shallow marine environments (Swift, 1968; Nummedal & Swift, 1987). During the transgression off Sendai, a sand sheet was deposited in contemporary water depths of more than 50 m in the offshore area from 14 to 9 ka (Saito *et al.*, 1989b) and in depths of 20–30 m from 9 to 6 ka. (Figure 13 shows the sea-level curve and the depth range over which the sand sheet of unit 2 was deposited. The depositional area shifted landward with rising sea level (Figs 13, 14).

To characterize the transgressive surface, the shelf and upper slope areas can be divided into four regions. Firstly the upper slope (more than 145 m in this case): sediments here have been deposited continuously through the fall and subsequent rise in sea level (Fig. 15), and a condensed layer or submarine hiatus is being formed at present. The second is the outer-shelf region (105–145 m), which is limited in depth by the lowest sea level (95 m) plus the maximum depth of shoreface erosion. Sediments have been deposited continuously from the fall to the rise in sea level except for the period of the early transgression. The third is the inner- to middle-shelf

**Fig. 9.** (b) Photographs of the RC-181 core obtained from the upper slope. Each slab is 5 cm wide and 25 cm long.

region (shallower than 105 m), where an erosional transgressive surface is produced during the transgression. The third region is divided into two parts with respect to the sediments on the transgressive surface: one is the middle-shelf region (c. 50–105 m), where transgressive sand has been deposited, and the other is the inner-shelf region (shallower than 50 m), where surficial shelf sediments composed of residuals are reworked by storm waves during the subsequent stand in sea level.

Viewed in relation to the actual sea level, the divisions outlined above are separated by (1) the deepest limit of shoreface erosion, and (2) the landward limit of redeposited sediments as a transgressive sand sheet. The former depth ranges from 10 to 30 m below sea level, and the latter depth ranges from 20 to 50 m in the Sendai area. During transgression, therefore, the sea bed at depths of 0 to about 10 m is erosional, that from 10 to 50 m is transitional from erosional, non-depositional to depositional, and that more than 50 m deep is depositional if sediments are supplied. The sand sheet is preserved below the estimated storm wave-base in this region.

## Preservation potential

In the survey area, marine strata formed during the early to middle Pleistocene have not been recognized

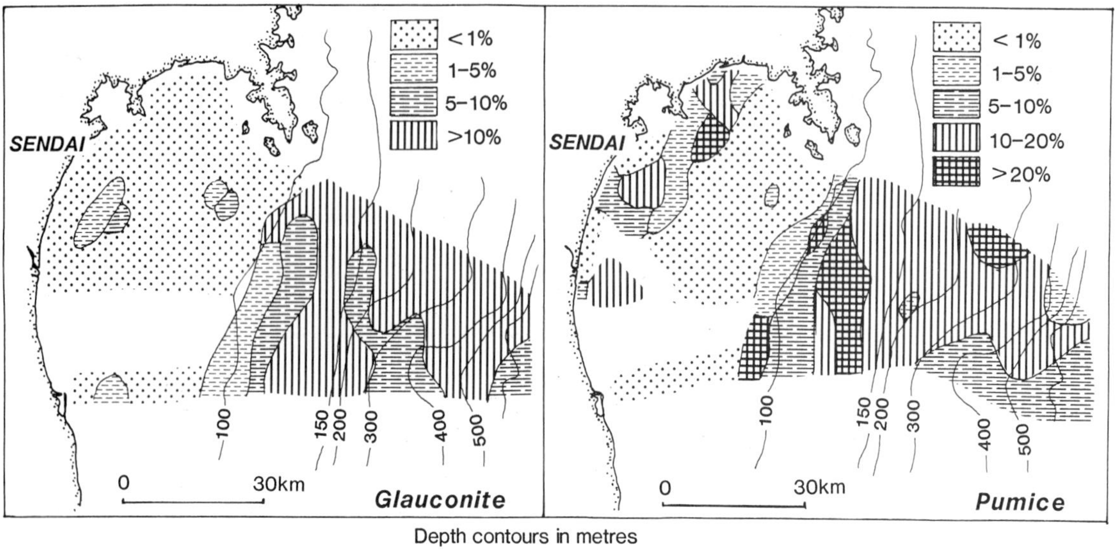


**Fig. 10.** Distribution of glauconite and pumice in the surface sediments expressed as a percentage of the sand fraction (after Hattori, 1967).

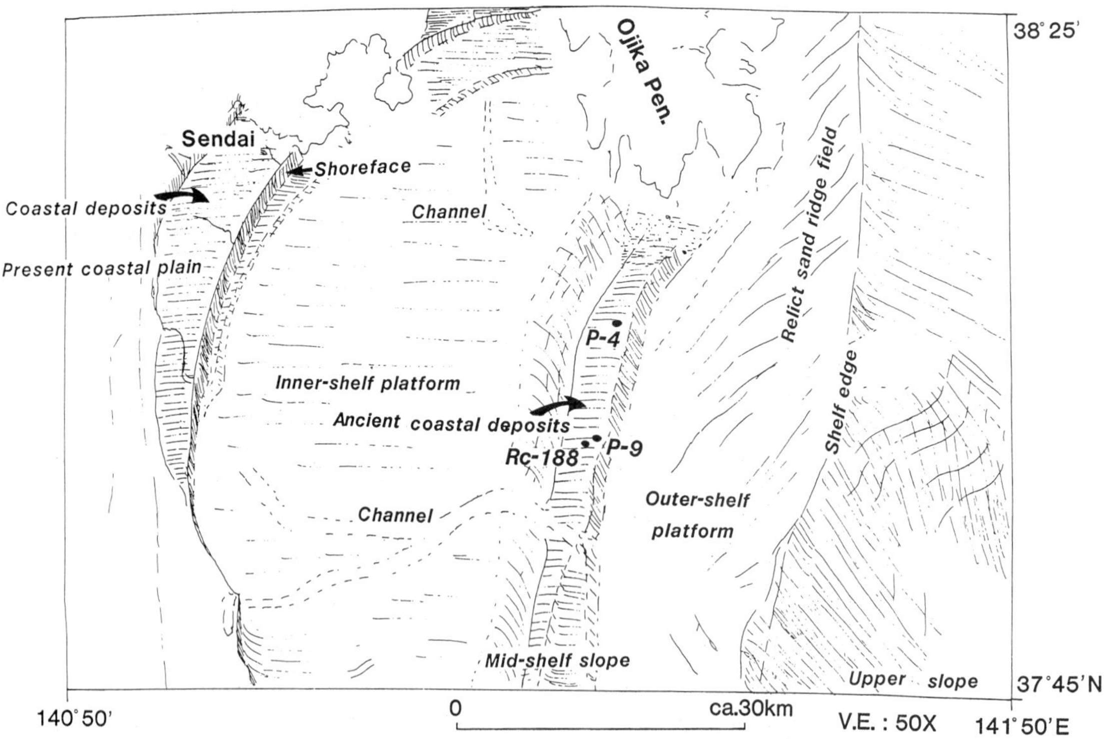


**Fig. 11.** Schematic present shelf topography and its interpretation, made from shaded relief map (after Saito, 1989d).

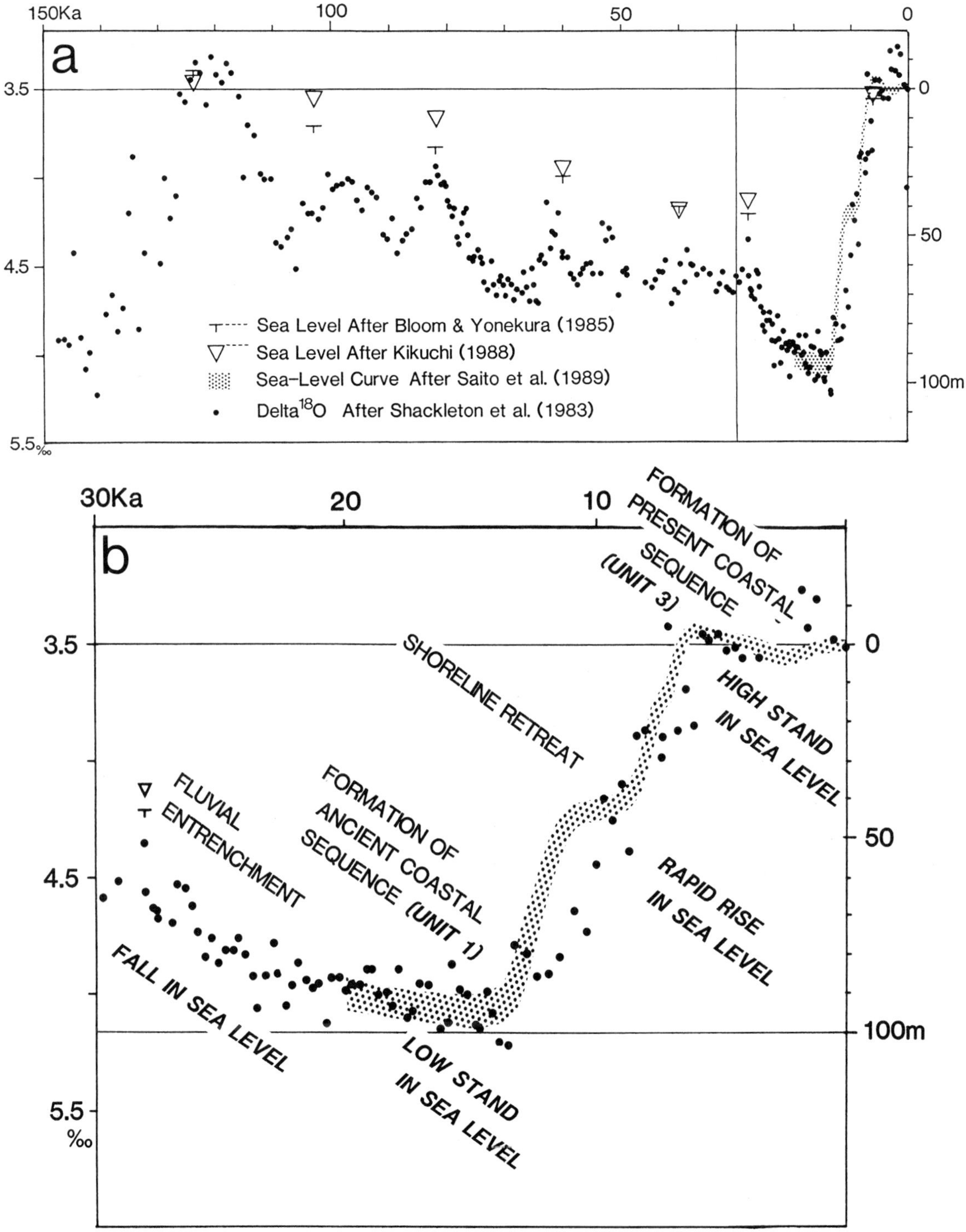


**Fig. 12.** (a) Sea-level curve and oxygen-isotope records during the last 150 ka based on the data of Kikuchi (1988), Bloom & Yonekura (1985), Saito *et al.* (1989b) and Shackleton *et al.* (1983). (b) Sea-level curve and coastal to shelf processes during the last 30 ka.

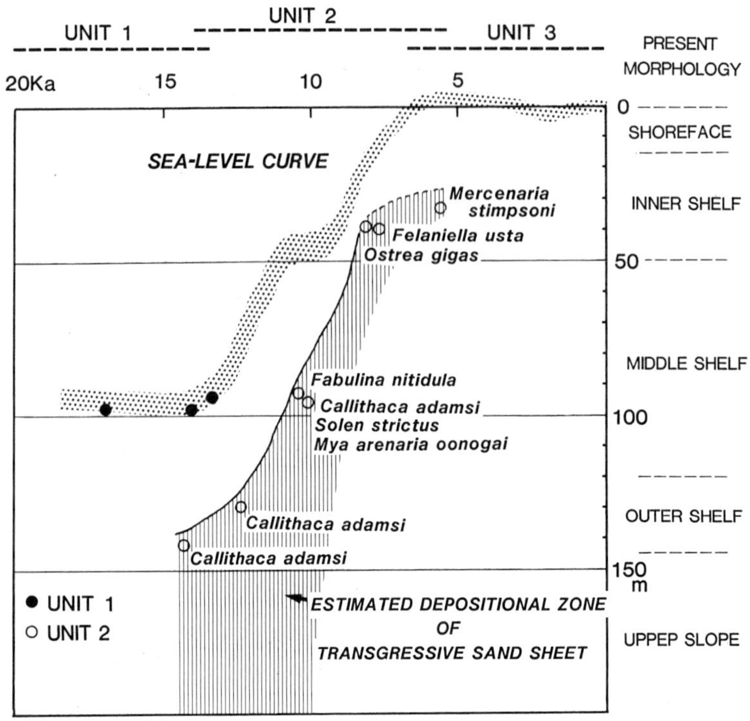


**Fig. 13.** The relationship between the sea-level curve and the transgressive sand sheet. Note that the sand sheet formed in palaeo-water depths of greater than 50 m during 14–9 ka, and of 20–30 m during 8–6 ka. Modified after Saito *et al.* (1989b) using the data of Hattori (1967).

on the shelf and coastal areas. If the same depositional processes occurred in this period as those typical of the last 30 ka, stacked or accumulated shelf sequences would be expected beneath the shelf. However, these marine sequences have not been found. This means that if early to middle Pleistocene sequences were formed on the shelf, then these sequences would have been eroded during later periodic sea-level changes. If this estimation is right and these phenomena occurred, the detritus that formed ancient coastal and shelf sequences would be redeposited on to the slope and into the adjacent fore-arc basin areas. Former sea-level stands are only recorded as condensed sections in slope and basin sequences.

## CONCLUSIONS

A depositional sequence, transgressive surface, and sand sheet on the shelf and upper slope off Sendai, northeast Japan, were formed in response to sea-level changes during the last 30 ka. The sediments on the shelf and in the adjacent coastal area form a depositional sequence that consists of three units overlying the latest Pleistocene erosion surface (Fig. 16). Unit 1 is an ancient coastal sequence, connected through a channel system to outer-shelf sand ridges, that formed during the fall and the last lowstand in sea level. Unit 2 consists of a sand sheet and also partly an onlap sequence that were deposited on the shelf and upper slope during the transgression from 14 to 6 ka. Unit 3, the modern coastal sequence, was deposited at present sea level during the last 6 ka.

The boundary between units 1 and 2 is a transgressive surface and partly associated erosional surface. The surface, which is sharp on the shelf, is defined by an erosional and ravinement surface. To characterize the transgressive surface, the shelf and upper slope areas can be divided into four regions; inner, middle and outer shelf, and upper slope. Moreover, based on contemporary water depth, shallow marine environments are divided into three regions. Where palaeo-water depth was shallower than about 10 m, the surface consists of a ravinement surface; where water depth always exceeded a storm wave-base of 50 m, the surface is a conformable surface; where water depth was between these two extremes, the surface is transitional from an unconformity to a conformity.

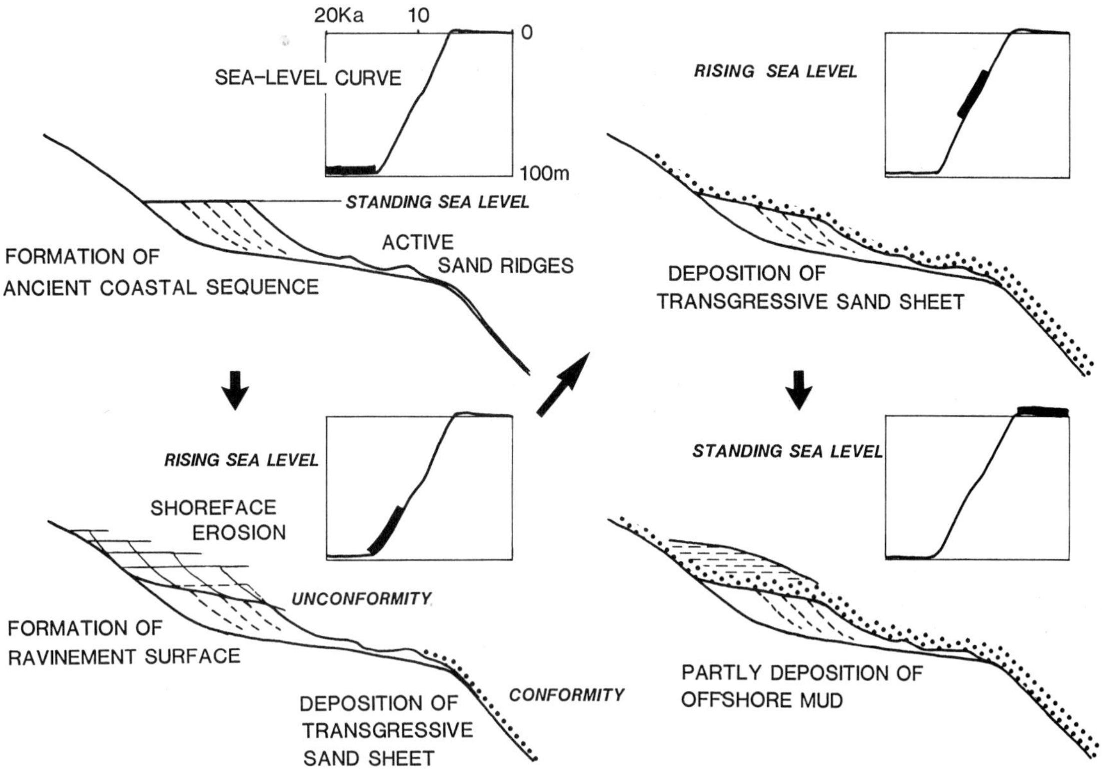


**Fig. 14.** Schematic depositional and erosional history in the middle to outer shelf areas in relation to sea-level changes during the last 20 ka.

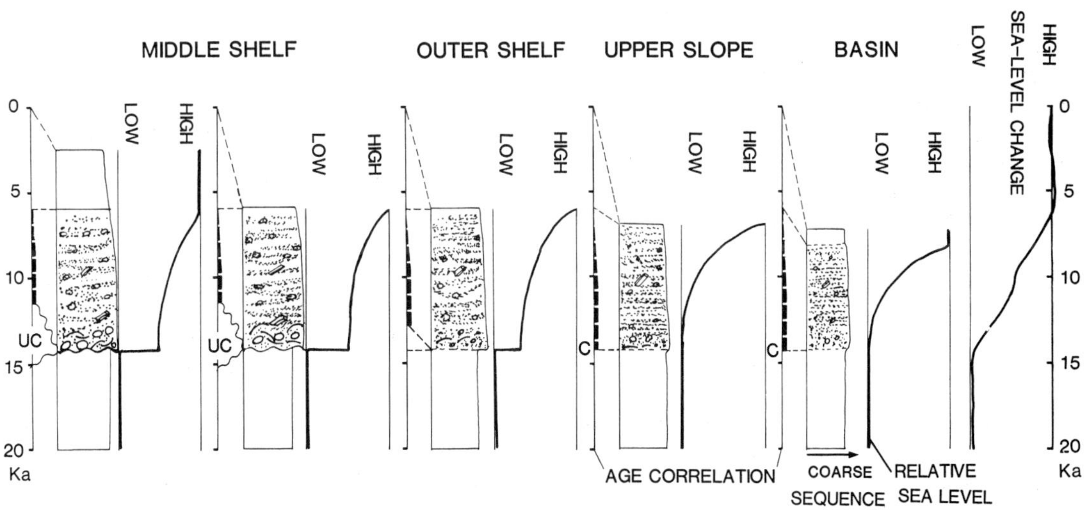


**Fig. 15.** Idealized sequences on the shelf to upper slope areas, in response to the relative sea-level curves. Mode of transgressive surface: UC, unconformity; C, conformity.

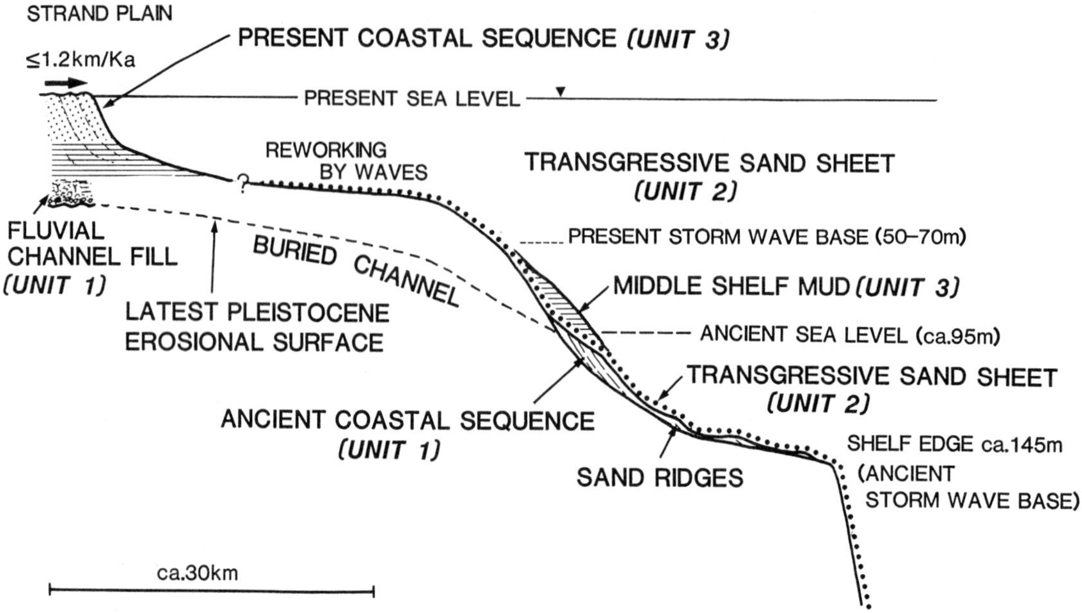


**Fig. 16.** Schematic model to explain shelf topography and shelf sequence.

The transgressive sand sheet that widely overlies the transgressive surface is commonly 0.5–1.5 m thick, and is composed of very poorly sorted, muddy and pebbly sand including mixed-habitat molluscan-shell fragments, glauconite, and volcanic grains. This sand sheet is diachronous across the shelf as an onlap pattern in seismic records. At present, the condensed layer or submarine hiatus, which is recognized by the distribution of relict sediment above the sand sheet, is generated in response to a very low sedimentation rate in this offshore area. The main part of the condensed section of Vail *et al.* (1984) is correlative with the transgressive sand in the outer shelf to upper slope areas, with the presence of scattered pebbles or faunal mixing horizons within the condensed section being indicative of its relationship with the transgressive sand sheet. In the inner shelf, shallower than 50–70 m, the sand sheet is reworked by present-day waves.

## ACKNOWLEDGEMENTS

I am indebted to Dr David Macdonald of the British Antarctic Survey, UK, for encouraging me to submit my paper for this special publication, and for his helpful reviews of the manuscript. I thank Dr Richard N. Hiscott of the Department of Earth Sciences, Memorial University of Newfoundland, Canada and Professor Kiyotaka Chinzei of Kyoto University, Japan for their critical readings of an early version of the manuscript, and acknowledge the reviews by Dr A.D. Reynolds of British Petroleum and Dr. S.P. Hesselbo of the University of Oxford.

## REFERENCES

Aziz, B.H. (1986) Coarse fraction analysis of the Sendai Bay sea-bottom sediments: a preliminary study. *Technical Report: Group Training Course in Offshore Prospecting*, pp. 71–88. Japan International Co-operation Agency, Tokyo.

Belknap, D.F. & Kraft, J.C. (1985) Influence of antecedent geology on stratigraphic preservation potential and evolution of Delaware's barrier systems. *Mar. Geol.* **63**, 235–262.

Berger, W.H., Killingley, J.S. & Vincent, E. (1987) Time scale of the Wisconsin/Holocene transition: oxygen isotope record in the western equatorial Pacific. *Quaternary Res.* **28**, 295–306.

Bloom, A.L. & Yonekura, N. (1985) Coastal terraces generated by sea-level change and tectonic uplift. In: *Models in Geomorphology* (Ed. Woldenberg, M.J.) pp. 139–154. Allen & Unwin, Boston, MA.

Bloom, A.L., Broeker, W.S., Chappell, J.M.A.,

Matthews, R.K. & Mesolella, K.J. (1974) Quaternary sea level fluctuations on a tectonic coast: New $^{230}Th/^{234}U$ dates from the Huon Peninsula, New Guinea. *Quaternary Res.* **4**, 185–205.

Bruun, P. (1962) Sea level rise as a cause of shoreface erosion. *Proc. Am. Soc. Civ. Eng., J. Waterways Harbors Div.* **88**, 117–130.

Clark, J.A., Farrell, W.E. & Peltier, W.R. (1978) Global changes in post-glacial sea level: a numerical calculation. *Quaternary Res.* **9**, 25–287.

Demarest, J.M. & Kraft, J.C. (1987) Stratigraphic record of Quaternary sea levels: implications for more ancient strata. In: *Sea-Level Fluctuation and Coastal Evolution* (Eds Nummedal, D., Pilkey, O.H. & Howard, J.D.). Soc. Econ. Paleontol. Mineral., Spec. Publ. 41, 223–240.

Fujii, S. & Nasu, N. (Eds) (1988) *Submerged Forest (Kaitei-rin)*. Tokyo University Press, Tokyo, 163 pp (in Japanese).

Hands, E.B. (1983) The Great Lakes as a test model for profile responses to sea level changes. In: *CRC Handbook of Coastal Processes and Erosion* (Ed. Komer, P.D.) pp. 167–189. CRC Press, Boca Raton, FL.

Haq, B.U., Hardenbol, J. & Vail, P.R. (1987) Chronology of fluctuating sea levels since the Triassic. *Science* **235**, 1156–1167.

Hattori, M. (1967) Recent sediments of Sendai Bay, Miyagi Prefecture, Japan. *Sci. Rep. Tohoku Univ. Ser. 2 (Geol.)* **39**, 1–61.

Honza, E., Inoue, E., Arita, M. & Ishihara, T. (1982) Activities of marine geological survey around Japan. *Chishitsu News* **331**, 13–35 (in Japanese).

Imbrie, J., Hays, J.D., Martinson, D.G., MacIntyre, A., Mix, A.C., Morley, J.J., Pisias, N.G., Prell, W.L. & Shackleton, N.J. (1984) The orbital theory of Pleistocene climate; support from a revised chronology of the marine delta $^{18}O$ record. In: *Milankovitch and Climate, Part 1* (Eds Berger, A.L., Imbrie, J., Hays, J., Kukla, G. & Saltzman, B.) pp. 269–305. D. Reidel, Dordrecht.

Itoh, H. (1989) *Recent ostracods from Sendai Bay, northeast Japan.* Graduation thesis, Shizuoka University, 83 pp.

Kikuchi, T. (1977) Pleistocene sea level changes and tectonic movements in the Boso Peninsula, central Japan. *Geogr. Rep., Tokyo Metro. Univ.* **12**, 77–103.

Kikuchi, T. (1988) Paleo sea-levels since the last interglacial — a mathematical method estimating for uplift rate of marine terraces and paleo sea-levels. *Trans. Jpn. Geomorphol. Union* **9**, 81–104 (in Japanese with English abstract).

Kraft, J.C., Chrzasowski, M.J., Belknap, D.F., Toscano, M.A. & Fletcher, C.H. III (1987) The transgressive barrier–lagoon coast of Delaware: morphostratigraphy, sedimentary sequences and responses to relative rise in sea level. In: *Sea-Level Fluctuation and Coastal Evolution* (Eds Nummedal, D., Pilkey, O.H. & Howard, J.D.). Soc. Econ. Paleontol. Mineral., Spec. Publ. 41, 129–143.

Leatherman, S.P. (1987) Beach and shoreface response to sea-level rise: Ocean City, Maryland, U.S.A.. *Prog. Oceanogr.* **18**, 139–149.

Lee, H.J., Jeong, K.S., Han, S.J. & Bahk, K.S. (1988) Heavy minerals indicative of Holocene transgression in the southeastern Yellow Sea. *Continent. Shelf Res.* **8**, 255–266.

Loutit, T.S., Hardenbol, J., Vail, P.R. & Baum, G.R. (1988) Condensed sections: the key to age dating and correlation of continental margin sequences. In: *Sea-Level Changes — an Integrated Approach* (Eds Wilgus, C.K., Hastings, B.S., Kendall, C.G.St.C., Posamentier, H.W., Ross, C.A. & van Wagoner, J.). Soc. Econ. Paleontol. Mineral., Spec. Publ. 42, 183–213.

Maenaka, K., Yokoyama, T. & Ishida, S. (1977) Paleomagnetic stratigraphy and biostratigraphy of the Plio-Pleistocene in the Kinki district, Japan. *Quaternary Res.* **7**, 341–362.

Maldonado, A., Swift, D.J.P., Young, R., Han, G., Nittrouer, C.A., Demaster, D.J., Rey, J., Palomo, C., Accosta, J., Ballester, A. & Castellvi, J. (1983) Sedimentation on the Valencia continental shelf: preliminary results. *Continent. Shelf Res.* **2**, 195–211.

Matsumoto, H. (1981) Sea-level changes during Holocene and geomorphic developments of the Sendai coastal plain, northeast Japan. *Geogr. Rev. Jpn.* **54**, 72–85 (in Japanese with English abstract).

Matsumoto, H. (1984) Beach ridges on Holocene coastal plains on northeast Japan. *Geogr. Rev. Jpn. Ser. A* **57**, 720–738 (in Japanese with English abstract).

Matsumoto, H. (1988) Morphology of the base of Chusekiso in the Sendai Coastal Plain. In: *The Study on the Formation of Alluvial Plain and Alluvium in Japan and the Relationship with Natural Environments in the Late Quaternary* (Ed. Iseki, H.) pp. 53–58. Ministry of Education, Tokyo, Survey no. (A) 61302084 (in Japanese).

Matsushima, Y. (1987) Holocene relative sea-level changes in the Tama and Tsurumi River lowlands. In: *The Geological Studies on the Alluvial Deposits in the Tama and Tsurumi River Lowlands, Kawasaki* (Ed. Matsushima, Y.) pp. 125–132. Kawasaki City, 145 pp (in Japanese).

Mitchum, R.M. Jr., Vail, P.R. & Sangree, J.B. (1977) Seismic stratigraphy and global changes of sea level, part 6: stratigraphic interpretation of seismic reflection patterns in depositional sequences. In: *Seismic Stratigraphy – Applications to Hydrocarbon Exploration* (Ed. Payton, C.E.). Mem., Am. Assoc. Petrol. Geol. 26 117–133.

Moriwaki, H. (1978) Problems concerning Holocene sea-level changes. *Geogr. Rep., Tokyo Metro. Univ.* **13**, 49–64.

Nummedal, D. & Swift, D.J.P. (1987) Transgressive stratigraphy at sequence-bounding unconformities: some principles derived from Holocene and Cretaceous examples. In: *Sea-Level Fluctuation and Coastal Evolution* (Eds Nummedal, D., Pilkey, O.H. & Howard, J.D.). Soc. Econ. Paleontol. Mineral., Spec. Publ. 41, 241–260.

Nummedal, D., Pilkey, O.H. & Howard, J.D. (Eds) (1987) *Sea-Level Fluctuation and Coastal Evolution*. Soc. Econ. Paleontol. Mineral., Spec. Publ. 41, 267 pp.

Okamura, Y. (1989) Multi-layered progradational sequences in the shelf and shelf slope of the Southwest Japan Forearc. In: *Sedimentary Facies in the Active Plate Margin* (Eds Taira, A. & Masuda, F.) pp. 295–317. Terra Sci. Publ., Tokyo.

PENLAND, S. & SUTER, J.R. (1983) Transgressive coastal facies preserved in barrier island arc retreat paths in the Mississippi River delta plain. *Trans. Gulf Coast Assoc. Geol. Soc.* **XXXIII**, 367–382.

SAITO, Y. (1987) Depositional models of Holocene coastal marine deposits. *Gekkan Chikyu (Earth Monthly)* **9**, 533–541 (in Japanese).

SAITO, Y. (1988) On the relationships among coastal topography and sediments, and wave base. *Gekkan Chikyu (Earth Monthly)* **10**, 458–466 (in Japanese).

SAITO, Y. (1989a) Modern storm deposits in the inner shelf and their recurrence intervals, Sendai Bay, northeast Japan. In: *Sedimentary Facies in the Active Plate Margin* (Eds Taira, A. & Masuda, F.) pp. 331–344. Terra Sci. Publ., Tokyo.

SAITO, Y. (1989b) Storm-built sand ridges in the inner shelf of Kashima-Nada, northeast Japan. In: *Sedimentary Facies in the Active Plate Margin* (Eds Taira, A. & Masuda, F.) pp. 319–330. Terra Sci. Publ., Tokyo.

SAITO, Y. (1989c) Shelf sediments off Akita, Northeast Japan. In: *Submarine Topography and Sediments Offshore Akita: The Report of Anthropogenic Influence of the Sedimentary Regime of an Open Type Bay, 1987FY Study, off Akita*, pp. 14–70. Geological Survey Japan, Tokyo, 100 pp (in Japanese).

SAITO, Y. (1989d) Late Pleistocene coastal sediments, drainage patterns, and sand ridge systems on the shelf off Sendai, northeast Japan. *Mar. Geol.* **89**, 229–244.

SAITO, Y., KASHIMA, K. & MATSUMOTO, E. (1989a) Sea level of the last glaciation maximum based on in-place sediments on the shelf off Sendai, northeast Japan. *Quaternary Res. Jpn. (Daiyonki-Kenkyu)* **28**, 111–119.

SAITO, Y., NISHIMURA, A. & MATSUMOTO, E. (1989b) Transgressive sand sheet covering the shelf and the upper slope off Sendai, northeast Japan. *Mar. Geol.* **89**, 245–258.

SHACKLETON, N.J. & OPDYKE, N.D. (1973) Oxygen isotope and paleomagnetic stratigraphy of equatorial Pacific core V28–238: oxygen isotope temperature and ice volume on a $10^5$ year scale. *Quaternary Res.* **3**, 39–55.

SHACKLETON, N.J., IMBRIE, J. & HALL, M.A. (1983) Oxygen and carbon record of East Pacific core V19–30: implications for the formation of deep water in the late Pleistocene North Atlantic. *Earth Planet. Sci. Lett.* **65**, 233–244.

SUGAWARA, K., KOBUNE, K., SASAKI, H., HASHIMOTO, N., KANEYAMA, Y. & NARITA, A. (1986) Wave statistics with 15-years data in the coastal wave observation network. *Tech. Note, Port Harbor Res. Inst.* **554**, 872 pp (in Japanese with English abstract).

SUNAMURA, T. (1987) Coastal cliff erosion in Nii-Jima Island, Japan. In: *International Geomorphology 1986, Part 1* (Ed. Gardinar, V.) pp. 1192–1212. Wiley, New York.

SUZUKI, T. & SAITO, Y. (1988) Heavy mineral composition and provenance of marine sediments in Sendai Bay, northeast Honshu, Japan. *Bull. Geol. Surv. Jpn.* **39**, 643–660 (in Japanese with English abstract).

SWIFT, D.J.P. (1968) Coastal erosion and transgressive stratigraphy. *J. Geol.* **7**, 444–456.

SWIFT, D.J.P., STANLEY, D.J. & CURRAY, J.R. (1971) Relict sediments on continental shelves: a reconstruction. *J. Geol.* **79**, 322–346.

VAIL, P.R., MITCHUM, R.M. JR. & THOMPSON, S. III (1977a) Seismic stratigraphy and global changes of sea level, part 3: relative changes of sea level from coastal onlap. In: *Seismic Stratigraphy – Applications to Hydrocarbon Exploration* (Ed. Payton, C.E.). Mem., Am. Assoc. Petrol. Geol. 26, 63–81.

VAIL, P.R., MITCHUM, R.M. JR. & THOMPSON, S. III (1977b) Seismic stratigraphy and global changes of sea level, part 4: global cycles of relative changes of sea level. In: *Seismic Stratigraphy – Applications to Hydrocarbon Exploration* (Ed. Payton, C.E.). Mem., Am. Assoc. Petrol. Geol. 26, 83–97.

VAIL, P.R., HARDENBOL, J. & TODD, R.G. (1984) Jurassic unconformities, chronostratigraphy, and sea-level changes from seismic stratigraphy and biostratigraphy. In: *Interregional Unconformities and Hydrocarbon Accumulation* (Ed. Schlee, J.S.). Mem., Am. Assoc. Petrol. Geol. 36, 129–144.

WALCOTT, R.I. (1972) Past sea levels, eustasy and deformation of the earth. *Quaternary Res.* **2**, 1–14.

WILGUS, C.K., HASTINGS, B.S., KENDALL, C.G.St.C. POSAMENTIER, H., ROSS, C.A. & VAN WAGONER, J. (Eds) (1988) *Sea-Level Changes – an Integrated Approach*. Soc. Econ. Paleontol. Mineral., Spec. Publ. 42, 407 pp.

YAJIMA, K., IMAI, M. & NISHIZAWA, K. (1982) 1:1,000,000 bathymetric charts 'Hokkaido', 'North-East Nippon' and submarine topography. *Rep. Hydrogr. Res.* **17**, 93–162 (in Japanese with English abstract).

YANAGISAWA, Y., NAKAMURA, K., SUZUKI, Y., SAWAMURA, K., YOSHIDA, S., TANAKA, Y., HONDA, Y. & TANAHASHI, M. (1989) Biostratigraphy and hiatuses of the subsurface Tertiary sequence in Futaba area, Fukushima Prefecture, Japan. *Bull. Geol. Surv. Jpn.* **40**, 405–467 (in Japanese with English abstract).

*Spec. Publs. int. Ass. Sediment.* (1991) **12**, 151–163

# Interplay between arc tectonics and sea-level changes as revealed by sedimentation patterns in the Aleutians

MAXWELL R. DOBSON[†], DAVID W. SCHOLL[‡] *and* ANDREW J. STEVENSON[‡]

[†] *Institute of Earth Studies, University College of Wales, Aberystwyth, UK; and* [‡] *US Geological Survey, Menlo Park, California, USA*

## ABSTRACT

Intra-oceanic convergent margin evolution is broadly related to the thermal character of the subducting plate such that it is considered the principal control both in terms of volcanic arc expansion and subsequent decline. Patterns of arc sedimentation are similarly influenced by the subducting plate although sea-level oscillations, with their impact on arc erosion, transportation, and deposition of sediment, are of comparable importance. In the dynamic environment of a volcanic arc, being able to recognize eustatic signals depends on the availability of thick, well-preserved sedimentary sequences. Aleutian arc-flanking sediment sequences, seen as the Lower and Middle Series, provide only facies indications for sea-level changes, as they accumulated during a phase of arc expansion and rapid uplift, whilst the Upper Series, having largely been deposited in fore-arc basins during a period of arc subsidence, has the potential to retain detailed evidence. GLORIA images, for the Aleutian Terrace, reveal the styles of Upper Series sedimentation and intimate that glacio-eustatic activity has been the principal control.

## INTRODUCTION

Intra-oceanic convergent margins owe much of their overall tectonic form and igneous activity to the nature, complexity, and orientation of plate subduction. Response to subduction is, to a significant degree, related to the age and thermal history of the downgoing plate. Indeed, primary subduction response occurs in the form of vertical motions involving the volcanic arc, the surface manifestation of which is usually limited to the location and amount of magmatic activity. However, when considering the geological evolution of a convergent margin as a whole, and not just the volcanic arc, patterns of erosion, transport, and sedimentation are also seen to be strongly influenced by subduction response. Yet, these vertical oscillations, which are a function of the thermal character of a downgoing plate, are not the only influence on sedimentation processes, for vertical sea-level oscillations have a similar role. Any process with an oscillation period measured in millions of years has the potential to generate large-scale regional signals. Onlap-determined first-order global eustatic sea-level changes are both comparable, in time-scale terms, with subduction-driven vertical motions and have the potential to generate regional signals. These two activities, although able to operate independently of each other, might be expected to cause observably similar phenomena, such as arc-wide unconformities and discrete sediment packages. Globally significant, second and even third-order sea-level oscillations, with their shorter time-scales, might generate smaller, though potentially sharper, signatures in a convergent margin rock record. Theoretically, it should be possible to separate the effects that these two independent variables might impose on arc geology. To examine this further, one of the world's largest arcs, the Aleutians, has been selected for analysis. The high-latitude location of this convergent margin and proximity to permanent icefields is especially pertinent, for it would appear there has been a close correspondence, throughout the Cenozoic, between climatic change and sea-level movements.

The Aleutian convergent margin, a 2200 km long, 55 Ma old, mostly intra-oceanic arc, extends from the Alaska Peninsula westwards to within a few hundred kilometres of Kamchatka. It consists of a volcanic-crested cordillera 100 km wide, together with a fore-arc and trench slope varying in width from 60 to 110 km. Total relief, from the Vsevidof volcano to the trench floor, exceeds 9300 m (Fig. 1).

## EVOLUTION OF THE ARC

Precise timing of arc inception is not known, but the inferred increase in Kula – Pacific plate velocity at 56 Ma is considered by Engebretson *et al.* (1984) and Scholl *et al.* (1975) to be associated with a jump in the locus of subduction from a trench north of the region now called the Bering Sea to the present Aleutian Arc. If the age of inception of the arc can only be inferred, Kula – Pacific plate history, over the last 60 Ma, may be modelled using synchronous changes in the motions of plate pairs. Engebretson *et al.* (1984) identified several of these simple patterns. More recently, Lonsdale (1988) reconstructed the Paleogene history of the Kula – Pacific plate, whilst Scholl *et al.* (1987) used the relative motions between oceanic and continental plates, in the northern Pacific basin, to model the timing of associated phenomena, such as terrane accretion and continental growth.

The concept that subduction of increasingly younger sea floor promotes arc uplift and subduction of progressively older plate encourages subsidence was applied to the Aleutians by Delong *et al.* (1976) and adapted by Scholl *et al.* (1987). Adaptation was necessary, for although the concept of downgoing plate age influencing arc activity and elevation was an important advance in understanding the kinematics of arc behaviour, the actual ridge of the Kula – Pacific plate did not start subducting until 5 – 10 Ma ago, 35 Ma after spreading activity had ceased. Scholl *et al.* (1987) further speculated that subduction of progressively older, colder, and denser oceanic crust, following subduction of the dead ridge crest, probably dampened arc-wide volcanic activity and contributed to regional subsidence.

The direct effect on arc volcanism of subduction of the dead Kula Ridge is clouded by the roughly coeval changes in subduction direction and speed and by the initial injections of trench turbidite sediment down the Benioff zone in concert with growth of the subduction complex. The response of the Aleutian Arc to the subducting plate's physical character has been summed up elegantly by Scholl *et al.* (1987) in their fig. 9, reproduced here (Fig. 2). Several points are especially relevant (see also Table 1).

**1** The Kula – Pacific plate formed and commenced rapid spreading (20 cm/yr) 85 Ma ago.

**2** Subduction from arc inception to about 43 Ma involved old, relatively cold, dense lithosphere of Late Jurassic age.

**3** A rapid subduction rate (20 cm/yr) for the Kula – Pacific plate compensated for the low temperature, Jurassic-aged ocean lithosphere, and allowed the arc to expand areally.

**4** At 43 Ma Kula – Pacific plate spreading terminated; subduction slowed while the relative motion changed direction from north to westnorthwest. Subducted crust age jumped from Late Jurassic to Late Cretaceous.

**5** Slowing of subduction rate continued. Oblique underthrusting with subduction of increasingly younger oceanic crust of Late Cretaceous age induced confinement of magmatic activity to the arc summit line and promoted elevation through increased volcanic activity.

**6** After 43 Ma the arc volcanic summit line was above sea level. This is only a tentative date based partly on plate subduction calculations (see Engebretson *et al.*, 1984).

**7** Dead Kula Ridge reached the subduction zone after about 10 Ma and probably closer to 5 Ma; when the subduction rate was 8 cm/year. Thereafter, the subducting plate became increasingly older. A small remnant of the dead ridge has yet to be subducted (Lonsdale, 1988).

**8** Arc complex subsided following subduction of the Kula Ridge. Ridge subduction migrated westwards along the arc.

**9** Present age of subducting lithosphere is Paleocene.

## PATTERNS OF EROSION TRANSPORTATION AND SEDIMENTATION

Three rock series defined, from seismic reflection data, by major unconformities have been recognized throughout the Aleutian Arc (Scholl *et al.*, 1987). The unconformities, which may be diachronous along the arc, are seen largely as extended interruptions in sedimentation or, in the case of the

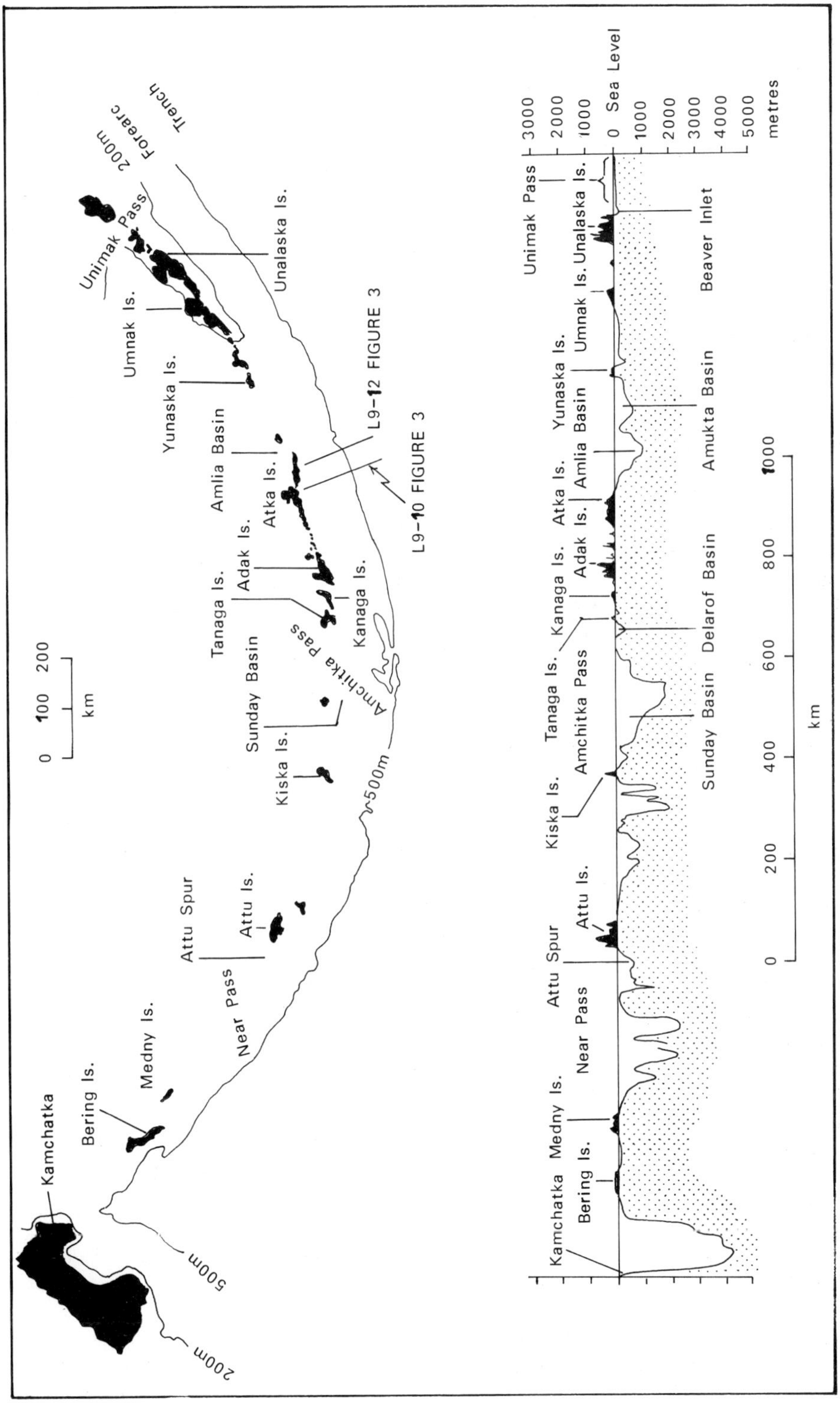


**Fig. 1.** The Aleutian Arc together with a profile along the summit to reveal the nature of the bevel.

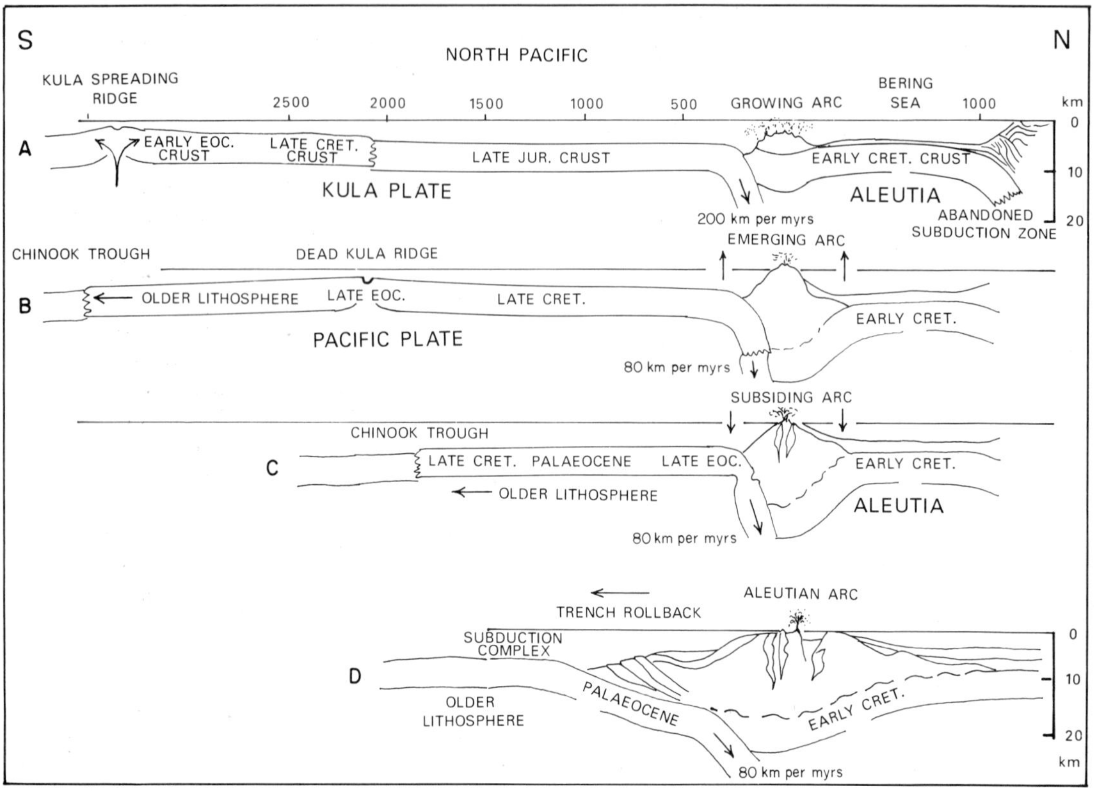


**Fig. 2.** A model of the evolution of the Aleutian Ridge based on a series of north south profiles. (A) Formation of the arc about 55 Ma. (B) End of spreading at the Kula–Pacific Ridge; this occurred about 40 Ma. (C) Subduction of the Kula–Pacific Ridge commenced after 10 Ma and continues to present. (D) Present configuration of the Aleutian Arc.

Middle–Upper Series boundary, related to the introduction of the subduction complex. Certainly, unconformities do not appear to represent periods of arc deformation by obduction as no ophiolites have been reported. Limited exposure and lack of lateral continuity of the rock series makes any attempt to erect a chronostratigraphy, for the Aleutian Arc as a whole, suspect. Table 1 should, therefore, be treated with caution and used only as a general framework reference.

### Lower Series (pre-Oligocene)

Rocks of this series range in age from inception of the arc (?55 Ma) to 37 Ma or approximately the end of the Eocene (Scholl *et al.*, 1970,1987). Most of the reliable dates for this series were obtained from lavas of the Finger Bay Volcanics of Adak, where Ar/Ar dating has provided an age greater than 50 Ma (Rubenstone, 1984), and the 35 Ma rhyolites of the Komandorsky Islands (Borsuk & Tsvetkov, 1982). Thin fossiliferous horizons of middle Eocene age have also been identified (Schmidt, 1978).

Lower Series rocks accumulated only on the crest and flanks of the expanding arc and in restricted summit basins of Paleogene age, as the fore-arc lacked both major depositional basins and a tectonically attached accretionary prism (Ryan & Scholl, 1989). This 'basement' series contains extrusive igneous sequences, notably pillow lavas, volcaniclastic suites, mostly agglomerates, and interbedded sedimentary material. The series exhibits a wide range of metamorphic grades induced by the emplacement of calc-alkaline plutons. Mixed clastic and volcaniclastic sediments become dominant in the upper part of the series. Prior to arc emergence most of the sedimentary material included laminates of chert and silt with occasional limestones; tuffaceous greywackes were also deposited. After about 43 Ma and subaerial exposure of the arc crest, sequences such as the Andrew Lake Formation, characterized by alternations of greywackes and biogenic-rich hemipelagites, were deposited in periodically sediment-starved summit basins in water depths between 200 and 500 m. Episodic turbidity current activity carried clastic material to those parts of the summit basins located below wave influence. Seismic profiles confirm that these Lower Series sediments lack lateral reflector continuity and have been truncated by the erosional surface of the summit platform (Fig. 1). The end of Lower Series time is broadly coincident with a sudden decline in eruptive activity, particularly along the flanks of the arc (Delong *et al.*, 1978).

### Middle Series (Oligocene–Miocene)

Rocks included in the Middle Series have been placed in the Oligocene and Miocene. Reliable dates can only be obtained from intrusive sills and small plutons and range from about 33–30 Ma (Citron *et al.*, 1980; Hein & McLean, 1979) to 5 Ma. (Marlow *et al.*, 1973; Perfit & Lawrence, 1979; Perfit *et al.*, 1980). The oldest known sedimentary rocks are of early Oligocene age. These data are derived from rock samples collected offshore Atka and Amlia islands (Scholl *et al.*, 1987). The Middle Series sediments were derived from the arc during a period of protracted emergence and erosion. Sequences preserved on the islands consist of shallow water deposits that are largely unmetamorphosed. They form dip-slope blanket sequences 1–3 km thick. The Nevidiskov Formation near the base of this series (Table 1) is interpreted to consist of prograding delta lobes that promoted large-scale slumps and proximal turbidites. Several other formations, including the Nilkol'ski, Chuniksak, and Chirikof, are shallow water, frequently coarse-grained marine sediments, with a shelly fauna and plant fragments. Debris flows and proximal turbidites interbedded with pelagic sediments characterize the Dutch Harbor Formation. An unconformity within the Middle Series at about 10 Ma, but recognized only on the Near Islands (Gates *et al.*, 1971), separates lower marine sediments from upper, mainly non-marine, fluviatile deposits. Non-marine deposits include the Upper Miocene Vodopad suite, the Massacre Bay and Faneto Formations of late Miocene age and the Upper Miocene Martin Harbor Formation. These widely developed, commonly red sediments, containing trunks of trees, are terminated by a major unconformity that marks the end of Middle Series sedimentation.

### Upper Series (Pliocene to Present)

This series, unlike the previous two, largely accumulated in well-developed fore-arc and summit basins that were created following major changes involving arc fragmentation and the emplacement of a subduction complex. Consequently, Upper Series sediments rest on the Middle Series with marked unconformity, except in those summit basins that were formed in the early Tertiary and escaped later deformation. The Upper Series consists almost exclusively of sediments that in the fore-arc basins are 3–4 km thick and in the summit basins are 1–3 km thick. Trench sediments, 2–3 km thick and Pacific-plate pelagic sediments 100 m thick, are included within the Upper Series as they both contribute to the 8 km thick subduction complex.

#### *Summit basins*

Geist *et al.* (1987) divided the Upper Series in the summit basins into three major units, each separated by depositional dip divergence. Individual units are composed of several distinct subunits; the basal unit appears to represent very rapid basin infilling. Most bedding appears crude and chaotic on seismic records and is poorly imaged on seismic records, inferring continuing rapid sediment influxes. The upper unit, is thick, well bedded, occasionally clinoform or progradational and may be divided into three further units, on the basis of wedge dip divergence, with the uppermost subunit seen as a pelagic veneer. The wedge-bedded units are thought to consist of turbidite–pelagic couplets.

#### *Fore-arc basins*

Evidence from DSDP boreholes 186 and 187 (Stewart, 1978) indicates that the basinal sequences consist of laminated sands (10%) and silts (40%) with a high pelagic, largely biogenic, component. Coarse sandy horizons are thought to represent distal turbidites. Analysis of the borehole sequences indicates that the fore-arc basinal zone, in common with all parts of the submerged arc, received and continues to receive a continuous pelagic sediment rain, seen as background sedimentation that is punctuated by turbidity current influxes and infrequent wind-transported volcanic ash (Fig. 3).

**Table 1.** Major arc events and their possible relationship to the global sea-level curve

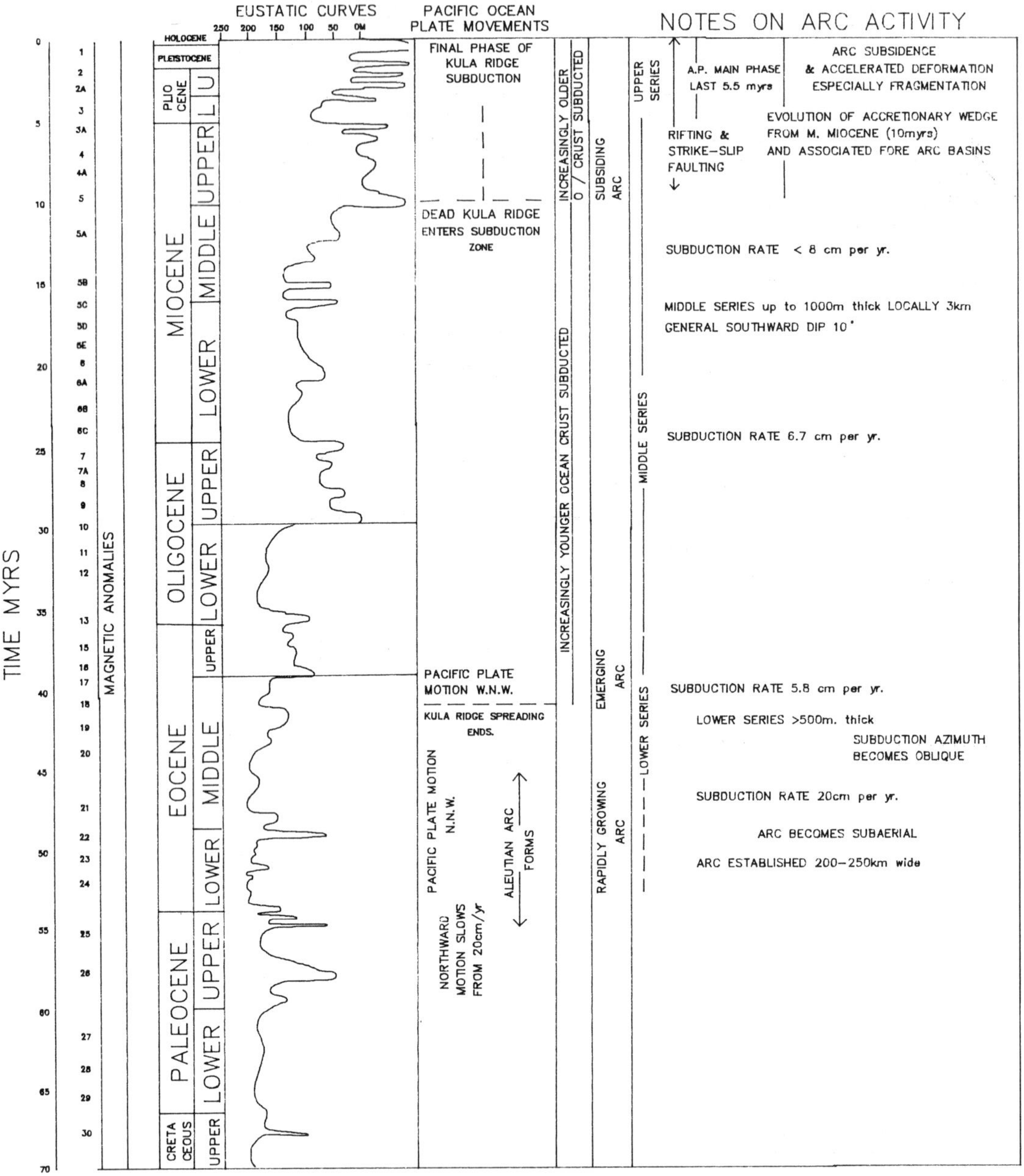

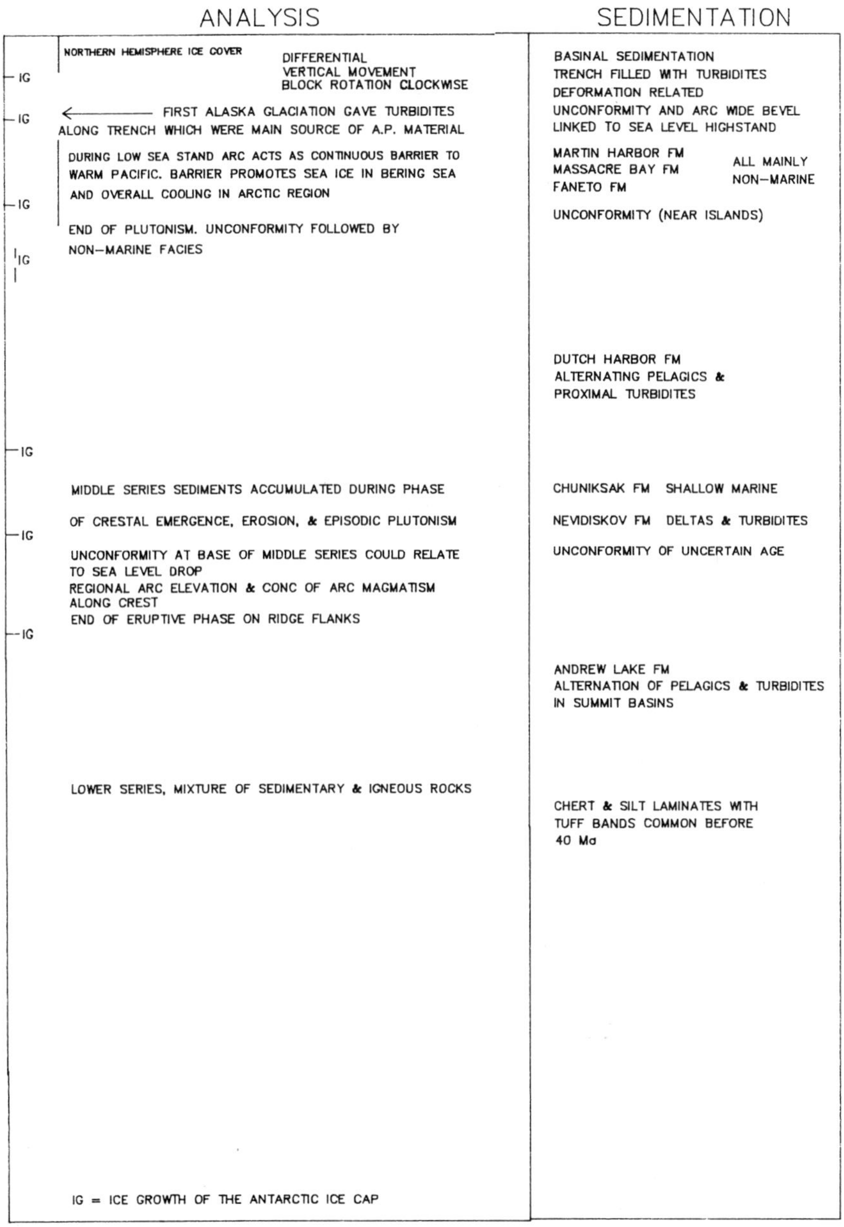
ANALYSIS
SEDIMENTATION
IG
IG
IG
IG
IG
IG
IG
NORTHERN HEMISPHERE ICE COVER
DIFFERENTIAL VERTICAL MOVEMENT BLOCK ROTATION CLOCKWISE
FIRST ALASKA GLACIATION GAVE TURBIDITES ALONG TRENCH WHICH WERE MAIN SOURCE OF A.P. MATERIAL
DURING LOW SEA STAND ARC ACTS AS CONTINUOUS BARRIER TO WARM PACIFIC. BARRIER PROMOTES SEA ICE IN BERING SEA AND OVERALL COOLING IN ARCTIC REGION
END OF PLUTONISM. UNCONFORMITY FOLLOWED BY NON-MARINE FACIES
MIDDLE SERIES SEDIMENTS ACCUMULATED DURING PHASE
OF CRESTAL EMERGENCE, EROSION, & EPISODIC PLUTONISM
UNCONFORMITY AT BASE OF MIDDLE SERIES COULD RELATE TO SEA LEVEL DROP
REGIONAL ARC ELEVATION & CONC OF ARC MAGMATISM ALONG CREST
END OF ERUPTIVE PHASE ON RIDGE FLANKS
LOWER SERIES, MIXTURE OF SEDIMENTARY & IGNEOUS ROCKS
IG = ICE GROWTH OF THE ANTARCTIC ICE CAP
BASINAL SEDIMENTATION
TRENCH FILLED WITH TURBIDITES
DEFORMATION RELATED
UNCONFORMITY AND ARC WIDE BEVEL LINKED TO SEA LEVEL HIGHSTAND
MARTIN HARBOR FM
MASSACRE BAY FM
FANETO FM
ALL MAINLY NON-MARINE
UNCONFORMITY (NEAR ISLANDS)
DUTCH HARBOR FM
ALTERNATING PELAGICS & PROXIMAL TURBIDITES
CHUNIKSAK FM SHALLOW MARINE
NEVIDISKOV FM DELTAS & TURBIDITES
UNCONFORMITY OF UNCERTAIN AGE
ANDREW LAKE FM
ALTERNATION OF PELAGICS & TURBIDITES IN SUMMIT BASINS
CHERT & SILT LAMINATES WITH TUFF BANDS COMMON BEFORE 40 Ma

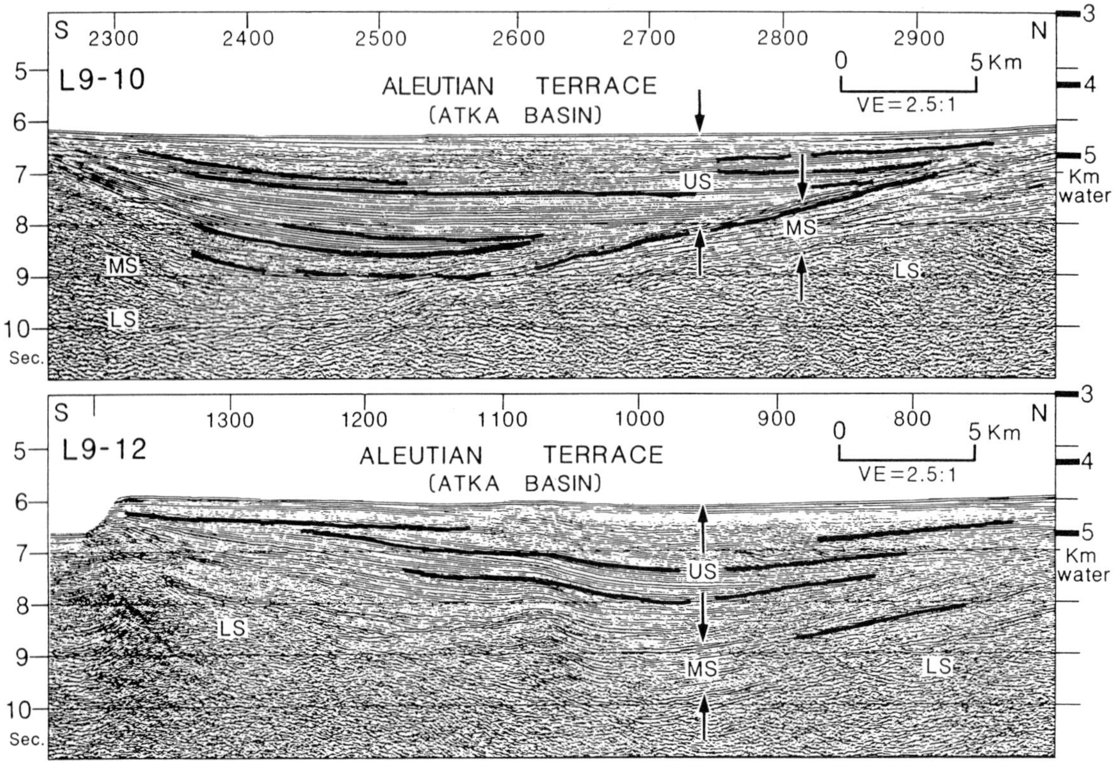


**Fig. 3.** Migrated multichannel seismic sections that traverse the Aleutian Terrace south of Amlia Island. For location see Fig. 1. Reprinted by permission of Circum-Pacific Council AAPG.

GLORIA surveys of the arc (Fig. 4), conducted by the US Geological Survey during 1987 and 1988, have revealed broad erosional zones along the shelf slope-break, seen as extensive canyon systems dominated by mass wasting. Large-scale sediment movement appears to be particularly active where bathymetric lows in the arc crestal zone provide conduits for releasing volcaniclastic material to the fore-arc (Fig. 4). Supply of sediment to the fore-arc is also a function of volcanic arc geology, seen as readiness to erode. Major plutonic masses provide little sediment to the fore-arc whereas adjacent sectors, composed largely of volcaniclastic material, are able to support major channel systems. Debris flows and slumps extend downslope to depositional zones dominated by submarine fans with braid channels and channel-overbank systems that terminate in ponded lobes. The routes taken by the channels are controlled by the location of the fore-arc basin depocentres. These bathymetric variations reflect, in part, earlier patterns of sedimentation and fault activity (Ryan & Scholl, 1989). Thus stacked and presently abandoned channel-overbank complexes occur as bathymetric highs. Lows or sites of active sedimentation appear locally to have suffered inversion in response to strike-slip faulting. Abandoned fans are frequently affected by slides, the failure sites of which become reoccupied by channel systems. Airgun profile analysis reveals that the channel zones exhibit high-amplitude signals with lateral continuity implying well-developed stratification separated by thinned chaotic units suggestive of rapid sedimentation. The channels are usually wide (>5 km) but short (<20 km), terminating in a series of fan lobes or sediment ponds.

Thus, in fore-arc basins, seismic profiles record no major erosion intervals or unconformities although gentle dip divergence is present along some basin margins. Dip divergence in the Atka Basin, for example, may reflect arc fragmentation caused by strike-slip tectonics. Sedimentation rates for the Upper Series in the Atka Basin (Fig. 3) appear to be

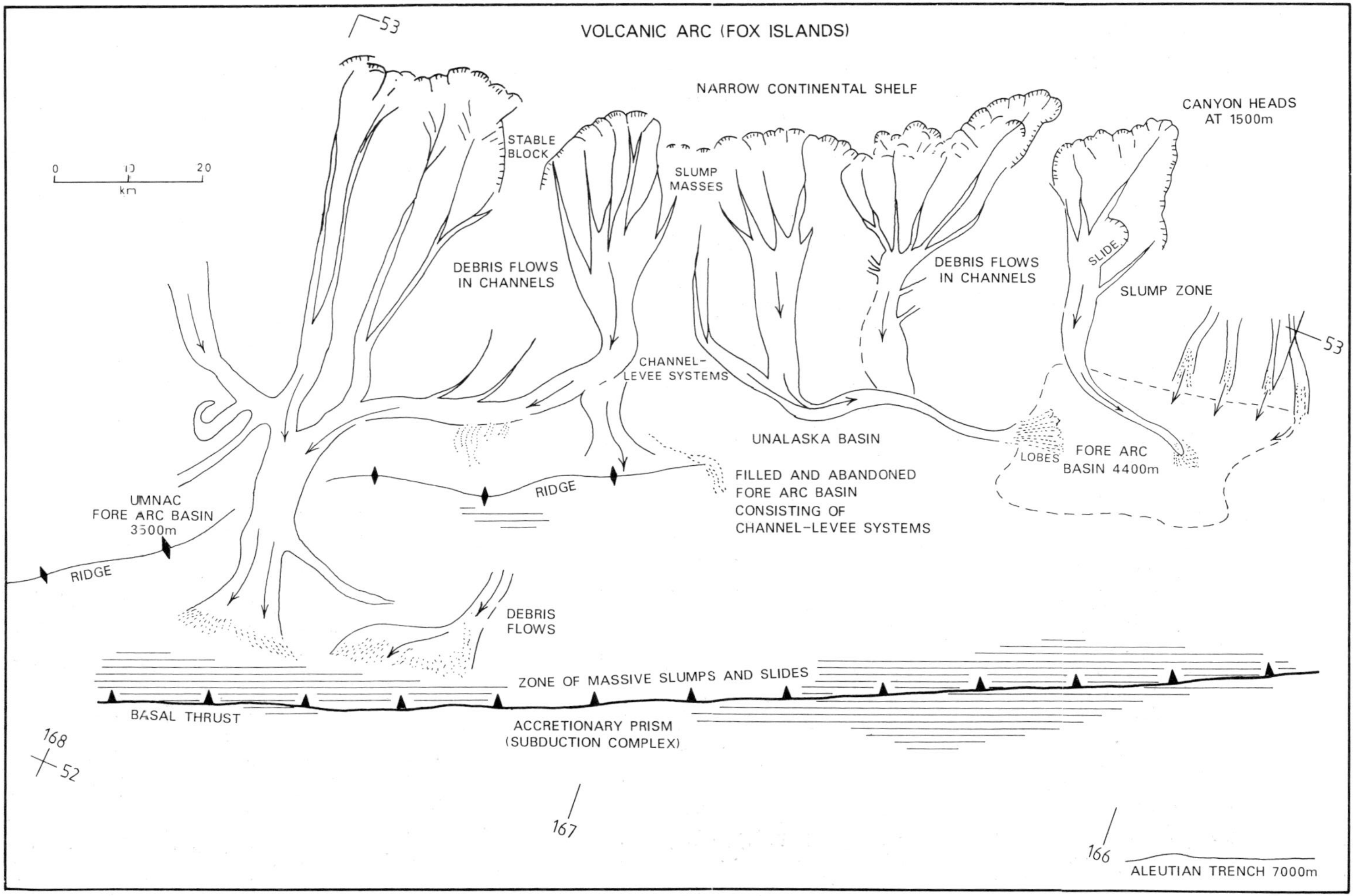


**Fig. 4.** Region of the Aleutian Terrace south of Unalaska and Umnak Islands. The interpretation is based on GLORIA data and single-channel reflection profiles. The area extends from the shelf of the volcanic arc to the Aleutian Trench.

particularly high. Based on analysis of seismic profiles the average rate for the last 5 Ma has been 72 cm/1000 yr. This compares with an average channel system thickness of 500 m, which at the above rate of sedimentation represents sustained deposition for more than 0.5 Ma.

### *Trench and Pacific plate sediments*

Migrated multichannel seismic profiles (see L5–5 line in Scholl *et al.*, 1987) reveal a sediment sequence 2–3 km thick in the Aleutian trench immediately south of Atka Island and the Atka fore-arc basin. The sequence appears to extend the whole length of the trench. Single-channel profiles collected during the GLORIA survey show laterally continuous parallel reflectors. Slumping into the trench is rare with only isolated events recorded from both the Pacific plate side and the accretionary wedge. GLORIA images reveal a uniformly dark grey reflection response. In detail, the trench sequences consist of packages of reflectors up to 500 m thick separated by transparent layers. Pelagic sediments on the Pacific plate show closely spaced clear reflectors up to 50 km from the trench, thereafter normal fault patterns associated with the outer trench high and flexing of the down-going plate create a chaotic sediment layer overlying the disrupted ocean floor.

### *The arc-wide bevel*

Much of the crestal zone of the volcanic arc is a platform 100 km wide and up to 200 m deep supporting late Cenozoic volcanic cones (Fig. 1). It is believed that the platform is of late Cenozoic age and was cut by prolonged wave-base erosion. Since it was cut, sections are thought to have subsided below 1500 m whilst other sections have been elevated above present sea level (Scholl *et al.*, 1987).

## ARC EVOLUTION IN RELATION TO VERTICAL MOTION AND SEA-LEVEL CHANGES

### Lower Series (Eocene)

The 20 Ma period from about 55 to 36 Ma was characterized by northward movement of the Kula–Pacific plate at 20 cm/yr. It is this rapid plate motion that is considered directly responsible for the generation of a volcanic arc 5 km high and 100 km wide. At about 43 Ma the arc reached the sea surface, the plate vector shifted to westnorthwest and Kula–Pacific plate spreading ended. The subduction rate also slowed. The arc build-up rate was probably slow: 5 km over 20 Ma translates into an average rate of 25 cm/1000 yr implying that sustained global sea-level changes should register in any preserved sediment sequence. An unconformity, recognized throughout the Aleutians as a major sedimentation break, marks the end of the Lower Series. Haq *et al.* (1987) show a global lowstand of only about 50 m at 36 Ma, which generated fan deposits; it is interpreted as a medium sequence boundary. The nearest, in time, major sequence boundary occurs at 30 Ma; this is the principal Oligocene eustatic event involving a sea-level drop of 175–200 m. As the 36 Ma age is only an approximation, for the Lower–Middle Series boundary, and with the oldest pluton dates for the Middle Series close to 30 Ma, it is not unreasonable to link the unconformity with the principal Oligocene sea-level fall.

### Middle Series (Oligocene–Miocene)

This 30 Ma long phase witnessed the continuing subduction of the Pacific plate. The Kula Plate effectively 'died' at 42 Ma; its dead spreading axis reached the arc by 5 Ma ago at the latest. Thereafter, cooler, older plate was subducted. However, it is not possible to make assumptions about elevation that involves the whole arc, as subduction of the dead Kula Ridge was diachronous with a small part of the old spreading axis yet to be subducted (Lonsdale, 1988). The first 20 Ma of this period is seen as one of protracted arc emergence and erosion. During the Oligocene, and particularly following the low sea-level stand at 30 Ma, there was a persistent, though fluctuating global sea-level rise up to 15 Ma ago. As continuing arc uplift coincided with this global rise of sea level, any eustatic impact on the sediment sequence would be confused or even nullified. From 14 to 10 Ma ago sea level fell sharply. The coincidence of continuing arc uplift at a time of rapidly falling sea level resulted in the accumulation of shallow-water sediments. This marked facies indicator only persisted whilst arc uplift was maintained. In fact, an unconformity is locally recorded at about 10 Ma — this may be the first sedimentation break associated with a global regression of major proportions. With arc uplift stabilized at a time of world-wide sea-level lowering, non-marine, red, wood-bearing sediments accumulated over a 5 Ma period. Significantly, subduction of the Kula – Pacific spreading axis was largely completed by about 5 Ma and arc uplift broadly ceased.

The major unconformity at the end of Middle Series time, essentially at the end of the Miocene, although seen as separating the Middle from the Upper Series, occurred during a period of change that affected the whole arc. In particular, the volcanic islands were extensively bevelled (see Fig. 1). It may be assumed that bevelling of the summit platform represents planation during a period of eustatic highstand. Certainly modern planations, by the post-glacial rising sea level, are recorded around all the arc islands (Black, 1980). Dating major erosion events such as the regional bevel is notoriously difficult, but field evidence points towards the end of the Miocene. On the Haq *et al.* (1987) curve a significant global sea-level rise is recorded for 5 Ma, which was sustained for at least 1 Ma. If the end Middle Series unconformity and bevel are similar in age with the latter, owing to a sea-level highstand, it is reasonable to conclude that the former is related to tectonic activity.

### Upper Series (Pliocene to Present)

Arc subsidence, in association with the active strike-slip tectonics caused by oblique subduction that also increasingly affected the evolving subduction complex, characterizes the last 5 Ma (Ryan & Scholl, 1989). This last 5 Ma has been a period of declining volcanic activity, regional subsidence and strong earthquakes at a time of major global sea-level oscillations. Whilst the chaotic basal unit of the Upper Series may have resulted from debris eroded during the bevelling phase it is not possible to link directly identified units or subunits to sea-level oscillations or, further, to relate them in any way to glacial eustatic events. Clearly these clinoform accumulations must reflect episodes of rapid sediment influx due, in all probability, to an enigmatic mix of tectonism, volcanism, and glacially induced oscillating sea levels. GLORIA and seismic profiles clearly point to sedimentation events that are probably directly due to outbuilding on the continental slope at times of lowstands followed by a sea-level rise, quiescence, and pelagic sediment dominance. Certainly, the presence of discrete channel-levee sediment packages remains the strongest potential evidence for eustatic control of sedimentation, which, with adequate dating, should provide a record of sea-level influence during the last 5 Ma of arc history.

## DISCUSSION AND ANALYSIS

The long period of sustained arc uplift, partly promoted by the younging of the subducting plate, cannot be viewed simply as a function of volume expansion and increased elevation through intrusion emplacement. The Lower Series, especially, reveals a dominance of extrusive activity. The likelihood that this extrusive activity was pulsed rather than continuous precludes relating rates or styles of sediment deposition to other influences, such as sea-level changes. Therefore, at this stage of study of the arc, the sediment packages in the Lower Series cannot be linked directly to global sea-level oscillations. Moreover, as strike-slip tectonics have been a prominent feature of the arc since the Oligocene, it is reasonable to conclude that arc fragmentation as a function of oblique subduction has strongly influenced patterns and loci of sedimentation over the last 30 Ma at least. The unconformity between the Middle and Lower Series, despite uncertainty as to its age, is a strong candidate for a mixed origin involving both global sea-level regression during the early Oligocene, and arc-wide tectonic and volcanic activity. It is not until the significant global regression at about 10 Ma, with evidence of an arc-wide facies change to a largely non-marine sedimentary sequence preserved on the islands, that any confidence can be placed in a sustained link between arc activity and sea-level changes. Thick non-marine sequences may be seen as the result of two independent variables reinforcing each other. As these non-marine deposits have yet to be mapped offshore it is not possible to determine where sea level was at the time of deposition; at the same level as today or lower. The age of the arc-wide bevel, generally placed at the Mio-Pliocene boundary, is too poorly constrained to link it with confidence to the marked sea-level rise at 5 Ma, although judging from the Haq *et al.* (1987) curve this would appear to be the only reasonable candidate. Nevertheless, this is perhaps the single most important erosional feature recognized throughout the arc and its scale implies a phase of wave cutting that took several million years to achieve. Moreover, the degree of erosion involved indicates that a relationship between sea-level changes and sedimentation must be sought. Further, this arc-wide bevel is a valuable indicator of differential subsidence and uplift that has occurred throughout the Aleutians since cutting took place. Differential subsidence is probably due to several factors. Strike-slip fragmentation may have encouraged cooling of the arc massif. Subsidence might equally be a manifestation of strain release associated with accelerated arc fragmentation (H. Ryan, personal communication, 1989). Again the underlying mantle wedge could well have been affected by the faster underthrusting of older and colder crust during the last 5 Ma. Perhaps crestal subsidence should be related to fundamental subcrustal factors. Clearly any subduction erosion by thinning of the base of the arc could equally cause regional subsidence, especially across the fore-arc region. Mapping evidence strongly supports the view that surface loads imposed by volcanic centres have contributed little to regional subsidence. By contrast, localized uplift might be related to possible isostatic adjustment following the melting of Aleutian icecaps. The unconformity, which is widely developed, appears to be tectonic in origin; certainly seismic evidence points to profound changes throughout the arc.

In marked contrast to the initial 50 Ma of arc evolution the period of the Upper Series preserves a detailed record of arc-related activity that has yet to be resolved. It is assumed that activities associated with climatic changes dominate the last 5 Ma of arc history. These activities probably included periodic and pronounced glacial expansion in Alaska, which episodically discharged sediment into the Aleutian trench (Dobson *et al.*, in press) through turbidite processes. These thick trench sequences, in turn, fed the subduction complex. As the bulk of the arc and the subduction complex rapidly expanded an effective barrier to deep, cold-ocean circulation was created. The between-island passes admitted and continued to admit warm surface water to the Bering Sea. The Bering Sea thus remains a wintertime evaporation basin, which can supply moisture to the glaciated mountain drainages of the Gulf of Alaska. This is a climate forcing influence promoting more extensive glacial phases.

Upper Series units and subunits suggest at least five phases or pulses of sediment injection that may relate to glacio-eustatic events. Certainly the GLORIA data allow recognition of the processes at work. However, the activity of mass wasting, whilst a clear indicator of the collapse of too rapid outbuilding along steep margins typical of passive margins (such as those off New England), in the Aleutians, where there are well-documented sectors in which some of the world's largest earthquakes occur, sediment injection may not necessarily be due entirely to rapid sediment outbuilding. Single-channel seismic records point to repeated episodes of sediment injection along well-defined channel systems that could just as easily be linked to arc instability as to glacially induced eustatic changes.

A programme of sampling, dating, and seismic profiling will do much to identify second-order sea-level changes, even though the generation of a subduction complex and creation of fore-arc basins, at a time of accelerating arc fragmentation induced by strike-slip tectonics, severely reduces the opportunity to recognize eustatic influences. Inspection of a relatively young arc, such as the Aleutians, as a model for studying ancient arcs suggests that the early evolving phase, which might last up to 30 Ma, will not readily record sea-level signals that can be recognized easily. Subsequent episodes of arc activity point to the prospect of first-order eustatic signals, with unconformities, planations, and major facies changes. The opportunity to examine basinal sequences in detail might reveal second-order patterns in the sediments. Generally, vertical motion of an arc massif is due to too many interrelated processes that invariably stem from subduction activity for unequivocal eustatic patterns to be easily discerned.

## REFERENCES


Black, R.F. (1980) Isostatic and eustatic movements of sea-level in the Aleutian Islands, Alaska. In *Earth Rheology, Isostacy and Eustacy* (Ed. Nils-Axel Morner, N.-A.). Wiley, New York, 599 pp.

Borsuk, A.M., & Tsvetkov, A.A. (1982) Magmatic associations of the western part of the Aleutian Island arc. *Int. Geol. Rev.* **24**, 317–327.

Citron, G.P., Kay, R.W., Kay, S.M., Snee, W.L. & Sutter, J.F. (1980) Tectonic significance of early Oligocene plutonisim on Adak Island, Central Aleutian Islands, Alaska. *Geology* **8**, 375–379.

DeLong, S.E., Fox, P.J. & McDowell, F.W. (1978) Subduction of the Kula Ridge at the Aleutian Trench. *Geol. Soc. Am. Bull.* **89**, 83–95.

Dobson, M.R., Huggett, Q., Bruns, T., Carlson, P. & Stevenson, A.J. (1991) Placer lag deposits in submarine channels in the Gulf of Alaska. *Circum-Pacific Coun. Energy Mineral Resour. Earth Sci. Ser.* **7**, Houston, Texas (in press).

Engebretson, D.C., Cox, A. & Gordon, R.G. (1984) Relative motions between ocean plates of the Pacific Basin. *J. Geophys. Res.* **89**(10), 291–310.

Gates, O., Powers, H.A., Wilcox, R.E. & Schafer, J.P. (1971) Geology of the Near Islands, Alaska. *U.S. Geol. Surv. Bull.* **1028-U**, 709–822.

GEIST, E.L., CHILDS, J.R. & SCHOLL, D.W. (1987) Evolution and petroleum geology of Amlia and Amukta intra-arc summit basins, Aleutian Ridge, *Mar. Petrol. Geol.* **4**, 334–352.

HAQ, B.U., HARDENBOL, J. & VAIL, P.R. (1987) The new chronostratigraphic basis of Cenozoic and Mesozoic sea-level cycles. *Cushman Found. Foraminiferal Res. Spec.* Publ. **24**, 7–13.

HEIN, J.R. & MCLEAN, H. (1979) Paleogene sedimentary and volcanogenic rocks from Adak Island, central Aleutian Islands, Alaska. *U.S. Geol. Surv. Prof. Pap.* **1126-A-J**, 16 pp.

LONSDALE, P. (1988) Paleocene history of the Kula plate: offshore evidence and onshore implications. *Geol. Soc. Am. Bull.* **100**, 733–754.

MARLOW, H.S., SCHOLL, D.W., BUFFINGTON, E.C. & ALPHA, T.R. (1973) Tectonic history of the central Aleutian Arc. *Geol. Soc. Am. Bull.* **84**, 1555–1574.

PERFIT, M.R. & LAWRENCE, J.R. (1979) Oxygen isotopic evidence for meteoric water interaction with the Captains Bay pluton, Aleutian Islands. *Earth Planet. Sci. Lett.* **45**, 16–22.

PERFIT, M.R., HRUECKNER, B., LAWRENCE, H. & KAY, R.W. (1980) Trace elements and isotopic variations in a zoned pluton and associated rocks, Unalaska Island, Alaska: a model for fractionation in the Aleutian calc-alkaline suite. *Contrib. Mineral. Petrol.* **73**, 69–87.

RUBENSTONE, J.L. (1984) *Geology and geochemistry of early Tertiary submarine volcanic rocks of the Aleutian Islands and their bearing on the development of the Aleutian Island Arc.* Unpublished PhD thesis, Cornell University, New York, 349 pp.

RYAN, H. & SCHOLL, D.W. (1989) The evolution of forearc structures along an oblique convergent margin, Central Aleutian Arc. *Tectonics* **8**, 497–516.

SCHOLL, D.W., GREENE, H.G. & MARLOW, M.S. (1970) Eocene age of the Adak Paleozoic(?) rocks, Aleutian Islands, Alaska. *Geol. Soc. Am. Bull.* **81**, 3583–3592.

SCHOLL, D.W., BUFFINGTON, E.C. & MARLOW, M.S. (1975) Plate tectonics and the structural evolutions of the Aleutian–Bering Sea region. *Geol. Soc. Am. Spec. Pap.* **151**, 1–31.

SCHOLL, D.W., VALLIER, T.L. & STEVENSON, A.J. (1987) Geologic evolution and petroleum geology of the Aleutian Ridge. In: *Geology and Resource Potential of the Continental Margins of Western North America and Adjacent Ocean Basins – Beaufort Sea to Baja California*. Circum-Pacific Coun. Energy Mineral Resour. Earth Sci. Ser. 6, Houston, Texas. 123–156.

SCHMIDT, O.A. (1978) Tectonics of the Komandorsky Islands and the structure of the Aleutian Ridge. *Acad. Naut. Geol. Inst. Moscow* **320**, 100.

STEWART, R.J. (1978) Neogene volcaniclastic sediments from Atka Basin, Aleutian Ridge. *Bull. Am. Assoc. Petrol. Geol.* **62**, 87–97.

*Spec. Publs. int. Ass. Sediment.* (1991) **12**, 165–186

# Neogene interaction of tectonic and glacial processes at the Pacific margin of the Antarctic Peninsula

R.D. LARTER *and* P.F. BARKER

*British Antarctic Survey, Natural Environment Research Council, High Cross, Madingley Road, Cambridge CB3 0ET, UK*

## ABSTRACT

The Antarctic Peninsula was a magmatic arc subducting Pacific Ocean floor throughout the Mesozoic. During the Cenozoic, subduction ceased at each of a series of ridge-crest–trench collisions, which migrated northeast along the margin. Multichannel seismic profiles across the Pacific margin of the Antarctic Peninsula show evidence of post-collision uplift, followed by subsidence. During Pliocene–Pleistocene time, ice sheets have grounded out to the shelf edge at times of glacial maximum, transporting sediment that has extended the outer shelf. Subsidence of the margin has preserved a unique sedimentary record of ice-sheet advances, which provides the opportunity of looking closely at the hypothesized relationship between change in grounded ice volume and global sea-level change.

## INTRODUCTION

The exposed geology of the Antarctic Peninsula indicates that it was a magmatic arc subducting oceanic lithosphere at its Pacific margin from at least the Triassic onwards (Rex, 1976; Pankhurst, 1982; Storey & Garrett, 1985). Studies of marine magnetic anomalies have revealed a series of collisions during the Cenozoic between ridge-crest sections of the Antarctic–Phoenix (ANT–PHO) plate boundary and the trench at the Pacific margin (Herron & Tucholke, 1976; Barker, 1982). The earliest of these ridge-crest–trench (RC–T) collisions took place about 50 Ma ago at the portion of the margin adjacent to southern Alexander Island (Fig. 1). Later collisions occurred progressively farther to the northeast, the most recent being to the west of Smith Island (Fig. 1) during the Pliocene. New marine magnetic data (Fig. 2) have enabled us to define more closely the age and extent of the Neogene RC–T collisions, between the Tula and Hero fracture zones (FZ). In each collision zone, sea-floor spreading and subduction both stopped at the time of collision, trench basement topography was eliminated (Tucholke & Houtz, 1976), and the margin became passive. That subduction stopped at the time of collision was an inevitable consequence of the unusual arrangement of plates involved. Throughout the series of collisions, the trailing flank of the ridge crest and the overriding lithosphere at the trench were both part of the Antarctic plate (Fig. 3), so even before collision there was no relative motion between them.

On the three sections of the ANT–PHO ridge northeast of the Hero FZ (Fig. 1), spreading stopped or decreased to a very slow rate about 4 Ma ago, before the spreading centre had reached the trench (Barker, 1982). Subduction is thought to have continued at the South Shetland Trench, driven by the excess weight of the subducted slab. Roll-back of the hinge of subduction of the oceanic lithosphere probably enabled the opening of the Bransfield Strait back-arc basin (Barker, 1982).

During the period spanning the Antarctic Peninsula RC–T collisions there has been a progressive cooling and glaciation of the polar regions, punctuated by periods of more rapid climatic change (Savin *et al.*, 1975; Shackleton & Kennett, 1975; Kennett, 1977; Barker *et al.*, 1988). In particular there is evidence of more rapid cooling and of increases in intensity of Antarctic glaciation in the middle Miocene, earliest Pliocene, and late Pliocene (about 14 Ma, 4.8 Ma, and 2.4 Ma ago, respectively).

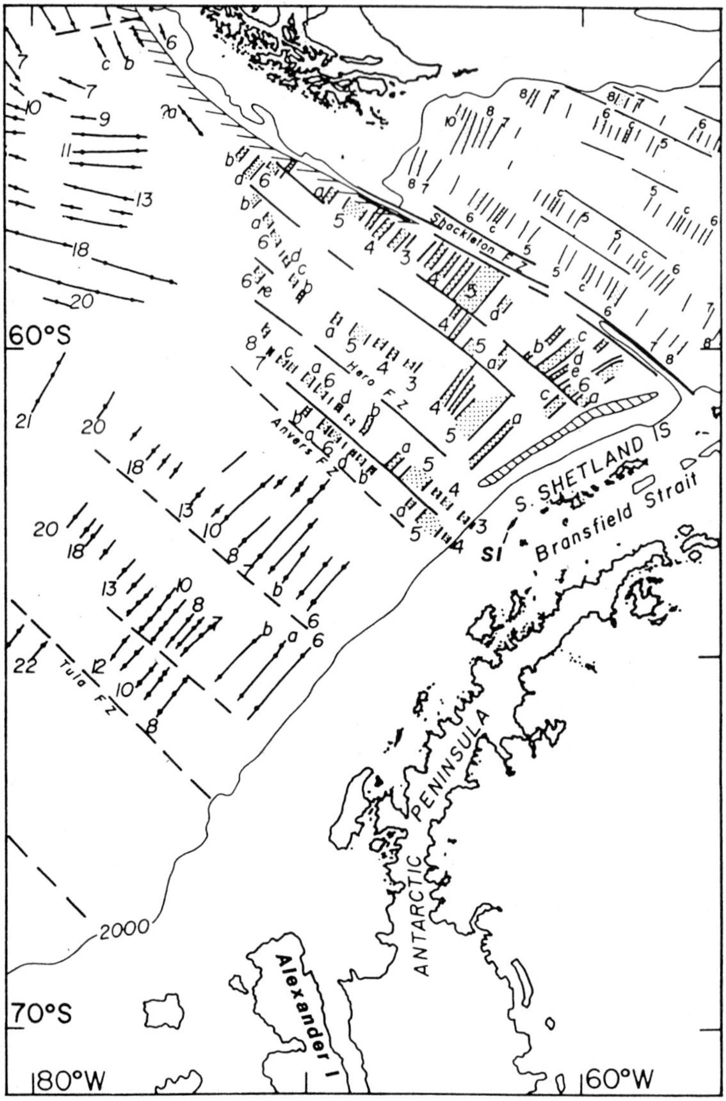


**Fig. 1.** Magnetic anomaly identifications in the southeast Pacific. Active subduction zones are cross-hatched (South Chile and South Shetland trenches); SI=Smith Island (after Barker, 1982). Reproduced by permission of the Geological Society.

During the last few million years sedimentation on the Antarctic Peninsula Pacific margin has been controlled by glacial processes rather than sea-level change (Larter & Barker, 1989). Many authors consider fluctuations in the volume of continental ice sheets to be the dominant mechanism controlling sea-level change over periods of 10 Ma or less (Vail *et al.*, 1977b; Pitman, 1978; Donovan & Jones, 1979; Beard *et al.*, 1982), so studies of stratigraphic sequences produced directly by the action of ice sheets can make an important contribution to the understanding of global sea-level change.

The purpose of this paper is to use multichannel seismic reflection (MCS) profiles to examine Neogene RC–T collision effects and glacial processes on the Pacific margin of the Antarctic Peninsula, and present a model to explain how they may have interacted to produce the observed pattern of depositional sequences.

## Ridge-crest–trench collisions

Ridge-crest–trench collisions probably occur several times during the life cycle of most large ocean basins.

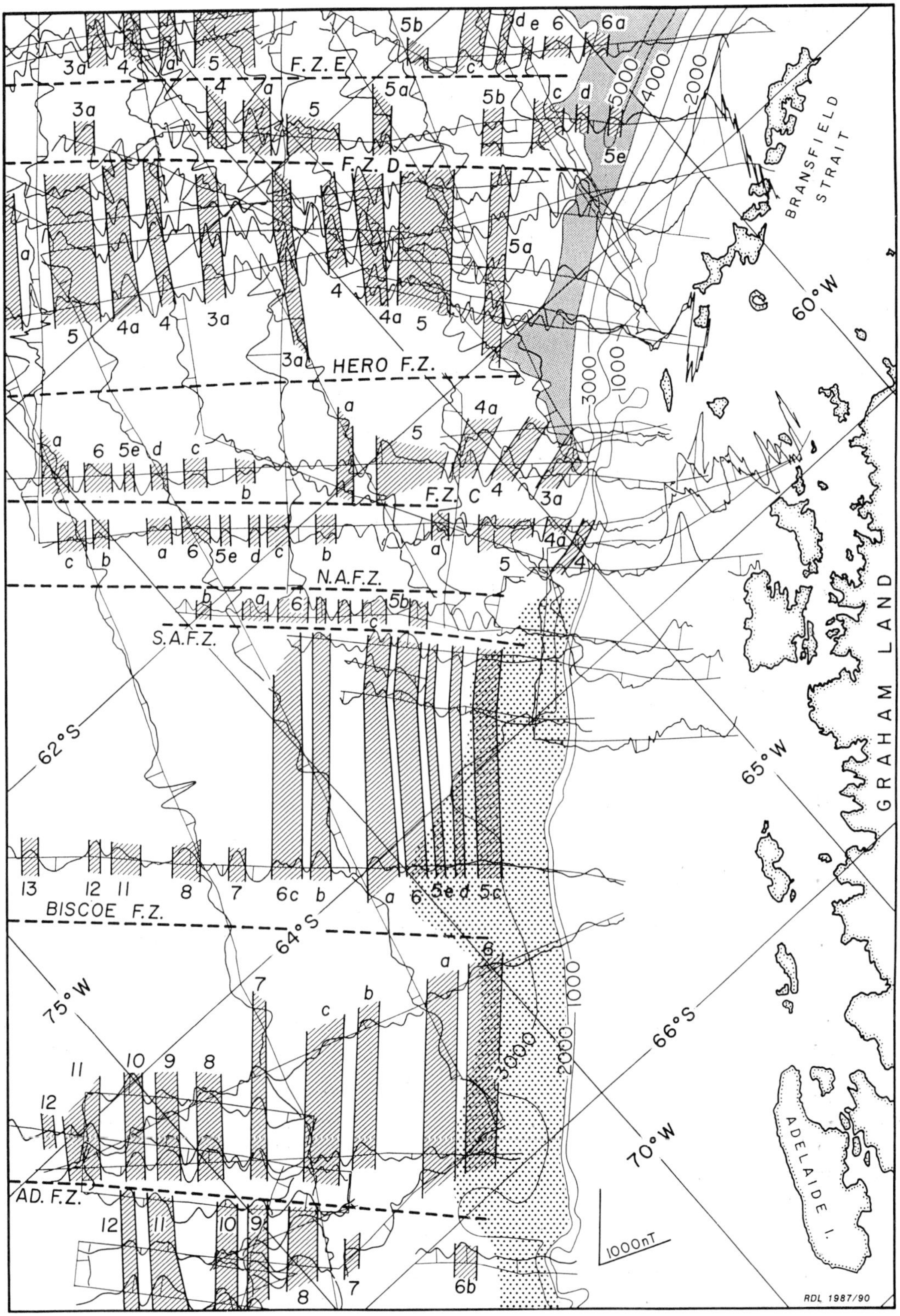


**Fig. 2.** Magnetic anomalies projected perpendicular to mean track directions and magnetic anomaly identifications (cross-hatched) near the Pacific margin of the Antarctic Peninsula. Bathymetric contours shown on margin at 1000 m intervals. Fine stipple is the South Shetland Trench (>4000 m). Coarse stipple shows the approximate extent of the magnetic quiet zone. N.A.F.Z.=North Anvers Fracture Zone, S.A.F.Z.=South Anvers Fracture Zone, AD.F.Z.=Adelaide Fracture Zone.

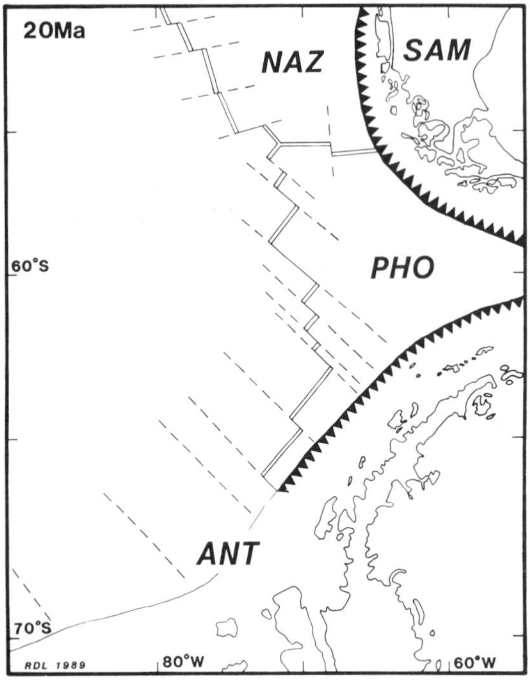


**Fig. 3.** Reconstruction of the southeast Pacific 20 Ma ago, in the Antarctic reference frame. South American–Antarctic rotation parameters from Barker & Lawver (1988).

However, in the world today there are only a small number of areas where RC–T collision is happening now, or has taken place in the last few million years.

On theoretical grounds many authors have predicted diverse effects of collisions on the overriding plate, and there seems little doubt that an RC–T collision represents a major crisis in the history of an active margin. The predicted effects of RC–T collision on the fore-arc and arc can be divided into:

**1** thermal effects, which may be expected to occur whenever a recently active spreading centre is subducted;

**2** other effects, which may occur in some RC–T collision events and not in others.

Thermal effects include thermal uplift and subsequent subsidence, regional thermal metamorphism, hydrothermal activity, and possibly also the hiatus in arc magmatism, which appears to be associated with RC–T collision (DeLong & Fox, 1977; DeLong *et al.*, 1979; Barker, 1982; Cande *et al.*, 1987). The second category includes tectonic erosion (Barker, 1982; Cande & Leslie, 1986; Cande *et al.*, 1987), ophiolite obduction (Christensen & Salisbury, 1975; Dewey, 1976; Forsythe *et al.*, 1986), anomalous magmatic activity close to the trench (Marshak & Karig, 1977; Forsythe *et al.*, 1986), and back-arc extension (Uyeda & Miyashiro, 1974).

The remainder of this paper is mainly concerned with the cycle of thermal uplift and subsidence. DeLong *et al.* (1978) suggested a simple way of modelling the uplift of a magmatic arc and fore-arc as an active spreading centre is subducted. They assumed that as the topographic profile of a spreading ridge is thermally induced and in isostatic equilibrium, it will maintain its form through RC–T collision. In this case uplift of the subducting margin is directly related to the age of ocean floor beneath. Applying this model to the collision of the Kula Ridge with the Aleutian Trench, DeLong *et al.* (1978) suggested an uplift–subsidence cycle of about 500 m amplitude over a period of about 50 Ma. Although more recent work now suggests that an actively spreading ridge may not have been subducted beneath the Aleutian Trench (Byrne, 1979; Lonsdale & Smith, 1986; Engebretson *et al.*, 1985), this does not invalidate the model for the general case of RC–T collision in which the ridge was active up to collision time and subduction continued after collision.

The same model can be applied to RC–T collisions such as those along the Antarctic Peninsula, where subduction ceased at collision time, if it is assumed that the oceanic lithosphere emplaced beneath the fore-arc cools and subsides like normal ocean floor. On this basis slower post-collision subsidence would be expected than in areas where subduction continued after collision. In fact the model predicts a post-collision subsidence history very similar to that of a young passive margin.

### Antarctic continental margins

Typically, Antarctic continental margins differ from temperate and low-latitude margins in several important respects.

**1** Antarctic continental shelves are much deeper, averaging about 450 m, and, in general, they slope gently towards the continent, not offshore. There has long been debate over whether this is principally a consequence of glacial erosion (e.g. Anderson *et al.*, 1983) or of depression of the continent by the ice load (e.g. Stagg, 1985).

**2** Seismic reflection profiles across Antarctic outer shelves commonly show truncation of inclined reflectors either at the sea floor or beneath a thin sequence (generally less than 400 m) of flat-lying sediments (Tucholke & Houtz, 1976; Stagg, 1985; Eittreim & Smith, 1987; Tanahashi *et al.*, 1987; Larter & Barker, 1989).

**3** Many Antarctic continental slopes are considerably steeper than the world-wide average (4.17° for the upper 1800 m according to Shepard (1973)). Seismic reflection profiles across the Wilkes Land (Eittreim & Smith, 1987; Tanahashi *et al.*, 1987), Prydz Bay (Stagg, 1985), and Antarctic Peninsula (Larter & Barker, 1989) margins show that the steep continental slopes (up to 17° on the Antarctic Peninsula Pacific margin) in these areas have been produced by shelf progradation. Despite the steep angle of the continental slopes, the reflection profiles show no evidence of major slope failure.

For the Antarctic Peninsula Pacific margin, Larter & Barker (1989) argue that these and other characteristics were produced by the action of icesheets grounded out to the shelf edge during Pliocene–Pleistocene glacial maxima. Similar groundings will have occurred on most Antarctic margins, probably starting earlier on East Antarctic margins (Barker *et al.*, 1988; Barron *et al.*, 1988; Barrett, 1989).

Similar characteristics (except the unusually steep continental slope) have been reported on some Arctic continental margins, and have also been interpreted to result from the action of grounded ice (Josenhans & Fader, 1989; Vorren *et al.*, 1989).

Since the first development of major marine icesheets, erosional unconformities on the Antarctic continental shelf are thought to have formed close to the equilibrium grounding depth of ice sheets rather than near sea level. This complicates the stratigraphic interpretation of seismic reflection profiles, as individual profiles may show both kinds of unconformity. Depositional sequences cannot be interpreted simply in terms of changes in 'relative sea level', as is done for temperate and low-latitude margins. To avoid making assumptions about the origin of individual unconformities we will use the term 'relative erosion level' (REL) instead of 'relative sea level' in the initial description and interpretation of reflection profiles.

## AGE AND EXTENT OF ANTARCTIC PENINSULA RIDGE-CREST–TRENCH COLLISIONS

In order to relate the uplift history of different parts of the Antarctic Peninsula Pacific margin to RC–T collisions, it is important to determine the age and extent of these collisions as precisely as possible. Over 4000 km of new magnetic profiles were acquired in the area to the west of the Antarctic Peninsula in 1985, aboard RRS *Discovery*. In 1988 another similar quantity of new magnetic data was acquired in the same area, also from RRS *Discovery*. Some of the new profiles have been plotted along track in Fig. 2, together with previously published profiles (Herron & Tucholke, 1976; Barker, 1982).

A collision is assumed to have taken place when the ridge crest met the base of the continental slope. Using all the available magnetic profiles the ages of the RC–T collisions north of the Tula FZ (Fig. 1) have been recalculated. For collision zones between the Tula FZ and the North Anvers FZ, the uncertainty in these calculations derives from three sources:

**1** the difficulty in identifying the exact location of the RC–T collisions beneath a continental slope that has subsequently prograded;

**2** the presence of a magnetic quiet zone (MQZ), which extends up to 95 km across the continental rise from the base of the slope, preventing identification of the key magnetic anomalies;

**3** uncertainties in a particular magnetic reversal time-scale.

Seismic reflection profiles indicate that progradation near the base of the continental slope has been minor, despite considerable oceanward advance of the shelf break since collision along most of this margin. The present 2500 m isobath is a reasonable approximation to the base of the slope at the time of collision. The times of RC–T collision have been estimated by extrapolation across the MQZ from the youngest confidently identified anomaly at spreading rates measured elsewhere on the ANT–PHO spreading system. The following RC–T collision-age estimates have been made assuming uncertainties of ±5 km in the locus of collision and ±10% in the spreading rates used to extrapolate across the MQZ, and using the magnetic reversal time-scale of LaBrecque *et al.* (1977).

The estimated ages of RC–T collisions between the Tula FZ and Adelaide FZ, and between the Adelaide FZ and Biscoe FZ are 19.8 ± 0.8 Ma and 16.5 ± 0.7 Ma, respectively. The estimated age of RC–T collision at the southwest part of the margin between the Biscoe FZ and the South Anvers FZ is 14.5 ± 0.6 Ma, but in the broad embayment in the margin centred about 64°15′S (Fig. 2), collision would not have taken place until about 1 Ma later. For the RC–T collision between the South Anvers FZ and North Anvers FZ the estimated age is 10.0 ± 0.4 Ma.

The MQZ is absent between the North Anvers FZ and Hero FZ, but the RC–T collisions in this region are complicated in a different way. A fan-shaped pattern of magnetic anomalies near the margin (Fig. 2) indicates rotation of the ridge crest during the last few million years before collision, so that the ridge crest entered the trench obliquely. Directly southwest of the C FZ, the magnetic anomaly identifications in Fig. 2 indicate that RC–T collision took place 6.0 Ma ago, then migrated southwest, reaching the North Anvers FZ about 5.6 Ma ago. In the most northeasterly RC–T collision zone, collision took place 5.5 Ma ago directly southwest of the Hero FZ, and may have reached the C FZ as late as 3.1 Ma ago (Larter & Barker, 1990, and in preparation).

## SEISMIC REFLECTION PROFILES

In 1985, aboard RRS *Discovery*, over 1000 km of 48-channel seismic reflection profiles were acquired in the area of the younger RC–T collision zones on the Antarctic Peninsula Pacific margin (Fig. 4). This

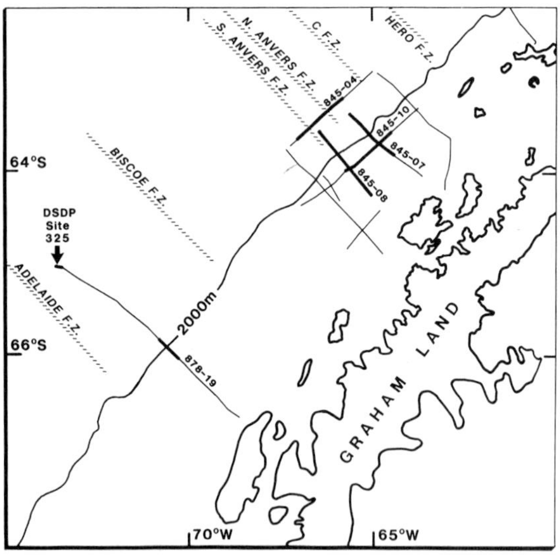


**Fig. 4.** Track chart showing the positions of multichannel seismic profiles acquired on RRS *Discovery* Cruise 154 (1985) and of Profile BAS 878-19, acquired on RRS *Discovery* Cruise 172 (1988). Thicker lines indicate parts of profiles reproduced or shown as line drawings in other figures.

survey included four margin-crossing profiles, each in a different RC–T collision zone and intersected by a strike profile along the outer shelf (Profile AMG 845-10). The seismic source consisted of four air guns with a total capacity of 8.5 l. This was fired at 50 m intervals, which in combination with the 50 m hydrophone group spacing gave 24-fold cover. Additional MCS data were acquired in 1988, also aboard RRS *Discovery*, in the same area and farther to the southwest. For most of the 1988 profiles a larger seismic source (four air guns, total capacity 15.8 l) was used, but with only a 16-channel hydrophone streamer. As before, shot and hydrophone group spacing were both 50 m, giving eight-fold cover. Parts of one profile from this more recent survey are presented here (see Figs 11, 12), but most of these data have yet to be processed. The track chart (Fig. 4) shows only this single profile from the 1988 survey.

All the MCS reflection profiles crossing the continental margin between the Adelaide FZ and C FZ show the same three main geological provinces on the middle and outer shelf: a prograded outer shelf is separated from a small mid-shelf sedimentary basin by a structural high (the 'Mid-shelf High' or MSH). Only two profiles continue farther inshore, on to the inner shelf, defined here as that area nearer to the Peninsula than a line joining the northwest coasts of the larger islands. In this area seismic refraction velocities combined with the absence of coherent subsea-bed reflections suggest that crystalline 'arc' rocks, similar to those exposed onshore (Tectonic Map, British Antarctic Survey, 1985), outcrop at the sea-bed. On all profiles the continental slope is unusually smooth and steep, and clearly has been constructed by depositional processes associated with progradation of the outer shelf. Mid-slope (1000–2000 m) angles range from 13 to 17°. Each profile shows an overall decrease in the thickness of continental-rise sediments overlying igneous oceanic basement with increasing distance from the continental slope. The thickness of continental-rise sediments near the foot of the slope also decreases with RC–T collision age, from up to 1.8 s two-way time (TWT) between the Adelaide FZ and South Anvers FZ to less than 0.5 s TWT northeast of the North Anvers FZ.

The shelf and margin farther to the northeast, opposite the youngest RC–T collision zone, are more complicated owing to their proximity to the active extensional basin of Bransfield Strait and to the South Shetland Trench. The neotectonics of

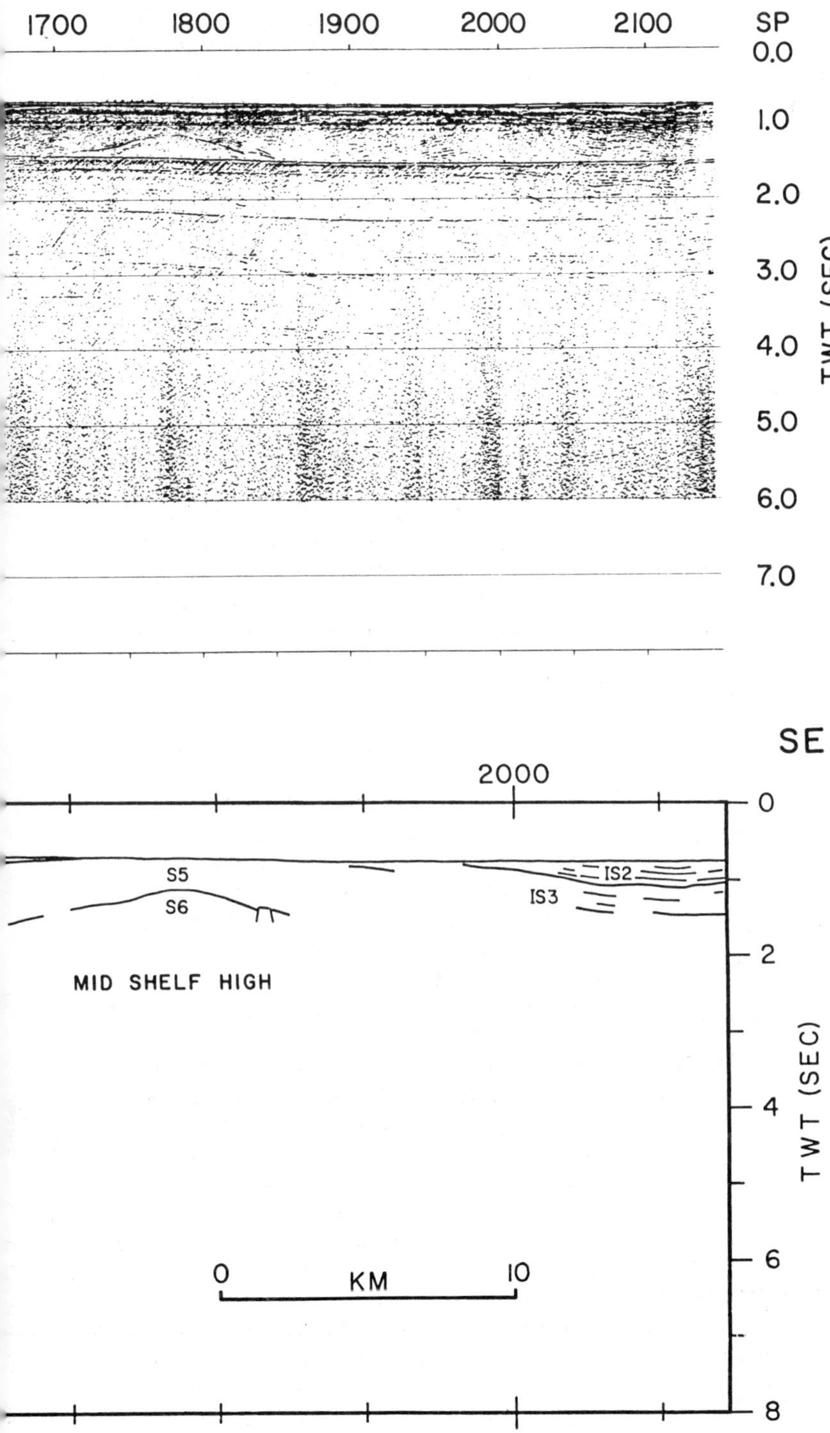


**Fig. 5.** Multichannel seismic reflection profile AMG 845-08 and interpretive line drawing. Vertical exaggeration in water 3.3:1.

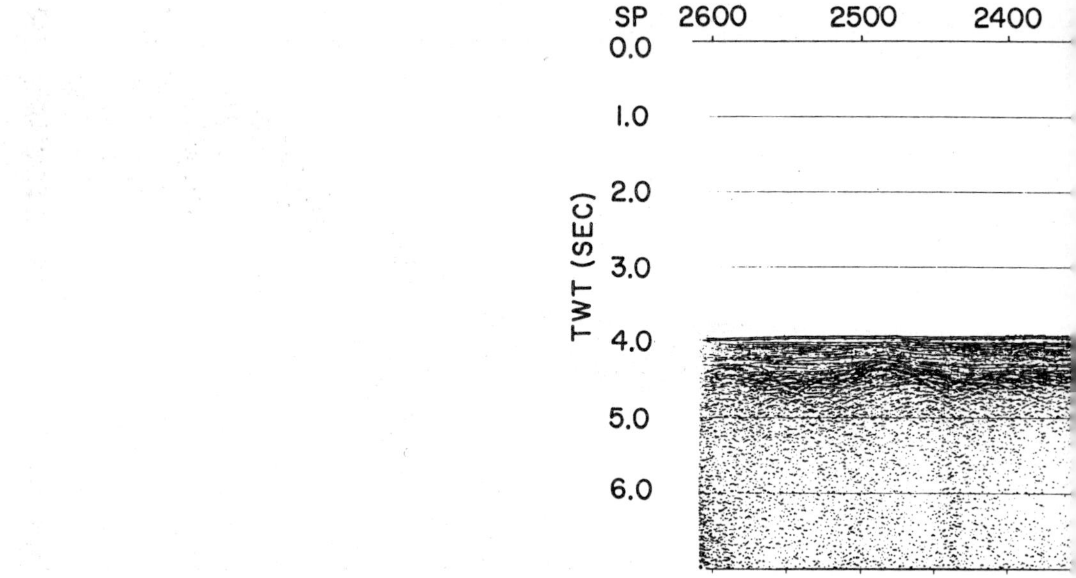

SP
2600
2500
2400
0.0
1.0
2.0
3.0
4.0
5.0
6.0
TWT (SEC)


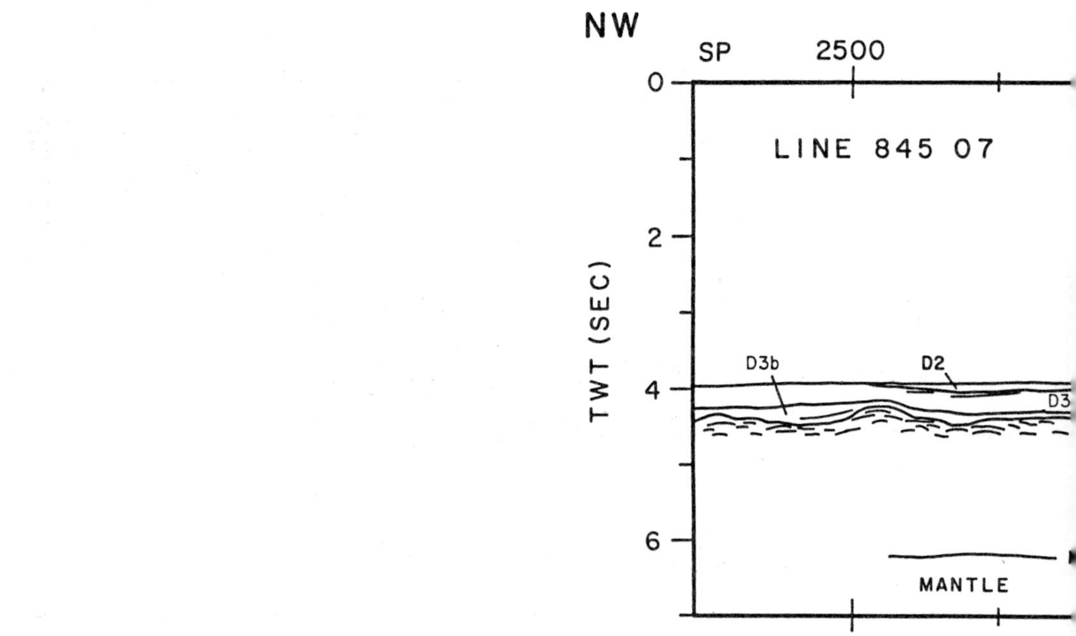

NW
SP
2500
0
2
4
6
TWT (SEC)
LINE 845 07
D3b
D2
D3
MANTLE

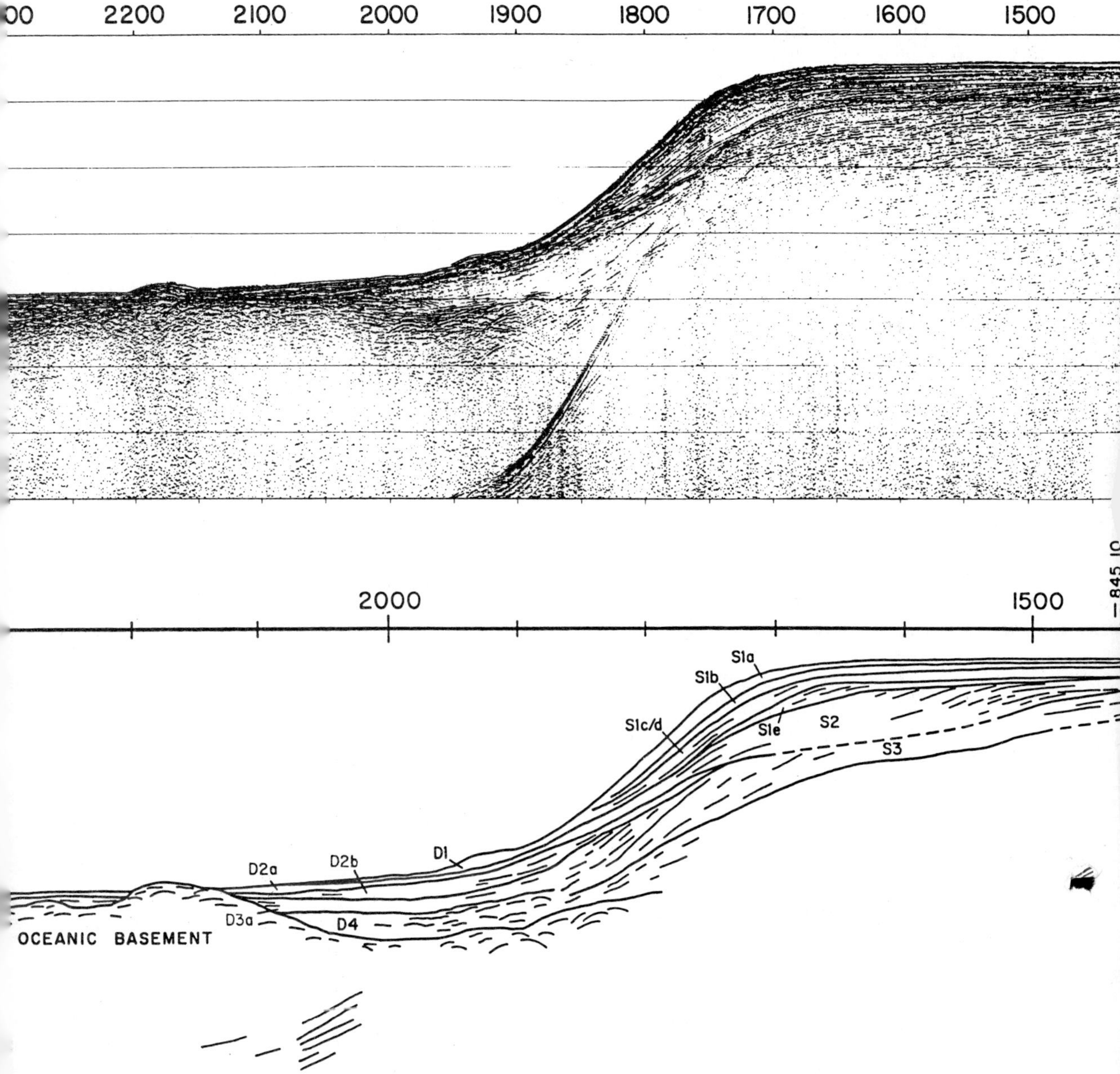
00
2200
2100
2000
1900
1800
1700
1600
1500
2000
1500
S1a
S1b
S1c/d
S1e
S2
S3
D1
D2a
D2b
D3a
D4
OCEANIC BASEMENT

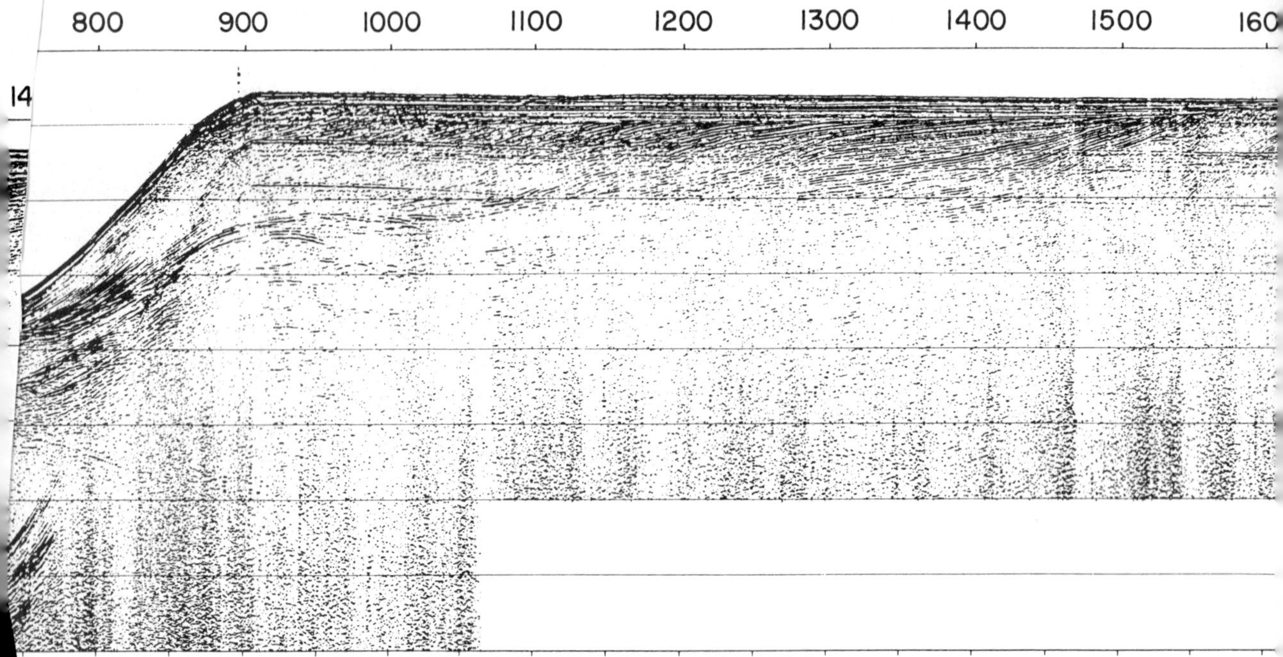
800
900
1000
1100
1200
1300
1400
1500
160
14

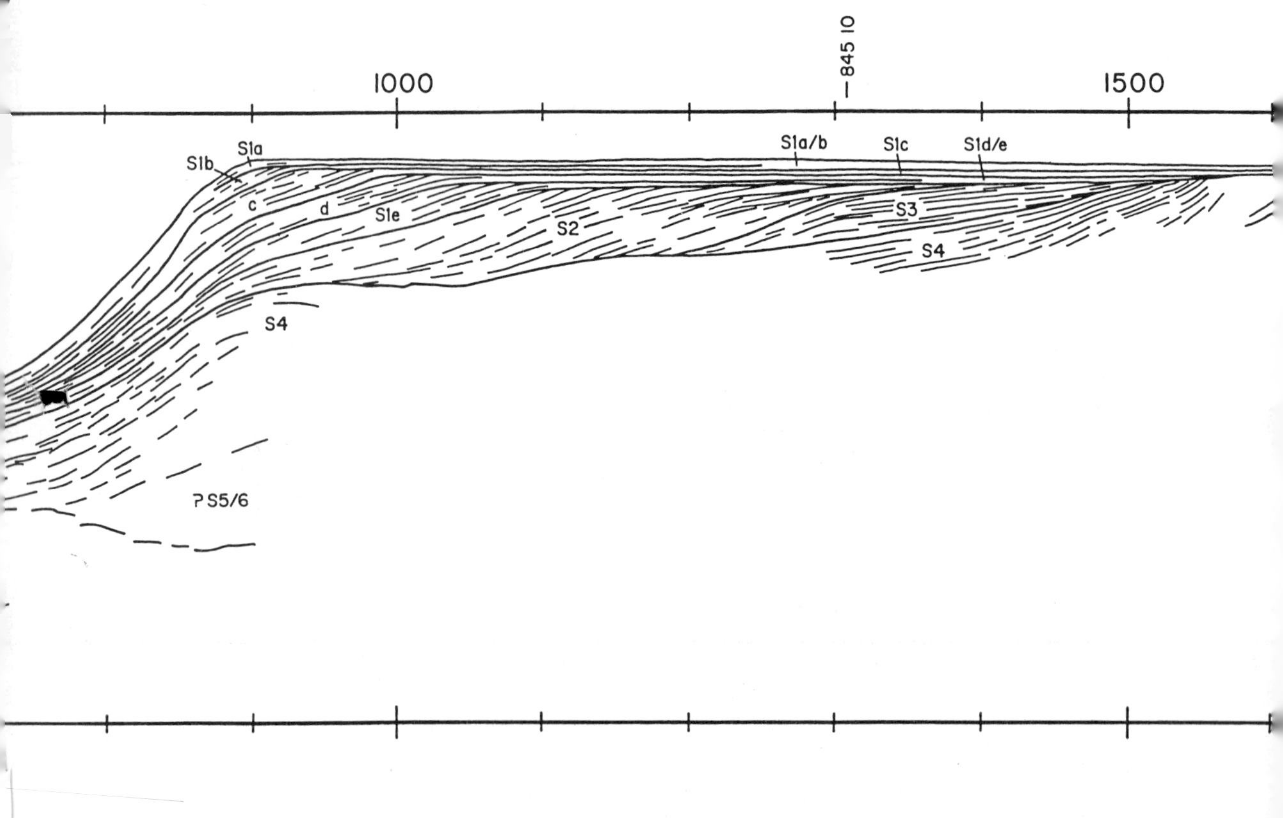
845 10
1000
1500
S1a
S1b
c
d
S1e
S2
S1a/b
S1c
S1d/e
S3
S4
S4
? S5/6

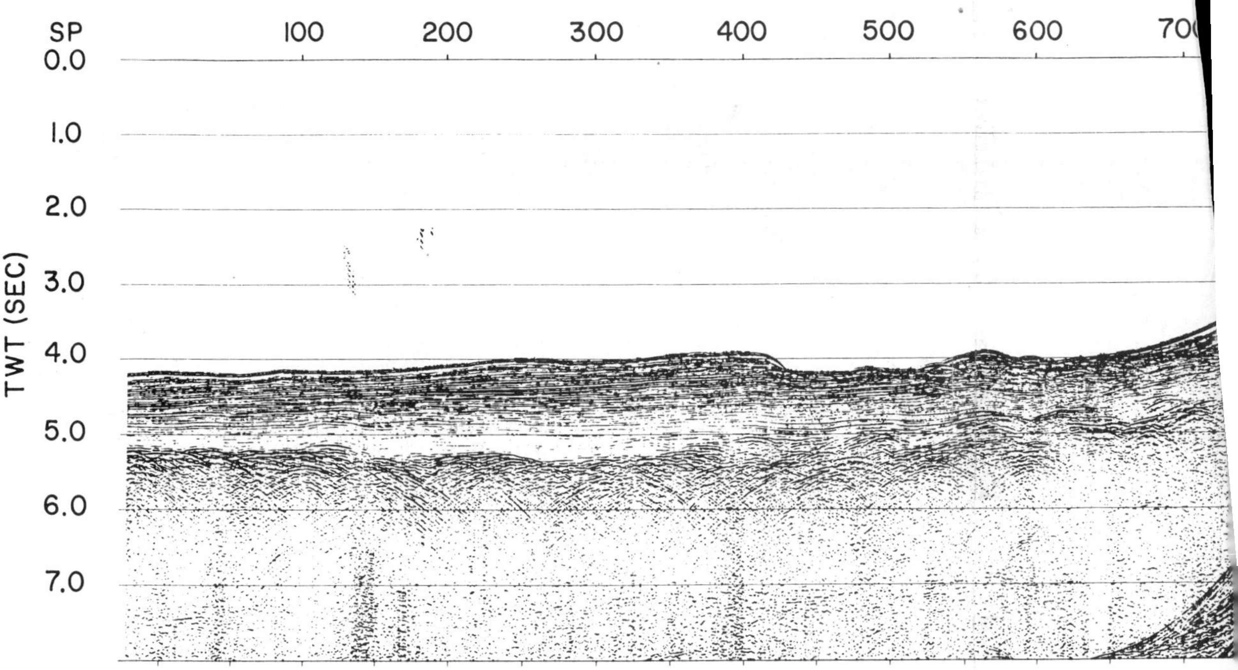

SP
100
200
300
400
500
600
0.0
1.0
2.0
3.0
4.0
5.0
6.0
7.0
TWT (SEC)


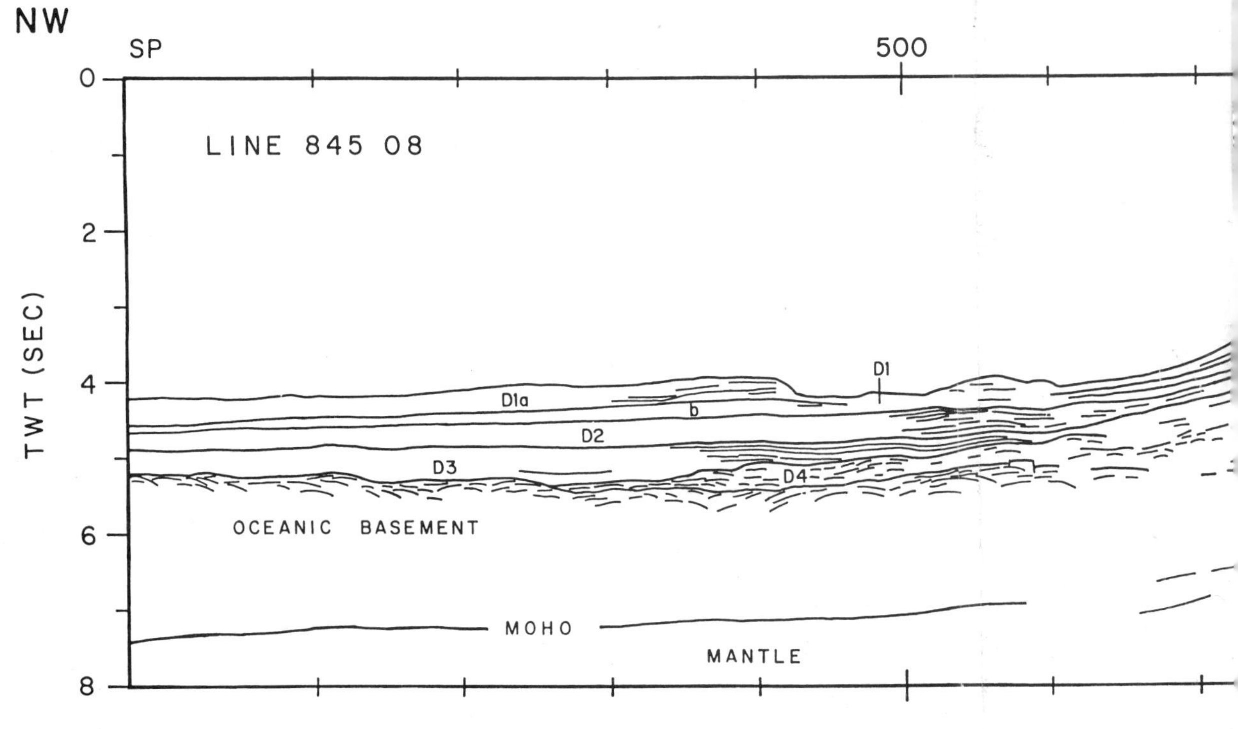

NW
SP
500
0
2
4
6
8
TWT (SEC)
LINE 845 08
D1
D1a
b
D2
D3
D4
OCEANIC BASEMENT
MOHO
MANTLE

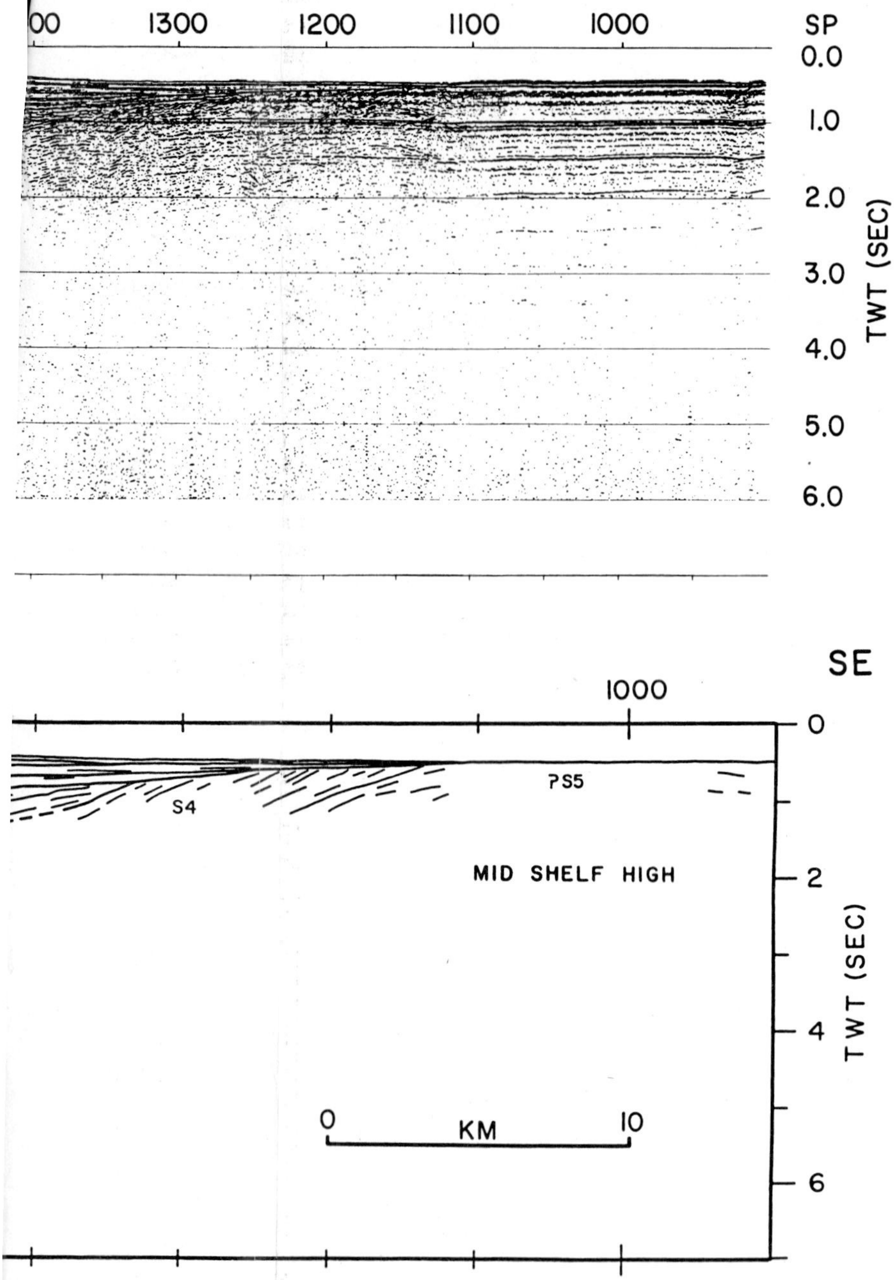


**Fig. 8.** Part of MCS profile AMG 845-07 and interpretive line drawing. Vertical exaggeration in water 3.3:1.

Bransfield Strait and its environs is beyond the scope of this paper.

The following sections describe the depositional sequences observed on the outer shelf, slope, and rise on individual profiles from the 1985 survey. Following the interpretation principles described by Mitchum *et al.* (1977), six major sequences or sequence groups (S1–S6) have been identified on the outer shelf and slope, while four principal sequences (D1–D4) have been recognized within the sedimentary pile overlying oceanic basement on the continental rise. On two margin-crossing profiles (AMG 845-08 and AMG 845-07), records of changes in REL have been determined from the outer shelf sequences and crude chronostratigraphic control has been derived by correlation to oceanic basement of age known from marine magnetic anomalies. The methods used to determine changes in REL are essentially the same as those described by Vail *et al.* (1977a) for determining relative change of sea level.

**Profile AMG 845-08**

Beneath the outermost shelf and slope on Profile AMG 845-08 (Fig. 5, opposite p. 170) the sequence boundary at the base of the prograded sediment wedge (the S4–S2 boundary) is a very prominent reflection event. This sequence boundary can be followed down the slope into the sediments on the continental rise, where it correlates with the D3–D2 boundary. Interpretation of marine magnetic anomalies (Fig. 2) suggests an age of 10.0 ± 0.4 Ma for oceanic basement beneath the lower continental slope, which sets an upper limit for the age of the younger parts of S4 and the overlying sequences. This age and a similar correlation with oceanic basement on Profile AMG 845-07 provide a crude time-scale for a chronostratigraphic section of the outer shelf on Profile AMG 845-08 (Fig. 6).

Towards the MSH it can be seen that S4 was erosionally truncated before deposition of S3. Reflectors in the upper part of S4 converge and show increasing dip as they are followed towards the MSH. Above the unconformity, reflectors within S3 begin to onlap at about shot point (SP) 1350. After correction for the dip of reflectors within S3, this indicates a drop in REL of at least 350 m before deposition of S3. As the uppermost part of S4 has been removed by erosion near the MSH, and as no correction has been made for compaction, the real change in REL was probably considerably greater than this. However, even a change of 350 m is twice the magnitude of the largest changes in REL that Haq *et al.* (1987) attribute to eustatic sea-level fluctuations. We interpret the combination of local erosional truncation of S4, the convergent reflection

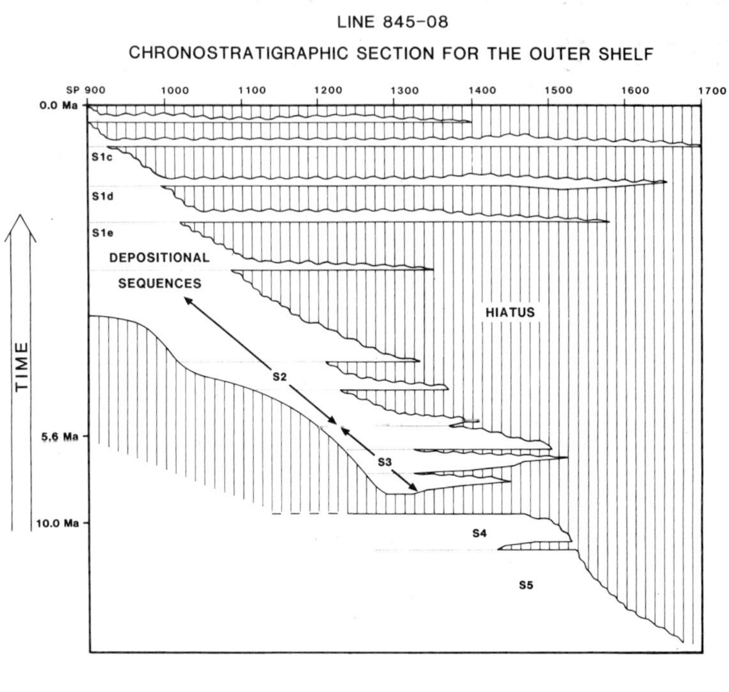


**Fig. 6.** Chronostratigraphic section for the outer shelf (SP 900–1700) of Profile AMG 845-08 (Fig. 5). The time-scale was constructed by assuming a constant time increment per reflection wavelength and so is expected to be non-linear. The labelled ages represent correlations to oceanic basement of age estimated from marine magnetic anomalies.

configuration and lateral change of dip within S4, and the major change in REL across the S4–S3 sequence boundary to indicate tectonic uplift, beginning during the deposition of S4 and centred on the MSH. From the age control provided by the relationship of S4 to oceanic basement of known age, it appears that this uplift began at about the time of the RC–T collision along this section of the margin.

No age control is available for the sequences of the mid-shelf sedimentary basins. On Profile AMG 845-08, sequences IS2 and IS3 show parallel reflection configuration patterns and increasing dip towards the MSH. From this it is inferred that they pre-date the episode of tectonic uplift.

The pronounced onlap seen at the base of S3 indicates a large rise in REL during the deposition of this sequence. Measurement of aggradation on the onlapping reflectors indicates a rise of about 410 m. Above the onlapping reflectors, still within S3, is an additional thickness of topset deposits of about 290 m for which onlap relationships are not seen because of subsequent erosion, but the accumulation of which implies a further rise in REL. The locus of a palaeoshelf-break is seen within S3, and retreats towards the MSH with time. Combined with the limited extent of S3 along the profile (Figs 5, 6), this indicates a period of restricted sediment supply to the outer shelf. We infer that the MSH was a positive topographic feature and constituted a barrier to cross-shelf sediment transport in this area during the period represented by S3.

The S2–S1 sequence boundary is a major erosional surface. Significant downward tilting of the underlying S3 and S2 topset deposits away from the MSH took place during the hiatus represented by this unconformity (Fig. 5). Consequently the total rise in REL since the start of deposition of S3 changes by about 60 m/km along this part of the profile. At SP 1400 a total thickness of about 290 m of topset deposits overlies S3, indicating an additional net rise in REL by this amount. This makes a total REL rise at this location, before any correction for compaction, of about 990 m since the start of deposition of S3. As in the case of the preceding fall in REL, a change of this magnitude is far too large to be explained by a glacio-eustatic mechanism, and tectonic subsidence is implied. In this usage the term 'tectonic' encompasses local vertical motion of a depositional surface owing to a variety of causes (thermal effects, isostatic compensation, lithospheric flexure, compaction of underlying sequences, etc.). The observed tilting of S2 and S3 topsets beneath the S2–S1 sequence boundary indicates an episode of differential vertical movement between the MSH and the outer shelf, in which the latter has subsided further. Slight thickening of S1 away from the MSH suggests that slow differential subsidence continued after the S2–S1 hiatus, and indeed may be continuing at the present day.

## Profile AMG 845-10

Profile AMG 845-10 (Fig. 7) was shot along the strike of the outer shelf to intersect all of the 1985 MCS profiles crossing the margin (Fig. 4). It allows correlation of the outer shelf sequences observed on these other lines. Of particular interest here are the correlations between Profiles AMG 845-08 and AMG 845-07, which allow the relative timing of uplift and subsidence on these two profiles to be determined. Also of interest is the termination of the outer shelf sediment wedge about 15 km northeast of the intersection of Profile AMG 845-07 with AMG 845-10 (SP 2537). In this area reflectors within S3 and S4 show parallel reflection configuration patterns and increasing southwest dip as they are followed to the northeast. They are erosionally truncated by the sequence boundary at the base of S2, and S3 pinches out entirely in this way. These reflection configurations suggest tectonic uplift of the margin to the northeast after deposition of S3. The absence of S1 and S2 over this region suggests that it remained uplifted during their deposition. This uplift may be related to the most recent RC–T collision, which concluded directly to the northeast of where the C FZ meets the margin (Fig. 2), perhaps as recently as 3.1 Ma ago (Larter & Barker, 1990, and in preparation).

## Profiles AMG 845-07 and AMG 845-04

From marine magnetic data (Fig. 2), RC–T collision is estimated to have taken place about 5.6 Ma ago on the part of the margin crossed by Profile AMG 845-07 (Fig. 4). Despite a 4.4 Ma difference in the age of RC–T collision, Profile AMG 845-07 (Fig. 8, opposite p. 171) shows very similar general characteristics to Profile AMG 845-08 (Fig. 5). Beneath the outer-most shelf and slope the sequence boundary at the base of S2 is once again a very prominent reflection event. However, on AMG 845-07 it is underlain by S3, which continues beneath the continental slope and appears to correlate with part of sequence D3a on the continental rise. On Profile

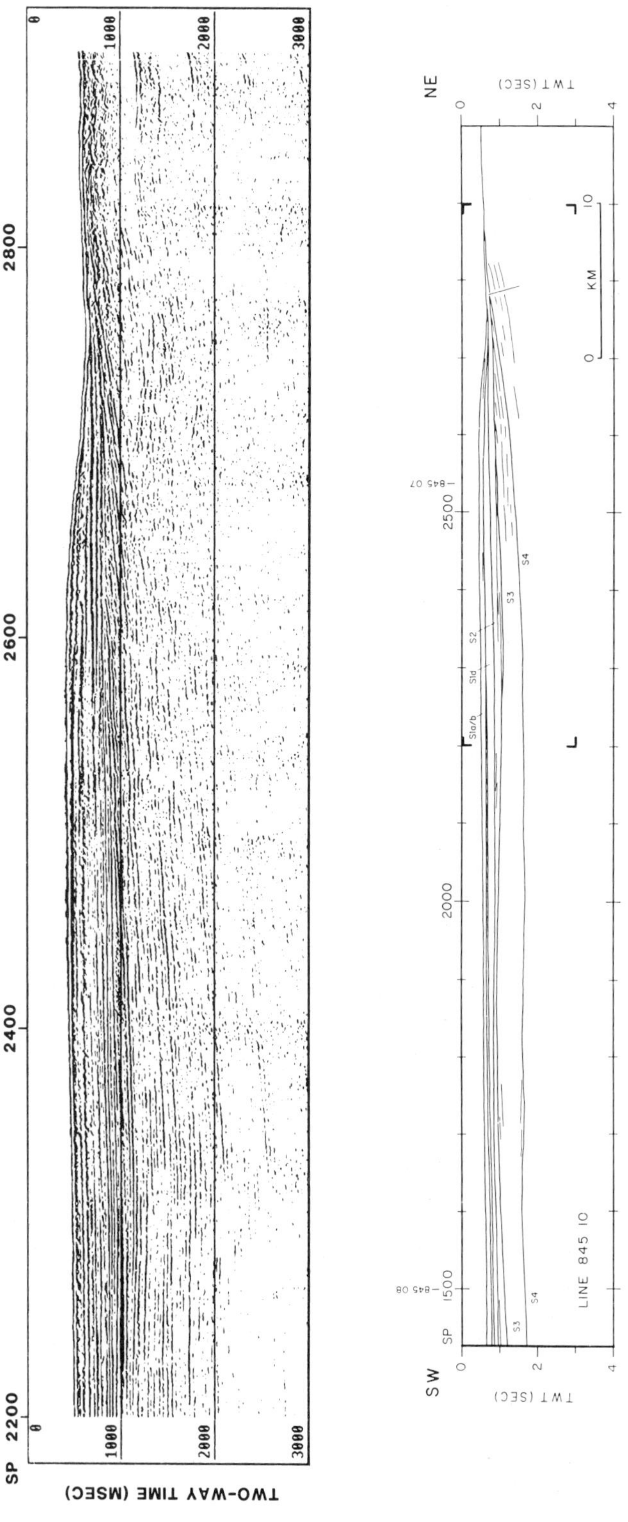


**Fig. 7.** Interpretive line drawing of part of MCS profile AMG 845-10, with seismic data from area enclosed by corner marks shown above. Vertical exaggeration in water 3.3:1.

AMG 845-08, there appears to be no lateral equivalent of S3 beneath the continental slope or rise, implying that D3a is absent or too thin to be resolved, and 'D3' on this profile is equivalent to D3b on AMG 845-07. If this is correct, D3a should pinch out as it is traced southwestward through the continental rise. Although Profile AMG 845-04 does not quite intersect the other profiles discussed here (Fig. 4), a tentative correlation of sequence boundaries has been attempted. The interpreted pattern of sequences on AMG 845-04 (Fig. 9) is consistent with D3a thinning and pinching out in the vicinity of the North Anvers FZ.

The unit labelled D4 on the line drawings in Figs 5, 8, and 9 is identified as a seismic facies unit, but is not considered to be a true sequence of chronostratigraphic significance. It contains high amplitude, discontinuous reflectors and occurs near the margin and along fracture-zone troughs, directly overlying oceanic basement. It probably represents rapid terrigenous sedimentation overflowing the ridge crest shortly before and during RC–T collision. Disruption by faulting in the underlying oceanic basement and hypabyssal intrusion by mid-ocean ridge magmas could explain its seismic character and apparent diachroneity along the margin. However, if burial of the ridge crest by sediment is the cause of the MQZ (Irving, 1970; Vogt *et al.*, 1970; Levi & Riddihough, 1986), then the lower part of D3 must also represent sediment that overflowed the ridge crest in some areas.

The continuity of sequence S3 beneath the continental slope and the presence of an equivalent sequence on the continental rise on Profile AMG 845-07 suggests that this section of the margin did not experience the restriction of sediment supply hypothesized to have occurred during this interval in the area to the southwest. Other evidence also points to a later uplift of the shelf along Profile AMG 845-07 than on AMG 845-08.

**1** There is no clear evidence of erosional truncation or of onlap at the S4–S3 sequence boundary on Profile AMG 845-07, but both S3 and S4 reflectors are erosionally truncated by the sequence boundary at the base of S2.

**2** The truncation of reflectors within D3a at the base of D2b and of reflectors within S3 at the base of S2 on the palaeoslope on Profile AMG 845-07 indicates a hiatus, and an episode of marine erosion. Sequence D2b on Profile AMG 845-07 is interpreted to consist of locally redeposited products of marine erosion (i.e. it is a 'lowstand systems tract slope fan' and 'basin-floor fan' in the terminology of van Wagoner *et al.*, 1988). This implies a restriction of sediment supply to the margin after the deposition of S3. Evidence of a similar restriction in sediment supply is seen on Profile AMG 845-08, but in this case it occurred earlier, during deposition of S3.

An upper limit for the age of the unconformity between S3 and S2 is provided by the 5.6 Ma age of the oceanic basement near the margin on Profile AMG 845-07. Since the S3–S2 sequence boundary can be followed into the continental-rise sediments overlying this basement, the younger parts of S3 and the overlying sequences must be less than 5.6 Ma old. The unconformity thus post-dates RC–T collision along this section of the margin.

The topset deposits of S2 and S1 show pronounced onlap on to the eroded surface of S3 and S4. Aggradation measurements indicate a net rise in REL close to the MSH of about 400 m since the start of deposition of S2. However, the overall thickness of topset deposits indicates a net rise in REL of about 590 m at SP 1440, implying an increase in total REL rise of about 20 m/km with increasing distance from the MSH (none of these estimates includes any correction for compaction). Neither the lateral variation nor the overall magnitude of this REL rise can be explained by a glacio-eustatic mechanism; they must indicate gradual tectonic subsidence, which increased with increasing distance from the MSH. We suggest that the S3–S2 unconformity on Profile AMG 845-07 was produced in the same way as the S4–S3 unconformity on Profile AMG 845-08: tectonic uplift of the middle and outer shelf resulting from RC–T collision was centred on the MSH, producing erosional truncation over the MSH itself, and restricting sediment supply to the margin.

### Deep Sea Drilling Project Site 325

Site 325 was drilled on the central continental rise between the Adelaide FZ and Biscoe FZ, about 170 km from the base of the Antarctic Peninsula continental slope (Fig. 4). A single hole was drilled to a sub-bottom depth of 718 m but failed to reach volcanic basement (Hollister *et al.*, 1976), which is estimated from magnetic data (Fig. 2) to be about 24 Ma old (anomaly 6c). The hole was spot-cored and the total length of core recovered was only 34.4 m. However, the biostratigraphy of these cores adequately defines the major variations in sedimentation rate (Fig. 10). The recovered sediments have a high terrigenous component (turbidite and ice-

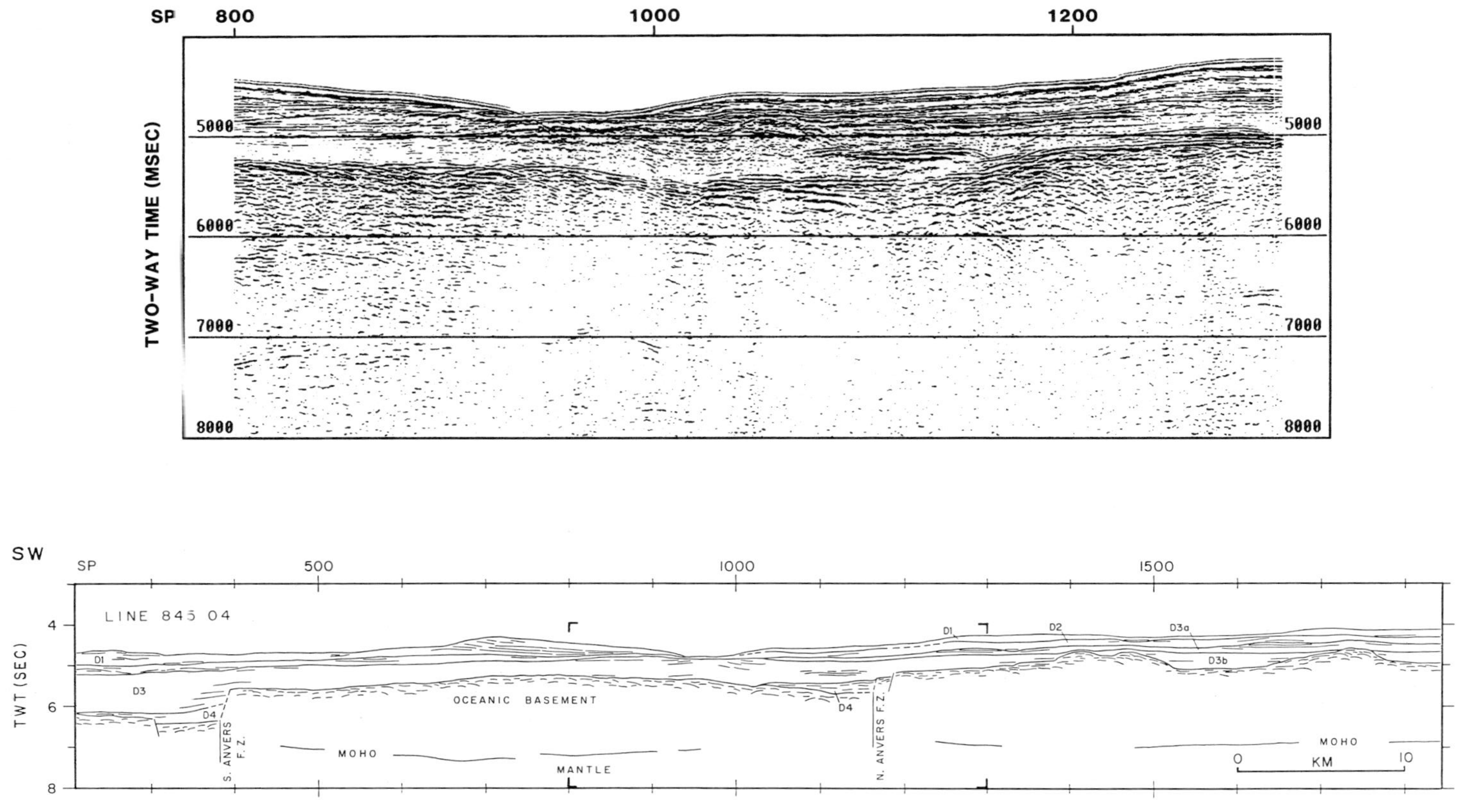


**Fig. 9.** Interpretive line drawing of part of MCS profile AMG 845-04, with seismic data from area enclosed by corner marks shown above. Vertical exaggeration in water 3.3:1.

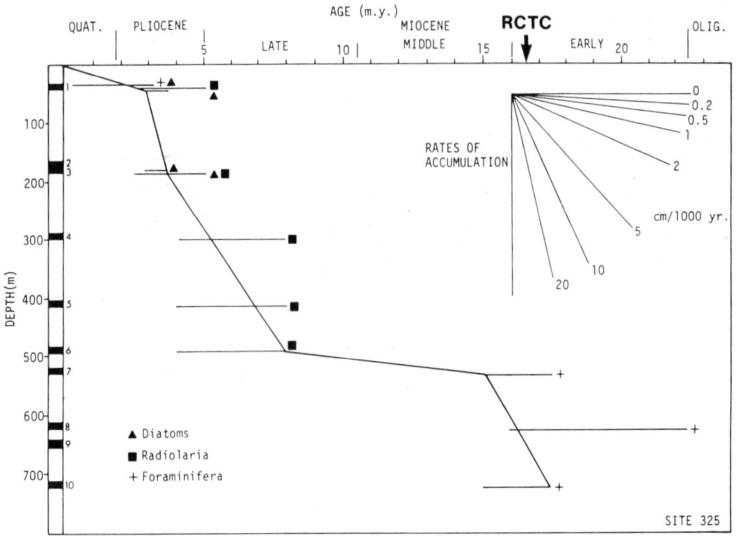


**Fig. 10.** Rates of sediment accumulation at DSDP Site 325, showing reduction in sedimentation rate following 16.5 Ma ridge-crest–trench collision (RCTC) (after Hollister, Craddock *et al.*, 1976).

rafted). Thus the severely reduced sedimentation rate (or hiatus) that started 15 Ma ago or slightly earlier, and continued at least until 8 Ma, is interpreted to indicate a reduction or cessation in the supply of terrigenous sediment to the Site.

Ridge-crest–trench collision at the continental margin opposite Site 325 is estimated to have taken place 16.5 ± 0.7 Ma ago. Thus, terrigenous sediment supply to the site, across the ridge crest, was continuous before collision, but stopped less than 2.2 Ma after collision for several million years, before being renewed.

The middle Miocene is a time of widespread unconformity in deep-sea sediments, probably owing to changes in deep-water circulation (e.g. Davies *et al.*, 1975). However, there is no evidence of a break in sedimentation at Site 322 or 323 (630 km to the north-northeast and 1170 km to the west of Site 325, respectively) on the Bellingshausen Abyssal Plain during this interval, so deep circulation alone is unlikely to explain the condensed section/hiatus at Site 325.

## Profile BAS 878-19

Profile BAS 878-19 was shot across the continental shelf and slope from near northern Adelaide Island and continued across the continental rise to pass through Site 325 (Fig. 4). Although individual sequences cannot be correlated directly with those on other profiles farther to the northeast, the general character of the shelf is similar: a MSH separates a small sedimentary basin inshore from a prograded outer shelf wedge.

In the vicinity of Site 325, Profile BAS 878-19 shows a prominent reflector 460 ms below the sea floor, which also marks a change in seismic facies (Fig. 11). As this reflector is followed towards the continental margin, occasional reflector terminations against it confirm its status as a sequence boundary and the prime candidate for correlation with the middle–late Miocene hiatus or condensed interval at Site 325. This correlation requires an average velocity of 2130 m/s from the sea floor down to 490 m, which is considerably higher than the velocity predicted by the empirical relationship of Carlson *et al.* (1986) for this sub-bottom depth. A sonobuoy profile recorded near Site 325 does suggest velocities above the norm for deep-sea sediments (Tucholke *et al.*, 1976). Another consideration, in view of the drilling technique ('washing') used between cores at Site 325, is that the small amount of material recovered (0.3 m) in the core ending at 490 m (core 6) may have been derived from higher in the hole (Gieskes & Lawrence, 1976; Tucholke *et al.*, 1976), which would reduce the interval velocity implied by correlation of the hiatus with the reflector. This consideration does not cast doubt on the existence of a middle-late Miocene condensed section or hiatus at Site 325. Even if core 6 is from higher in the hole, the biostratigraphic data still adequately define a period of reduced sedimentation (Fig. 10).

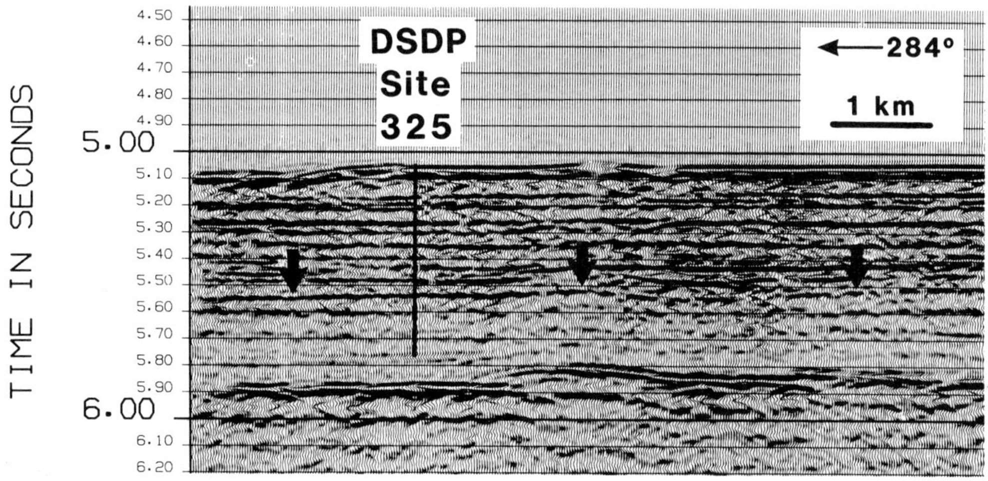


**Fig. 11.** Part of MCS profile BAS 878-19 (eightfold stack) near DSDP Site 325. Arrows mark reflector correlated with mid-Miocene hiatus (or period of slow sedimentation). High-amplitude reflector at 5.8–5.9 s TWT is interpreted as oceanic basement. Vertical exaggeration in water 3.3:1.

The sequence boundary 460 ms below the sea floor at Site 325 can be traced with confidence to the continental margin and beneath the continental slope. Onlap on to it on the upper continental rise indicates a systematic decrease in the age of the oldest sediments above it as the margin is approached. There is no indication of systematic change in the age of the sediments below the sequence boundary. Beneath the continental slope this sequence boundary is a very prominent reflector and separates sequences of markedly different seismic character (Fig. 12). In terms of its position, its character, the character of the sequences above and below it, and the age of the youngest sediments

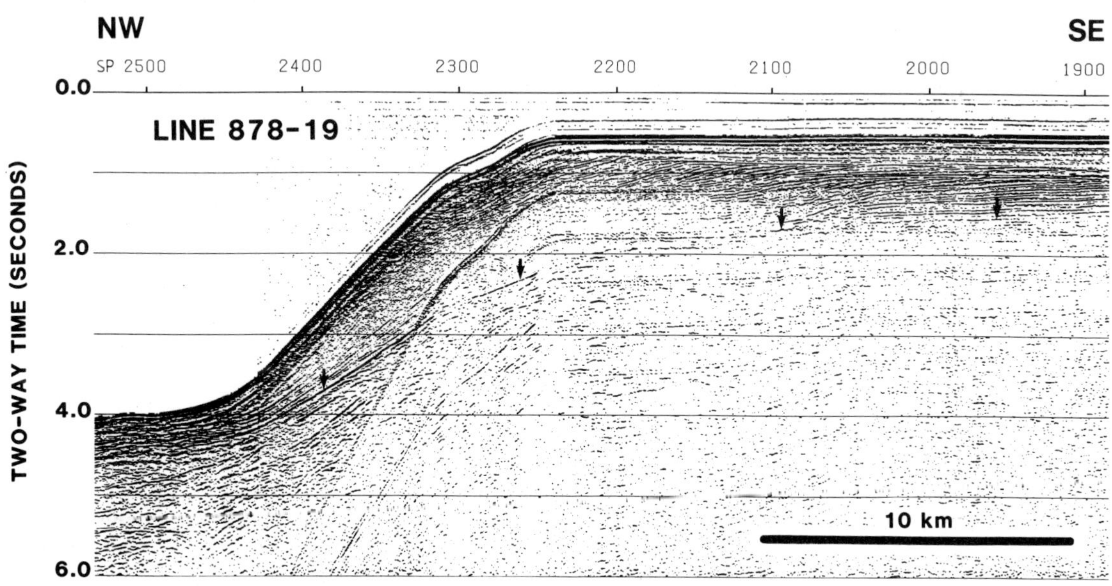


**Fig. 12.** Part of MCS profile BAS 878-19 (eightfold stack) crossing continental slope. Arrows mark sequence boundary correlated with mid-Miocene hiatus (or period of slow sedimentation) at DSDP Site 325. The 'reflector' above and parallel to the sea-floor reflection on the shelf and upper slope is a processing artefact. Vertical exaggeration in water 3.3:1.

below it relative to RC–T collision (based on correlation with Site 325), this sequence boundary is very similar to the S4–S2 boundary on Profile AMG 845-08 and the S3–S2 boundary on Profile AMG 845-07. Once again, it is thought that this is an unconformity produced by thermal uplift of the middle and outer shelf areas following RC–T collision.

Near the base of the slope on Profile BAS 878-19, the sequence boundary overlies about 1 km of undisturbed sediments, which in turn overlie a D4-type unit. The estimated age of the of RC–T collision (16.5 ± 0.7 Ma) and the age of the sediments directly below the sequence boundary (15–17 Ma, by correlation with DSDP Site 325) imply that this 1-km-thick sequence accumulated in less than 2.2 Ma following collision, and therefore at an extremely high sedimentation rate (greater than 45 cm/1000 yr). The existence of a magnetic quiet zone 50 km wide adjacent to the margin in the vicinity of this profile (Fig. 2) is thought to indicate that sedimentation rates were also high before RC–T collision. Severely diminished magnetic anomaly amplitudes over highly sedimented spreading centres have been observed in several areas and have been attributed to reduction of the amount of rapidly quenched, very highly magnetized pillow basalts (Irving, 1970; Vogt *et al.*, 1970) and, more recently, to the effects of hydrothermal reactions (Levi & Riddihough, 1986).

Sea-floor multiple reflections incompletely removed during processing make it difficult to trace the sequence boundary seen beneath the continental slope into the shelf sequences. However, in Fig. 12 a tentative correlation has been made with a shelf sequence boundary that is overlain by about 1280 m of topset deposits. If correct, this indicates a net rise in REL by this amount since the deposition of the first sediments above this sequence boundary. As on Profiles AMG 845-07 and AMG 845-08, such a large rise in REL cannot be explained by a glacio-eustatic mechanism and must indicate local tectonic subsidence.

Thus on every seismic profile crossing the margin between the Adelaide FZ and C FZ, similar outer shelf and slope sequences that record similar, but systematically diachronous, histories of vertical motion are observed. Following RC–T collision, uplift of the continental shelf centred on the MSH appears to cut off the supply of terrigenous sediment to the margin. A few million years after collision, uplift ends and a period of long-term subsidence begins.

The present state of the margin opposite the youngest collision zone (northeast of the C FZ, 3.1–5.5 Ma ago) is consistent with this interpretation. We deliberately avoided detailed consideration of the shelf and margin of this youngest section, because of the likelihood of thermal and tectonic complication from Bransfield Strait extension, in the northeast. However, at least part of the slope in this area is sediment-starved (British Antarctic Survey, unpublished dredge data), and the shelf is divided by two NE–SW-trending bathymetric highs 50 km long that lie along the line of the MSH. Smith Island, which exposes part of the Scotia Metamorphic Complex (British Antarctic Survey, 1985), is situated on one of these highs.

### Pliocene–Pleistocene shelf and slope sequences

On MCS Profiles AMG 845-07 (Fig. 8) and AMG

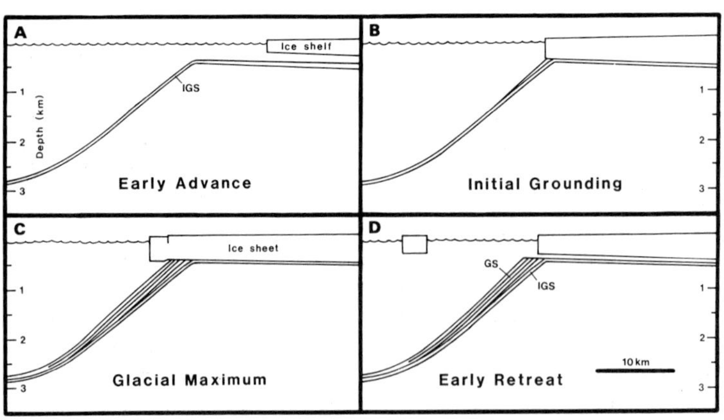


**Fig. 13.** Model for development of a single glacial margin sequence. (A) Early stage of ice-sheet advance; intergrounding subsequence (IGS) is being deposited on shelf and slope. (B) Initial grounding. (C) Glacial maximum; ice sheet is eroding and compacting shelf and is depositing grounding subsequence (GS) on slope. (D) Early stage of glacial retreat; hiatus on outer shelf has ended and deposition of next IGS has begun.

845-08 (Fig. 5), eight distinct oblique progradational sequences can be identified within S1 and S2. As pointed out previously, the stratigraphic relationship of S2 on Profile AMG 845-07 to ocean floor of age known from magnetic anomalies establishes a maximum age for S2 of 5.6 Ma. The five sequences within S1 can be correlated, via Profile AMG 845-10, across a wide area of the outer shelf, and have been labelled 'S1a' to 'S1e' on the line drawings. At least three similar sequences are present within S2, but the available data do not allow these to be correlated confidently from profile to profile.

These sequences are characterized by steeply dipping foresets and a depocentre on the upper continental slope, so that the slope has advanced oceanward and steepened with time. The most recent slope sediments on the mid-slope of Profile AMG 845-08 have a depositional dip of about 17°. The sequence boundaries are most easily recognized where these steeply dipping slope sediments are truncated by an unconformity at the base of the overlying sequence. In the case of the youngest sequence this unconformity is at the present-day sea floor, or closer to it than the limit of resolution of the MCS data (5–10 m). The present-day sea floor on the outer shelf is mostly 400–500 m deep and becomes steadily shallower towards the shelf break (the gradient in Fig. 5 is 1 in 430). The present-day shelf break is generally quite sharp, and this appears to have been the case at earlier sequence boundaries as well.

Despite the very steep angle of the continental slope, the reflection profiles show no evidence of major slope failure during deposition of S2 and S1. Shallow seismic refraction data and interval velocities from CDP gathers show that outer shelf sediments directly below the sea floor have velocities in excess of 2000 m/s.

### Grounded ice-sheet model

Larter & Barker (1989) argue that the presence of Pliocene–Pleistocene erosional unconformities in such deep water, the morphology of the shelf, the steepness and stability of the continental slope, and the high seismic velocity of the uppermost outer shelf sediments cannot be explained in terms of sedimentary processes common at low latitudes. It is proposed instead that the S1 and S2 sequences were produced by the action of ice sheets grounded out to the shelf edge at times of glacial maxima. Supporting evidence of recent ice-sheet striation of the sea-bed comes from sidescan sonar, 3.5 kHz profiles, and piston cores on the shelf (Barker *et al.*, in preparation).

The model of sequence deposition proposed by Larter & Barker (1989) is illustrated by Fig. 13. During an interglacial (Fig. 13A), an intergrounding subsequence (IGS) of unconsolidated glacial marine and pelagic sediment would be deposited on the shelf and slope. At some stage during the next cooling, an advancing ice sheet would ground on the shelf (Fig. 13B). Depending on changes in sediment supply from inshore, ice-sheet thickness, and basal temperature (Sugden, 1978; Hughes, 1981), a grounded ice sheet might erode part of the IGS or deposit (and consolidate) a basal till above it. More certainly, it would transport unsorted material in a basal debris zone to the shelf edge and upper slope to form the grounding subsequence (GS), steepening the slope (Fig. 13C). Thus a sequence begins, and ends, at the start of a glacial recession, with the floating of a previously grounded icesheet, or shortly beforehand if a thinning but still grounded ice sheet becomes everywhere depositional (Fig. 13D).

Note that the identified sequence boundaries probably represent only the most major ice advances. A particularly thick icesheet could erode all of the previous IGS on the shelf: since the sequence boundary is recognized by truncation of the upper ends of the foresets at the base of the topsets of the overlying IGS, such erosion would combine sedimentation within the previous and the current glacial–interglacial cycle into one sequence. Similarly, a rather weak glacial maximum, in which the ice sheet did not ground out to the shelf edge, would produce a less extensive and perhaps barely detectable unconformity on the shelf, and a less distinctive GS lithology on the slope. Thus the sedimentary record needs careful interpretation.

This model explains the particular bathymetric and seismic character of the shelf. The high seismic velocities observed in near-surface sediments are thought to be due to compaction resulting from the overburden of grounded ice around glacial maximum. That MCS profiles do not show evidence of major slope failure, despite the extremely steep angle of the continental slope, may be owing to the textural composition of the slope sediments. The model presented above predicts that these will be largely unsorted glacial marine diamictons (Powell, 1984). Laboratory tests on sand and gravel (Hoek & Bray, 1981) show that the friction angle (an indicator of slope stability) decreases with increased sorting. The particle surface roughness of much glacial marine diamicton will also tend to increase the friction angle (Rowe, 1962). Another factor influencing submarine slope stability is pore pressure (Prior & Coleman, 1984); overcompaction of the shelf sediments will increase stability in the region of the shelf break.

## DISCUSSION

### Vertical motion

The glacial sequences produced in the manner described above are particularly well preserved at the Pacific margin of the Antarctic Peninsula because of steady subsidence during deposition (Larter & Barker, 1989). As discussed earlier, theoretical models predict that thermal subsidence should follow uplift associated with RC–T collision. The interpretation of Profile BAS 878-19 in this paper, combined with data from DSDP Site 325, suggests that the onset of the period of restricted sediment supply to the margin, which is presumed to have preceded maximum uplift, occurred less than 2.2 Ma after the RC–T collision in that area. However, the onlap at the base of S3 on Profile AMG 845-08 is thought to indicate subsidence during its deposition. Elsewhere, the stratigraphic relationship of the base of S3 to ocean floor of age known from magnetic anomalies suggests that it is at least 5.6 Ma old, in which case subsidence was in progress in the area of Profile AMG 845-08 about 4 Ma after the local RC–T collision. Thus, if we assume that the shelf adjacent to each RC-T collision zone responded in a similar manner, the limited stratigraphic control available suggests that maximum uplift occurred about 1–4 Ma after RC–T collision.

On profiles AMG 845-07 and AMG 845-08 on the outer shelf the measured net rise in REL since maximum uplift (presumed to be due mainly to thermal subsidence) was found to increase with increasing distance from the MSH. On AMG 845-07 the net rise in REL was about 400 m close to the MSH, increasing towards the shelf edge at a rate of about 20 m/km. On AMG 845-08 the net rise in REL was about 990 m close to the MSH, increasing towards the shelf edge at a rate of about 60 m/km. The net rise in REL since maximum uplift on the outer shelf on profile BAS 878-19 was about 1280 m at a point about 17 km from the shelf break, with no evidence of significant lateral variation.

The simplest theoretical model for predicting the pattern of thermal subsidence following maximum uplift is to assume that the oceanic lithosphere emplaced beneath the fore-arc cools and subsides like normal young ocean floor. Although in these RC–T collisions, convergence stopped at collision time, it is probable that a subducted slab several hundred kilometres long still existed (Jarrard, 1986) and continued to sink into the mantle, requiring continued separation of it from the Antarctic plate, and some sort of spreading process beneath the fore-arc. In this case, the lithosphere beneath the outer shelf will have been thermally rejuvenated in the 1–2 Ma following RC–T collision as the spreading axis migrated beneath it. This, combined with the slow transfer of heat by conduction (DeLong *et al.*, 1979), probably explains the delay in the time of maximum uplift relative to RC–T collision time observed near the MSH. It also provides a possible explanation of the differential subsidence between the outer shelf and MSH observed on some profiles, since subsidence would have started later with increasing distance from the margin. If it is assumed that the average thermal age of the lithosphere beneath the outer shelf is the length of time since RC–T collision less 2 Ma, and that this lithosphere follows the same subsidence curve as young ocean floor (Trehu, 1975), the predicted thermal subsidence on Profile AMG 845-07 (5.6 Ma collision) is about 640 m, on AMG 845-08 (10.0 Ma collision) it is about 960 m, and on BAS 878-19 (16.5 Ma collision) about 1310 m.

Thus the amounts of subsidence predicted by the simple oceanic model agree remarkably well with the estimated rises in REL on these three profiles. However, this agreement may be fortuitous since several important considerations have been ignored.

**1** The estimates of REL change would have been greater if corrected for subsequent compaction within the measured parts of S3, S2, and S1.

**2** With deepening glaciation there has been a downward shift of erosion level relative to sea level of several hundred metres. True subsidence exceeds rise in REL by the shift in erosion level relative to sea level since the time of maximum uplift.

**3** The thermal component of subsidence is probably less than is predicted by the simple oceanic model, since the blanket of fore-arc material overlying the young lithosphere will smooth the peak of thermal uplift and retard its subsequent cooling (Fig. 14).

**4** Additional subsidence will have taken place beneath the outer shelf by isostatic response to post-

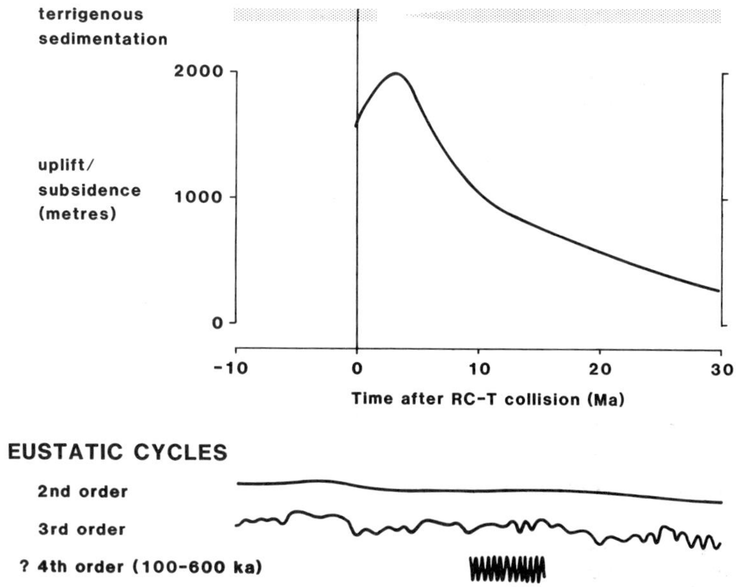


**Fig. 14.** Schematic diagram showing the interpreted temporal relations between ridge-crest–trench collision, uplift, and supply of terrigenous sediment to the margin. Portions of the 'long-term' (second order) and 'short-term' (third order) eustatic sea-level curves of Haq *et al.* (1987) are shown at the same scale for comparison, together with a crude representation of the more rapid fluctuations (fourth order) thought to have been important in producing S2 and S1.

collision sediment load and by compaction of pre-collision sequences under this load.

That the model predictions correspond well to the estimated REL changes suggests that errors in the estimates resulting from (1) and (2) combined with errors in the model predictions resulting from (3) are largely offset by additional subsidence due to sedimentary loading (4).

Sequences S1 and S2, which form the major part of the sediment load on the outer shelf, have been deposited by glacial action within the last 5.6 Ma. Both compaction of older sediments and isostatic compensation for S1 and S2 are expected to produce greater subsidence near the shelf edge (where the bulk of the load is located) than near the MSH. Too little is known about the precollision accretionary complex to model its compaction response, but the outer parts of some accretionary prisms (e.g. the Barbados fore-arc, Mascle *et al.*, 1988) are little more consolidated than undeformed ocean-floor sediments, so could compact substantially under subsequent load.

The isostatic response to load of the rejuvenated lithosphere beneath the outer shelf should be similar to that of young ocean floor, i.e. local, becoming more distributed with age (Turcotte & Schubert, 1982). Totally local isostatic compensation for the load of S1 and S2 would produce a maximum of about 1 km of additional subsidence near the shelf break on Profile AMG 845-08 (Crough, 1983). Even for an RC–T collision of this age (10.0 Ma) the response to loading by S1 and S2 will be distributed to some extent and will be more so in the older collision zones, so the maximum local response to loading will be less.

Detailed numerical modelling of vertical motion history at the margin through RC–T collision is likely to prove informative, but is beyond the scope of this paper.

## Sediment supply to the margin

Profile AMG 845–08 shows a sustained period of restricted supply of terrigenous sediment to the outer shelf following RC–T collision (during the deposition of S3). From this it has been inferred that the MSH continued to be a barrier to cross-shelf sediment transport even after considerable regional subsidence. It appears that sediment supply to the outer shelf and margin remained low from the time of uplift until the start of deposition of S2. The duration of the hiatus (or period of slow sedimentation) at Site 325, extrapolated to the continental slope on Profile BAS 878–19, suggests a similar barrier on the shelf off Adelaide Island (Fig. 2), despite the much longer interval between uplift there and the start of deposition of S2. It is inferred that in each RC–T collision zone the MSH became a positive topographic feature during the uplift that followed collision, and that it maintained a greater elevation than the surrounding shelf areas during later subsidence. Thus sediments derived from the Antarctic Peninsula would either have been trapped on the inner/middle shelf or diverted to an adjoining part of the margin where RC–T collision had not yet taken place. It seems that the early grounded ice-sheet advances of S2 age overrode the residual relief on the MSH, eroding it as they did so, and brought the first major influx of sediments to the outer shelf areas since their various times of uplift. Sediments trapped on the inner and middle shelf during this interval have probably been largely or wholly removed by subsequent shelf overdeepening.

## Importance to sequence stratigraphy

The basic concept of sequence stratigraphy (Vail *et al.*, 1977b; Haq *et al.*, 1987) is that global sea-level fluctuations have produced depositional sequences with synchronous boundaries on continental margins in many parts of the world. Change in the volume of continental ice sheets is often cited as a cause of global change in sea level (e.g. Vail *et al.*, 1977b; Beard *et al.*, 1982) and the difficulty in explaining such widespread and synchronous change otherwise has been a major source of interest in the early history of Antarctic glaciation. However, there is not a close agreement between the curves of global sea-level change deduced from sequence stratigraphy (Haq *et al.*, 1987) and the oxygen-isotope record (Miller *et al.*, 1987; Shackleton, 1987; Prentice & Matthews, 1988), both of which are presumed to record fluctuations in the volume of continental ice.

It has been argued (Larter & Barker, 1989, and this paper) that the S1 and S2 seismic sequences observed on the Pacific margin of the Antarctic Peninsula result directly from the action of grounded ice sheets. Under the model presented here (Fig. 13), the bounding unconformities of these sequences on the continental shelf coincide with glacial maxima and their succeeding deglaciations. Although the Antarctic Peninsula will carry only a small proportion of Antarctic ice at glacial maximum, glacial–interglacial fluctuations there are fairly simple and of large amplitude, are in phase with East Antarctic fluctuations, and, owing to rapid subsidence of the outer shelf, leave a particularly complete sedimentary record. Uplift and subsidence resulting from RC–T collision are of much longer period than glacial–interglacial fluctuations, so the tectonic and glacial components of REL change can easily be separated (Fig. 14).

On low-latitude continental shelves, in general terms, erosional unconformity is most widespread during sea-level lowstands, and the depocentre migrates offshore; rising sea level causes renewed sedimentation inshore, with a reduction in erosional unconformity. Thus, if global sea-level lowstands are assumed to be an indirect result of the accumulation of continental ice sheets, then high-latitude (as represented by the Antarctic Peninsula) and low-latitude sequences will be approximately in phase.

The global cycle chart of Haq *et al.* (1987) shows six sequences in the last 5.5 Ma. On the Pacific margin of the Antarctic Peninsula we have identified eight sequences that we are confident have been deposited in the last 5.6 Ma or less. Additionally, we recognize that the seismic evidence of some ice advances has probably been removed by later ice sheets which eroded all the intervening topset deposits, merging the adjacent sequences into one. Oxygen-isotope studies on deep-sea cores indicate that major climatic fluctuations have occurred over much shorter periods than hypothesized by sequence stratigraphers, at least during the last 700 ka (e.g. Shackleton, 1987). It is most likely that seismic stratigraphic investigations on both high and low-latitude margins have aliased a climatic signal which, in part, is beyond the resolution of the method. Support for this view comes from recent detailed studies of well logs, cores, and outcrops in parts of the USA (e.g. van Wagoner & Mitchum, 1989).

Direct correlation between the Antarctic Peninsula sequences and those on low-latitude margins, or with the oxygen-isotope curve, will require detailed stratigraphic data, which can only be obtained by drilling. Drilling of the outer shelf and slope sequences, together with high-resolution seismic profiling, would make possible a detailed study of the interaction between ice-volume fluctuations and global sea-level change. High-resolution stratigraphic data from palaeoslope deposits would be particularly interesting, since the model presented in this paper predicts the existence of whole glacial–interglacial sequences from which only the foresets are preserved. With the present, limited stratigraphic control it is only possible to highlight potential correlations with major, independently attested palaeoclimatic changes. Palaeoclimatic evidence (e.g. Hodell *et al.*, 1986; Barker *et al.*, 1988) suggests that the most likely time since 6 Ma ago for the onset of major ice-sheet grounding cycles on the Antarctic Peninsula shelf was either 2.4 or 4.8 Ma ago. If the former, then there is a crude similarity to the sequences described by Beard *et al.* (1982). If the latter, then the S1–S2 sequence boundary, which truncates foresets over a wide area of the shelf, could coincide with the further cooling 2.4 Ma ago.

## CONCLUSIONS

**1** All multichannel seismic reflection profiles crossing the Pacific margin of the Antarctic Peninsula to the southwest of the C FZ show the same three main geological provinces on the middle and outer shelf: a prograded outer shelf is separated from a small mid-shelf basin by the Mid-shelf High (MSH).

**2** A magnetic quiet zone, which extends up to 95 km across the continental rise from the base of the slope, is thought to be a consequence of burial of sections of the ANT–PHO ridge by terrigenous sediment before collision with the margin. The distribution of the quiet zone indicates vigorous terrigenous sedimentation continuing until at least 10 Ma ago, opposite some parts of the margin. The absence of a quiet zone opposite the two most recent RC–T collision zones indicates a change in the sedimentary regime, which is probably related to progressive climatic cooling.

**3** The Neogene RC–T collisions at the Pacific margin of the Antarctic Peninsula were followed by major uplift of the opposing section of the margin in the 1–4 Ma following each collision.

**4** In each case the greatest uplift occurred along the MSH, which would have been close to the shelf edge at the time. The MSH became a barrier to sediment transport from the Antarctic Peninsula to the continental slope and ocean floor, and the MSH itself was truncated by erosion. The unconformities thus produced are similar in character on seismic reflection profiles from several collision segments, but the time intervals they represent become progressively later from southwest to northeast along the margin, as would be expected from their inferred relationship to collision.

**5** Uplift has been followed by a period of steady subsidence, which has continued to the present day and is probably mainly thermal in origin. Estimates of subsidence based on seismic reflection profiles correspond well to the predictions of a simple thermal cooling model. However, deficiencies are apparent in both the estimates and the model. The close agreement between the two may indicate that errors produced by these deficiencies are cancelled by additional subsidence due to isostatic response to sedimentary load, and compaction of precollision margin sediments.

**6** During the last few million years ice sheets have grounded out to the shelf edge at times of glacial maximum, erosionally overdeepening the middle shelf and transporting glacial debris to the margin to produce a prograded outer shelf. Such action started later than 5.6 Ma ago, and possibly at one of the rapid climatic changes at 4.8 or 2.4 Ma ago.

**7** The prograded sequences produced on the outer shelf and slope by the action of these grounded ice sheets have a seismic character that is quite different from sequences on low-latitude margins. They have steep foresets (up to 17°) that are sharply truncated beneath the topset deposits of the overlying sequence. The planar erosion surface is usually a prominent reflector and ends at a sharp palaeoshelf-break. These sequences have high seismic velocities for young sediments, probably owing to compaction from the overburden of grounded ice at glacial maximum. The apparent stability of the continental slope, despite its very steep angle, may be due to the coarse, unsorted composition of the foreset deposits.

**8** Steady subsidence of the outer shelf, following uplift related to RC–T collision, has preserved a particularly complete sedimentary record of grounded ice-sheet advances. Eight sequences of the type described above are recognized (all less than 5.6 Ma old), but ice advances probably outnumber these, the recognized sequence boundaries in the topset beds representing only the most major advances. In contrast, the foreset deposits are expected to contain a virtually complete record of all ice advances.

**9** Short-term eustatic sea-level fluctuations are generally thought to be an indirect result of the waxing and waning of continental ice sheets. If so, the prograded sequences on the outer shelf and slope of the Antarctic Peninsula Pacific margin, which we contend result directly from the action of grounded ice sheets, will correlate closely with sequences identified on low-latitude, passive margins. However, precise stratigraphic correlation of these sequences with low-latitude margin sequences can only be achieved by drilling. Drilling, combined with high-resolution seismic profiling, on the Pacific margin of the Antarctic Peninsula would permit a detailed study of the interaction between ice-volume fluctuations and global sea-level change.

## ACKNOWLEDGEMENTS

We would like to thank the officers and crew who sailed on RRS *Discovery* cruises 154 and 172, and the staff of Research Vessel Services, Barry for invaluable technical support. MCS data acquired in 1985 were processed under contract by GECO (UK), and Profile BAS 878-19 was processed by BP Petroleum Development. We thank Rob Kidd for his comments, and Richard Hobbs for reproducing Profiles AMG 845-07 (Fig. 8) and AMG 845-08 (Fig. 5) on the Bullard Laboratories Versatec plotter. This research was supported in part by British Antarctic Survey research contract F60/E1/07 to Birmingham University, Department of Geological Sciences.

## REFERENCES

ANDERSON, J.B., BRAKE, C., DOMACK, E.D., MYERS, N. & SINGER, J. (1983) Sedimentary dynamics on the Antarctic continental shelf. In: *Antarctic Earth Science* (Eds Oliver, R.L., James, P.R. & Jago, J.B.) pp. 387–389. Australian Academy of Science, Canberra.

BARKER, P.F. (1982) The Cenozoic subduction history of the Pacific margin of the Antarctic Peninsula: ridge crest–trench interactions. *J. Geol. Soc. London* **139**, 787–801.

BARKER, P.F. & LAWVER, L.A. (1988) South American–Antarctic plate motion over the past 50 Myr, and the evolution of the South American–Antarctic Ridge. *Geophys. J.* **94**, 377–386.

BARKER, P.F., KENNETT, J.P. & SCIENTIFIC PARTY (1988) Weddell Sea palaeoceanography: preliminary results of ODP leg 113. *Palaeogeogr. Palaeoclimatol. Palaeoecol.* **67**, 75–102.

BARRETT, P.J. (1989) Antarctic Cenozoic history from the CIROS-I drillhole, McMurdo Sound. *Dept. Sci. Indus. Res. Bull. Wellington* **245**, 251 pp.

BARRON, J., LARSEN, B. & SCIENTIFIC PARTY (1988) Early glaciation of Antarctica. *Nature* **333**, 303–304.

BEARD, J.H., SANGREE, J.B. & SMITH, L.A. (1982) Quaternary chronology, paleoclimate, depositional sequences and eustatic cycles. *Bull., Am. Assoc. Petrol. Geol.* **66**, 158–169.

BRITISH ANTARCTIC SURVEY (1985) *Tectonic Map of the Scotia Arc, 1:3 000 000*. BAS, Cambridge, BAS (Misc) 3.

BYRNE, T. (1979) Late Paleocene demise of the Kula-Pacific spreading center. *Geology* **7**, 341–344.

CANDE, S.C. & LESLIE, R.B. (1986) Late Cenozoic tectonics of the southern Chile trench. *J. Geophys. Res.* **91**, 471–496.

CANDE, S.C., LESLIE, R.B., PARRA, J.C. & HOBART, M. (1987) Interaction between the Chile Ridge and Chile Trench: geophysical and geothermal evidence. *J. Geophys. Res.* **92**, 495–520.

CARLSON, R.L., GANGI, A.F. & SNOW, K.R. (1986) Empirical reflection travel time versus depth and velocity versus depth functions for the deep sea sediment column. *J. Geophys. Res.* **91**, 8249–8266.

CHRISTENSEN, N.I. & SALISBURY, M.H. (1975) Structure and constitution of the lower oceanic crust. *Rev. Geophys. Space Phys.* **13**, 57–86.

CROUGH, S.T. (1983) The correction for sediment loading on the seafloor. *J. Geophys. Res.* **88**, 6449–6454.

DAVIES, T.A., WESER, O.E., LUYENDYK, B.P. & KIDD, R.B. (1975) Unconformities in the sediments of the Indian Ocean. *Nature* **253**, 15–19.

DELONG, S.E. & FOX, P. (1977) Geological consequences of ridge subduction. In: *Island Arcs, Deep Sea Trenches and Back-arc Basins* (Eds Talwani, M. & Pitman, W.C.). Am. Geophys. Union, Maurice Ewing Ser. 1, 221–228.

DELONG, S.E., FOX, P. & MCDOWELL, R.W. (1978) Subduction of the Kula Ridge at the Aleutian Trench. *Geol. Soc. Am. Bull.* **89**, 83–95.

DELONG, S.E., SCHWARZ, W.M. & ANDERSON, R.M. (1979) Thermal effects of ridge subduction. *Earth Planet. Sci. Lett.* **44**, 239–246.

DEWEY, J.F. (1976) Ophiolite obduction. *Tectonophysics* **31**, 93–120.

DONOVAN, D.T. & JONES, E.J.W. (1979) Causes of world-wide changes in sea-level. *J. Geol. Soc. London* **136**, 187–192.

EITTREIM, S.L. & SMITH, G.L. (1987) Seismic sequences and their distribution on the Wilkes Land margin. In: *The Antarctic Continental Margin: Geology and Geophysics of Offshore Wilkes Land* (Eds Eittreim, S.L. & Hampton, M.A.) Circum-Pacific Coun. Energy Mineral Resour., Earth Sci. Ser. Houston, TX, 5A, 15–43.

ENGEBRETSON, D.C., COX, A. & GORDON, R.G. (1985) Relative motions between oceanic and continental plates in the Pacific basin. *Geol. Soc. Am. Spec. Pap.* **206**, 59 pp.

FORSYTHE, R.D., NELSON, E.P., CARR, M.J., KAEDING, M.E., HERVE, M., MPODOZIS, C., SOFFIA, J.M. & HARAMBOUR, S. (1986) Pliocene near-trench magmatism in southern Chile: a possible manifestation of ridge collision. *Geology* **14**, 23–27.

GIESKES, J.M. & LAWRENCE, J.R. (1976) Interstitial water studies, Leg 35. In: *Initial Reports of the Deep Sea Drilling Project*, Vol. 35, pp. 407–424. US Government Printing Office, Washington, DC.

HAQ, B.U., HARDENBOL, J. & VAIL, P.R. (1987) Chronology of fluctuating sea levels since the Triassic. *Science* **235**, 1156–1167.

HERRON, E.M. & TUCHOLKE, B.E. (1976) Sea floor magnetic patterns and basement structure in the southeastern Pacific. In: *Initial Reports of the Deep Sea Drilling Project,* Vol. 35, pp. 263–278. US Government Printing Office, Washington, DC.

HODELL, D.A., ELMSTROM, K.M. & KENNETT, J.P. (1986) Latest Miocene benthic $\delta^{18}O$ changes, global ice volume, sea level and the "Messinian salinity crisis". *Nature* **320**, 411–414.

HOEK, E. & BRAY, J.W. (1981) *Rock Slope Engineering*, 3rd edn. Institution of Mining and Metallurgy, London, 358 pp.

HOLLISTER, C.D., CRADDOCK, C. ET AL. (1976) *Initial Reports of the Deep Sea Drilling Project*, Vol. 35. US

Government Printing Office, Washington, DC, 930 pp.
Hughes, T.J. (1981) Numerical reconstruction of paleo-ice sheets. In: *The Last Great Ice Sheets* (Eds Denton, G.H. & Hughes, T.J.) pp. 221–261. Wiley, New York.
Irving, E. (1970) The Mid-Atlantic Ridge at 45°N XIV. Oxidation and magnetic properties of basalt: review and discussion. *Can. J. Earth Sci.* **7**, 1528–1538.
Jarrard, R.D. (1986) Relations among subduction parameters. *Rev. Geophys.* **24**, 217–284.
Josenhans, H.W. & Fader, G.B.J. (1989) A comparison of models of glacial sedimentation along the eastern Canadian margin. *Mar. Geol.* **85**, 273–300.
Kennett, J.P. (1977) Cenozoic evolution of Antarctic glaciation, the circum-Antarctic ocean, and their impact on global paleoceanography. *J. Geophys. Res.* **82**, 3843–3860.
LaBrecque, J.L., Kent, D.V. & Cande, S.C. (1977) Revised magnetic polarity time scale for late Cretaceous and Cenozoic time. *Geology* **5**, 330–335.
Larter, R.D. & Barker, P.F. (1989) Seismic stratigraphy of the Antarctic Peninsula Pacific margin: a record of Pliocene–Pleistocene ice volume and paleoclimate. *Geology* **17**, 731–734.
Larter, R.D. & Barker, P.F. (1990) A propagating subduction zone at the Antarctic Peninsula Pacific margin (abstract). *Geophys. J. Int.* **101**, 277.
Levi, S. & Riddihough, R. (1986) Why are marine magnetic anomalies suppressed over sedimented spreading centers? *Geology* **14**, 651–654.
Lonsdale, P. & Smith, D. (1986) Kula Plate not kula. *Eos (Trans. Am. Geophys. Union)* **67**, 1199.
Marshak, R.S. & Karig, D.E. (1977) Triple junctions as a cause for anomalously near-trench igneous activity between the trench and volcanic arc. *Geology* **5**, 233–236.
Mascle, A., Morre, J.C. et al. (1988) *Proceedings Ocean Drilling Project, Init. Rep. (Part A)*, Vol. 110. Ocean Drilling Program, College Station, TX, 603 pp.
Miller, K.G., Fairbanks, R.G. & Mountain, G.S. (1987) Tertiary oxygen isotope synthesis, sea-level history and continental margin erosion. *Paleoceanography* **2**, 1–19.
Mitchum, R.M., Vail, P.R. & Thompson, S. (1977) Seismic stratigraphy and global changes of sea level, part 2: the depositional sequence as a basic unit for stratigraphic analysis. In: *Seismic Stratigraphy – Applications to Hydrocarbon Exploration* (Ed. Payton, C.E.). Mem., Am. Assoc. Petrol. Geol. 26, 53–62.
Pankhurst, R.J. (1982) Rb–Sr geochronology of Graham Land, Antarctica. *J. Geol. Soc. London* **139**, 701–711.
Pitman, W.C., III (1978) Relationship between eustacy and stratigraphic sequences of passive margins. *Geol. Soc. Am. Bull.* **89**, 1389–1403.
Powell, R.D. (1984) Glacimarine processes and inductive lithofacies modelling of ice shelf and tidewater glacier sediments based on Quaternary examples. *Mar. Geol.* **57**, 1–52.
Prentice, M L. & Matthews, R.K. (1988) Cenozoic ice volume history: development of a composite oxygen isotope record. *Geology* **16**, 963–966.
Prior, D.B. & Coleman, J.M. (1984) Submarine slope stability. In: *Slope Instability* (Eds Brunsden, D. & Prior, D.B.) pp. 419–455. Wiley, Chichester.
Rex, D.C. (1976) Geochronology in relation to the stratigraphy of the Antarctic Peninsula. *Br. Antarct. Surv. Bull.* **43**, 49–58.
Rowe, P.W. (1962) The stress-dilatancy relation for static equilibrium of an assembly of particles in contact. *Proc. R. Soc. London, Ser. A* **269**, 500–527.
Savin, S.M., Douglas, R.G. & Stehli, F.G. (1975) Tertiary marine paleotemperatures. *Geol. Soc. Am. Bull.* **86**, 1499–1510.
Shackleton, N.J. (1987) Oxygen isotopes, ice volume and sea-level. *Quaternary Sci. Rev.* **6**, 183–190.
Shackleton, N.J. & Kennett, J.P. (1975) Paleotemperature history of the Cenozoic and the initiation of Antarctic glaciation: oxygen and carbon isotope analyses in DSDP sites 277, 279 and 281. In: *Initial Reports of the Deep Sea Drilling Project*, Vol. 29 (Eds Kennett, J.P., Houtz, R.E., *et al.*) pp. 743–755. US Government Printing Office, Washington, DC.
Shepard, F.P. (1973) *Submarine Geology*, 3rd edn. Harper & Row, New York, 517 pp.
Stagg, H.M.J. (1985) The structure and origin of Prydz Bay and MacRobertson Shelf, East Antarctica. *Tectonophysics* **114**, 315–340.
Storey, B.C. & Garrett, S.W. (1985) Crustal growth of the Antarctic Peninsula by accretion, magmatism and extension. *Geol. Mag.* **122**, 5–14.
Sugden, D.E. (1978) Glacial erosion by the Laurentide ice sheet. *J. Glaciol.* **20**, 367–91.
Tanahashi, M., Saki, T., Oikawa, N. & Sato, S. (1987) An interpretation of multichannel seismic reflection profiles across the continental margin of the Dumont d'Urville Sea, off Wilkes Land, East Antarctica. In: *The Antarctic Continental Margin: Geology and Geophysics of Offshore Wilkes Land* (Eds Eittreim, S.L. & Hampton, M.A.). Circum-Pacific Coun. Energy Mineral Resour. Earth Sci. Ser., Houston, TX, 5A, 1–13.
Trehu, A.M. (1975) Depth versus $(age)^{1/2}$: a perspective on mid-ocean rises. *Earth Planet. Sci. Lett.* **27**, 287–304.
Tucholke, B.E. & Houtz, R.E. (1976) Sedimentary framework of the Bellingshausen Basin from seismic profile data. In: *Initial Reports of the Deep Sea Drilling Project*, Vol. 35, pp. 197–227. US Government Printing Office, Washington, DC.
Tucholke, B.E., Edgar, N.T. & Boyce, R.E. (1976) Physical properties of sediments and correlations with acoustic stratigraphy: Leg 35, Deep Sea Drilling Project. In: *Initial Reports of the Deep Sea Drilling Project*, Vol. 35, pp. 229–249. US Government Printing Office, Washington, DC.
Turcotte, D.L. & Schubert, G. (1982) *Geodynamics – Applications of Continuum Physics of Geological Problems*. Wiley, New York, 450 pp.
Uyeda, S. & Miyashiro, A. (1974) Plate tectonics and the Japanese Islands: a synthesis. *Geol. Soc. Am. Bull.* **85**, 1159–1170.
Vail, P.R., Mitchum, R.M. & Thompson, S. (1977a) Seismic stratigraphy and global changes of sea level, part 3: relative changes of sea level from coastal onlap. In: *Seismic Stratigraphy – Applications to Hydrocarbon Exploration* (Ed. Payton, C.E.). Mem., Am. Assoc. Petrol, Geol. 26, 63–82.
Vail, P.R., Mitchum, R.M. & Thompson, S. (1977b) Seismic stratigraphy and global changes of sea level, part 4: global cycles of relative changes of sea level. In:

*Seismic Stratigraphy – Applications to Hydrocarbon Exploration* (Ed. Payton, C.E.). Mem., Am. Assoc. Petrol. Geol. 26, 83–98.

VAN WAGONER, J.C. & MITCHUM, R.M. (1989) High frequency sequences and their stacking patterns. *Abstracts, 28th International Geological Congress*, Washington **3**, 284.

VAN WAGONER, J.C., POSAMENTIER, H.W., MITCHUM, R.M., VAIL, P.R., SARG, J.F., LOUTIT, T.S. & HARDENBOL, J. (1988) An overview of the fundamentals of sequence stratigraphy and key definitions. In: *Sea-level Changes – an Integrated Approach* (Eds Wilgus, C.K., Hastings, B.S., Kendall, C.G.St.C., Posamentier, H.W., Ross, C.A. & van Wagoner, J.). Soc. Econ. Paleontol. Mineral. Spec. Publ. 42, 39–45.

VOGT, P.R., ANDERSON, C.N., BRACEY, D.R. & SCHNEIDER, E.D. (1970) North Atlantic magnetic smooth zones. *J. Geophys. Res.* **75**, 3955–3968.

VORREN, T.O., LEBESBYE, E., ANDREASSEN, K. & LARSEN, K.-B. (1989) Glacigenic sediments on a passive continental margin as exemplified by the Barents Sea. *Mar. Geol.* **85**, 251–272.

# Ancient Arc and Transform Margins

*Spec. Publs. int. Ass. Sediment.* (1991) **12**, 189–200

# Relative importance of regional tectonics and eustasy for the Mesozoic of the Andes

A. HALLAM
*School of Earth Sciences, University of Birmingham, Birmingham B15 2TT, UK*

## ABSTRACT

In various parts of the Andes there is a good marine record of Mesozoic deposits from the Norian to the Maastrichtian. These were laid down with associated continental deposits in a series of basins formed by extensional tectonics and thermal subsidence, the best known of which are the back-arc Andean and Magallanes Basins. Of several episodes of regional tectonics and associated igneous activity during this period, causing widespread uplift and regression of the sea, the most important took place in the southern Andes in the Mid-Cretaceous, and the second most important in Late Jurassic times. Both of these events were probably associated with eastward shifts of the subduction zone related to the opening of the South Atlantic. During the much longer intervals of tectonic quiescence, the pattern of marine sedimentation substantially reflected global events, suggesting that eustasy was a significant controlling factor.

## INTRODUCTION

Substantial thicknesses of marine Mesozoic strata are found throughout extensive tracts of the Andes, with sufficient ammonites occurring to allow correlation with classic sections in Europe and North America to at least substage level. The large distance (several thousand km) from the North Atlantic margins, where most of the data was obtained for the successive Exxon sea-level curves (Vail *et al.*, 1977,1984; Haq *et al.*, 1987) allows, among other things, a test of Watts' (1982) claim that Vail and his colleagues have confused eustasy with regional tectonic events associated with the opening of the North Atlantic.

The presence of significant quantities of calc-alkaline volcanic rocks intercalated with both marine and continental sediments, and associated granitoid plutons, signifies a more or less continuous history of subduction along the Andean margin since the late Palaeozoic. With regard to northern Chile, in the early Mesozoic, volcanicity was substantially confined to the western zones, such as the Coastal Cordillera of Chile, but the principal locus of igneous activity subsequently shifted eastwards, most notably from Mid-Cretaceous times onwards, such that at the present day the currently or recently active volcanoes are situated well inland from the Pacific margin. High sedimentation rates in some localities are signified by thicknesses of more than a kilometre for individual stratigraphic stages. A large proportion of the coarser clastic debris is evidently volcanic in origin, and the same probably applies to many of the finer clastic rocks as well.

Tight folding in many areas signifies compressive tectonic activity at various times, but there is also considerable evidence of extensive regional tensional tectonics. It is to episodes of tension that the principal depositional basins of the Andes owe their origin, with initial horst-and-graben fault-controlled collapse being succeeded by more widespread thermal subsidence. Figure 1 shows the major present-day basins that preserve considerable thicknesses of Cretaceous strata, according to Macellari (1988). Macellari's Neuquén Basin is better termed the Andean Basin (Malumian *et al.*, 1983), with the name Neuquén being reserved for an embayment in the southeast. The Andean Basin is valid also for the earlier Mesozoic but the Magallanes Basin did not develop before the latest Jurassic. There was extensive deposition of Triassic and Early Jurassic marine strata in Peru, but the distribution of basins is

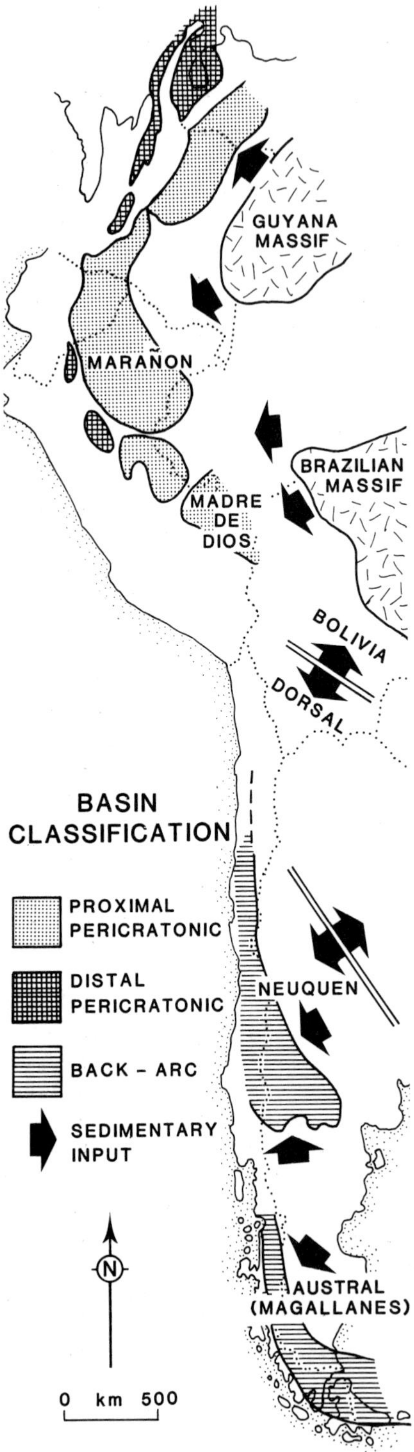


**Fig. 1.** Cretaceous basins of the Andes. Simplified from Macellari (1988, fig. 2).

unclear because of considerable post Mid-Jurassic erosion. North of Peru most pre-Cretaceous sediments are continental and in consequence poorly dated. In contrast to cordilleran North America, displaced terranes are apparently rare in the Andes, being confined to the Western Cordillera of Colombia and the western margin of southern Chile (e.g. McCourt *et al.*, 1984; Hallam, 1986).

## TRIASSIC AND JURASSIC

Regional accounts of Triassic and Jurassic stratigraphy are provided by Mégard (1978), Geyer (1973,1980), Riccardi (1983), Stipanicic (1983) and Gröschke & Hillebrandt (1985). It is convenient to make a distinction between the southern Andes, embracing cordilleran Argentina and Chile, and the northern Andes, embracing Peru, Ecuador, Colombia, and Venezuela.

### Southern Andes

The best region for study is the Andean Basin, which occupies an extensive area in a back-arc setting with respect to the Coastal Cordillera of Chile (Malumian *et al.*, 1983; Macellari, 1988). The Magallanes or Austral Basin did not develop until near the end of Jurassic times, and the Jurassic stratigraphy is still not precisely defined. The deposits of the Coastal Cordillera of Chile have a high proportion of volcanic rocks and the limited amount of intervening sediment is usually poorly dated. In consequence these regions will not be considered further here.

Apart from beds with a few Anisian fossils on the north central Chilean coast (Stipanicic, 1983) the oldest marine Triassic is late Norian (including Rhaetian) in age. Together with deposits of the basal Jurassic (Hettangian) stage, these rocks occur only in the northern part of the Andean Basin, in and around the pre-cordilleran Domeyko Range, where the thickness of the combined Triassic and Jurassic strata may reach 3000 m (Gröschke & Hillebrandt, 1985). Above a series of continental clastic rocks, including conglomerates, with horizons of Triassic plants, shelly marine limestones pass up into ammonite-bearing laminated black shales with extremely sparse benthos, which pass up without change of facies into Hettangian. The lithology thereafter is rather uniform, with a series of thin-bedded silty shales with subordinate sandstone, and horizons of calcitic concretions. Ammonites are moderately common but benthic fossils extremely rare, apart from horizons of the bivalve *Bositra*. In the thicker sequences of the Domeyko Range, turbidites are common. The author shares Gröschke & Hillebrandt's (1985) opinion that the strata were probably deposited in the deeper neritic zone below the depth at which sea-level fluctuations would be clearly recorded in the succession.

Away from the basin centre a number of marine transgressions can be recognized, subsequent to the late Norian transgression that caused the sea to enter the northern part of the Andean Basin. Marine Hettangian rests on continental Triassic in the Coastal Cordillera of northern Chile. A widespread late Sinemurian transgression caused the sea to spread both eastwards towards the High Cordillera of northern Chile and southwards into the Neuquén Embayment, where up to 2500 m of Lower and Middle Jurassic strata are recorded (Riccardi, 1983). North of the mid-Domeyko region (24°S), marine Pliensbachian rests on continental Triassic (Gröschke & Hillebrandt, 1985). During the Early Jurassic there was in the High Cordillera a more or less gradual transgression eastwards (Hillebrandt, 1971,1973). In the Cerro Jaspé, northeast of Calamá (22°S), marine Bajocian limestones rest directly on Palaeozoic granodiorite. Transgressive beds of various ages on the basin margins are characteristically rich in benthos, principally a moderate diversity of bivalves, gastropods, and brachiopods, in marked contrast to the more basinal deposits.

A number of regressive deposits are also recognizable. There is a widespread gypsum deposit, up to 200 m thick, stratigraphically located at about the Oxfordian–Kimmeridgian boundary, which is traceable for over 800 km throughout the Andean Basin (Riccardi, 1983). In the Domeyko Range it passes up into marine Kimmeridgian, but farther north the gypsum is followed by continental deposits. This is the case also farther south, with thin marine Kimmeridgian passing into a thick unit of coarse continental volcaniclastics overlain by marine Tithonian, up to 3 km thick in the Volcán Valley southeast of Santiago (Hallam *et al.*, 1986). Eastwards into Argentina this unit becomes thinner and finer grained, with minor intercalations of marine Kimmeridgian deposits. In the Rio Atuel region of Argentina, close to the Chilean border southeast of Santiago (34°S), a thin gypsum unit occurs within marine strata containing Callovian fossils. Over an extensive region of northern and central Chile, Lower Callovian or uppermost Bathonian strata rest non-sequentially on upper Lower Bajocian strata, indicating a widespread regressive event in the late Bejocian or early Bathonian (Riccardi, 1983).

The continental Kimmeridgian deposits of central Chile and adjacent Argentina are abruptly overlain by fully marine Tithonian limestones and shales with abundant ammonites and bivalves, which pass up into the Cretaceous without facies charge or stratigraphic break (Hallam *et al.*, 1986). Further north, marine Tithonian can be recognized intermittently and in the Asientos region of northern Chile can be seen resting non-sequentially on Callovian (Perez, 1982).

### Northern Andes

A marine sequence ranging in age from the Upper Norian to Middle Jurassic occurs over extensive areas of the Peruvian Andes, where it is non-sequentially overlain by non-marine Lower Cretaceous sandstones; in the more northerly areas these overlie Lower Jurassic (Mégard, 1978; Westermann *et al.*, 1980; Loughman & Hallam, 1982; Prinz, 1985). The Norian Chambará Formation, composed of limestones and dolomites with a rich molluscan, brachiopod, and coral fauna, passes up into the basal Jurassic Aramachay Formation consisting mainly of laminated black silty shales rich in ammonites but extremely poor in benthos; locally this has a transgressive base, with Lower Sinemurian strata overlying Triassic. As in northern Chile there is no apparent hiatus or facies change between the Triassic and Jurassic. The deeper water Hettangian and Sinemurian deposits pass up gradually into a shallower water limestone sequence of Pliensbachian and Toarcian age, the Condorsinga Formation. Ammonites are too sparse in this formation to determine stage or substage boundaries precisely. Passing northwards into Ecuador and Colombia, marine transgressive units of Norian, Sinemurian, and Tithonian age can be detected in a thick continental or marginal marine volcaniclastic sequence, deposits of the first two stages being readily traceable into, respectively, the Chambará and Aramachay Formations of Peru (Geyer, 1974,1980). Because of the paucity of fully marine deposits this large region is unsuitable for comprehensive eustatic analysis.

### Comparison of sea-level curves

By comparing a latest Triassic to early Jurassic rela-

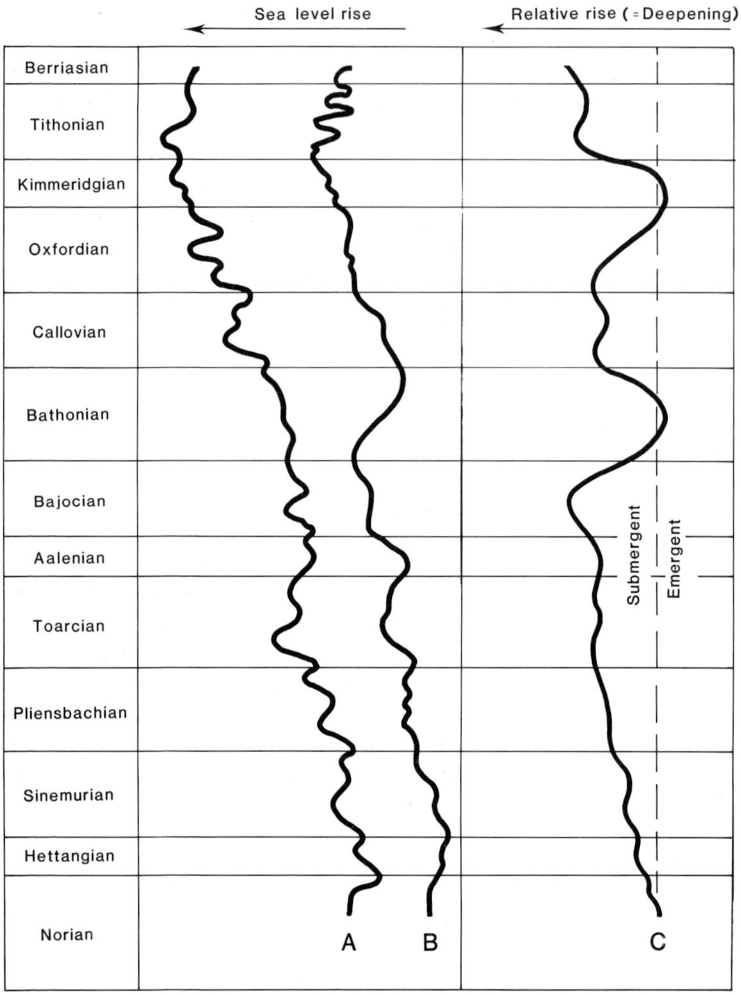


**Fig. 2.** Comparison of latest Triassic to Jurassic eustatic curves (A) and (B) with relative sea-level curve (C) for the Andean Basin. (A) After Hallam (1988, fig. 10); (B) after Haq *et al.* (1987, figs 4, 5).

tive sea-level curve established for Andean South America with two independently derived eustatic curves it should be possible to distinguish regional from global events at least to a first order of approximation. For the reason outlined above it is necessary to concentrate attention on the Andean Basin of Chile and Argentina.

The curve of Fig. 2C indicates a more or less gradual rise from the late Norian to the Bajocian when, according to Riccardi (1983), the Jurassic seas reached their greatest extent. Within this general trend a number of pulses of sea-level rise can be recognized, marked by marine transgressions over continental deposits and/or deepening events within the marine sequence. The first of these is obviously the late Norian, with the initiation of marine deposition in the Andean Basin and areas to the north extending as far as Colombia. The next is in the early Sinemurian, marked by spread of a distinctive black shale facies in Peru (Loughman & Hallam, 1982) and similar *Arnioceras*-rich deposits sandwiched between non-marine facies in Ecuador and Colombia (Geyer, 1980). A late Sinemurian transgressive event is also clearly signalled within the Andean Basin, with a major spread of sea to the east and south. A mid- to late Pliensbachian transgression is recorded locally (Gröschke & Hillebrandt, 1985), which evidently corresponds to a deepening pulse within the major depocentre that continued into the early Toarcian. This is clearly shown in the Rio

Transito region of Chile east of Vallenar (28°S), where the Early Jurassic deposits are of generally shallower water character than the region further north (Hillebrandt & Schmidt-Effing, 1981). Here the late Pliensbachian and early Toarcian is represented by benthos-poor organic-rich silty shales sandwiched between Upper Sinemurian to lower Pliensbachian and Middle Toarcian limestones rich in benthic fossils including the large bivalves *Weyla, Gryphaea*, and *Lithiotis*, which were without doubt deposited in the shallow neritic zone. The mid-Toarcian shallowing event, recognizable also in Peru, was followed by a late Toarcian deepening marked by the widespread development of *Phymatoceras*-rich shales. An Aalenian shallowing event is clearly recorded in the Rio Transito region by the occurrence of bioclastic limestones extremely rich in a diversity of bivalves dominated by *Gryphaea*, *Lopha*, and trigoniids. A widespread late early Bajocian transgression is recorded by benthos-rich *Stephanoceras*-bearing limestones overstepping rocks of a variety of ages down to Palaeozoic (as in the Cerro Jaspé).

After the Bajocian an extensive regression took place over a large area of the Andean Basin, followed by a transgression traditionally dated as early Callovian, but more likely in fact to have started in latest Bathonian times (Riccardi, 1983; Westermann *et al.*, 1984). Much depends on the precise age assignation of the *Neuqueniceras* fauna. Westermann *et al.* (1984) date it as latest Bathonian but in a more recent paper it is referred to the early Callovian (Riccardi *et al.*, 1989). Not everywhere, however, do the *Neuqueniceras* beds rest on Bajocian strata, and a possibly complete Bathonian sequence is present locally, suggesting that the post-early Bajocian regression was regional even within the Andean Basin. A minor regressive event within the Callovian is recorded locally by the gypsum horizons in the Rio Atuel and Chacay Melehue sections in Argentina (Riccardi, 1983; Riccardi *et al.*, 1989), and a more regionally significant one, affecting the whole Andean Basin, at or near the end of the Oxfordian, with a major episode of gypsum deposition. The Kimmeridgian is mainly regressive, being developed for the most part in continental volcaniclastic facies, and is sharply overlain by marine Tithonian; two transgressive/deepening pulses can be inferred, in the early and late Tithonian (Hallam *et al.*, 1986). The Tithonian transgression is a major event through much of the Andes, being recognizable both in Colombia and over a wide region of Chile and Argentina. The absence of Jurassic marine deposits in the northern Andes between Bajocian and Tithonian indicates an extensive regional regression, which cannot, however, be precisely dated. It could correspond either to the Bathonian or late Oxfordian to Kimmeridgian regressive events recognizable in the Andean Basin, to a compound of both or perhaps to some other event. Jurassic marine conditions in Venezuela and northern Colombia only developed after the opening of the Caribbean. Before that time only continental facies (La Quinta, Giron Formations) were deposited within rapidly subsiding grabens.

Figures 2A, 2B are tentative eustatic curves proposed respectively by Hallam (1988) and Haq *et al.* (1987). Figure 2A is a modification of an earlier curve (Hallam, 1978) based on a combination of facies analysis of stratigraphic sections and percentage flooding of the continents in successive stages, as determined from palaeogeographic maps inferred from the spread of marine sediments. Figure 2B is a modification of two earlier efforts by the Exxon group based on seismic stratigraphy. Vail & Todd (1981) put forward a eustatic curve based entirely on seismic sequence analysis of North Sea data, which carries the clear implication that analysis of sequences in one region will record a global signature. The Jurassic part of the eustatic curve of Vail *et al.* (1984) differs little from that of Vail & Todd except that sea level is no longer confused with coastal onlap.

Subsequently the Exxon group attempted to correlate their data derived from analysis of seismic sequences on the continental shelf with information from a few classic sequences exposed onshore in western Europe, principally in Dorset and Yorkshire, England. This led to a number of modifications to their curve. Thus the lowest Jurassic sea-level stand is now seen as end-Hettangian rather than early Sinemurian, and a gradual rise of sea level is recognized through the Sinemurian and Pliensbachian rather than there having been virtually no change through this interval. The early Toarcian rise and early Aalenian fall become more pronounced and a Bathonian fall is recognized, to a low near the end of the stage, rather than a more or less stable stand. Otherwise their two curves are essentially similar, most obviously for the Tithonian, which is seen as a time of considerable sea-level instability. With regard to the amount of change, a c. 100 m rise through the Jurassic in Vail *et al.* (1984) becomes a c. 200 m rise in Haq *et al.* (1987), but the reasons for adopting this change are not explained.

Hallam (1988) criticized the Exxon group's work on two grounds, that the facies model adopted was inapposite to the classic European sequences deposited on a relatively stable craton at modest rates of subsidence, and that insufficient attention had been paid to the distorting effects of regional tectonics. Thus the pronounced Aalenian and Bathonian regressions, and the strongly oscillatory Tithonian part of the curve, primarily reflect events in and around the North Sea. A number of other differences are recognizable from comparison of Fig. 2A and 2B, and not unnaturally the author favours the curve in Fig. 2A as better reflecting the world picture, but there is agreement on other points, such as notable Sinemurian, early Toarcian, and early Bajocian rises. Of more general significance is that the long-term trend is similar, namely a sea-level rise from its lowest value at or close to the start of the Jurassic, checked somewhat in the mid-Jurassic, to a Kimmeridgian to early Tithonian highstand, before reversing later in the Tithonian.

Turning back to the Andean curve (Fig. 2C), it is important to note that biostratigraphically well-studied sections, such as those in the Domeyko Range, were deposited in water too deep to record minor fluctuations of sea level (cf. Gröschke & Hillebrandt, 1985) and correlation between Europe and South America can often not be achieved at a level more refined than substage. The general trend from the beginning of the Jurassic to the Bajocian is, however, remarkably similar to Figs 2A, 2B. Major regressions in Bathonian and late Oxfordian/Kimmeridgian times had the effect of severely restricting the Jurassic seas in the Andes, which were barely restored to their early mid-Jurassic level by the extensive Tithonian transgression. On a more refined scale it is noteworthy that there were important transgressive pulses in the Hettangian, early and late Sinemurian, late Pliensbachian to early Toarcian, early Bajocian, early Callovian, and early Tithonian. These all correspond to important episodes of sea-level rise elsewhere in the world. As regards the Late Triassic, the significant sea-level rise in the Andes marked by the extensive late Norian transgression is not obviously reflected globally. It is difficult to be precise about the time of inception of marine conditions, but it probably predates slightly the Rhaetian transgression clearly recognizable in northern Europe and some other parts of the world (Loughman & Hallam, 1982; Hillebrandt, 1990). Possibly the deepening phase exhibited in northern Chile and Peru shortly before the end of the Triassic corresponds with this latest Triassic event.

## CRETACEOUS

It is conventional among South American geologists to include informally the Tithonian with the Cretaceous (Malumian *et al.*, 1983; Macellari, 1988). Macellari (1988) has recently provided a valuable review of Andean Cretaceous deposits in relationship to major depositional cycles and degree of marine inundation. He distinguishes three regions with their own individual history that need to be discussed separately: the northern region from Venezuela to Peru; the region of northern and central Chile and Argentina, known variously as the Andean or Neuquén Basin; the region embracing the southern limits of Chile and Argentina, known as the Magallanes or Austral Basin (Fig. 1). In each case Macellari relates the depositional history to basin development.

### Northern Andes

Five major cycles are recognized. Cycle 1 (Tithonian–Valanginian) is restricted to NE–SW-trending grabens originated at an early rifting stage, with no break being recognizable across the Jurassic–Cretaceous boundary (a phenomenon similar to and already noted for the Andean Basin). During cycle 2 (Valanginian to Upper Aptian) the pattern of subsidence changed from being more confined and tectonically driven to being more widespread and thermally driven. During cycle 3 (Upper Aptian to mid-Cenomanian), generalized thermal subsidence continued and a high sea-level stand resulted in the flooding of several previously emergent palaeohighs. In cycle 4 (Upper Cenomanian? to Lower Campanian) the maximum flooding of North and South America was achieved, followed by regression during cycle 5 (Upper Campanian to Maastrichtian).

### Andean Basin

The Mesozoic stratigraphy of this region, with up to 6000 m of Jurassic and Cretaceous sediments, is relatively well known both in outcrop and subsurface because of extensive exploration for petroleum (see paper by Legaretta & Uliana). Two major unconformities related to episodes of regional tectonic disturbance have been recorded: the 'Araucanian' (Kimmeridgian) and 'Peruvian' (Mid-Cretaceous) phases. Early Mesozoic extension was followed by thermal subsidence in a back-arc setting during the Jurassic and Early Cretaceous. The palaeogeographic map of Fig. 3 illustrates this setting for the Tithonian, and Fig. 4 indicates the principal regional facies changes through the latest Jurassic to Early Cretaceous marine interval sandwiched between thick continental volcaniclastic sequences in the western part of the basin. Eastwards into Argentina the proportion of volcanogenic sediments and volcanics diminishes with the overall stratigraphic thickness. Macellari (1988) recognizes six cycles, the earliest two of which are illustrated in Fig. 4 (Tithonian to Lower Valanginian, Upper Valanginian to Lower Hauterivian). Cycles 3 and 4 (Upper Hauterivian to mid-Cenomanian) are represented by beds showing more restricted marine influence and cycle 5 (mid-Cenomanian to Lower Campanian?) has exclusively continental deposits. Cycle 6 (Upper Campanian? to Paleocene) is marked by a renewal of marine deposits signifying a widespread latest Cretaceous transgression.

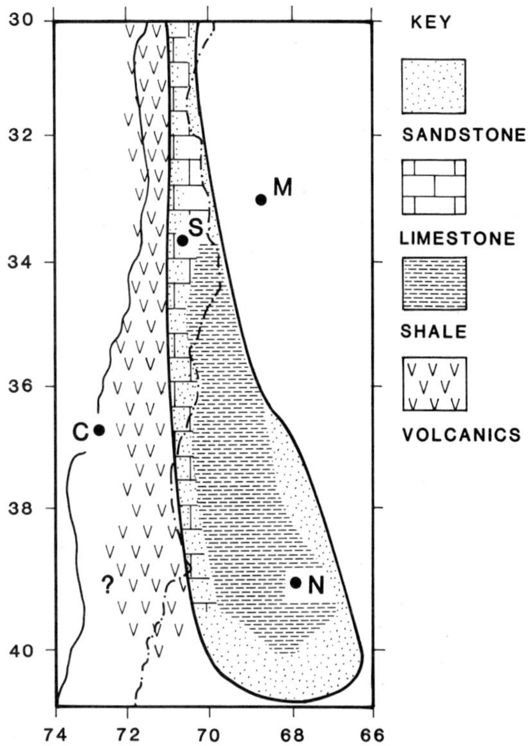


**Fig. 3.** Palaeogeographic sketch map for the southern part of the Andean Basin in the Tithonian. S, Santiago; C, Concepcion; M, Mendoza; N, Neuquén. After Hallam *et al.* (1986, fig. 5).

### Magallanes Basin

This originated in a latest Jurassic extensional episode associated with the formation of a marginal basin to the west, followed as in the other basins farther north by thermal subsidence. By Late Cretaceous times a thick pile of sediments accumulated in a foreland setting following Andean uplift and the formation of a foredeep adjacent to the uplifted orogen. The boundaries of the five cycles recognized by Macellari are more poorly defined than for the other basins because of insufficient biostratigraphic control. It is apparent, however, that maximum flooding took place in the late Hauterivian and that there was significant regression thereafter, associated with the Mid-Cretaceous closure of the western marginal basin and formation of the 'Palaeoandes'. A major hiatus exists between the Upper Cenomanian and Lower Coniacian.

Macellari's scheme of cycles for the three regions is shown in Fig. 5. Only two unconformities have a continent-wide distribution, the intra-Cenomanian and intra-Campanian, and only the latter corresponds with a major unconformity in the Haq *et al.* (1987) global scheme. Precise correlation with the Haq *et al.* sequences is difficult because of the poor dating of regressive facies, and because the earliest transgressive deposits are often poorly fossiliferous. (On a more detailed scale, the same problem of insufficient biostratigraphic control limits the possibility of precise correlation of the sequences recognized in the chapter by Legaretta & Uliana in the Andean Basin with the Haq *et al.* sequences.) Reliable dates provided by marine intervals in northern South America indicate that the contacts between cycles 1 and 2, 2 and 3, and 4 and 5 coincide with major unconformities in the Haq *et al.* scheme. For the Andean Basin, only cycles 1 and 2 and 4 and 5 bear some resemblance to their scheme.

An alternative approach to sea-level analysis is to plot a curve indicating the estimated percentage of area flooded through time. This was done for South America by Macellari (1988) and is reproduced in Fig. 6. The curve for northern South America indicates a maximum flooding in the Mid-Cretaceous, with a mid-Cenomanian sea-level fall intervening

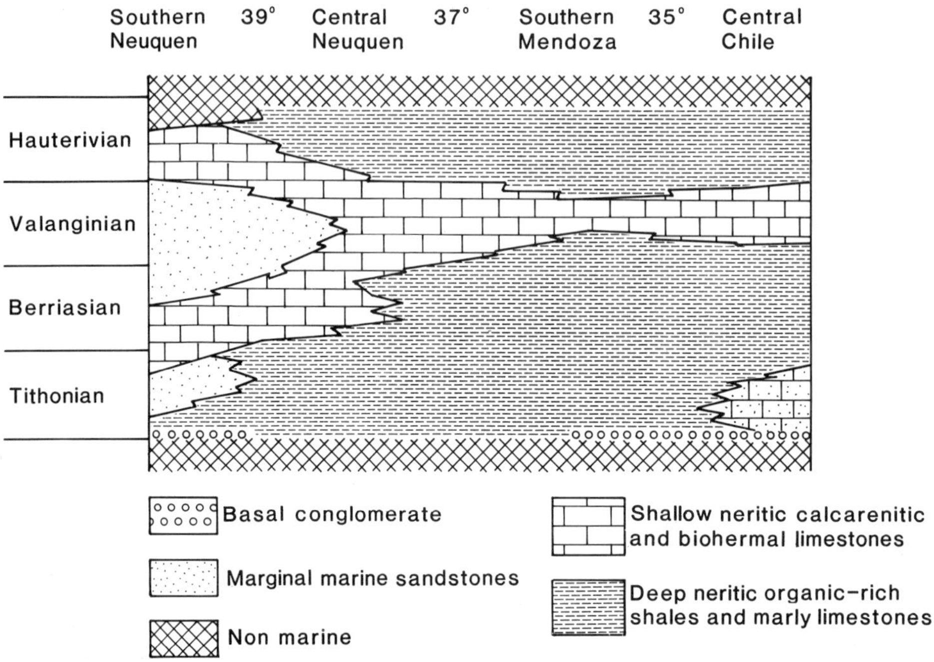


**Fig. 4.** Major facies patterns for the marine Tithonian–Neocomian of the southern part of the Andean Basin. After Hallam *et al.* (1986, fig. 4).

between late Albian and Turonian highstands (Macellari, 1988). It is broadly similar to the Haq *et al.* (1987) curve of long-term sea-level change and remarkably similar to the Sliter (1976) curve based on estimated flooding of the continents. On the other hand, the curve for southern South America is quite different, indicating Mid-Cretaceous regression and Late Cretaceous transgression. This

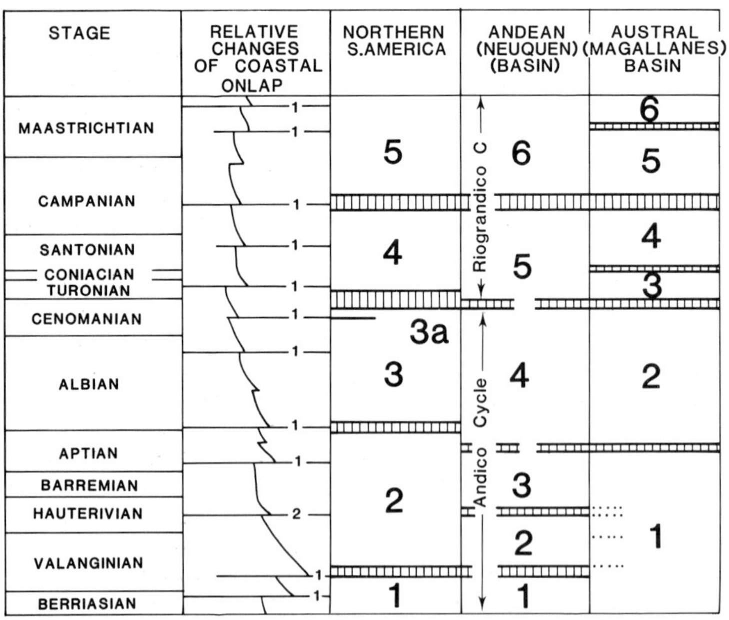


**Fig. 5.** Comparison of Cretaceous cycles recognized in the northern Andes and in the Andean and Magallanes Basins with the global chart of coastal onlap of Haq *et al.* (1987). Simplified from Macellari (1988, fig. 26).

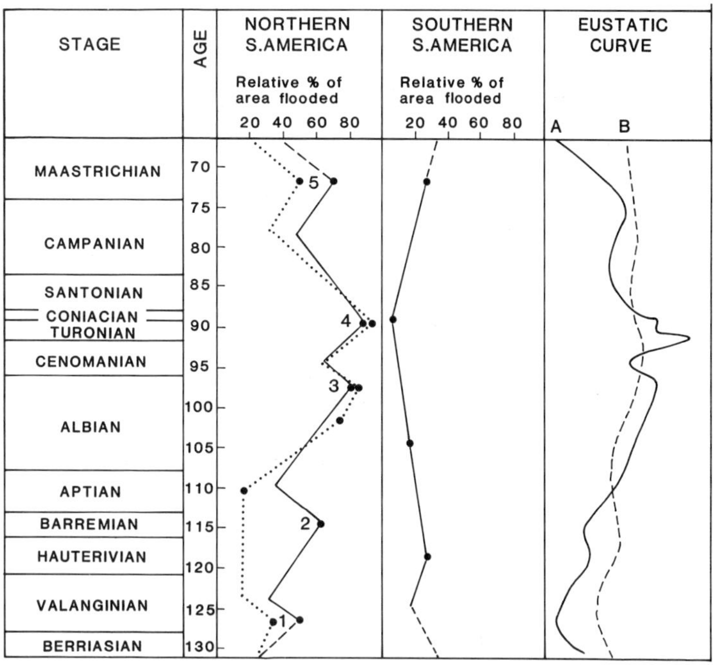


**Fig. 6.** Comparison of variation in percentage of area flooded in northwestern South America (solid line, Venezuela–Colombia; dotted line, Equador–Peru; numbers indicate major flooding events) and southern South America, with eustatic fluctuations and percentages of area flooded (solid line after Sliter (1976), broken line after Haq *et al.* (1987)). Simplified from Macellari (1988, fig. 27).

is clearly the consequence of regional tectonics, because both in the Andean and Magallanes Basins there is evidence of increased igneous activity and tectonic disturbance in the Mid-Cretaceous, with angular unconformities in the stratigraphic sequences, associated with an early phase of uplift of the Andes and concomitant destruction or severe modification of the back-arc basins.

## CONCLUSIONS

That there are some notable differences between the marine Mesozoic stratigraphy of Andean South America and Europe is readily apparent if one considers the system boundaries. While in Europe the Triassic–Jurassic and Jurassic–Cretaceous boundaries are located at intervals of widespread regression there is no obvious facies change across these boundaries where they are well exposed in South America. Indeed, because one of the principal Andean regressive intervals precedes the Tithonian, this stage is normally classed by South American geologists with the Cretaceous, with which it is intimately associated. Similarly, without evidence from other parts of the world it would be quite logical for the same geologists to class the Norian with the Jurassic.

There were several important episodes of regional uplift and associated sedimentary regression and/or erosional truncation in the Andes, affecting areas of hundreds of thousands of square kilometres, if not substantially more, thereby ruling out merely local tectonics. In the northern Andes there were one or more episodes within the Mid- or Late Jurassic. More precise dating is possible further south: late Bajocian to Bathonian, late Oxfordian to Kimmeridgian, and, most important of all, Mid-Cretaceous. However, during the longer periods of relative tectonic and igneous quiescence, Andean South America experienced a pattern of marine inundation that can be matched quite closely to the world picture, suggesting the operation of eustatic control. It is almost always easier in eustatic analysis to recognize signals of marine deepening/transgression than shallowing/regression. Successive Norian, Hettangian, Sinemurian, Pliensbachian–Toarcian, Bajocian, Callovian, Tithonian, and Albian events correspond more or less precisely with contemporary events in other parts of the world. With regard to shallowing events, at least those recognized in the Aalenian and Valanginian (Fig. 4) are likely to reflect eustasy. Remarkable as it may seem, because of their association with tectonic instability, even convergent plate margins with long histories of subduction may record eustatic signals with greater or lesser fidelity through long periods of their history.

As regards regional tectonics it has been claimed by South American geologists that the spectacular Mid-Cretaceous regression in the southern Andes is a consequence of uplift associated with the inception of a compressive orogenic phase and increased igneous activity, which is probably bound up with the initial opening of the South Atlantic (Frutos, 1981; Coira *et al.*, 1982; Malumian & Ramos, 1984). This opening event apparently caused an eastward shift of the magmatic arc towards the High Cordillera, thereby bringing to an end the back-arc basin that had existed since the latest Triassic (Fig. 7). The late Jurassic regressive phase has received less attention. Bearing in mind the evidence, from igneous rock dating, that the volcanic arc has shifted progressively eastwards through the course of the Mesozoic and Cenozoic (Coira *et al.*, 1982) it may be tentatively suggested that this phase marks a sudden eastward extension of the volcanic arc as the Chilean Coastal Cordillera grew dramatically in size, causing huge quantities of coarse erosional debris to be carried eastwards to fill the Andean Basin. This may perhaps relate to the earliest stage of opening of the South Atlantic, because palaeobiogeographic data suggest that a direct marine connection between the southern Andes and southern Africa had become established by Tithonian times (Hallam *et al.*, 1986). The scenario as outlined implies compression in northern and central Chile, but it has also been argued that the Late Jurassic and Early Cretaceous were times of tension in this region, with the for-

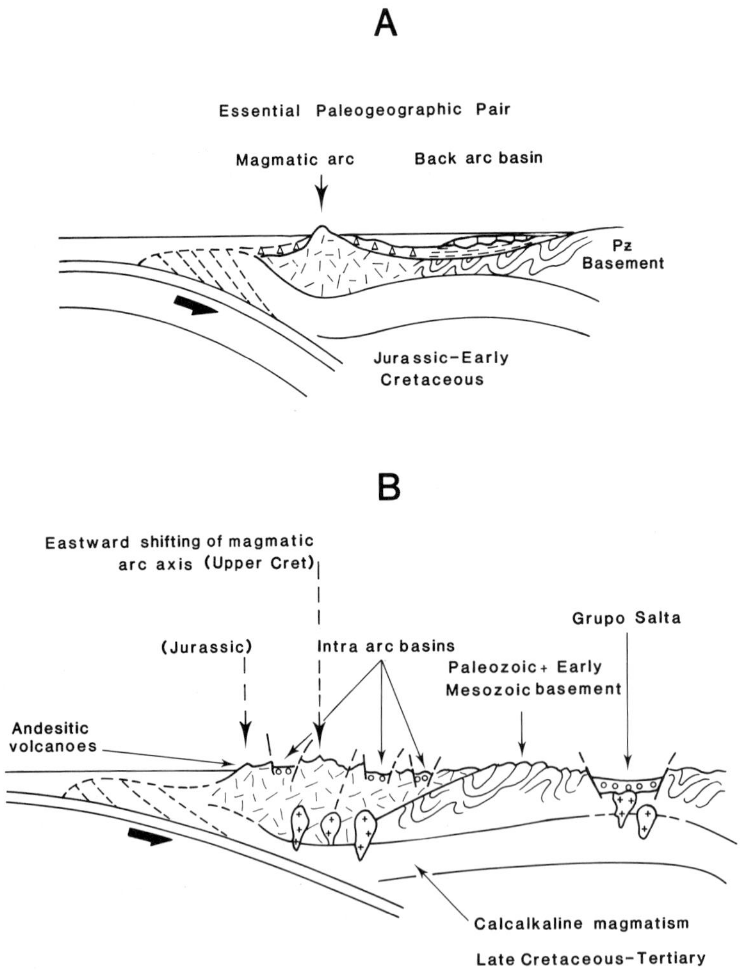


**Fig. 7.** Sketch diagrams to illustrate Mesozoic migration of the subduction zone and magmatic arc in the southern Andes. (A) The magmatic arc as developed since the latest Triassic to Early Jurassic and continuing until the end of the Early Cretaceous; (B) eastward shift of the arc in Mid to Late Cretaceous. After Coira *et al.* (1982, figs 7, 8).

mation of the Central Valley as an aborted back-arc rift (Pankhurst *et al.*, 1988). The two models must somehow be reconciled.

## ACKNOWLEDGEMENTS

Fieldwork in Chile in 1989 was made possible by financial support from the Royal Society and the Comision Nacional de Investigacion Cientifica y Tecnologia. I am also very grateful for the logistic support and collaboration of Professor G. Chong and Dr A. Quinzio of the Universidad del Norte, Antofagasta and Sr E. Perez of the Servicio Nacional de Geologia y Mineria.

## REFERENCES

Coira, B., Davidson, J., Mpodozis, C. & Ramos, V. (1982) Tectonic and magmatic evolution of the Andes of northern Argentina and Chile. *Earth Sci. Rev.* **18**, 303–332.

Frutos, J. (1981) Andean tectonics as a consequence of sea-floor spreading. *Tectonophysics* **72**, T21–32.

Geyer, O.F. (1973) Das präkretazische Mesozoikum von Kolumbien. *Geol. Jahrb. Reihe B*, **5**, 3–155.

Geyer, O.F. (1974) Der Unterjura (Santiago-Formation) von Ekuador. *Neues Jahrb. Geol. Paläont. Monatsh.* **1974**(H9), 525–541.

Geyer, O.F. (1980) Die mesozoische Magnafazies-Abfolge in den nördlichen Anden (Peru, Ekuador, Kolumbien). *Geol. Rundsch.* **69**, 875–891.

Gröschke, M. & Hillebrandt, A.V. (1985) Trias und Jura in der mittleren Cordillera Domeyko von Chile (23°31′–24°30′). *Neues Jahrb. Geol. Paläontol. Abh.* **170**, 129–166.

Hallam, A. (1978) Eustatic cycles in the Jurassic. *Palaeogeogr. Palaeoclimatol. Palaeoecol.* **23**, 1–32.

Hallam, A. (1986) Evidence of displaced terranes from Permian to Jurassic faunas around the Pacific margins. *J. Geol. Soc. London* **143**, 209–216.

Hallam, A. (1988) A re-evaluation of Jurassic eustasy in the light of new data and the revised Exxon curve. In: *Sea-Level Changes – an Integrated Approach* (Eds Wilgus, C.K., Hastings, B.S., Kendall, C.G.St.C., Posamentier, H.W., Ross, C.A. & van Wagoner, J.). Soc. Econ. Paleontol. Mineral. Spec. Publ. 42, 261–273.

Hallam, A., Biro-Bagoczky, L. & Perez, E. (1986) Facies analysis of the Lo Valdés Formation (Tithonian–Hauterivian) of the High Cordillera of Central Chile, and the palaeogeographic evolution of the Andean Basin. *Geol. Mag.* **123**, 425–435.

Haq, B.U., Hardenbol, J. & Vail, P.R. (1987) Chronology of fluctuating sea levels since the Triassic. *Science* **235**, 1156–1167.

Hillebrandt, A.V. (1971) Der Jura in der chilenisch–argentinisch Hochkordillera (25° bis 32°30′S). *Muenstersche Forsch. Geol. Palaeontol.* **20/21**, 63–87.

Hillebrandt, A.V. (1973) Neue Ergebnisse über den Jura in Chile und Argentinien. *Muenstersche Forsch. Geol. Palaeontol.* **31/32**, 167–199.

Hillebrandt, A.V. (1990) The Triassic–Jurassic boundary in northern Chile. *Cahiers Univ. Cathol. Lyon, sér. Sci.* **3**, 27–53.

Hillebrandt, A.V. & Schmidt-Effing, R. (1981) Ammoniten aus dem Toarcium (Jura) von Chile (Südamerika). *Zitteliana* **6**, 3–74.

Loughman, D.L. & Hallam, A. (1982) A facies analysis of the Pucará Group (Norian to Toarcian carbonates, organic-rich shale and phosphate) of central and northern Peru. *Sediment. Geol.* **32**, 161–194.

Macellari, C. (1988) Cretaceous paleogeography and depositional cycles of western South America. *J. S. Am. Earth Sci.* **1**, 373–418.

Malumian, N. & Ramos, V.A. (1984) Magmatic intervals, transgression–regression cycles and oceanic events in the Cretaceous and Tertiary of southern South America. *Earth Planet. Sci. Lett.* **67**, 228–237.

Malumian, N., Nullo, F.E. & Ramos, V.A. (1983) The Cretaceous of Argentina, Chile, Paraguay and Uruguay. In: *The Phanerozoic Geology of the World, II: the Mesozoic, B* (Eds Moullade, M. & Nairn, A.E.M.) pp. 265–304. Elsevier, Amsterdam.

McCourt, W.J., Aspden, J.A. & Brook, M. (1984) New geological and geochronological data from the Colombian Andes: continental growth by multiple accretion. *J. Geol. Soc. London* **141**, 831–845.

Megard, F. (1978) Etude geologique des Andes du Perou Central. *Mém. Off. Rech. Sci. Techn. Outre Mer.* **86**, 1–310.

Pankhurst, R.J., Hole, M.J. & Brook, M. (1988) Isotopic evidence for the origin of Andean granites. *Trans. R. Soc. Edinburgh* **79**, 123–133.

Perez, E. (1982) Bioestratigrafia del Jurásico de Quebrada Asientos, norte de Potrerillos, region de Atacama. *Serv. Nac. Geol. Min. Chile Bol.* **37**, 1–149.

Prinz, P. (1985) Stratigraphie und Ammonitenfauna der Pucará-Gruppe (Obertrias-Unterjura) von Nord-Peru. *Palaeontographica* **A188**, 153–197.

Riccardi, A.C. (1983) The Jurassic of Argentina and Chile. In: *The Phanerozoic Geology of the World, II: the Mesozoic, B* (Eds Moullade, M. & Nairn, A.E.M.) pp. 201–264. Elsevier, Amsterdam.

Riccardi, A.C., Westermann, G.E.G. & Elmi, S. (1989) Les zones d'ammonites du Bathonien–Callovien (Jurassique moyen) des Andes argentino–chiliennes. *Geobios* **22**, 553–597.

Sliter, W.V. (1976) Cretaceous foraminifers from the southwest Atlantic Ocean, Leg 36, Deep Sea Drilling Project. In: *Initial Reports of the Deep Sea Drilling Project*, Vol. 36 (Eds Barker, P.F., Dalziel, I.W.D., *et al.*) pp. 519–573. US Government Printing Office, Washington, DC.

Stipanicic, P.N. (1983) The Triassic of Argentina and Chile. In: *The Phanerozoic Geology of the World, II: the Mesozoic, B* (Eds Moullade, M. & Nairn, A.E.M.) pp. 181–200. Elsevier, Amsterdam.

Vail, P.R. & Todd, R.G. (1981) Northern North Sea Jurassic unconformities, chronostratigraphy and sea-level changes from seismic stratigraphy. In: *Petroleum Geology of the Continental Shelf of North-West Europe*

(Eds Illing, L.V. & Hobson, G.) pp. 216–235, Heyden, London.

Vail, P.R., Mitchum, R.M. & Thompson, S. (1977) Seismic stratigraphy and global changes of sea level. In: *Seismic Stratigraphy – Applications to Hydrocarbon Exploration* (Ed. Payton, C.E.). Mem., Am. Assoc. Petrol. Geol. 26, 63–97.

Vail, P.R., Hardenbol, J. & Todd, R.G. (1984) Jurassic unconformities, chronostratigraphy and sea-level changes from seismic stratigraphy and biostratigraphy. In: *Interregional Unconformities and Hydrocarbon Accumulation* (Ed. Schlee, J.S.). Mem., Am. Assoc. Petrol. Geol. 6, 229–144.

Watts, A.B. (1982) Tectonic subsidence, flexure and global changes of sea-level. *Nature* **297**, 469–474.

Westermann, G.E.G., Riccardi, A.C., Palacios, O. & Rangel, C. (1980) Jurásico Medio en el Peru. *Inst. Geol. Min. Metal. Peru, Bol.* **9**, 1–47.

Westermann, G.E.G., Corona, R. & Carasco, R. (1984) The Andean Mid-Jurassic *Neuqueniceras* ammonite assemblage of Cualac, Mexico. *Spec. Pap. Geol. Soc. Can.* **27**, 99–112.

*Spec. Publs. int. Ass. Sediment.* (1991) **12**, 201–215

# Cyclic sedimentation in three Neogene basins in California

J. ROBERT DODD[†] *and* ROBERT J. STANTON, JR[‡]

[†] *Department of Geology, Indiana University, Bloomington, Indiana 47405, USA; and* [‡] *Department of Geology, Texas A & M University, College Station, Texas 77843, USA*

## ABSTRACT

Neogene marine deposition in California largely took place in basins that formed in middle to late Miocene time, reached maximum depth by early Pliocene time, and then progressively filled and shoaled to non-marine conditions during Pleistocene time. Cyclic patterns are poorly developed in the transgressive phase of basin history but are well developed in some basins in the later part of the regressive phases, when shelf conditions were attained. The late Miocene to Pleistocene cycles in the San Joaquin (about 12 cycles), Merced (about 30 cycles), and Eel River (about 4 cycles) Basins reflect differences in the history and palaeogeographic settings of the individual basins. Apparent eustatic cycles are best developed in the Merced Basin, where subsidence and depositional rates are nearly balanced. A larger tectonic component may be responsible for the apparently longer cycles in the other basins. These basins particularly illustrate the fact that for eustatic cycles to be well developed, the depositional surface should cross either of the datum levels of shoreline, or storm wave-base. If deposition is entirely below wave base or above sea level, even though both of these datums fluctuated during eustatic cycles, cyclicity in the sedimentology and fossil record will be poorly developed. If deposition is entirely between the two datums, cyclicity may be well developed, but this depends to a large extent on the intrinsic basin characteristics of sediment supply and wave and current action.

## INTRODUCTION

For a number of reasons, Neogene strata are particularly appropriate subjects for the study of cyclic sedimentation, i.e. cyclic variation in the position of the environment of deposition relative to sea level. (1) Major changes in sea level during Neogene time are especially well documented. (2) Fine details of sea-level history can be inferred from the oxygen isotope recorded in planktonic organisms in deep-sea cores (Prentice & Matthews, 1988). (3) The tectonic setting of Neogene deposition is well known. In many cases the present tectonic setting of an area is the same as it was in Neogene time (Ingersoll, 1988). (4) Conditions of sedimentation under which the strata were deposited were often very similar to modern conditions existing in the vicinity at the present time (Clifton, 1988). This maximizes confidence in interpretation of sedimentary features and sedimentation rates by comparison with modern analogues. (5) Many of the fossil genera and even some species in Neogene strata are still extant, allowing considerable confidence in palaeoecological interpretations of depositional environments (Dodd & Stanton, 1981).

Cyclic sedimentation has been described in Neogene strata of at least three basins in California: the San Joaquin Basin (Stanton & Dodd, 1970,1972), the Merced Basin (Hunter *et al.*, 1984; Clifton *et al.*, 1988; Clifton & Leithold, 1991), and the Eel River Basin (Bourgeois & Leithold, 1983; Leithold, 1989; Leithold & Bourgeois, 1989; Clifton & Leithold, 1991; and in this paper). Cyclic sedimentation probably occurs in other California Neogene basins as well. The extent of development of cycles differs considerably between the Merced, San Joaquin, and Eel River Basins. In this paper we will describe the cycles in each of these basins and explain why they differ as a result of differences in eustasy, tectonics, and sedimentation.

Eustasy, or variation in world-wide sea level, an obvious possible cause of cyclic sedimentation, has received a great deal of attention recently in large part owing to the work of Vail *et al.* (1977) and more recently Haq *et al.* (1987). Haq *et al.* (1987) show five major sea-level cycles during the last 6 Ma. However, there have clearly been many more large (up to about 170 m, Prentice & Matthews (1988)) changes in sea level that occurred during Neogene time in response to the waxing and waning of glaciers and the polar ice caps. The periodicity of these cycles is probably correlated with variation in Earth's orbital parameters (Milankovich cycles, Hayes *et al.* (1976)), and must be taken into account in any interpretation of Neogene cyclic sedimentation.

Effects of tectonic cycles have been described (e.g. Blair & Bilodeau, 1988) but are less well understood than eustatic cycles. The tectonic component may be either positive (i.e. variation in uplift of the sediment source area affecting sedimentation rate and sediment properties) or negative (i.e. variation in subsidence in the depositional area affecting depth of deposition). A tectonic component in any event is necessary to account for cyclic sedimentation, because the depositional basin must subside to make room for the sediment being deposited. In most cases we assume that this subsidence occurs at a uniform rate, but we usually have little direct evidence about what the rates may have been. Geological studies and neotectonic data in many areas suggest that deformation proceeds at a fairly uniform rate, but abrupt movements associated with stick-slip motion on major faults indicates large short-term variability that might affect both local sources and depositional sites (Wang & Burnett, 1990). Cyclic patterns of subsidence could produce sedimentation cycles that may be identical to cycles caused by eustasy.

Sedimentation rates are also important in producing sedimentary cycles. Cyclic variation in sedimentation rate in a uniformly subsiding basin could also produce sedimentary cycles. In order to determine what really caused a given cycle in the sedimentary record, the relative influence of eustasy, tectonics, and sedimentation must be evaluated.

## GEOLOGICAL SETTING

Neogene sediments in California were deposited in a number of more or less independent basins (Fig. 1). Each basin has a slightly different tectonic setting resulting in a correspondingly different sedimentary record. Correlation between the basins has long been a problem because of these differences in tectonics and sedimentation. Sediments that were deposited in relatively deep water can be correlated with some confidence based on fossils, such as planktonic forams and diatoms, which were relatively little affected by tectonic variations. Shallow-water strata have been more difficult to correlate from basin to basin because local tectonic differences resulted in contemporaneous differences in the depositional environment and thus in the biota.

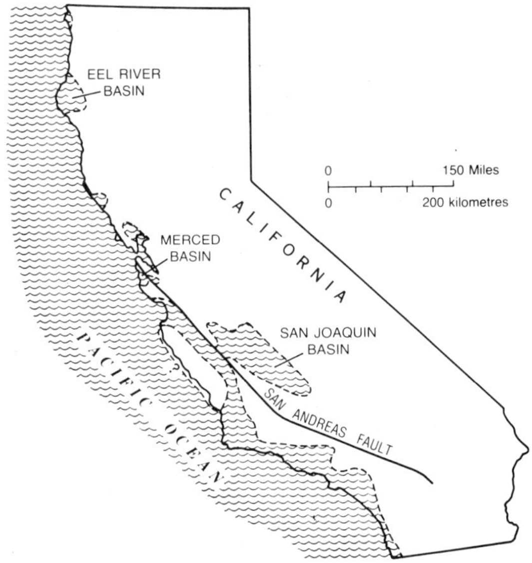


**Fig. 1.** Approximate position of the Pliocene shoreline in California showing position of sedimentary basins included in this study. Not corrected for probable right-lateral slip on San Andreas Fault of about 40 km.

The overall history of each basin is similar, beginning with apparent rapid subsidence to maximum water depth in mid- to late Miocene time. Deposition of deep-water diatomaceous sediments began and continued into Pliocene time (the Monterey, Reef Ridge, McClure, and Pullen Formations are examples of such deposits). Continued deposition resulted in progressively shallower water sediments through Pliocene and Pleistocene time, culminating in non-marine sediments in the onland basins.

### San Joaquin Basin

Neogene sediments (late Miocene through to Pleistocene in age), which are well exposed in the Kettleman

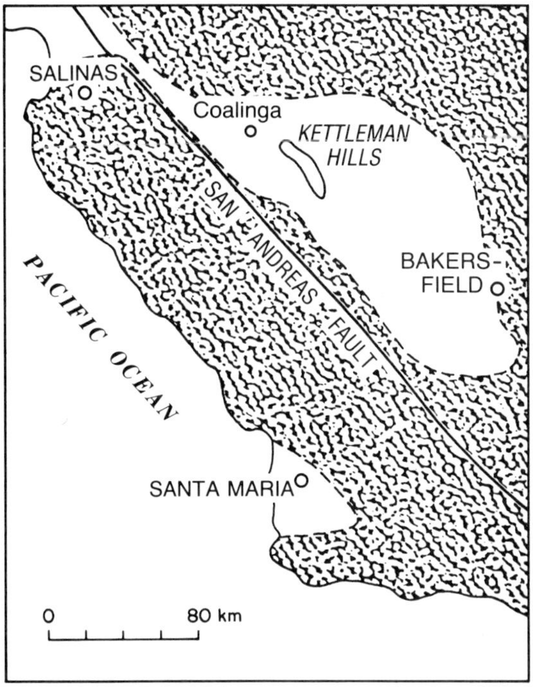


**Fig. 2.** Late Pliocene palaeogeography of central California showing seaway connecting the San Joaquin Basin with the open Pacific Ocean. The patterned area was land. White area was sea. Modified from Galehouse (1967).

Hills on the west side of the central valley of California (Fig. 2), were deposited as a late phase of filling of the San Joaquin Basin (Stanton & Dodd, 1970). The Kettleman Hills are approximately 25 miles (40 km) east of the San Andreas Fault. The stratigraphy and palaeontology of these strata have been extensively studied for many years. The cyclical nature of a portion of these strata was noted by us (Stanton & Dodd, 1970) and subsequently more extensively studied (Stanton & Dodd, 1972; Dodd & Stanton, 1975).

The San Joaquin Basin was formed during Eocene time as a fore-arc basin related to the convergent plate boundary between the Pacific and North American plates. It was initially a deep basin containing a thick section of siliciclastic sediments. Beginning some 15 Ma ago, the Mendocino Triple Junction began to migrate northwestward along the western boundary of the basin and the western boundary changed from a convergent to a transform margin (DeCelles, 1988). During this latter phase of development of the basin, rapid subsidence continued, but water depth shoaled considerably because of rapid sedimentation. Thus most of the Neogene sediments filling the west side of the basin were deposited in a shallow marine to terrestrial environment. Deposition during Pleistocene time was entirely non-marine (Stanton & Dodd, 1970).

### *Stratigraphy*

The oldest Neogene sediments in the San Joaquin Basin crop out to the west of the Kettleman Hills along the west flank of Reef Ridge, especially in Jacalitos Canyon (Stanton & Dodd, 1976). The lower part of the section consists of deeper water deposits of the McClure and Reef Ridge Formations. The overlying sediments become coarser upwards and were deposited, in general, in progressively shallower water. These sediments, of the Santa Margarita and Jacalitos Formations, have not been studied in as much detail as the younger strata exposed in the Kettleman Hills, and cyclic deposition patterns, although noted, have not been carefully analysed (Stanton & Dodd, 1976; Loomis, 1988).

During deposition of the Etchegoin and San Joaquin Formations (Fig. 3), the units we have investigated in most detail, the San Joaquin Basin was reduced to a broad but largely shallow inland sea or embayment connected to the open Pacific both to the north and the south (Galehouse, 1967). Later the southern connection was blocked and the only connection was to the north in the Monterey Bay area. In the Kettleman Hills region many lines of evidence indicate that marine conditions were more prevalent during most of this time to the northwest and that restricted, brackish conditions or non-marine conditions predominated to the southeast (Stanton & Dodd, 1970). During early Etchegoin time, deposition was entirely shallow marine; later in Etchegoin time and during San Joaquin time, deposition fluctuated between marine and non-marine. During Tulare time, deposition was almost entirely non-marine with only a few brackish intervals.

The age of these Neogene units was determined largely on the basis of molluscs and echinoids, making precise correlation with the better dated, deeper water units difficult. The Miocene–Pliocene boundary (c. 5 Ma) has been placed at the San Joaquin–Etchegoin contact (Armentrout & Echols, 1981), and the Pliocene–Pleistocene boundary is usually placed at the San Joaquin–Tulare contact

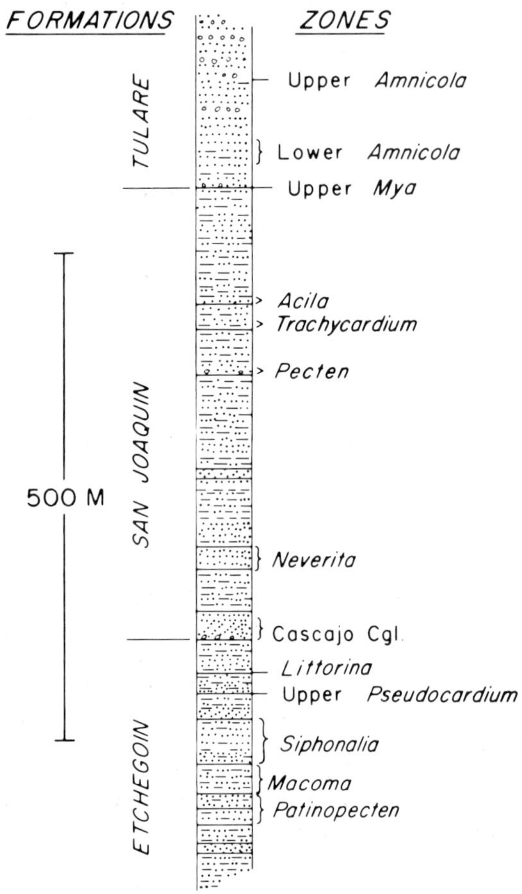


**Fig. 3.** Stratigraphic section exposed in the Kettleman Hills of central California showing formations and fossil zones. After Woodring *et al.* (1940).

(Roth, 1979) or within the lower Tulare (Armentrout & Echols, 1981). Loomis (1988) reviewed more recent work utilizing dating of ash beds, magnetostratigraphy, and $^{87}Sr/^{86}Sr$, and concluded that the Pliocene–Pleistocene boundary probably occurs within the Tulare Formation. The age of the base of the Echegoin Formation is not well defined but could be as old as late Miocene.

## Sedimentology

The Etchegoin and San Joaquin Formations consist of conglomerate, sandstone, siltstone, mudstone, and even a few carbonate beds (Stanton & Dodd, 1970). Much of the sediment is texturally and compositionally immature owing to deposition into a rapidly subsiding basin adjacent to the source area. Sorting ranges from poor to good, reflecting the shallow marine, beach, and deltaic depositional settings.

A typical sedimentation cycle (Fig. 4) begins with a scoured base overlain by well-sorted sand, commonly containing pebbles and abundant marine fossils. Woodring *et al.* (1940) called these very fossiliferous intervals 'zones', which they named for the dominant or conspicuous genera occurring within each. We have used these names to identify the individual cycles, which usually contain a single zone at their base. Thick, cross-bedded sandstone, sometimes containing abundant sand dollars (clypeasteroid echinoids), is common in the lower part of the cycle. Finer grained sediment increases upward in the cycle, and the concentration of marine fossils usually decreases. The upper part of each typical cycle is predominantly rather massive unfossiliferous siltstone and mudstone containing common plant fragments and local freshwater molluscs and vertebrate bones.

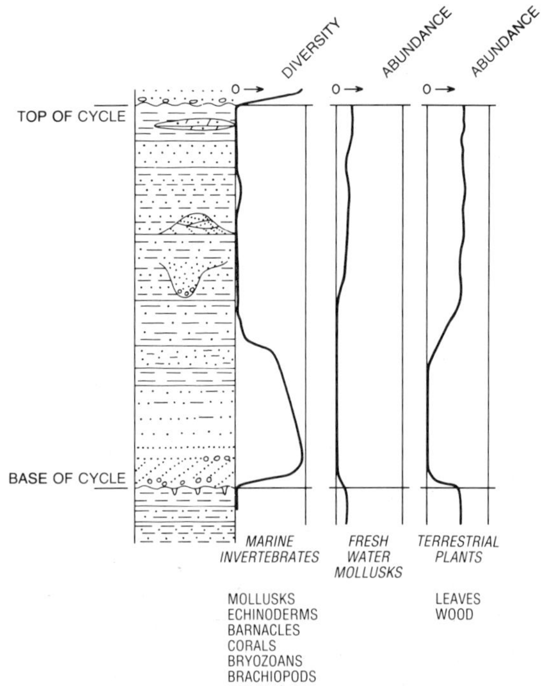


**Fig. 4.** Idealized section of *Pecten* zone of the San Joaquin Formation showing lithology, diversity of marine fossils, and abundance of freshwater molluscs and probable terrestrial plant material.

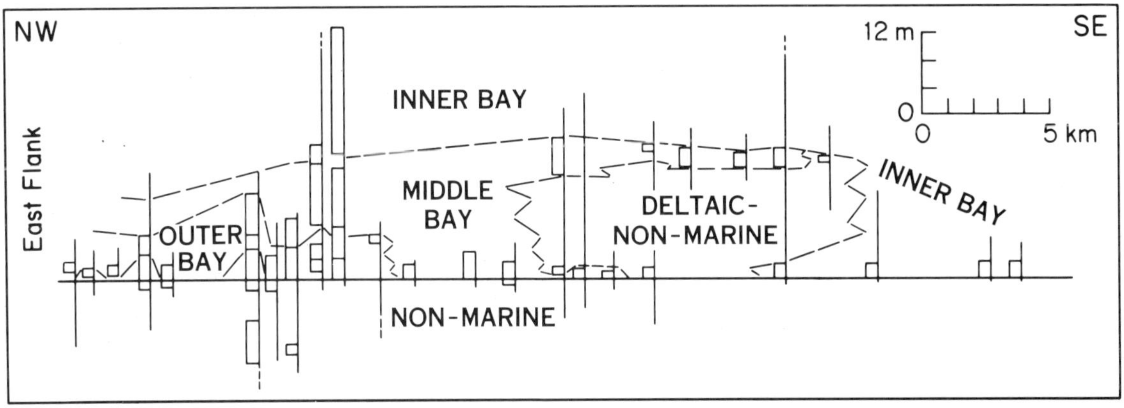


**Fig. 5.** Stratigraphic sections of the *Pecten* zone along the east flank of the Kettleman Hills. The datum line is the base of the *Pecten* zone. Vertical lines indicate the extent of the stratigraphic section examined. The vertical rectangles indicate the distribution of fossil communities. Modified from Dodd & Stanton (1975).

Lateral variation can occur within a cycle (Stanton & Dodd, 1970). Cycles as described above are best developed in the northwest Kettleman Hills. The thickness of coarse-grained sediments within any cycle generally is greatest there and decreases southeast as the proportion of siltstone and mudstone correspondingly increases. Fossils indicating open marine conditions are also most common in the northwest, but decrease southeast as brackish water and non-marine fossils increase in abundance (Fig. 5, from Dodd & Stanton, 1975). These lithologic and biotic gradients reflect an environmental gradient at the base of the cycles from more open marine and higher energy conditions to the northwest and more restricted marine to freshwater and deltaic non-marine conditions to the southeast, farther from the entrance to the bay.

These typical characteristics of vertical and lateral variation are especially well developed in the *Pecten* zone (Figs 4, 5). The sediment underlying the base of the *Pecten* zone is siltstone and claystone with abundant plant fragments, probably deposited in a very restricted marine to freshwater swamp environment. The base of the *Pecten* zone is a scoured and extensively burrowed (or bored) surface. Immediately above the scoured base is a thin, muddy sandstone bed containing rare worn and reworked horse teeth and other terrestrial vertebrate fossils. This basal unit is overlain in turn by coarse, locally cross-bedded sand containing pebble layers and abundant open marine fossils, such as brachiopods, corals, and molluscs, that are diagnostic of open marine conditions. Some massive sands contain only sand dollars, suggesting a shifting, unstable environment that could not support a more diverse fauna. The overlying sediment is increasingly muddy upward, and fossil diversity declines. The muddy sandstone and siltstone of the upper part of the section contain abundant plant fossils and scattered vertebrate fossil remains.

Fossils in the Kettleman Hills are well preserved and widely distributed, both geographically and stratigraphically within the Neogene strata (Stanton & Dodd, 1970). In addition, most of the fossil genera and some of the species are still extant in the northeast Pacific Ocean, and we know a great deal about their ecology. Consequently palaeoecological analysis provides valuable detail in the cyclic depositional model. We have used a number of palaeoecological methods (Dodd & Stanton, 1981). The most useful approach is palaeocommunity analysis supplemented by other methods, such as biogeochemical and taphonomic analysis (Dodd & Stanton, 1975). Within the *Pecten* zone, eight fossil communities can be recognized that are very similar to modern communities that we have identified in San Francisco Bay and the adjacent open ocean. Differences in the composition, diversity, and taphonomic characteristics of the communities correspond to differences in salinity, substrate, and degree of restriction or isolation from the open ocean. Oxygen and carbon isotopic composition of the fossils indicate a salinity gradient both vertically and in a NW–SE direction within the zone (Dodd & Stanton, 1975).

The most open, normal marine communities occur

at the base of the zone, in the northwest Kettleman Hills (Fig. 5; Dodd & Stanton, 1975). This is the area nearest the opening of the bay to the Pacific Ocean. The communities are characteristic of increasingly restricted, brackish to non-marine environments both upward in the section and to the southeast, away from the entrance to the bay.

In summary, the typical cycle in the San Joaquin Basin was initiated by a transgression of the sea over paludal to non-marine sites producing relatively open marine conditions (Fig. 6). This transgression was accompanied by probably minor erosion and reworking of the underlying sediment. After the initial transgression, relative sea level gradually fell resulting in decreased marine influence in the embayment. Eventually sea level fell below the entrance to the bay, and the bay then became a freshwater lake.

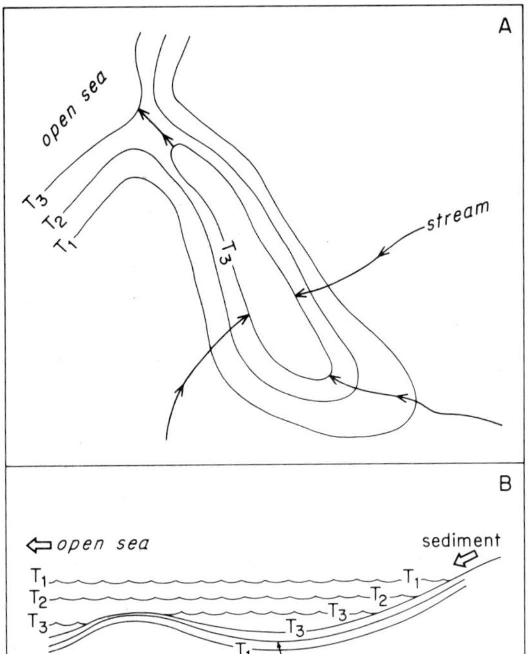


**Fig. 6.** Schematic representation of the San Joaquin Basin at three times during a eustatic sea-level cycle. $T_1$ shows sea level at the time of maximum transgression. Open communication is established with the Pacific Ocean. $T_3$ shows sea level at the time of maximum regression. The basin is cut off from the Pacific Ocean and is in part filled with a freshwater lake. $T_2$ shows sea level at an intermediate stage with restricted circulation between the embayment and the Pacific Ocean. (A) Map view, (B) section view.

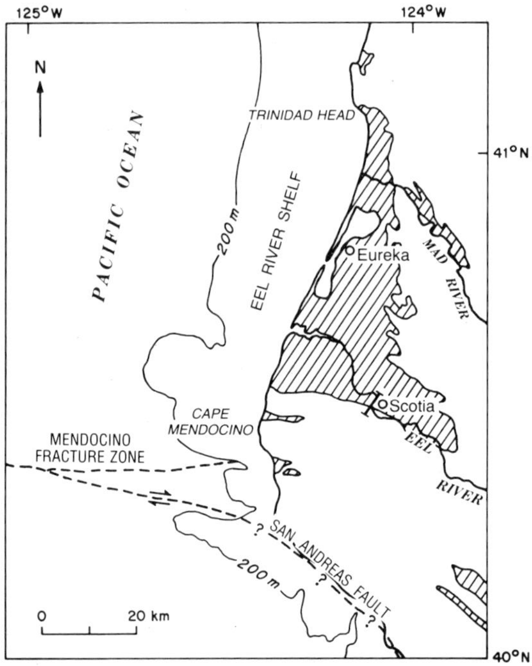


**Fig. 7.** Onshore portion of the Eel River Basin. Areas with diagonal lines are underlain by Neogene strata. Line at Scotia shows position of measured section in Fig. 8. After Dodd *et al.* (1984).

Lacustrine, paludal, and deltaic deposits comprise the upper part of the cycles.

## Eel River Basin

Deposition in the Eel River Basin of northern California during this time interval occurred in a tectonic setting very different from that for the San Joaquin Basin (Fig. 7; Dodd *et al.*, 1984). The Eel River Basin is located just north of the Cape Mendocino triple junction near the subducting plate margin. The Eel River deposits form a broad wedge at the continental margin extending as far north as the Oregon border (Field *et al.*, 1980) rather than a localized basin fill in the more usual sense. The eastern (onshore) portion of this wedge was deformed during Pleistocene time into the structural Eel River Basin, while sedimentation is continuing offshore (Leithold, 1989). The sediments of the Eel River Basin range in age from late Miocene through to Pleistocene based on Foraminifera (Keller & Ingle, 1981), diatoms (Burckle *et al.*, 1980), magnetic polarity reversals (Dodd *et al.*, 1977; Kodama, 1979), and absolute age determination of ash beds (Sarna-Wojcicki *et al.*, 1987). Thus these strata are considerably more securely dated but essentially contemporaneous with those of the San Joaquin Basin.

### *Stratigraphy*

The lowermost strata in the Eel River Basin are sandstones and conglomerates of the lower Pullen Formation deposited on an eroded surface of the underlying Eocene Yager Formation (Fig. 8). These basal sediments are overlain by siltstones and mudstones (in part diatomaceous) of the Pullen, Eel River, and lower Rio Dell Formations. The sand content increases in the upper Rio Dell Formation, and sandstone is the predominant lithology in the Scotia Bluffs Formation. The Carlotta Formation consists of sandstone, conglomerate, and mudstone.

The macrofossil fauna in the lower Pullen Formation is relatively diverse and abundant, consisting largely of shallow water molluscs. The mudstones and associated sandstones of the Pullen, Eel River, and lower Rio Dell Formations contain few macrofossils but a relatively diverse microfossil biota. The microfossils have been interpreted to indicate that the basin subsided rapidly in late Miocene and earliest Pliocene time to a depth of about 2000 m and then progressively shoaled (Piper *et al.*, 1976). Molluscs and echinoids become more abundant in the upper part of the Rio Dell Formation and indicate deposition at shelf depths. Molluscs and echinoids in the Scotia Bluffs Formation are indicative of shallow marine environments. The Carlotta Formation is largely unfossiliferous except for a few horizons containing marine molluscs (Morrison, 1980), suggesting brief periods when brackish water conditions occurred during largely non-marine deposition.

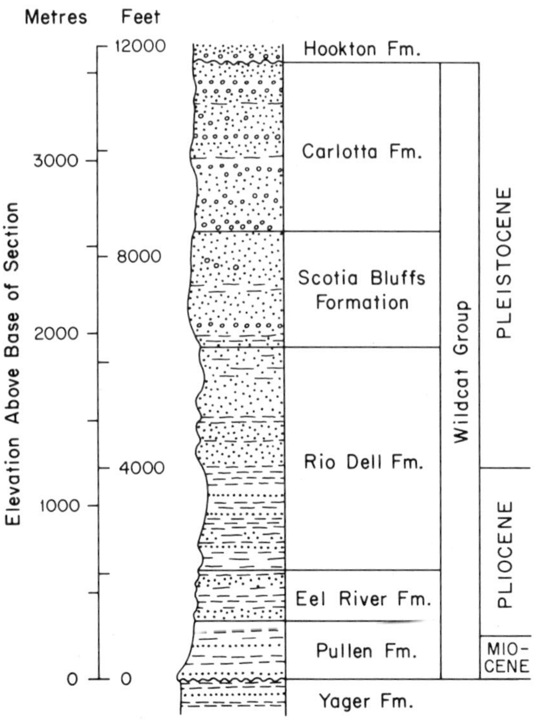


**Fig. 8.** Generalized stratigraphic section exposed along the Eel River at Scotia and Rio Dell, California.

### *Sedimentology*

The sediments in the Eel River Basin form a single thick transgressive–regressive cycle. Within this cycle, smaller scale cycles are evident only in sediments deposited when the depth was shallower than storm wave-base. The lower part of the section contains a thin interval of coarse sands and conglomerates, which suggest two or three poorly developed cycles. The strata in the upper Pullen, Eel River, and lower Rio Dell Formations are predominantly mudstones interbedded with turbidite sandstones.

In the upper half of the Rio Dell Formation there are four to five sand-rich intervals about 30–60 m thick, which alternate with intervals about 30–150 m thick of the normal mudstone lithology. Strata within the sandy intervals are characterized by (1) a sharp scoured base, (2) laminated sandstone at the base grading up to increasingly burrowed muddy sandstone and then to mudstone, and (3) larger scale hummocky cross-stratification. Bivalves are the predominant fossil. In the basal part of the unit, they are clearly reworked and oriented concave down, but are articulated and more or less in living position in the upper, burrowed part of the bed.

These cycles, each a metre to several metres thick, were formed by normal shelf sedimentation processes below fair-weather wave-base. These processes were interrupted by storms reworking and scouring the shelf sediments and bringing coarser sediment out on to the shelf. The thicker intervals of siltstone alternating with these sandier intervals probably resulted from variation in water depth. With sea-level fall, the sea floor came into the depth range to be affected by storms; with sea-level rise, it was below wave-base and only mud was deposited.

Leithold (1989) and Leithold & Bourgeois (1989) have recognized four sedimentary cycles in the Rio Dell Formation exposed on Centerville Beach. These cycles are rather subtle, consisting largely of mud-rich sediment deposited on a relatively deep shelf. They interpreted water depth as ranging between approximately 50–175 m in the lower cycle. Bourgeois & Leithold (1983) described two cycles in the lower Scotia Bluffs Formation exposed along the Eel River. They interpreted the cycles as beginning with muddy sands deposited near storm wave-base and terminating with conglomerates deposited in a stream-mouth bar or fluviatile environment. Most of the Scotia Bluffs sediments were deposited in shallow water with such persistent wave action and rapid sediment supply that neither short-term variation in sedimentary features nor expected trace fossils, such as *Ophiomorpha*, are preserved. Marine to non-marine shoreline fluctuations are present at the Scotia Bluffs–Carlotta Formation boundary, but do not appear to form any simple cyclic pattern.

### Merced Basin

The Merced Formation, a section 2000 m thick of poorly consolidated sediments on the San Francisco Peninsula (Fig. 9), has been extensively studied by Hunter *et al.* (1984), Clifton & Hunter (1987), and Clifton *et al.* (1988). The Merced Formation is primarily preserved in a structural trough along the coast southwest of San Francisco, but was deposited over a considerably larger area as indicated by strata of similar age and lithology in scattered discontinuous outcrops in the Coast Ranges north of San Francisco. Strata that crop out south of San Francisco were deposited along the open Pacific Coast. The formation is predominantly Pleistocene in age, ranging from perhaps latest Pliocene in the lower 100 m to late Pleistocene (c. 100 000 years) at the top (Clifton *et al.*, 1988). Thus it is correlative with the upper part of the San Joaquin Formation and the Tulare Formation in the San Joaquin Basin and probably the upper Rio Dell, Scotia Bluffs, and a portion of the Carlotta Formation in the Eel River Basin section.

Based on a very detailed study of sedimentary textures and structures, Clifton *et al.* (1988) have identified approximately 30 sedimentation cycles (Fig. 10). A complete cycle begins with a scoured surface on which sediment was deposited at depths as deep as outer shelf. The cycles shoal upward and commonly end with deposition in alluvial, aeolian, or paludal conditions. Many cycles do not contain all of the above components. Clifton *et al.* interpret these cycles as being due to eustatic sea-level changes in a relatively rapidly and uniformly subsiding basin. An important reason for considering these to be largely eustatic and to discount tectonic effects is because the number of cycles approximately equals the number of major sea-level cycles in the Pleistocene, as indicated by oxygen isotopic data from Foraminifera in deep-sea cores (Fig. 11). Unfortunately, time control is not good enough to attempt to match individual cycles.

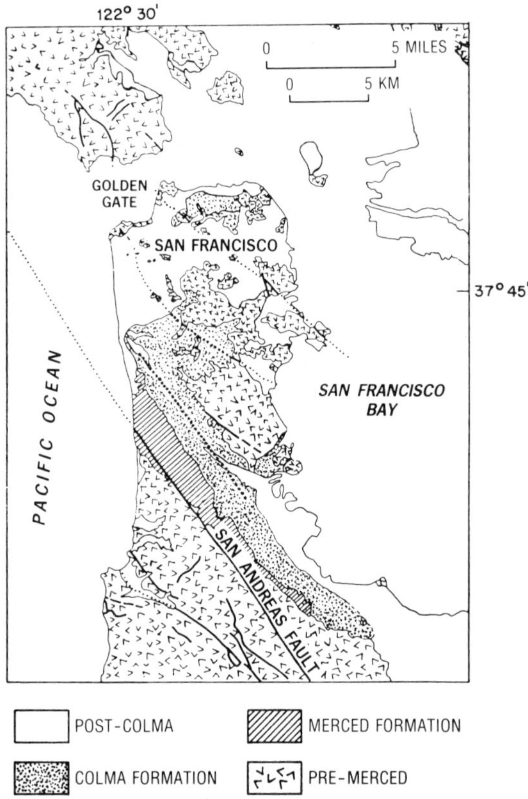


**Fig. 9.** Distribution of the Merced Formation and other units in the San Francisco area. Modified from Clifton *et al.* (1988).

## ORIGIN OF CYCLES

We believe that the cycles observed in all three basins surely had a strong eustatic component.

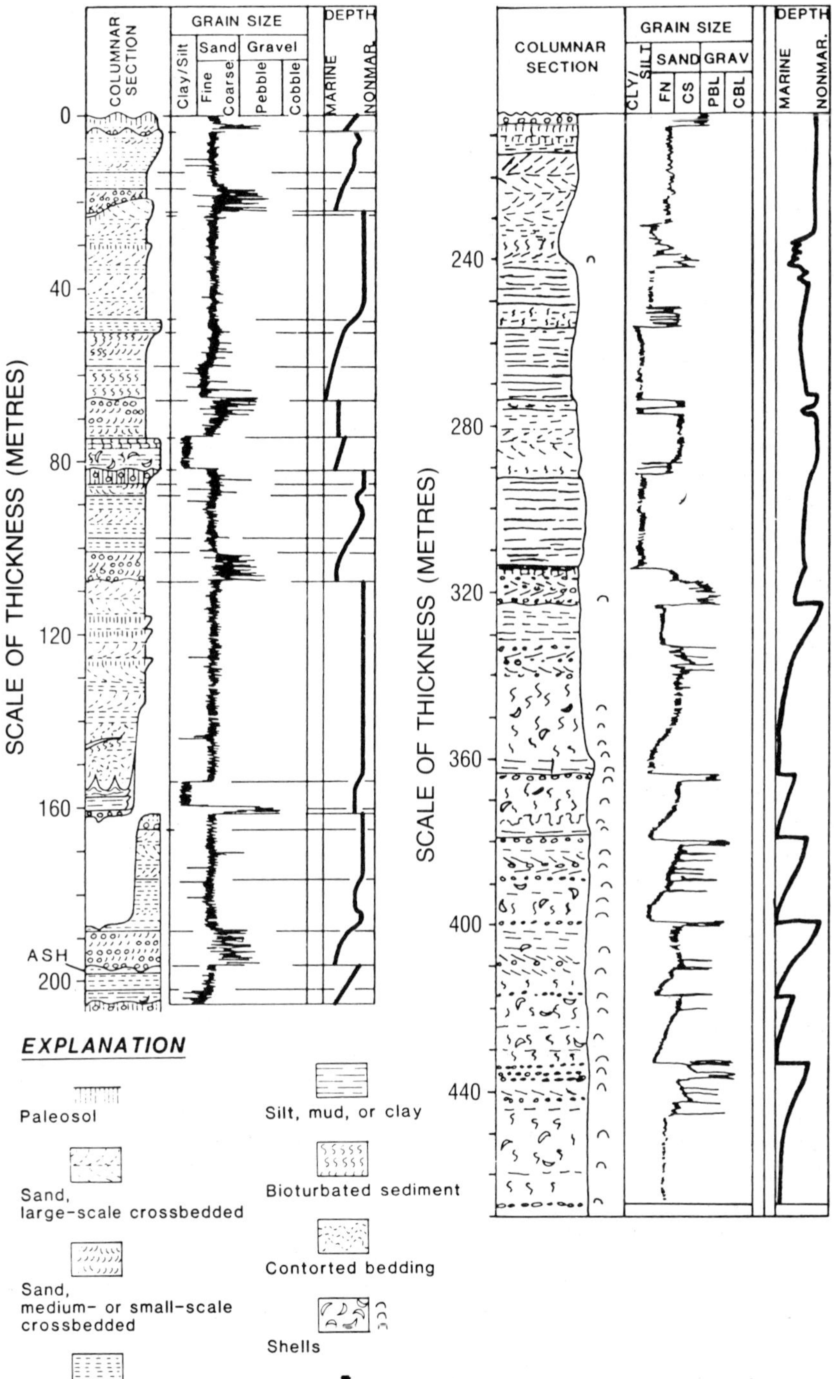


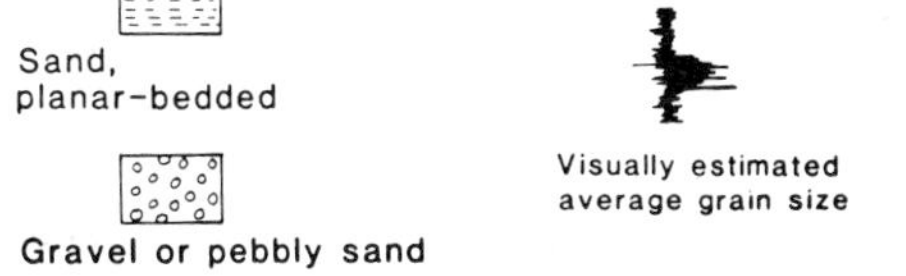


**Fig. 10.** Stratigraphic section of a portion of the upper Merced Formation exposed on the west coast of the San Francisco Peninsula. Modified from Hunter *et al.* (1984).

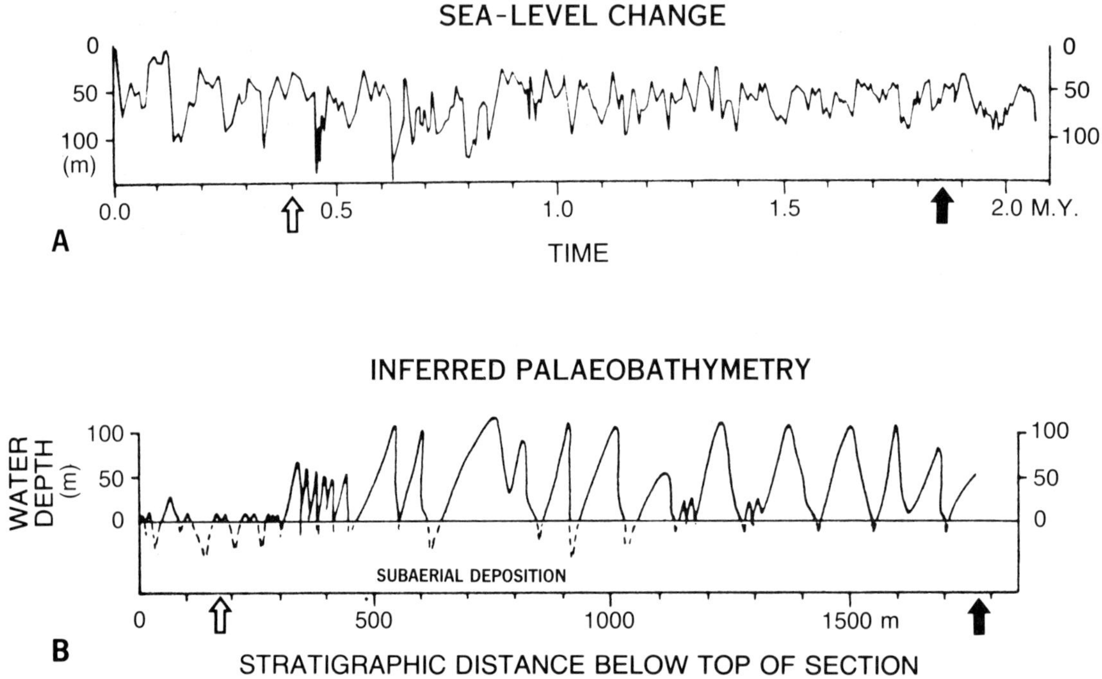


**Fig. 11.** Comparison of sea-level during Pleistocene time as determined from oxygen isotopic composition of planktonic Foraminifera from deep-sea core V28–239 (A) with water depth inferred from sedimentary features and fossils in the Merced Formation (B). The open and closed arrows show correlation points on the two curves. After Clifton *et al.* (1988).

Oxygen isotopic data from deep-sea cores suggest that extensive glaciers have existed throughout the Neogene and that they waxed and waned in a cyclic fashion (Prentice & Matthews, 1988). This growth and contraction of glaciers must be reflected in sea-level position during this time. The sea-level variation surely must affect shallow-water sediments on active as well as passive margins.

On the other hand, tectonic events in areas as active as these three basins are quite likely to be reflected in sedimentation patterns. Certainly each basin has subsided thousands of metres. We commonly assume that subsidence is uniform, but observation of modern elastic rebound movement on faults indicates that tectonic events are not uniform in the short term (Wang & Burnett, 1990). Episodic subsidence in an area of uniform sedimentation or episodic uplift of an adjacent source area affecting sedimentation rate could also produce cycles, as Blair & Bilodeau (1988) have documented from largely non-marine settings in which eustasy clearly was not a factor.

Finally, variation in sedimentation rate, whether related to tectonics or to other factors, such as changing drainage or current patterns, can also produce cycles in a basin that is uniformly subsiding.

All three of these factors, eustasy, tectonics, and sedimentation, were likely involved in producing the cycles observed in the Neogene basins. In these three examples of contemporaneous Neogene deposition in very different tectonic settings but within a short distance of one another, sorting out the magnitude of the effect of the three parameters is difficult.

Although the cycles in the three sections have many similarities, they also differ, especially in terms of age and apparent duration (Fig. 12). Although the age of the strata in the San Joaquin Basin is not well constrained, the cycles are mostly older than those in either the upper Eel River Basin section or the Merced Basin section. The ages of the Eel River Basin cycles and the Merced cycles overlap, but the upper cycles in the Merced Basin strata are probably younger than those in the Eel River Basin. The approximately 12 cycles identified in the San Joaquin

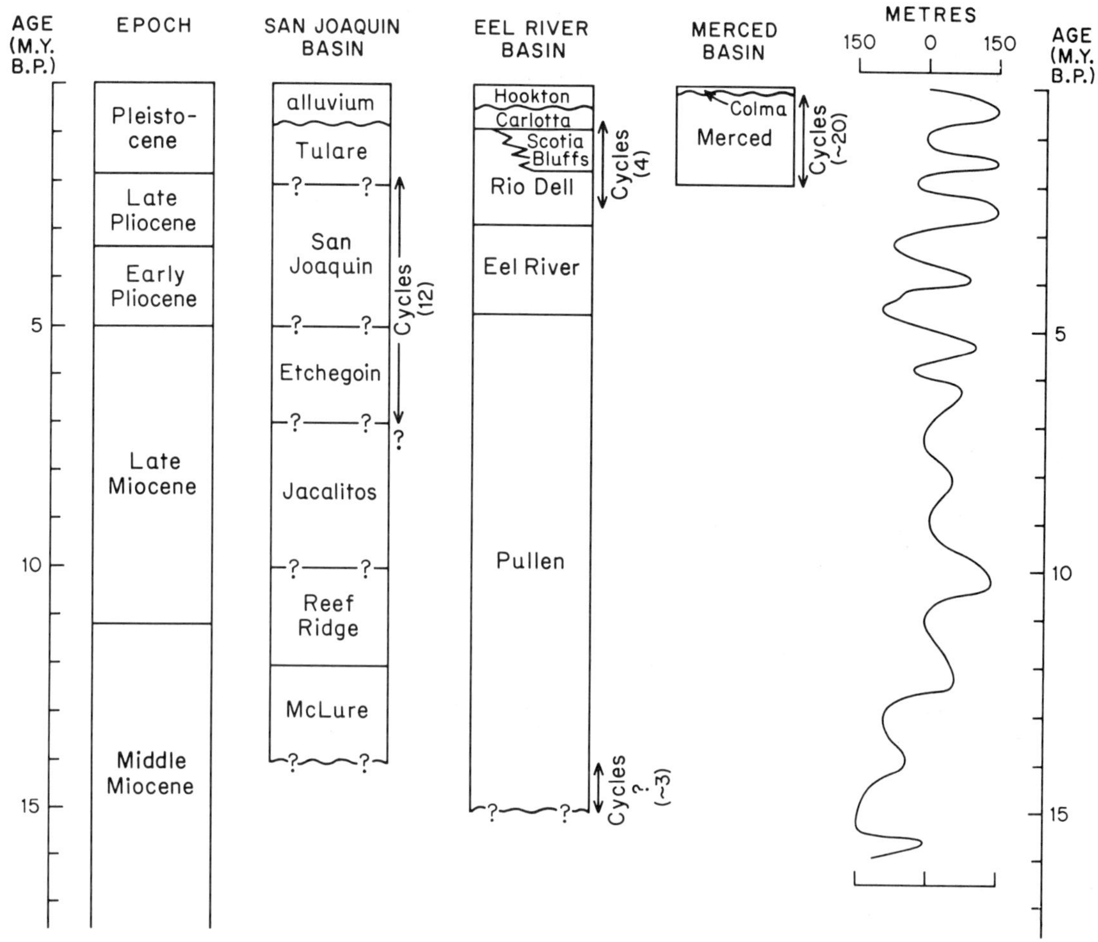


**Fig. 12.** Correlation between the stratigraphic sections in the San Joaquin, Eel River, and Merced Basins. Vertical scale is time in years before present. Correlation is largely from Armentrout & Echols (1981). Sea-level curve (curve on right) taken from Haq *et al.* (1987). Double headed arrows show position and approximate number of cycles in the three sections.

section formed over a time period of perhaps 5 Ma, giving an average duration per cycle of about 400 ka. The four cycles in the upper Eel River Basin probably were deposited over a timespan of about 0.75 Ma (Clifton & Leithold, 1991). This gives a periodicity of about 190 ka (the upper cycle appears to be about 400 ka long according to Clifton & Leithold). As in the case of the San Joaquin cycles, these cycles are considerably longer than those in the Merced Basin sediments. About 30 cycles were formed in the Merced Basin sediments during somewhat less than 2 Ma, indicating an average cycle duration of 60–70 ka.

Although a great deal of uncertainty about the origin of the cycles remains, some conclusions can be drawn.

**1** In both the San Joaquin Basin and the Eel River Basin, a single major, tectonically controlled Neogene transgressive–regressive cycle is observed. This corresponds to the cycle present throughout the North Pacific (Ingle, 1981), which must be the result of large-scale, long-term plate interactions. The section of the Merced Formation is too short to reveal this cycle, but it was probably deposited during the late stage of regression.

**2** Three lines of reasoning suggest that cyclic deposition in the Merced Basin probably was controlled largely by eustatic fluctuations.

(a) Many types of evidence, especially oxygen isotopic data from deep-sea cores, indicate that sea level fluctuated throughout Neogene time in response to glacial waxing and waning. The principle of Occam's Razor suggests that we should accept the most simple explanation that sea level in the basin varied in response to these worldwide, glacially controlled sea-level changes.

(b) The predominantly eustatic cause of the fluctuations in the Merced Basin is inferred from the fact that the number of cycles approximates the previously independently established number of eustatic cycles for that time interval.

(c) The magnitude of inferred change in water depth in all the basins, to the extent that it can be determined, is reasonable for previously postulated glacial lowering of sea level (Prentice & Matthews, 1988).

**3** Several explanations are possible for the apparently longer duration of the cycles in the San Joaquin and Eel River Basins.

(a) These cycle lengths are determined on the assumption that sedimentation was continuous and that there are no gaps in the stratigraphic record. The age of different parts of the sections is not well enough constrained to determine if such gaps do indeed exist, but physical evidence does not suggest that large gaps in these sections are significantly greater than in the Merced Basin.

(b) All fluctuations in Neogene sea level were not of the same magnitude (Fig. 11; Prentice & Matthews, 1988). Correspondingly, the sedimentary cycles that resulted from these fluctuations were not equally well developed (e.g. the Merced cycles, Figs 10, 11). Smaller cycles may not be clearly developed in the Eel River and San Joaquin Basins, giving the impression that the frequency of sea-level fluctuations is less than it actually was. If this is so, more detailed study may reveal additional poorly developed cycles.

(c) The eustatic sea-level curve for late Neogene time published by Haq *et al.* (1987; Fig. 12) shows cycles that are much longer than those suggested by isotopic studies. These cycles are even longer than those described from the Eel River and San Joaquin Basins.

The cause of sea-level changes producing the different cycle lengths in the three basins may be different patterns of tectonically determined variations in uplift, subsidence, and sediment supply. Cycle duration may have been tectonically controlled by cyclic variation in tectonism, as proposed by Blair & Bilodeau (1988). However, the cycles they describe range from 2.4 to 15 Ma in length. We know of no data on tectonic cycles at the time-scale of the cycles described in this paper.

**4** The cyclic pattern in the San Joaquin Basin strata may have been more affected by tectonic process because of its more inland location, separated from the ocean by the tectonically active belt of the California Coast Ranges. Cycles might reflect not so much a eustatic cause as tectonic activity that varied the extent of water circulation between the open ocean and the inland sea. At this stage of our analysis of sedimentary history in the San Joaquin Basin, we are not able adequately to evaluate that possibility.

**5** In the upper Rio Dell Formation, episodic storm events were recorded when the basin gradually shoaled so that the depositional surface was above storm wave-base but below fair-weather wave-base. These high-frequency depositional events are distinguished from the longer term periodicity due to eustatic or tectonic mechanisms.

The interpretation of sedimentation cycles depends upon an understanding of the conditions in which eustatic cycles are most likely to form and be evident. These three examples point out clearly that in clastic strata, eustatic cycles are most evident when at least part of the deposition occurs between sea level and storm wave-base (Fig. 13; Clifton,

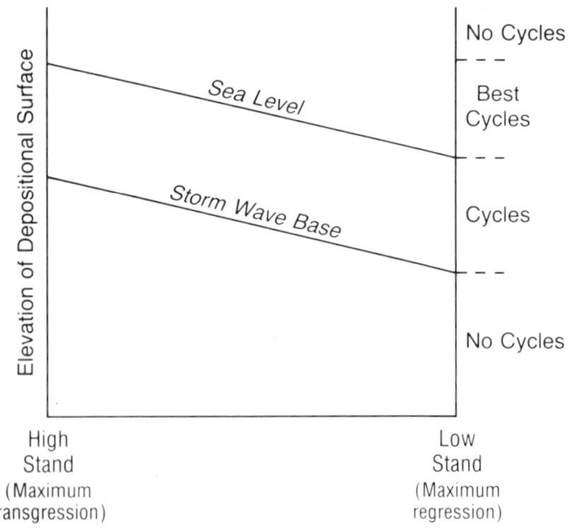


**Fig. 13.** Schematic representation of sea-level position during highstand and lowstand parts of a eustatic cycle. Sedimentary cycles develop when the depositional surface is between sea level and storm wave-base for at least a portion of the cycle.

1988). This is not to say that cyclic sedimentation does not occur at greater depths and in non-marine environments; but cycles formed in these settings are likely to be subtle and more easily overlooked. When sedimentation occurs entirely below storm wave-base, both during highstands and lowstands, conspicuous cycles would only result from variation in the rate of turbidite input and related effects. Likewise when sedimentation occurs entirely above sea level, both during highstands and lowstands, no obvious eustatically controlled cycles will be produced. The clearest cycles will be produced when highstand deposition is marine and lowstand deposition is non-marine.

In the Merced Basin, long-term tectonics (subsidence) and sedimentation rates appear to have been relatively uniform and counterbalanced so that eustatic cycles were well expressed throughout the section. Deposition within most cycles was above storm wave-base and extended in many of the cycles to above sea level. Deposition of the Merced sediments was also essentially entirely during Pleistocene time, when the magnitude of glacially controlled sea-level cycles was greatest (Prentice & Matthews, 1988). Conditions were thus optimal for depositing and preserving eustatically controlled sedimentary cycles.

In the Eel River Basin, much of the section was either deposited below storm wave-base or above sea level. Thus, only the lower Pullen and upper Rio Dell to lower Scotia Bluffs intervals have potential for containing clear, probably eustatically controlled, cycles. The cycles that occur in these units are not as conspicuous as those in either the Merced or San Joaquin sediments for several reasons. (1) Subsidence was very rapid during the transgressive phase of the major cycle. Thus the depositional site was in the critical zone for formation of sedimentation cycles for a very short time. (2) Relatively little sediment reached the depositional site during early Pullen time because most sediment was trapped in drowned estuaries closer to the sediment source area. (3) The upper Rio Dell cycles are subtle because they involve sedimentation above and below storm wave-base with no part of the cycle being above sea level. (4) Scotia Bluffs cycles contain shallow-marine sediments and probably some non-marine sands and gravels; however, determining whether these gravels are marine bar deposits or fluviatile deposits is difficult (Bourgeois & Leithold, 1983). Erosion may have removed most of the non-marine portions of these cycles (Bourgeois & Leithold, 1983). Lack of non-marine fossils in this section inhibits recognition of cycles.

The San Joaquin Basin situation was intermediate between these end-members. Much of the Etchegoin and San Joaquin Formations were deposited in the environmental window in which part of each cycle formed across the storm wave-base or sea-level transition. The cycles in the San Joaquin Formation are perhaps best developed because they contain marine and non-marine portions. Older parts of the San Joaquin section do not contain clear cycles because they were deposited below storm wave-base. The upper part of the section (Tulare Formation) is entirely non-marine and does not contain obvious cycles. The San Joaquin cycles are especially well preserved because they were deposited in a relatively low energy environment within a bay. The abundance of both marine and non-marine fossils in this section greatly aids cycle recognition.

## CONCLUSIONS

**1** Sedimentary cycles occur in the Neogene San Joaquin, Eel River, and Merced Basins of California.
**2** Each basin has a similar Neogene history, as do all the basins in the region, of rapid downwarping in middle to late Miocene time, maximum water depths in latest Miocene and earliest Pliocene time, and subsequent filling and shoaling to the present.
**3** The depositional setting ranged from approximately 2000 m below sea level to above sea level. Eustatic cycles were best developed during the regressive limb of the overall Neogene cycle.
**4** The cycles are probably largely the result of eustatic sea-level variation. Tectonics are very important in controlling subsidence rate and depth of deposition. Sedimentation rate is also important in influencing water depth in the subsiding basins. Rate of sediment supply and wave and current action at the depositional site appear to have strong effects on development of sedimentary cycles.
**5** Clear cycles can only form when at least part of the cycle is deposited above storm wave-base but below sea level. The cyclic pattern is best developed if the depositional surface crosses from marine to non-marine and back again during the cycle. All of the Merced section meets these requirements, allowing formation of many cycles. Much of the San Joaquin section falls in this range. A relatively smaller portion of the Eel River section was deposited in the critical range.

## REFERENCES


Armentrout, J.M. & Echols, R.J. (1981) Biostratigraphic–chronostratigraphic scale of the northeastern Pacific Neogene. *Proceedings IGCP-114 International Workshop on Pacific Neogene Biostratigraphy*, Osaka, 1981, pp. 7–27. Osaka Museum of Natural History, Osaka.

Blair, T.C. & Bilodeau, W.L. (1988) Development of tectonic cyclothems in rift, pull-apart, and foreland basins: sedimentary response to episodic tectonism. *Geology* **16**, 517–520.

Bourgeois, J. & Leithold, E.L. (1983) Sedimentation, tectonics and sea-level change as reflected in four wave-dominated shelf sequences in Oregon and California. In: *Cenozoic Marine Sedimentation, Pacific Margin, U.S.A.* (Eds Larue, D.K. & Steel, R.J.) pp. 1–16, Pacific Section, Soc. Econ. Paleontol. Mineral., Los Angeles, CA.

Burckle, L.H., Dodd, J.R. & Stanton, R.J., Jr. (1980) Diatom biostratigraphy and its relationship to paleomagnetic stratigraphy and molluscan distribution in the Neogene Centerville Beach section, California. *J. Paleontol.* **54**, 664–674.

Clifton, H.E. (1988) Sedimentologic approaches to paleobathymetry, with applications to the Merced Formation of central California. *Palaios* **3**, 507–522.

Clifton, H.E. & Hunter, R.E. (1987) The Merced Formation and related beds: a mile-thick succession of late Cenozoic coastal and shelf deposits in the seacliffs of San Francisco, California. In: *Geol. Soc. Am. Centennial Field Guide — Cordilleran Section*, pp. 257–262.

Clifton, H.E. & Leithold, E.L. (1991) Quaternary coastal and shallow-marine facies sequences, northern Californian and the Pacific Northwest. In: *Quaternary Non-glacial Geology: Conterminous United States* (Ed. R. Morrison). *Decade of North American Geology*, Vol. K–2. Geol. Soc. Am. (in press).

Clifton, H.E., Hunter, R.E. & Gardner, J.V. (1988) Analysis of eustatic, tectonic, and sedimentologic influences on transgressive and regressive cycles in the upper Cenozoic Merced Formation, San Francisco, California. In: *New Perspectives in Basin Analysis* (Eds Kleispehn, K.L. & Paola, C.) pp. 109–128. Springer-Verlag, New York.

DeCelles, P.G. (1988) Middle Cenozoic depositional, tectonic, and sea level history of southern San Joaquin Basin, California. *Bull., Am. Assoc. Petrol. Geol.* **72**, 1297–1322.

Dodd, J.R. & Stanton, R.J., Jr. (1975) Paleosalinities within a Pliocene bay, Kettleman Hills, California: a study of the resolving power of isotopic and faunal techniques. *Geol. Soc. Am. Bull.* **86**, 51–64.

Dodd, J.R. & Stanton, R.J., Jr. (1981) *Paleoecology, Concepts and Applications.* Wiley-Intersciences, New York, 559 pp.

Dodd, J.R., Mead, J. & Stanton, R.J., Jr. (1977) Paleomagnetic stratigraphy of Pliocene Centerville Beach section, northern California. *Earth Planet. Sci. Lett.* **34**, 381–386.

Dodd, J.R., Stanton, R.J., Jr. & Johnson, M. (1984) Oxygen isotopic composition of Neogene molluscan fossils from the Eel River Basin of California. *Geol. Soc. Am. Bull.* **95**, 1253–1258.

Field, M.E., Clarke, S.H., Jr. & White, M.E. (1980) Geology and geologic hazards of offshore Eel River Basin, northern California continental margin. *U.S. Geol. Surv. Open File Rep.* **80–1080**, 80 pp.

Galehouse, J.S. (1967) Provenance and paleocurrents of the Paso Robles Formation, California. *Geol. Soc. Am. Bull.* **78**, 951–978.

Haq, B.U., Hardenbol, J. & Vail, P.R. (1987) Chronology of fluctuating sea levels since the Triassic. *Science* **235**, 1156–1167.

Hayes, J.D., Imbrie, J. & Shackleton, N.J. (1976) Variations in the Earth's orbit: pacemaker of the ice ages. *Science* **194**, 1121–1132.

Hunter, R.E., Clifton, H.E., Hall, H.E., Csaszar, G., Richmond, B.M. & Chin, J.L. (1984) Pliocene and Pleistocene coastal and shelf deposits of the Merced Formation and associated beds, northwestern San Francisco Peninsula, California. In *Field Trip Guidebook No. 3, 1984 Midyear Meeting, San Jose, California*, pp. 1–29. Soc. Econ. Paleontol. Mineral., Tulsa, OK.

Ingersoll, R.V. (1988) Tectonics of sedimentary basins. *Geol. Soc. Am. Bull.* **100**, 1704–1719.

Ingle, J.C. (1981) Origin of Neogene diatomites around the North Pacific rim. In: *The Monterey Formation and Related Siliceous Rocks of California* (Eds Garrison, R.E., Douglas, R.G., Pisciotto, K.E., Isaacs, C.M. & Ingle, J.C.) pp. 159–179. Pacific Section, Soc. Econ. Paleontol. Mineral., Los Angeles, CA.

Keller, G. & Ingle, J.C., Jr. (1981) Planktonic foraminiferal biostratigraphy, paleoceanographic implications, and deep-sea correlation of the Pliocene–Pleistocene Centerville Beach section, northern California. In: *Pacific Northwest Cenozoic Biostratigraphy* (Ed. Armentrout, J.M.). Geol. Soc. Am. Spec. Pap. 184, 127–135.

Kodama, K.P. (1979) New paleomagnetic results from the Rio Dell Formation, California. *Geophys. Res. Lett.* **6**, 253–256.

Leithold, E.L. (1989) Depositional processes on an ancient and modern muddy shelf, northern California. *Sedimentology* **36**, 179–202.

Leithold, E.L. & Bourgeois, J. (1989) Sedimentation, sea-level change, and tectonics of an early Pleistocene continental shelf, northern California. *Geol. Soc. Am. Bull.* **101**, 1209–1224.

Loomis, K.B. (1988) Paleoenvironmental and paleoclimatic interpretation of upper Miocene–Pliocene lithofacies and macrobiota of the Etchegoin Group, Jacalitos Canyon, San Joaquin Valley, California. In: *Studies of the Geology of the San Joaquin Basin* (Ed. Graham, S.A.) pp. 303–318. Pacific Section, Soc. Econ. Paleontol. Mineral., Los Angeles, CA.

Morrison, S.D. (1980) Late Neogene biostratigraphy of the Humboldt Basin, northern California: progress report (abstr.). *Geol. Soc. Am., Abstr. Prog.* **12**, 143.

Piper, D.J.W., Normark, W.R. & Ingle, J.C., Jr. (1976) The Rio Dell Formation: a Plio-Pleistocene basin slope deposit in northern California. *Sedimentology* **23**, 309–328.

Prentice, M.L. & Matthews, R.K. (1988) Cenozoic ice-volume history: development of a composite oxygen isotope record. *Geology* **16**, 963–966.

Roth, B. (1979) *Late Cenozoic Marine Invertebrates from Northwest California and Southwest Oregon.* Unpublished PhD thesis, University of California, Berkeley, 803 pp.

Sarna-Wojcicki, A.M., Morrison, S.D., Meyer, C.E. & Hillhouse, J.W. (1987) Correlation of upper Cenozoic tephra layers between sediments of the western United States and eastern Pacific Ocean, and comparison with biostratigraphic magnetostratigraphic age data. *Geol. Soc. Am. Bull.* **98**, 207–223.

Stanton, R.J., Jr. & Dodd, J.R. (1970) Paleoecologic techniques — comparison of faunal and geochemical analyses of Pliocene paleoenvironments, Kettleman Hills, California. *J. Paleontol.* **44**, 1092–1121.

Stanton, R.J., Jr. & Dodd, J.R. (1972) Pliocene cyclic sedimentation in the Kettleman Hills, California. In: *Geology and Oil Fields, West Side Central San Joaquin Valley* (Ed. Rennie, E.W., Jr.) pp. 50–58. Guidebook. Pacific Sections, Am. Assoc. Petrol. Geol., Soc. Explor. Geol., Soc. Econ. Paleontol. Mineral. Los Angeles, CA.

Stanton, R.J., Jr. & Dodd, J.R. (1976) Pliocene biostratigraphy and depositional environments of the Jacalitos Canyon area, California. In: *The Neogene Symposium* (Eds Fritsche, A.E., TerBest, H., Jr. & Wornardt, W.W.) pp. 85–94. Pacific Section, Soc. Econ. Paleontol. Mineral., Los Angeles, CA.

Vail, P.R., Mitchum, R.M., Jr. & Thompson, S., III (1977) Seismic stratigraphy and global changes in sea level. In: *Seismic Stratigraphy – Applications to Hydrocarbon Exploration* (Ed. Payton, C.E.). Mem., Am. Assoc. Petrol. Geol. 26, 53–62.

Wang, C. & Burnett, W.C. (1990) Holocene mean uplift rates across an active plate-collision boundary in Taiwan. *Science* **248**, 204–206.

Woodring, W.P., Stewart, R. & Richards, R.W. (1940) Geology of the Kettleman Hills oil field, California. *U.S. Geol. Surv. Prof. Pap.* **195**, 170 pp.

*Spec. Publs. int. Ass. Sediment.* (1991) **12**, 217–240

# Anatomy of an evolving island arc: tectonic and eustatic control in the south Central American fore-arc area

HARTMUT SEYFRIED, ALLAN ASTORGA, HUBERT AMANN, CLAUDIO CALVO, WOLFGANG KOLB, HANNELORE SCHMIDT *and* JUTTA WINSEMANN

*Institut für Geologie und Paläontologie, Universität Stuttgart, Böblinger Strasse 72, D 7000 Stuttgart 1, Germany*

## ABSTRACT

The southern part of the Central American isthmus is the product of an island arc. It evolved initially as a ridge of primitive island-arc tholeiites at a collision zone between the Farallón plate and proto-Caribbean crust (Albian–Santonian). During the Campanian, a major tectonic event (most probably subduction reversal) caused décollement of different units of the former plate margin. The resulting structural high was covered by a carbonate platform. From Maastrichtian to Eocene times continuous subduction produced a stable morphotectonic configuration (trench-slope–outer-arc–fore-arc–calcalkaline-arc). Fore-arc sedimentation was controlled by volcaniclastic input and tectonic activity along the outer arc's inner margin. Eustatic control is essentially recognized through lowstand signals such as extensive turbidite sand lobes. Steady accretionary uplift of the outer arc gradually closed the bypasses between fore-arc and trench slope. Eustatic control is verified by lowstand signals (sands) on the trench slope and highstand signals on the outer arc (carbonate ramps). During the Oligocene another major tectonic event affected the entire system: accretion ceased, segments decoupled, and regional compression resulted in general uplift and erosion. From latest Oligocene to Pliocene times, three episodes of tilting created a series of fault-angle depressions. Subsidence varies enormously among these basins, but sedimentation is largely shallow marine. Facies architecture reflects complex interactions between tectonic processes, changes in volcaniclastic sediment supply, and eustasy. Subsequently, very strong explosive volcanic activity resulted in excessive sediment input that overfilled most basins. The history of the island arc shows that tectonic processes largely controlled composition, distribution, and geometry of the major sedimentary units. Eustatic signals do indeed occur when they are expected, but may be considered as an overprint rather than a dominating factor.

## INTRODUCTION

Island arcs are considered highly mobile tectonic belts. The aim of this paper is to evaluate whether control on island arc sedimentation is predominantly tectonic or whether other factors, such as volcanogenic sediment supply or eustatic changes, are important. The paper comprises: (1) a compilation of the present-day structural data, (2) a summary of the magmatic evolution, and (3) a step-by-step reconstruction of the evolutionary stages of the south Central American island arc to its present-day mature stage, with concomitant evaluation of data. In some cases it has proved almost impossible to present 'proper' references, as most of the data we work with are still in unpublished Costa Rican 'Licenciatura' theses and German PhD theses. In this context, it should be mentioned that this paper summarizes the results of very successful joint field work of Costa Rican and German workers, which started in 1985 with three small groups and now comprises 18 persons.

## OUTLINE OF THE TECTONIC SITUATION

### Fore-arc area

Central America is situated at the western margin of the Caribbean plate near the triple junction of the Cocos, Nazca, and Caribbean plates (Fig. 1). Along the Middle America Trench, the Cocos plate is currently being subducted beneath the North American and Caribbean plates. Two aseismic ridges (Tehuantepec Ridge and Cocos Ridge) intersect the Middle America Trench off southern Mexico and southern Costa Rica, respectively. The Cocos Ridge marks the southern end of the Middle America Trench.

#### *Southern Costa Rica*

In southern Costa Rica, the arrival and initial subduction of the Cocos Ridge during the Quaternary (Lonsdale & Klitgord, 1978) has affected the arc–trench system in many ways.

*Fracture zones.* The plate boundary between the Nazca and Cocos plates jumped from the Coiba Fracture Zone to the Panamá Fracture Zone (which now borders the Cocos Ridge towards the east), shearing off the Coiba Ridge (van Andel *et al.*, 1971; Lonsdale & Klitgord, 1978; see Fig. 1). Recent seismic activity suggests that the plate boundary might displace again towards the west and establish in the Quepos area (Burbach *et al.*, 1984; Morales, 1985). A fracture zone along the western margin of the Cocos Ridge ('East Nicoya Fracture Zone'; Corrigan *et al.*, 1990) slightly displaces the trench towards the northeast (Burbach *et al.*, 1984; Morales, 1985; see Fig. 2). This fracture zone extends towards mainland Costa Rica, where it joins a system of braided transcurrent faults in the Valle Central area (Montero & Dewey, 1982), which has been active since at least the Oligocene (Rivier & Calvo, 1988; Astorga *et al.*, 1989). Further to the east, it merges in the Limón thrust belt, thus dissecting the isthmus into two tectonically independent parts (Astorga *et al.*, 1989; compare Figs 1, 2). In this paper, this composite structural feature is referred to as the 'Trans-Isthmic Fault System'; the area affected is called 'Dissected Zone of Central Costa Rica'.

*Isostatic adjustment.* Isostatic compensation produced uplift and deformation in the fore-arc (Corrigan *et al.*, 1990). In cross-section (Fig. 2), this appears as a series of arcward-tilted uplifts and fault-angle depressions (Seyfried *et al.*, 1987). According to Gardner *et al.* (1987), Wells *et al.* (1988), Corrigan *et al.* (1990) and Adamek *et al.* (1987), a major fault ('longitudinal fault', see Fig. 2), most probably with a strike-slip component, indicates tectonic decoupling between the outer and inner fore-arc areas.

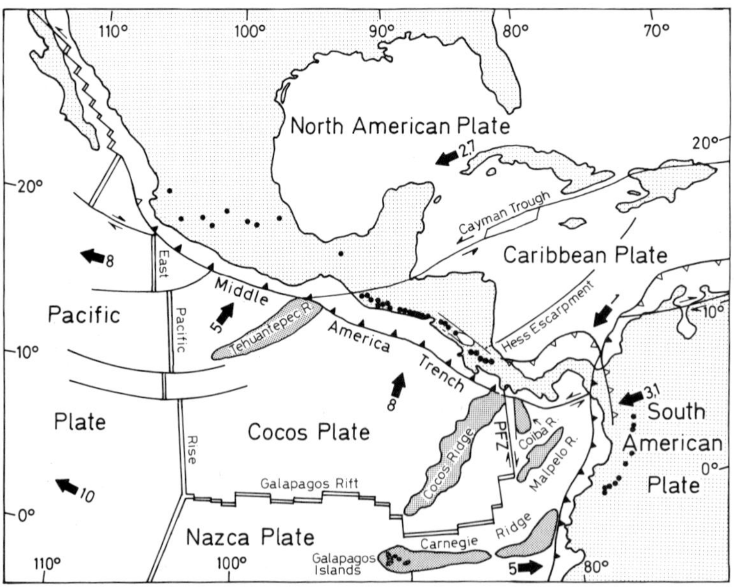


**Fig. 1.** Tectonic setting of Central America. Redrawn after Drummond (1981). Black arrows indicate drift directions; numbers represent absolute movements in cm/ yr; black dots are active volcanoes.

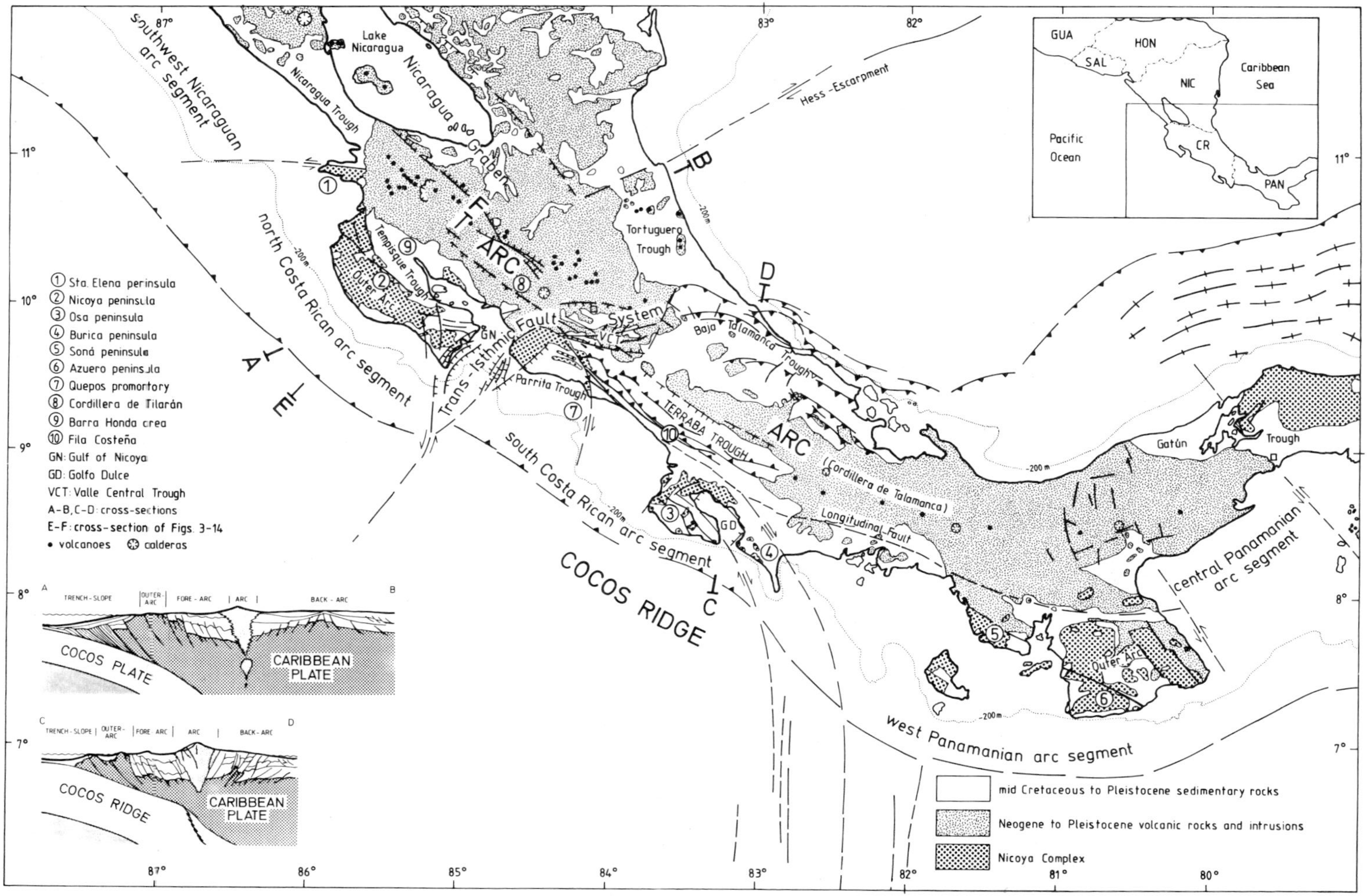


**Fig. 2.** Structural map of southern Central America, showing major tectonic elements, main tectonostratigraphic units, large sedimentary basins, and localities mentioned in text. Synthesized after Case & Holcombe (1980), Dengo (1962), Campos (1987), Fernandez (1987), Gardner *et al.* (1987), Berrangé & Thorpe (1988), de Boer *et al.* (1988), Denyer & Montero (1989), Astorga *et al.* (1989), Berrangé *et al.* (1989), and unpublished mapping by the authors.

*Uplift history.* According to Morales (1985) and McGeary *et al.* (1985), spatial gaps in arc volcanism indicate collision with oceanic plateaux. In southern Costa Rica, the width of the Cocos Ridge corresponds almost exactly to the breadth of the summit level of the Cordillera de Talamanca, which is the volcanically inactive segment of the arc. In the Chirripó area (with a present maximum altitude of 3820 m), mid-Miocene shallow-water volcaniclastics, pyroclastic flows, lahars, and intrusive granodiorites (G. Calvo, 1987; Wunsch, 1988), as well as late Pleistocene moraines (Weyl, 1956), indicate the maximum timespan for the uplift. Detrital quartz appears for the first time in upper Miocene shallow-water deposits of the back-arc basin (Campos, 1987); granodioritic boulders follow in lower Pliocene fan-delta deposits (Bottazzi, personal communication, 1989; F. Alvarado, 1984). Thus, the uplift of the Cordillera de Talamanca must have started as early as the late Miocene. This early uplift may have been a side-effect of the collision between the Panamá block and northwest South America (Wadge & Burke, 1983; Mann & Burke, 1984; Pindell *et al.*, 1988). According to S. Mora (1979), Rivier (1985), and Rivier & Seyfried (1985), the imbrication of the Fila Costeña and Limón thrust belts along the flanks of the Cordillera de Talamanca (see Fig. 2) also can be placed into the late Miocene timespan. Subsequent arcward tilting of both thrust belts also affected late Pliocene to early Pleistocene alluvial fans (Campos *et al.*, 1984) and thus can be dated as late Pleistocene. It is considered to be an effect of the compression exerted by the approaching Cocos Ridge (Heywood, 1984; Corrigan *et al.*, 1990).

### *Northern Costa Rica*

North of the Trans-Isthmic Fault System, the neotectonic situation is less complicated than in southern Costa Rica. In a cross-section from SW to NE (Fig. 2), it appears that tilting is also arcward. There are only minor trenchward thrusts in the inner fore-arc region prior to tilting ('Barra Honda' area; see Fig. 14); imbrication occurred more or less simultaneously with the Fila Costeña and Limón thrusts (C. Calvo, 1987). The trench slope is rather uniform and reaches the Middle America Trench without any indications of an approaching aseismic ridge (Mora, 1985; Crowe & Buffler, 1985). The present volcanic arc has been very active since at least Pliocene times (Tournon, 1984; G. Alvarado, 1985).

The present-day neotectonic features of the fore-arc area in southern Central America are the result of a rather complex structural history. The present arc is divided into the following segments (Fig. 2): southwest Nicaraguan arc segment, north Costa Rican arc segment, south Costa Rican arc segment, west Panamanian arc segment, and central Panamanian arc segment. Each segment has an individual structural, magmatic, and sedimentary history. This study deals mainly with the north and south Costa Rican arc segments and the dissected zone in between.

## Back-arc area

From a structural point of view, the back-arc area is far less complicated than the fore-arc area. Since the Late Cretaceous, a thick (almost 8 km) shallowing upward sedimentary sequence has been laid down more or less continuously (Fig. 3). The final transition from marine to terrestrial deposits started in the mid-Miocene (Obando, 1986; Calvo & Bolz, 1987); in places, marine ingressions are recorded until the early Pliocene (Astorga *et al.*, 1989). Seismic data reveal that an 'outer' structural high was emergent during the late Eocene.

The Trans-Isthmic Fault System divides the back-arc area into two structurally different regions: the Baja Talamanca Trough in the south and the Tortuguero Trough and Nicaragua graben in the north (Fig. 2). In the Baja Talamanca Trough (Rivier & Seyfried, 1985), late Miocene thrusting has been related to the initial uplift of the Cordillera de Talamanca and to the collision of the Panamá block with northwest South America, respectively. The Limoń Thrust Belt merges in the 'Panamá Deformed Belt' (Heywood, 1984; Vitali, 1985; see Fig. 2). Huge, extremely coarse Pliocene to lower Pleistocene clastic wedges were trapped between the imbricated sheets of the Limón Thrust Belt. Late Pleistocene arcward tilting was caused by the collision with the Cocos Ridge.

The Tortuguero Trough (see Fig. 2) is the undeformed northern complement of the Talamanca Trough; extensive coastal swamps indicate that the basin is still subsiding. According to Astorga *et al.* (1989), the Nicaragua graben started to subside in the late Miocene; there is sedimentological evidence that the graben was still active at the beginning of the late Pliocene (Kolb, 1990). It extends from the Honduran Pacific coastal borderlands through Nicaragua (Lake Managua and Lake Nicaragua) to northern Costa Rica, where the southeastern termin-

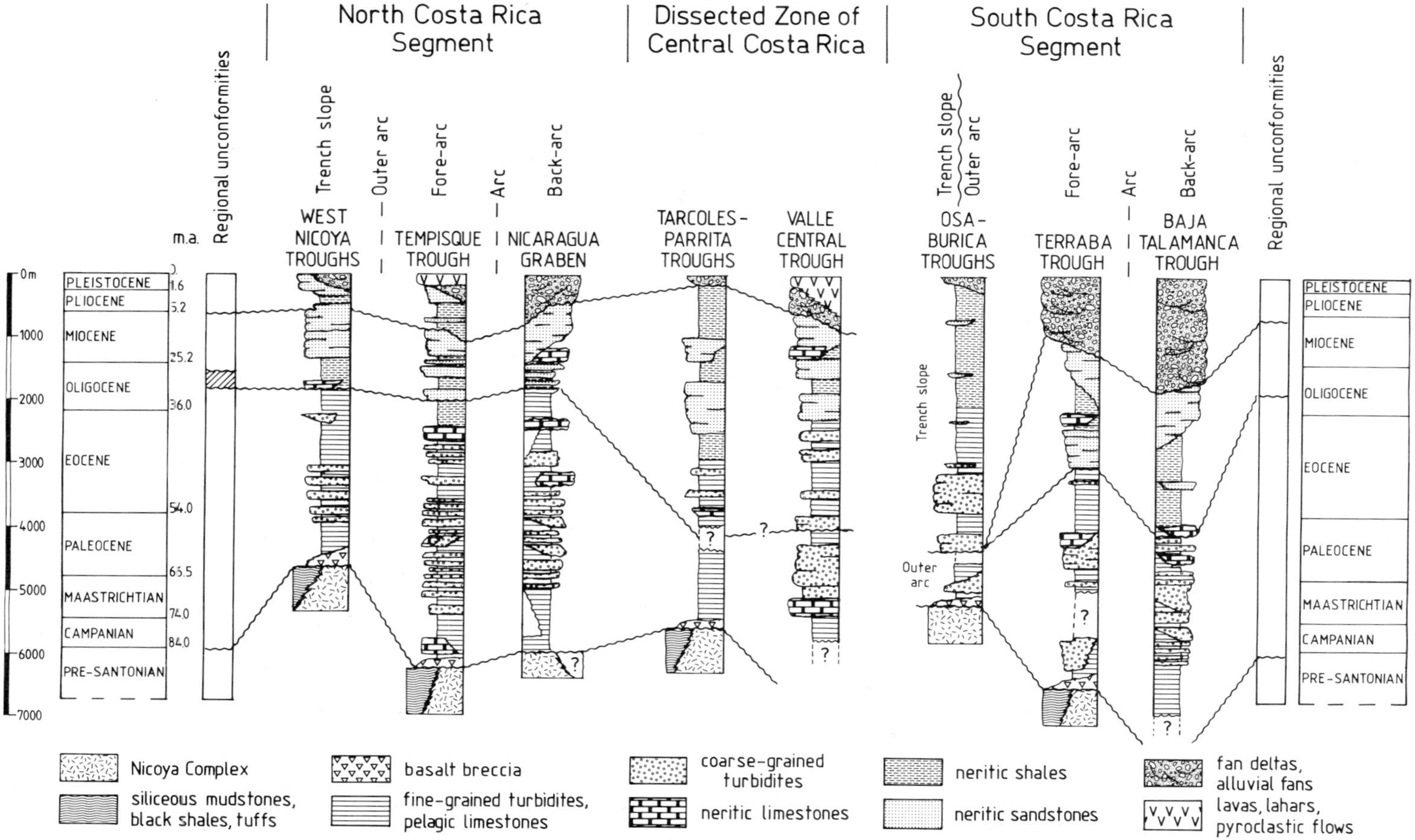


**Fig. 3.** Generalized stratigraphic sections along and across the Costa Rican part of the south Central American isthmus. For location, compare with Fig. 2. Compiled after Astorga *et al.* (1989), Rivier & Calvo (1988), Corrigan (1986), and unpublished data from the authors.

ation is obscured by Quaternary volcanics (see Figs 1, 2). Existing geological maps suggest (but provide no definite evidence) that the master faults cut across an older, ENE–WSW-striking structure, which is considered to represent the onshore prolongation of the Hess Escarpment (Burke *et al.*, 1984; Astorga *et al.*, 1989; compare Figs 1, 2). Most probably, it is this same fault that reappears underneath Quaternary volcanics at the Pacific side of Costa Rica in the area of the Santa Elena Peninsula (Tournon & Azéma, 1980; Burke *et al.*, 1984; see Fig. 2). Calvo (1987), Astorga (1988), and Winsemann & Seyfried (in their chapter) demonstrate that this fault (or one of its predecessors) has been active at least from Late Cretaceous to early Eocene times; in contrast, the geological situation at the Caribbean side (see Fig. 2) suggests that it may have been active even until the Pliocene.

## THE MAGMATIC SUITE

### Outer arc

#### *Northern Costa Rica*

In northern Costa Rica, the basement of the outer arc is formed by harzburgites and lherzolites (and their serpentinized equivalents), dunites, gabbros, plagiogranites, normal and transitional 'mid-ocean ridge' basalts (N-MORB, T-MORB), within-plate alkalibasalts and within-plate tholeiites (WPA, WPT), and 'island-arc' tholeiites (IAT) (Jager, 1977; Kuypers, 1980; Tournon & Azéma, 1980; Tournon, 1984; Wildberg, 1984,1987; Bourgois *et al.*, 1984; Azéma *et al.*, 1985; Desmet & Rocci, 1988; Meschede *et al.*, 1988; Sick, 1989; Appel, 1990). Discontinuous units of radiolarites found among the basalts have been assigned ages ranging from early Liassic until Santonian (de Wever *et al.*, 1985; Schmidt-Effing, 1979,1980; H.J. Gursky & Schmidt-Effing, 1983; Baumgartner, 1984,1987; Baumgartner *et al.*, 1984; Azéma *et al.*, 1985; H.-J. Gursky, 1988). Dengo (1962) named this ensemble the 'Nicoya Complex'. Schmidt-Effing (1979), Wildberg (1984), M.M. Gursky (1986), and H.-J. Gursky (1988,1989) made a further subdivision into 'Lower Nicoya Complex' (LNC), which comprises the oceanic suite, and 'Upper Nicoya Complex', which contains island-arc tholeiites and their associated sediments as well as within-plate basalts, which Wildberg (1984) related to the Santonian 'Caribbean flood basalt event' *sensu* Donnelly (1975,1985).

No radiometric age determinations older than 85 Ma (late Santonian) have been obtained so far from the Nicoya Complex. According to G. Alvarado (written communication), most determinations concentrate between 65 and 55 Ma (Paleocene). Recently, Appel (1990) also obtained K–Ar ages between 65 and 54 Ma, with samples coming from both the Lower and Upper Nicoya Complexes. He draws the conclusion that these ages, which are not in agreement with biostratigraphic and sedimentological data, represent alteration ages due to a thermal event. Kussmaul (1987) inferred that widespread crustal heating is needed to create the extensive granitic batholiths that penetrated the Cordillera de Talamanca during the Miocene. Because of the location (calc-alkaline arc) and age of these intrusions, this possibility should be discarded as a source of Paleocene heating in the outer arc. Marshak & Karig (1977), Hill *et al.* (1981), and Forsythe *et al.* (1986) argued that short-lived pulses of thermal and magmatic activity in the fore-arc area might be related to the subduction of an oceanic spreading centre. The WPT, which have recently been recognized throughout the outer arc area, are mostly attributed a Paleocene age, but their geochemistry does not correlate with an MOR derived magma. In addition, it is still not clear whether their emplacement was hot (intrusion) or cold (obduction).

#### *Southern Costa Rica*

*Osa and Burica Peninsulas.* This area shows basement rocks similar to the outer arc of northern Costa Rica (mainly T-MORB, N-MORB, IAT, and radiolarites; Lew, 1983; Tournon, 1984; Obando, 1986; Berrangé & Thorpe, 1988; Berrangé *et al.*, 1989; Sick, 1989; Appel, 1990). Biostratigraphically and radiometrically obtained ages range from Early Cretaceous to mid-Eocene (Azéma *et al.*, 1981; Lew, 1983; Tournon, 1984; Baumgartner, 1987; Berrangé *et al.*, 1989). Recently, WPA and WPT have also been identified in the Osa–Burica area (Berrangé & Thorpe, 1988; Sick, 1989; Appel, 1990). The spatial and genetic relations of these intra-plate basalts with the MORB and IAT units are still unclear. They might correspond to the younger magmatic events (Paleocene, mid-Eocene) that have been identified by Azéma *et al.* (1981), Tournon (1984), and Berrangé *et al.* (1989).

*Quepos promontory and Parrita Basin.* Wildberg (1984), Sick (1989), and Appel (1990) reported extensive outcrops of WPB (which commonly display high contents of vesicles) and IAT (with almost no vesicles). According to Azéma *et al.* (1979a) and Baumgartner *et al.* (1984), interpillow sediments assign these basalts an early Paleocene age. In contrast, Appel (1990) obtained K–Ar ages that spread around 46 Ma (mid-Eocene). Azéma *et al.* (1979a) and Sick (1989) describe both cold and hot contacts with overlying sediments. The presence of IAT and the occurrence of boulders of radiolarite and pebbles of Late Cretaceous limestones (Henningsen, 1966) among lower Paleocene conglomerates, which overlie the basalts in places, point to the existence of an older, Nicoya Complex type, basement in this area. These data allow two explanations: (1) cold emplacement — the intra-plate basalts are sheared-off remnants of a subducted seamount that obducted on to the trench slope and subsequently became uplifted (Sick, 1989; Appel, 1990); (2) hot emplacement — as in the Osa–Burica area, a two-phase magmatic evolution might explain the superposition of WPB over IAT. The intrusion of an alkaline magma into an island arc, however, requires a back-arc setting. This in turn means a temporary reversal of subduction polarity in the south Costa Rica arc segment. A back-arc origin has already been seriously discussed by Berrangé & Thorpe (1988) for the basalts of the Osa–Burica area. However, apart from the Paleocene heating event there are no further unambiguous facts that would point to subduction reversal. In the following, we shall adapt the obduction hypothesis as a working base. The question of the Paleocene thermal event must remain open for the time being.

### *Palaeomagnetic data*

Sick (1989) reports counter-clockwise rotations for the different units of the Nicoya Complex ranging from 90° to 37°. The different amounts of rotation are explained as a result of Campanian shear tectonics. The Nicoya Complex must have originated at near-equatorial latitudes. The tracks of relative motion of the Nicoya Complex and South America are almost exactly parallel from Late Jurassic until Oligocene times. This implies that an origin at a suture within the Pacific plate (as proposed by Burke *et al.* (1984) and Ross & Scotese (1988)) must be discarded. Post-Campanian rocks of the north Costa Rican arc segment do not show rotations, whereas their equivalents from the southern sector register slight differential rotations.

## The calc-alkaline arc

### *Palaeogene volcanics*

In northern Costa Rica, the first appearance of calc-alkaline volcanic detritus in the fore-arc sedimentary record dates back to the Maastrichtian (Tournon, 1984; Astorga, 1988; see chapter by Winsemann & Seyfried). Measured transport directions and proximality estimations throughout the Palaeogene fore-arc sequence (Astorga, 1987) clearly indicate that a calc-alkaline island arc must have grown rapidly behind the outer arc suite at a distance that more or less corresponds to the present-day distance between outer arc and arc. In the present situation, however, the greater part of the Paleogene arc is deeply buried underneath younger volcanics and hardly accessible to direct observation. Thus the only evidence of Paleogene magmatic activity comes from the detritus that has been shed into the fore-arc basin. Owing to strong alteration, geochemical analyses of boulders taken from different stratigraphic levels have so far failed to give more insight into possible geochemical trends in arc evolution. According to Astorga (1988) and the chapter by Winsemann & Seyfried, the presence of autoclastic flow breccias and boulders of rhyolitic tuffs among coarse-grained late Eocene deep-water channel-fill deposits indicates that magma differentiation must have reached a first climax during the Eocene.

In southern Costa Rica, parts of the Paleogene calc-alkaline arc are exposed along the Cordillera de Talamanca. Weyl (1957) and Pichler & Weyl (1975) mainly report andesites. So far, rhyolitic rocks have not been reported from this area.

As to Oligocene magmatic rocks, there are almost no data. It seems probable that there was little activity throughout the entire isthmus.

### *Miocene volcanics and intrusions*

Miocene calc-alkaline rocks have been studied in much detail by Pichler & Weyl (1973,1975), Kussmaul *et al.* (1982), Tournon (1984), Kussmaul (1987), and Appel (1990). An assemblage of plutonic rocks ranging from gabbros to alkali granites was intruded as laccoliths or batholiths mainly into the Cordillera de Talamanca and the Cordillera de Tilarán area (Fig. 2). K–Ar age determinations spread around 19 Ma (gabbros; Appel, 1990) and 11.5–8 Ma (granitoids; Tournon, 1984; Appel, 1990). Volcanic rocks, mainly of andesitic composition, are reported from both early and late Miocene times (Tournon, 1984). Parts of a large late Miocene volcano (with intercalated shallow marine sediments) are still preserved in the Chirripó area (G.Calvo, 1987; Wunsch, 1988).

### *Pliocene and Quaternary volcanics*

According to Pichler & Weyl (1975), Kussmaul *et al.* (1982), Tournon (1984), G. Alvarado (1985), Kussmaul (1987), Nyström *et al.* (1988), and Appel (1990), the Pliocene to Quaternary volcanics of the isthmus area cover a wide spectrum ranging from basaltic andesites to rhyolites. Most volcanoes correspond to the pyroclastic cone and the caldera type (O. Mora, 1989). Geochemically, the volcanic rocks can be subdivided into an alkaline and a calc-alkaline group. The alkaline group is mostly concentrated in the back-arc area, but sporadic occurrences in the arc region are also recorded. The calc-alkaline group still shows transitions towards IAT and displays extreme differentiations, which according to Pichler & Weyl (1975) and Kussmaul (1987) reveal that the island arc had entered into the 'mature' stage.

## THE SEDIMENTARY SUITE

Figure 3 shows eight highly generalized, synthesized stratigraphic columns that correspond to the major tectonostratigraphic units of Costa Rica (for location, compare with Fig. 2). These columns reveal that the way in which sequences developed is comparable throughout the entire isthmus:

**1** the overall pattern is shallowing upward (SU) (in the Osa–Burica Trough, there are two SU sequences);

**2** during the late Paleocene, mid- to late Eocene, and latest Oligocene, carbonate ramps appear simultaneously along the flanks of both outer arc and arc;

**3** major unconformities, which appear in the earliest Campanian, the late Oligocene, and the earliest Pliocene, are correlatable on a regional scale (900 km from southwestern Nicaragua to mid-western Panamá; see Fig. 3).

Biostratigraphic control is commonly very poor owing to strong diagenetic alteration, especially in fine sandstones (which account for approximately 70% of the total sediment volume of the island arc). Good control has been obtained with (1) pelagic and neritic limestones, (2) turbidites containing larger Foraminifera as grains, (3) some of the radiolaria-bearing sediments, and (4) some of the neritic shales. Again, this accounts for, at best, 5% of the total sediment volume. Fortunately, there are some regional marker horizons, for example an amber layer in the early Pliocene (Schmidt, 1989; Kolb, 1990). Further stratigraphic details are given in the papers by Winsemann & Seyfried, Schmidt & Seyfried, and Kolb & Schmidt.

Figures 4–14 summarize the main stages of development of the fore-arc area in the north Costa Rican arc segment from the Mid-Cretaceous until the present (for location of the cross-sections, see Fig. 2). At this stage of our investigations, the data base for an extension of these cross-sections into the back-arc area, or even into the south Costa Rican arc segment, is still insufficient. The sections were drawn mostly perpendicular to the present trench and are not to scale. The main structural elements and their internal facies architecture have been reconstructed on the basis of field observations by the authors, published sedimentological and biostratigraphical data, Costa Rican geological maps and 'licenciatura' theses (Escuela Centroamericana de Geología, University of Costa Rica), and seismic profiles (mainly Crowe & Buffler (1985), Shipley & Moore (1986), and Astorga *et al.* (1989)). As to the incorporation of geochemical data, it must be emphasized that there are still some open questions. Thus, in the course of drawing the cross-sections we had to adopt working hypotheses that are still subject to further proof.

### **Albian to early late Campanian** (Figs 4, 5)

#### *Observations*

The first sedimentary unit that shows a genetic relationship to incipient island-arc magmatism ('primitive' IAT) is composed of black shales, siliceous mudstones, and tuffs. Azéma *et al.* (1979b) and Baumgartner (1987) assigned this formation an age between Albian and Santonian; Astorga (1987) provided sedimentological details and named the unit 'Loma Chumico Formation'.

The next sedimentary unit is a predominantly basaltic sedimentary breccia; the contact with the underlying sedimentary and magmatic rocks is a regional unconformity (cf. Fig. 3). This breccia

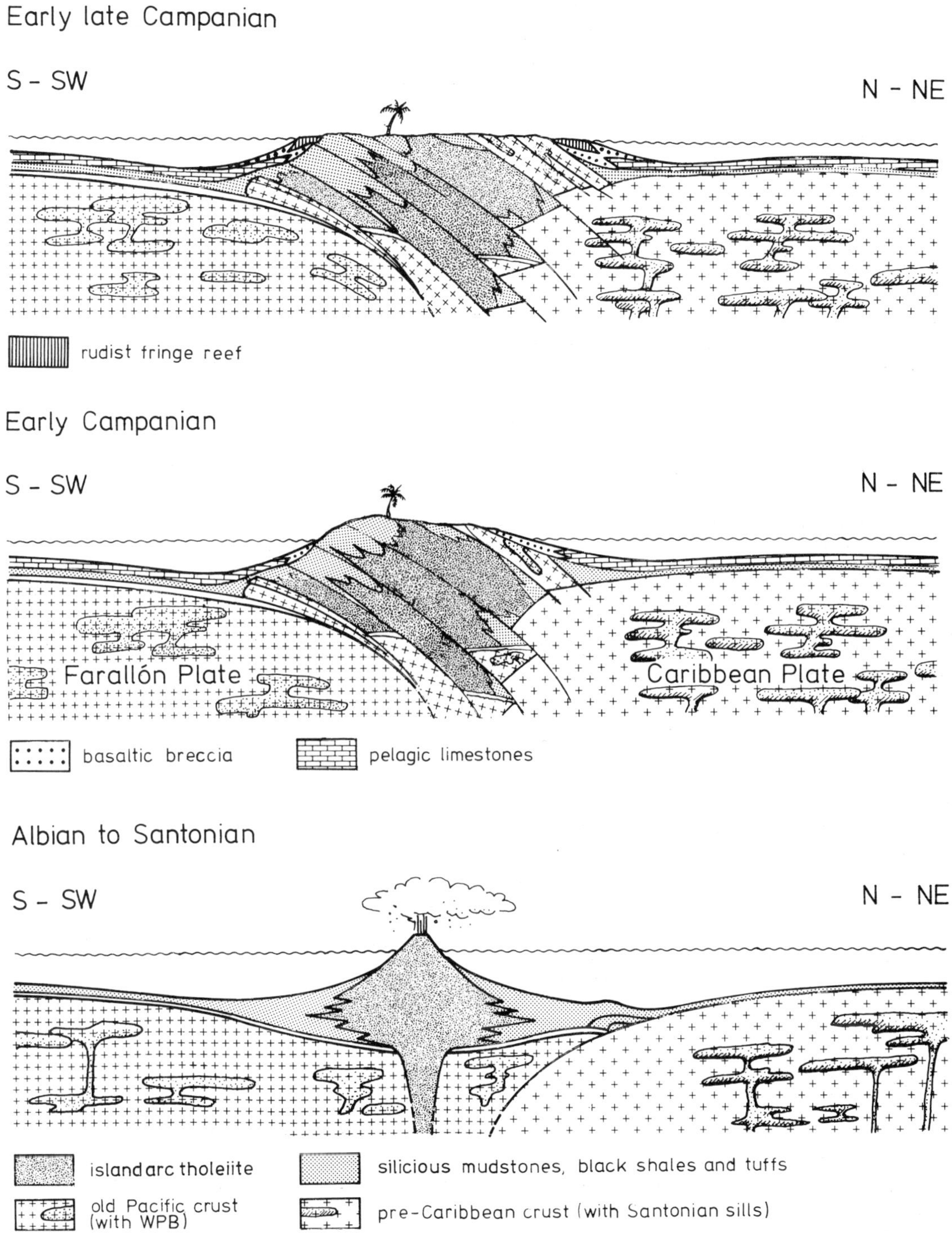


**Fig. 4.** Alternative A. Subduction reversal model as an explanation for the major tectonic event that affected the initial island arc during the early Campanian. The originally S–SW-directed subduction was reversed as the proto-Caribbean crust became unsubductable after the Santonian 'sill-event'. The wedge between the reversing subduction zones was telescoped by décollement. During the early late Campanian, the first rudist fringe reefs established on the structural high.

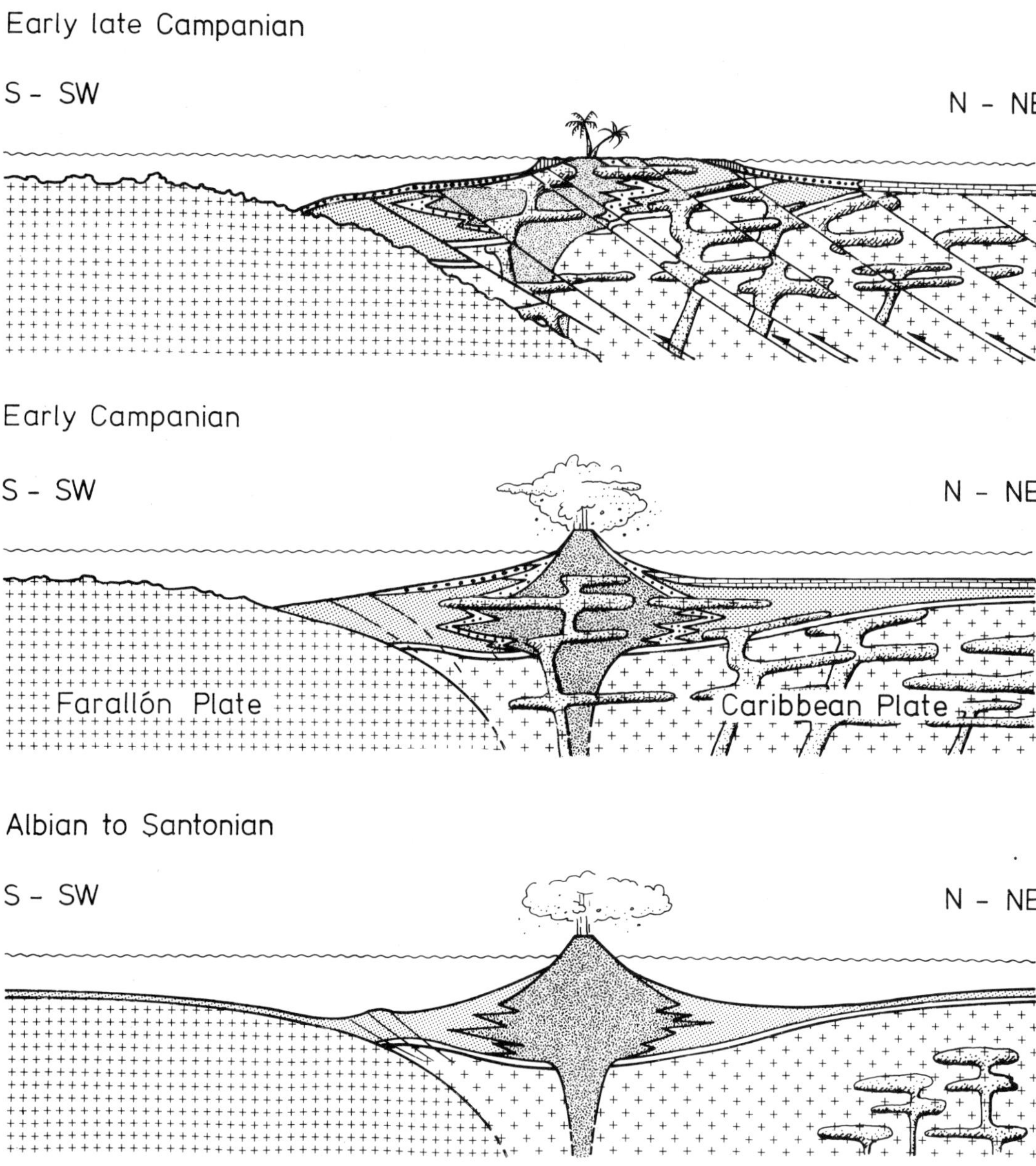


**Fig. 5.** Alternative B. Continuous subduction model as an alternative explanation for the major tectonic event that affected the initial island arc during the Campanian. In contrast to alternative A, continuous N–NE-directed subduction is assumed. Initial subduction would have been triggered by thickening of the proto-Caribbean crust during the Santonian 'sill-event'. Décollement is explained by strong plate coupling due to the arrival and initial subduction of an oceanic plateau. Symbols are the same as in Fig. 4.

spreads over both the 'Loma Chumico' unit and almost all units of the Nicoya Complex, even the oldest ones (Baumgartner *et al.*, 1984; Sprechmann, 1984). Seyfried & Sprechmann (1986) found that in some places the basal unit of the breccia consists entirely of reddish weathered, basalt rubble and gravel, which obviously formed a debris apron around islands. The upper unit of the breccia contains less-altered, probably cliff-derived basalt detritus and shows intercalations of rudist debris, which points to initial colonization by rudist fringe reefs. An ammonite (*Canadoceras* sp.) from one of these intercalations allowed this unit to be dated as late Campanian. On a regional scale, these brecciated deposits laterally interfinger with pelagic limestones, which yielded both early and late Campanian ages (Schmidt-Effing, 1979; Baumgartner *et al.*, 1984; Astorga, 1988).

### *Interpretation*

The indiscriminate exposure of both Lower and Upper Nicoya Complex units below the above-mentioned regional unconformity, the appearance of a widespread basaltic breccia above it, and the first appearance of upheaved islands indicate that a major tectonic process must have affected the area during the early Campanian. There are at least two possibilities.

*Alternative A* (subduction reversal model; Fig. 4). The edifice of primitive IAT was created by S–SW-directed subduction of proto-Caribbean crust underneath Pacific crust. Such a 'stationary' palaeogeographic model would be in accordance with the above-mentioned palaeomagnetic results (Sick, 1989). Crustal thickening during the Santonian 'sill event' rendered the Caribbean crust unsubductable. The ensuing reversal in subduction polarity telescoped the wedge in between, thus producing a stack of former plate-margin slices with the 'deep' units (Lower Nicoya Complex) on top and the 'shallow' units (Upper Nicoya Complex) underneath — which is exactly the situation found today in northern Costa Rica.

This model implies that (1) the entire Nicoya Complex, including radiolarites and WPB, has a Pacific origin, but now lies upon Caribbean crust — this would account for the anomalously thick crust underneath Costa Rica (40 km; Matumoto *et al.*, 1977); (2) the Caribbean sill event would not have had any direct effects upon the arc; (3) IAT volcanism would have stopped from the Santonian onward. Baumgartner *et al.* (1984) and Astorga (1988), however, report some minor volcanic activity from the early Campanian. The strong point of this model is that it explains satisfactorily the drastic spatial, temporal, and genetic separation between the Nicoya Complex and the calc-alkaline arc.

*Alternative B* (continuous subduction model; Fig. 5). This model starts with the assumption that the edifice of primitive IAT was created by N–NE-directed subduction of Pacific crust underneath proto-Caribbean crust. Such a suture could have developed elsewhere in the Pacific. The palaeogeographic model is not necessarily stationary and thus fits with the scenarios proposed by Burke *et al.* (1984) and Ross & Scotese (1988). The shortening of the overriding plate margin and stacking of slices of Lower Nicoya Complex over slices of Upper Nicoya Complex requires very strong plate coupling and deep-reaching transmission of compressive forces into the Caribbean plate. Such a situation could have developed during the initial collision and subduction of an oceanic plateau. According to the observations made in present-day southern Costa Rica (collision with the Cocos Ridge — see earlier section on tectonics), initial uplift should then be followed by isostatic relaxation (which indeed happened during the next episode).

The model implies that (1) the Upper Nicoya Complex formed at the Caribbean side of the suture upon the Lower Nicoya Complex, which originally was Pacific crust; (2) initial subduction was triggered by thickening of proto-Caribbean crust in the Santonian (evidence from the Loma Chumico Formation suggests, however, that IAT volcanism must have started much earlier); (3) continuous subduction allowed continuous IAT production until the early Campanian; (4) this in turn sets the telescoping event into the early late Campanian. It implies further that the Nicoyan WPB are either the product of the Caribbean sill event or represent accreted slices from a subducted oceanic plateau. Wildberg (1984) and Sick (1989) favour an origin as Caribbean sills; the latter even provided an indirect palaeomagnetic age determination (33r isochron = Campanian). In our opinion, however, it is difficult to imagine how these Caribbean sills should have penetrated during the entire Santonian to early Campanian timespan right into the leading edge of the Caribbean plate, where they interfered with active IAT volcanism.

As matters stand, neither of the alternatives presents unambiguous interpretations. In the following, we shall adapt model A as a working hypothesis.

## **Terminal Campanian to early late Maastrichtian** (Fig. 6)

### *Observations*

Seyfried & Sprechmann (1986) reported that the

first deposits that can be found upon the abovementioned upheaved islands are thin patchy veneers of extremely coarse basaltic and ultramafic debris. Outcrops on the Santa Elena Peninsula and in the Tempisque Basin provide excellent evidence that large areas were overgrown by rudist reefs. Obviously, the biodetritus supplied by these reefs helped smooth the pre-existing relief. In places, the reefs even grew over rocky relief. A considerable portion of the former islands remained emerged, still exposing Nicoya Complex and rocks of the Loma Chumico Formation.

Subsequently, the entire relief was covered by a carbonate platform ('Barra Honda Formation'). Bio-

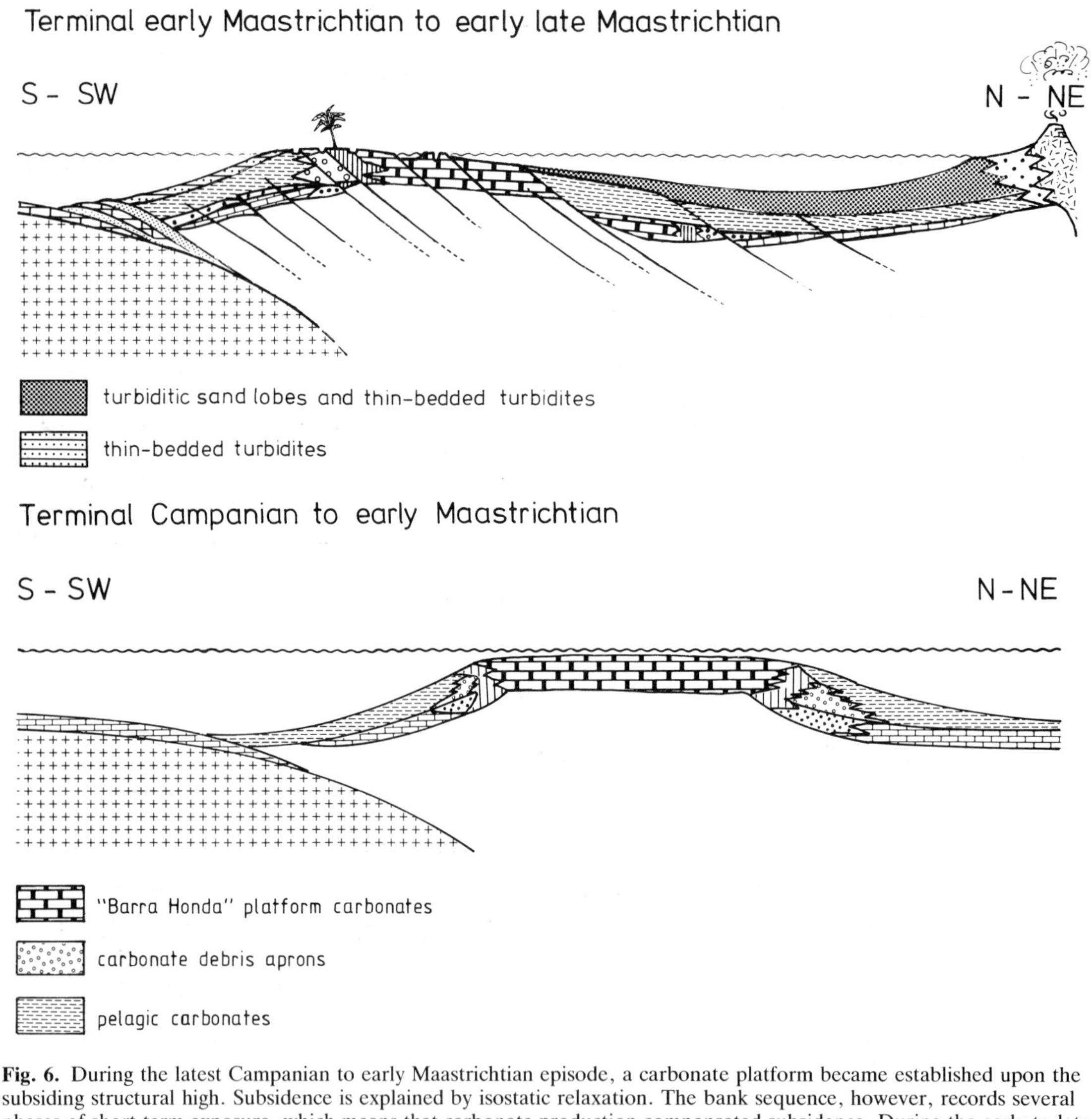


**Fig. 6.** During the latest Campanian to early Maastrichtian episode, a carbonate platform became established upon the subsiding structural high. Subsidence is explained by isostatic relaxation. The bank sequence, however, records several phases of short-term exposure, which means that carbonate production compensated subsidence. During the early to late Maastrichtian transition, a final karstification occurred, which most probably represents a eustatic overprint over steady uplift due to incipient accretion. During that episode, the activity of a calc-alkaline arc becomes increasingly evident. By now, the classical subdivision into trench slope, outer arc, fore-arc, and calc-alkaline arc is clearly established. Connections between the fore-arc and the trench slope, however, must have existed as sand lobes obviously correlate with thin-bedded turbidite systems.

stratigraphical data (C. Calvo, 1987; Sprechmann *et al.*, 1987) prove that this platform was initiated in the latest Campanian and ended shortly before the end of the early Maastrichtian. Microfacies analysis (C. Calvo, 1987) revealed that micritic facies types predominate; towards the windward side, they were protected by fringing coral reefs. The bank sequence records several minor meteoric diagenetic events and a final karstification episode.

In basinal areas, during the early Maastrichtian, turbiditic sand lobes and thin-bedded turbidite systems, composed of andesitic material, gradually replaced carbonate oozes. Thin-bedded turbidites, which are also typical of the trench slope area, have been observed to prograde diachronically over the eastern parts of the 'Barra Honda' platform.

*Interpretation*

The gradual overgrowth, firstly by fringing rudist reefs, and then by a carbonate platform, of the formerly upheaved islands clearly points to subsidence by isostatic relaxation, probably aided by a relative rise of sea level. Several minor meteoric diagenetic events recorded in the bank sequence might correspond to small-scale, short-term sea-level fluctuations. The final karstification event could be explained by uplift of the outer arc. Seismic data reveal (though not very clearly) that accretion must have started more or less during this episode and thus could have provoked the uplift.

During later development (see Figs 7, 8) the most uplifted parts of the platform became the nucleus of the outer arc. The less uplifted eastern parts subsided and were rapidly buried underneath thick units of fore-arc turbidites, derived from a rapidly rising calc-alkaline arc. Thus, the differences in tectonic style between outer arc and fore-arc must have started to develop during this episode. In addition to this purely tectonic explanation it must be emphasized, however, that a significant drop of sea level could also have produced the karstification event. In our opinion, a eustatic overprint over steady uplift in the outer arc and continuing subsidence in the fore-arc would best explain the given situation.

**Latest Maastrichtian to early late Paleocene** (Fig. 7)

*Observations*

At the beginning of this episode, the pattern of sedimentation was still dominated by turbidites spreading over the entire area. During the early Paleocene, however, parts of the former trench-slope turbidite sequence were removed along deep incisions. In the earliest late Paleocene these features were being filled with debris deriving from both the arc and the former 'Barra Honda' platform (Baumgartner *et al.*, 1984; Astorga, 1988). Simultaneously, a carbonate ramp began to develop discontinuously along both sides of the outer arc, shedding grainflows of larger Foraminifera into the adjacent basins. In the fore-arc trough, turbidite systems became increasingly coarser.

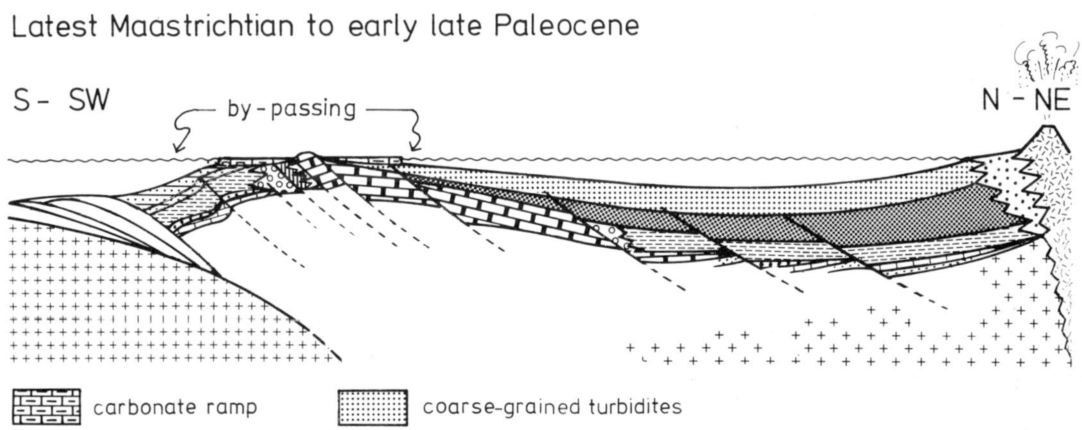


**Fig. 7.** During this episode, the outer arc was first overgrown by a carbonate ramp and then undercut by canyons. The growth of the ramp is considered as a highstand signal, whereas the canyon cuts are believed to represent a lowstand signal. The uplift of the outer arc is attributed to accretion.

*Interpretation*

The incisions at the edge of the outer arc can be interpreted as canyon cuts. Their filling proves both a direct connection to the fore-arc and exposure of parts of the 'Barra Honda' platform in the outer arc. Thus, it may be assumed that the outer arc was tilted towards the arc and cross-cut by canyons that allowed bypassing of debris coming from the arc. Concomitant with tilting, carbonate ramps started to form along the edges of the outer arc. Boulders from that ramp are found in the canyon fill debris, where they occur with boulders from the 'Barra Honda' platform. As to the causes of tilting, again a combination of continuing accretion and eustatic overprint could have produced the observed situation. The occurrence of carbonate megabreccias along the inner edge of the outer fore-arc, however, might be considered as a strong tectonic signal indeed (C. Calvo, 1987; Seyfried *et al.*, 1987; Astorga, 1988; for further details, see paper by Winsemann & Seyfried).

**Terminal late Paleocene** (Fig. 8)

*Observations*

The carbonate ramp continued to grow during this episode. In the fore-arc, turbidites became again predominantly thin-bedded. Along the trench slope, sedimentation turned to siliceous limestones.

*Interpretation*

No obvious structural changes are observed during this episode. The continuing growth of the ramp, together with the onset of siliceous limestone deposition on the trench slope, and the return, in the fore-arc basin, from coarse-grained to fine-grained turbidites clearly indicate that the situation was controlled by rising sea level. The overall trend signals an increasing separation between fore-arc and trench slope (Lundberg, 1982; Astorga, 1988; see paper by Winsemann & Seyfried). Continuing accretion in the trench is indicated by seismic data (Crowe & Buffler, 1985).

**Early to mid-Eocene** (Fig. 9)

*Observations*

In the trench slope area, sedimentation continued with siliceous limestones. Intercalations of allodapic limestones are reported from the mid-Eocene (Baumgartner *et al.*, 1984; C. Mora, 1985; C. Calvo, 1987). The carbonate ramp on the outer arc continued growing along the inner flank. During the mid-Eocene, it extended on to the outer flank (C. Calvo, 1987). Sedimentation in the fore-arc was still characterized by thin-bedded turbidites. Towards the mid-Eocene, volcaniclastic influx increased gradually, creating minor lobe systems in the inner fore-arc. There is some field evidence

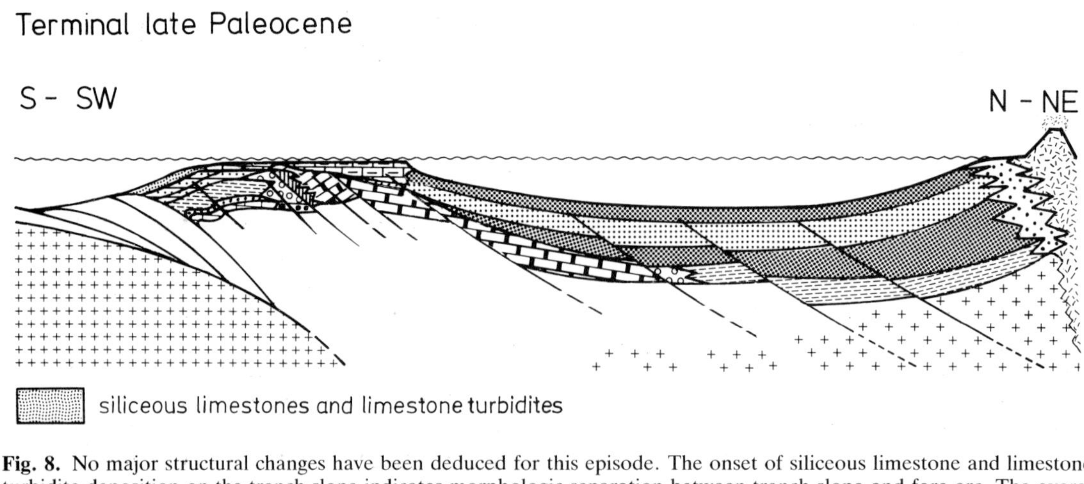


**Fig. 8.** No major structural changes have been deduced for this episode. The onset of siliceous limestone and limestone turbidite deposition on the trench slope indicates morphologic separation between trench slope and fore-arc. The overall situation suggests control by a rising sea level, with the growing carbonate ramps shedding fore-arc detritus.

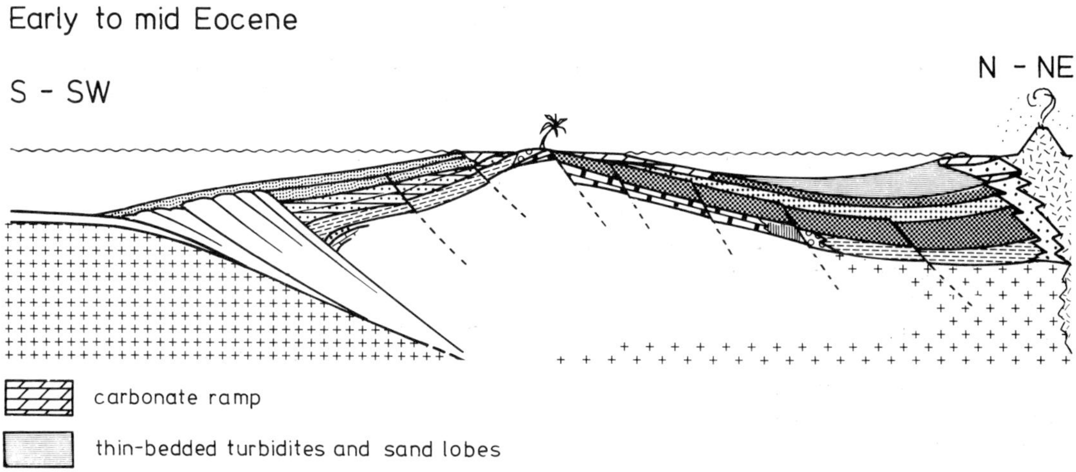


**Fig. 9.** The outer arc continued to be uplifted (erosion of parts of the previous sequence); consequently, separation between trench slope and fore-arc increased. The outer arc was again overgrown intermittently by carbonate ramps.

(larger foraminiferal grainflows) that carbonate ramps must have developed intermittently along the inner margin of the fore-arc basin.

*Interpretation*

The more or less continuous growth of the carbonate ramp along the inner edge of the outer arc can be explained by differential movements between outer arc (uplift by accretion) and fore-arc (subsidence by loading). Seismic data (Crowe & Buffler, 1985) indicate that accretion continued. The extension, during the mid-Eocene, of the carbonate ramp on to the outer edge of the outer arc may be interpreted as a strong signal of rising sea level.

**Late Eocene to early Oligocene** (Fig. 10)

*Observations*

During the late Eocene, both flanks of the outer arc were again overgrown by a carbonate ramp (Baumgartner *et al.*, 1984; Ċ. Mora, 1985; C. Calvo, 1987). There is little information on sedimentation in the trench slope; seismic data (Crowe & Buffler, 1985) show that sedimentation continued. Late Eocene fore-arc sedimentation is characterized by very coarse-grained channel-lobe systems. Submarine canyons were deeply incised into the volcanic flanks (or small internal clastic/carbonate ramp shelves) of the arc. Volcaniclastic flows with enormous andesite boulders moved directly into canyon heads (Astorga, 1988; see paper by Winsemann & Seyfried). Finally, the entire fore-arc basin was filled up by prograding slope apron systems. There is some evidence that compressional deformation of the fore-arc area started during this episode (Kuypers, 1980; Rivier, 1983; Denyer *et al.*, 1987).

Little information is available on early Oligocene sedimentation. The outer arc and the outer margin of the fore-arc basin do not show early Oligocene deposits at all. Recent field observations made by one of us (Astorga) suggest that sedimentation may have persisted along the inner margin, but no biostratigraphic proof has been obtained so far.

*Interpretation*

The existence of carbonate ramps on the outer arc in the late Eocene, their demise during that substage, and the lack of early Oligocene deposits in this area suggest that the outer arc became emergent during the early Oligocene. The gradual filling (with very coarse clastic detritus), shoaling, and narrowing of the fore-arc basin, as well as the indications of compressional deformation indicate that the fore-arc area became uplifted as well. Seismic data provided by Crowe & Buffler (1985) clearly prove that accretion stopped towards the end of the Eocene. In this context, the overall uplift and compressional deformation could be interpreted as a consequence

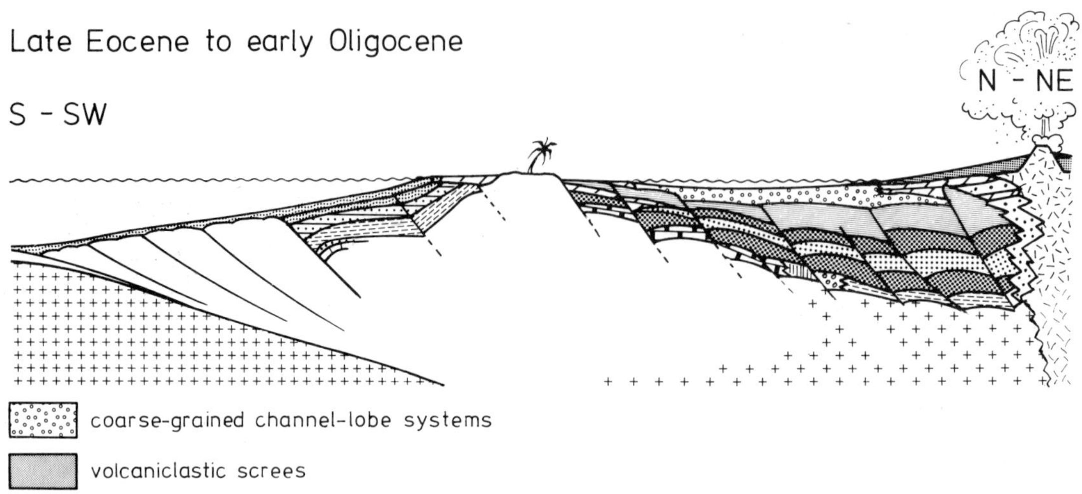


**Fig. 10.** During this episode a major tectonic event affected the entire island arc. Volcanism increased, accretion stopped, and both the outer arc and the fore-arc were subjected to compressional stress. In the course of the late Eocene, the fore-arc basin became filled up with coarse volcaniclastic debris. In the more protected areas of the external outer arc, a carbonate ramp developed during the late Eocene. The almost complete lack of early Oligocene sediments suggests that the entire area became uplifted during this time.

of increased plate coupling. If strong enough, this process would stop further accretion and transmit considerable compressional forces into the fore-arc. The present-day situation in southern Costa Rica demonstrates that the arrival and initial subduction of an oceanic plateau produces precisely such a situation.

Alternatively, according to Astorga *et al.* (1989), compression could also have been caused by the rotation of the southern segments with respect to the northern one (Sick, 1989), which in turn may have been a result of the tectonic rearrangements that affected the Caribbean plate during the episode in question (Burke *et al.*, 1984; Mann & Burke, 1984; Pindell *et al.*, 1988).

## Late Oligocene to early Miocene (Fig. 11)

### *Observations*

Late Oligocene to early Miocene deposits have only been found so far along the outer edge of the outer arc. Seismic data (Crowe & Buffler, 1985) prove they extend downslope towards the trench. On the peninsula of Nicoya, the sequence starts with mixed carbonate–clastic coastal bar systems which rest unconformably on different older rock units ranging from Paleocene trench-slope deposits to late Eocene carbonate ramps. The coastal deposits in turn are overlain by delta deposits (Baumgartner *et al.*, 1984; C. Mora, 1985; Seyfried *et al.*, 1987; Laurito, 1988; Schmidt, 1989; see paper by Schmidt & Seyfried). Field observations recently made by one of us (Astorga) suggest that sediments of this episode might also exist in the inner fore-arc but no biostratigraphic proof has been obtained so far.

### *Interpretation*

The fact that sedimentation started on an angular unconformity along the outer edge of the outer arc suggests previous uplift and subsequent subsidence. As large parts of the outer arc and fore-arc remained emergent there can be no doubt that subsidence was tectonic. Isostatic relaxation after the late Eocene to early Oligocene compressive event could be an explanation of this phenomenon. As with previous uplift events (Campanian, Paleocene), the overprint of a eustatic signal can be detected again, with the coastal deposits representing transgressive and the delta deposits indicating regressive conditions; poor biostratigraphic control does not allow more detailed conclusions. It must be emphasized, however, that the basal angular unconformity is regional (cf. Fig. 3). Evidence from adjacent basins puts onset of

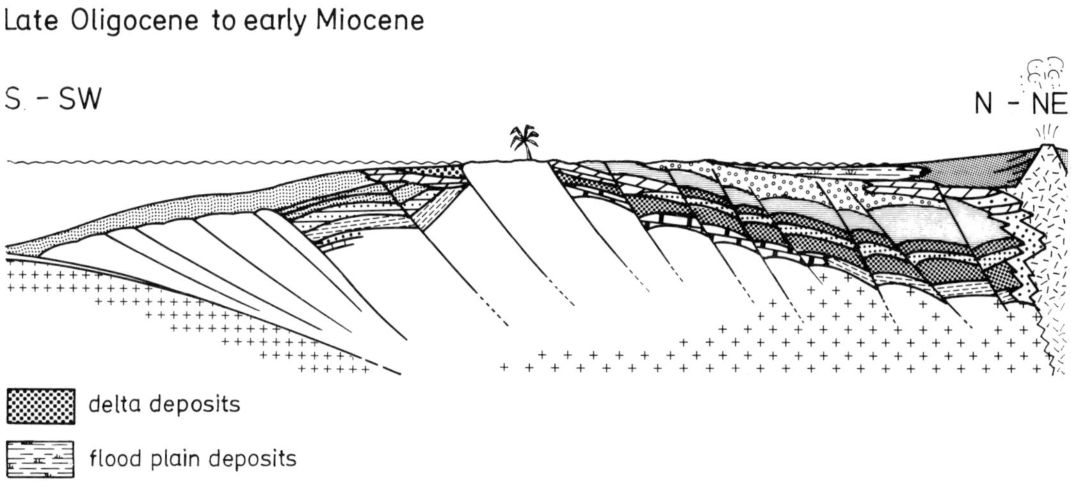


**Fig. 11.** Sedimentation started again in the latest Oligocene on a regional unconformity produced by the previous uplift. The transgressive episode, however, was of short duration; early Miocene delta and floodplain deposits soon overfilled the small sedimentary basins, prograding most of their detritus on to the trench slope.

sedimentation into the latest Oligocene; several pulses of deltaic sand supply are recorded from early Miocene shallow-shelf sequences (Schmidt, 1989).

**Mid-Miocene** (Fig. 12)

*Observations*

At the outer edge of the outer arc, mid-Miocene sediments are separated from early Miocene deposits by a disconformity. The sequence is made up by some few tens of metres of estuarine, fan-delta, and shallow-shelf deposits (Schmidt, 1989). Similar facies types could be observed in the inner fore-arc (Kuypers, 1979; Schmidt, 1989); there, they interfinger with tephritic layers.

*Interpretation*

The combination of estuarine, fan-delta, and shallow-shelf deposits is typical of the coastal embayments along the present-day Pacific coastline of Costa Rica.

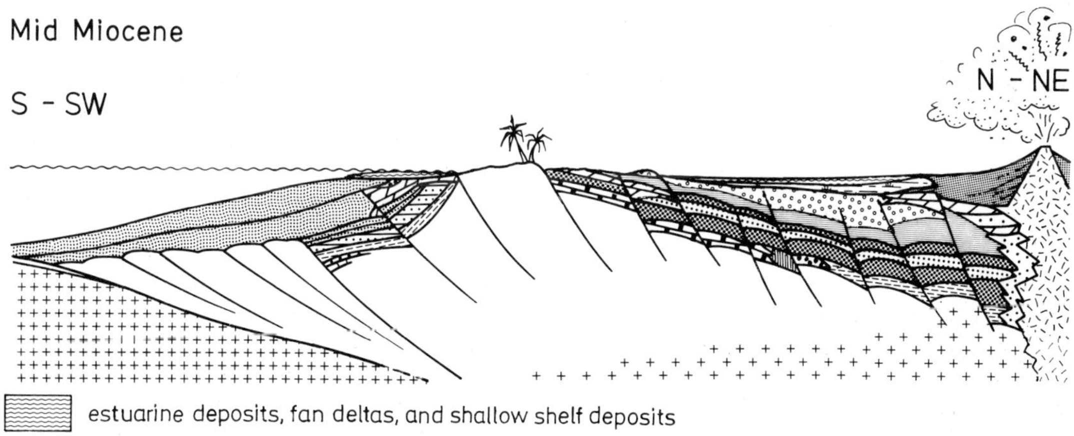


**Fig. 12.** The outer arc continued rising, whereas fore-arc and trench slope subsided. The basins depicted in this cross-section contain estuarine deposits, which means that highstand deposits were preserved by local subsidence.

Structurally, these basins are fault-angle depressions. Estuarine and paralic facies are located along the hinge zone, whereas fan-deltas and cliff-bound deposits accumulate along the opposite fault-scarped coasts, with the shelf deposits (mainly fine sands) lying in between. As this facies pattern can be recognized in most of the Neogene basins, an analogous interpretation has been proposed by Seyfried *et al.* (1987). Detailed palaeoecological and sedimentological studies confirmed this hypothesis (Seyfried *et al.*, 1985; Laurito, 1988; Schmidt, 1989). In contrast to most other Neogene troughs, the basins shown in Fig. 12 did not accumulate large thicknesses of sediment (some tens of metres in comparison to almost 1 km in the Parrita Basin). Under these circumstances, sequences are expected to register densely stacked eustatic signals. Field studies recently carried out by one of us (Amann) in the Tarcoles Basin indeed revealed the existence of four shallowing upward parasequence sets. Seismic profiles (Crowe & Buffler, 1985) indicate that there was no accretion during this episode. Both seismic data and Deep Sea Drilling Project results (von Huene *et al.*, 1985) verified the existence of a thick carbonate–clastic turbidite sequence that overlies the ancient accretionary prism (see Fig. 12). This points to trench-slope subsidence through continuing tectonic erosion in the subduction zone.

**Late Miocene to early Pliocene** (Fig. 13)

*Observations*

Late Miocene marine deposits are rare in the area. At the outer edge of the outer arc, some tens of metres of braid-delta deposits occur together with shoreface sandstones (Schmidt, 1989). In the inner fore-arc, a flood-plain–braid-delta sequence 100 m thick, which contains abundant volcanic debris and shows intercalations of lahars, has been assigned a late Miocene age (Fischer, 1981). Strong volcanic activity has also been reported from other basins (Astorga *et al.*, 1989; Kolb, 1990). Compressional phenomena within the outer arc (overthrusts and karstification of uplifted slices of 'Barra Honda' limestones) have been dated as late Miocene by Dengo (1962), Kuypers (1980), C. Calvo (1987), and Denyer *et al.* (1987). Accordingly, Silver *et al.* (1985) relate seismically observed 'duplex' phenomena to compression in the trench (cf. Fig. 13).

The boundary between upper Miocene and lower Pliocene deposits is again a regional angular unconformity (cf. Fig. 3). In the outer arc area, lower Pliocene deposits have only been recognized from one small basin ('Montezuma Basin' in southwest Nicoya), which filled with a sequence 35 m thick composed of cliff-derived debris, fan-delta deposits, shoreline and floodplain sands, and shallow-shelf sediments (Chinchilla, 1989). These deposits rest unconformably on both Nicoya Complex and Paleocene trench-slope sediments. Palaeoecological aspects point to initial deepening followed by shoaling. From the inner fore-arc area, only borehole data are available. Astorga *et al.* (1989) described a Plio-Pleistocene sequence 600 m thick composed of shallow-shelf, shoreface, floodplain or lagoonal, and lacustrine facies types that rests unconformably on upper Miocene rocks. Similar sequences are known from adjacent basins (see papers by Kolb & Schmidt and Schmidt & Seyfried). A common feature of these sequences is a basal shell-lag deposit, which in places contains high concentrations of amber pebbles. An enormously thick Plio-Pleistocene section (3 km) is known from the Burica Peninsula (Corrigan *et al.*, 1990). The filling pattern is comparable to the above-mentioned Montezuma Basin (although thicknesses differ by a factor of 100): the sequence starts with cliff-derived debris and channel-lobe systems and passes through slope apron deposits to shallow-shelf and estuarine sediments (cf. Fig. 3).

*Interpretation*

The predominance of braid-delta and flood-plain deposits as well as the occurrence of karstification phenomena in the late Miocene indicate general uplift of the entire outer-arc–fore-arc system. This is corroborated by compression observed both in the outer arc and in the trench. The lower Pliocene basin-fill sequences, in contrast, suggest the reappearance of the same type of fault-angle depressions as occurred in the mid-Miocene. Such a succession of tectonic events recalls the situations that existed during the Campano-Maastrichtian and Oligo-Miocene episodes: compression is followed by relaxation; uplift, emersion, and erosion precede the creation of new basins, which display the fault-angle depression pattern. From this point of view, the three major regional unconformities shown in Fig. 3 testify that something unusual happened to the subduction zone. A further common characteristic is the fact that eustatic changes overprint the tectonic signal only when isostatic relaxation allowed

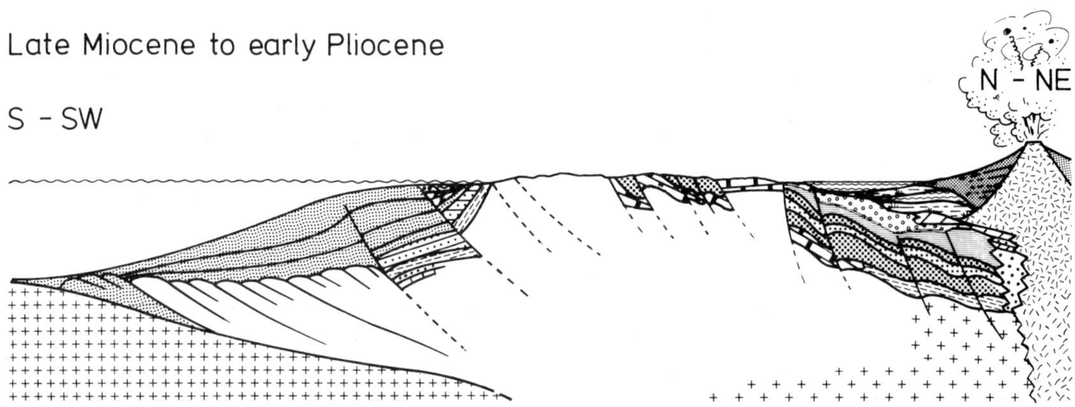


**Fig. 13.** During the late Miocene, strong regional compression led to exposure and erosion of large areas. The third and last regional unconformity shown in Fig. 3 developed during this episode. The last episode of basin formation started in the early Pliocene. In the area covered by the present cross-section, there is only one comparatively small basin in the inner fore-arc. In neighbouring regions, Pliocene troughs subsided by as much as 3 km. Seismic information indicates small 'duplex' structures, which could mean temporary resumption of accretion.

them to do so. In the early Pliocene, this seems to be particularly true for the basal shell-lag concentrations that contain high amounts of amber pebbles.

**Late Pliocene to Recent** (Fig. 14)

*Observations*

The only record of sedimentation relative to this episode comes from a borehole in the Tempisque area (inner fore-arc) where upper Pliocene floodplain and lacustrine deposits (diatomites) have been described by Astorga *et al.* (1989). These deposits are overlain by $>300$ m of Quaternary ignimbrites, which relate to the volcanic chain of the Guanacaste and Tilarán cordilleras (O. Mora, 1989). According to Crowe & Buffler (1985), there is no further accretion, but sedimentation continues along the trench slope. Geomorphologic criteria, such as uplifted wave-cut terraces and rivulets falling into active cliffs, point to ongoing uplift of the outer arc (Hare & Gardner, 1985).

*Interpretation*

The late Pliocene situation evolved into the present-day situation by steady uplift of the outer arc. The portion of the inner fore-arc shown in the cross-section of Fig. 14 most probably subsided until the late Pleistocene and became uplifted during the Holocene. This can be deduced from the geomorphology of the Tempisque valley, which shows inselbergs (= heads of thrust units) sticking out of a floodplain, and only recently being eroded along river banks. The recent Gulf of Nicoya, situated southeast of the cross-section (compare with Fig. 2) represents an inner fore-arc fault-angle basin inherited from Pliocene times. The fact that it still remains open may be due to the intersection with the Trans-Isthmic Fault System (cf. Fig. 2).

## CONCLUSIONS

**Albian to early late Campanian**

The southern part of the Central American isthmus was created at a suture between the Farallón plate and proto-Caribbean crust. From the Albian to the Santonian, this intra-oceanic suture produced a ridge of IAT that was partly covered by siliceous mudstones, black shales, and tuffs.

During the Campanian, a major tectonic event destroyed this configuration. We conclude that subduction reversal was responsible for the décollement tectonics that produced the stacking of different units of the former plate margin. This model explains the drastic spatial, temporal, and genetic separation between the older Nicoya Complex (which subsequently became the basement of the outer arc) and the younger calc-alkaline arc.

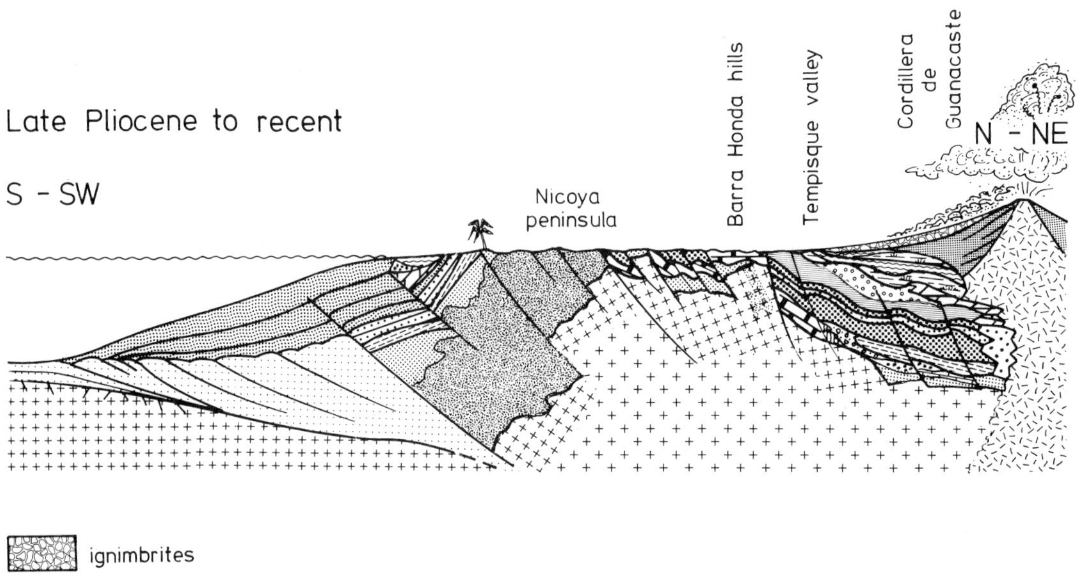


**Fig. 14.** This figure shows the present-day situation in the Costa Rican fore-arc–outer-arc area, which by now is fully emerged and evolved from the early Pliocene configuration by general uplift and strong explosive volcanic activity, which produced large amounts of volcaniclastic debris, overfilling the remnants of the former basins.

## Late Campanian to late Eocene

During the late Campanian to early Maastrichtian, a carbonate platform established on the gradually subsiding tectonic high. From the late Maastrichtian onward, continuous subduction produced both accretion and a continuous growth of the calc-alkaline arc. This resulted in a relatively stable morphotectonic configuration (trench-slope/outer-arc/fore-arc/calcalkaline-arc), which persisted until the late Eocene.

Sedimentation on the trench slope was largely controlled by the morphologic evolution of the outer arc. Thin-bedded turbidite systems or coarse-grained channel-lobe systems developed when canyons cut across the outer arc, allowing the bypass of arc debris towards the trench slope. Siliceous limestones (or calcareous turbidites) predominated during episodes of relative isolation. They correlate with carbonate ramps that repeatedly overgrew the outer arc. Thus, it may be concluded that though outer arc uplift (and morphology) controlled the general pattern of sedimentation in the trench slope, eustatic changes left a remarkable overprint by overdeepening and back-cutting pre-existing depressions during lowstands and by allowing carbonate ramps to grow and shed fore-arc clastics during highstands.

Sedimentation in the fore-arc was strongly influenced by volcanic activity and by tectonic activity along the outer margins of the basin (which most probably had a transcurrent component). Volcanic activity is expressed by the occurrence of coarse-grained channel-lobe systems, which may contain andesite boulders with diameters of up to 4 m as well as autoclastic flow breccias. Tectonic activity is mainly indicated by debris flows with carbonate ramp components. Eustatic control is expressed in the style of turbiditic sedimentation: during lowstands, huge amounts of first-cycle sands were mobilized along internal clastic shelf areas and ponded into the basin; during highstands, thin-bedded turbidite systems prevailed. The predominant eustatic signal, however, is the lowstand signal (for more details, see the paper by Winsemann & Seyfried). Towards the end of the Eocene, the fore-arc basin became almost entirely filled, mainly as a result of increased volcanic activity of the calc-alkaline arc.

## Oligocene to Recent

At the beginning of this episode, two important events affected the area in question: accretion stopped and the north and south Costa Rican arc segments began to decouple along the Trans-Isthmic Fault System. During most of the Oligocene, large parts of the outer arc and fore-arc of the northern block became uplifted and eroded. As to the process responsible for this widespread uplift, there are two possibilities: (1) collision with an oceanic plateau (with general initial uplift and subsequent differential isostatic relaxation); (2) compression as a consequence of rotation of the southern blocks, which in turn may have resulted from tectonic rearrangements within the Caribbean plate.

From the latest Oligocene until the Pliocene, differential tilting and (to a minor degree) strike-slip faulting created several fault-angle depressions along both the outer arc and the fore-arc area. Each basin-fill sequence starts on an unconformity, which in most cases is correlatable on a regional scale. Cliff-bound deposits, fan-deltas, and braid-deltas are characteristic of the fault-scarped coastlines of these basins, whereas estuarine facies, mixed carbonate–clastic coastal bars, and floodplain deposits are common along the hinge zone. Characteristically, the shallow-shelf sands which commonly lie in between show bimodal compositions: first-cycle components (feldspars, calc-alkaline volcanic rock fragments) are intricately mixed with strongly altered 'Nicoya Complex grains' derived from weathered structural highs.

In most basins, subsidence was fairly strong, but water depth seldom fell below shallow-shelf depths. Some few basins, especially those in the outermost fore-arc, display extremely complex subsidence/uplift histories. The present-day situation is the result of a general slight uplift of the entire island arc. Strong explosive volcanic activity during the Pleistocene–Holocene also led to shoaling of most of the pre-existing basins.

The cycle: compression – uplift – erosion – unconformity – subsidence (tilting) – basin-filling has occurred three times since the late Eocene (Oligocene to early Miocene, latest early Miocene to early late Miocene, latest Miocene to Pleistocene). These repetitive phases of compression might be related to changes in the intra-plate stress field, which in turn are the result of such factors as convergence of the North and South American plates and changing spreading rates, respectively.

## ACKNOWLEDGEMENTS

This study was supported by grants from the German Research Foundation (DFG; Se 490/1-1, Se 490/1-2, Se 490/1-3), from the German Academic Exchange Service (DAAD; Seyfried, Amann, Astorga, Bolz, Calvo, Kolb, Winsemann), and from the universities of Mainz (Schmidt) and Stuttgart (Amann). The ideas expressed in this paper benefited from discussions with Gerhard Wörner and Holger Appel (University of Mainz), Rudolf Fischer (University of Hannover) and Peter Sprechmann (University of Uruguay), as well as with members of the exploration department of the Costa Rican Oil Company (RECOPE) and the staff of the Escuela Centroamericana de Geología (University of Costa Rica). Transport facilities and generous support in the field were provided by RECOPE, the University of Costa Rica, and the Instituto Nicaragüense de la Minería (INMINE). To all these colleagues and institutions we express our sincere gratitude. We also thank Neil Lundberg, Philip Nell, and David Macdonald for encouragement and their most thorough reviews, which sharpened our ideas and much improved the final text.

## REFERENCES

Adamek, S., Tajima, F. & Wiens, D.A. (1987) Seismic rupture associated with subduction of the Cocos Ridge. *Tectonics* **6**, 757–774.

Alvarado, F. (1984) Formación Suretka. In: *Manual de Geologia de Costa Rica* (Ed. Sprechmann, P.) pp. 255–257. Editorial de la Universidad de Costa Rica, San José.

Alvarado, G.E. (1985) Consideraciones Petrológicas de los Estratovolcanes de Costa Rica. *Rev. Geol. Am. Central* **3**, 103–128.

Appel, H. (1990) *Geochemie und K/Ar-Datierung an Magmatiten in Costa Rica, Zentralamerika*. Unpublished Diplomarbeit, University of Mainz, 149 pp.

Astorga, A. (1987) *El Cretácico Superior y el Paleógeno de la vertiente Pacifica de Nicaragua meridional y Costa Rica septentrional: origen, evolución y dinámica de las cuencas profundas relacionadas al margen convergente de Centroamérica*. Unpublished tesis de Licenciatura, University of Costa Rica, San José, 250 pp.

Astorga, A. (1988) Geodinámica de las cuencas del Cretácico Superior-Paleógeno de la región 'forearc' del Sur de Nicaragua y Norte de Costa Rica. *Rev. Geol. Am. Central* **9**, 1–40.

Astorga, A., Fernández, J., Barboza, G., Campos, L., Obando, J., Aguilar, A. & Obando, L. (1989) Cuencas sedimentarias de Costa Rica: evolución Cretácico Superior-Cenozoica y potencial de hidrocarburos. *Transactions of the Fourth Circum-Pacific Council for Energy and Mineral Resources Conference*, San José (Costa Rica). Abstr. vol., pp. 1–18.

Azéma J., Glaçon, G., Tournon, J. & Vila, J.-H. (1979a) Precisiones acerca del Paleoceno de Puerto Quepos y sus alrededores, Provincia de Puntarenas, Costa Rica. *Inst. Geogr. Nac., Inf. Semestr. Julio-Dic.* **1979**, 77–88.

Azéma, J., Sornay, J. & Tournon, J. (1979b) Découverte d'Albien supérieur à ammonites dans le matériel volcano-sedimentaire du 'Complexe de Nicoya' (province de Guanacaste, Costa Rica). *C.R. Somm. Seances Soc. Geol. Fr.* **1979**, 129–132.

Azéma, J., Glaçon, J. & Tournon, J. (1981) Nouvelles données sur le Paléocène a Foraminifères planctoniques de la bordure pacifique du Costa Rica (Amérique Centrale). *C.R. Somm. Seances Soc. Geol. Fr.* **1981**, 85–88.

Azéma, J., Bourgois, J., Tournon, J., Baumgartner, P.O. & Desmet, A. (1985) L'orogène pré-sénonien supérieur de la marge pacifique du Costa Rica (Amérique Centrale). *Bull. Soc. Géol. Fr.* **1985**, 173–179.

Baumgartner, P.O. (1984) El Complejo ofiolítico de Nicoya (Costa Rica): modelos estructurales analizados en función de las edades de los radiolarios (Calloviense a Santoniense). In: *Manual de Geología de Costa Rica* (Ed. Sprechmann, P.) pp. 115–123. Editorial de la Universidad de Costa Rica, San José.

Baumgartner, P.O. (1987) Tectónica y sedimentación del Cretácico Superior en la zona pacífica de Costa Rica (América Central) In: *El Cretácico de México y América Central* (Eds Barbarin, J.M., Gursky, H.-J. & Meiburg, P). Actas Fac. Cienc. Tierra, Univ. Autónoma Nueva León, Linares **2**, 251–260.

Baumgartner, P.O., Mora, C.R., Butterlin, J., Sigal, J., Glaçon, J., Azéma, J. & Bourgois, J. (1984) Sedimentología y paleogeografía del Cretácico y Cenozoico del Litoral Pacífico de Costa Rica. *Rev. Geol. Am. Central* **1**, 57–136.

Berrangé, J.P. (1989) The Osa Group: an auriferous Pliocene sedimentary unit from the Osa Peninsula, southern Costa Rica. *Rev. Geol. Am. Central,* **10**, 67–93.

Berrangé, J.P. & Thorpe, R.S. (1988) The geology, geochemistry and emplacement of the Cretaceous–Tertiary ophiolitic Nicoya Complex of the Osa Peninsula, southern Costa Rica. *Tectonophysics* **147**, 193–220.

Berrangé, J.P., Bradly, D.R. & Snelling, N.J. (1989) K–Ar age dating of the ophiolic Nicoya Complex of the Osa Peninsula, southern Costa Rica. *J. S. Am. Earth Sci.* **2**, 49–59.

Bourgois, J., Azéma, J., Baumgartner, P.O., Tournon, J., Desmet, A. & Aubouin, J. (1984) The geologic history of the Carribean–Cocos plate boundary with special reference to the Nicoya Complex (Costa Rica) and D.S.D.P. results (Legs 67 and 84 off Guatemala): a synthesis. *Tectonophysics* **108**, 1–32.

Burbach, G., Frohlich, C., Pennington, W. & Matumoto, T. (1984) Seismicity and tectonics of the subducted Cocos Plate. *J. Geophys. Res.* **81**, 4885–4896.

Burke, K., Cooper, C., Dewey, J.F., Mann, P.M. & Pindell, J.L. (1984) Caribbean tectonics and relative plate motion. In: *The Caribbean–South American Plate Boundary and Regional Tectonics* (Eds Bonini, W.E., Hardgraves, R.B. & Shagman, R.). Geol. Soc. Am. Mem. 162, 31–63.

Calvo, C. (1987) *Las calizas neríticas de la vertiente Pacífica del norte de Costa Rica y sur de Nicaragua: épocas y sistemas, asociados con la apertura y evolución del margen convergente de la América Central Meridional.* Unpublished tesis de Licenciatura, University of Costa Rica, San José, 165 pp.

Calvo, C. & Bolz, A. (1987) La secuencia de Venado, un estuario lagunar tropical del Mioceno Medio, San Carlos, Costa Rica. *Rev. Geol. Am. Central* **6**, 1–24.

Calvo, G. (1987) *Geologia del Macizo del Chirripó, Cordillera de Talamanca, Costa Rica.* Unpublished Inf. Campaña Geol. 1987, University of Costa Rica, San José, 37 pp.

Campos, L. (1987) *Geología de la Fila Asunción y zonas aledañas, Atlántico Central, Costa Rica.* Unpublished tesis de Licenciatura, University of Costa Rica, San José, 250 pp.

Campos, L., Montalto, F. & Solano, C. (1984) Estratigrafía del sur de Costa Rica y Fila Costeña. In: *Manual de Geología de Costa Rica* (Ed. Sprechmann, P.) pp. 175–213. Editorial de la Universidad de Costa Rica, San José.

Case, J.E. & Holcombe, T.L. (1980) Geologic–tectonic map of the Caribbean region, Scale 1:2 500 000. *U.S. Geol. Surv. Misc. Invest. Ser.* **Map I-1100**.

Chinchilla, A.L. (1989) *Estudio geológico de la Formación Montezuma (Plio-Pleistoceno, Peninsula de Nicoya, Costa Rica).* Unpublished tesis de Licenciatura, University of Costa Rica, San José, 148 pp.

Corrigan, J. (1986) *Geology of the Burica Peninsula, Panama–Costa Rica: Neotectonic implications for the southern Middle America convergent margin.* Unpublished MA thesis, University of Texas, Austin, 152 pp.

Corrigan, J., Mann, P. & Ingle, J.C. (1990) Forearc response to subduction of the Cocos Ridge, Panama–Costa Rica. *Geol. Soc. Am. Bull.* **102**, 628–652.

Crowe, J.C. & Buffler, R.T. (1985) The Middle America Trench, Costa Rica — detached sediments (décollement tectonics and active margins). In: *Middle America Trench off western Central America* (Eds Ladd, J.W. & Buffler, R.T.). Ocean Margin Drilling Program, Regional Data Synthesis Series, Atlas 7, Woods Hole (Marine Science International).

de Boer, J.Z., Defant, M.J., Steward, R.H., Restrepo, J.F., Clark, L.F. & Ramírez, A.H. (1988) Quaternary calc-alkaline volcanism in western Panamá: regional variation and implications for the plate tectonic framework. *J. N. Am. Earth Sci.* **1**, 275–291.

Dengo, G. (1962) *Estudio Geológico de la Región de Guanacaste, Costa Rica.* Instituto Geográfico Nacional, San José IX + 112 pp.

Denyer, P. & Montero, W. (1989) *Mapa geo-estructural y sismos del Valle Central, Costa Rica, Scale 1:100 000.* Instituto Geográfico Nacional, San José, Costa Rica.

Denyer, P., Montero, W., Soto, G.J., Quezada, A., Leandro, L., Pérez, C. & Rodríguez, D. (1987) Geología tectónica de la margen oriental del Golfo de Nicoya, Coast Rica. *Rev. Cienc. Tecnol.* **11**, 17–31.

Desmet, A. & Rocci, G. (1988) Les dolérites et les ferrobasaltes du complexe ophiolitique de Santa Elena (Costa Rica): relations, géochimie et contexte géodynamique. *Bull. Soc. Geol. Fr.* **1988**, 479–487.

de Wever, P., Azéma, J. & Desmet, A. (1985) Découverte de matériel océanique du Lias-Dogger inférieur dans la péninsule de Santa Elena (Costa Rica, Amérique Centrale). *C.R. Acad. Sci. Paris* **1985**, 759–764.

Donnelly, T.W. (1975) The geological evolution of the Caribbean and Gulf of Mexico. Some critical problems and areas. In: *The Ocean Basins and Margins*, Vol. 3: *the Gulf of Mexico and the Caribbean* (Eds Nairn, A.E.M. & Stehli, F.G.) pp. 663–689. Plenum Press, New York.

Donnelly, T.W. (1985) Mesozoic and Cenozoic plate evolution of the Caribbean region. In: *The Great American Biotic Interchange* (Eds Stehli, F.G. & Webb, S.D.) pp. 89–121. Plenum Press, New York.

Drummond, K. (1981) *Plate Tectonic Map of the Circum-Pacific Region, Northeast Quadrant, Scale 1:10 000 000*. Am. Assoc. Petrol. Geol., Tulsa, OK.

Fernández, J.A. (1987) *Geología de la hoja topográfica Tucurrique (1:50 000 I.G.N.C.R., Nr.34451)*. Unpublished tesis de Licenciatura, University of Costa Rica, San José, 206 pp.

Fischer, R. (1981) Die Herausformung des mittelamerikanishchen Isthmus im Miozan Costa Ricas. *Zentralbl. Geol. Palaeontol. Teil I* **3/4**, 210–221.

Forsythe, R.D., Nelson, E.P., Carr, M.J., Kaeding, M.E., Herve, M., Mpodozis, C., Soffia, J.M. & Harambour, S. (1986) Pliocene near-trench magmatism in southern Chile: a possible manifestation of ridge collision. *Geology* **14**, 23–27.

Gardner, T.W., Back, W., Bullard, T.F., Hare, P.W., Kesel, R.H., Lowe, D.R., Menges, C.M., Mora, S.C., Pazzagalia, F.J., Sasowsky, I.D., Troester, J.W. & Wells, S.G. (1987) Central America and the Caribbean. In: *Geomorphic systems of North America* (Ed. Graf, W.L.). Geol. Soc. Am., Centennial Spec. Vol. 2, 343–402.

Gursky, H.-J. (1988) Gefüge, Zusammensetzung und Genese der Radiolarite im ophiolitischen Nicoya-Komplex (Costa Rica). *Muenstersche Forsch. Geol. Palaeontol.* **68**, 1–189.

Gursky, H.-J. (1989) Presencia y origen de rocas sedimentarias en el basamento ofiolítico de Costa Rica. *Rev. Geol. Am. Central* **10**, 19–66.

Gursky, H.-J. & Schimdt-Effing, R. (1983) Sedimentology of radiolarites within the Nicoya ophiolite complex, Costa Rica, Central America. In: *Siliceous Deposits in the Pacific Region: Developments in Sedimentology* (Eds Iijima, A., Hein, J.R. & Siever, R.) pp. 127–142. Elsevier, Amsterdam.

Gursky, M.M. (1986) *Tektonische und thermische Deformationen im ophiolitischen Nicoya-Komplex und seinem sedimentären Auflager (Nicoya-Halbinsel, Costa Rica) und ihre Bedeutung für die geodynamische Entwicklung im südlichen Zentralamerika*. PhD thesis, University of Marburg, 394 pp.

Hare, P.W. & Gardner, T.W. (1985) Geomorphic indicators of vertical tectonism along a converging plate margin. In: *Tectonic Geomorphology. Proceedings of the 15th Geomorphology Symposium, Birmingham* (Eds Hack, J. & Morisawa, M.) pp. 76–104.

Henningsen, D. (1966) Die pazifische Küstenkordillere (Cordillera Costeña) Costa Ricas und ihre Stellung innerhalb des süd-zentralamerikanischen Gebirges. *Geotektonische Forsch.* **23**, 3–66.

Heywood, C.E. (1984) *Forearc deformation in Southern Costa Rica: a consequence of the collision of the aseismic Cocos Ridge*. Unpublished MSc thesis, University of California, Santa Cruz, 104 pp.

Hill, M., Morris, J. & Whelan, J. (1981) Hybrid granodiorites intruding the accretionary prism, Kodiak, Shumagin, and Sanak Islands, south-west Alaska. *J. Geophys. Res.* **86**, 10569–10590.

Jager, G. (1977) *Geología de las mineralizaciones de cromita al este de la Península de Santa Elena, Prov. de Guanacaste, Costa Rica*. Unpublished tesis de Licenciatura, University of Costa Rica, San José, 136 pp.

Kolb, W. (1990) *Die neogene Schelfsedimentation in Südwest-Nicaragua: Ablagerungssequenzen und paläogeographische Szenarien*. Unpublished PhD thesis, University of Stuttgart, 163 pp.

Kussmaul, S. (1987) Petrología de las rocas intrusivas neógenas de Costa Rica. *Rev. Geol. Am. Central* **7**, 83–112.

Kussmaul, S., Paniagua, S. & Gaínza, J. (1982) Recopilación, clasificación e interpretación petroquímica de las rocas de Costa Rica. *Inst. Geogr. Nac., Inf. Semestr. Julio-Dic.* **1982**, 17–79.

Kuypers, E.P. (1980) The geologic history of the Nicoya Ophiolite Complex, Costa Rica, and its geotectonic significance. *Tectonophysics* **68**, 233–255.

Laurito, C. (1988) Análisis sedimentológico de las Formaciones El Fraile, en la costa suroeste de Nicaragua, Carmen y Punta Carballo, en la costa Pacífica de Costa Rica Unpublished tesis de Licenciatura, University of Costa Rica, San José, 117 pp.

Lew, L. (1983) *The Geology of the Osa Peninsula, Costa Rica: observations and speculations about the evolution of a part of the outer arc of the Southern Central American Orogen*. Unpublished MSc thesis, Pennsylvania State University, 128 pp.

Lonsdale, P. & Klitgord, K.D. (1978) Structure and tectonic history of the eastern Panamá Basin. *Geol. Soc. Am. Bull.* **89**, 981–999.

Lundberg, N. (1982) Evolution of the slope landward of the Middle America Trench, Nicoya Peninsula, Costa Rica. In: *Trench–Forearc Geology* (Ed. Leggett, J.K.). Geol. Soc. London, Spec. Publ. 10, 133–147.

Mann, P.M. & Burke, K. (1984) Neotectonics of the Caribbean. *Rev. Geophys. Space Phys.* **22**, 309–362.

Marshak, K.S. & Karig, D.E. (1977) Triple junctions as a cause for anomalously near-trench igneous activity between the trench and volcanic arc. *Geology* **5**, 233–236.

Matumoto, T., Ohtake, M., Latham, G. & Umaña, J.J. (1977) Crustal structure of southern Central America. *Bull. Seismol. Soc. Am.* **67**, 121–134.

McGeary, S., Nur, A. & Ben-Avraham, Z. (1985) Spatial gaps in arc volcanism: the effect of collision on subduction of oceanic plateaus. *Tectonophysics* **119**, 195–221.

Meschede, M., Frisch, W. & Sick, M. (1988) Interpretación geodinámica de los complejos ofiolíticos de Costa Rica. *Rev. Geol. Am. Central* **8**, 1–17.

Montero, W. & Dewey, J. (1982) Shallow-focus seismicity, composite focal mechanism, and tectonics of the Valle Central, Costa Rica. *Bull. Seismol. Soc. Am.* **71**, 1611–1626.

Mora, C.R. (1985) *Sedimentología y geomorfología del Sur de la Península de Nicoya (Provincia de Puntarenas, Costa Rica)*. Unpublished tesis de Licenciatura, University of Costa Rica, San José, 148 pp.

Mora, O. (1989) *Estudio geológico–petrológico de las piroclastitas en los alrededores de Bagaces, Provincia de Guanacaste, Costa Rica*. Unpublished tesis de Licenciatura, University of Costa Rica, San José, 61 pp.

Mora, S. (1979) *Estudio geológico de una parte de la región sureste del Valle del General*. Unpublished tesis de Licenciatura, University of Costa Rica, San José, 210 pp.

Morales, L.D. (1985) Las zonas sismicas de Costa Rica. *Rev. Geol. Am. Central*, **3**, 69–101.

Nyström, J.O., Levy, B., Troeng, B., Ehrenborg, J. & Carranza, G. (1988) Geochemistry of volcanic rocks in a traverse through Nicaragua. *Rev. Geol. Am. Central* **8**, 77–109.

OBANDO, J.A. (1986) *Sedimentología y tectónica del Cretácico y Paleógeno de la región de Golfito, Península de Burica y Península de Osa, Provincia de Puntarenas Costa Rica*. Unpublished tesis de Licenciatura, University of Costa Rica, San José, 210 pp.

PICHLER, H. & WEYL, R. (1973) Petrochemical aspects of Central American magmatism. *Geol. Rundsch.* **62**, 357–396.

PICHLER, H. & WEYL, R. (1975) Magmatism and crustal evolution in Costa Rica (Central America). *Geol. Rundsch.* **64**, 457–475.

PINDELL, J.L., CANDE, S.C., PITMAN, W.C., III, ROWLEY, D.B., DEWEY, J.F., LA BRANQUE, J.L. & HAXBY, W. (1988) A plate-kinematic framework for models of Caribbean evolution. In: *Mesozoic and Cenozoic Plate Reconstructions* (Eds Scotese, C.R. & Sager, W.W.). Tectonophysics 155, 121–138.

RIVIER, F. (1983) Sintesis geologica y mapa geologico del area del Bajo Tempisque, Guanacaste, Costa Rica. *Inst. Geog. Nac. Inf. Semestr.* Enero-Junio, **1983**, 7–30.

RIVIER, F. (1985) Sección geológica del Pacífico al Atlántico a través de Costa Rica. *Rev. Geol. Am. Central* **2**, 23–32.

RIVIER, F. & CALVO, C. (1988) Terciario del sur del Valle Central: sección estratigráfica del Cerro Caraigres, Provincia de San José, Costa Rica. *Rev. Geol. Am. Central* **9**, 61–77.

RIVIER, F. & SEYFRIED, H. (1985) Genèse des bassins molassiques de part et d'autre de la Cordillera de Talamanca (Costa Rica, Amérique Centrale). *Abstr. Symp. Foreland Basins, Fribourg*, p. 110.

ROSS, M.I. & SCOTESE, C.R. (1988) A hierarchical tectonic model of the Gulf of Mexico and Caribbean region. In: *Mesozoic and Cenozoic Plate Reconstructions* (Eds Scotese, C.R. & Sager, W.W.). Tectonophysics 115, 139–168.

SCHMIDT, H.E. (1989) *Sequenzstratigraphie des Neogenen Inselbogen-Schelfes im 'Forearc'-Bereich Costa Ricas und Nicaraguas*. Unpublished PhD thesis, Technische Universität Berlin, 250 pp.

SCHMIDT-EFFING, R. (1979) Alter und Genese des Nicoya-Komplexes, einer ozeanischen Paläokruste (Oberjura bis Eozän) im Zentralamerika. *Geol. Rundsch.* **68**, 457–494.

SCHMIDT-EFFING, R. (1980) Radiolarien der Miltel-Kreide aus dem Santa-Elena Massiv von Costa Rica. *Neues Jahrb. Geol. Palaeontol.* **160**, 241–57.

SEYFRIED, H. & SPRECHMANN, P. (1986) Über die Frühgeschichte (Campan bis Eozän) der südlichen mittelamerikanischen Landbrücke. *Neues Jahrb. Geol. Palaeontol. Monatsh.* **1986**, 38–55.

SEYFRIED, H., SPRECHMANN, P. & AGUILAR, T. (1985) Sedimentología y paleoecología de un estuario del litoral Pacífico del istmo Centroamericano primordial (Mioceno Medio, Costa Rica). *Rev. Geol. Am. Central* **3**, 1–68.

SEYFRIED, H., ASTORGA, A. & CALVO, C. (1987) Sequence stratigraphy of deep and shallow water deposits from an evolving island arc: the upper Cretaceous and Tertiary of Central America. *Facies* **17**, 203–214.

SHIPLEY, T. & MOORE, G.F. (1986) Sediment accretion, subduction, and dewatering at the base of the trench slope off Costa Rica: a seismic reflexion view of the décollement. *J. Geophys. Res.* **91**, 2019–2028.

SICK, M. (1989) Paleomagnetism of the Ophiolite Complexes from the southern Middle American land-bridge (Costa Rica and Panamá). *Tübinger Geowiss. Arb.* **4**, 1–108.

SILVER, E.A., ELLIS, M.J., BREEN, N.A. & Shipley, T.H. (1985) Comments on the growth of accretionary wedges. *Geology* **13**, 6–9.

SPRECHMANN, P. (1984) Estratigrafía del Complejo de Nicoya. In: *Manual de Geología de Costa Rica* (Ed. Sprechmann, P.) pp. 99–123. Edit. Univ. Costa Rica, San José.

SPRECHMANN, P., ASTORGA, A., BOLZ, A. & CALVO, C. (1987) Estratigrafía del Cretácico de Costa Rica. In: *El Cretácico de México y América Central* (Eds Barbarin, J.M., Gursky, H.-J. & Meiburg, P.). Actas Fac. Cienc. Tierra, Univ. Autonoma Nuevo León, Linares 2, 251–260.

TOURNON, J. (1984) Magmatismes du Mésozoique a l'actuel en Amérique Centrale: l'exemple de Costa Rica, des ophiolites aux andesites. Unpublished thèse d'Etat, Univ. Pierre et Marie Curie, 335 pp.

TOURNON, J. & AZÉMA, J. (1980) Sobre la estructura y petrología del macizo ultrabásico de Santa Elena (Provincia de Guanacaste, Costa Rica). *Inst. Geogr. Nac., Inf. Semestr., Enero-Julio* **1980**, 17–54.

VAN ANDEL, T.H., HEATH, G.R., MALFAIT, B.T., HEINRICHS, D.F. & EWINF, J.L. (1971) Tectonics of the Panamá Basin, eastern Equatorial Pacific. *Geol. Soc. Am. Bull.* **82**, 1489–1508.

VITALI, C. (1985) *Etude morphostructurale des prismes tectoniques de Panamá, de Colombie et du Venezuela: leurs relations avec les domaines océaniques et continentaux proches*. Unpublished thèse Docteur 3e cycle, Univ. Pierre et Marie Curie, 224 pp.

VON HUENE, R., AUBOUIN, J., *et al.* (1985) Site 565. *Initial Reports of the Deep Sea Drilling Project*, Vol. 84, pp. 21–77. US Government Printing Office, Washington, D.C.

WADGE, G. & BURKE, K. (1983) Neogene Caribbean plate rotation and associated Central American tectonic evolution. *Tectonics* **2**, 633–643.

WELLS, S.G., BULLARD, T.F., MENGES, P.G., DRAKE, P.G., KARAS, P.A., KELSON, K.I., RITTER, J.B. & WESLING, J.R. (1989) Regional variations in tectonic geomorphology along a segmented convergent plate boundary, Pacific coast of Costa Rica. *Geomorphology* 1–33.

WEYL, R. (1956) Eiszeitliche Gletscherspuren in Costa Rica (Mittelamerika). *Zeitschr. Gletscherkde. Glazialgeol.* **III**, 317–325.

WEYL, R. (1980) *Geology of Central America*. Gebr. Borntraeger, Berlin-Stuttgart, 371 pp.

WILDBERG, H.G.H. (1984) Der Nicoya-Komplex, Costa Rica, Zentralamerika: Magmatismus und Genese eines polygenetischen Ophiolith-Komplexes. *Muenstersche. Forsch. Geol. Palaeontol.* **62**, 1–123.

WILDBERG, H.G.H. (1987) High level and low level plagiogranites from the Nicoya ophiolite complex, Costa Rica, Central America. *Geol. Rundsch.* **76**, 285–301.

WUNSCH, O. (1988) *Geologie der alpinen Zone des Chirripó-Massives in Costa Rica, Mittelamerika*. Unpublished Diplom-Kartierungsbericht, Freie University, Berlin, 39 pp.

*Spec. Publs. int. Ass. Sediment.* (1991) **12**, 241–258

# Depositional sequences and sequence boundaries in fore-arc coastal embayments: case histories from Central America

HANNELORE SCHMIDT *and* HARTMUT SEYFRIED

*Institut für Geologie und Paläontologie, Universität Stuttgart, Böblinger Strasse 72, D 7000 Stuttgart 1, Germany*

## ABSTRACT

From Oligocene to Recent times a series of tectonically controlled coastal embayments formed on the Pacific fore-arc side of the southern Central American island-arc system. Each of these basins shows characteristic stratal geometries and facies distributions reflecting the complex interaction of changes of sea level, volcaniclastic input, and tectonic activity (subsidence, uplift).

Sequential stratigraphic correlations based on sequence analysis and discontinuity surfaces indicate that eustatic sea-level changes control architecture, geometry, and facies distribution of depositional sequences at the level of second-order cycles. Owing to the particular tectonic position, complete sets of systems tracts are seldom developed. Voluminous sediment supply, especially during episodes of strong volcanic activity, may overcompensate transgressions. Strong uplift may annihilate any sedimentary documentation or reduce sequences to a strongly condensed package with random preservation of estuarine or deltaic facies. The bounding unconformities (sequence boundaries) are traceable on a regional scale and are related at the level of second-order sequence boundaries to tectonic events that repeatedly affected the island arc during the episode in question.

## INTRODUCTION

The southern Central American land bridge was formed at a suture zone between the Farallón Plate and proto-Caribbean crust. According to the chapter by Seyfried *et al.*, the following stages of evolution can be distinguished: (1) formation of a primitive tholeiitic ridge (Albian–Santonian), (2) compression of the overriding plate margin (Campanian), (3) development of a mature island-arc system ranging from calc-alkaline arc through fore-arc and outer arc to trench slope (Maastrichtian–Eocene), (4) regional compressive uplift and subsequent repetitive opening of fault-angle depressions (Oligocene–Recent).

In the early Paleocene, the outer arc became a morphological barrier between the fore-arc and trench slope, exerting a direct control on trench slope sedimentation. The strong middle Oligocene (Haq *et al.*, 1988) fall of sea level could have enhanced the effects of Oligocene compressive uplift. As a result, extensive parts of the north and south Costa Rican fore-arc and outer arc areas were deeply eroded. In parts of the northern and central areas, marine sedimentation started again in the latest Oligocene; in the inner fore-arc of southern Costa Rica as well as in southwestern Nicaragua, however, Oligocene sedimentation was obviously continuous and most probably predominantly marine.

From the latest Oligocene until the Pliocene, a series of short-lived coastal embayments formed along both fore-arc and outer arc. The geometry of these coastal embayments was controlled by a combination of normal faults and strike-slip fault systems with dominantly transtensile regimes. From a structural point of view, the following categories of embayments can be recognized (for location, see Fig. 1):

**1** coastal embayments bounded by normal faults (Malpaís Basin, Dominical Basin);

**Fig. 1.** Location map of Neogene fore-arc coastal embayments. (1) Malpaís Basin, (2) Gulf of Nicoya, (3) Punta Carballo Basin, (4) Punta Leona–Tarcolitos Basin, (5) Parrita Basin, (6) Dominical Basin.

**2** coastal embayments created by fault-angle depressions (Punta Carballo Basin, the actual Gulf of Nicoya);

**3** coastal embayments formed by transtensional strike-slip movements (Punta Leona–Tarcolitos Basin, Parrita Basin).

Each of these basins displays an individual structural and sedimentary history. Several tilting phases characterize the Malpaís Basin, resulting in the formation of strongly reduced shallow-marine sequences separated by angular unconformities. The Punta Carballo Basin existed for a relatively short period of time and is characterized by tide-influenced sedimentation. In the Punta Leona–Tarcolitos Basin there are two distinctive stages of basin development: a uniform basal unit formed in an open-shelf–beach system is followed by a higher unit characterized by differentiation in various sedimentary environments. In the Parrita Basin, transgressive–regressive sequence couplets are well developed on a background of strong subsidence. The Dominical Basin shows a simple development from an undifferentiated open shelf system to a small embayment.

The purpose of this paper is to demonstrate the extent to which sequence-stratigraphic concepts can be applied to siliciclastic shallow-water deposits from a convergent plate margin.

The general geological and tectonic setting of the Central American island-arc system and the sequence-stratigraphic interpretation of the deep-water deposits will be discussed in the papers by Seyfried *et al.* and Winsemann & Seyfried.

## Conceptual model

According to the concept of sequence stratigraphy (cf. Vail *et al.*, 1977), sequences and their bounding unconformities are interpreted as forming in response to cycles of eustatic sea-level fluctuation. The fundamental stratigraphic unit is the depositional sequence, which is bounded by unconformities and their correlative conformities.

The bounding unconformities (sequence boundaries) are inferred to be produced by eustatic falls of sea level. Vail & Todd (1981) distinguish between type 1 and type 2 sequence boundaries. A type 1 sequence boundary is characterized by both subaerial and submarine erosion and is formed when the rate of eustatic sea-level fall exceeds the rate of subsidence at the depositional shoreline break (Posamentier & Vail, 1988). A type 2 sequence boundary is characterized by widespread subaerial erosion of the exposed shelf and a change within the sedimentary succession from progradational to aggradational stacking patterns. Type 2 sequence boundaries are formed when the rate of eustatic sea-level fall is less than subsidence rate at the depositional shoreline break (Posamentier & Vail, 1988). A depositional sequence comprises a succession of three systems tracts. Type 1 sequences consist of lowstand, transgressive, and highstand systems tracts and are bounded at the base by type 1 sequence boundaries. Type 2 sequences consist of shelf-margin, transgressive, and highstand systems tracts and are bounded at the base by type 2 sequence boundaries. Both type 1 and type 2 sequences can be bounded at the top by either a type 1 or a type 2 sequence boundary. Haq *et al.* (1988) give a sequence-stratigraphic hierarchy and subdivide sequence-stratigraphic units into first-order cycles (about 250 Ma long), second-order cycles (about 9–10 Ma long), third-order cycles (about 1– 2 Ma long), and fourth-order or higher cycles (400–20 ka long).

Systems tracts are defined by their position within the sequence, their bounding surfaces, and by the stacking patterns of their constituent parasequences and parasequence sets. Systems tracts are inferred to be associated with a specific part of the eustatic cycle. Lowstand systems tracts will form during a time of rapid eustatic fall and early eustatic rise of sea level; shelf-margin systems tracts will form during a time of slow eustatic fall and early eustatic rise of sea level; transgressive systems tracts will form during a time of a rapid eustatic sea-level rise; highstand systems tracts will form during a time of a late eustatic rise, a eustatic stillstand, and a time of an early eustatic fall (Posamentier *et al.*, 1988).

Parasequences are defined as a succession of genetically related depositional systems with progradational, retrogradational, or aggradational stacking pattern. Two or more parasequences form a parasequence set (van Wagoner *et al.*, 1988).

Depositional sequences, systems tracts, parasequences and sets of parasequences are each bounded below and above by surfaces of regional or local extent, which provide a chronostratigraphic framework for correlating and mapping sedimentary rocks (van Wagoner *et al.*, 1988).

## THE NEOGENE COASTAL EMBAYMENTS

During latest Oligocene to Recent times, a series of tectonically controlled coastal embayments formed on the Pacific fore-arc side of the southern Central American island-arc system (Fig. 1). Each of these embayments shows characteristic sequences of shallow-marine strata, which reflect the complex interaction of (1) changes of sea level, (2) subsidence, (3) tectonic uplift, and (4) variable sediment supply (mostly volcanoclastic).

Two of these coastal embayments, each characterized by distinctive controlling factors, will be discussed in more detail. The Malpaís Basin shows a prevailing tectonic control, especially uplift and block faulting. The Parrita Basin reflects strong subsidence, probably in response to abundant sediment supply. Both document opposing extremes in basin development, with sedimentation being controlled either by tectonics or by eustasy.

### The Malpaís Basin

The Malpaís Basin is located in the south of the Nicoya Peninsula (Fig. 1). The uppermost Oligocene to lower Pliocene shallow-marine sediments of the Malpaís Basin make up the upper 250 m of an upper trench slope succession, 5000 m thick in total (cf. paper by Seyfried *et al.*).

Middle to upper Eocene shallow marine, carbonate ramp deposits (Calvo, 1987) mark the onset of shallow water sedimentation in the area. These carbonates are separated from the underlying upper trench slope deposits by an angular unconformity that cuts across slightly folded and intensely slumped strata, thus indicating previous compressive uplift. This set is again cross-cut by another angular unconformity, which dissects an intensely sheared substratum, exposing both ramp carbonates and older units in random order. Thus, it may be concluded that strong compressive uplift affected the entire outer arc and fore-arc area (cf. paper by Seyfried *et al.*). During early and mid-Oligocene times no sedimentation is documented. Mixed carbonate–clastic coastal bar systems of latest Oligocene age (Calvo, 1987; Schmidt, 1989) form the first deposits lying on the unconformity. They are followed by shallow-marine siliciclastic sequences, separated by major unconformities. The unconformities cut across tilted blocks, which each expose the previously formed stratigraphic sets. Regional compression must be inferred as a cause for tilting.

The sedimentary succession was described for the first time by Dengo (1962). Further workers (Schmidt-Effing, 1979; Lundberg, 1982; Baumgartner *et al.*, 1984; Sprechmann, 1984) established several formal rock units; Mora (1985) presented the first sedimentological study. Astorga (1987), Calvo (1987), and Schmidt (1989) provided detailed facies analysis and interpretations of sequence dynamics of the deep-water deposits, carbonate systems, and siliciclastic shallow-water sediments, respectively. A preliminary summary of the tectono-sedimentary development of the Malpaís Basin was given by Seyfried *et al.* (1987).

#### *Biostratigraphic remarks*

Owing to the poor preservation of planktonic Foraminifera, biostratigraphic control is sparse. Available palaeontological data assign a mid to late Eocene age to the shallow-marine carbonates (Baumgartner *et al.*, 1984; Mora, 1985; Calvo, 1987). Age determination of the siliciclastic shallow-marine deposits range from late Oligocene to early Miocene (Lundberg, 1982; Baumgartner *et al.*, 1984; Mora, 1985) for the basal part of the succession, probably middle Miocene (personal communication,

D. Pizarro, Micropaleontological Laboratory, Recope) for the middle part of the succession, and late Miocene to early Pliocene (Fischer, 1981; Goudkoff & Porter, in Sprechmann, 1984) for the upper part.

## Facies descriptions and depositional environment

Within the siliciclastic shallow-marine series, seven facies types were identified in the outcrops examined. It should be pointed out that grain size is commonly 1–2 size classes coarser than in comparable environments described in the literature (the same observation can be made for the Parrita Basin). A representative but idealized vertical section is shown in Fig. 2.

### *Facies 1 (prodelta)*

This facies consists of several centimetre to decimetre-thick interbeds of siltstones and fine-grained sandstones. Shell debris and plant remains are rare. Frequency and thickness of the sandstone beds increase upwards slightly. The sandstones exhibit sparse planar lamination and normal grading. Occasionally basal contacts of the sandstones are scoured and show load structures. Soft-sediment deformation (e.g. flame structures, convolute lamination) is common. The siltstones show no visible sedimentary structures. Siltstone and sandstone beds show varying degrees of bioturbation. Trace fossils include *Chondrites*, *Helminthopsis*, *Planolites*, *Terebellina*; *Ophiomorpha* and *Thalassinoides* occur only in sandstone beds.

Facies 1 is interpreted as having been deposited in a prodelta position under varying energy conditions and high rates of sediment supply. The bioturbated siltstones are considered to have been deposited by suspension fall-out during fair weather conditions, representing 'normal' conditions of sedimentation. The graded and laminated sandstone beds can be related to suspension-laden currents under decreas-

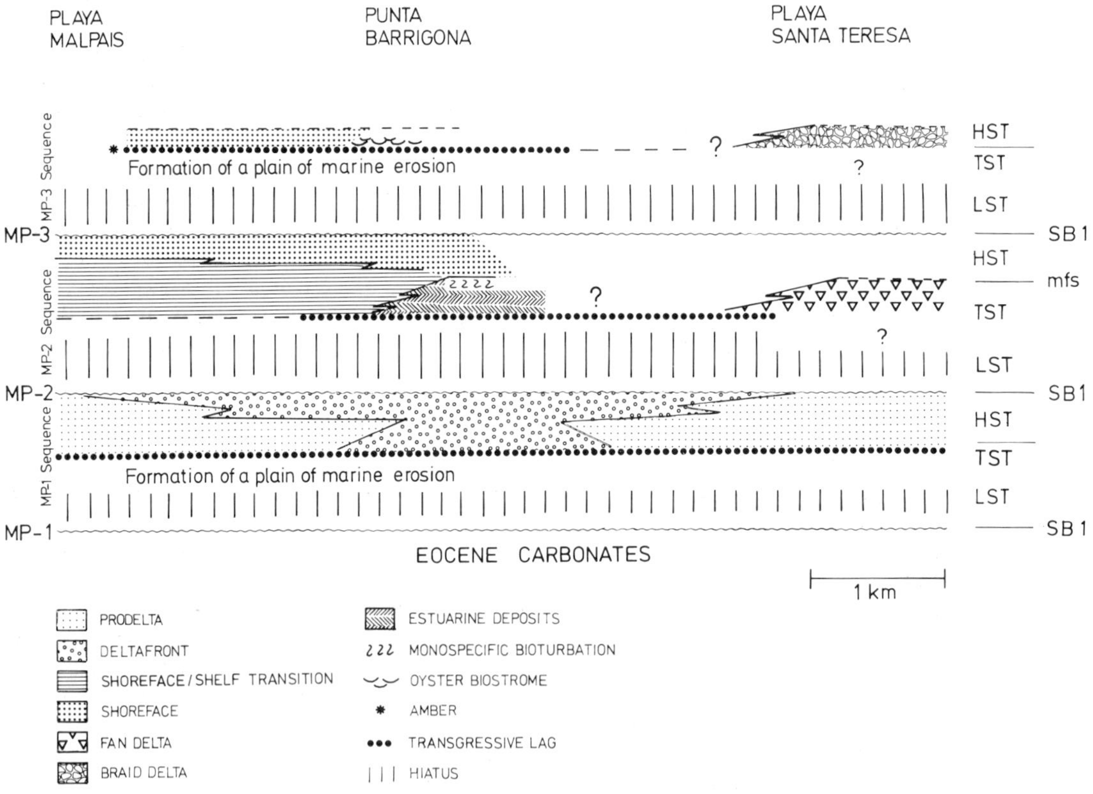


**Fig. 2.** Idealized vertical succession and sequence development of the late Oligocene to early Pliocene Malpaís Basin. The left column refers to nomenclature of sequence analysis; the right column refers to sequence-stratigraphic interpretation (modified after Schmidt, 1989).

ing velocities. The sedimentary features and occurrence of scoured bases suggest sedimentation under storm weather conditions.

*Facies 2 (delta front)*

This facies consists of fine to coarse-grained centimetre to decimetre-thick bedded sandstones and may contain shell fragments. The sandstones exhibit horizontal lamination, low-angle planar cross-stratification, current ripples, and synsedimentary deformation structures, such as flame structures, convolute lamination, and load casts. Basal bedding contacts are generally sharp and occasionally erosional. Some beds are graded and show erosional bases and rip-up clasts. The sandstones show moderate to strong and partly complete bioturbation with *Ophiomorpha*, *Thalassinoides*, escape structures, *Chondrites*, *Helminthopsis*, *Terebellina*, *Teichichnus*, and *Planolites*. This facies also exhibits massive sandstone beds without any visible stratification or bioturbation, and massive sandstone beds with primary bedding structures completely destroyed by biogenic reworking. The vertical succession suggests increasing energy levels and presumably upward shallowing, indicated by increase of grain size and increase of primary sedimentary structure.

This facies originated from suspension fall-out and bedload in a high energy delta-front position. Planar lamination can result from upper flow regime currents. Planar cross-stratified beds have been considered to form under similar conditions. The coarse-grained nature of the sandstones and current ripples suggest moderately strong tractional bottom currents. The massive sandstones lacking physical or biogenic structures are considered to be deposited from suspension fall-out in a more proximal delta-front position. The high density of burrows in some sandstone beds suggests a more distal position where bioturbation kept pace with sedimentation. Graded beds with erosional bases and rip-up clasts are interpreted as representing tempestites.

*Facies 3 (shelf–shoreface transition zone)*

Facies 3 is lithologically similar to facies 1, but the grain size is somewhat coarser and soft-sediment deformation is less common. Sandstone beds predominate within this facies. They show sparse parallel lamination and some beds are characterized by low-angle cross-stratification. Bioturbation is rare and restricted to *Ophiomorpha* and *Thalassinoides*. This bioturbation mainly affects the tops of sandstone beds, which in turn commonly show sharp erosional basal contacts and normal grading. Concretion horizons are common.

Facies 3 is considered to represent a shelf–shoreface transitional environment. The massive siltstones and sandstones represent deposition from suspension fall-out during fair weather conditions or below storm wave-base. The laminated and planar cross-stratified sandstones are deposited under higher energy conditions, indicating a slightly shallower environment. The graded, sharp-based sandstone beds are interpreted as indicating episodic storm events.

*Facies 4 (shoreface)*

This facies consists of centimetre to decimetre bedded, fine to coarse-grained sandstones. Physical sedimentary structures include parallel lamination, low-angle planar cross-stratification, and small-scale trough cross-bedding. Graded beds with sharp-based erosional contacts are common. Mollusc shells are common and often concentrated in small lenses. Bioturbation is moderate to strong. *Ophiomorpha* and *Thalassinoides* are predominant; *Chondrites* and *Helminthopsis* are less common.

Facies 4 exhibits sedimentary structures typical of a high-energy environment. Planar and trough cross-stratification have been considered to form under high-energy oscillatory flow conditions in the near-shore setting. The graded sharp-based beds together with small lenticular shell concentrations are considered to be tempestites.

*Facies 5 (estuary)*

Facies 5 is characterized by extremely fossiliferous medium to coarse-grained sandstones, which interdigitate laterally with deposits of the shelf–shoreface transition zone. The lower part of the succession exhibits lenticular mollusc concentration and/or gravel lags. The upper part shows parallel lamination, low-angle planar cross-stratification, sigmoidal foresets, and bidirectional cross-stratification. Grain size decreases in an upward direction. Within these deposits a horizon 70 cm thick occurs that displays monospecific bioturbation by *Thalassinoides*.

This facies is interpreted to represent the estuarine–marine portion of a bar-built estuary. Lenticular mollusc concentrations and gravel lags are a product of concentration and reworking by currents, probably at the bottom of a small runnel. The sedimentary structures reflect moderate to high energy conditions in a shallow environment, and tidal influence is indicated by bidirectional cross-stratification suggesting deposition on a shoal.

*Facies 6 (fan-delta)*

This facies consists of decimetre-thick beds and lenses of poorly sorted, fine to coarse-grained breccias, matrix-supported conglomerates, clast-supported conglomerates with interbedded sandy layers, pebbly sandstones, and fine to coarse-grained sandstones. Sedimentary structures include low-angle planar cross-stratification, trough cross-stratification, parallel lamination, and herring-bone cross-stratification. Some beds exhibit normal grading. Basal contacts are generally sharp and erosional. The beds show varying degrees of bioturbation, which is restricted to *Ophiomorpha*.

Facies 6 is considered to represent fan-delta deposits. The poorly sorted breccias are interpreted as sheet-flood deposits; the matrix-supported conglomerates have been deposited as debris flows. The sedimentary structures indicate reworking by marine processes in a high energy, shallow marine environment.

*Facies 7 (braid-delta)*

Lithologically, this facies consists of centimetre to decimetre-thick siltstones, which are characterized by a high content of organic material (plants, wood) and medium to coarse-grained sandstones with lenticular intercalations of conglomerates. The beds show low-angle planar cross-stratification, parallel lamination, trough cross-stratification, herring-bone cross-stratification, flaser bedding, and high-angle planar foresets.

This facies is interpreted as representing braid-delta deposits. Physical sedimentary structures indicate a high energy, delta-front environment with intercalated small distributary channels and mouth bar deposits. The siltstones with plant and wood remains represent low-energy braidplain deposits.

## Sequence analysis and sea-level history

The siliciclastic shallow-marine series of the Malpaís Basin are characterized by the presence of laterally traceable discontinuity surfaces, which subdivide the succession into the three depositional sequences (Fig. 2). The surfaces are numbered successively MP-1, MP-2, and MP-3. Depositional sequences are identified by the same number as their basal discontinuity surface.

The MP-1 discontinuity surface is identified as an angular unconformity by the truncation of underlying sequences. The MP-1 surface represents a wave-cut platform showing pronounced microrelief and *Gastrochaenolites* borings. The MP-1 sequence starts with a conglomeratic lag deposit. This lag contains mainly cobbles of the underlying sequences, as well as basalt and radiolarite from Nicoya Complex exposures, and less common shells, rhodoliths, and shark teeth. Some of the components have borings filled with detritus of the overlying sediments. This is interpreted as having been deposited as a transgressive lag during a period of a relative sea-level rise. The overlying sequence comprises an offshore–onshore facies tract from prodelta to delta-front environments. The overall coarsening upward and shallowing upward trend reflects a prograding deltaic system, thus implying a relative fall of sea level.

The MP-2 discontinuity surface has been formed as a marine erosional surface, characterized by *Thalassinoides* burrows, filled with overlying coarser material. A basal conglomerate horizon is locally present; it contains cobbles derived from the underlying shallow-marine sequences, basalt, radiolarite, and quartz components, as well as abundant shells and shell fragments. Basinwards, this surface merges into a correlative conformity. The MP-2 sequence exhibits an offshore–onshore facies tract and comprises laterally interfingered fan-delta, estuarine, and shelf–shoreface transition zone environments, which gradually pass into shoreface environments. The vertical succession indicates a transgressive–regressive couplet.

The MP-3 boundary is similar to the MP-1 discontinuity surface. The major distinguishing feature is the occurrence of amber within the overlying conglomeratic lag deposit. This lag is interpreted as a transgressive lag due to a relative sea-level rise. The MP-3 sequence is only poorly preserved and comprises shoreface and braid-delta deposits.

## Evolution of the Malpaís Basin

Based on detailed measurement and description of sections, available palaeontological data, and sequence analysis, the tectono-sedimentary develop-

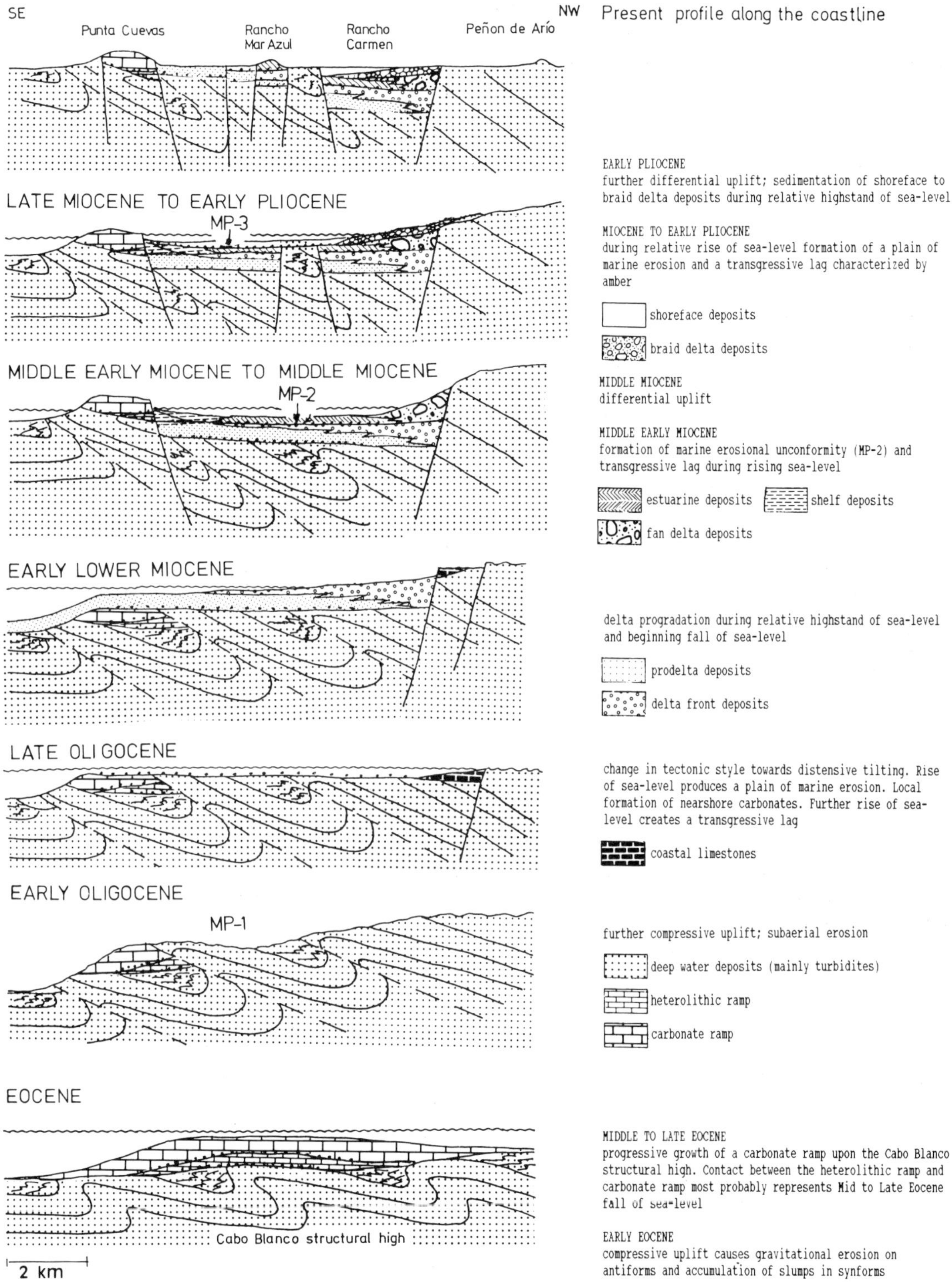


**Fig. 3.** Tectono-sedimentary development of the Malpaís Basin (modified after Seyfried *et al.* (1987) and Schmidt (1989)). See text for further explanation.

ment of the Malpaís Basin can be reconstructed as follows (Fig. 3).

The shallow-marine history of the Malpaís Basin starts in middle to late Eocene times, with progradational growth of a carbonate and heterolithic ramp system after previous compressive uplift, which transformed the former upper trench slope into an external outer arc.

Between the end of the late Eocene and the terminal Oligocene, another compressive event occured, which caused strong shearing and finally led to widespread erosion.

From the latest Oligocene onward, isostatic relaxation produced tilting of blocks parallel to the outer edge of the outer arc. Thus, younger sequences start with an angular unconformity (MP-1), which in places appears as a wave-cut platform bored by *Gastrochaenolites* and covered with a thin veneer of coarse clastics. A concomitant relative sea-level rise is suggested by the formation of mixed carbonate–clastic depositional systems that cover near-shore areas (Nosara region, Calvo, 1987). Carbonate detritus derived from these shoals has been reworked among the lower Miocene deltaic sediments, evidencing continuous tectonic activity (Schmidt, 1989). The prograding delta sequence should have formed during a relative highstand of sea level and one, or several, episodes of falling sea level.

In the outcrops examined, the MP-2 marine erosional unconformity is not angular and thus might correspond to a relative fall of sea level rather than to tectonic uplift (which always follows the tilting style). The basement components in the following conglomeratic lag, however, prove tilting revival. Correspondingly, the marine unconformity should be considered a correlative conformity. Tilting fits well with the fact that the overlying mid-Miocene sediments were deposited in a completely different environment: small coastal embayment where estuarine deposits closely interfinger with fan-delta and shallow-shelf deposits. The estuarine sequence is topped by a horizon with monospecific bioturbation by *Thalassinoides*.

After another phase of tilting the MP-3 angular unconformity was produced. Middle Miocene to lower Pliocene deposits start with an amber-bearing transgressive lag, which in turn covers a wave-cut platform. In contrast to the former depositional sequences, the present one does not display facies types that unambiguously prove a highstand signal in the basin-fill.

## The Parrita Basin

The Parrita Basin is located on the Pacific Coast of central Costa Rica (Fig. 1). Basin geometry is defined by a strike-slip fault system (Astorga *et al.*, 1989; Schmidt, 1989). More than 1300 m of shallow-marine deposits accumulated during the mid-Miocene and probably early Pliocene on a continually subsiding block. Facies types reflect a wide environmental spectrum, ranging from deeper shelf environments through tidal-flat deposits to paralic conditions. The basement of the sedimentary succession is known from geological mapping in the surrounding areas and is represented by basaltic rocks of the Nicoya Complex. The Neogene sequence is topped by the alluvial deposits of the modern Parrita Valley.

The sedimentary suite of the Parrita Basin was described for the first time by Dengo (1962). Shallow-marine deposits are included among the Miocene 'Punta Carballo Formation' (Kuypers, 1979; Fischer, 1981; Sprechmann, 1984). Geological studies concerning sedimentology, environmental, and sequential dynamics, as well as palaeoecological aspects, have been carried out in the last few years by Seyfried & Aguilar (1984), Seyfried *et al.* (1985), Seyfried & Sprechmann (1985), and Schmidt (1989).

### *Biostratigraphic remarks*

Kuypers (1979) and Sprechmann (1984) claim a mid-Miocene age for the Punta Carballo Formation. Fischer (1981), however, argued that the age of the formation ranges from middle Miocene to late Miocene. F. Rivier (Escuela Centroamericana de Geología, Universidad de Costa Rica, personal communication) emphasizes an early Pliocene age for the upper part of the succession. Planktonic Foraminifera of the basal part of the sequence belong to zone N9 (after Blow (1969,1979), cited in Seyfried *et al.*, 1985).

## Facies description and depositional environments

The type section at Punta Judas (Fig. 4) shows 12 facies types. The mainly litharenitic volcaniclastic sediments characteristically show an intricate mixing of two grain populations: relatively well-preserved first-cycle grains, such as feldspars and calc-alkaline volcanic rock fragments, are almost equal in abundance to heavily altered 'Nicoya Complex' grains (mostly basalt fragments). It is obvious that a 'fresh' influx of calc-alkaline arc grains had un-

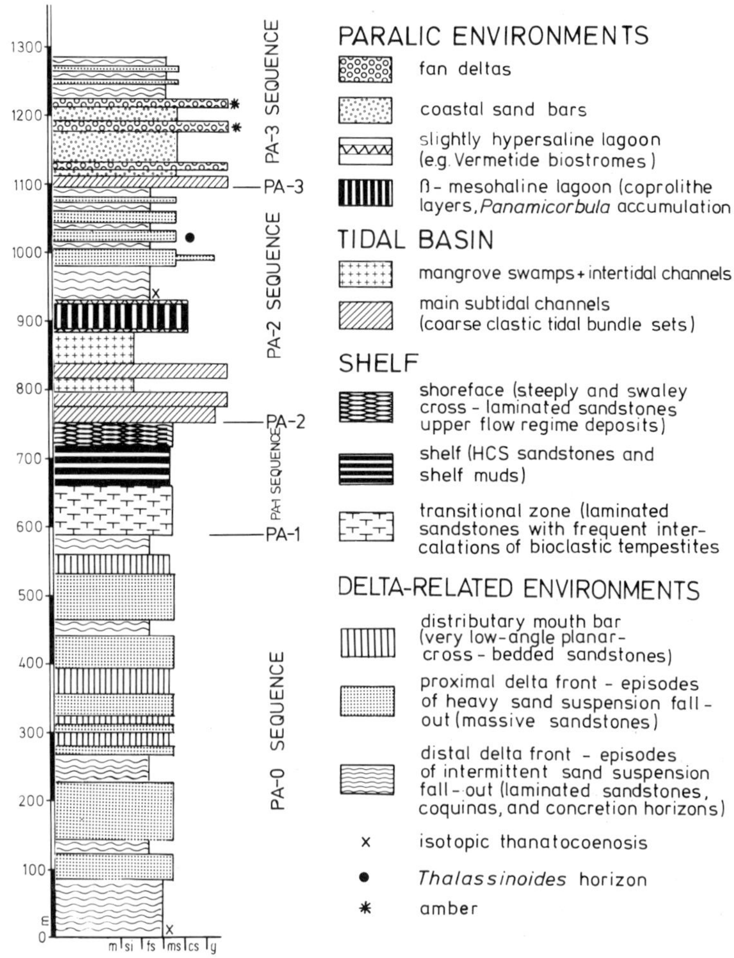


**Fig. 4.** Generalized type section (Punta Judas Section) of the shallow-marine sequences of the middle Miocene to early Pliocene(?) Parrita Basin (modified after Seyfried *et al.* (1987) and Schmidt (1989)). The vertical succession comprises four transgressive–regressive cycles (PA-0 to PA-4 sequences).

hindered access to the basin, whereas the Nicoya Complex grains already had a history; they are considered to derive from adjacent weathered uplifts.

Most facies are extremely fossiliferous and exhibit varied trace fossil assemblages, so that sedimentological interpretation can be thoroughly controlled and completed by palaeoecological data.

### *Facies 1 (distal delta-front)*

This facies consists of alternating fine to medium-grained, decimetre thick, well bedded, massive sandstones, laminated sandstones, and highly bioturbated sandstones containing a high diversity of molluscs (Seyfried *et al.*, 1985). The sandstones are characterized by abundant regularly spaced concretion levels, parautochthonous shell accumulations, and monospecific assemblages of molluscs (e.g. *Atrina* sp., *Semele* sp.) in life position (biocoenosis, cf. Seyfried *et al.*, 1985).

Facies 1 is interpreted as having been deposited within a deltaic environment in a distal delta-front position. The massive sandstones most probably originated from suspension fall-out. The bioturbated sandstones reflect slightly diminished sediment influx, so that bioturbation was able to keep pace with sedimentation. Laminated sandstones and particularly the occurrence of biocoenoses reflect reduced sediment input or even short episodes of no net sedimentation.

*Facies 2 (proximal delta-front)*

Lithologically, this facies consists of fine to medium-grained homogeneous sandstones without physical sedimentary structures and with very scarce bioturbation. The sandstones may contain rare mollusc shells and isolated concretions.

This facies is considered to reflect a position in a proximal delta-front environment. The lack of any structure within the sandstones and the low content of molluscs reflect a high sedimentation rate, attributed to intense suspension fall-out.

*Facies 3 (distributary mouth bar)*

Facies 3 is made up of decimetre thick, bedded, fine to medium-grained sandstones, with common concretion horizons, and laminated fine to medium-grained sandstones, which exhibit low-angle planar cross-stratification. Bioturbation with *Ophiomorpha* and *Thalassinoides* is moderately common.

This facies clearly reveals higher energy conditions than facies 2. Planar lamination and low-angle planar cross-stratification are considered to result from upper flow regime currents. The sandstones of this facies type are therefore interpreted to represent mouth bar deposits within a deltaic environment.

*Facies 4 (shelf)*

Facies 4 consists of alternations of parallel laminated, fine-grained sandstones, fine-grained sandstones with primary bedding structures that have been completely destroyed by bioturbation, and fine to medium-grained, decimetre thick, bedded sandstones displaying sharp erosional basal contacts, hummocky cross-stratification, and *Ophiomorpha* burrows, which mainly affect the tops of the sandstone beds. Intercalations of shell beds and lenticular shell accumulations are common.

Facies 4 is interpreted to represent a shelf–shoreface transitional environment. The parallel-laminated and bioturbated fine-grained sandstones reflect sedimentation from suspension fall-out within a relatively tranquil environment and probably represent fair weather conditions. Conversely, the sedimentary structures of the fine to medium-grained sandstones indicate conditions of waning flow and appear to be storm deposits. Shell beds and lenticular shell accumulations are densely packed with *Anadara* (*Scapharca*) sp., a mesohaline bivalve, and hence must be considered as tempestites (Seyfried & Sprechmann, 1985).

*Facies 5 (shelf)*

This facies is characterized by alternations of laminated or massive argillaceous siltstones and mudstones with fine to medium-grained sandstones that alternatingly exhibit horizontal and low-angle cross-stratification (interpreted as hummocky cross-stratification).

This facies is attributed to a shelf environment between fair weather wave-base and storm wave-base. The silt- and mudstones of this facies probably represent the deepest deposits of the section, with sandy intercalations indicating occasional wave-induced turbulence.

*Facies 6 (shoreface)*

This facies type consists of well-bedded, fine to medium-grained sandstones. Physical sedimentary structures include parallel lamination, small-scale trough cross-stratification, steeply inclined planar foresets and swaley cross-stratification. Moderate to strong bioturbation is restricted to *Ophiomorpha*.

The predominance of physical sedimentary structures formed under high-energy conditions verify the sedimentation in a shoreface position. This facies is dominated by both bidirectional wave motion and unidirectional currents.

*Facies 7 (main tidal channels)*

Lithologically, this facies is made up of centimetre to decimetre-thick, bedded, medium-sand grained to pebble deposits. The beds exhibit thin horizontal bedding, small-scale and large-scale trough cross-stratification, steep planar foresets, and decimetric sigmoidal foresets with well-developed toesets. These foresets exhibit a marked grading from medium pebble size to medium sand size and are separated by double mud drapes, thus proving their origin as tidal bundles.

Sedimentary structures and the occurrence of tidal mud drapes verify this facies type to be originated in subtidal and intertidal channels.

*Facies 8 (mangrove swamps)*

The sediments of this facies type are composed of laminated, wavy-bedded sets of argillaceous siltstones, and medium to coarse-grained sandstones. Some thicker sandstone beds show grading and lenticular channel fills with basal rip-up clasts. Wood debris is often concentrated in thin sheets; leaves are common among argillaceous layers. *Thalassinoides* and *Skolithos* burrows are filled with medium to coarse-grained sand. Pholas borings have been observed mainly in the silty facies.

This facies is assigned to a mangrove swamp environment of a tropical tidal flat. The sandstone intercalations and foresetted mud-chip accumulations are interpreted as landward tempestites.

### *Facies 9 (β-mesohaline lagoon)*

The deposits of this facies are fine to medium-grained sandstones with densely packed accumulations of *Corbula* (*Panamicorbula*) sp. as well as several coprolite layers.

The dominance of *Corbula* (*Panamicorbula*) sp., a mesohaline bivalve (Seyfried *et al.*, 1985), suggests an interpretation as hyposaline lagoonal deposits.

### *Facies 10 (hypersaline lagoon)*

This facies is characterized by biostromes of the vermetid gastropod *Petaloconchus sculpturatus* (Seyfried *et al.*, 1985). Giant bivalves and gastropods are common accessory faunal elements.

The vermetide biostromes originated in an ephemeral, extremely shallow, and most probably hypersaline lagoon.

### *Facies 11 (coastal sand bars)*

Lithologically, this facies is similar to facies 6. Additional sedimentary structures comprise convolute lamination, escape structures, climbing ripples, and wave ripples reflecting general high sediment supply. This facies contains abundant 'sand dollars' (*Melitta* sp.), a sea urchin, which is typical for the breaker zone (Seyfried *et al.*, 1985).

The sedimentary structures and the occurrence of *Melitta* sp. place this facies in an extremely shallow marine setting under high-energy oscillatory flow conditions. This is characteristic for coastal sand bars.

### *Facies 12 (fan-delta)*

Facies 12 exhibits the same lithological and sedimentary features as facies 6 of the Malpaís Basin.

## Sequence analysis and sea-level history

The designation of discontinuity surfaces and sequences follows the same pattern as used for the Malpaís Basin. The surfaces and corresponding sequences are numbered PA-1, PA-2, and PA-3 (Figs 4, 6). The PA-1 discontinuity surface is characterized by a change from distal delta-front deposits to sediments of the shelf–shoreface transition zone. This surface is interpreted as having been produced under submarine conditions during a transgression, which was preceded by a relative fall of sea level. The PA-2 surface is characterized by the erosional incision of subtidal channels into shoreface deposits. The development of such discontinuity features is interpreted to be related to periods of low sea level (Shanmugam, 1988). The PA-3 discontinuity surface is an angular unconformity produced by tectonic uplift, which in turn is proved by the overlying fan-delta deposits, and probably concomitant with a period of relative fall of sea level.

The vertical sequence development, indicated by the facies framework and stratal pattern within the sedimentary record, points out four environmental scenarios. These scenarios reflect different sequence development according to variable interaction between rates of eustasy, subsidence, and sediment supply (Fig. 5).

### *Scenario A (PA-1 sequence)*

A thick succession of delta-front deposits characterizes this situation. Sedimentological and biostratonomic features indicate three different facies types, which formed during a time of relatively rising sea-level, controlled by a continuing subsidence and variable sediment input.

**1** Massive sandstones reflect high sedimentation rates during episodes of heavy sand suspension fall-out. Local conditions were: uplift of the hinterland to produce strong erosion, and/or a short-term drop of sea level to produce prograding delta systems.

**2** Alternations of laminated, commonly bioturbated sandstones, parautochthonous shell accumulations, and biocoenoses reflect relatively low sedimentation rates. During a short-term rise of sea level, the delta became an estuary that served as a sediment trap and little sediment could reach the depocentre. Alternations of facies types 1 and 2 were produced by fluctuations of sediment influx owing to episodic shifting of the delta system — producing a local

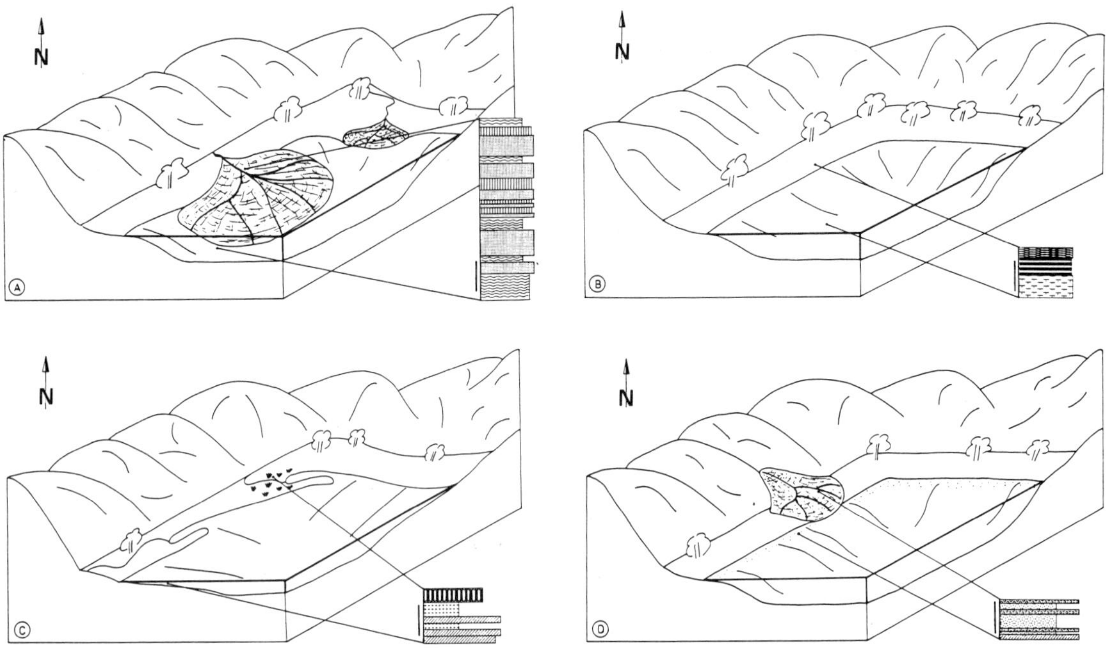

**Fig. 5.** Environmental scenarios for the sedimentary development of the Parrita Basin, reflecting different sequence development according to variable interaction between rates of eustasy, subsidence, and sediment supply. See text for further explanation. Key for columns corresponds to Fig. 4; bars equal 100 m.

relative sea-level rise, climate changes, or short-term transgressive episodes.

**3** Interbeds of very low-angle planar cross-bedded sandstones, massive sandstones, and concretion horizons indicate relatively low sedimentation rates from continuing suspension fall-out. Concretion horizons, strong bioturbation, and small-scale lamination are characteristic for low rates of sedimentation. Massive sandstones, containing a sparse fauna, indicate continuing sedimentation from suspension fall-out. To have produced such a facies, a highstand of sea level and a high rate of sediment input would have been needed.

*Scenario B (PA-2 sequence)*

This situation reveals a transgressive–regressive couplet, which is superimposed by a general shallowing upward trend. The succession starts with sediments of the shelf–shoreface transition zone, is followed by shelf sediments, and then grades upwards into shoreface sediments. This pattern must have been produced by a rise and highstand of sea-level and the start of a fall of sea level.

*Scenario C (PA-3 sequence)*

During a relative lowstand of sea level, subtidal deposits formed in nearshore areas, and the marginal regions are characterized by deposits of mangrove swamps and lagoons.

*Scenario D (PA-3 sequence)*

This scenario also reflects lowstand of sea level. In contrast to scenario C, fan-deltas interfingering with coastal sand bars indicate contemporaneous tectonic activity.

## SEQUENCE-STRATIGRAPHIC INTERPRETATION

Sequence analysis of the Malpaís succession indicates three transgressive–regressive sequences separated by discontinuity surfaces. In terms of sequence stratigraphy these sequences correspond to depositional sequences bounded below and above by unconformities and their correlative conformities.

The MP-1 and MP-3 sequence boundaries are developed as angular unconformities and correspond to type 1 sequence boundaries, which formed during a rapid fall of relative sea level. Truncation of underlying sequences gives evidence for simultaneous local uplift and/or tilting of blocks. The MP-2 sequence boundary is also developed as a type 1 sequence boundary and formed as a marine erosional surface owing to a relative fall of sea level.

### *The MP-1 sequence*

A slow relative rise of sea level produced a plain of marine erosion. With continuing rise of sea level a conglomerate bed developed. The plain of marine erosion and the overlying lag deposit are interpreted as representing the transgressive systems tract of this depositional sequence. The following prodelta sediments reflect the deepest sedimentary environment within the MP-1 sequence and are considered as having been deposited during a period of relative highstand of sea level. Regressive conditions, caused by a relative fall of sea level, are indicated by delta-front deposits, which prograde over the prodelta deposits. This succession of prograding deltaic deposits is considered as representing the highstand systems tract.

### *The MP-2 sequence*

The MP-2 sequence boundary developed owing to a relative fall of sea level. This fall produced a marine erosional surface, which is traceable basinwards into a correlative conformity. A basal conglomerate, overlain by laterally interdigitating estuarine deposits and deposits of the shelf–shoreface transition zone, is interpreted as having been formed during a relative rise of sea level. This succession represents the transgressive systems tract. The highstand systems tract contains prograding shoreface deposits. The stratigraphic position of the fan-delta deposits within this sequence is not clear owing to outcrop conditions.

A horizon with extremely strong monospecific bioturbation by *Thalassinoides* at the top of the estuarine deposits indicates hardground conditions (Kidwell, 1984) with low or even no net sedimentation. This horizon is interpreted as a flooding surface. This surface (maximum flooding surface, cf. Baum & Vail, 1987) marks the boundary between the transgressive systems tract and the highstand systems tract.

### *The MP-3 sequence*

Uplift and block faulting produced an angular unconformity (MP-3 sequence boundary). A plain of marine erosion, covered by an amber-bearing conglomerate, is interpreted as the transgressive systems tract. This interpretation implies a preceding fall of sea level. The transgressive succession is overlain by an oyster biostrome and shoreface deposits; the stratigraphic position of braid-delta deposits within the sequence is unclear.

The following general conclusion can be drawn for the shallow-marine series of the Malpaís Basin.

**1** Lowstand systems tracts are missing.

**2** Transgressive systems tracts are represented by wave-cut platforms, transgressive lags, and mixed carbonate–clastic marginal-marine deposits.

**3** Highstand systems tracts reflect various sedimentary environments depending on interaction of controlling factors. High sediment influx produced prograding delta systems.

## Parrita Basin

The shallow-marine Parrita succession (Fig. 6) can be subdivided into four depositional sequences bounded by discontinuity surfaces, interpreted to represent sequence boundaries.

### *The PA-0 sequence*

Owing to outcrop conditions the basal boundary of this sequence cannot be observed. The PA-0 sequence consists of a succession of thick deltaic deposits. Within this succession the vertical arrangement of deltaic depositional systems forms genetically related units, interpreted to represent parasequences. The parasequences consist of combinations of facies type 1 (distal delta-front, characterized by episodes of intermittent sand suspension fall-out; numbering according to facies description of the Parrita Basin), facies type 2 (proximal delta-front, characterized by episodes of heavy sand suspension fall-out), and facies type 3 (distributary mouth bar, characterized by very low-angle planar cross-bedded sandstones). Two general combinations of these facies types can be distinguished forming parasequence sets.

*Parasequence set 1* shows an overall aggrading stacking pattern. Such a pattern is only possible if there is an equilibrium between sediment influx, subsidence, and relative change of sea level (Vail *et*

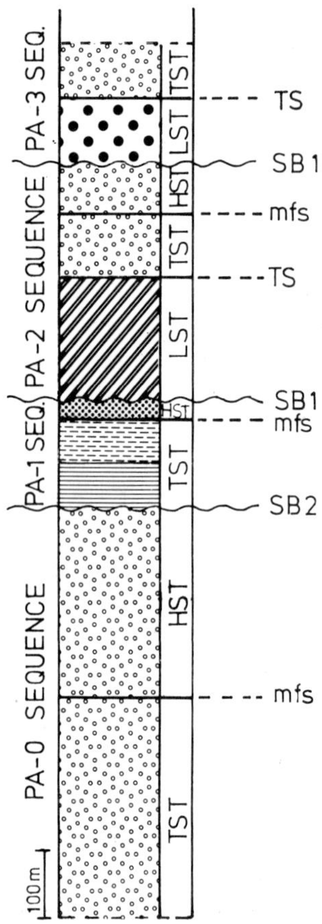


**Fig. 6.** Sequence-stratigraphic interpretation of the Parrita Basin (key as for Fig. 7).

*al.*, 1977). *Parasequence set 2* contains parasequences with an overall prograding pattern, reflecting regressive conditions.

Parasequence set 1 forms the lower part of the PA-0 sequence. The aggrading stacking pattern of set 1 is interpreted as having been established during rising sea level, high sediment influx, and continuing subsidence, representing the transgressive systems tract of the PA-0 sequence. Biocoenoses, characterized by *Atrina* sp. and *Semele* sp. in life position, occur within the basal parts. This feature indicates periods of reduced sediment influx due to flooding events during the relative rise of sea level. Therefore, these horizons are considered to be marine-flooding surfaces.

The upper part of the PA-0 sequence consists mainly of parasequence set 2. The overall prograding pattern of this part of the sequence indicates regressive conditions. It is interpreted as having been deposited during an early fall of sea level representing the highstand systems tract of the PA-0 sequence.

### *PA-1 sequence*

The boundary between the PA-0 and PA-1 sequence is formed by a marine erosional surface and is characterized by the change from delta-front deposits to deposits of the shelf–shoreface transition zone. Since no obvious evidence of subaerial erosion can be observed, this surface is interpreted as having been formed in response to a slow fall of relative sea level and represents a type 2 sequence boundary. The transgressive–regressive couplet of the PA-1 sequence consists of the following components.

**1** Deposits of the shelf–shoreface transition zone exhibit a retrograding pattern which is characteristic for transgressive conditions. This part of the sequence is correlated with the transgressive systems tract, possibly containing elements of the shelf-margin systems tract.

**2** A coarsening and shallowing upwards sequence indicates a relative highstand of sea level and a subsequent relative fall in sea level. This part is considered to represent the highstand systems tract of the PA-1 sequence.

### *PA-2 sequence*

The PA-2 sequence boundary is represented by erosional tidal channels in shoreface sediments. This surface is correlated with the development of a type 1 sequence boundary during a relative fall of sea level. The lower part of the PA-2 sequence is made up of tidal-flat deposits, which were deposited after the relative fall of sea level. Their stratigraphic position within the sequence permits correlation with the lowstand systems tract. A clear facies change from tidal to deltaic deposits reflects the transition to transgressive conditions. The occurrence of a biocoenosis (*Semele* sp. in life position) represents a transgressive surface. As with the PA-1 sequence, a succession of deltaic sediments was deposited during transgressive and regressive times. The bounding surface between the transgressive systems tract and the highstand systems tract (maximum flooding surface) is marked by a *Thalassinoides* horizon (compare with MP-2 sequence of the Malpaís Basin).

### *PA-3 sequence*

The PA-3 sequence boundary is developed as an angular unconformity. This surface was produced by tectonic uplift and probably reinforced by a relative fall of sea level. It represents a type 1 sequence boundary. In response to the tectonic movement, fan-delta sediments were deposited and form the lower part of the PA-3 sequence. These deposits represent part of the lowstand systems tract and the transgressive systems tract. The transgressive character of the upper fan-delta deposits is supported by the occurrence of amber (compare with MP-3 sequence of the Malpaís Basin). The discontinuity surface between the transgressive and the highstand systems tract reflects a clear change in facies to delta-front sediments, which were deposited during the highstand systems tract.

## Interaction of controlling factors

The sequence-stratigraphic interpretation of the shallow-marine Malpaís and Parrita Basins clearly reveals similarities of systems tract architecture within these basins. The statements given in this section are valid for the Malpaís and Parrita Basins and can be applied to all fore-arc coastal embayments of the southern Central American island-arc system (Schmidt, 1989; see paper by Kolb & Schmidt).

The systems tracts show characteristic stratal patterns and facies architecture. Lowstand systems tracts are characterized by non-marine deposits (e.g. coastal plain deposits, fan-delta deposits), marginal marine deposits (e.g. estuarine deposits, tidal deposits) or a hiatus. Transgressive systems tracts and highstand systems tracts reflect various sedimentary environments, including the formation of a plain of marine erosion and/or a transgressive lag. This indicates that not only changes of sea level, but also various interactions of sediment influx, uplift, and subsidence control formation of facies types. The facies associations within the transgressive and highstand systems tracts reveal the following general relations between uplift/subsidence, rate of sedimentation, and changes of sea level.

**1** Transgressive systems tract:
(a) uplift > rise of sea level = formation of fan-deltas, interdigitated with shelf deposits;
(b) subsidence = rise of sea level + high rate of sedimentation – formation of thick prograding or aggrading delta systems;
(c) subsidence < rise of sea level – formation of transgressive shelf deposits;
(d) subsidence > rise of sea level – formation of *Thalassinoides* horizon.

**2** Highstand systems tract:
(a) uplift > fall of sea level – formation of coastal plains with braid-delta systems;
(b) subsidence = fall of sea level + high rate of sedimentation – thick aggrading shoreface deposits;
(c) subsidence < fall of sea level – formation of prograding shelf–beach systems.

The following surfaces and sequence boundaries to define depositional sequences and systems tracts can be recognized.

### *Type 1 sequence boundary* (cf. Vail & Todd, 1981)

Regional angular unconformities as well as subaerial and submarine erosional surfaces are interpreted to be type 1 sequence boundaries. They are characterized by subaerial exposure and erosion or submarine erosion, the formation of root horizons, and the deposition of non-marine or very shallow marine (e.g. estuarine or lagoonal) facies types about the sequence boundary.

### *Type 2 sequence boundary* (cf. Vail & Todd, 1981)

Surfaces without obvious evidence of subaerial or submarine erosion, but characterized by a significant facies change are interpreted as type 2 sequence boundaries.

### *Transgressive surfaces*

Marine flooding surfaces above the lowstand systems tract, indicated by a veneer of conglomerates, a shell lag or a biocoenosis, are interpreted as transgressive surfaces.

### *Maximum flooding surface*

A horizon of monospecific bioturbation by *Thalassinoides*, which indicates hardground conditions, is interpreted as a maximum flooding surface.

## Basin to basin correlation

The reconstruction of sedimentary basin development and basin dynamics of the Neogene coastal embayments demonstrates a widely varied lithology and facies distribution (Schmidt, 1989). Therefore a comprehensive correlation of different basins is often problematic. The application of sequence-stratigraphic criteria, however, gives a more detailed insight in basin-wide sedimentary dynamics and may be helpful for basin to basin correlation (Fig. 7).

The Malpaís Basin and the Parrita Basin are two examples reflecting the high diversity of basin development in a tectonic realm, such as the fore-arc setting of an island-arc system. In the Malpaís Basin repetitive tilting phases clearly control sequence development and eustatic changes only contribute an overprint. The Parrita Basin reflects strong subsidence and primary eustatic control over sequence development.

Two marker beds correlate between these two basins:

**1** an amber-bearing conglomerate;

**2** a horizon with extreme *Thalassinoides* burrows. Both marker beds occur only once in each basin and therefore are regarded to be unique events.

The amber-bearing conglomerate occurs above discontinuity surfaces developed as angular unconformities which are interpreted as type 1 sequence boundaries (MP-3 and PA-3 surfaces). Biostratigraphic data indicate late Miocene to early Pliocene ages for the MP-3 sequence. The same age can be assumed for the PA-3 sequence. Within both basins, the *Thalassinoides* horizon seems to form the discontinuity surface between highstand and transgressive systems tracts of the underlying sequence (MP-2 and PA-2 sequences). Age determination within the Malpaís Basin reveals a probable middle Miocene age for the MP-2 sequence. This coincides with palaeontological data of the Parrita Basin.

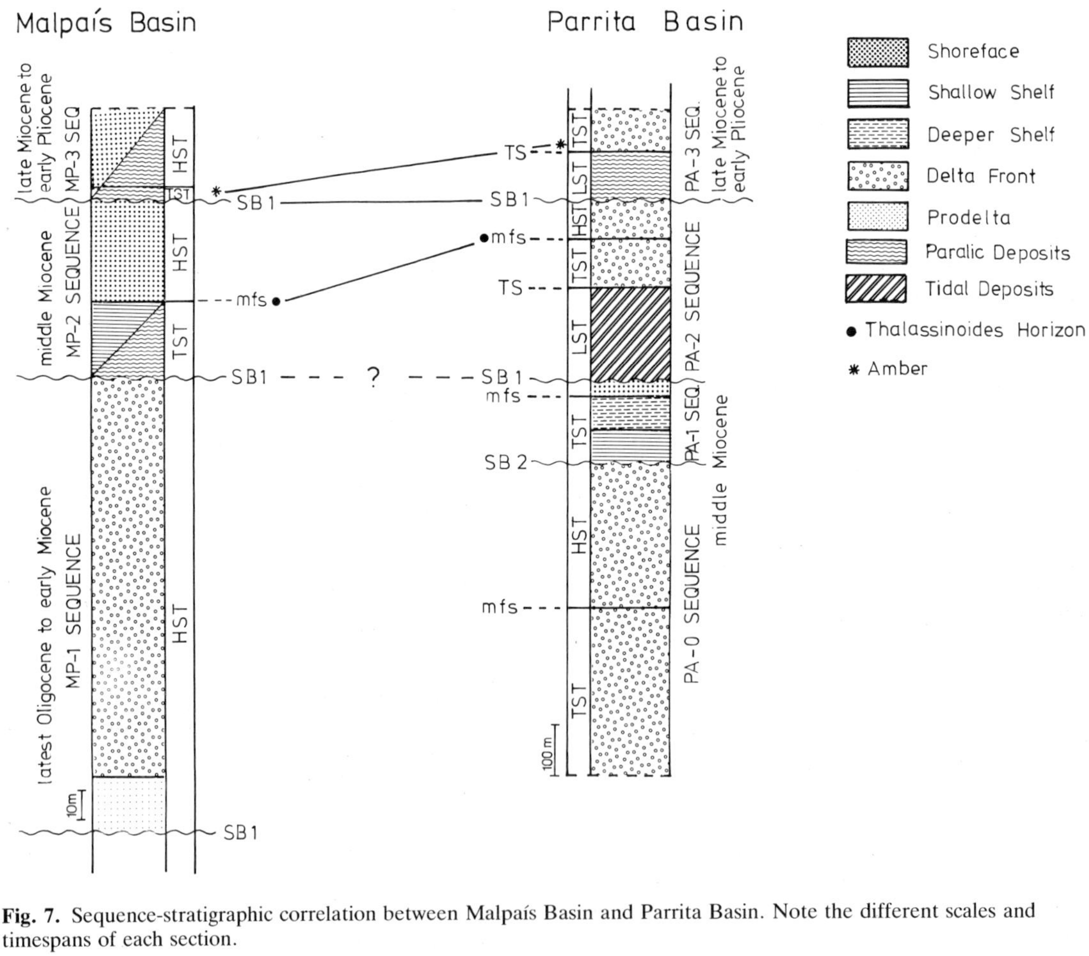


**Fig. 7.** Sequence-stratigraphic correlation between Malpaís Basin and Parrita Basin. Note the different scales and timespans of each section.

The correlation between MP-2 and PA-2 sequence boundaries is uncertain but seems probable with respect to sequence analysis. Correlation between MP-1 and PA-1 sequence boundaries is not possible. Age determination within the MP-1 sequence indicates a late Oligocene to early Miocene age, whereas planktonic Foraminifera of the basal part of the PA-1 sequence belong to zone N9 (early middle Miocene). This implies a hiatus associated with the MP-2 sequence boundary.

## CONCLUSIONS

Taking into account the thickness of the shallow-marine successions (Malpaís Basin, c. 250 m; Parrita Basin, c. 1300 m), sequence development, sequence-stratigraphic interpretation, and correlation, as well as biostratigraphic data and correlation with published global sea-level charts, the following conclusions can be made.

The sequence boundaries developed within the relatively thin late Oligocene to early Pliocene Malpaís succession most probably represent sequence boundaries of second-order cycles according to Haq *et al.* (1988). Sequence boundaries within the middle Miocene to early Pliocene(?) succession of the Parrita Basin in contrast are considered as sequence boundaries of third-order cycles according to Haq *et al.* (1988). A basin to basin correlation is reliable at the level of second-order sequence boundaries.

Comparison with the global sea-level charts of Haq *et al.* (1988) implies a possible correlation of the MP-3 and PA-3 sequence boundary with the TB2/TB3 sequence boundary, and a correlation of the MP-1 sequence boundary with the TA4/TB1 supercycle sequence boundary, and the MP-2 sequence boundary with the TB1/TB2 supercycle boundary. The PA-1 and PA-2 sequence boundaries seem to correlate with sequence boundaries of middle Miocene third-order cycles.

Sequence boundaries of second-order cycles coincide with angular unconformities, implying a controlling function of tectonic events to the formation of second-order sequence boundaries within the fore-arc setting of the southern Central American island-arc system.

The application of sequence-stratigraphic criteria for interpretation and correlation of basin dynamics turned out to be a useful tool for basin analysis, even in such an inhomogeneous tectonic realm as an island-arc system. The examples presented show that sequences with similar stratal patterns developed at the same time in widely varied coastal embayments. This suggests a globally effective control, such as eustatic sea-level change, on sedimentation and basin development at the level of systems tracts, third-order depositional sequences, and parasequences.

## ACKNOWLEDGEMENTS

This study was benefited by a grant from the Nachwuchsförderungsgesetz (Berlin) for one of the authors (H. Schmidt). The 'Deutsche Forschungsgemeinschaft' (DFG) provided financial support for field trips (Se 490/1-1, Se 490/1-2). The research in Costa Rica was supported through the close collaboration of RECOPE (national oil company of Costa Rica and the 'Escuela Centroamericana de Geología' (University of Costa Rica). S. Brown, R.H. Dott, Jr. and D.I.M. Macdonald reviewed the manuscript and made many helpful suggestions.

## REFERENCES

Astorga, A. (1987) *El Cretácico Superior y el Paleógeno de la vertiente Pacifica de Nicaragua meridional y Costa Rica septentrional: orígen, evolución y dinámica de las cuencas profundas relacionadas con el margen convergente de Centroamérica.* Unpublished tesis de Licenciatura, University of Costa Rica, San José, 250 pp.

Astorga, A., Fernandez, J.A., Barboza, G., Campos, L., Obando, J., Aguilar, A. & Obando, L.G. (1989) *Cuencas Sedimentarias de Costa Rica: Evolución y Potential de Hidrocarburos.* Transactions of the Fourth Circum-Pacific Council for Energy and Mineral Resources Conference, Abstr. Vol., San José (Costa Rica), pp. 1–18.

Baum, G.R. & Vail, P.R. (1987) Sequence stratigraphy, allostratigraphy, isotope stratigraphy and biostratigraphy — putting it all together in the Atlantic and Gulf Paleocene. *8th Research Conference SEPM/GCS (Innovative Biostratigraphic Approaches to Sequence Analysis: New Exploration Opportunities)* Houston, Texas, pp. 15–23.

Baumgartner, P.O., Mora, C., Butterlin, J., Sigal, J., Clacon, G., Azema, J. & Bourgois, J. (1984) Sedimentología y paleogeografía del Cretacico y Cenozoico del litoral Pacífico de Costa Rica. *Rev. Geol. Am. Central* **1**, 57–136.

Calvo, C. (1987) *Las calizas neriticas de la vertiente Pacífica del norte de Costa Rica y sur de Nicaragua: epocas y sistemas de sedimentación asociadas, con la apertura y evolución del margen convergente de la América Central Meridional.* Unpublished tesis de Licenciatura, University of Costa Rica, San José, 165 pp.

Dengo, G. (1962) *Estudio geológico de la región de Guanacaste, Costa Rica.* Inst. Geogr. Nac., San José, ix + 112 pp.

Fischer, R. (1981) Die Herausformung des mittelamerikanischen Isthmus im Miozän Costa Ricas. *Zentral. Geol. Palaeontol. Teil I* **3/4**, 219–221.

Haq, B.U., Hardenbol, J. & Vail, P.R. (1988) Mesozoic and Cenozoic chronostratigraphy and cycles of sea-level. In: *Sea-level Changes: an Integrated Approach* (Eds Wilgus, C.K., Hastings, B.S., Kendall, C.G.St.C., Posamentier, H.W., Ross, C.A. & van Wagoner, J.C.). Soc. Econ. Paleontol. Mineral., Spec. Publ. 42, 71–108.

Kidwell, S.M. (1984) Outcrop features and origin of basin margin unconformities in the Lower Cheasapeake Group (miocene) Atlantic Coastal Plain. In: *Interregional Unconformities and Hydrocarbon Accumulation* (Ed. Schlee, F.S.). Mem., Am. Assoc. Petrol. Geol. 36, 37–58.

Kuypers, E.P. (1979) Analisis sedimentológico de la Formación Punta Carballo (Mioceno), Costa Rica. *Inst. Geogr. Nac., Informe Semest.* **julio–diciembre**, 77–94.

Lundberg, N. (1982) Evolution of the slope landward of the Middle America Trench, Nicoya Peninsula, Costa Rica. In: *Trench–Forearc Geology* (Ed. Legget, J.K.). Geol. Soc. London, Spec. Publ. 10, 133–147.

Mora, R.C. (1985) *Sedimentología y geomorphología del Sur de la Peninsula de Nicoya (Provincia de Puntarenas, Costa Rica).* Unpublished tesis de Licenciatura, University of Costa Rica, San José, 148 pp.

Posamentier, H.W. & Vail, P.R. (1988) Eustatic controls on clastic deposition, II — sequence and systems tract models. In: *Sea-level Changes: an Integrated Approach* (Eds Wilgus, C.K., Hastings, B.S., Kendall, C.G.St.C., Posamentier, H.W., Ross, C.A. & van Wagoner, J.). Soc. Econ. Paleontol. Mineral., Spec. Publ. 42, 125–154.

Posamentier, H.W., Jervey, M.T. & Vail, P.R. (1988) Eustatic controls on clastic deposition, I — conceptual frameword. In: *Sea-level Changes: an Integrated Approach* (Eds Wilgus, C.K., Hastings, B.S., Kendall, C.G.St.C., Posamentier, H.W., Ross, C.A. & van Wagoner, J.). Soc. Econ. Paleontol. Mineral., Spec. Publ. 42, 109–124.

Schmidt, H. (1989) *Sequenzstratigraphie des neogenen Inselbogen-Schelfes im 'Forearc'-Bereich Costa Ricas und Nicaraguas.* Unpublished PhD dissertation, Technische Universität Berlin, 250 pp.

Schmidt-Effing, R. (1979) Alter und Genese des Nicoya-Komplexes, einer ozeanischen Paläokruste (Oberjura bis Eozän) im südlichen Mittelamerika. *Geol. Rundsch.* **68** (2), 467–494.

Seyfried, H. (1986) Strukturgeschichte und Sedimentation an einem konvergenten Plattenrand: Das Beispiel der südlichen mittelamerikanischen Landbrücke. *Berliner Geowiss. Abh., (A), Sonderbd. X. Geowiss. Lateinamer.- Koll.*, 7–13.

Seyfried, H. & Aguilar, T. (1984) Sedimentologie und Paläoökologie jungtertiärer Sandwatt-Ablagerungen Costa Ricas. *Tagungsh. 9. Geowiss. Lateinamer.-Koll.*, 156–159.

Seyfried, H. & Sprechmann, P. (1985) Sedimentology and paleoecology of a perilittoral estuarine siliciclastic sequence from the Middle Miocene of Costa Rica (Central America). *Abstracts, Symposium of Modern and Ancient Clastic Tidal Deposits, Utrecht*, pp. 116–119.

Seyfried, H., Sprechmann, P. & Aguilar, T. (1985) Sedimentología y paleoecología de un estuario del litoral Pacífico del Istmo Centroamericano Primordial (Mioceno Medio, Costa Rica). *Rev. Geol. Am. Central* **3**, 1–68.

Seyfried, H., Astorga, A., Calvo, C. & Laurito, C. (1987) Sequence response (cyclicity, biostratinomy, ichnofacies) to subsidence, sea-level fluctuations, and exceptional events in Cenozoic fore-arc basins of southern Central America. *8th Research Conference SEPM/GCS (Innovative Biostratigraphic Approaches to Sequence Analysis: New Exploration Opportunities)*, Houston, TX. pp. 131–141.

Shanmugan, G. (1988) Origin, recognition, and importance of erosional unconformities in sedimentary basins. In: *New Perspectives in Basin Analysis* (Eds Kleinspehn, K.L. & Paola, C.) pp. 83–108. Springer-Verlag, New York.

Sprechmann, P. (Ed.) (1984) *Manual de Geología de Costa Rica, 1: Estratigrafía.* Editorial de la Universidad de Costa Rica, San José, 324 pp.

Vail, P.R., Mitchum, R.M., Jr., Todd, R.G., Widmier, J.M., Thompson, S., III, Sangree, J.R., Dubb, J.N. & Haslid, W.G. (1977) Seismic stratigraphy and global changes of sea-level (parts 1–6). In: *Seismic Stratigraphy — Applications to Hydrocarbon Exploration* (Ed. Payton, C.E.). Mem. Am. Assoc. Petrol. Geol. 26 49–205.

Vail, P.R. & Todd, G.R. (1981) North Sea Jurassic unconformities, chronostratigraphy and sea-level changes from seismic stratigraphy. In: *Petroleum Geology of the Continental Shelf, Northwest Europe* (Eds Cling, J.V. & Hobson, C.D.) pp. 216–235. Institute of Petroleum, London.

Van Wagoner, J.C., Posamentier, H.W., Mitchum, R.M., Vail, P.R., Sarg, J.F., Loutit, T.S. & Hardenbol, J. (1988) An overview of the fundamentals of sequence stratigraphy and key definitions. In: *Sea-level Changes: an Integrated Approach* (Eds Wilgus, C.K., Hastings, B.S., Kendall, C.G.St.C., Posamentier, H.W., Ross, C.A. & van Wagoner, J.). Soc. Econ. Paleontol. Mineral., Spec. Publ. 42, 39–46.

Wildberg, H. (1984) Der Nicoya-Komplex, Costa Rica, Zentralamerika: Magmatismus und Genese eines polygenetischen Ophiolith-Komplexes. *Muensterche Forsch. Geol. Palaeontol.* **62**, 1–123.

*Spec. Publs. int. Ass. Sediment.* (1991) **12**, 259–272

# Depositional sequences associated with equilibrium coastlines in the Neogene of southwestern Nicaragua

WOLFGANG KOLB *and* HANNELORE SCHMIDT

*Institut für Geologie und Palaontologie, Universitat Stuttgart, Böblinger Strasse 72, D 7000 Stuttgart 1, Germany*

## ABSTRACT

The Neogene shallow-marine 'El Fraile' basin in southwestern Nicaragua reveals three depositional sequences, each comprising shallow-shelf to shoreface deposits. In places, coarse clastic fan-delta complexes occur owing to local tectonic uplift combined with volcanic activity. The depositional sequences are the product of a complex interaction between eustatic sea-level fluctuation and tectonic activity. Systems tracts and their internal facies association, however, are mainly controlled by local parameters: local sediment supply, subsidence/uplift, and relative sea-level changes.

## INTRODUCTION

Throughout the latest Eocene to early Oligocene, uplift of the Central American island arc resulted in extended emergent stretches within the northern and central parts of the fore-arc area. The 30 Ma sea-level lowering (Vail *et al.*, 1977) enhanced this effect, so that sedimentation did not resume until the latest Oligocene.

During the Miocene, several shallow-water sedimentary basins formed in the fore-arc area bounded by normal or strike-slip faults. In Costa Rica, the coast consisted of shallow-marine embayments during the Neogene (Schmidt, 1989), while in Nicaragua the Miocene depositional sequences indicate an open shoreline. Previous workers (Zoppis & Del Giudice, 1958; Paz Rivera, 1964; Kuang, 1971; Parson Corporation, 1972) mainly described these sediments in terms of lithostratigraphy. The present study analyses facies development of the Miocene shallow-water sedimentary environments in southwestern Nicaragua with regard to the sequence-stratigraphic concept of Vail *et al.* (1977), Vail & Posamentier (1988), and Haq *et al.* (1988).

## GEOLOGICAL SETTING

McBirney & Williams (1965) subdivide southwestern Nicaragua into three morphotectonic units (Figs 1, 2). The *Interior Highland* consists of Tertiary lavas and tuffs, which are genetically related to similar Tertiary volcanics in El Salvador, Honduras, and Guatemala. The *Nicaragua Depression* is a NW–SE-striking graben structure extending from the Gulf of Fonseca in the north to the Costa Rican Caribbean Coastal Plain in the south. The Pacific Coastal Plain cuts across the c. 6000 m of Upper Cretaceous to Tertiary sedimentary fill of the *Nicaragua Trough* (Fig. 1). The trough extends from the Nicaragua Depression to the Pacific coast and continues on to the shelf. The strata strike NW–SE and dip 5–25° to the southwest. The sediments are not deformed, except for slight folding in the Rivas anticline. According to Weyl (1980), the Nicaragua Trough represents a tilted fault-block. The NW–SE-striking fault systems extend into Pleistocene deposits; some are still active.

In the Nicaragua Trough, sedimentation commenced with turbiditic deep-water sediments of Late Cretaceous to Eocene age. Previous workers refer to them as the Rivas and Brito Formations. Basin configuration and sedimentation dynamics of these sediments are described by Astorga (1987). The turbidite complexes are overlain by shelf deposits of probably early Oligocene age (J. Winsemann, personal communication, 1989) (Masachapa Formation; cf. Wilson, 1941; Auer, 1942). Nearshore siliclastic

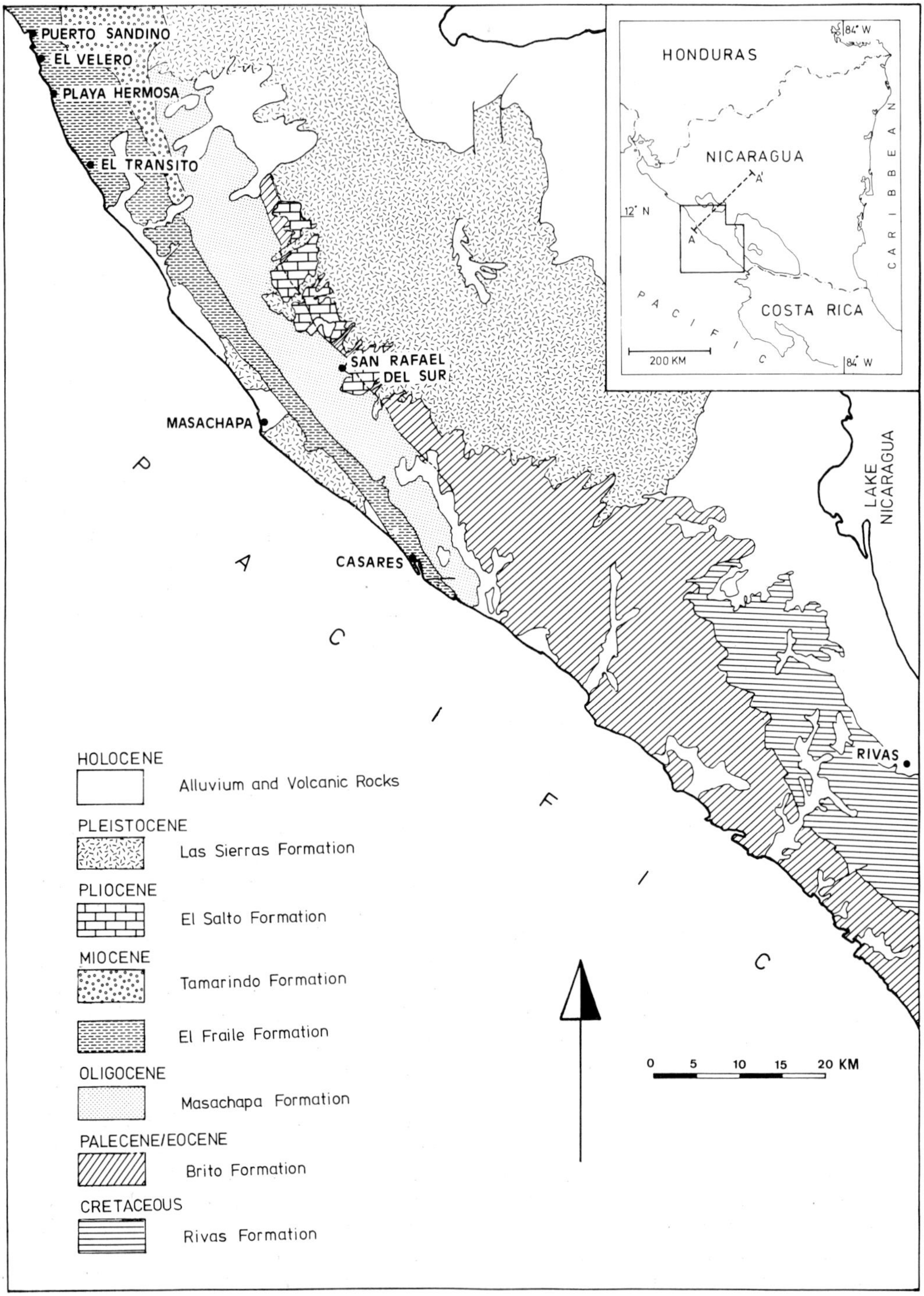


**Fig. 1.** Geological map of the Nicaragua Trough (modified after Weyl, 1980). (Line A – A′ in the inset indicates location of cross-section in Fig. 2.)

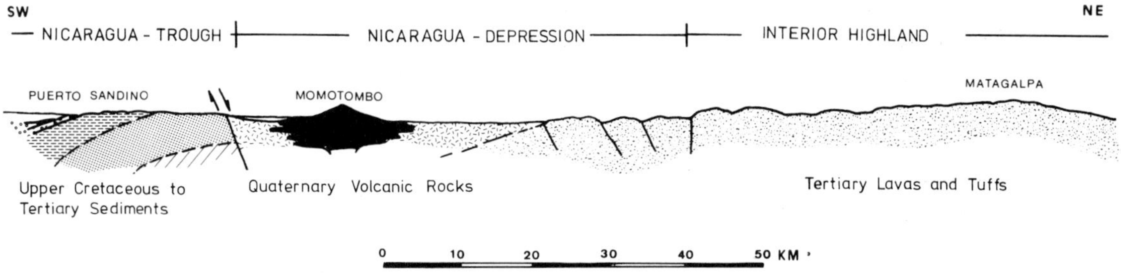


**Fig. 2.** Generalized geological section through southwestern Nicaragua (redrawn from McBirney & Williams, 1965).

deposits first appear in the Miocene (El Fraile Formation; cf. Wilson, 1941). The upper part of the El Fraile Formation seems to reach the Pliocene, based on key-bed correlation with similar deposits that crop out along the Pacific coast of Costa Rica (Schmidt, 1989; see paper by Schmidt & Seyfried). In the northeastern part of the Nicaragua Trough, these sediments interfinger laterally (Weyl, 1980) with continental pyroclastic deposits and tuffs (Tamarindo Formation; cf. Wilson, 1941). The deposits of siliciclastic shallow-water sedimentary environments dominate in the middle Miocene to early Pliocene of the fore-arc area of southwestern Nicaragua. In the late Pliocene, the flooding of a very shallow half-graben, probably related to the start of subsidence in the Nicaragua Depression (Kolb, 1990), produced a succession of mixed carbonate–clastic systems, which consist of calcarenites, marls, coquinas, and oyster reefs (El Salto Formation; cf. Wilson, 1941).

The Nicaragua Trough became emergent after the deposition of the shallow-marine sediments of the El Fraile and El Salto Formations. The Pleistocene cover consists of tuffs, lahars, ignimbrites, and lavas (Las Sierras Formation; cf. Wilson, 1941). According to Weyl (1980), these rocks are related to the Quaternary volcanism of the Nicaragua Depression.

## Study area

The present study deals with the siliciclastic shallow-marine El Fraile Formation, which formed during the Miocene and early Pliocene in the fore-arc area of the southern Central American island arc. Outcrops occur along the Pacific coast between Puerto Sandino (12°10′N) and Casares (11°36′N) (cf. Fig. 1). The strata strike almost parallel to the shoreline, dipping 10–25° to the southwest. El Fraile Formation consists of 100 m of fine to coarse-grained sandstones, tuffaceous siltstones, conglomerates, and volcaniclastic breccias. Facies development reveals depositional environments ranging from shallow shelf to shoreface. Pronounced prograding–retrograding patterns within this succession account for repeated regressive–transgressive intervals. An idealized section gives an outline of the overall facies architecture (Fig. 3); individual outcrop areas along the wave-cut platform are summarized in three general sections (Figs 5–8).

## Facies relationships

### *Coastal plain deposits*

In the Playa Hermosa section, sedimentation of the El Fraile Formation begins with 3 m of alternating lenticularly bedded, inversely graded, matrix-supported conglomerates. The deposits mainly consist of subangular volcanic rock fragments, tuff clasts, and scoria. Individual sets are characterized by non-erosive contacts. The overall structures, as well as the general lack of fossils or biomicritic intercalations, suggest that the deposits are subaerial high-viscosity debris flows, which are mainly described from middle and upper levels of alluvial fans (Hooke, 1967; Steel *et al.*, 1977). The conglomerates pass upward into 2 m of alternating sets of channelized coarse-grained sandstones — each individual set comprising a conglomeratic channel-floor lag — and parallel laminated fine to medium-grained sandstones with occasional clay drapes. This succession is interpreted as meandering channel fills with intercalated crevasse splay deposits.

The basal part of El Velero section is made up of 2–5 m of pyroclastic deposits extending laterally for

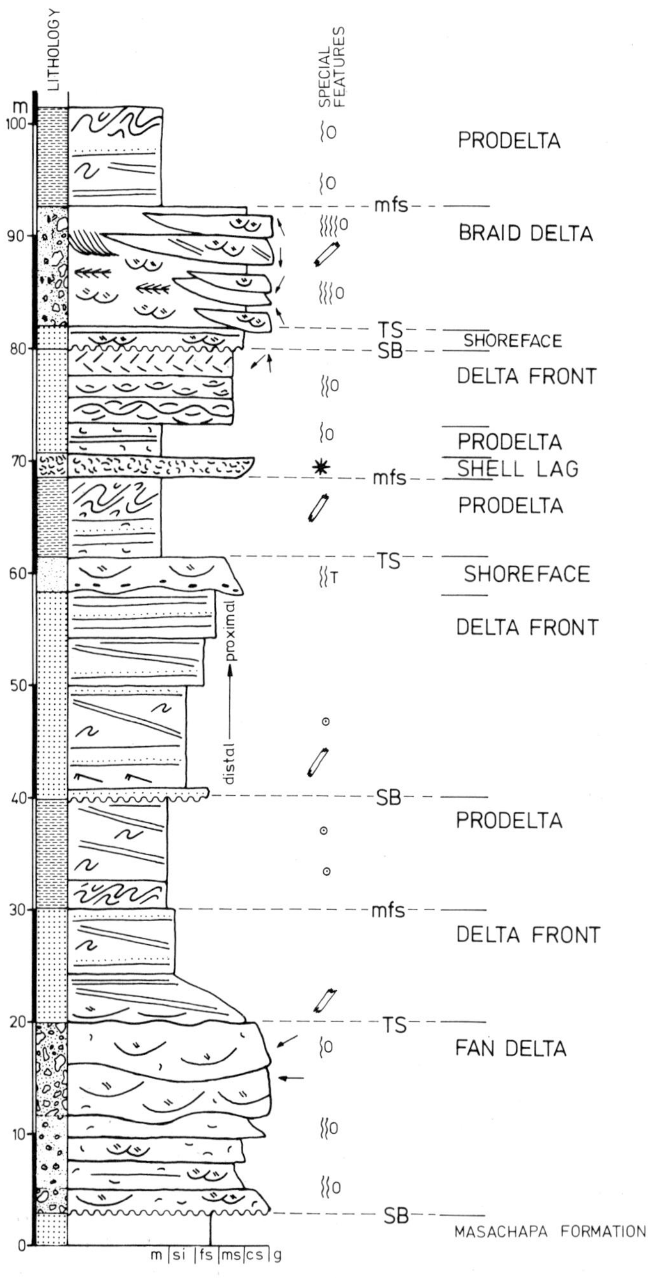


**Fig. 3.** Generalized section of the El Fraile Formation (key: see Fig. 4).

about 300 m. Volcaniclastic and igneous rock fragments, volcanic glass, and chips of charred wood are dispersed in a fine to coarse-grained matrix. Massive beds are common, some of them exhibit convolute lamination, normal, and inverse graded bedding. Both the structures and the type of stratification suggest a pyroclastic flow rather than a lahar origin (Fisher & Schmincke, 1984; Cas & Wright, 1987). According to Cas & Wright (1987), the occurrence of charred wood and scoria points to a high emplacement temperature (>400°).

### *Fan-delta and braid-delta deposits*

Sedimentation of the El Fraile Formation within the Casares section starts with a fan-delta complex, un-

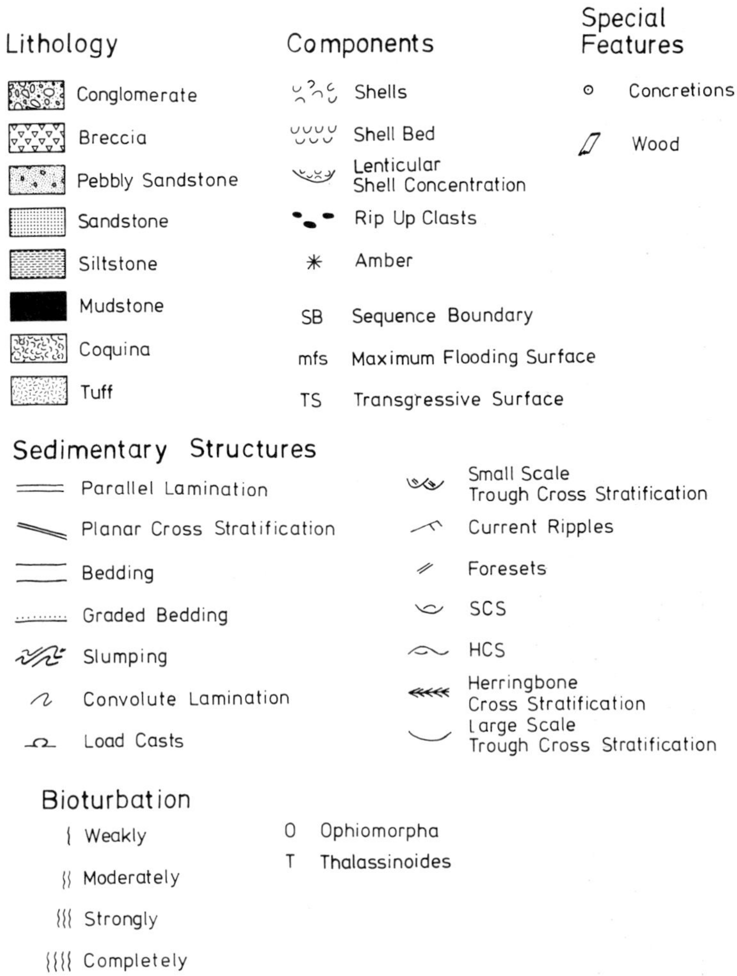


**Fig. 4.** Key to Figs 3, 6, 7 and 8.

conformably resting upon tuffaceous siltstones of the Oligocene Masachapa Formation. The fan-delta deposits mainly consist of a fan-shaped succession of unstratified clast-supported coarse-grained conglomerates with intercalated coarse-grained pebbly sandstones, made up mainly of subangular volcanic rock fragments. Such sediments are probably the deposits of alternating sheet flows and debris flows (cf. Rust & Koster, 1984; McPherson *et al.*, 1988). Thin intercalations of low-angle, planar and trough cross-stratified, coarse-grained sandstones within this succession are probably related to processes of shallow marine reworking (McPherson *et al.*, 1988). The submarine character of these deposits is documented by the occurrence of *Tridacna* sp., *Ostrea* sp., *Pecten* sp., *Conus* sp., *Oliva* sp., *Strombus* sp., corals, and hydrozoans, which may have colonized shallow-marine areas of the fan-delta complex. Bioturbation in the coarse-grained sandstone units is attributed to *Ophiomorpha nodosa*. The fan-delta complex extends laterally for about 2–3 km.

Within the upper part of the Playa Hermosa section, a succession of conglomeratic channel fills several metres wide containing large trunks of silicified wood and coarse-grained sandstones is interpreted as a braid-delta complex. These vertically and horizontally interfingering channel fills consist of well-rounded, fine to medium-grained volcaniclastic conglomerates, which erosively cut into the underlying sandstone units. The predominant sedimentary structures within the channel fills are high-angle planar cross-stratified and normally graded conglomerates; foresets dip to SW–NW. The strongly diverging palaeocurrent directions may be related to tide-influenced sedimentary processes. Strong influence of tide-generated currents is documented by herringbone cross-stratification and tidal bundles within the upper portions of the coarse-grained sandstone units. The strongly channelized character of the conglomerates as well as the geometry of the depositional body point to a braid-delta complex rather than a fan-delta environment (Kerr, 1984; Rust & Koster, 1984; McPherson *et al.*, 1987,1988).

### *Deltaic deposits*

Within all measured sections of the El Fraile Formation, deltaic deposits occur with distinct prograding and retrograding patterns. None of the deltaic successions reveals delta-plain facies. This could be a result of outcrop conditions, but since the majority of deltaic successions within the Neogene fore-arc area of the southern Central American island arc lack delta topset facies (Schmidt, 1989), we suppose that most of these deltaic environments never reached a subaerial stage. Hence we suggest that these deltaic environments (including the deltaic successions within the El Fraile Formation) have to be regarded as estuarine deltas. Delta-front sediments within the El Fraile succession are composed of well-sorted medium to coarse-grained sandstones, mainly consisting of volcaniclastic rock fragments and isolated subangular quartz grains. Prodelta deposits mainly consist of tuffaceous siltstone.

The lower portion of the El Velero section displays a fining upward succession of 10 m of delta-front sediments. The lower part of these deposits is composed of well-sorted, coarse-grained sandstones. Massive beds are common, whilst decimetre to metre-thick layers display low-angle planar cross-stratification, trough cross-stratification, parallel lamination, and tempestite intercalations. Upsection, this facies changes into poorly sorted medium to fine-grained sandstones with convolute lamination. This succession could be the result of a retrogradation from proximal to distal delta-front environment (Coleman & Prior, 1982; Wright, 1985). A sequence of 20 m of coarsening upward delta-front deposits can be observed at the 24-m level in the El Velero section. This succession provides a mirror image of the deltaic environment described above. Here the lower portions consist of poorly sorted, parallel-laminated, medium-grained sandstones, which pass upsection into well-sorted, trough cross-stratified, coarse-grained sandstones. These deposits are interpreted as a prograding delta-front environment. Another shallowing upward delta-front environment can be observed in the Playa Hermosa section. Layers within the lower portion of this sequence display hummocky cross-stratification (HCS), accounting for shallow-marine processes between storm wave-base and fair-weather wave-base (Hamblin & Walker, 1979; Walker, 1984). Within the upper portions of this succession there are swaley cross-stratified sandstones (SCS), suggesting deposition in water depths above fair-weather wave-base (Leckie & Walker, 1982; Elliot, 1986). In the proximal part of the delta front, strata bundles (*sensu* Nummedal *et al.*, 1986) point to bimodal transport directions. These structures suggest the existence of migrating longshore bars on the lower shoreface (B. Flemming, personal communication, 1989).

Prodelta environments within the El Fraile Formation are made up of several tens of metres of thick monotonous tuffaceous siltstones and fine-grained sandstones. Except for occasional thin tempestite layers, these strata reveal parallel lamination, load casts, slumps, and minor angular unconformities.

### *The shell lag*

Within all measured sections of the El Fraile Formation a biodetrital conglomeratic layer 0.5–1 m thick occurs, which can also be observed in other Neogene sedimentary basins within the southern Central American fore-arc area farther south (Costa Rica) (see paper by Schmidt & Seyfried). In all sections the layer consists of medium to coarse-grained sandstone containing a high concentration of crustaceans, gastropods, and bivalves (*Architectonica nobilis, Ficus carbacea, Melongena consors, Turritella altilira, Turritella gatunensis, Xancus* cf. *volidus hysterus, Conus* sp., *Cancellaria* sp., *Cymia* sp., *Phos (Cymatophos)* sp., *Clementia dariena, Pitar* sp., *Tellina* sp., *Ostrea* sp.). The most outstanding feature of this layer is a large quantity of amber cobbles.

In a sequence-stratigraphic context, biodetrital layers are often interpreted as condensed sections atop a transgressive systems tract ('maximum flooding surface'; *sensu* Loutit *et al.*, 1988; Vail & Sangree, 1988), which accumulate owing to low net sedimentation during transgression ('hiatal shell concentration'; *sensu* Kidwell *et al.*, 1986). According to Kidwell & Aigner (1985), Kidwell *et al.* (1986), and Beckvar & Kidwell (1988), a high percentage of densely packed disarticulated bivalve shells, as well as good sorting of matrix, indicates a condensed section rather than a storm rip accumulation. We interpret this amber-bearing shell concentration as a maximum flooding surface. According to D. Schlee (personal communication, 1989) the large quantity of amber cobbles can be traced to the fact that a high sea-level probably led to the flooding of coastal forests; the trees affected (e.g. by sea water and roots washed bare of soil) reacted by increasing resin production.

### *Shoreface deposits*

Shoreface deposits within the El Fraile Formation consist of centimetre to decimetre-scale beds of fine to coarse-grained sandstone and pebbly sandstone. Sedimentary structures are composed of parallel lamination, low-angle planar cross-stratification, and small-scale trough cross-stratification. Graded beds with sharp-based erosional contacts and lenticular shell concentrations are interpreted as tempestite layers and storm rip accumulations. According to Niedoroda *et al.* (1985) low-angle planar cross-stratification and small-scale trough cross-stratification suggest deposition under a high-energy oscillatory flow regime in the nearshore zone. Sediments are often intensely bioturbated, revealing *Thalassinoides* and *Ophiomorpha nodosa* burrows.

## The sedimentary suite

### *El Velero section*

The basal part of this section (Fig. 6) is made up of 2–5 m of pyroclastic deposits extending laterally for about 300 m south of Puerto Sandino. These deposits are probably related to Miocene ignimbrites and tuffs of the Tamarindo Formation, which interfinger laterally with the shallow-marine strata of the El Fraile Formation just north of Puerto Sandino. At the 5-m level the pyroclastic deposits are capped by a well-defined erosive unconformity, overlain by a fining upward series of 10 m of retrogradational delta-front siltstones and sandstones. The delta-front sandstones are capped by a monotonous series of thin-bedded tuffaceous siltstones commonly with synsedimentary deformation (mud diapirs, convolute lamination, load casts, slumps, and minor angular unconformities), which are interpreted as prodelta deposits. At 24 m these siltstones are capped by an erosive unconformity. The overlying strata consist of poorly sorted, fine-grained laminated sandstones exhibiting convolute lamination. These deposits gradually pass into well-sorted, medium to coarse-grained sandstones with low-angle planar cross-stratification, trough cross-stratification, and occasional storm layers. This sequence is interpreted as progradational delta-front environment. The overlying sediments consist of tuffaceous prodelta siltstones capped by amber-bearing shell lag. The top of the El Velero section is made up of fine-grained prodelta sandstones and coarse-grained parallel-laminated shoreface deposits.

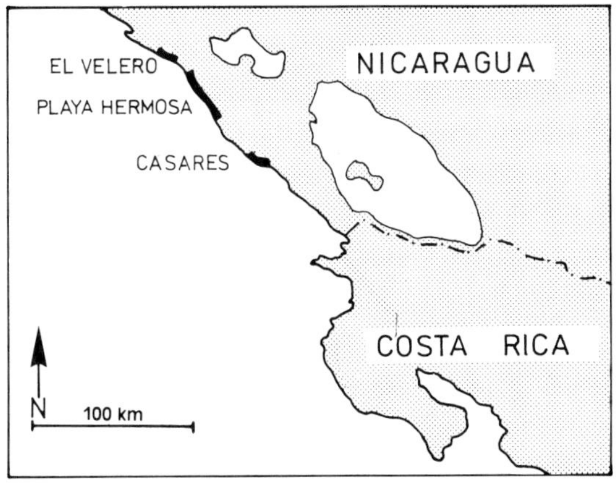


**Fig. 5.** Location of sections (Figs 6–8). (Black bars indicate adjoining outcrop areas along the wave-cut platform.)

### *Playa Hermosa section*

The Playa Hermosa section (Fig. 7) starts with coastal plain deposits. Trough cross-stratified and horizontally laminated sands, and normally graded conglomerates are interpreted as conglomeratic, braided channel fills alternating with interchannel deposits. The top of this series exhibits fine-grained sandstones containing wood fragments and occasional clay drapes representing floodplain and crevasse splay deposits. This unit is separated by an unconformity from the overlying sediments, which are made up of medium to coarse-grained upper shoreface sediments. The overlying tuffaceous prodelta siltstones contain the amber-bearing shell lag.

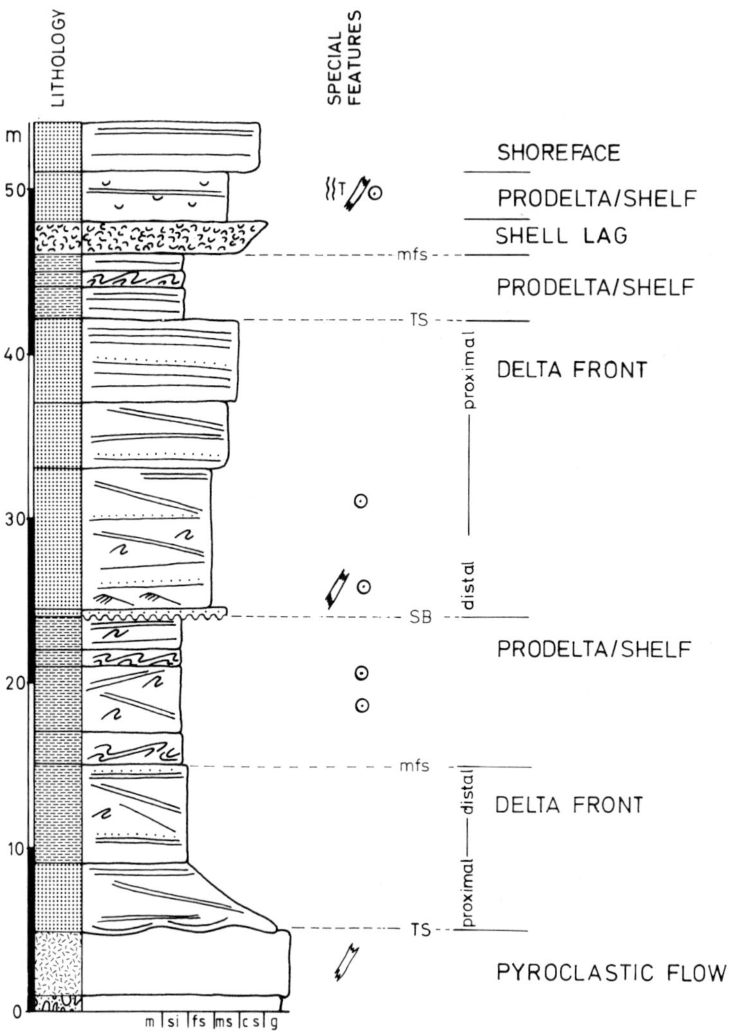


**Fig. 6.** El Velero section (outcrops extend over 5 km along the wave-cut platform 2 km south of Puerto Sandino). Middle Miocene to early Pliocene, El Fraile Formation, southwestern Nicaragua. See Fig. 4 for key.

The prodelta siltstones are topped by a progadational delta-front environment revealing a shallowing upward trend with HCS, SCS, and low-angle planar cross-stratification suggesting decreasing water depth up to fair-weather wave-base. Separated by a well-defined erosive unconformity, the overlying strata consist of coarse-grained parallel-laminated shoreface sediments. The overlying braid-delta complex consists of alternating conglomeratic channel fills and medium to coarse-grained sandstones. The sandstones display herringbone cross-stratification as well as coarse clastic tidal-bundles related to major subtidal channels. The upper portion of the braid-delta succession displays densely packed *Ophiomorpha nodosa* burrows. The braid-delta complex is overlain by monotonous prodelta siltstones revealing convolute lamination, minor angular unconformities, and occasional tempestite layers.

### *Casares section*

In the southernmost part of the basin, sedimentation of the El Fraile Formation starts with coarse clastic conglomerates of a fan-delta complex, which rest unconformably upon sediments of the Oligocene Masachapa Formation (Fig. 8). The exposed basal part of the section represents a conglomeratic in-

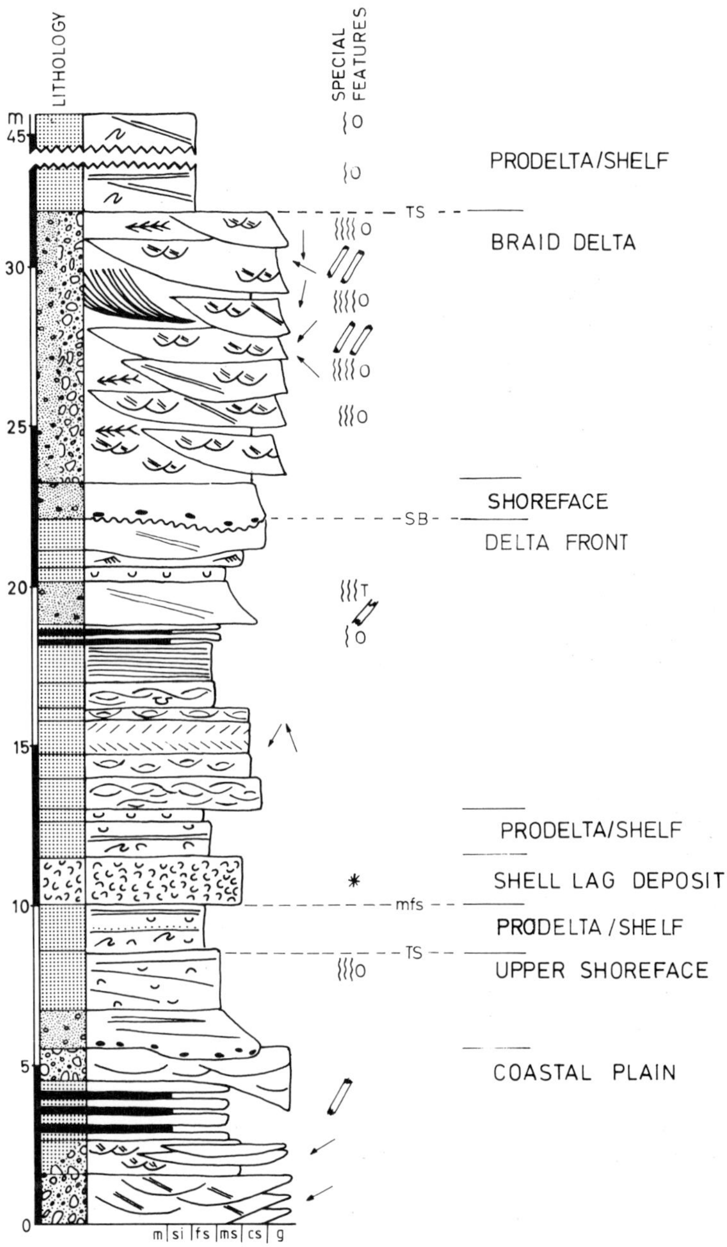


**Fig. 7.** Playa Hermosa section (outcrops extend over 20 km along the wave-cut platform near the village of El Transito). Middle Miocene to early Pliocene, El Fraile Formation, southwestern Nicaragua. See Fig. 4 for key.

fill of one of the major channels. The overlying succession reveals a deepening upward sequence ranging from proximal to distal delta-front deposits, topped by tuffaceous prodelta siltstones with slumps and convolute lamination. An erosive unconformity separates these prodelta deposits from overlying coarse-grained delta-front sediments and bioturbated shoreface deposits. Tuffaceous prodelta siltstones are repeatedly intercalated. The amber-bearing shell lag is encountered 55 m above the section base. This key bed is overlain by aggradational coarse-grained shoreface deposits exhibiting parallel lamination, occasional storm layers, and lenticular shell concentrations.

## SEQUENCE-STRATIGRAPHIC INTERPRETATION AND CORRELATION

The stratal relationship and stacking pattern within the sedimentary succession of the El Fraile Formation suggest an organization into three sequences. These sequences are bounded by laterally traceable discontinuity surfaces. In terms of sequence stratigraphy the sequences and discontinuity surfaces will correspond to depositional sequences and sequence boundaries, respectively (Fig. 9).

The measured sections were correlated using the amber-bearing shell lag. This lag appears only once in each section and therefore is considered to represent a unique event. Observations in contemporary fore-arc coastal embayments of Costa Rica confirm this assumption (Schmidt, 1989; see chapter by Schmidt & Seyfried). The comparison with the Costa Rican basins permits a more precise chronostra-

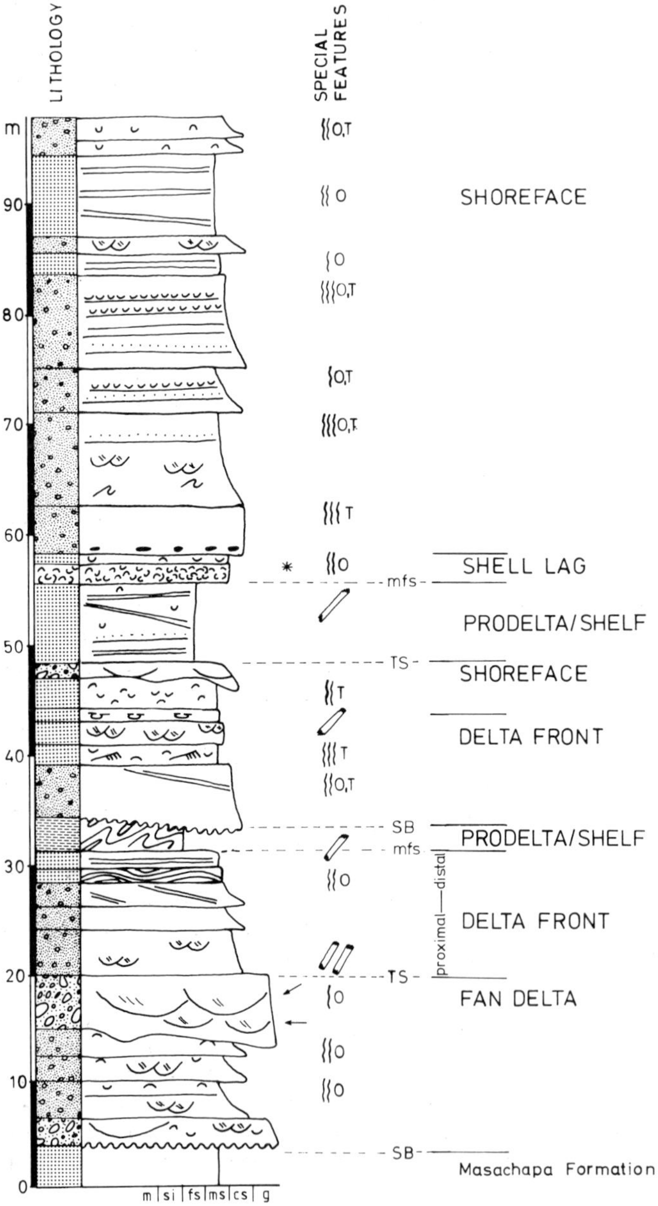


**Fig. 8.** Casares section (outcrops extend over 8 km along the wave-cut platform near the village of Casares). Middle Miocene to early Pliocene, El Fraile Formation, southwestern Nicaragua. See Fig. 4 for key.

tigraphic subdivision of the El Fraile succession. Age determinations in the Costa Rican deposits reveal a late Miocene to early Pliocene age for the sequences upsection of this key bed. We presume the same age for the corresponding Nicaraguan deposits.

**Depositional sequence I**

The basal sequence boundary of this sequence is well displayed in the Casares section: conglomeratic channel fills of a fan-delta complex rest unconformably upon tuffaceous siltstones of the Oligocene Masachapa Formation.

*Lowstand systems tract*

Sea-level lowstand and local tectonic uplift in the southern part of the basin (Casares section) resulted in conglomeratic fan-delta deposits, which erosively cut into the underlying sediments. In the northern part of the basin (El Velero section), the lowstand systems tract consists of pyroclastic deposits that extend over parts of the coastal plain.

*Transgressive systems tract*

The boundary between lowstand systems tract and transgressive systems tract is marked by a transgressive surface. This unconformity was formed by erosion and reworking of sediments during the transgression and is well displayed in the El Velero and Casares sections. A fining upward trend within delta-front sediments is obvious in both these sections, documenting an increase in water depth. In both sections these sediments are interpreted as retrogradational delta-front environments.

*Highstand systems tract*

A sudden change in grain size characterizes the boundary between transgressive systems tract and highstand systems tract in the El Velero and Casares sections (maximum flooding surface); medium to coarse-grained delta-front deposits are overlain by silty prodelta–shelf sediments.

**Depositional sequence II**

The sequence boundary between depositional sequences I and II is marked by an abrupt facies change from prodelta–shelf sediments to coarse-grained shoreface and delta-front deposits. This boundary appears throughout the basin as an erosional surface suggested to have formed due to a rapid sea-level fall.

*Lowstand systems tract*

In the El Velero and Casares sections this lowstand systems tract consists of prograding delta-front and shoreface sediments deposited during sea-level lowstand. These sections reflect more or less offshore positions within the basin, whereas more marginal conditions are encountered in the Playa Hermosa section. Here the lowstand systems tract first consists of coastal plain deposits, which pass upsection into upper shoreface sandstones owing to an onset of sea-level rise.

*Transgressive systems tract*

The succeeding transgression deposited silty prodelta–shelf sediments that rest upon coarse-grained delta front and shoreface deposits of the lowstand systems tract. The transgressive surface is evident in all three sections as an erosional unconformity.

*Highstand systems tract*

Sea-level rise caused reworking of nearshore sediments and generated the amber-bearing shell-lag deposit (maximum flooding surface) found in all three sections. The positive interaction between tectonic uplift, subsidence, and sediment input is illustrated by the great thickness of the highstand systems tract in the Playa Hermosa and Casares sections. The Playa Hermosa section reveals a prograding delta system in spite of a high sea level. This is supposed to be created by a short-term increase of sediment input, most probably caused by tectonic uplift in the hinterland. However, in the southern part (Casares section), presumably accelerated subsidence compensates for this increased sediment input, thus resulting in the accumulation of aggradational shoreface deposits.

**Depositional sequence III**

Depositional sequence III is incomplete and crops out only in the Playa Hermosa section. The sequence boundary is indicated by an erosional unconformity separating shoreface sediments from underlying delta-front sandstones. Sea-level lowstand resulted

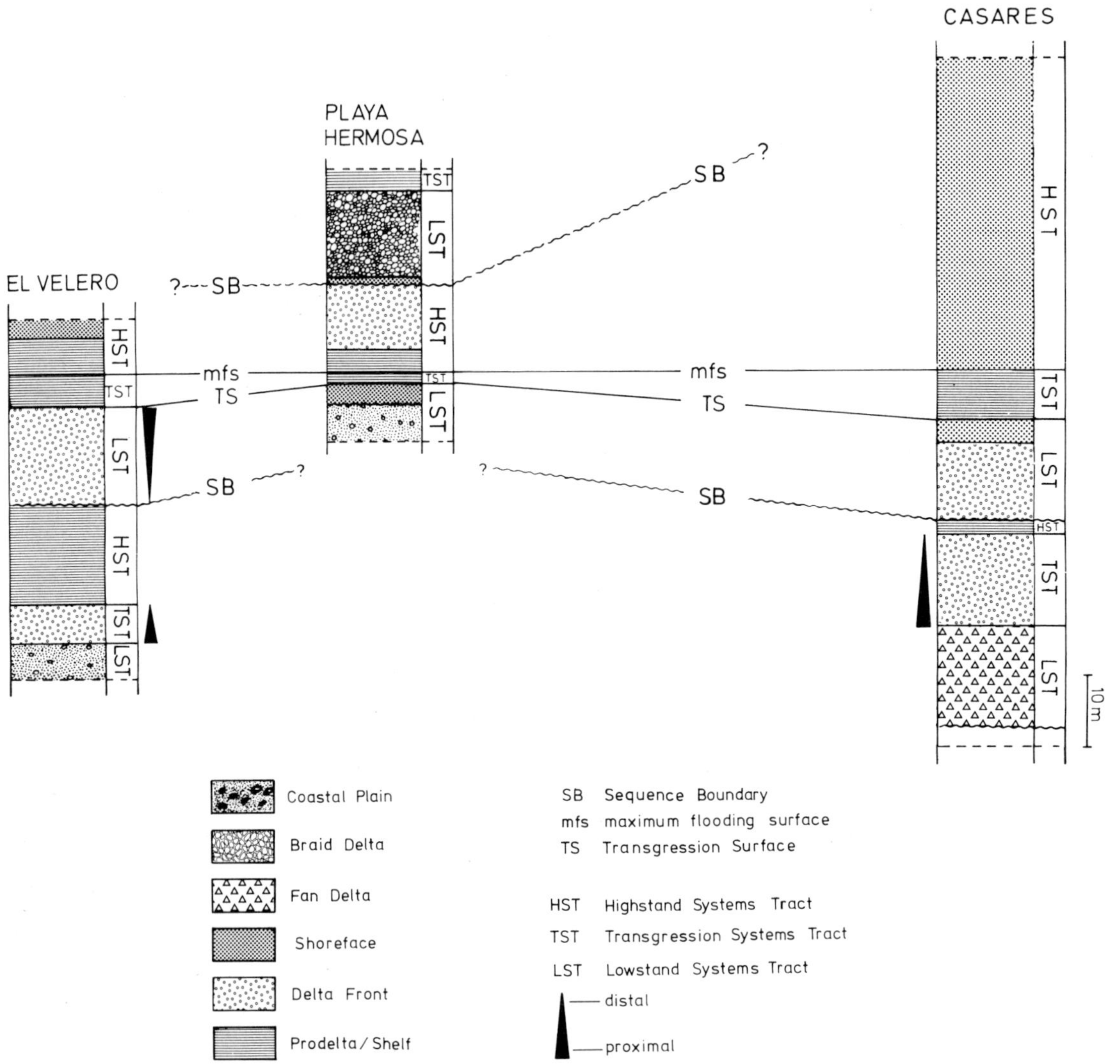


**Fig. 9.** Sequence-stratigraphic correlation of sections within the El Fraile Formation (correlation is based on the amber-bearing key-bed; 'maximum flooding surface' in depositional sequence II).

in the deposition of coarse-grained shoreface sediments overlain by a braid-delta complex. Intensely bioturbated (*Ophiomorpha nodosa*) upper portions of the braid-delta complex are probably related to episodes of low net-sedimentation during the following transgression. The braid-delta complex is overlain by tuffaceous prodelta siltstones probably deposited within the transgressive systems tract.

## Morphotectonic evolution

Early Miocene (21 Ma) sea-level fall led to shelf exposure. Fan deltas evolved in tectonically uplifted areas extending over truncated Oligocene sediments. Pyroclastic and alluvial deposits spread across the exposed shelf. Increased sediment input resulted in prograding delta systems despite a sea-level rise. In the late middle Miocene, sea-level attained its maximum and the coastline retrogressed far into the hinterland. Prograding delta systems developed during sea-level fall at the boundary from middle to upper Miocene. The subsequent lowstand initiated the deposition of a braid-delta complex.

## SUMMARY

The El Fraile Formation reflects the shallow-marine history of the Nicaraguan section of the Central American fore-arc area during Miocene to early Pliocene(?) times. The combined occurrence of coastal plain, shoreface, delta, and prodelta deposits point to a more or less equilibrated open shoreline system, which prevailed at least during the middle Miocene. This may be explained by a change in tectonic style along the plate margin, which is observed just north of the Costa Rican border: highly convergent in Costa Rica and convergent–extensional in Nicaragua and northward (Azéma & Tournon, 1979).

The El Fraile succession is organized into three depositional sequences. The bounding surfaces are interpreted as sequence boundaries (*sensu* Vail *et al.*, 1984). The comparison with contemporaneous fore-arc coastal embayments of Costa Rica implies a possible correlation of sequence boundaries with the TB1/TB2 and TB2/TB3, second-order supercycle boundaries of Haq *et al.* (1988).

Sequence development of the El Fraile succession clearly reveals that depositional sequences mainly respond directly to sea-level fluctuations. Systems tracts and their internal facies associations, however, are mainly controlled by the combined effects of variations in sediment supply, subsidence, and sea-level changes. Clearly recognizable tectonic signals occur only episodically and are restricted to a few areas with tectonic uplift.



## ACKNOWLEDGEMENTS

The field trips were sponsored by grants of the Deutscher Akademischer Austauschdienst (DAAD) and the Deutsche Forschungsgemeinschaft (DFG). Field work in Nicaragua would have been impossible without the help of Dr Hugo Moreno and his experienced staff of the Instituto Nicaragüense de la Mineria (INMINE), Managua. Fossils were determined by R. Fischer (University of Hannover) and Teresita Aguilar (University of Costa Rica). We would like to thank D.I.M. Macdonald (BAS), H.E. Clifton (USGS), and P.B. Wignall (University of Leeds) who provided helpful comments on the manuscript.


## REFERENCES


Astorga, A. (1987) *El Cretácico Superior y el Paleógeno de la vertiente Pacífica de Nicaragua meridional y Costa Rica septentrional: orígen, evolución y dinámica de las cuencas profundas relacionadas con el margen convergente de Centroamérica*. Unpublished tesis de Licenciatura, University of Costa Rica, San José, 250 pp.

Auer, W.F. (1942) Summary report: Paleontology of the Pacific coast area. Cited in Parson Corporation (1972) The geology of Western Nicaragua. In: *Nicaragua Tax Improvement and Natural Resources Inventory Project, Final Technical Report*, Vol. IV, Parson Corporation, Managua, 220 pp.

Azéma, J. & Tournon, J. (1979) The Guatemalan margin, the Nicoya complex, and the origin of the Caribbean plate. In: *Initial Reports of the Deep Sea Drilling Project*, Vol. 67 (Eds Aubouin, J., von Huene, R., *et al.*) pp. 739–745. US Government Printing Office, Washington, DC.

Beckvar & Kidwell (1988) Hiatal shell concentrations, sequence analysis, and sea-level history of a Pleistocene coastal alluvial fan, Punta Chueca, Sonora. *Lethaia* **21**, 257–270.

Cas, R.A.F. & Wright, J.V. (1987) *Volcanic Successions, Modern and Ancient*. Allen & Unwin, London, 528 pp.

Coleman, J.M. & Prior, D.B. (1982) Deltaic environments of deposition. In: *Sandstone Depositional Environments* (Eds Scholle, P.A. & Spearing, D.). Mem., Am. Assoc. Petrol. Geol. 31, 139–178.

Elliott, T. (1986) Siliciclastic shorelines. In: *Sedimentary Environments and Facies*, 2nd edn (Ed. Reading, H.G.) pp. 155–188. Blackwell Scientific, Oxford.

Fisher, R.V. & Schmincke, H.-U. (1984) *Pyroclastic Rocks*. Springer-Verlag, Berlin, 472 pp.

Hamblin, A.P. & Walker, R.G. (1979) Storm-dominated shallow marine deposits: the Fernie–Kootenary (Jurassic) transition, southern Rocky Mountains. *Can. J. Earth Sci.* **16**, 1673–1690.

Haq, B.U., Hardenbol, J. & Vail, P.R. (1988) Mesozoic and Cenozoic chronostratigraphy and eustatic cycles. In: *Sea-level Changes: an Integrated Approach* (Eds Wilgus, C.K., Hastings, B.S., Kendall, C.G. St. C., Posamentier, H.W., Ross, C.A. & van Wagoner, J.). Soc. Econ. Paleontol. Mineral., Spec. Publ. 42, 71–108.

Hooke, R. LeB. (1967) Processes on arid region alluvial fans. *J. Geol.* **75**, 438–460.

Kerr, D.R. (1984) Early Neogene continental sedimentation in the Vallecito and Fish Creek Mountains, western Salton Trough, California. *Sediment. Geol.* **38**, 217–246.

Kidwell, S.M. & Aigner, T. (1985) Sedimentary dynamics of complex shell beds: implications for ecologic and evolutionary patterns. In: *Sedimentary and Evolutionary Cycles* (Eds Bayer, U. & Seilacher, A.) pp. 382–395. Springer-Verlag, Berlin.

Kidwell, S.M., Fürsich, F.T. & Aigner, T. (1986) Conceptual framework for the analysis and classification of fossil concentrations. *Palaios* **1**, 228–238.

Kolb, W. (1990) *Die neogene Schelfsedimentation in Südwest-Nicaragua: Ablagerungssequenzen und paläo-*

*geographische Szenarien.* Unpublished PhD dissertation, University of Stuttgart, 163 pp.

Kuang, S.J. (1971) Estudio geológico del Pacífico de Nicaragua. *Catastro de Inventario de Recursos Naturales, División de Geología, Inform. Geol.* **3**, 1–101.

Leckie, D.A. & Walker, R.G. (1982) Storm- and tide-dominated shorelines in Cretaceous Moosebar–lower gates interval–outcrop equivalents of deep basin gas trap in Western Canada. *Bull., Am. Assoc. Petrol. Geol.* **66**(2), 138–157.

Loutit, T.S., Hardenbol, J., Vail, P.R. & Baum, G.R. (1988) Condensed sections: the key to age dating and correlation of continental margin sequences. In: *Sea-level Changes: an Integrated Approach* (Eds Wilgus, C.K., Hasting, B.S., Kendall, C.G.St.C., Posamentier, H.W., Ross, C.A. & van Wagoner, J.). Soc. Econ. Paleontol. Mineral., Spec. Publ. 42, 183–213.

McBirney, A.R. & Williams, H. (1965) Volcanic history of Nicaragua. *Univ. Calif. Publ. Geol. Sci.* **55**, 1–65.

McPherson, J.G., Shanmugam, G. & Moiola, R.J. (1987) Fan-deltas and braid deltas: varieties of coarse-grained deltas. *Geol. Soc. Am. Bull.* **99**, 331–340.

McPherson, J.G., Shanmugam, G. & Moiola, R.J. (1988) Fan deltas and braid deltas: conceptual problems. In: *Fan Deltas: Sedimentology and Tectonic Settings* (Eds Nemec, W. & Steel, R.J.) pp. 14–22. Blackie & Son, London.

Niedoroda, A.W., Swift, D.J.P. & Hopkins, T.S. (1985) The shoreface. In: *Coastal Sedimentary Environments*, 2nd edn (Ed. Davis, R.A., Jr.) pp. 533–615. Springer-Verlag, New York.

Nummedal, D., Swift, D.J.P. & Wright, R. (1986) Depositional sequence and shelf sandstones in Cretaceous strata of the San Juan Basin, New Mexico. *Field Guide 7th Ann. Res. Conf. GCS/SEPM*, 277 pp.

Parson corporation (1972) The geology of Western Nicaragua. *Nicaragua Tax Improvement and Natural Resources Inventory Project, Final Technical Report,* Vol. IV. Parson Corporation, Managua, 220 pp.

Paz Rivera, N. (1964) Reconocimiento geológico de la costa del Pacífico de Nicaragua. *Bol. Serv. Geol. Nac. Nicaragua* **8**, 69–87.

Rust, B.R. & Koster, E.H. (1984) Coarse alluvial deposits. In: *Facies Models*, 2nd edn (Ed. Walker, R.G.). Geosci. Can. Reprint Ser. 1, 53–69.

Schmidt, H. (1989) *Sequenzstratigraphie des neogenen Inselbogen-Schelfes im 'Forearc'-Bereich Costa Ricas und Nicaraguas.* Unpublished PhD dissertation, Technische Universität Berlin, 250 pp.

Steel, R.J., Maehle, S., Nilsen, H., Roe, S.L. & Spinnangr, A. (1977) Coarsening-upward cycles in the alluvium of Hornelen Basin (Devonian), Norway: sedimentary response to tectonic events. *Geol. Soc. Am. Bull.* **88**, 1124–1134.

Vail, P.R. & Posamentier, H.W. (1988) Principles of sequence stratigraphy. In: *Sequences, Stratigraphy, Sedimentology: Surface and Subsurface* (Eds James, D.P. & Leckie, D.A.). Mem., Can. Soc. Petrol. Geol. 15, 572.

Vail, P.R. & Sangree, J.B. (1988) Sequence stratigraphy concepts. *Short course in Conjunction with the 'Sequence, Stratigraphy, Sedimentology: Surface and Subsurface' Conference, Course Notes.* Can. Soc. Petrol. Geol., Calgary, September 1988.

Vail, P.R., Mitchum, R.M., Jr., Todd, R.G., Widmier, J.M., Thompson, S., III, Sangree, J.R., Dubb, J.N. & Hatlelid, W.G. (1977) Seismic stratigraphy and global changes of sea-level. In: *Seismic Stratigraphy – Applications to Hydrocarbon Exploration* (Ed. Payton, C.E.). Mem., Am. Assoc. Petrol. Geol. 26, 49–205.

Vail, P.R., Hardenbol, J. & Todd, R.G. (1984) Jurassic unconformities, chronostratigraphy and sea-level changes from seismic stratigraphy and biostratigraphy. In: *Interregional Unconformities and Hydrocarbon Accumulation* (Ed. Schlee, F.S.). Mem., Am. Assoc. Petrol. Geol. 36, 129–144.

Walker, R.G. (1984) Shelf and shallow marine sands. In: *Facies Models*, 2nd edn (Ed. Walker, R.G.). Geosci. Can., Reprint Ser. 1, 141–170.

Weyl, R. (1980) *Geology of Central America.* 371 S., Borntraeger, Berlin, 371 pp.

Wilson, T.C. (1941) *Structures and Stratigraphy of the San Rafael–El Carmen Area.* Department of Managua, Nicaragua, 53 pp.

Wright, L.D. (1985) River Deltas. In: *Coastal Sedimentary Environments*, 2nd edn (Ed. Davis, R.A., Jr.) pp. 1–76, Springer-Verlag, New York.

Zoppis, B.L. & Del Giudice, D. (1958) Geología de la costa del Pacífico de Nicaragua. *Bol. Serv. Geol. Nac. Nicaragua* **1**, 29–34.

*Spec. Publs. int. Ass. Sediment.* (1991) **12**, 273–292

# Response of deep-water fore-arc systems to sea-level changes, tectonic activity and volcaniclastic input in Central America

JUTTA WINSEMANN *and* HARTMUT SEYFRIED

*Institut für Geologie und Paläontologie, Universität Stuttgart, Böblinger Strasse 72, D-7000 Stuttgart 1, Germany*

## ABSTRACT

The incipient island-arc system of southern Central America (Cretaceous – early Oligocene) is characterized by thick turbidite systems, which mainly filled inner fore-arc troughs. Outcrop data show four, second-order depositional sequences in the deep-water sediments. The formation of these depositional sequences is strongly related to the morphotectonic evolution of the island-arc system. Each depositional sequence reflects the complex interaction between global sea-level fluctuations, sediment supply, and tectonic activity.

Strong marginal uplift and high volcaniclastic sediment supply during early to late Paleocene and late Eocene times caused the formation of coarse-grained channel-lobe systems. During late Paleocene and mid-Eocene times, fine-grained, thin-bedded turbidite systems were deposited, owing to regional subsidence and a decrease in volcanic supply. Uplift and subsidence of sediment-source areas acted as major controls on deposition of basinal cycles.

## INTRODUCTION

The fore-arc area of southern Central America belongs to the middle America arc–trench system, which extends along southern Mexico and the Central American isthmus. The Cocos plate is being subducted beneath the North American and Caribbean plates (Burbach *et al.*, 1984) and is currently being subducted beneath Costa Rica at 9 cm/yr (see Fig. 1 in paper by Seyfried *et al.*)

Subduction and related processes in southern Central America started in Late Cretaceous times. The incipient island-arc system is characterized by narrow structural basins in the inner fore-arc area, an outer arc, and a trench slope system (Fig. 1).

The sediments of the ophiolitic basement (Nicoya Complex) are mainly pelagic deposits and consist of radiolarian cherts, black shales, and deep-water limestones of Late Jurassic to Late Cretaceous age (Schmidt-Effing, 1979; Lundberg, 1982; Baumgartner *et al.* 1984; Gursky, 1984; Astorga, 1987).

From Maastrichtian to Eocene times, thick successions of volcaniclastic turbidites were deposited in the inner fore-arc basins, while thin-bedded, calcareous turbidites and slump deposits prevailed in the outer fore-arc basins. The shape of these sedimentary basins was probably controlled by a major strike-slip system, which transected the island arc perpendicularly in the region of central Costa Rica. A broader discussion of the regional geological setting is given in the chapter by Seyfried *et al.*

Because modern and ancient submarine fans show great variability in shape and facies relationship, the type and timing of turbidite events have been discussed recently by many authors with regard to sediment supply, tectonic activity, and sea-level fluctuation (Barnes & Normark, 1985; Mutti, 1985; Shanmugam *et al.*, 1985; Stow *et al.*, 1985; Mutti & Normark, 1987; Kolla & Macurda, 1988; Shanmugam & Moiola, 1988). Classifications of submarine fans have been compiled, based on their tectonic setting (Mutti & Normark, 1987; Shanmugam & Moiola, 1988).

According to the sequence-stratigraphic model, the development of submarine fans is mainly related to global sea-level lowstands and they form part of

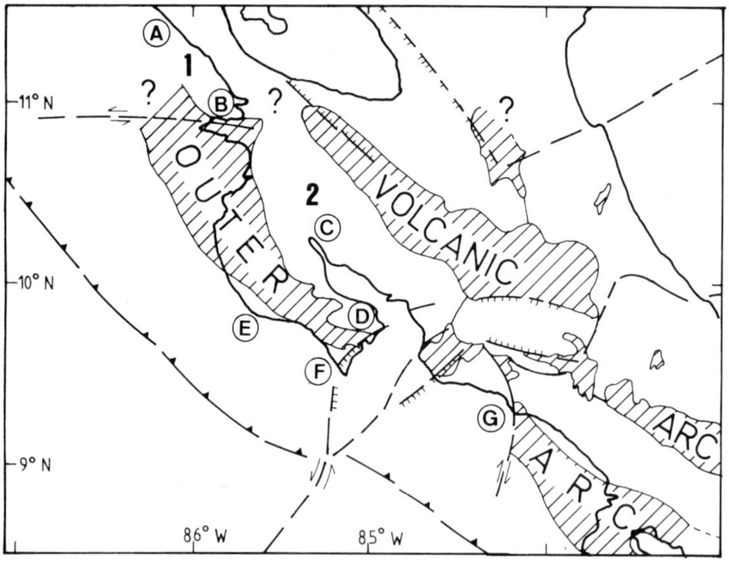


**Fig. 1.** Tectonic setting and location map of Upper Cretaceous to upper Eocene turbidite systems of the fore-arc area of Costa Rica and southwest Nicaragua (modified after Astorga *et al.*, 1989). Inner fore-arc basins: (1) Nicaragua Trough (A, southwest Nicaragua; B, Bahia Salinas); (2) Tempisque Trough (C, Barbudal; D, Curú). Outer fore-arc basins: E, Sámara Basin; F, Cabo Blanco Basin; G, Quepos Basin. For location of figure within Central America, compare with Fig. 2 of the paper by Seyfried *et al.*, this volume.

the lowstand systems tract. This lowstand systems tract can be divided into a lowstand fan (basin-floor fan) and a lowstand wedge (Posamentier & Vail, 1988). The lowstand fan is characterized by the deposition of sand lobes, whereas the subsequent lowstand wedge is dominated by channel-fill deposits and finer grained, wedge-shaped slope deposits. The lowstand wedge has also been described as slope fan, slope-front fill, and wedge and cone (Mitchum, 1985; Vail & Sangree, 1988). In contrast, carbonate platform environments suggest that the sedimentary response to relative sea-level changes in these systems is opposite to that observed in siliciclastic systems (Sarg, 1988; Dolan, 1989). During highstands, mud- and sand-size carbonate sediment is shed off the platform and transported into the adjacent basin, whereas conversely, during sea-level lowstands, the production capability of shallow water carbonate is strongly reduced and subaerial exposure of platforms may result in meteoric cementation and karstification (Sarg, 1988). In hybrid siliciclastic–carbonate basinal cycles, carbonate deposition therefore reflects conditions of relative highstands of sea level, whereas lowstands of sea level are documented by siliciclastic sedimentation. In contrast, small to moderate falls of sea level in ramp settings may expose only the inner part of the ramp, resulting in lowstand deposition of mixed siliciclastic–allodapic carbonate sediments (Dolan, 1989).

Another model was presented by Mutti (1985), who distinguished three types of turbidite system, type I, type II, and type III (Fig. 2), which more recently were related to the long-term stability of basins (Mutti & Normark, 1987). Active-margin fans are considered usually to be small, coarse-grained systems, owing to proximal sources, narrow shelves, and coastal plains (Mutti & Normark, 1987; Shanmugam & Moiola, 1988). However, in the study area, small, sand-rich fans only occur in lower Paleocene and upper Eocene deposits, whereas fine-grained, thin-bedded turbidite systems formed from late Paleocene to mid-Eocene times.

## DEPOSITIONAL SEQUENCES IN THE INNER FORE-ARC TROUGHS

The inner fore-arc area comprises two major basin-fill systems: (1) Upper Cretaceous to upper Eocene turbidite systems of the Nicaragua Trough and (2) Upper Cretaceous to upper Eocene turbidite systems of the Tempisque Trough (Fig. 1).

Outcrop data show four second-order depositional sequences in the deep-water sediments, which range considerably in size, geometry and location. Each depositional sequence is characterized by a particular association of turbidite systems, which reflect the prevailing local conditions of sediment supply, tectonic activity, and global sea-level fluctuations during Late Cretaceous to Eocene times (Fig. 3).

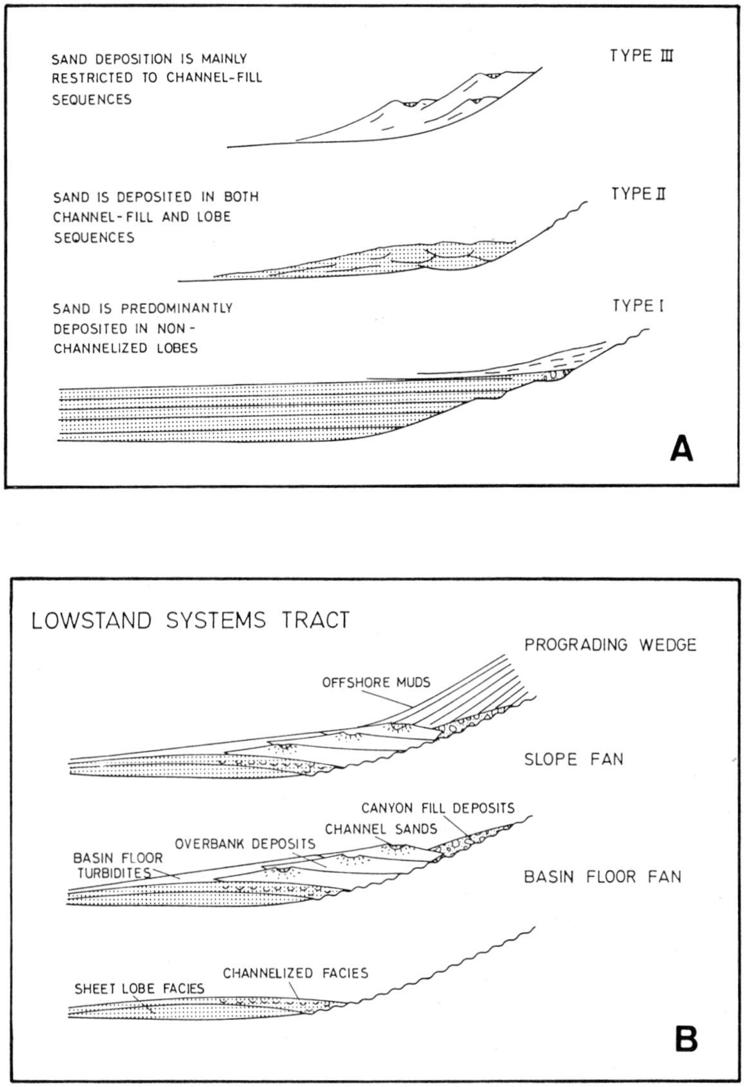


**Fig. 2.** Comparison of Mutti's depositional systems and the lowstand systems tract of the sequence-stratigraphic model. The type I turbidite system corresponds to the basin-floor fan, the type II and type III systems correspond to the lowstand wedge (also called slope fan or prograding wedge). (A) Simplified after Mutti & Normark (1987). (B) Simplified after Vail & Sangree (1988).

## Nicaragua Trough

The deposits of the Nicaragua Trough are mainly exposed along the Pacific coast of southwest Nicaragua and northern Costa Rica and were recorded in two major sections (A, B, Fig. 1).

Large-scale subsidence of the southern basin margin is recorded by thickness variations and facies development of turbidite systems from Late Cretaceous to Eocene times. The greatest thickness of the basinal sequence occurs in the northern part of the Nicaragua Trough (section A), where almost 6000 m of deep-water sediments are recorded. Thickness decreases strongly to the south and is only about 3000 m in section B.

### *Section A*

In southwest Nicaragua, Upper Cretaceous to Paleocene rocks crop out only in poor inland exposures and are mainly recorded from well data (Anonymous, 1972, cited in Weyl, 1980). According to Weyl (1980) these deposits mainly consist of about 3500 m of siliciclastic turbidites and tuffaceous shales.

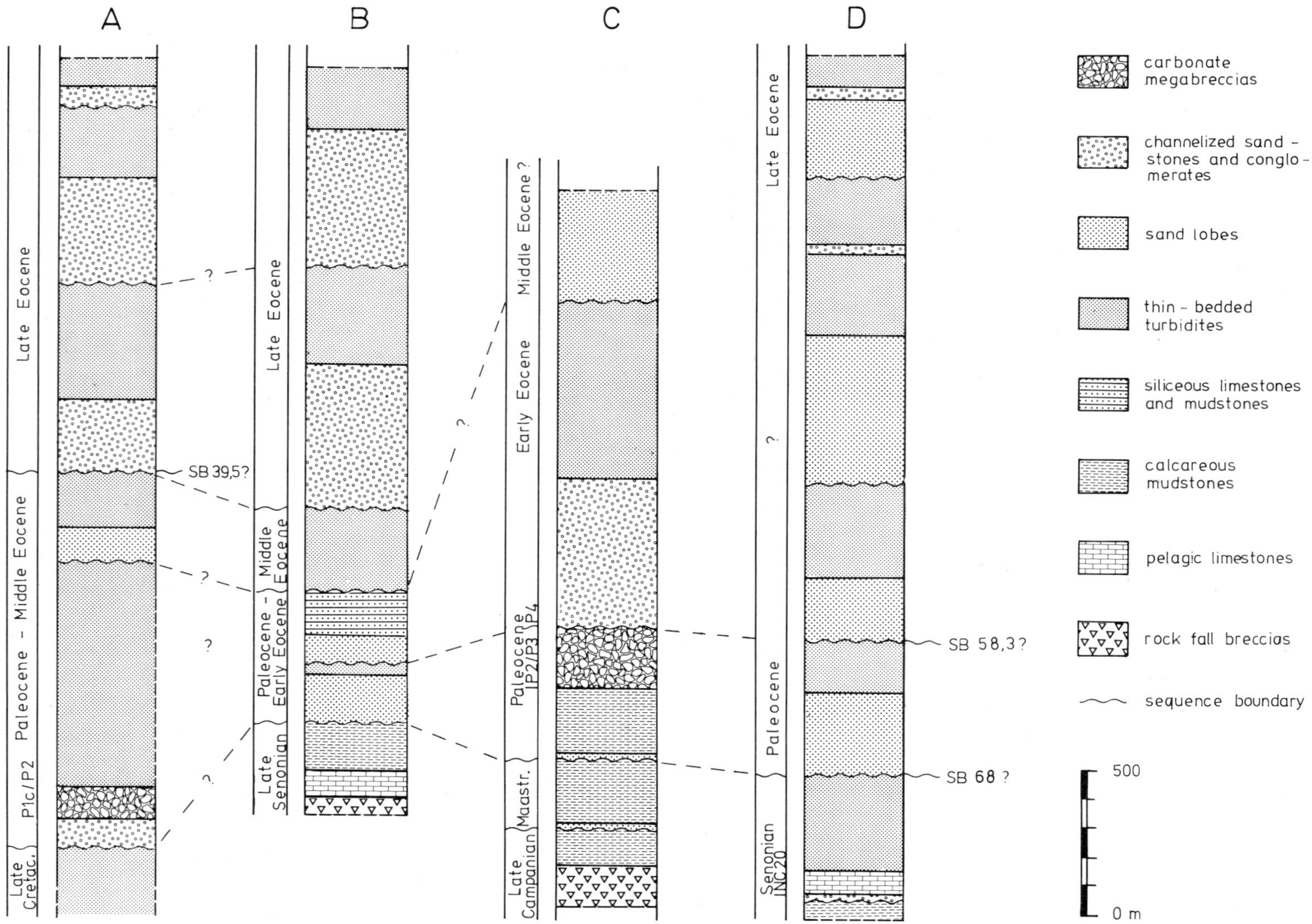


**Fig. 3.** Basin-wide correlation of depositional sequences in deep-water sediments of the inner fore-arc area of southwest Nicaragua and Costa Rica. Location of columns are shown in Fig. 1. Age determinations from Schmidt-Effing (1979), Rivier (1983), Baumgartner *et al.* (1984), Seyfried & Sprechmann (1986), Sprechmann *et al.* (1987), Hagn (personal communication, 1989), and C. Calvo (personal communication, 1990).

Large-scale slump and cobbly mudstone facies in the uppermost Cretaceous section indicate a slope or base of slope environment. Cobbly mudstones contain clasts of well-rounded basalts, radiolarites, and sandstones, indicating erosion of upheaved basaltic islands (see, for comparison, the chapter by Seyfried *et al.*). Locally, this slump apron is overlain by volcanic and carbonate breccias of early Paleocene age (Weyl, 1980; Hofherr, 1983; Calvo, 1987; Sprechmann *et al.*, 1987). The volcanic breccias are up to 50 m thick and consist of dominantly pebble-sized angular volcanic rock fragments and minor constituents of neritic carbonates, hemipelagic calcilutites, and cherts (Hofherr, 1983). Up section, these volcanic breccias are overlain by carbonate megabreccias with huge blocks of shallow-water limestones, which reach maximum diametres of about 50 m (Calvo, 1987). The matrix is made up of calcilutites and volcanic detritus, most probably derived from reworking of periplatform deposits and erosion of the underlying strata. According to Calvo (1987), the neritic carbonates represent allodapic platform carbonates of Late Cretaceous age and are similar to the Costa Rican 'Barra Honda' platform carbonates (see chapter by Seyfried *et al.*). Foraminiferal biostratigraphy indicates that initial allodapic limestone deposition occurred during early Paleocene times (P1c/P2; Sprechmann *et al.*, 1987). Upper Paleocene to middle Eocene deposits are poorly exposed and mainly consist of thin to medium-bedded turbidites. A lobe complex, which occurs within these sedimentary units, probably indicates the transition from early to middle Eocene deposits, but no biostratigraphic data are available. These lobe deposits are up to 80 m thick and consists of thick-bedded to massive, amalgamated sandstones, which are rich in volcanic rock fragments. Up section, they are overlain by 300 m of thin-bedded volcaniclastic turbidites of probably middle Eocene age (Weyl, 1980; Astorga, 1987).

The upper Eocene deposits are well exposed along the Pacific coast and consist of multicycle fan complexes. This sedimentary unit shows three cycles ranging in thickness from about 250 to 600 m and is characterized by well-developed channel overbank systems. Each cycle consists of a basinal unit of channelized conglomerates/sandstones, overlain by an interval of thin-bedded turbidites, interpreted as channel-associated overbank systems and overbank wedges (Mutti & Normark, 1987). The channelized facies are commonly characterized by an upward decrease in grain size and bed thickness. Within each cycle, the channel overbank systems may consist of one or several nested channels that formed as a result of channel avulsion.

The channelized deposits are characterized by different facies:

**1** Basinal minor channelized sand lobes, very rich in isolated larger Foraminifera and other shelf-derived fossils.

**2** Channel-fill deposits rich in resedimented neritic carbonates and bioclasts of larger Foraminifera, corals, molluscs, and echinoids.

**3** Very coarse-grained channel-fill deposits, which consist of volcaniclastic conglomerates and breccias with boulders up to 4 m across. Debris flow deposits and high-density turbidites contain crystal-rich, juvenile volcanic detritus, indicating rapid redeposition of subaerial pyroclastic flows by subaqueous debris flows or high-density turbidity currents.

**4** Sand-rich channel-fill deposits, which are associated with overbank complexes and thick overbank wedges. These sand-rich deposits consist of amalgamated sand bodies with very common loading and dewatering structures and overlie the coarser grained volcaniclastic conglomerates and/or carbonate breccias. Channel-associated overbank deposits consist of fine-grained, thin-bedded turbidities, which may be intensively current laminated. The overlying overbank wedges are characterized by thicker bedded turbidites with common progradational cycles.

### *Section B*

An almost complete section of Upper Cretaceous to upper Eocene deposits crops out along the northwestern Pacific coast of Costa Rica. These sedimentary units were deposited close to the southern basin margin of the Nicaragua Trough and overlie a pre-Santonian basement (Baumgartner *et al.*, 1984; see chapter by Seyfried *et al.*).

Sedimentation started with rock-fall breccias of late Campanian age (Baumgartner *et al.*, 1984). These breccias are rich in resedimented neritic carbonate clasts and pass up-section into pebbly sandstones that contain large amounts of larger Foraminifera. The breccias are overlain by 80 m of interbedded hemipelagic limestones and thin-bedded sandstones. The siliciclastic influx increases up section and sandstone beds become thicker. Unchannelized, thick-bedded turbidites abruptly overlie the calcareous unit. These lobe deposits consist of two major progradational cycles, separated by a small interval of thin to mediun-bedded turbidites. Up section the thick-bedded turbidites gradually pass into siliceous limestones and mudstones of late Paleocene to early Eocene age (Baumgartner *et al.*, 1984). The siliceous limestones are rich in Radiolaria and planktonic Foraminifera and are similar to siliceous limestones, which formed simultaneously in the outer fore-arc region of Sámara (section E) and Cabo Blanco (section F). The middle Eocene deposits consist of thin to medium-bedded turbidites, rich in volcanic ash. The transition from the underlying lower Eocene deposits is gradational. Owing to the poor exposures of these deposits, no further data are available.

The upper Eocene deposits make up the greatest part of the sedimentary record. These sedimentary units consist, as in the northern basin area, of multicycle fan complexes and are characterized by well-developed channel overbank systems. Two cycles are recorded, ranging in thickness from about 600 to 700 m. As in section A, each cycle consists of a basal unit of channelized conglomerates/sandstones, overlain by an interval of thin-bedded turbidites.

*Interpretation*

In the Nicaragua Trough, four second-order depositional sequences have been identified. Two third-order depositional sequences could be determined within the upper Eocene sequence of section A.

*Sequence 1* is mainly exposed in the southern Nicaragua Trough (section B) and consists entirely of unchannelized lobe deposits. The sequence boundary is indicated by the abrupt onset of sand lobes during the latest Maastrichtian at approximately the 68 Ma sea-level fall (Haq *et al.*, 1988). The sand lobes are interpreted to be part of the lowstand systems tract and to have formed during a rapid fall of sea level when sediment excavated from the slope and shelf were transported via submarine canyons into deep water (Posamentier & Vail, 1988). The small thicknesses of these lobe sediments indicate that they probably were deposited close to a structural high, which deflected the bulk of the sediment from this area. The interval of thin to medium-bedded turbidites are interpreted to have formed during the late lowstand systems tract (Vail & Sangree, 1988). Condensed sections of the transgressive/highstand systems tracts could not be determined and most probably were eroded during a subsequent fall of sea level (cf. Weimer, 1990).

In the northern basin area, sequence 1 is poorly exposed. The sequence boundary may be indicated by the occurrence of large-scale slump features and cobbly mudstone facies. However, there is no age control and debris aprons and slides have been reported also from highstand deposits and therefore may form at any time independent of the position of sea level (Weimer, 1990). The deposition of the volcanic and carbonate megabreccias during early Paleocene times is attributed to a major volcanic/tectonic event that affected the southeastern basin margin. A previous interpretation of the emplacement of the carbonate megabreccias has been given by Calvo (1987), that these blocks detached from steep submarine fault-scarps of the Hess Escarpment. However, the association with volcanic breccias indicates that, most probably, volcanic activity and accompanying basin-margin fault activity resulted in the break up of carbonate shelf rocks, which subsequently were transported into the basin.

*Sequence 2* is only exposed in section B. The sequence boundary between sequence 1 and sequence 2 most probably developed during late Paleocene times and is indicated by the onset of a new progradational lobe complex. These sand lobes are interpreted to be part of a new lowstand systems tract (Posamentier & Vail, 1988). Up section the sandstones gradually pass into siliceous limestones and mudstones of late Paleocene to early Eocene age, indicating a rapid sea-level rise with a drastic decrease in siliciclastic sediment supply. These deposits are interpreted to be part of the transgressive and highstand systems tracts.

*Sequence 3* mainly consists of thln to medium-bedded, ash-rich turbidites. In section B, the sequence boundary between sequence 2 and sequence 3 is indicated by the transition from siliceous limestones to siliciclastic turbidites, which occurred during the middle Eocene (Baumgartner *et al.*, 1984) and may be correlatable with the 49.5 Ma sea-level fall (Haq *et al.*, 1988). Progradational and retrogradational cycles within these sequences probably indicate lowstand, transgressive, and highstand deposition.

In the northern basin area, the sequence boundary is indicated by the formation of unchannelized lobe deposits, which overlie thin to medium-bedded turbidites. These lobe deposits are interpreted as the lowstand systems tract. The overlying thin to medium-bedded volcaniclastic turbidites most probably form part of the transgressive and highstand systems tracts, but no age control has been obtained.

*Sequence 4* is characterized by multicycle fan complexes with channelized lobes, very coarse-grained channel-fill deposits, and thick overbank wedges. The sequence boundary between sequences 3 and 4 is indicated by the abrupt onset of coarse-grained, channelized deposits over thin-bedded turbidites, reflecting initiation of erosion and a deep-marine expression of a type 1 sequence boundary (Posamentier & Vail, 1988).

Cycles most probably developed in response to third-order eustatic sea-level changes and are interpreted as cyclic stacks of individual lowstand systems tracts, consisting dominantly of the lowstand wedge (Posamentier & Vail, 1988).

The high content of larger Foraminifera within channelized lobe deposits most probably reflects the resedimentation of carbonate shoals during the initial fall of sea level at the shelf margin (Shanmugam *et al.*, 1985). Exposure of arc-fringing reefs or carbonate shelfs during the continuous fall of sea level led to the destruction and resedimentation of these systems into the basin, forming allodapic carbonate breccias, which commonly were deposited in channels.

Deposition of very coarse-grained channel-fill deposits of the lowstand systems tract occurred when major tectonic/volcanic pulses increased the supply of siliciclastic detritus and submarine canyons were deeply incised into the small shelves and/or volcanic flanks. Neritic carbonate blocks up to 10 m across within channels suggest the repeated destruction of basin-margin areas. Channel-overbank systems formed when lowstand deltas began to prograde during the period of lowstand wedge deposition. As the depositional slope of the delta became oversteepened and unstable, sediment failure began and large amounts of fine-grained sediments were transported into the basin, forming extensive channel overbank systems and overbank wedges. Sections of the transgressive/highstand systems tract could not be determined and were most probably eroded as sea level began to drop again and erosion and rapid sedimentation occurred in the deep water (cf. Weimar, 1990).

## Tempisque Trough

The deposits of the Tempisque Trough are mainly recorded from the Barbudal area (section C) and the Curú area (section D) where a sedimentary unit of almost 3500 m is recorded. The development of sequences within this basin is strongly influenced by local parameters of volcanic and tectonic activity.

### *Section C*

The oldest sediments of the Barbudal area consist of tuffs and siliceous limestones of probable early Campanian age (Seyfried & Sprechmann, 1986). Up section, coarse-grained basaltic breccias and conglomerates occur. The basal unit of these breccias consists of weathered basalt rubble and gravel, indicating erosion of basaltic islands (see paper by Seyfried *et al.*). The upper units contain less-altered, probably cliff-derived basalt detritus and show intercalations of resedimented rudist debris, echinoids, oysters, corals, and larger Foraminifera. Resedimentation of these breccias occurred during the late Campanian (UC11–12; Seyfried & Sprechmann, 1986). The breccias are overlain by a unit of thin to medium-bedded mudstones, which reach a thickness of 730 m and contain two intercalations of calcarenite beds, rich in neritic fossils. Deposition of these dominantly fine-grained facies persisted from terminal late Campanian until early Paleocene times (Rivier, 1983). Up section, 200 m of carbonate breccias and calcarenites occur. These breccias and calcarenites form small debris apron-like features and were deposited in early Paleocene times (P2/P3A; Rivier, 1983). Deposition of the carbonate sediments terminated abruptly with the onset of coarse-grained channelized conglomerates and sandstones during late Paleocene times (P4; Rivier, 1983). These channel-fill deposits consist of coarse-grained volcaniclastic conglomerates and are commonly not associated with overbank deposits. Up section they gradually pass into finer grained sandstones and calcareous mudstones. The transition from coarse-grained channelized deposits to the fine-grained calcareous turbidites occurred during the latest Paleocene (Rivier, 1983) and the lower and central parts of this fine-grained facies have been dated as early Eocene (Rivier, 1983). Up section the siliciclastic influx markedly increased and minor lobe systems formed. These lobe deposits consist of thin to medium-bedded sandstones with intercalated ash beds and pebble-sized volcanic conglomerates. Fragments of larger Foraminifera are commonly observable. The onset of these dominantly volcaniclastic deposits most probably occurred during the early middle Eocene (A. Astorga, personal communication, 1990).

### *Section D*

In the Curú area, sedimentation started in the early

Campanian with siliceous mudstones and beds of volcanic ash, which unconformably overlie the ophiolitic basement or pre-Campanian deep-water sediments (Schmidt-Effing, 1979; Lundberg, 1982; Astorga, 1987). Intercalated channelized debris-flow deposits consist of resedimented hemipelagic mudstones, locally derived neritic fossils, and minor contents of recycled material from the Nicoya Complex. Upwards, Foraminifera-rich calcareous mudstones occur. These limestones are late Campanian in age (Calcarata zone, Schmidt-Effing, 1979) and grade up-section through argillaceous limestones into thin-bedded turbidites, reflecting an increasing sediment input during the early Maastrichtian. These interbedded sandstone and mudstone facies are overlain by thick lobe systems, which initially formed in the latest Maastrichtian and persisted until the late Paleocene (Lundberg, 1982; Baumgartner *et al.*, 1984). The sand lobes consist of thick-bedded to massive sandstones, which are up to 12 m thick and interbedded with thin-bedded turbidites and shales. Two major progradational cycles are separated by a larger interval of thin-bedded turbidites and shales. The lobe deposits are overlain by fine-grained, thin-bedded turbidite systems, which probably were deposited during latest Paleocene and early Eocene times (Astorga, 1987). During middle and late Eocene times, thick amalgamated sandstone complexes are absent and instead medium to thick-bedded turbidites developed, usually forming minor thickening and coarsening upward cycles. These lobe systems are rich in larger Foraminifera and are overlain by small channelized conglomerates, composed of resedimented neritic carbonates. The top of the section is composed of thin-bedded turbidites and slump deposits.

### *Interpretation*

In the Tempisque Trough, four depositional second-order sequences have been identified. However, sequences vary in size, geometry, and location.

*Sequence 1* is mainly exposed in the southern basin area (section D) and consists entirely of unchannelized lobe deposits. As in section B, the sequence boundary is indicated by the abrupt onset of sand lobes during the latest Maastrichtian at approximately the 68 Ma sea-level fall (Haq *et al.*, 1988). These sand lobes are interpreted to be part of the lowstand systems tract. The large thickness of these lobe sediments indicates that very high sedimentation rates occurred during early Paleocene times in this area. In addition, palaeocurrent data suggest deposition by axial flows down an elongate basin, most probably supporting the development of lobe deposits. The overlying thin-bedded turbidites and shales are interpreted to have been formed during the late lowstand systems tract and probably are also parts of the transgressive and highstand systems tracts. Condensed sections could not be identified.

In the northern basin area, thick lobe deposits are absent and instead mainly fine-grained mudstones were deposited. The sequence boundary may be indicated by a thick calcarenite bed, rich in neritic fossils, reflecting a relative lowstand of sea level with erosion of shelfal sediments. Deposition of this calcarenite bed occurred during the latest Maastrichtian (Rivier, 1983) at approximately the 68 Ma sea-level fall (Haq *et al.*, 1988). A major volcanic/tectonic event is indicated by the formation of carbonate breccias and calcarenites on top of the sequence. According to the biostratigraphic data, deposition of these allodapic limestones could have been controlled also by the brief eustatic drop that occurred at the 63 Ma sea-level fall (Haq *et al.*, 1988). However, the 63 Ma sea-level fall is also related to major global tectonic events (Schwan, 1980), and the high content of volcanic detritus calls rather for a major control on deposition by volcanic events.

In the southern Tempisque Trough, the boundary between sequence 1 and *sequence 2* is indicated by the onset of a new progradational lobe complex. These sand lobes are interpreted to be part of a new lowstand systems tract (Posamentier & Vail, 1988) that formed during the late Paleocene, probably at approximately the 58.3 Ma sea-level fall (Haq *et al.*, 1988). The overlying interval of thin-bedded turbidites and shales indicates a drastic decrease in siliciclastic sediment supply, most probably owing to a rapid sea-level rise. These deposits are interpreted to be part of the late lowstand systems tract as well as of the transgressive and highstand systems tract. In the northern basin area, the sequence boundary between sequence 1 and sequence 2 is indicated by the occurrence of very coarse-grained, volcaniclastic channel-fill deposits. Biostratigraphic data suggest that deposition of these channelized conglomerates occurred at approximately the late Paleocene 58.3 Ma sea-level fall (Haq *et al.*, 1988). This channel complex is interpreted to be part of the lowstand wedge (Posamentier & Vail, 1988). The overlying late Paleocene to early Eocene fine-grained calcareous mudstones and calcarenites are interpreted to have been formed during the subsequent sea-level rise and to be part of the transgressive/highstand systems tract.

The boundary between sequence 2 and *sequence 3* is indicated by the onset of lobe deposits, both in section C and section D. In section C the onset of lobe deposits most probably occurred during early mid-Eocene times (A. Astorga, personal communication, 1990) and may be related to the 49.5 Ma sea-level fall (Haq *et al.*, 1988). In section D, no biostratigraphic control is available. The lobe deposits are interpreted to be part of the early lowstand systems tract, whereas the overlying thin-bedded turbidites most probably were deposited during the late lowstand systems tract and the transgressive and highstand systems tract. In section C, lower sedimentation rates can be deduced from minor sediment thicknesses and the finer grained nature of the deposits.

*Sequence 4* is only recorded in section D. The vertical facies arrangement of these upper Eocene sediments indicates a typical lowstand systems tract. The sequence boundary is marked by the onset of a new lobe complex, indicating sedimentation during the early lowstand systems tract. Overlying channelized conglomerates are interpreted to have been deposited during the early stage of lowstand-wedge formation. The slump and debris apron on top of the sequence are interpreted to be a part of the later stage of lowstand wedge formation, when sedimentation within channels had stopped, and sediment failure in the slope was caused by rapid progradation of lowstand deltas (Posamentier & Vail, 1988).

## DEPOSITIONAL SEQUENCES IN TRENCH SLOPE BASINS

Several small slope basins developed on the seaward margin of the outer structural high. Deposits of these trench slope basins are documented in the area of Sámara, Cabo Blanco, and Quepos (Figs 1, 4). Facies associations within the sedimentary record vary considerably and strongly depended on local conditions, such as size and location of submarine canyons as well as differential uplift/subsidence of the outer fore-arc area.

### *Section E*

The oldest sediments of the Sámara area consist of siliceous mudstones and limestones of early to late Campanian age (Baumgartner *et al.*, 1984), which unconformably overlie the basaltic basement or its cover of basaltic breccias.

From late Maastrichtian to late Paleocene times a coarse-grained submarine fan complex developed. Foraminiferal biostratigraphy indicates that the transition from calcareous limestones and mudstones to siliciclastic deposition occurred during the terminal late Maastrichtian (NC22/NC23; Schmidt-Effing, 1979).

A basal lobe complex consists of unchannelized, medium to thick-bedded sandstones, which are rich in resedimented neritic fossils. These lobe deposits are overlain by channelized, chaotic mass flow deposits, including large, well-rounded andesitic boulders up to 3 m across and huge blocks of both Upper Cretaceous and Paleocene platform carbonates (Calvo, 1987). Up section, a further lobe complex consists of channelized, massive to thick-bedded sandstones. This lobe system initially formed during the late Paleocene (P4; Lehfeld, 1989) and is overlain by coarse-grained channel-fill deposits, which are very rich in resedimented neritic fossils and carbonate debris. These conglomerates and chaotic mass flow deposits are overlain by thin-bedded turbidites, which pass upwards into siliceous limestones and mudstones of late Paleocene to early Eocene age (Schmidt-Effing, 1979; Baumgartner *et al.*, 1984; Lehfeld, 1989).

### *Interpretation*

*Sequence 1* is built up by a typical lowstand systems tract with basinal lobe deposits and an overlying lowstand wedge. The sequence boundary is marked by the onset of sand lobes, indicating sedimentation during the early lowstand systems tract. The transition from mainly biogenic sedimentation to siliciclastic deposition occurred during the terminal Maastrichtian at approximately the 68 Ma sea-level fall (Haq *et al.*, 1988). The overlying channelized conglomerates and chaotic mass flow deposits are interpreted to be part of the lowstand wedge. Sections of the transgressive and highstand systems tract do not occur.

The boundary between sequence 1 and *sequence 2* is indicated by the deposition of a new lobe system. The onset of these lobe deposits occurred during the late Paleocene at approximately the 58.3 Ma sea-level fall (Haq *et al.*, 1988) and reveals sedimentation in an early stage of the lowstand systems tract. The

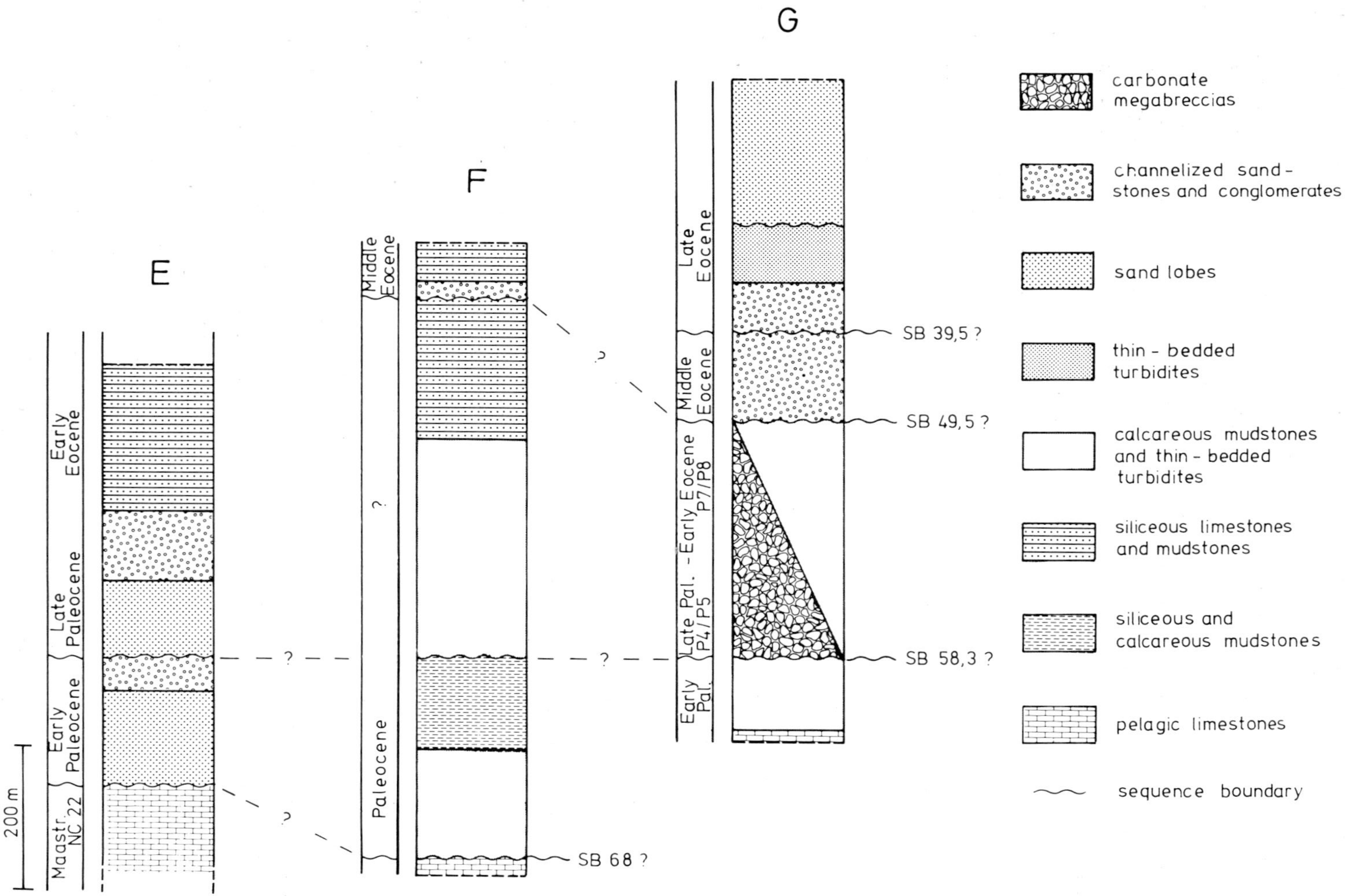


**Fig. 4.** Basin-wide correlation of depositional sequences in deep-water sediments of the outer fore-arc area of Costa Rica. Location of columns are shown in Fig. 1. Age determination from Azéma *et al.* (1979), Schmidt-Effing (1979), Baumgartner *et al.* (1984), and Lehfeld (1989).

overlying channel-fill deposits are interpreted to have been formed during the lowstand wedge. The increase in reef and shelf-derived shallow-water fossils indicates a relative rise of sea level, which caused a re-establishment of shallow-water production during the late lowstand wedge. The fine-grained siliceous limestones and mudstones on top of the sequence reflect a rapid rise of sea level with a drastic decrease in sediment supply and are interpreted to have been formed during the transgressive and highstand systems tracts.

*Section F*

In the area of Cabo Blanco, thick siliciclastic turbidite successions are absent and biostratigraphic control is very poor. Instead, fine-grained calcareous mudstones and thin-bedded turbidites overlie the oceanic basement and Upper Cretaceous biogenic rocks (Lundberg, 1982; Baumgartner *et al.*, 1984; Astorga, 1987).

The basinal section consists of Paleocene thin to medium-bedded siliciclastic turbidites, which are intercalated with fine-grained siliceous and calcareous mudstones (Lundberg, 1982; Baumgartner *et al.*, 1984). Turbidites show a thickening upward trend and are overlain by siliceous and calcareous mudstones. These mudstones are thin to medium-bedded and individual beds show subtle grading of foraminiferan tests, representing locally derived hemipelagic material that was redeposited by minor currents. Up section, a new interval of interbedded mudstones and sandstones occurs. This interval is characterized by a progradational and retrogradational cycle. The uppermost section consists of siliceous limestones and mudstones of probably late Paleocene to middle Eocene age (Baumgartner *et al.*, 1984). On top of the section, minor channelized calcarenites and conglomerates occur. These channel-fill deposits are middle Eocene in age (Baumgartner *et al.*, 1984) and consist of resedimented neritic carbonates, isolated larger Foraminifera and volcanic rock fragments.

*Interpretation*

The absence of thick sand-rich turbidites suggests the occurrence of smaller submarine canyons downslope of the outer structural high that did not receive a direct supply of terrigenous detritus.

The *sequence 1* boundary is indicated by the onset of Paleocene siliciclastic turbidites over Upper Cretaceous biogenic sediments. This transition most probably occurred during the clearly defined 68 Ma fall of sea level (Haq *et al.*, 1988), when coarser terrigenous sediments were also funnelled into smaller canyons of the trench slope, leading to the formation of intercalated siliciclastic turbidites and hemipelagic mudstones. The overlying fine-grained hemipelagic mudstones are interpreted to have been formed during the subsequent sea-level rise and be part of the transgressive and highstand systems tracts.

The boundary between sequence 1 and *sequence 2* is marked by the new onset of siliciclastic turbidites, which probably occurred during late Paleocene times. The occurrence of these siliciclastic turbidites is related to lowstand deposition, whereas the overlying siliceous limestones and mudstones are interpreted as highstand deposits.

The boundary between sequence 2 and *sequence 3* is indicated by the occurrence of channelized calcarenites and conglomerates. Lowstand deposition is marked by the appearance of significant sand- and pebble-sized volcanic material as well as the resedimentation of large amounts of larger Foraminifera, which indicate erosion of carbonate shoals during an initial fall of sea level (Sarg, 1988).

*Section G*

The Quepos section has unusual facies. Specific characteristics of this basin-fill system show close similarities with wrench-fault basins, described from the Californian borderlands (Howell *et al.*, 1980) and the eastern Betics (Montenat *et al.*, 1987). The depositional regime was characterized by repeated fault activity, resulting in deepening of the basin, followed by rapid erosion of borderlands and sedimentation of very coarse clastic sediments on a narrow coastal fringe of fan-deltas leading into submarine fans. Facies development is divided into coarse-grained megabreccias and small conglomeratic cones along the basin margin and a finer grained turbiditic facies along the basin axis.

Basal slope strata overlie probably accreted alkaline basalts (Sick, 1989) and basaltic sedimentary breccias. The basaltic breccias are of early Paleocene age (P1b/P1c; Schmidt-Effing, 1979) and contain radiolarites and Late Cretaceous limestones (Azéma *et al.*, 1979), which point to the existence of an older basement in this area (see, for comparison, the paper by Seyfried *et al.*). Pelagic carbonates, which subsequently formed (P2/3A; cf. Azéma *et al.* 1979;

Schmidt-Effing, 1979), indicate a very low sediment input up to the early late Paleocene.

The overlying strata consist of siliceous mudstones and thin-bedded sandstones and are very similar to the deposits of the Cabo Blanco area. The mudstones are rich in Radiolaria and planktonic Foraminifera with an up-section increase of calcareous oozes. Sandstones are mainly composed of volcaniclastic rock fragments. Some sandstones also contain basaltic and radiolaritic rock fragments, which were recycled from the Nicoya Complex.

Biostratigraphic data indicate an early Paleocene to early Eocene age (P2/P3B – P7/P8; Azéma *et al.*, 1979; Schmidt-Effing, 1979). These dominantly fine-grained deposits are associated with chaotic mass flow deposits of late Paleocene to early Eocene age (Schmidt-Effing, 1979). The chaotic mass flow deposits consist of megabreccias with huge blocks of siliceous limestones, up to 100 m across. Resedimented neritic carbonates are rich in larger Foraminifera of late Paleocene age and make only a minor contribution to the components. The matrix is composed of siliceous and calcareous mudstones, similar to the underlying deposits. Biostratigraphic ages fall within planktonic foraminiferal biozone P4/5 (Schmidt-Effing, 1979).

Rapid uplift in middle Eocene times resulted in the formation of small, very coarse-grained braided channel systems, which most probably were related to fan-deltas. The clast composition is still dominated by recycled material from the Nicoya Complex (basalts, cherts, and pelagic carbonates), but also includes coarse volcanic detritus, derived from the calc-alkaline arc. Age determination from resedimented larger Foraminifera indicates deposition in early middle Eocene times (Azéma *et al.*, 1979). The coarse-grained middle Eocene deposits are overlain by upper Eocene sand-rich turbidites. These upper Eocene deposits start with coarse-grained, channelized conglomerates, which pass up-section into sand-rich channel-fill deposits. These sand-rich channel fills are associated with overbank deposits and are overlain by a thick overbank wedge. The uppermost section is composed of lobe deposits, which consist of medium to thick-bedded turbidites, rich in larger Foraminifera. Sand lobes are overlain by an interval of thin to medium-bedded turbidites.

### *Interpretation*

*Sequence 1* is not completely exposed and consists of intercalated thin-bedded turbidites and hemipelagic mudstones. Similar to sequence 1 of section E, this deposition may indicate a lowstand during early Paleocene times, when small-sized submarine canyons received a larger amount of terrigenous material.

The boundary between sequence 1 and *sequence 2* is indicated by deposition of megabreccias. These megabreccias initially formed during the late Paleocene at approximately the 58.3 Ma sea-level fall (Haq *et al.*, 1988). As deposition of breccias continued until early Eocene times it is suggested that tectonic control on deposition owing to uplift and tilting of the outer fore-arc was the dominant controlling factor. However, well-rounded clasts of neritic carbonates as well as basalts and radiolarites indicate erosion of neritic environments, which probably occurred because of the 58.3 Ma sea-level fall (Haq *et al.*, 1988) and interfered with tectonism in this area.

The boundary between sequence 2 and *sequence 3* is marked by the onset of well-organized, braided channel systems. Well-rounded clasts of conglomerates and high contents of larger Foraminifera may indicate a major eustatic control on deposition.

The internal facies construction of late Eocene deposits indicates a clear eustatic control on deposition, resulting in the formation of typical lowstand systems tracts. The boundary between sequence 3 and *sequence 4* is marked by the deposition of large-scale channel-fill deposits and probably occurred because of the late Eocene 39.5 Ma sea-level fall (Haq *et al.*, 1988). The vertical development of coarse-grained channelized conglomerates and sand-rich, channel overbank complexes are interpreted to be part of a lowstand wedge. Lobe deposits, which overlie the channel overbank complex, indicate a third-order sequence boundary and are interpreted to have been formed during a rapid fall of sea level and are therefore interpreted to be part of a new lowstand systems tract. The final interval of thin to medium-bedded turbidites, which overlies the lobe deposits, formed during the subsequent lowstand-wedge stage.

## EFFECTS OF TECTONIC UPLIFT/ SUBSIDENCE, VOLCANIC SUPPLY, AND EUSTATIC FLUCTUATIONS ON DEPOSITION

At the end of the late Maastrichtian, sedimentation

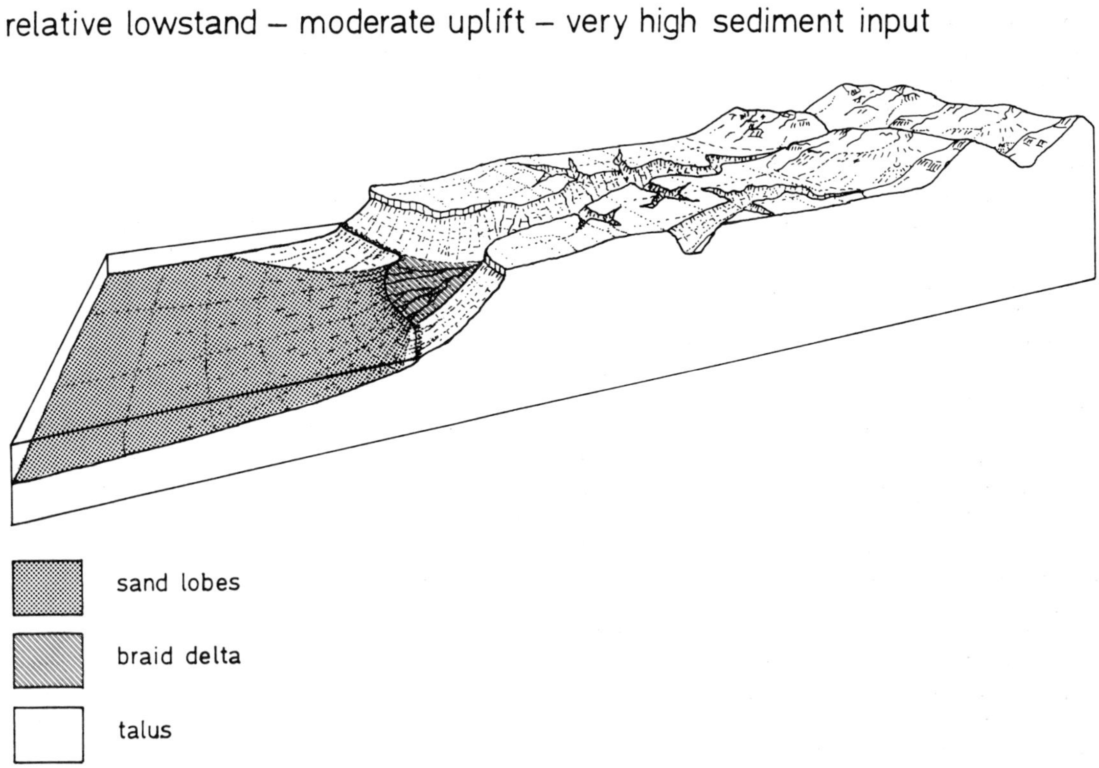


**Fig. 5.** Very high sedimentation rates and a relative lowstand of sea-level caused the formation of thick lobe systems in the southern Tempisque Trough. This preferred development of sand-lobe systems from early Paleocene to Eocene times was most probably favoured by a small, elongate basin shape in this area, and very high sedimentation rates during this time may have outpaced the rate of sea-level rise during early Paleocene times.

changed markedly, and sand-rich turbidite systems formed both in the inner and outer fore-arc troughs owing to the 68 Ma sea-level fall (Haq *et al.*, 1988). In the southern Tempisque Trough, extensive sand lobe systems formed. For the first time in the evolution of the island arc, there must have been a neritic system, which was able to store large amounts of sand. Lobe sedimentation persisted during the entire Paleocene with an up-section increase in grain size. The overall thickening and coarsening upward cycles within these Paleocene lobe systems indicate that a very high sediment influx, delivered from the rising calc-alkaline island arc, was able to compensate the effects of sea-level rises during Paleocene times (Fig. 5). In addition, the preferred development of sand lobe systems may have been favoured by an elongate basin shape which funnelled axial flows in this area.

In the early Paleocene, a major tectonic/volcanic event affected the northern parts of the fore-arc area. This event led to the destruction of Upper Cretaceous platform carbonates, which were deposited in both the northern Tempisque Trough and the southeastern Nicaragua Trough. The subsequent 58.3 Ma sea-level fall (Haq *et al.*, 1988) interfered with an episode of major tectonic/volcanic activity. As a result, a coarse-grained turbidite system developed (Fig. 6). The concomitant uplift and tilting of the outer fore-arc area (see paper by Seyfried *et al.*) caused back-cutting of submarine canyons that deeply incised across the outer arc and headed on to the narrow shelf of the calc-alkaline arc. At the mouth of these canyons, very coarse-grained submarine fans formed in the outer fore-arc area (Fig. 7). Where smaller submarine canyons only headed on to the slope of the outer arc, locally derived mass flow deposits and thin-bedded turbidites prevailed.

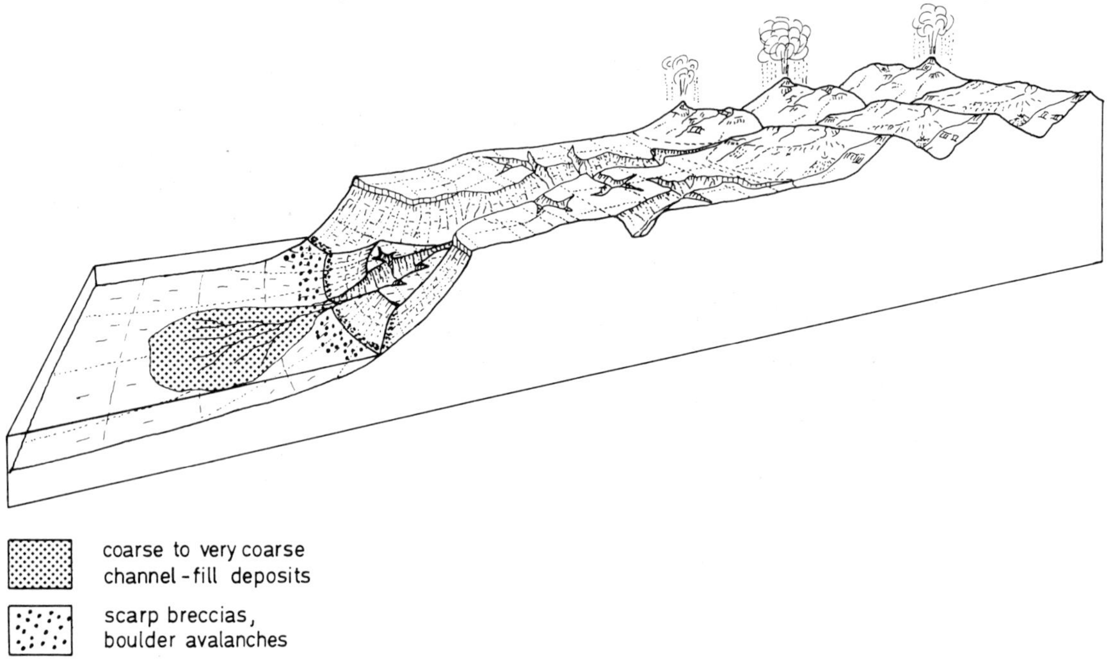


**Fig. 6.** Strong uplift, high sedimentation rates, and a relative lowstand of sea level caused the formation of channelized lobes and very coarse-grained channel-fill deposits in the inner fore-arc area of southwest Nicaragua and northwest Costa Rica during the late Paleocene and late Eocene.

The subsequent sea-level rise in the latest Paleocene and the early Eocene is mainly documented by retrogradational cycles. In addition, the calc-alkaline arc was largely inactive throughout late Paleocene to early Eocene times and contributed only minor volcaniclastic sediments to the basins. In the inner fore-arc troughs, thin-bedded turbidites formed where siliciclastic neritic systems persisted, whereas siliceous mudstones and limestones developed when carbonate ramps were establised (Fig. 8). The uplift of the outer structural high during the late Paleocene closed the Sámara bypass area (see paper by Seyfried *et al.*) and subsequently only fine-grained siliceous mudstones and limestones were deposited in the outer fore-arc basins. Carbonate ramps, concomitantly developed on the outer structural high, periodically delivered neritic carbonate material to the lower slope (Fig. 9). In the southern outer arc, tectonic uplift and major fault activity persisted until the middle Eocene and caused the formation of chaotic megabreccias and small-scale braided fan systems, which were related to fan-deltas.

In middle Eocene times, volcanic activity increased again and ash-rich sediments were deposited in the inner fore-arc basins. Common resedimented larger Foraminifera indicate the persistence of neritic carbonate systems upon the shelf. Small to moderate drops of sea level are documented by the formation of minor siliciclastic slope fans, rich in resedimented larger Foraminifera (Fig. 10).

During the early late Eocene, volcanic/tectonic activity intensified and very coarse-grained submarine fan systems formed. Fragmentation of the basin margin caused the destruction and redeposition of patch reefs, which were resedimented into the adjacent basins (Fig. 6). The formation of repetitive channel overbank complexes in this depositional sequence is interpreted to be the result of third-order eustatic sea-level fluctuations when the strong volcanic/tectonic activity had decreased (Fig. 11).

## CONCLUSIONS

The formation of depositional sequences in the deep-water sediments of southern Central America is

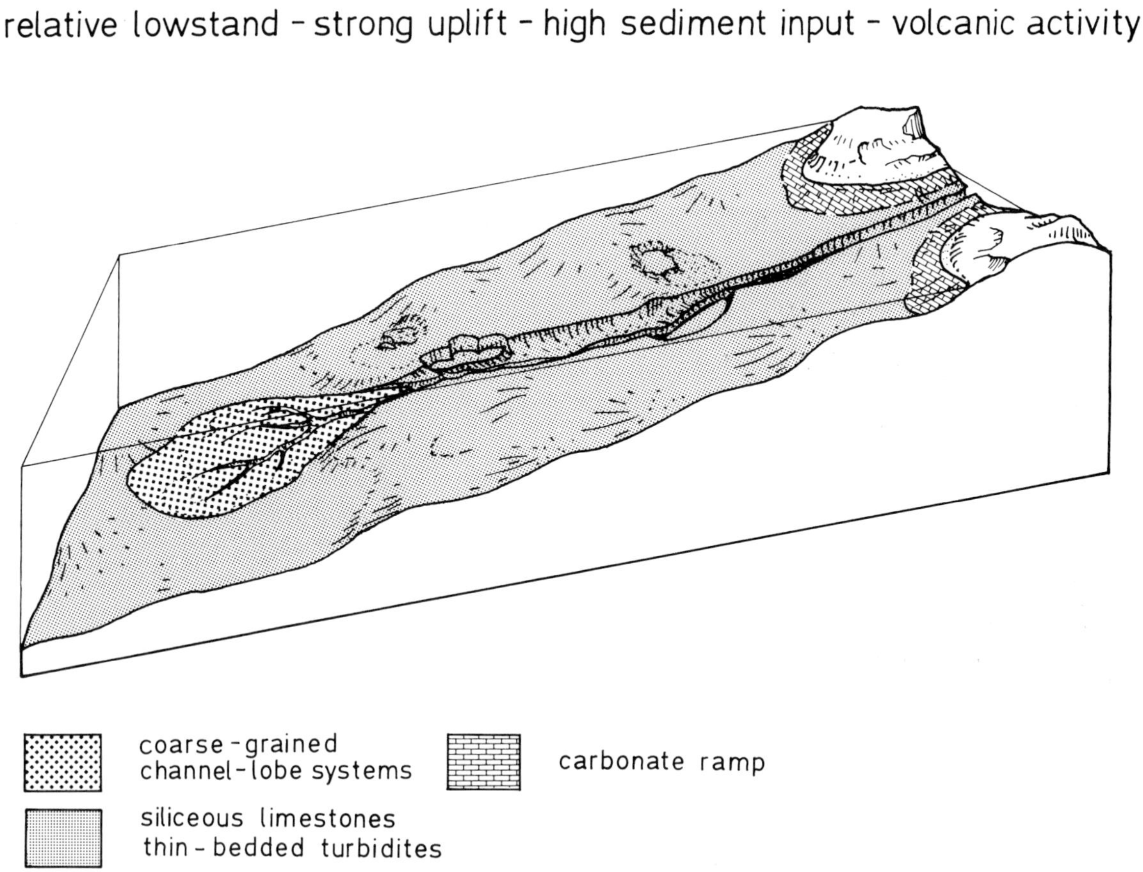


**Fig. 7.** A model for submarine fan sedimentation in the outer fore-arc during the early Paleocene and earliest late Paleocene. Strong uplift of the fore-arc area, high sedimentation rates, and a relative lowstand of sea level caused the formation of sand lobes and very coarse-grained channel-fill deposits in the outer fore-arc area, where submarine canyons deeply incised into the outer arc and headed on to the narrow shelf of the volcanic arc. Where smaller submarine canyons only headed on to the slope of the outer arc, locally derived mass flow deposits and thin-bedded turbidites prevailed.

strongly related to the morphotectonic evolution of the island-arc system. Each depositional sequence reflects the complex interaction between global sea-level fluctuations, sediment supply, and tectonic activity. Sediment supply and tectonic activity overprinted the eustatic effects and enhanced or lessened them. If large supplies of clastics or uplift overcame the eustatic effects, deep marine sands were also deposited during highstand of sea level, whereas under conditions of low sediment input, thin-bedded turbidites were deposited even during lowstands of sea level.

The internal facies architecture of depositional sequences can be reduced strongly, and recognition of transgressive and highstand systems tracts may be uncertain, where outcrop data do not allow determination of the larger scale stratal pattern or potential condensed sections are eroded. Third-order depositional sequences could be determined only when high rates of sedimentation and subsidence occurred.

Though lowstands of sea level favour the development of turbidite or fan systems, the amount and type of sediments will significantly influence the type and timing of turbidite events. Very coarse channel-lobe systems will only form in mature island-arc systems when simultaneous tectonic uplift and high volcanic activity occur in addition to sea-level lowering. During periods of tectonic and volcanic quiescense, fine-grained, thin-bedded turbidite systems prevail even during lowstands of sea level.

Remarkable changes in sedimentation occurred

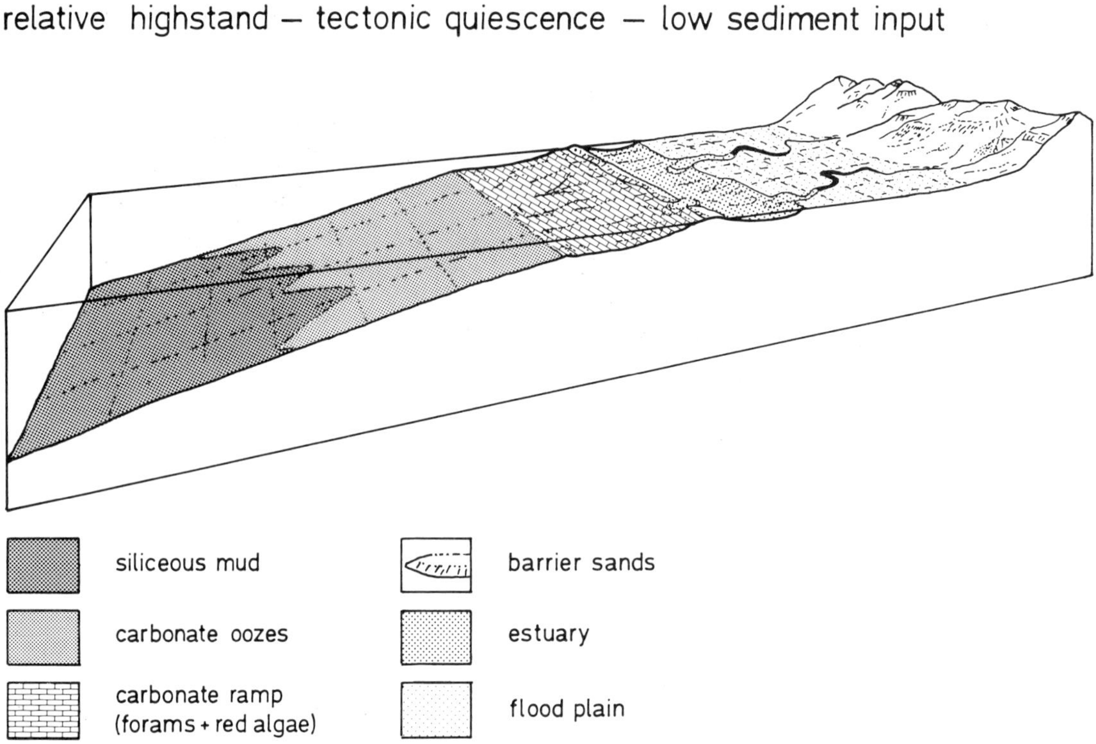


**Fig. 8.** A relative highstand of sea level, tectonic quiescense, and low sedimentation rates during late Paleocene to early Eocene times caused the formation of siliceous mudstones and limestones in the inner fore-arc basins, where carbonate ramps became established. Where siliciclastic neritic systems persisted, thin-bedded turbidites formed in the adjacent basin areas.

when tectonic/volcanic pulses and second-order cycles of eustatic fluctuation interfered. Therefore, uplift and subsidence of sediment-source areas acted as major controls on deposition of basinal cycles. Alternate immersion and exposure of the carbonate shelf that supplied the shallow-water carbonate sediments into the fore-arc basins caused long-term hybrid siliciclastic–carbonate basin cycles. The abruptness of lithologic transitions between carbonate breccias and siliciclastic deposits suggests that resedimentation of carbonate breccias mainly occurred owing to tectonic/volcanic events. Siliciclastic sedimentation indicates that most detritus reached the deep sea during relative sea-level lowstands, when the sediment bypassed subaerially exposed shelves. Basinal siliciclastic sedimentation decreases when most of the detritus was trapped in inner shelf environments. During relative highstands, mud- and sand-sized carbonate sediment was shed off neritic carbonate systems and transported into the fore-arc basins, resulting in increased carbonate contents in the basinal sediments. Conversely, during sea-level lowstands, subaerial exposure of platforms caused meteoric cementation and karstification and, as a result, only a little mud- and sand-sized carbonate shelf detritus reached the deep sea. In ramp settings a sea-level fall exposes only the upper parts of carbonate ramp systems and production of shallow-water carbonate sediment continues in the outer ramp environment (Shanmugam & Moiola, 1985; Sarg, 1988; Dolan, 1989). Such moderate or initial lowstands are documented by grainflow deposits and turbidites, rich in larger Foraminifera.

The most striking feature is the almost simultaneity of tectonic/volcanic activity and second-order cycles of eustatic sea-level fluctuations during Late Cretaceous, Paleocene, and mid-late Eocene times.

Many causes have been proposed for changes in global sea levels, by changing either the total volume of sea water or the total volume of the oceanic

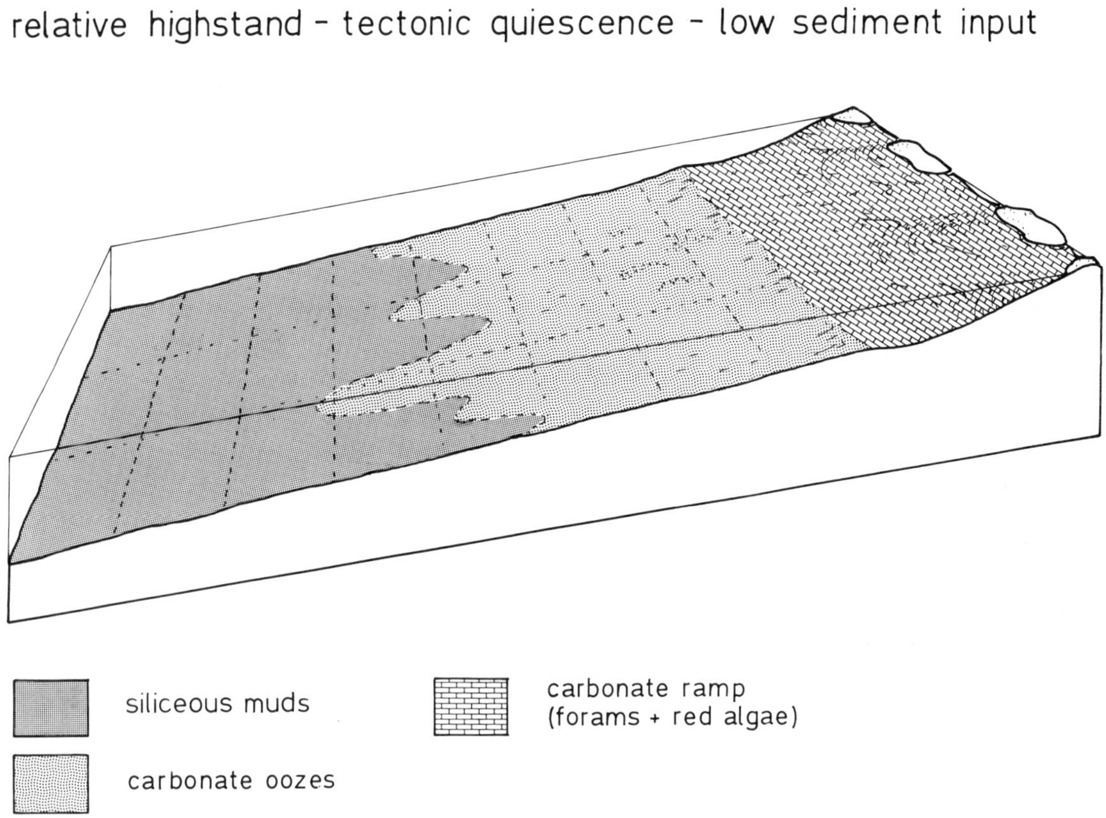


**Fig. 9.** A relative highstand of sea level, tectonic quiescense, and low sedimentation rates caused the formation of siliceous mudstones and limestones in the outer fore-arc basins during late Paleocene to middle Eocene times. Carbonate ramps, which concomitantly developed on the outer structural high, periodically delivered neritic carbonate material to the lower slope.

basins (Pitman, 1979; Howell & von Huene, 1980; Miall, 1984). Cloetingh (1988) proposed a tectonic mechanism for apparent sea-level fluctuations with a maximum magnitude of a few hundred metres. According to this model, tectonically induced vertical motions of the lithosphere may cause third-order cycles. Larger fluctuations in apparent sea level could be related to major reorganizations of lithospheric stress fields owing to rifting and fragmentation of plates, dynamic changes at convergent plate boundaries, or collision processes. Schwan (1980) compiled data showing that world-wide orogenic activity in the Late Jurassic to Recent, including the development of unconformities and compressional structures, metamorphic and plutonic episodes, could be correlated with major readjustments in the world plate patterns.

Late Cretaceous to late Eocene drift rates calculated from the North Atlantic Ocean correlate closely with global orogenic phases which in turn coincide with crustal movements in Central America. These episodes of major global tectonic/volcanic activity occurred at 80–75, 63, 55–53, and 42–38 Ma and agree with major tectonic and volcanic pulses in Central America (see paper by Seyfried *et al.*). This corroborates that global tectonic processes play an important role in generating major sea-level fluctuations and hence in the formation of depositional sequences.

## ACKNOWLEDGEMENTS

This study was supported by the Deutsche Forschungsgemeinschaft (DFG; Se 490/1-1, Se 490/1-3). The field trips were sponsored by grants of the Deutscher Akademischer Austauschdienst (DAAD). The research in Costa Rica was further

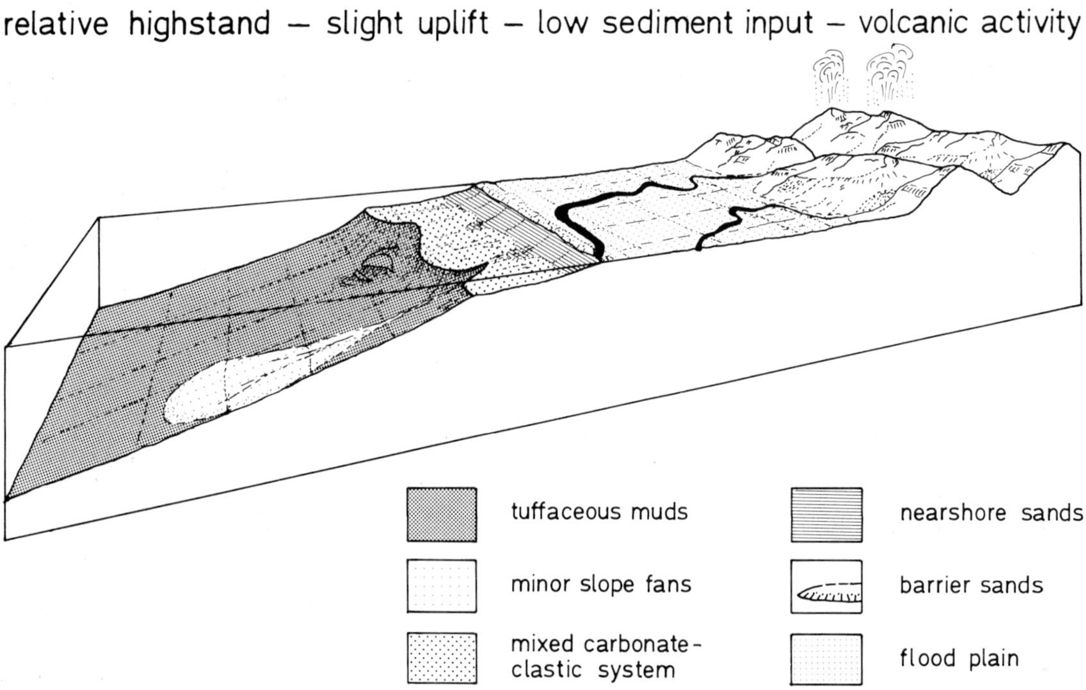


**Fig. 10.** A relative highstand of sea level, slight uplift, and low sedimentation rates during the middle Eocene led to the formation of tuffaceous mudstones as well as thin-bedded turbidites. Smaller slope fans developed as a result of a minor sea level lowstand or a local increase in volcaniclastic supply.

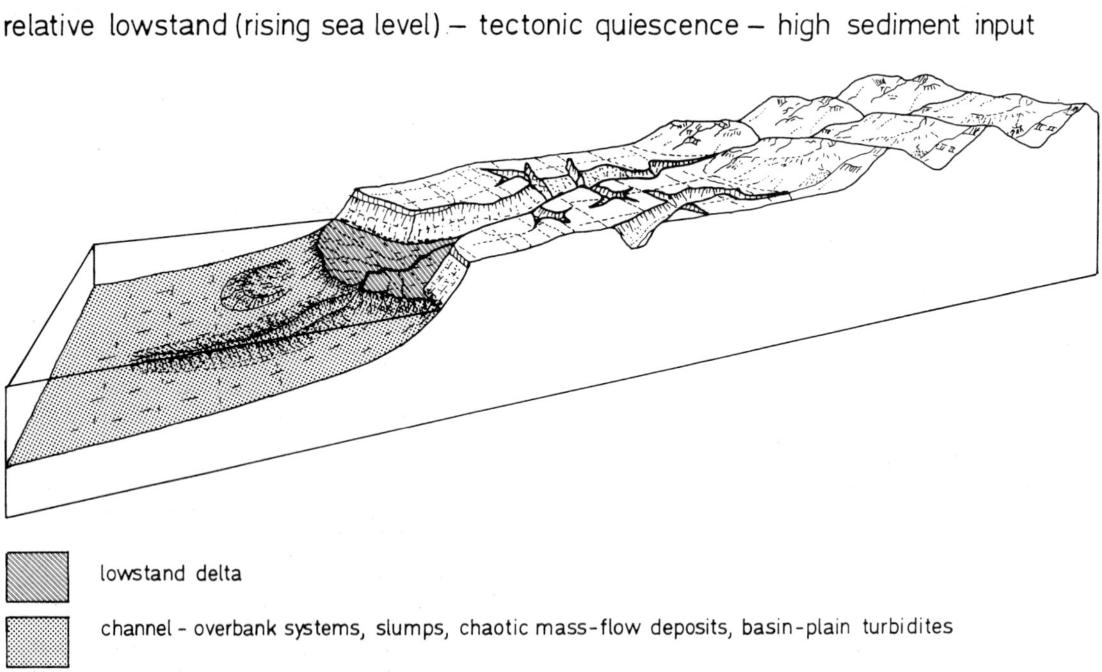


**Fig. 11.** A relative rise of the sea level in late Eocene times changed the coarse-grained sedimentation shown in Fig. 6 and channel overbank systems developed. Overbank deposits and overbank wedges are rich in organic material and must have been related to lowstand deltas.

supported by the close cooperation with the national oil company of Costa Rica (RECOPE) and the Department of Geology, University of Costa Rica. Transport facilities and field support were further provided by the Instituto Nicaragüense de la Minería (INMINE). We would like to thank David Macdonald (British Antarctic Survey) for his helpful comments and review of this paper.

## REFERENCES


ASTORGA, A. (1988) Geodinámica de las cuencas del Cretácico Superior-Paleógeno de la region 'forearc' del sur de Nicaragua y norte de Costa Rica. *Rev. Geol. Am. Central* **9**, 1–40.

ASTORGA, A., FERNANDEZ, J., BARBOZA, G., CAMPOS, L., OBANDO, J., AGUILAR, A. & OBANDO, L. (1989) Cuencas sedimentarias de Costa Rica: evolución Cretácico Superior-Cenozoica y potencial de hidrocarburos. *Transactions of the Fourth Circum-Pacific Council for Energy and Mineral Resources Conference*, Abstr. Vol., San José, pp. 1–18.

AZÉMA, J., GLACON, G., TOURNON, J. & VILA, J.-H. (1979) Precisiones acerca del Paleoceno de Puerto Quepos y sus alrededores, Provincia de Puntarenas, Costa Rica. *Inst. Geogr. Nac., Inf. Semestr. Julio-Dic.* **1979**, 77–88.

BARNES, N.E. & NORMARK, W.R. (1985) Diagnostic parameters for comparing modern submarine fans and ancient turbidite systems. In: *Submarine Fans and Related Turbidite Systems* (Eds Bouma, A.H., Normark, W.R. & Barnes, N.E.) pp. 13–14. Springer-Verlag, New York.

BAUMGARTNER, P.O., MORA, C.R., BUTTERLIN, J., SIGAL, J., GLACON, G., AZÉMA, J. & BOURGOIS, J. (1984) Sedimentación y paleogeografia del Cretácico y Cenozoico del litoral pacífico de Costa Rica. *Rev. Geol. Am. Central* **1**, 57–136.

BURBACH, G., FROHLICH, C., PENNINGTON, W. & MATUMOTO, T. (1984) Seismicity and tectonics of the subducted Cocos Plate. *J. Geophys. Res.* **81**, 31–63.

CALVO, C. (1987) *Las calizas neriticas de la vertiente Pacífica del norte de Costa Rica y sur de Nicaragua: epocas y sistemas de sedimentación asociadas, con la apertura y evolución del margen convergente de la América Central meridional*. Unpublished tesis de Licenciatura, University of Costa Rica, San José, 165 pp.

CLOETINGH, S. (1988) Intraplate stresses: a tectonic cause for third-order cycles in apparent sea-level? In: *Sea-level Changes: an Integrated Approach* (Eds Wilgus, C.K., Hastings, B.S., Kendall, C.G.St.C., Posamentier, H.W., Ross, C.A. & van Wagoner, J.C.) Soc. Econ. Paleontol. Mineral.,Spec. Publ. 42, 18–29.

DOLAN, J.F. (1989) Eustatic and tectonic controls on deposition of hybrid siliciclastic/carbonate basinal cycles: discussion with examples. *Bull., Am. Assoc. Petrol. Geol.* **73**, 1233–1246.

GURSKY, H.-J. (1984) *Die Sedimentgesteine im ophiolithischen Nicoya-Komplex (Oberjura bis Alt-Tertiär von Costa Rica): ihre Verbreitung, Fazies und geologische Geschichte mit besonderer Berücksichtigung der Radiolarite*. Unpublished dissertation, University of Marburg, 394 pp.

HAQ, B.U., HARDENBOL, J. & VAIL, P.R. (1988) Mesozoic and Cenozoic chronostratigraphy and eustatic cycles. In: *Sea-level Changes: an Integrated Approach* (Eds Wilgus, C.K., Hastings, B.S., Kendall, C.G. St. C., Posamentier, H.W., Ross, C.A. & van Wagoner, J.). Soc. Econ. Paleontol. Mineral., Spec. Publ. 42, 71–108.

HOFHERR, G. (1983) Geología de las calizas de Sapoá. *Inf. Final Proy. 05. Unidad Geol. COMNOMET* (unpublished). Instituto Nicaragüense de la Minería (INMINE), Managua, 53 pp.

HOWELL, D.G. & VON HUENE, R. (1980) Tectonics and sediment along active continental margins. *Soc. Econ. Paleont. Miner., Short Course*, San Francisco, **1980**.

HOWELL, D.G., CRAICH, J.K., GREENE, H.G., MCCULLOCH, D.S. & VEDDER, J.G. (1980) Basin development along the late Mesozoic and Cainozoic California margin: a plate tectonic margin of subduction, oblique subduction and transform tectonics. In: *Sedimentation in oblique-slip Mobile Zones* (Eds Ballance, P.F. & Reading, H.G.) Spec. Publ. Int. Ass. Sediment. 4, 43–62.

KOLLA, V. & MACURDA D.B., JR. (1988) Sea-level changes and timing of turbidity-current events in deep-sea fan systems. In: *Sea-level Changes: an Integrated Approach* (Eds Wilgus, C.K., Hastings, B.S., Kendall, C.G. St. C., Posamentier, H.W., Ross, C.A. & van Wagoner, J.). Soc. Econ. Paleontol. Mineral., Spec. Publ. 42, 381–392.

LEHFELD, H. (1989) *Biostratigraphie und taxonomische Untersuchungen an planktonischen Foraminiferen aus dem Alt-Tertiär-Profil von Garza bei Samara (Prov. Guanacaste, C.R.)*. Unpublished Diplomarbeit, University of Marburg, 68 pp.

LUNDBERG, N. (1982) Evolution of the slope landward of the Middle America Trench, Nicoya Peninsula, Costa Rica. In: *Trench–Forearc Geology* (Ed. Leggett, J.K.). Geol. Soc. London, Spec. Publ. 10, 133–147.

MIALL, A.D. (1984) *Principles of Sedimentary Basin Analysis*. Springer-Verlag, New York, 490 pp.

MITCHUM, R.M. (1985) Seismic stratigraphic expression of submarine fans. In: *Seismic Stratigraphy, II: an Integrated Approach to Hydrocarbon Exploration* (Eds Berg, O.R. & Woolverton, D.G.). Mem., Am. Assoc. Petrol. Geol. 39, 117–136.

MONTENAT, C., OTT D'ESTEVOU, P., LAROUZIERE, F.D. & BEDU, P. (1987) Originalité géodynamique des bassins néogènes du domaine bétique oriental. *Notes Mém., Total Co. Fr. Pétrole* **21**, 11–49.

MUTTI, E. (1985) Turbidite systems and their relations to depositional sequences. In: *Provenance of Arenites* (Ed. Zuffa, G.G.) pp. 65–93. D. Reidel, Dordect.

MUTTI, E. & NORMARK, W.R. (1987) Comparing examples of modern and ancient turbidite systems: problems and concepts. In: *Marine Clastic Sedimentology* (Eds Leggett, J.K. & Zuffa, G.G.) pp. 1–38. Graham & Trotman, London.

PITMAN, W.C., III (1979) The effect of eustatic sea level changes on stratigraphic sequences at Atlantic margins. *Mem., Am. Assoc. Petrol. Geol.* **29**, 453–460.

POSAMENTIER, H.W. & VAIL, P.R. (1988) Eustatic controls

on clastic deposition, II — sequence and systems tract models. In: *Sea-level Changes: an Integrated Approach* (Eds Wilgus, C.K., Hastings, B.S., Kendall, C.G. St. C., Posamentier, H.W., Ross., C.A. & van Wagoner, J.). Soc. Econ. Paleontol. Mineral., Spec. Publ. 42, 155–181.

Rivier, F. (1983) Síntesis geológica y mapa geológico del área del bajo tempisque, Guanacaste, Costa Rica. *Inst. Geogr. Nac., Inf. Semestr. Enero-Junio*, **1983**, 7–30.

Sarg, J.F. (1988) Carbonate sequence stratigraphy. In: *Sea-level Changes: an Integrated Approach* (Eds Wilgus, C.K., Hastings, B.S., Kendall, C.G. St. C., Posamentier, H.W., Ross, C.A. & van Wagoner, J.). Soc. Econ. Paleontol. Mineral., Spec. Publ. 42, 155–181.

Schmidt-Effing, R. (1979) Alter und Genese des Nicoya-Komplexes, einer ozeanischen Paläokruste (Oberjura bis Eozän) im südlichen Zentralamerika. *Geol. Rundsch.* **68**(2), 457–494.

Schwan, W. (1980) Geodynamic peaks in Alpino-type orogenesis and changes in ocean-floor spreading during Late Jurassic–Late Tertiary time. *Bull., Am. Assoc. Petrol. Geol.* **64**, 359–373.

Seyfried, H. & Sprechmann, P. (1986) Über die Frühgeschichte (Campan bis Eozän) der südlichen mittelamerikanischen Landbrücke. *Neues Jahrb. Geol. Palaeontol. Monatsh.* **1986**, 38–55.

Shanmugam, G. & Moiola, R.J. (1988) Submarine fans: characteristics, models, classification, and reservoir potential. *Earth Sci. Rev.* **24**, 383–428.

Shanmugam, G., Moiola, R.J. & Damuth, J.E. (1985) Eustatic control of submarine fan development. In: *Submarine Fans and Related Turbidite Systems* (Eds Bouma, A.H., Normark, W.R. & Barnes, N.E.) pp. 23–28. Springer-Verlag, New York.

Sick, M. (1989) Paleomagnetism of the Ophiolite Complexes from the southern Middle America land-bridge (Costa Rica and Panamá). *Tübinger Geowiss. Arb.* **4**, 1–108.

Sprechmann, P., Astorga, A., Bolz, A. & Calvo, C. (1987) Estratigrafía del Cretácico de Costa Rica. In: *El Cretácico de México y América Central* (Eds Barbarin, J.M., Gursky, H.-J. & Meiburg, P.). Actas Fac. Cienc. Tierra, Univ. Autonoma Nuevo León, Linares 2, 251–260.

Stow, D.A.V., Howell, D.G. & Nelson, C.H. (1985) Sedimentary, tectonic, and sea-level controls. In: *Submarine Fans and Related Turbidite Systems* (Eds Bouma, A.H., Normark, W.R. & Barnes, N.E.) pp. 15–22. Springer-Verlag, New York.

Vail, P.R. & Sangree, J.B. (1988) Sequence stratigraphy concepts. *Short course in conjunction with the 'Sequence, Stratigraphy, Sedimentology, Surface and Subsurface' Conference, Course Notes.* Can. Soc. Petrol. Geol., Calgary, September 1988.

Weimer, P. (1990) Sequence stratigraphy, facies geometries, and depositional history of the Mississippi Fan, Gulf of Mexico. *Bull., Am. Assoc. Petrol. Geol.* **74**, 425–453.

Weyl, R. (1980) *Geology of Central America*. Gebr. Borntraeger, Berlin, 371 pp.

*Spec. Publs. int. Ass. Sediment.* (1991) **12**, 293–305

# The role of tectonics and eustasy in the evolution of a marginal basin: Cretaceous–Tertiary Larsen Basin, Antarctica

D. PIRRIE[†] [*], A.G. WHITHAM[†] [**] *and* J.R. INESON[‡]

[†] *British Antarctic Survey, Natural Environment Research Council, High Cross, Madingley Rd, Cambridge CB3 OET, UK; and* [‡] *Geological Survey of Denmark, Thoravej 8, DK 2400, Copenhagen NV, Denmark*

## ABSTRACT

A 5–6 km thick, Cretaceous–Tertiary, marginal basin succession is exposed in the James Ross Island area (northern Antarctic Peninsula). Deep-marine fan facies and slope-apron facies (c. 2.5 km) are succeeded by shallow-marine shelf-deltaic deposits (c. 3 km). The deep-marine sedimentary rocks were deposited adjacent to a tectonically active basin margin and show little evidence of relative sea-level change. Intra- and extra-basinal tectonics were the major control on sedimentation; eustatic events have not been recognized. During the (?)Coniacian a peak in proximal volcanism increased sediment supply to the basin in the James Ross Island area. Continued tectonic uplift and passive basin filling led to partial shallowing within the basin. Widespread shallow-marine sedimentation did not develop until the Santonian–Campanian, coinciding with basin uplift, widening of the shelf, and decreased tectonic activity. Post-Santonian sedimentation was partially controlled by minor base-level changes, possibly eustatically driven, although the current biostratigraphic resolution does not allow direct correlation.

## INTRODUCTION

Differentiation between the effects of eustasy and tectonics in the sedimentary record has taxed generations of stratigraphers and sedimentologists (see Miall, 1984). Despite the enthusiastic response in recent years to sequence stratigraphy, as presented by Vail and co-workers (e.g. Vail *et al.*, 1977), the basis for the global eustatic curve has been disputed on a number of grounds. The role of local tectonics and regional intra-plate stresses in producing base-level variation has been emphasized by a number of workers (e.g. Cloetingh, 1988; Nemec, 1988), even within so-called passive margins (e.g. Hiscott *et al.*, 1990). The problem of biostratigraphic accuracy has also been highlighted (Parkinson & Summerhayes, 1985; Miall, 1986; Hubbard, 1988) since proving global synchroneity is considered a key point in the eustatic argument. If differentiating eustatic from tectonic effects is proving contentious in passive margin successions, greater ambiguity can be expected in active tectonic settings. The aim of this paper is to present a case study — the Cretaceous to Tertiary Larsen Basin, a back-arc basin that lies to the east of the Antarctic Peninsula (Fig. 1). The data base for this paper is mainly from outcrop studies in the James Ross Island area, which represents only the northern part of the Larsen Basin.

The Larsen Basin has been studied in some detail over the last decade (Elliot & Trautman, 1982; Ineson, 1985,1989; Macellari, 1988a; Pirrie, 1989), but is known primarily from surface exposures since no seismic data have been obtained to date, and the geophysical data base is restricted to regional gravity and aeromagnetic surveys. Furthermore, exposure within the basin is insufficient to construct the three-dimensional (or even well-constrained two-dimensional) transects that facilitate the recognition of the geometry of individual depositional tracts and

* Present Address: Camborne School of Mines, Redruth, Cornwall, TR15 3SE, UK.

** Present Address: Cambridge Arctic Shelf Programme, Gravel Hill, Huntingdon Road, Cambridge CB3 ODJ, UK.

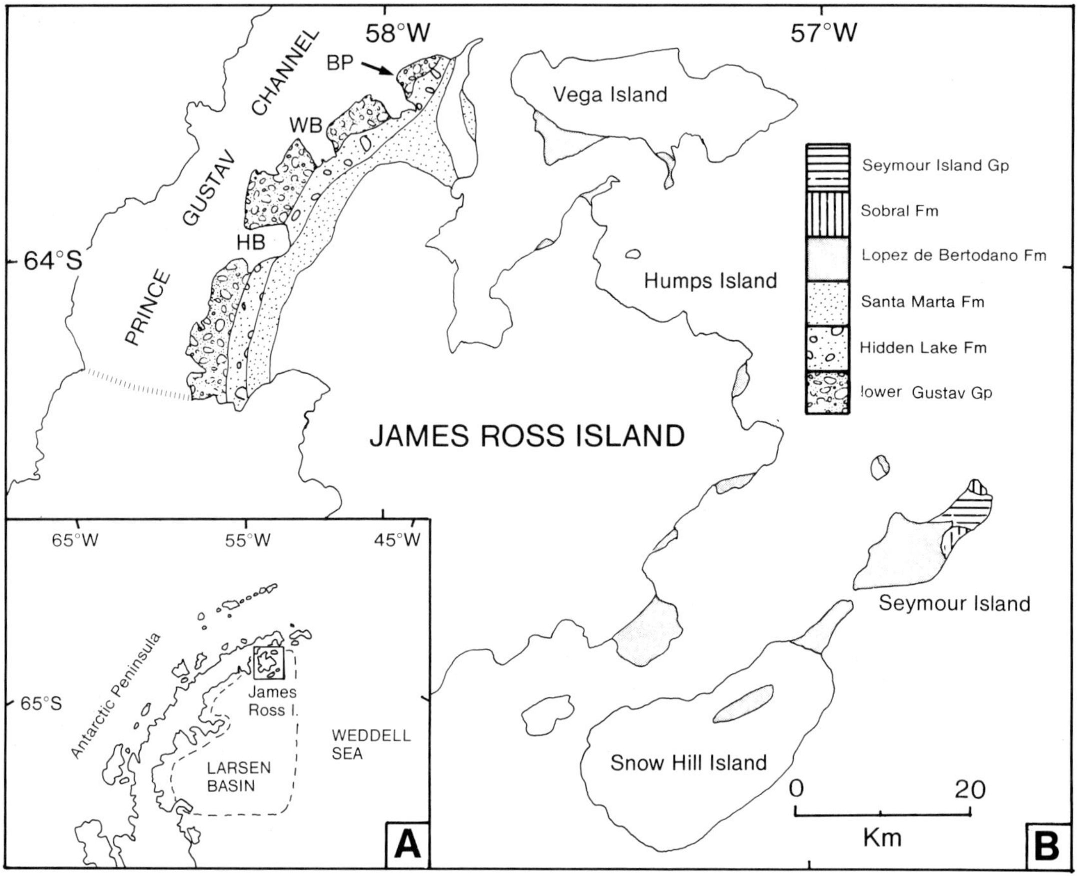


**Fig. 1.** Geological sketch map of the James Ross Island area, showing the distribution of (?)Barremian–(?)Oligocene sedimentary strata. Blank areas represent older or younger rocks or permanent ice cover. Inset map A shows the location of map B and the outline of the Larsen Basin. BP, Bibby Point; HB, Holluschickie Bay; WB, Whisky Bay.

sequences. The aim, therefore, is to describe the evolution of the basin based on classical outcrop techniques, and to discuss the problems and ambiguities of recognizing eustatic signatures in an active tectonic setting.

## GEOLOGICAL SETTING

The Antarctic Peninsula represents the eroded roots of a Mesozoic to Tertiary volcanic arc that formed during the subduction of proto-Pacific ocean crust. During the Cretaceous–Tertiary, a large sedimentary basin, termed the Larsen Basin by Macdonald *et al.* (1988), developed in the back-arc region to the east of the Antarctic Peninsula. The Cretaceous–Tertiary strata that crop out in the James Ross Island area represent part of the fill of this sedimentary basin.

The tectonic setting of the Larsen Basin is complex since both subduction-related processes and the opening of, and spreading within, the Weddell Sea affected its development. During Gondwana break-up, the Weddell Sea region underwent stretching and subsidence, with Cretaceous sea-floor spreading along an E–W-trending ridge. The orientation of this ridge rotated to NE–SW before subduction beneath the Scotia Arc (Barker *et al.*, in press). Storey & Nell (1988) suggested that the opening of the Weddell Sea may have led to strike-slip movement or oblique extension, recognized along the eastern margin of the Antarctic Peninsula (e.g. Whitham & Storey, 1989), and thus may have controlled the development of sedimentary basins along this margin. Although the Larsen Basin developed in a back-arc setting with its sediment fill derived wholly from the volcanic arc, the influence of back-arc tectonic processes during basin development is poorly known. However, the initiation of the basin coincided with a period of arc uplift and extensive volcanism, possibly related to the increased rate of subduction.

The Antarctic Peninsula is dominated by calc-alkaline arc volcanic and plutonic rocks that were intruded within a basement of deformed Permian–Triassic sedimentary rocks (the Trinity Peninsula Group (TPG)). Macdonald *et al.* (1988) inferred that the Larsen Basin is floored by either stretched TPG or earlier granitic basement. This basement is thought to be overlain by Upper Jurassic to Lower Cretaceous interbedded mudstones and tuffs (the Nordenskjöld Formation; Whitham & Doyle, 1989). Macdonald *et al.* (1988) considered this unit to represent the oldest strata within the basin. In the James Ross Island area a succession of Cretaceous–Tertiary sedimentary rocks crop out, with an exposed thickness in excess of 5 km (Fig. 2). This sequence has been subdivided into three major groups, the Gustav Group ((?)Barremian–Santonian), the Marambio Group (Santonian–Paleocene), and the Seymour Island Group (Paleocene–(?)Oligocene) (Elliott & Trautman, 1982; Ineson *et al.*, 1986; Olivero *et al.* 1986). Unconformably overlying this succession is a major suite of Miocene–Pliocene alkaline volcanics, the James Ross Island Volcanic Group. The extrusion of these volcanics has been related to a phase of extensional intra-plate volcanism (Smellie, 1987).

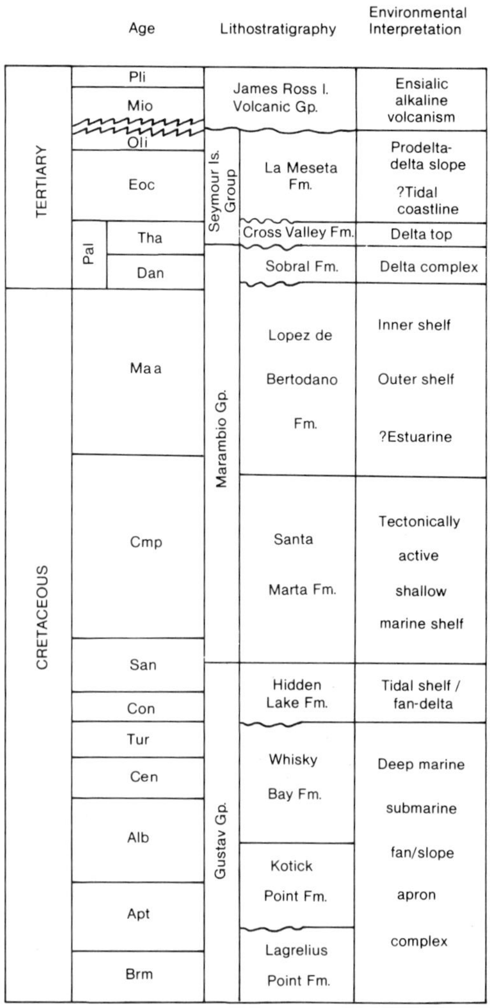


**Fig. 2.** Age, lithostratigraphy, and environmental interpretation of the James Ross Island succession.

## STRUCTURAL EVOLUTION OF THE JAMES ROSS ISLAND SUCCESSION

Throughout much of the James Ross Island area the sedimentary strata show a consistent dip towards the southeast. The most intense deformation is in northwest James Ross Island where the Gustav Group is exposed in a monocline. Dips are steepest along a line paralleling the coast between Lagrelius Point (8 km north of Holluschickie Bay) and Bibby Point (Fig. 1), where the dip ranges from subvertical to 45°. Southeast of this line, dips decrease progressively over a distance of approximately 4 km to 12°. Localized unconformities occur within this marginal zone of deformation, but do not represent significant breaks in sedimentation. There is much less structural deformation in the Marambio and Seymour Island groups, which crop out to the south and east of this zone of marginal deformation. Over much of the area covered by these units dips are < 10° and for the most part average 5–7°. The direction of dip is predominantly southeastward, although locally northeast dips are found (cf. Sadler, 1988). In a section on Seymour Island, low-angle dips show a progressive decrease up-section in a manner similar to that seen near the northwest margin of the basin.

Small-scale normal and reverse faults occur throughout the succession with displacements generally < 1 m; some of these faults are synsedimentary

(Doktor *et al.*, 1988; Pirrie & Riding, 1988; Sadler, 1988). Synsedimentary faulting occurs throughout the James Ross Island succession. There are also rare large-scale faults, such as that seen near Lagrelius Point where the James Ross Island Volcanic Group is down-thrown against sediments of the Lagrelius Point Formation. Another large-scale fault is postulated to cause the repetition of Marambio Group strata seen on the southeast coast of James Ross Island.

Although the structural evolution of the James Ross Island area is still poorly understood, it is becoming clear that the sedimentary successions were disturbed by several phases of deformation. Recent work by Whitham & Marshall (1988), combining vitrinite reflectance measurements and sedimentological observations, has shown that the monocline in which the Gustav Group is exposed formed during a protracted period of syndepositional deformation between the (?)Barremian and (?)Coniacian. The low dips in the strata away from this marginal zone may be a reflection of the fact that structural deformation was restricted to the basin margin. However, this cannot be stated with certainty since this is the only part of the region where pre-Santonian strata are exposed. The relative lack of deformation in post-Santonian strata coupled with the typical low vitrinite reflectance values of these sediments (Whitham & Marshall, 1988) is evidence of decreased tectonic activity. The presence of synsedimentary faulting implies some syndepositional tectonism; however, many faults of this nature are purely the result of sedimentary processes. The rare large-scale faults, and some small-scale faults, that transect the alkaline basalts of the James Ross Island Volcanic Group are of Tertiary age and related to the period of extension that accompanied the emplacement of these rocks.

## SEDIMENTOLOGICAL EVOLUTION OF THE JAMES ROSS ISLAND SUCCESSION

The Cretaceous–Tertiary succession is a varied sequence of siliciclastic sediments. Overall it is a regressive megasequence, which shows a transition from predominantly deep marine conditions (Gustav Group) to shallow marine conditions (Marambio and Seymour Island Groups). Three main stages are recognized and are described below.

### Stage 1: lower Gustav Group

The lower Gustav Group (Lagrelius Point, Kotick Point, and Whisky Bay Formations) form a heterogeneous succession of marine conglomerates, sandstones, and mudstones, c. 2000 m thick and of (?)Barremian to (?)Coniacian age (Fig. 3) (Ineson *et al.*, 1986). They crop out along the northwest coast of James Ross Island parallel to, and 5–15 km distant from, the basin margin, inferred from geophysical and geological data (Farquharson *et al.*, 1984). Conglomerate-dominated portions of this succession form three major composite bodies represented by the Lagrelius Point Formation ((?)Barremian–Aptian), the middle member of the Kotick Point Formation (late Aptian to early Albian) and the bulk of the Whisky Bay Formation (middle Albian to (?)Coniacian). Although the large-scale geometry of the first two conglomerate units is poorly known, the Whisky Bay Formation conglomerates form a composite lenticular body up to 700 m thick and c. 40 km long parallel to the basin margin. Clast- and matrix-supported conglomerates and pebbly sandstones dominate and typically show grading (normal and inverse), scoured bed contacts, imbrication, and parallel stratification. These facies were deposited by a range of sediment gravity flow processes in a relatively deep marine (substorm wave-base) setting; details of facies and inferred processes are given in Ineson (1989). The Whisky Bay Formation conglomerate body shows a marked radial palaeocurrent pattern, which, together with the lenticular morphology parallel to the basin margin and evidence of well-developed channel systems, indicates depositon on a submarine fan. Ineson (1989) recognized three distinct assemblages of facies representing axial fan channels, channel terraces or marginal ephemeral channels, and levee or interchannel settings. The vertical alternation of these assemblages defines a large-scale cyclicity (70–250 m) within the Whisky Bay Formation, which is attributed to autocyclic processes, such as fan channel switching or lateral migration (Ineson, 1989). Although more poorly exposed, the internal construction and facies development of the Lagrelius Point Formation and Kotick Point Formation conglomerate bodies are broadly similar to the Whisky Bay Formation and are interpreted to represent comparable gravelly submarine fan environments.

The conglomerate bodies are vertically separated by, and wedge out laterally into, monotonous successions of thinly interbedded graded sandstone tur-

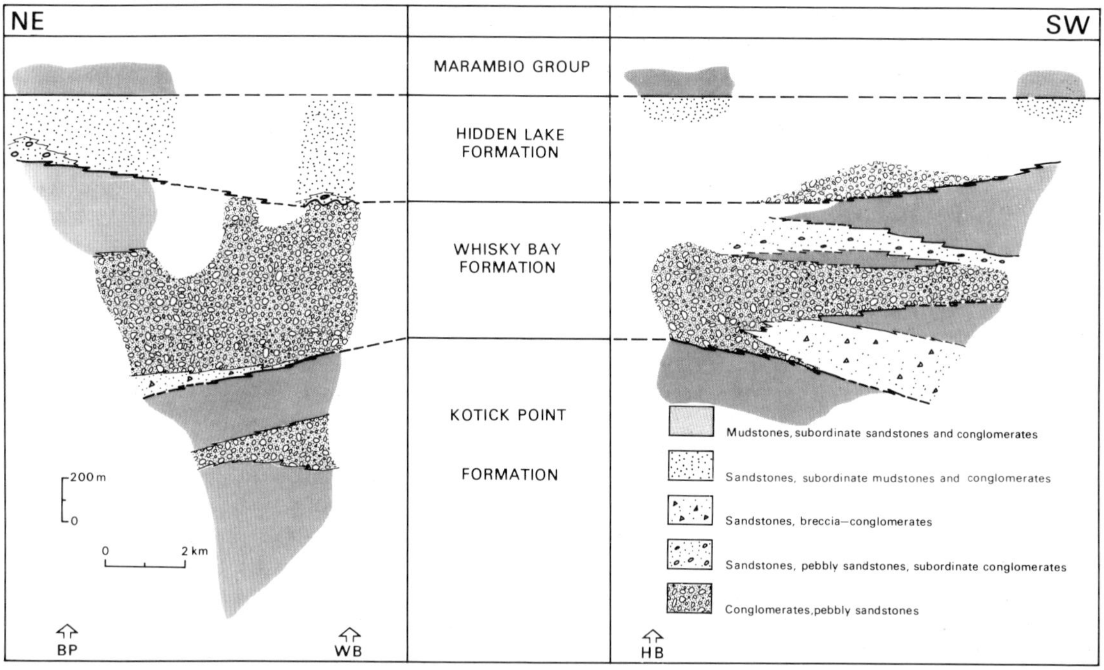


**Fig. 3.** NE–SW transect along the northwest coast of James Ross Island showing the distribution of the major sedimentary units in the Gustav Group; abbreviations as in Fig. 1 (modified from Ineson *et al.*, 1986).

bidites and bioturbated silty mudstones. These deposits are interspersed with thick mud-rich matrix-supported conglomerates (debris flow deposits), slump sheets, and sand-rich intervals with lenticular sandy breccia scour fills (see detailed analysis in Ineson (1989)). These mud-rich deposits are interpreted to represent a slope apron that flanked the basin margin and was locally interrupted by the gravelly submarine fan systems described previously. Slide blocks of the Jurassic Nordenskjöld Formation (up to 800 m across, see Ineson (1985)), stacked slump sheets, debris flow deposits, and angular disconformities within these slope apron deposits attest to periods of instability, and suggest that the slope apron depositional patterns were dictated largely by tectonism along the fault-controlled basin margin (Ineson, 1989). This is supported by the occurrence of abundant Lower Cretaceous intra-basinal clasts, typically reworked concretions, that occur within mass-flow debris sheets and testify to uplift and erosion of intra-basinal strata along the basin margin. Similarly, abundant reworked vitrinite in the succession argues for 'cannibalization' of intra-basinal strata along a tectonically active basin margin (Whitham & Marshall, 1988).

The 2-km thick succession accumulated in a relatively deep marine setting (estimated water depths 300–1000 m, see Ineson (1989)), and shows no inherent sedimentological evidence of relative sea-level change. This does not, however, preclude a link between sediment supply to the deep basin and eustatic sea-level variation, as proposed by many workers (e.g. Mutti, 1985). Sequence stratigraphy predicts that submarine fan development is favoured during sea-level lowstands, when sediment bypasses, or is eroded from, the narrow or exposed shelf. Such a relationship is not unique, however, particularly in basins in active tectonic settings that are undergoing active faulting, possess steep gradients, and are close to sources of coarse-grained sediments (Kolla & Macurda, 1988). In such settings, phases of coarse sediment supply and fan development may equally record discrete pulses of tectonic uplift in the source region (e.g. Surlyk, 1978). Indeed, Higgs (1990) recently suggested that deep-water gravel-rich fans (his 'deep-water fan deltas') may develop preferentially during sea-level highstands when alluvial fans emanating from continental highlands shed sediment directly into deep water.

Gravel-rich submarine fans developed in the proximal reaches of the Larsen Basin during the (?)Barremian–Aptian, the late Aptian to early Albian, and the middle Albian to (?)Coniacian. Given the uncertainty and overall ranges of these ages, correlation of these intervals with short-term global sea-level changes is clearly impossible and the role of eustasy in controlling basinal sedimentation patterns remains an open question. However, in view of the evidence outlined above, indicating a tectonically active basin margin and source area, it is most likely that the tectonic and magmatic evolution of the adjacent volcanic arc was the dominant factor controlling coarse sediment input and fan development.

**Stage 2: upper Gustav Group**

The uppermost formation of the Gustav Group, the Hidden Lake Formation, separates the relatively deep marine fan and slope apron strata from the overlying marine shelf strata of the Marambio Group (see stage 3; Pirrie, 1989). The Hidden Lake Formation is a distinctive suite of volcaniclastic sandstones, mudstones, and subordinate conglomerates, which is recognized along the length of northwest James Ross Island, parallel to the basin margin. Although facies vary along this transect, a characteristic feature of the formation is the abundance of volcanic detritus, such as pumice, accretionary lapilli, and reworked volcanic bombs. The sandstone petrography and microprobe analysis of clinopyroxene grains indicate derivation from an andesitic stratovolcanic source (Pirrie, 1991). The abundance of volcaniclastic detritus suggests a major peak of volcanism on the adjacent portion of the volcanic arc during the Coniacian.

The formation is 300–>400 m thick and shows a crude fining upward trend. The lower half to third of the formation consists of sandstones, pebbly sandstones, and matrix-supported conglomerates with inoceramid-bearing sandstones and mudstones. North of Bibby Point, thick-bedded graded and channelled pebbly sandstones are interbedded with thick (1–4 m) beds of parallel-stratified medium to coarse-grained sandstones. These facies are attributed to a variety of sediment gravity flow processes. Cross-stratified sandstones occur at regular intervals through the succession and are a characteristic feature of the lower Hidden Lake Formation as a whole, particularly north of Whisky Bay. The medium to large-scale cross-sets exhibit mudstone drapes on foresets and reactivation surfaces and indicate a consistent northwest-directed palaeocurrent, parallel to the basin margin. The succession in the area north of Bibby Point is interpreted as the toe-set of a fan delta, fed from a local source on the arc.

Farther south towards Whisky Bay (Fig. 2), the formation is generally finer grained, consisting of interbedded volcaniclastic sandstones and silty mudstones with localized slump sheets and cross-bedded intervals. Significantly, the slump sheets are mud-rich, in contrast to the sand-rich arc-derived sediment. Furthermore, slump fold vergence directions suggest derivation from the southeast. Evidence of intra-basinal topography and mass-wastage towards the arc is also observed in upper levels of the underlying Whisky Bay Formation 2 km south of Bibby Point, where a suite of syndepositional extensional normal faults downfault towards the northwest. These faults are associated with slumped strata and may represent the extensional zone of a major slide.

South of Holluschickie Bay, the lower interval of the formation consists of a poorly exposed succession of chaotic or graded, imbricated conglomerates, graded pebbly sandstone, and sandstones, comparable in many respects to the conglomeratic submarine fan deposits of stage 1. Poor exposure precludes detailed analysis of the depositional setting and a general 'base of slope' interpretation is preferred.

The upper Hidden Lake Formation is lithologically more uniform along its outcrop, typically consisting of thinly bedded, graded, laminated and cross-laminated volcaniclastic sandstones interbedded with bioturbated silty mudstones, isolated cross-sets, and thin slump sheets generally $< 1$ m.

Preliminary interpretations of the depositional setting relied heavily on the presence of cross-bedding of inferred tidal origin to propose a shallow marine environment for much of the formation (see Macdonald *et al.*, 1988; Ineson, 1989). Recent work on the sedimentology suggests that the evidence argues for a marine shelf setting below storm wave-base, in water depths in excess of 200 m, i.e. outer shelf depths. Evidence in support of this conclusion includes the absence of storm-induced stratification (hummocky cross-stratification and swaly cross-stratification), the dominance of sediment gravity flow deposits indicating appreciable marginal relief, the restricted ichnofauna, and a limited macrofauna dominated by inoceramids. Shallow-marine fossils only occur in sediment gravity flows and were clearly derived from a shallower setting at the basin margin. The presence of cross-bedding of probable tidal origin is interpreted to reflect tectonic modification of the basin morphology in the late Turonian. Reactivation and local inversion of marginal and intra-basinal faults is indicated by the angular unconformity at the base of the formation in Whisky Bay and indirectly by mass flow deposits derived from the southeast. It is proposed that differential uplift in the proximal region of the basin resulted in the formation of localized intra-basinal highs that acted both as a source of mud-rich mass flows and resulted in amplified tidal currents parallel to the basin axis. By analogy with the modern Straits of Messina, Colella & d'Alessandro (1988) proposed a bathyal setting ($>350$ m) for Plio-Pleistocene tidal sand-waves, demonstrating that currents flowing within a constricted basinal configuration are capable of moving sand-sized particles at water depths of several hundred metres.

The overall upward-fining nature of the Hidden Lake Formation reflects decreasing supply of coarse volcanic detritus to the basin and probably records a decline in proximal arc volcanism. The upward-fining trend is accompanied by a decrease in the number of slump sheets, debris flow deposits, and cross-bedded sandstones. This trend may reflect the decreasing importance of active intra-basinal tectonism during the Coniacian and thus the progressive reduction of basinal relief and water depth owing to sediment ponding and drape. The absence of appreciable tilting of the formation and younger beds, and the reduced proportion of reworked vitrinite and intra-basinal clasts (Whitham & Marshall, 1988), suggests a reduction in the amount of syndepositional tectonism along the basin margin in post-Coniacian times.

In conclusion, the Hidden Lake Formation records a discrete volcano-tectonic event, during which the configuration of the basin was altered, possibly owing to transpression and partial basin inversion. This coincided with a peak in volcanism and an influx of sediment from volcanic sources close to the basin margin. Although previously regarded as representing a marked shallowing event (Macdonald *et al.*, 1988; Ineson, 1989), re-evaluation indicates a less significant relative sea-level change from the Whisky Bay to Hidden Lake Formations. This minor fall in relative sea level is most readily attributed to tectonic uplift, in view of the evidence of basin reorganization outlined above. The effect of the postulated large fall in global sea level in the late Turonian (Haq *et al.*, 1987) has not been isolated.

**Stage 3: Marambio and Seymour Island Groups**

The Marambio and Seymour Island Groups reflect a change to typically fine-grained sedimentation within the basin. The Marambio Group, of Santonian to Paleocene age, is approximately 2.4 km thick. The group is subdivided into three formations, the Santa Marta, Lopez de Bertodano, and Sobral. The overlying Seymour Island Group is 850 m thick and can be subdivided into two formations, the Cross Valley Formation and overlying La Meseta Formation.

*Marambio Group*

The basal unit of the Marambio Group (facies association 1 of the Santa Marta Formation, see Pirrie (1989)) is approximately 750 m thick and composed of interbedded volcaniclastic sandstones and pebbly sandstones, silty sandstones, mudstones and scarce volcaniclastic conglomerates, intra-basinal conglomerates, and accretionary lapilli horizons. The volcaniclastic sandstones, pebbly sandstones, and conglomerates are typically ungraded or normally graded, massive to parallel laminated, and dominantly composed of andesitic lithic-volcanic grains, plagioclase, glass shards, and pumice. These facies are interpreted as being deposited from a range of mass flow processes including debris flows, high and low-concentration turbidity currents, and possibly storm-driven density currents, below storm wave-base (see detailed analysis in Pirrie (1989)). Accretionary lapilli horizons and scarce shard-rich units reflect air-fall sedimentation. These volcaniclastic facies are interbedded with finer-grained, typically bioturbated, massive silty sandstones and mudstones. The finer grained facies yield an abundant and diverse faunal assemblage. Poorly sorted, ungraded intra-formational conglomerates, yielding a diverse macrofaunal assemblage, also occur, and are interpreted as deposited from debris flows evolved from synsedimentary slumps and slides (Pirrie, 1989). Pirrie (1989) considered these sediments to have been deposited in a mid- to outer shelf environment, below storm wave-base. The input of coarser grained volcaniclastic facies, deposited by mass flow processes, was thought to have been controlled directly by increased rates of sediment supply to the basin following coeval arc volcanism. The finer grained facies were related to shelf processes similar to those described by Leithold (1989). The macrofaunal assemblage within the finer grained facies supports the environmental interpretation (e.g. Richter & Ward, 1990).

This unit is overlain by a sandstone-dominated packet on northern James Ross Island (see facies association 2 of Pirrie (1989)). It is composed of clean, well-sorted quartz-rich sandstones, in some beds showing vague large-scale planar cross-bedding, interbedded with hummocky cross-stratified sandstones, coquinas, and scarce siltstones and mudstones (Pirrie, 1989,1990). Some of the tabular cross-bedded sandstones have thin silt drapes on foresets. The sandstones are quartz-rich with an increased proportion of K-feldspar relative to plagioclase, interpreted to indicate an apparent cessation in proximal arc volcanism (Pirrie, 1991). This unit yields a less diverse macrofauna, with more shallow-marine affinities than the underlying unit (cf. Richter & Ward, 1990). The hummocky cross-stratified sandstones and coquinas are interpreted as deposited from storm-generated combined flow processes (Pirrie, 1989,1990) and are typically interbedded with finer grained suspension sedimentation deposits. The tabular cross-bedded sandstones with silt drapes on foresets may represent tidal sedimentation. Pirrie (1989,1990) interpreted this unit as deposited within an inner shelf setting, with sedimentation above storm wave-base. The transition between these two units was thought to reflect a regression, possibly in the late Campanian (Fig. 4). This inner shelf sandstone unit has been correlated with a sequence of interbedded sandstones and mudstones, deposited away from the basin margin on Humps Island (Pirrie & Riding, 1988). The sandstone beds on Humps Island are ungraded to normally graded, and usually structureless, although some beds show indistinct planar lamination, wavy lamination, and dish structures. These beds were interpreted by Pirrie & Riding (1988) as deposited from either turbidity currents or storm-related offshore-directed density currents deposited below storm wave-base. On the basis of the associated faunal assemblage a substorm wave-base, outer shelf depositional setting was proposed. The deposition of the sandstone beds was related to a regression in the late Campanian to early Maastrichtian.

The overlying Lopez de Bertodano Formation is composed of mudstones, silty sandstones, and sandstones. The base of the formation is best exposed on Vega Island where approximately 500 m of late Campanian to Maastrichtian strata crop out. The

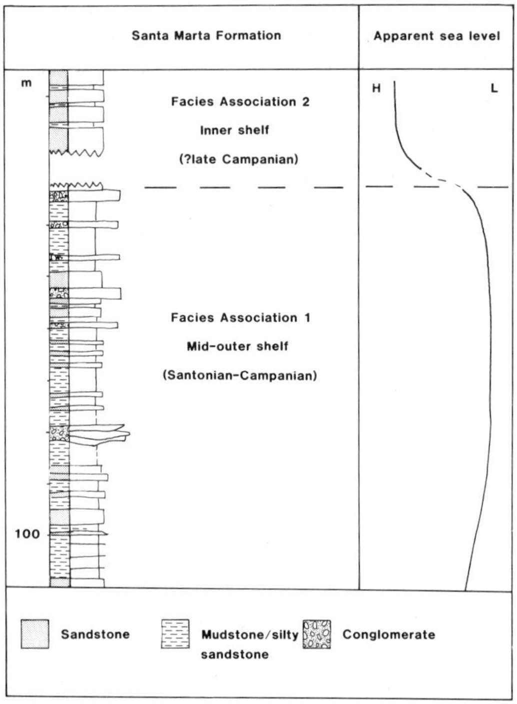


**Fig. 4.** Composite sedimentary log for the Santa Marta Formation. The transition from the mid- to outer shelf facies of association 1 to the inner shelf facies of association 2 is poorly exposed. This change is interpreted as representing a late Campanian regression.

underlying sandstone-dominated packet described above is overlain by fossiliferous, bioturbated muddy siltstones, which dominate the lower 260 m of the section, and coarsen up into a sandier dominated succession, erosionally overlain by a laterally persistent conglomerate horizon. This conglomerate horizon is overlain by fossiliferous silty mudstones and sandstones, in turn unconformably overlain by a sequence of coarse-grained volcaniclastic sandstones.

The uniform bioturbated silty mudstones represent suspension sedimentation in substorm wave-base water depths. The succession yields an abundant and diverse shallow-marine faunal assemblage indicative of a possible outer shelf depositional setting (cf. Macellari, 1988a). The coarsening upwards cycle from silty mudstones to silty sandstones and sandstones with a related decrease in the diversity and abundance of the associated faunal assemblage is comparable to sequences described by Leithold (1989). This coarsening upwards cycle is interpreted to reflect a period of regression. The erosionally overlying conglomerate horizon represents a hiatus, with a distinct change in palynomorph assemblages at this level. The conglomerate is directly comparable to the transgressive lag facies described by Plint & Walker (1987), and is overlain by silty mudstones and sandstones that are similar to the outer shelf sediments below. Thus this succession is interpreted as representing a minor regressive–transgressive cycle.

Younger Lopez de Bertodano Formation strata crop out on Seymour Island, to the southeast, away from the inferred basin margin. This section is again dominated by silty mudstones and silty sandstones, which, on the basis of palaeoecological data, have been interpreted by Macellari (1988a) as representing a transgressive through to regressive cycle, with deposition in a range of shelf settings. Macellari (1988a) interpreted the sequence as representing a transgression through the Maastrichtian, with a regression in the late Maastrichtian and in the early Paleocene. A late Maastrichtian regression is also indicated by the volcaniclastic sandstones that unconformably overlie the section on Vega Island. This section is composed of interbedded mudstones and sandstones rich in lithic-volcanics. Some sandstone beds show hummocky cross-stratification and wave rippling, and, at the top of the section, a thin rootlet horizon with dominantly non-marine palynomorphs indicates periodic emergence. Palynological studies have indicated a late Maastrichtian age for this unit. The latest Maastrichtian sediments on both Vega and Seymour Islands show increasing proportions of plagioclase and lithic-volcanic grains relative to quartz, implying a renewal of proximal coeval arc volcanism (Macellari, 1988a).

The Lopez de Bertodano Formation is unconformably overlain by the Sobral Formation on Seymour Island (Sadler, 1988). This unconformity has been dated as early Paleocene (Danian). The Sobral Formation is of Danian age and is composed of interbedded sandstones, silty sandstones, and mudstones. Tabular and trough cross-stratified sandstones are interbedded with bioturbated silty sandstones. Macellari (1988a) interpreted the Sobral Formation as the prograding facies of a delta complex, with deposition in prodelta slope, coastal barrier sands, and delta top environments, with intercalated marine and non-marine strata at the top of the section.

### *Seymour Island Group*

The unconformably overlying Seymour Island Group is divided into two main units, a basal sandstone-dominated, volcaniclastic-rich unit (the Cross Valley Formation) and the overlying finer grained, more quartzose, La Meseta Formation. The Cross Valley Formation is typified by coarse-grained tabular and trough cross-bedded volcaniclastic sandstones, and planar bedded pebbly sandstones. Scarce marine trace fossils and invertebrates occur, along with rare rootlet horizons implying periodic emergence. The sandstones are highly volcaniclastic with abundant pumice, glass shards, and andesitic lithic fragments. Elliot & Trautman (1982) interpreted the formation to have been deposited in distributary channels and interdistributary areas of a delta, with sedimentation related to a pulse of arc volcanism, supplying coarse detritus to the basin.

A major unconformity of Eocene age, with at least 70 m of erosional relief, marks the base of the overlying La Meseta Formation (Sadler, 1988). Sadler (1988) considered that the La Meseta Formation is a grossly lenticular unit, deposited within a 'large channel' cut into an emergent shelf. The sedimentary infill at the base of the unconformity appears to show onlap on to the margins of the unconformity. Reworked clasts from the Sobral and Cross Valley Formations also occur near the base of the formation. The La Meseta Formation is composed of a number of coarsening upward cycles, 100–150 m thick, with interbedded sandstones–mudstones at the base, and sandstone-dominated tops. Elliot & Trautman (1982) interpreted the formation as deposited in prodelta, tide-dominated delta top, and possibly lagoonal environments. Recent work by Doktor *et al.* (1988) suggested deposition in a tide influenced, prograding environment. Growth faulting, fault-bounded channels with slumped walls, and angular unconformities are attributed to syndepositional tectonic instability (Doktor *et al.*, 1988).

As described above, the Marambio and Seymour Island Groups reflect typically shallow-marine sedimentation in shelf-deltaic environments. Basin uplift (or decreased basin subsidence outpaced by the sedimentation rate) led to the development of a broad shallow shelf, possibly 50 km wide, with southeastward progradation of the basin margin. Syndepositional tectonic deformation was significantly reduced during this period of basin development. Small-scale synsedimentary faulting does occur in both groups, but is relatively minor. More intense synsedimentary deformation is only recorded from the La Meseta Formation, although much of this deformation can be related to the formation's deltaic depositional setting (Doktor *et al.*, 1988). In contrast to the underlying Gustav Group, there is little evidence for tectonically induced sedimentation, such as intra-basinal sediment reworking. Sediment reworking is only seen above major unconformities, for example reworking of Sobral and Cross Valley Formation clasts into the base of the La Meseta Formation. Although periodic coeval arc volcanism is shown by both petrographic and facies studies (e.g. the Cross Valley Formation), the volcanic arc was considerably less active proximal to the basin margin during the Late Cretaceous to Tertiary. Decreased volcanism led to a decreased supply of coarse-grained sediment to the basin.

As a consequence of the development of a broad stable shelf environment, decreased synsedimentary tectonic deformation, and decreased coeval arc volcanism, sedimentation was more directly controlled by small-scale changes in base level. A series of regressive–transgressive cycles can be recognized within these shallow-marine strata. The sandstone-dominated packet at the top of the Santa Marta Formation reflects a basin-wide late Campanian to early Maastrichtian regressive event. This is overlain by the outer shelf sediments of the basal Lopez de Bertodano Formation, which shows a minor regressive–transgressive cycle in the early Maastrichtian, with a late Maastrichtian regression coinciding with a pulse in coeval arc volcanism. Regressive cycles are also recognized in the early Paleocene with the major unconformity at the base of the La Maseta Formation dated as Ypresian. On the basis of both detailed facies studies and palaeoecology, the scale of the relative base-level changes recognized can be estimated to be in the order of 50–100 m, assuming a maximum outer shelf water depth of approximately 200 m. The frequency and scale of the base-level changes seen is comparable to that suggested by the Haq *et al.* (1987) sea-level curve.

## DISCUSSION

As noted by many authors (e.g. Cloetingh & Kooi, 1989), discrimination between the effects of tectonics and eustasy can be difficult, particularly where the biostratigraphic framework is limited in resolution. In the Cretaceous of the James Ross Island area many of the major stratigraphic packets have a possible error of ±5 Ma. With this level of accuracy, which is probably not atypical of many basins, correlation with sea-level cycles on the Exxon global eustatic curve is possible but highly equivocal. In any event, given the continuing debate concerning the validity and/or biostratigraphic accuracy of these cycles, it is perhaps unwise to correlate even a well-dated sequence boundary to the global sea-level curve purely on the basis of synchroneity. As discussed by Cloetingh & Kooi (1989), it is important to support or refute such a correlation by detailed study of stratigraphic criteria that aid differentiation of eustatic and tectonic effects.

Clearly, rigorous analysis of the controlling factors in basin evolution requires both a high-resolution biostratigraphy and a detailed knowledge of the structural and stratigraphic development of the basin (see e.g. Morton, 1989). Although our present knowledge of the Larsen Basin is insufficient to confidently differentiate the effects of tectonism from those of eustasy on sedimentation throughout the basin's history, there are some significant conclusions that can be drawn from this case study.

The three evolutionary stages described show a progressive decrease in the influence of tectonism on sedimentation through the Cretaceous and early Tertiary. The (?)Barremian to Turonian history of the basin (stage 1) is one of continued active tectonism along the basin margin, reflected in deformed marginal strata, local angular unconformities, and redeposition or 'cannibalization' of intra-basinal deposits. Discrete periods of submarine fan development were most probably linked to tectonic uplift in the source area; the influence of eustatic sea-level variation has not been recognized.

In the Coniacian (stage 2) an influx of coarse volcaniclastic sediment followed flexure and the development of a localized angular unconformity. Following flexure, the development of the marginal monocline ceased and progressive shallowing occurred. There is minor shallowing associated with this event which is attributed to partial basin inversion and tectonic reorganization of the proximal reaches of the basin. The eustatic sea level fall at 90 Ma (Haq *et al.*, 1987) may have contributed to this shallowing, but the imprint of magmatism and tectonism dominates.

During stage 3, in post-Coniacian strata, syndepositional flexure cannot be proven, suggesting that marginal and intra-basinal tectonism was on the decline. Decreased intra-basinal sediment reworking, and the development of a broad shallow shelf, both support the decreasing influence of synsedimentary tectonism on sedimentation. Sedimentation was more directly controlled by a series of relative base-level changes, with a magnitude of 50–100 m. Whether these base-level changes were tectonically or eustatically driven is poorly constrained; current biostratigraphic resolution within the basin does not allow direct correlation with the global cycle chart of Haq *et al.* (1987). For example, the regression dated as late Campanian to early Maastrichtian within the basin may correspond to the 75 Ma lowstand on the Haq *et al.* (1987) cycle chart. However, compressional synsedimentary faulting of this age implies a possible tectonic control on the base-level change. The repeated pattern of regressions/transgressions within the Late Cretaceous of the basin are broadly comparable to depositional cycles within the Austral Basin of southern South America (Macellari, 1988b; Macellari *et al.*, 1989). A late Campanian regressive sandstone packet followed by a transgressive cycle and subsequent Maastrichtian regression are recognized (Macellari, 1988b; Macellari *et al.*, 1989). Macellari (1988b), however, considered the base-level changes within the Cretaceous basins of southern South America to be due to the interplay of both tectonic and eustatic controls.

Within the Tertiary sequence on Seymour Island, the unconformities at the base of the Sobral and Cross Valley Formations do not correspond to lowstands on the global cycle chart. The major unconformity at the base of the La Meseta Formation does correspond to the documented late Ypresian (49.5 Ma) lowstand (Sadler, 1988). This well-dated major unconformity probably represents an inter-regional unconformity, and is the only major base-level change within the basin that can, at present, be correlated with any degree of confidence with the global cycle chart of Haq *et al.* (1987).

The changing style of sedimentation within the basin, from a tectonically active margin during stage 1, declining during stage 2 to an essentially quiescent margin during stage 3, corresponds well to the changing tectonic evolution of the region. By the Late Cretaceous, the major tectonic influences of subduction and opening of the Weddell Sea had decreased in significance. Subduction-related magmatism migrated northwestward away from the basin margin (Hole *et al.*, 1991). Furthermore, the Weddell Sea came to the end of its major period of active development. As a consequence of these tectonic changes, basin margin tectonism declined, with a corresponding increase in the width of the shelf by stage 3. In the initial stages of basin development, intra- and extra-basinal tectonics were the dominant controls on sedimentation, masking the possible influences of eustatic sea-level variation. Conversely the effects of eustatic base-level changes can tentatively be recognized during the latter stage of basin development as tectonic activity declined. However, the current biostratigraphic resolution does not allow these base-level changes to be correlated with the global cycle charts.

Despite the present uncertainties within the basin it is important to persevere with basin studies such as this, especially from the lesser-known continents, as the assimilation of independent data from tectonically unrelated basins is essential to test the 'globality' of the sea-level variations documented by the Exxon group.

## ACKNOWLEDGEMENTS

The authors would like to thank P.J. Butterworth and D.I.M. Macdonald for commenting on earlier drafts of this paper, and M.E. Field and M.G. Audley-Charles for their constructive reviews.

## REFERENCES

Barker, P.F., Dalziel, I.W.D. & Storey, B.C. (in press) Tectonic development of the Scotia Arc region. In: *The Geology of Antarctica* (Ed. Tingey, R.J.). Oxford University Press, Oxford.

Cloetingh, S. (1988) Intraplate stress: new perspective in basin analysis. In: *New Perspectives in Basin Analysis* (Eds Kleinspehn, K.L. & Paola, C.) pp. 205–230. Springer-Verlag, New York.

Cloetingh, S. & Kooi, H. (1989) Tectonic subsidence and sea-level change: a reappraisal. In: *Correlation in Hydrocarbon Exploration* (Ed. Collinson, J.D.) pp. 3–11. Graham & Trotman, London.

Colella, A. & d'Alessandro, A. (1988) Sand waves, *Echinocardium* traces and their bathyal depositional setting (Monte Torre Palaeostrait, Plio-Pleistocene, southern Italy). *Sedimentology* **35**, 219–237.

Doktor, M., Gazdzicki, A., Marenssi, S.A., Porebski, S.J., Santillana, S.N. & Vrba, A.V. (1988) Argentine–Polish geological investigations on Seymour (Marambio) Island, Antarctica, 1988. *Polish Polar Res.* **9**, 521–541.

Elliot, D.H. & Trautman, T.A. (1982) Lower Tertiary strata on Seymour Island, Antarctic Peninsula. In: *Antarctic Geoscience* (Ed. Craddock, C.) pp. 287–297. University of Wisconsin Press, Madison, WI.

Farquharson, G.W., Hamer, R.D. & Ineson, J.R. (1984) Proximal volcaniclastic sedimentation in a Cretaceous back-arc basin, northern Antarctic Peninsula. In: *Mar-*

*ginal Basin Geology* (Eds Kokelaar, B.P. & Howells, M.F.). Geol. Soc. London, Spec. Publ. 16, 219–229.

Haq, B.U., Hardenbol, J. & Vail, P.R. (1987) Chronology of fluctuating sea-levels since the Triassic. *Science* **235**, 1156–1167.

Higgs, R. (1990) Sedimentology and tectonic implications of Cretaceous fan-delta conglomerates, Queen Charlotte Islands, Canada. *Sedimentology* **37**, 83–103.

Hiscott, R.N., Wilson, R.C.L., Gradstein, F.M., Pujalte, V., Garcia-Mondejar, J., Boudreau, R.R. & Wishart, H.A. (1990) Comparative stratigraphy and subsidence history of Mesozoic rift basins of North Atlantic. *Bull., Am. Assoc. Petrol. Geol.* **74**, 60–76.

Hole, M.J., Pankhurst, R.J. & Saunders, A.D. (1991) Geochemical evolution of the Antarctic Peninsula magmatic arc; the importance of mantle–crust interactions during magma genesis. In: *Geological Evolution of Antarctica* (Eds Thomson, M.R.A., Crame, J.A. & Thomson, J.W.). Cambridge University Press, Cambridge.

Hubbard, R.J. (1988) Age and significance of sequence boundaries on Jurassic and Early Cretaceous rifted continental margins. *Bull., Am. Assoc. Petrol. Geol.* **72**, 49–72.

Ineson, J.R. (1985) Submarine glide blocks from the Lower Cretaceous of the Antarctic Peninsula. *Sedimentology* **32**, 659–670.

Ineson, J.R. (1989) Coarse-grained submarine fan and slope apron deposits in a Cretaceous back-arc basin, Antarctica. *Sedimentology* **36**, 793–819.

Ineson, J.R., Crame, J.A. & Thomson, M.R.A. (1986) Lithostratigraphy of the Cretaceous strata of west James Ross Island, Antarctica. *Cretaceous Res.* **7**, 141–159.

Kolla, V. & Macurda, D.B. (1988) Sea-level changes and timing of turbidity current events in deep-sea fan systems. In: *Sea-level Changes: an Integrated Approach* (Eds Wilgus, C.K., Hastings, B.S., Kendall, C.G.St.C., Posamentier, H.W., Ross, C.A. & van Wagoner, J.C.) Soc. Econ. Paleontol. Mineral., Spec. Publ. 42, 381–392.

Leithold, E.L. (1989) Depositional processes on an ancient and modern muddy shelf, northern California. *Sedimentology* **36**, 179–202.

Macdonald, D.I.M., Barker, P.F., Garrett, S.W., Ineson, J.R., Pirrie, D., Storey, B.C., Whitham, A.G., Kinghorn, R.R.F. & Marshall, J.E.A. (1988) A preliminary assessment of the hydrocarbon potential of the Larsen Basin, Antarctica. *Mar. Petrol. Geol.* **5**, 34–53.

Macellari, C.E. (1988a) Stratigraphy, sedimentology and paleontology of late Cretaceous/Paleocene shelf deltaic sediments of Seymour Island. In: *Geology and Paleontology of Seymour Island, Antarctic Peninsula* (Eds Feldmann, R.M. & Woodburne, M.O.). Geol. Soc. Am. Mem. 169, 25–53.

Macellari, C.E. (1988b) Cretaceous paleogeography and depositional cycles of western South America. *J. S. Am. Earth Sci.* **1**, 373–418.

Macellari, C.E., Barrio, C.A. & Manassero, M.J. (1989) Upper Cretaceous to Paleocene depositional sequences and sandstone petrography of southwestern Patagonia (Argentina and Chile). *J. S. Am. Earth Sci.* **2**, 223–239.

Miall, A.D. (1984) *Principles of Sedimentary Basin Analysis*. Springer-Verlag, New York, 490 pp.

Miall, A.D. (1986) Eustatic sea-level changes interpreted from sequence stratigraphy: a critique of the methodology with particular reference to the North Sea Jurassic record. *Bull. Am. Assoc. Petrol. Geol.* **70**, 131–137.

Morton, N. (1989) Jurassic sequence stratigraphy in the Hebrides Basin, N.W. Scotland. *Mar. Petrol. Geol.* **6**, 243–260.

Mutti, E. (1985) Turbidite systems and their relations to depositional sequences. In: *Provenance of Arenites* (Ed. Zuffa, G.G.). pp. 65–93. D. Reidel, Dordrecht.

Nemec, W. (1988) Coal correlation and intrabasinal subsidence: a new analytical perspective. In: *New Perspectives in Basin Analysis* (Eds Kleinspehn, K. & Paola, C.) pp. 161–188. Springer-Verlag, New York.

Olivero, E.B., Scasso, R.A. & Rinaldi, C.A. (1986) Revision of the Marambio Group, James Ross Island, Antarctica. *Contrib. Inst. Antárt. Argent.* **331**, 1–29.

Parkinson, N. & Summerhayes, C. (1985) Synchronous global sequence boundaries. *Bull., Am. Assoc. Petrol. Geol.* **69**, 685–687.

Pirrie, D. (1989) Shallow-marine sedimentation within an active margin basin, James Ross Island, Antarctica. *Sediment. Geol.* **63**, 61–82.

Pirrie, D. (1990) A new sedimentological interpretation for part of the Santa Marta Formation, James Ross Island. *Antarct. Sci.* **2**, 77–78.

Pirrie, D. (1991) Controls on the petrographic evolution of an active margin sedimentary sequence: The Larsen Basin, Antarctica. In: *Developments in Sedimentary Provenance Studies* (Eds Morton, A.C., Todd, S.P. & Haughton, P.D.W.) Geol. Soc. London, Spec. Publ. 57, 231–249.

Pirrie, D. & Riding, J.B. (1988) Sedimentology, palynology and structure of Humps Island, northern Antarctic Peninsula. *Br. Antarct. Surv. Bull.* **80**, 1–19.

Plint, A.G. & Walker, R.G. (1987) Cardium Formation 8. Facies and environments of the Cardium shoreline and coastal plain in the Kakwa field and adjacent areas, northwestern Alberta. *Bull. Can. Petrol. Geol.* **35**, 48–64.

Richter, M. & Ward, D.J. (1990) Fish remains from the Santa Marta Formation (Late Cretaceous) of James Ross Island, Antarctica. *Antarct. Sci.* **2**, 67–76.

Sadler, P.M. (1988) Geometry and stratification of uppermost Cretaceous and Paleogene units on Seymour Island, northern Antarctic Peninsula. In: *Geology and Paleontology of Seymour Island, Antarctic Peninsula* (Eds Feldmann, R.M. & Woodburne, M.O.). Geol. Soc. Am. Mem. 169, 303–320.

Smellie, J.L. (1987) Geochemistry and tectonic setting of alkaline volcanic rocks in the Antarctic Peninsula: a review. *J. Volcanol. Geotherm. Res.* **32**, 269–285.

Storey, B.C. & Nell, P.A.R. (1988) Role of strike-slip faulting in the tectonic evolution of the Antarctic Peninsula. *J. Geol. Soc. London* **145**, 33–337.

Surlyk, F. (1978) Submarine fan sedimentation along fault scarps on tilted fault blocks (Jurassic–Cretaceous boundary, East Greenland). *Bull. Grnl. Geol. Unders.* **128**, 1–108.

Vail, P.R., Mitchum, R.M., Todd, R.G., Widmier, J.M., Thompson, S., Sangree, J.B., Bubb, J.N. & Hatlelid, W.G. (1977) Seismic stratigraphy and global changes of sea-level, parts 1–6. In: *Seismic Stratigraphy – Applications to Hydrocarbon Exploration* (Ed. Payton, C.E.). Mem., Am. Assoc. Petrol. Geol. 26, 49–212.

WHITHAM, A.G. & DOYLE, P. (1989) Stratigraphy of the Upper Jurassic–Lower Cretaceous Nordenskjold Formation of eastern Graham Land, Antarctica. *J. S. Am. Earth Sci.* **2**, 371–384.

WHITHAM, A.G. & MARSHALL, J.E.A. (1988) Syn-depositional deformation in a Cretaceous succession, James Ross Island, Antarctica: evidence from vitrinite reflectivity. *Geol. Mag.* **125**, 583–591.

WHITHAM, A.G. & STOREY, B.C. (1989) Evidence for strike-slip deformation during Late Jurassic and Early Cretaceous times from the east coast of northern Graham Land, Antarctic Peninsula. *Antarct. Sci.* **1**, 269–278.

*Spec. Publs. int. Ass. Sediment.* (1991) **12**, 307–329

# The role of eustasy in the development of a regional shallowing event in a tectonically active basin: Fossil Bluff Group (Jurassic–Cretaceous), Alexander Island, Antarctica

P.J. BUTTERWORTH*

*British Antarctic Survey, Natural Environment Research Council, High Cross, Madingley Road, Cambridge, CB3 0ET, UK*

## ABSTRACT

A 7 km thick, Kimmeridgian–Albian fore-arc basin succession, unconformably overlying an accretionary complex, is exposed on the east coast of Alexander Island adjacent to the Antarctic Peninsula. The basin fill represents a single regressive megasequence, strongly controlled by active faulting along the arc-basin margin. This regressive megasequence comprises a large-scale (minimum 440 m thick) slope-collapse deposit overlain by a 2.2 km thick Tithonian–Berriasian sequence of proximal submarine fan deposits that record three tectonic pulses of coarse clastic sediment input. Intra-basinal synsedimentary tectonism was the primary control on the development of the overlying, 1 km thick, slope mudstone sequence (Valanginian–Aptian). The basin fill is capped by an Aptian–Albian shallow-marine and terrestrial sandstone sequence, 3.5 km thick, which is probably diachronous across the basin. A clearly defined petrographic arc-unroofing trend is documented by a shift in conglomerate clast composition and sandstone petrofacies, and indicates a substantial episode of arc uplift and dissection during Tithonian and Berriasian times.

This overall, 7 km thick, regressive megasequence is punctuated during the late Berriasian by a well-exposed, abrupt facies from inner fan channel, deep marine conglomerates to a 70–90 m thick, shallow marine, storm-dominated shelf sandstone unit. This synchronous, basin-wide event is interpreted as reflecting a sedimentary response, at least in part, to a short-term eustatically controlled sea-level fall. This event correlates with a third-order short-term sea-level fall from the Exxon coastal onlap cycle chart.

## INTRODUCTION

For many years it has been realized that global sea-level change (eustasy) can be a major control on sediment dispersal patterns and the evolution of many sedimentary basins (e.g. the Hampshire Basin of southern England: Stamp, 1921; Plint, 1988). Documented examples of eustatically controlled changes in sea level are, however, typically restricted to basins developed at passive margins, where the influence of local and regional-scale tectonics is of minor importance. This paper examines the evolution of a well-exposed Mesozoic sedimentary basin at an active margin, which records a prolonged (45 Ma) history of tectonic control on sedimentation. Within this basin-fill sequence there is a well-exposed, basin-wide abrupt facies shift from inner fan channel, deep marine conglomerates to a storm-dominated transgressive shelf sequence. Unravelling tectonic and eustatic controls on sedimentation within fore-arc basins poses many problems (Miall, 1984; see chapters by Fortuin & DeSmet, and Hallam). The purpose of this paper is to discuss a specific example from the Fossil Bluff Group (FBG)

* Present address: The Robertson Group plc, Llandudno, Gwynedd, LL30 1SA, UK.

fore-arc basin of Alexander Island, and to document the facies distribution, style of sedimentation, and proposed palaeobathymetric setting, which suggests that a primary control on basin shallowing at one time slice in basin evolution was eustasy. This single event is correlated with an event within the Exxon coastal onlap chart of Haq *et al.* (1987).

## Geological setting

During Mesozoic times the Antarctic Peninsula was the site of an active magmatic arc associated with the eastward subduction of Pacific and proto-Pacific oceanic crust beneath the western margin of Gondwana (Barker, 1982; Storey & Garrett, 1985). The Fossil Bluff Group (FBG) of eastern Alexander Island crops out as a belt 250 km long by up to 30 km wide (Butterworth *et al.*, 1988) (Fig. 1). The FBG is a thick (4–7 km), little deformed, Kimmeridgian–Albian fore-arc basin sequence, forming a perched basin most probably floored by rocks of an accretionary complex. Strata young to the south accompanied by a facies change from deep marine through slope to shelf/paralic deltaic in the extreme south (Butterworth *et al.*, 1988). The basin was structurally confined to the east by the faulted scarp margin of the arc-basin boundary, which is probably analogous to the Tertiary transtensional feature of George VI Sound (Fig. 1), which is thought to have exploited an older tectonic boundary (Maslanyj, 1988; Storey & Nell, 1988). This episode of Cenozoic extension along the west side of the magmatic arc gave rise to thinning of the continental crust, a gabbroic intrusive suite, and the formation of linear block faults, which obscure the exact relationship of arc to fore-arc in the southern Antarctic Peninsula region. To the west, the FBG rests in faulted and unconformable contact on the accretionary complex of the LeMay Group (Edwards, 1980; Fig. 2). This boundary is complex, and recent work has suggested that there was local emergence of the accretionary prism during Late Jurassic times as a semi-emergent

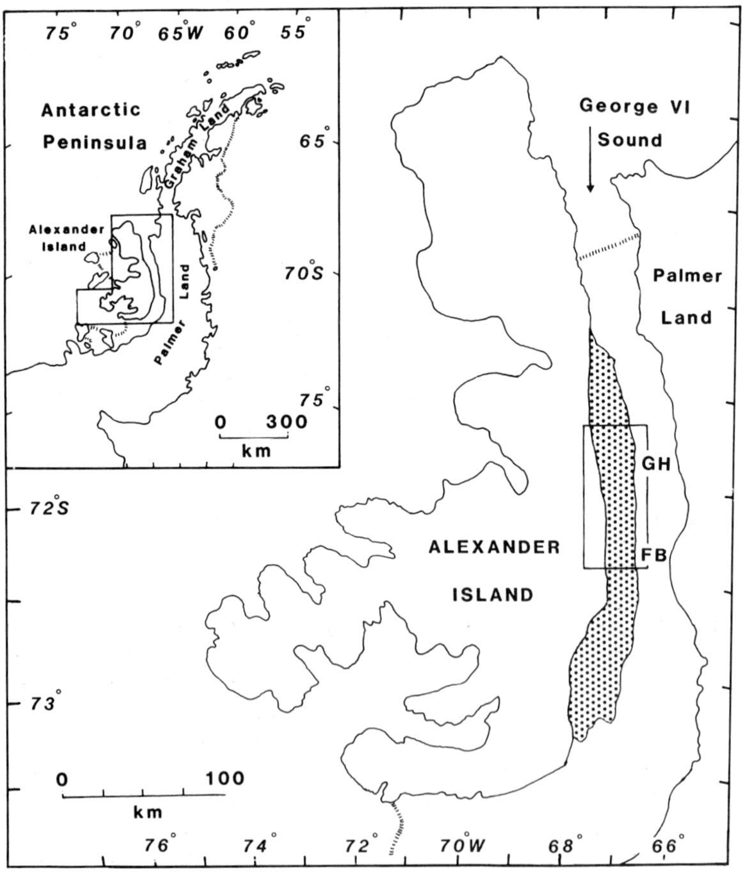


**Fig. 1.** Map of the Antarctic Peninsula and sketch map of Alexander Island, highlighting the elongate fore-arc basin of the Fossil Bluff Group (stippled). Palmer Land represents the remnant Mesozoic volcanic arc terrane, which is displaced from Alexander Island by George VI Sound, a product of Cenozoic transtension. The FBG fore-arc basin is in faulted and unconformable contact with the accretionary complex of the LeMay Group of central and northern Alexander Island. FB, Fossil Bluff; GH, Ganymede Heights. Outline shows area of map in Fig. 2.

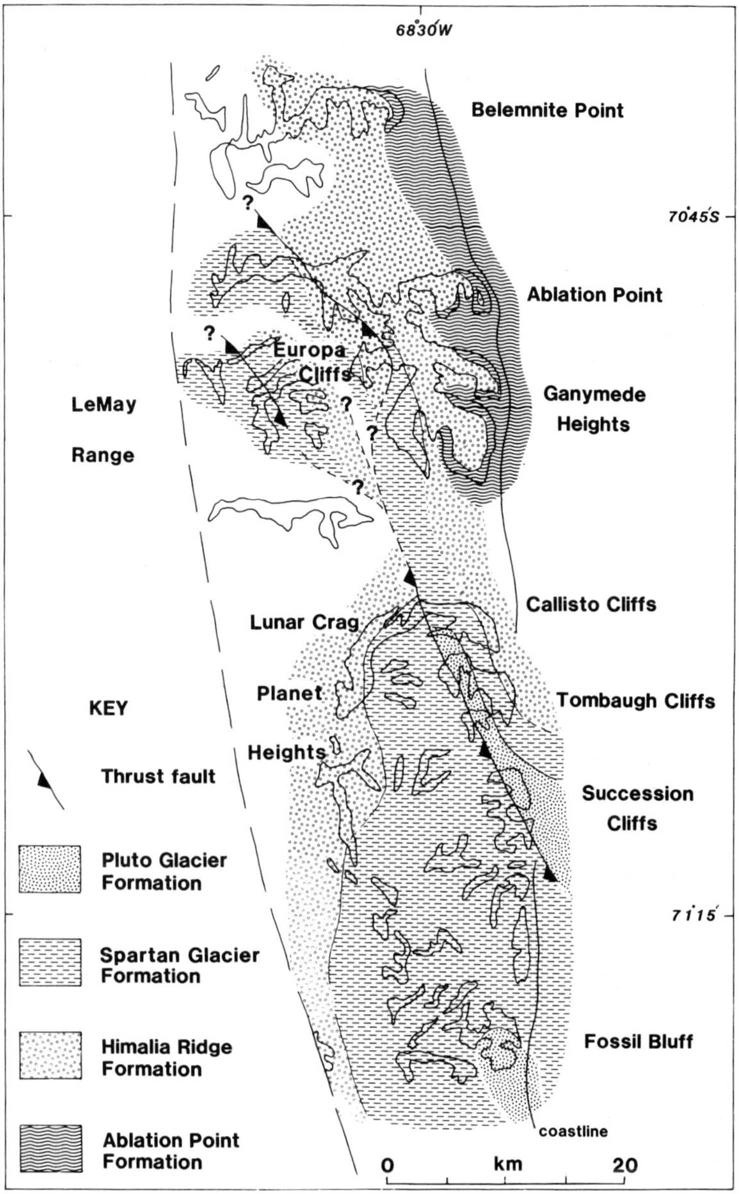


**Fig. 2.** Geological sketch map modified from Butterworth *et al.* (1988) of the northern area of the FBG between Fossil Bluff and Belemnite Point. Strata young to the south accompanied by a facies change from a deep-marine slope-collapse deposit (Ablation Point Formation) and submarine fan (Himalia Ridge Formation) through slope (Spartan Glacier Formation) to shelf/deltaic sequences in the south (Pluto Glacier Formation). The key localities discussed in the text are highlighted.

isolated chain of islands in an outer-arc setting. The presence of local Le May Group derived conglomerates along the bounding fault zone suggests that it may have marked the original edge of the fore-arc basin (Nell & Storey, in press).

The present-day Antarctic Peninsula largely corresponds to the Mesozoic arc (Thomson, 1982). The arc terrane comprises volcanic rocks of the Antarctic Peninsula Volcanic Group, intruded by co-genetic plutons. The timing of tectonic and volcanic events in the southern area of the Antarctic Peninsula (Palmer Land; Fig. 1) is poorly defined, but plutonism and volcanism span at least Early Jurassic until mid-Cretaceous times (Meneilly *et al.*, 1987). A widespread suite of basic sills, dykes, and plutons was emplaced in the arc terrane during Late Jurassic to earliest Cretaceous times (Palmer Land; Fig. 1) (Piercy & Harrison, 1991), which coincides with the earliest recorded sedimentation within the fore-arc basin. This major phase of Late Jurassic extension and magmatism correlates with several other well-documented tectonic and volcanic events along the western margin of Gondwana, including the opening of the back-arc basins in South Georgia (Storey & Macdonald, 1984) and South America (Dalziel *et al.*, 1975).

Intermediate-acid plutonism resumed in mid-Cretaceous times, accompanied by compression, uplift, and abundant volcanism (Piercy & Harrison, in press). Mid-Cretaceous uplift is also recorded in Graham Land (Fig. 1) by syndepositional deformation of sediments at the margin of the Larsen Basin (Whitham & Marshall, 1988), and in southern South America and South Georgia by uplift and deformation of the Rocas Verdes marginal basin (Dalziel *et al.*, 1975).

Dextral transcurrent motion in the Antarctic Peninsula region is well documented from at least mid-Cretaceous to Cenozoic times (Storey & Nell, 1988). The FBG basin fill is deformed by a series of NNW-trending open anticlines related to major ENE-directed thrusts (Nell & Storey, 1991) (Fig. 2). Abundant sandstone dykes, which also cut the sediments, trend east-northeast. There is no evidence to suggest that strike-slip processes were active during fore-arc basin sedimentation, although a phase of transpression must have occurred prior to lithification.

## BASIN EVOLUTION

The evolution of the 7 km thick, FBG basin-fill sequence is summarized here as a series of time slices; overall it is a regressive megasequence from deep-marine submarine fan to shallow marine–terrestrial conditions.

### Kimmeridgian

The Late Jurassic episode of extension and shearing within the arc corresponds to the earliest sedimentation within the presently exposed FBG. The oldest strata in the basin are part of a large Kimmeridgian slump unit (Ablation Point Formation), comprising folded and rafted blocks of sedimentary and (subordinate) volcanic rock in chaotic, laterally discontinuous units. This unit is interpreted as a seismically induced, slope-collapse feature involving an estimated 100–150 $km^3$ of sediment, and probably reflects break-up of a submarine fan complex (Macdonald & Butterworth, 1986; Butterworth, 1988).

### Tithonian–Berriasian

During this time, the proximal fringe of sediment flanking the arc (eastern) basin margin were deep marine muds accumulated during periods of low coarse clastic sediment supply into the basin. This low-energy environment is incised by two discrete deep marine fan depocentres, composed of large channel-fill complexes up to 170 m thick and 20 km wide (Fig. 3A). These two submarine fan complexes, centred on Ganymede Heights and Planet Heights, are separated laterally by mud-rich interchannel sediments (Fig. 4; Butterworth, 1988). Together these comprise the Himalia Ridge Formation (Fig. 2).

The development of distinct, isolated deep marine fans at the basin margin contrasts with many ancient examples of resedimented conglomerates. For example, Surlyk (1978,1984) documented a sequence of slope apron/coalescing fans from the Cretaceous Wollaston Foreland Group of East Greenland, and Winn & Dott (1979) have described a single linear channelled complex from the Cretaceous of southern Chile. A narrow shelf probably caused a localization of the input of coarse clastics into the deep marine basin of the Fossil Bluff Group via incised canyons, and aided rounding of clasts within either a fluvial or fan delta environment. Palaeocurrent and provenance studies clearly indicate derivation from the volcanic arc to the east (Butterworth, 1988). At the type section of the formation three distinct pulses of coarse-grained sediment input have been recognized, based on the association of channelled conglomerate-overbank and interchannel associations exposed in vertical section (Butterworth, 1988; Fig. 5). These three major pulses of coarse-grained sediment input reflect a tectonic allocyclic control at source related to rejuvenation and uplift of the hinterland. However, the subordinate cyclic alternation of axial channel-fill, channel abandonment, overbank, and quiet-water interchannel sedimentation within any one pulse (Fig. 5) reflects an aggradational style of fill related to the lateral juxtaposition and vertical stacking of braided, gravelly submarine channels. It is important to note that this cyclical alternation of conglomerate–sandstone–mudstone facies (100–300 m thick fining upward cycles) within any one pulse reflects an intrabasinal, autocyclic control

**Fig. 3.** (A) View of the Himalia Ridge Formation looking east on to part of the 250 m high cliffs at Lunar Crag. Two conglomerate channel-fill complexes comprise the two vertical scarps, separated by a synsedimentary slumped unit 70 m thick. Note the occurrence of internal channelization within each conglomerate complex. (B) View of the 500 m thick section at Fossil Bluff. A synsedimentary slump unit 40 m thick composed of thinly interbedded mudstone/sandstone blocks at the base of the section is overlain by slope mudstones of the Spartan Glacier Formation. The transition to the shallow-marine sandstones of the Pluto Glacier Formation crops out on the pyramidal peak in the background, picked out by the thick pale-weathered scarps. (C) Prominent angular unconformity within the Spartan Glacier Formation at Ganymede Heights, running approximately through the centre of the photograph. Total stratigraphic thickness is 700 m. A series of sandstone dykes cut the succession at this locality.

on channel switching due to avulsion (Butterworth, 1988), and not extrabasinal tectonism.

Dramatic arc uplift during Tithonian–Berriasian times is supported by a clearly defined shift in the composition of conglomerate clasts from volcanic to plutonic (and minor metamorphic) compositions (Fig. 5). This is similarily reflected in the sandstone petrofacies trends from undissected arc to dissected arc and basement uplift fields, utilizing the QFL plots of Dickinson *et al.* (1983) (Fig. 5). Similar arc unroofing trends have been documented from the Great Valley sequence of California (Ingersoll, 1978a,1983) and the Mesozoic of central Oregon (Dickinson *et al.*, 1979).

Thickness variations between the conglomeratic submarine fan and interchannel mud sequences across the basin (Fig. 4) suggest that subsidence and sedimentation rates were greatest at Ganymede Heights. An average sedimentation rate for submarine fan deposition is 250 m/Ma, which is com-

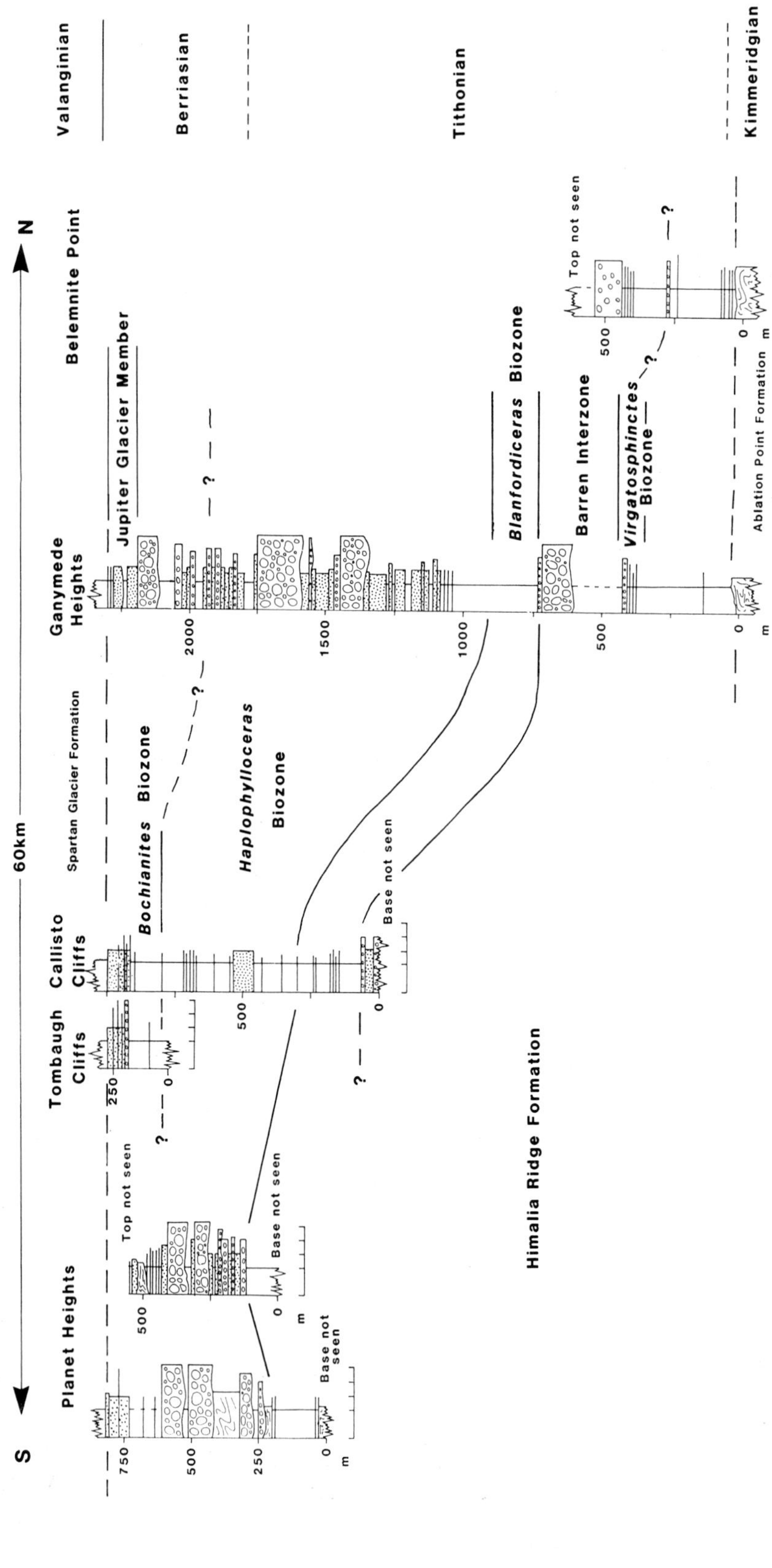
Valanginian
Berriasian
Tithonian
Kimmeridgian
S
N
60km
Planet Heights
Tombaugh Cliffs
Callisto Cliffs
Ganymede Heights
Belemnite Point
Jupiter Glacier Member
Spartan Glacier Formation
Bochianites Biozone
Haplophylloceras Biozone
Blanfordiceras Biozone
Barren Interzone
Virgatosphinctes Biozone
Ablation Point Formation
Himalia Ridge Formation
Top not seen
Base not seen
Base not seen
Top not seen
Base not seen
Base not seen
Top not seen
750
500
250
0
m
500
0
m
250
0
500
0
2000
1500
1000
500
0
m
500
0
m

**Fig. 4.** N–S transect through Kimmeridgian to Valanginian aged strata of the Fossil Bluff Group based on measured sections. The biozonation scheme of Howlett (1989) is used. Note the lateral juxtaposition of conglomerate channel-fill complexes and interchannel mudstones, and the laterally extensive sandstone unit (Jupiter Glacier Member) overlying the submarine fan deposits at all localities.

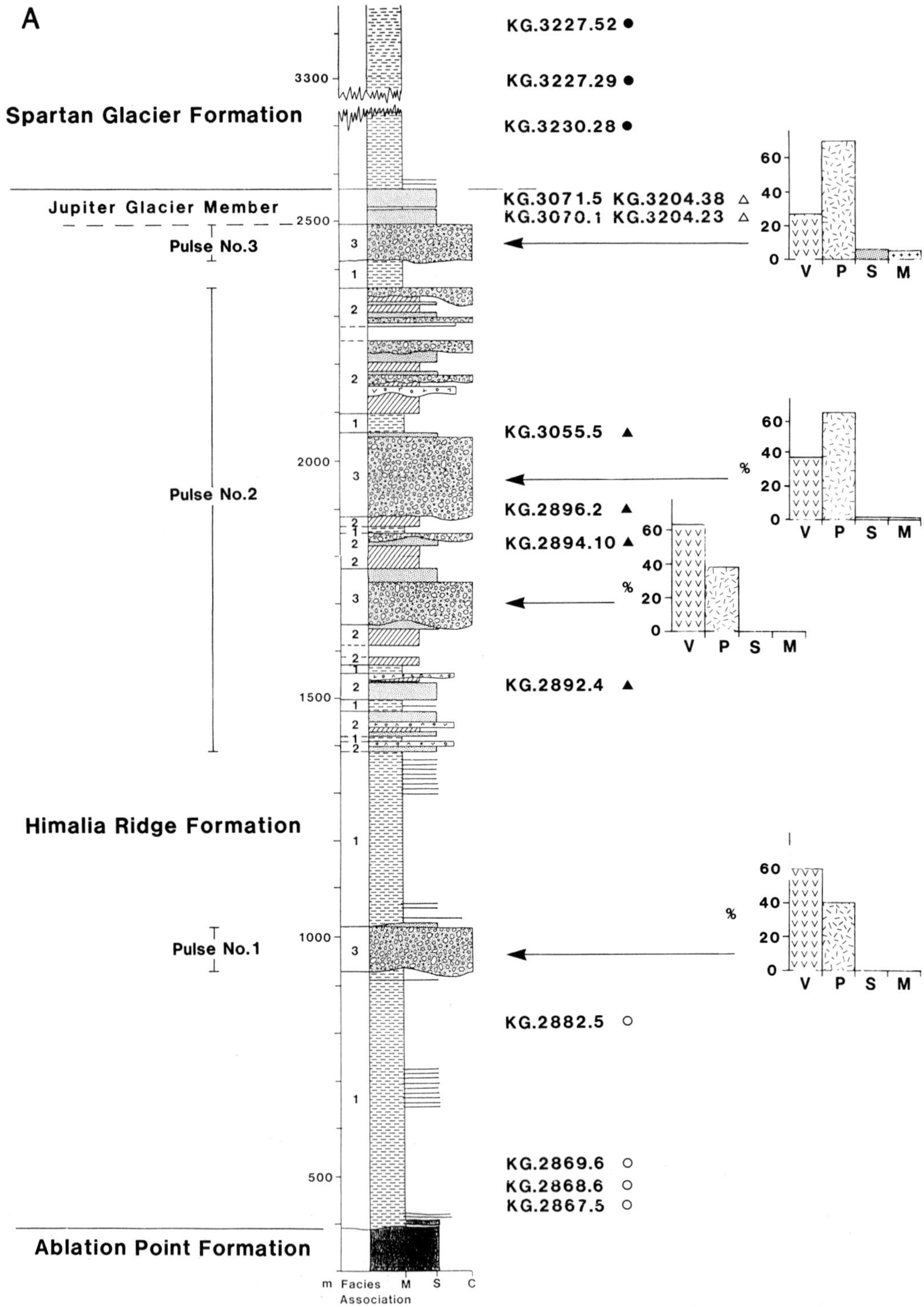

A
Spartan Glacier Formation
Jupiter Glacier Member
Pulse No.3
Pulse No.2
Himalia Ridge Formation
Pulse No.1
Ablation Point Formation
3300
2500
2000
1500
1000
500
m
Facies Association
M
S
C
KG.3227.52
KG.3227.29
KG.3230.28
KG.3071.5 KG.3204.38
KG.3070.1 KG.3204.23
KG.3055.5
KG.2896.2
KG.2894.10
KG.2892.4
KG.2882.5
KG.2869.6
KG.2868.6
KG.2867.5
%
V
P
S
M

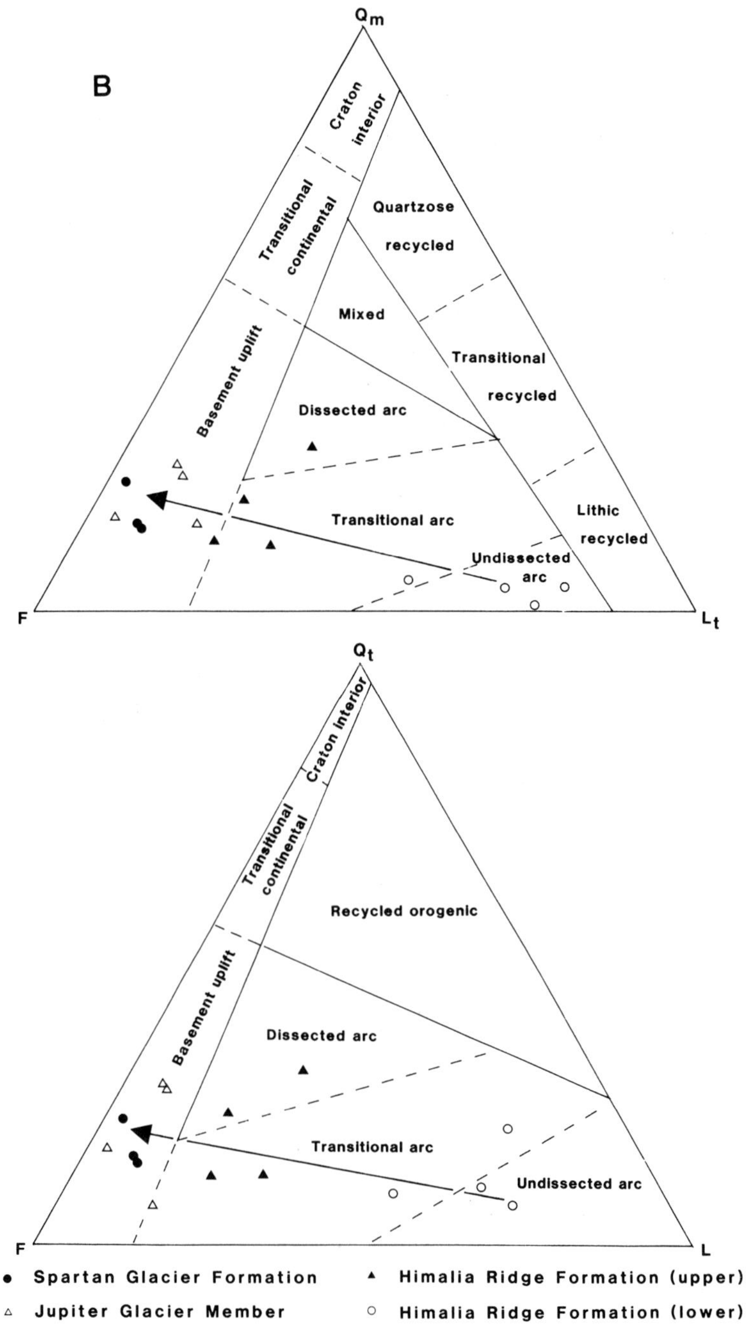


**Fig. 5.** (A) Stratigraphic position of frequency histograms of conglomerate clast compositions and sandstone detrital mode samples from the measured section at Ganymede Heights. Clast compositions are: V, volcanic; P, plutonic; S, sedimentary; M, metamorphic. Facies association refers to submarine fan subenvironment (Butterworth, 1988). 1, interchannel; 2, overbank; 3, channel fill. Three tectonic pulses are identified on the basis of the association of channel fill, overbank, and interchannel facies associations exposed in vertical section. Pulse 1 is encased in a thick sequence of interchannel mudstones. Pulse 2 represents an alternation of channel-fill/overbank subenvironments due to channel avulsion. Pulse 3 is separated from pulse 2 by a thick sequence of interchannel muds. A full interpretation of this measured section is given in Butterworth (1988). Sample numbers and symbols show levels for sandstone samples used in Fig. 5B. (B) $Q_tFL$ and $Q_mFL_t$ ternary plots of detrital sandstone modes from the Ganymede Heights section, based on the fields of Dickinson *et al.* (1983). Arrows indicate approximate trends.

parable to the sedimentation rate documented from the Late Cretaceous Great Valley fore-arc of California (Ingersoll, 1979). The occurrence of many minor slumped zones, up to 25 m thick, throughout the deep marine sequence suggests sedimentation on an unstable slope.

The laterally juxtaposed gravelly submarine fan deposits and interchannel muds of the Himalia Ridge Formation are all sharply overlain by a 70–90 m thick basin-wide blanket of fine-grained, laminated sandstones (Jupiter Glacier Member). This unit is discussed in detail below.

### Valanginian–Albian

The Jupiter Glacier Member is overlain by a 1000 m thick, Valanginian–Aptian aged monotonous sequence of mudstone and siltstone with subordinate thin, fine-grained sandstone interbeds (Spartan Glacier Formation; Butterworth *et al.*, 1988). The formation is characterized by slumps and synsedimentary melanges up to 120 m thick (Fig. 3B), swarms of sandstone dykes (Taylor, 1982), local angular unconformities (Fig. 3C), and synsedimentary faults. This basin-wide blanket of fine-grained sediment probably represents a series of individual mud lobes deposited within a slope–outer-shelf environment. Intrabasinal tectonic processes controlled sedimentation, at least in part (Butterworth & Macdonald, in press), and is similar to the Roberts Island slope system of the Great Valley sequence of California (Cherven, 1983).

During Aptian times, there is a transition to bioturbated, cross-bedded medium sandstones (Fig. 3B), probably deposited in a shelf environment (Pluto Glacier Formation) (Horne, 1969). These deposits are overlain in the southernmost exposure of the basin fill by delta plain facies where extensive fossil forests have been found (Jefferson, 1982). It is possible that the diachronous development of a shelf system across the basin at this time corresponds to an episode of volcanism, fault rejuvenation, and uplift documented from the arc during the mid-Cretaceous (Piercy & Harrison, 1991). Sandstone petrofacies compositions support an episode of active arc volcanism (Horne & Thomson, 1972), emphasizing the prolonged influence of the arc as a sediment source throughout basin evolution.

## SUBMARINE-FAN–SHELF TRANSITION: A LATE BERRIASIAN REGIONAL SHALLOWING EVENT

It is apparent that the tectonically active FBG basin-fill sequence represents an overall regressive megasequence from submarine fan to shelf during Kimmeridgian to Aptian times (c. 45 Ma). However, this simple facies trend is punctuated by a spectacular, abrupt, regional shallowing event of late Berriasian age. The sedimentology of this unit and the implications for a tectonic or eustatic control on relative sea-level change are detailed here.

A distinctive, buff-coloured unit of sandstone, up to 90 m thick, crops out as a resistantly weathered scarp overlying the topmost strata of the deep-marine fan sequence of the Himalia Ridge Formation in the northern area of the basin fill (Figs 4,6A). The Jupiter Glacier Member is formally defined here as an update to the previous stratigraphic scheme of the FBG (Butterworth *et al*; 1988) (Fig. 4); it is named after the Jupiter Glacier, as many of the exposures crop out adjacent to this feature. Lithofacies are dominated by fine- and very fine-grained sandstones with flat and low-angle/swaley stratification; there is a diverse faunal association including *in situ* infaunal bivalves (*Pinna, Myophorella, Pleuromya*) and epifaunal bivalves (*Grammatodon, Entolium*, pectinids). Both the lithofacies and fauna are in striking contrast to the turbiditic sandstones of the underlying deep-marine fan sequence.

Three vertically stacked facies associations are recognized on the basis of grain size, structure, fauna, and bioturbation (summarized in Table 1).

### Association A: transitional facies assemblage

Facies comprise cross-stratified medium sandstone to granulestone, complexly interbedded with scour-fill conglomerates and pebbly sandstones (Fig. 6B), plus flat-stratified sandstones, rare breccias, and pebbly mudstones (Table 1). All beds are typically less than 1 m thick, with variable bed contacts ranging from very sharp to diffuse over 10–20 m along the same bedding plane, and with high-angle planar cross-stratification (Fig. 6B).

### Association B: inner shelf assemblage

Facies are dominated by fine-grained, well-sorted, flat and low-angle/undulating stratified sandstones with lenses of trough cross-stratification (Figs 6C,7);

many beds are amalgamated. Individual sets range from 10 to 150 cm thick and may be laterally continuous for up to 5 m, typically comprising broad troughs that scour into each other, or into gently undulating flat lamination (Fig. 6C). The stratification consists of curving laminae, typically concave up, with inclinations of up to 10°. The laminae are overall fairly flat, although they commonly intersect at low angles and may display scattered dip directions or thicken laterally within any one set. Lenses of trough cross-stratified sandstone are common, with typically steeper laminae (up to 20°) (Fig. 6C). Many beds are cut by small scours and/or deformed by loading. Symmetrical wave ripples formed by vertical accretion are common both in cross-section and in plan view (Fig. 6D), with amplitudes up to 3 cm and wavelengths of 5–8 cm. Several beds contain mudflake intraclasts.

Interbedded with these sandstones are rare, thin, discontinuous pebble lags, and medium to thick-bedded, sharp-based, clast-supported, normally graded, scour-fill conglomerates (Fig. 7). Mudstone beds are very rare, and are typically restricted to thin, isolated horizons, which may be bioturbated (e.g. *Chondrites*) and/or drape low-angle bedforms. A very restricted fauna, limited to rare *in situ Myophorella* is present.

### Association C: outer shelf assemblage

This association is typically dominated by very fine-grained, massive, or more commonly extensively bioturbated, thick-bedded sandstones (Fig. 7). Rare beds may be diffusely parallel-stratified, with up to 20% silt and mud laminae (Fig. 6E). Infaunal and epifaunal bivalves, serpulid coquinas in lenses up to 40 cm thick by 3 m long, and wood fragments up to 30 cm long are very common (Fig. 7, Table 1). The definition of bed boundaries is often difficult owing to extensive bioturbation; bed thicknesses are correspondingly very variable (5–160 cm thick). The trace fossil assemblage includes *Chondrites*, *Zoophycos*, *Phycosiphon*, *Palaeophycus*, *Ophiomorpha*, and *Thalassinoides*, indicating a well-oxygenated environment (Ekdale *et al.*, 1984).

These sandstones are interbedded with subordinate thin-bedded fine-grained sandstones, thin-bedded mudflake breccias, and mudstones (Fig. 7).

### Facies distribution

The spatial distribution and internal organization of facies associations A–C within the Jupiter Glacier Member across the basin are relatively consistent (Fig. 8). Association A (0–23 m thick) is only found associated with underlying conglomerate channel complexes; it is not found overlying fine-grained interchannel areas (Fig. 8). Association B typically comprises up to 75% of Jupiter Glacier Member, and is a basin-wide interval of variable thickness (Fig. 8). Association C is typically 20–30 m thick, and is restricted to the top of the sand packet, although it is absent in the section at Callisto Cliffs. There is very little change in the total thickness of the Jupiter Glacier Member across the basin (Fig. 8).

### Discussion: a transgressive shelf sequence

The abrupt vertical transition within the sedimentary pile from a gravelly, mass-flow dominated, submarine fan sequence (Himalia Ridge Formation) to a shallow-marine shelf sequence (Jupiter Glacier Member; associations B and C) reflects a rapid, synchronous basin-wide shallowing event (relative sea-level fall). Association A represents a transitional facies most probably resulting from continued, episodic input and reworking of conglomerate and sandstone facies beneath turbidity currents, which are commonly confined within channels (cf. Hiscott & Middleton, 1979; Lowe, 1982). A possible modern analogue is the submarine sand dune field documented from Oceanographer Canyon on the continental margin of New England (Valentine *et al.*, 1984), which is interpreted as mixing of shelf sands with immobile gravel deposits within the confines of the canyon.

The very low-angle to undulating, stratified sandstones of association B (Fig. 6C), which comprise up to 75% of this sand packet, are directly comparable to both 'swaley cross-stratified sandstones' (Duke, 1985; McCrory & Walker, 1986; Rosenthal & Walker, 1987) and 'hummocky cross-stratified sandstones' (Hamblin & Walker, 1979; Bourgeois, 1980; Brenchley *et al.*, 1986) discussed from many ancient shelf sequences. Rosenthal & Walker (1987) suggest that there is probably a genetic relationship between 'swaley' and 'hummocky' cross-stratification, and most workers have inferred an origin involving storm-generated progressive surface gravity waves (e.g. Duke, 1985; Swift *et al.*, 1987). The presence of wave ripples (Fig. 6D) in conjunction with hummocky cross-stratification and swaley-stratified sandstones supports a shallow marine, storm-

**Fig. 6.** (A) View of the 1100 m thick section through the Himalia Ridge Formation at Callisto Cliffs. The prominent, white-coloured sandstone scarp immediately below the skyline is the Jupiter Glacier Member. Note the sharp contrast with the underlying interchannel mudstones of the submarine fan complex.

(B) Conglomerate overlain by cross-stratified, coarse-grained sandstone–granulestone, association A, Jupiter Glacier Member. Hammer shaft is 35 cm long.

(C) Flat and low-angle stratified fine-grained sandstone, association B, Jupiter Glacier Member. Note the lense of trough cross-stratified sandstone with relatively steep laminae. The mottled effect is a product of laumontitization. Lens cap is 5.5 cm in diameter.

(D) Wave-rippled fine-grained sandstone, association B, Jupiter Glacier Member. Lens cap is 5.5 cm in diameter.

(E) Bioturbated, massive and diffusely parallel-laminated very fine-grained sandstone, with thin silt and mud laminae, association C, Jupiter Glacier Member. Lens cap is 5.5 cm in diameter.

dominated inner shelf/shoreface setting. Planar-stratified sandstones may reflect sedimentation on a beach-shoreface subenvironment (Hamblin & Walker, 1979; Rosenthal & Walker, 1987), although a range of environmental settings is possible (Brenchley *et al.*, 1986).

Associated with these thick sequences of swaley- and low-angle, undulating-stratified, fine-grained sandstones are sporadic, thin-bedded pebble lags and thick-bedded conglomerates (Fig. 7). The occurrence of thin pebble lags can best be explained as a storm-generated feature, similar to those documented from the Miocene of southwest Oregon by Leithold & Bourgeois (1984). However, the presence of thick-bedded mass flow conglomerates confined to scours is more problematical. They are confined within high-energy, storm-wave dominated, shallow-marine deposits (swaley cross-stratified sandstones), and reflect relatively rare events (Fig. 7). It is possible that these reflect redeposition of shoreface deposits owing to storm reworking (Leithhold & Bourgeois, 1984), although similar linear conglomerate bodies from the Cardium Formation of Alberta (Bergman & Walker, 1987) have been interpreted as reflecting the interaction of small-scale fluctuations in sea-level and changes in sediment supply to the shelf.

The transition from association B to C reflects a deepening of the depositional environment from inner to outer shelf. The occurrence of very fine-

**Table 1.** Summary of the main characteristics of facies associations A–C recognized from the Jupiter Glacier Member shelf-sandstone unit. Order of facies indicates relative frequency

| Association | Thickness (m) | Facies | Bioturbation | Fauna | Depositional process | Environment |
|---|---|---|---|---|---|---|
| C | 10–75 | Very fine-grained:<br>Bioturbated sandstone<br>Massive sandstone<br>Diffusely flat-stratified sandstone<br>+ mudstone<br>sandstone<br>mudflake breccia | Variable — ranges from a little (rare) to very extensive:<br>*Chondrites*, *Zoophycos*, *Phycosiphon*, *Ophiomorpha*, *Palaeophycus*, *Thalassinoides* | Extensive (often *in situ*):<br>*Entolium*, *Grammatodon*, *Myophorella*, pectinids, *Pinna*, *Pleuromya*, *Belemnopsis*, *Haplophylloceras*, *Raimondiceras*, *Spiticeras* | Rare storm deposits<br>Increased suspension sedimentation | Outer shelf |
| B | 0–80 | Fine-grained:<br>Low-angle–undulating ('swaley') stratified sandstone<br>Flat-stratified sandstone<br>Trough cross-stratified sandstone<br>Massive sandstone<br>+ conglomerate<br>mudstone | Very limited:<br>*Chondrites* | Very limited:<br>*Myophorella*, *Belemnopsis* | Storm-dominated sedimentation of fine-grained sandstone ('swaley' cross-stratification) punctuated by rare gravelly mass flows<br>Suspension sedimentation minimal | Storm-dominated inner shelf |
| A | 0–23 | Stratified coarse-grained (granule) sandstone<br>Pebbly sandstone<br>+ mudflake breccia<br>pebbly mudstone | Very limited:<br>*Chondrites* | Very limited:<br>*Belemnopsis* | Deposition by gravelly and sandy high density turbidity currents, with substantial reworking of coarse-grained sandstones | Re-worked channel deposits (base of slope) transitional to the shelf environment |

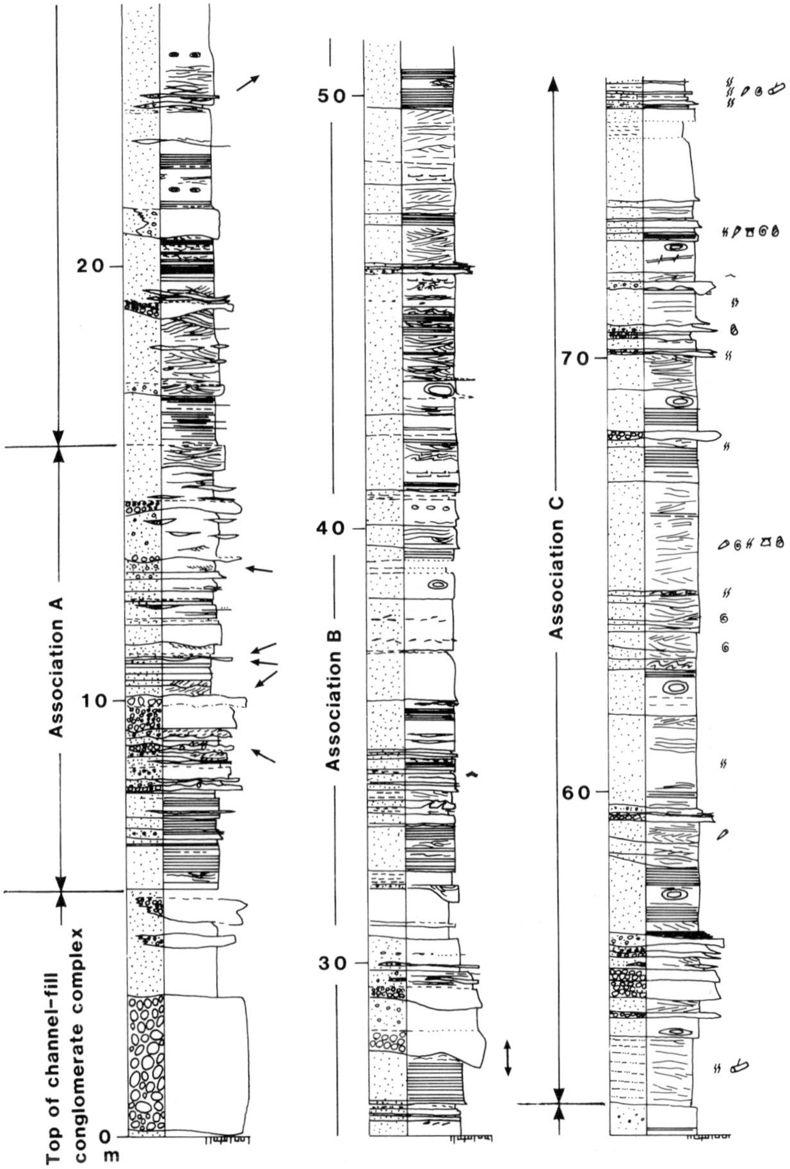


**Fig. 7.** Representative logged section through the Jupiter Glacier Member shallow-marine sandstone packet from Ganymede Heights. Associations A–C are highlighted. Key as for Fig. 4.

grained, parallel-laminated sandstones without swaley cross-stratification in association C may reflect the remobilization of sediment by storm-generated currents flowing offshore transporting sediment to an outer shelf setting below storm wave-base (e.g. Brenchley *et al.*, 1986). The abundant fauna (Table 1) is comparable to that classically interpreted as indicative of shelf depths (Kauffman, 1969; Wilson, 1982; Fursich, 1984). Many of the features are similar to those discussed by Bourgeois (1980) from the transgressive sequence within the Cape Sebastian sandstone of Oregon. Other trends within the Jupiter Glacier Member that are indicative of a transgressive sequence have been widely docu-

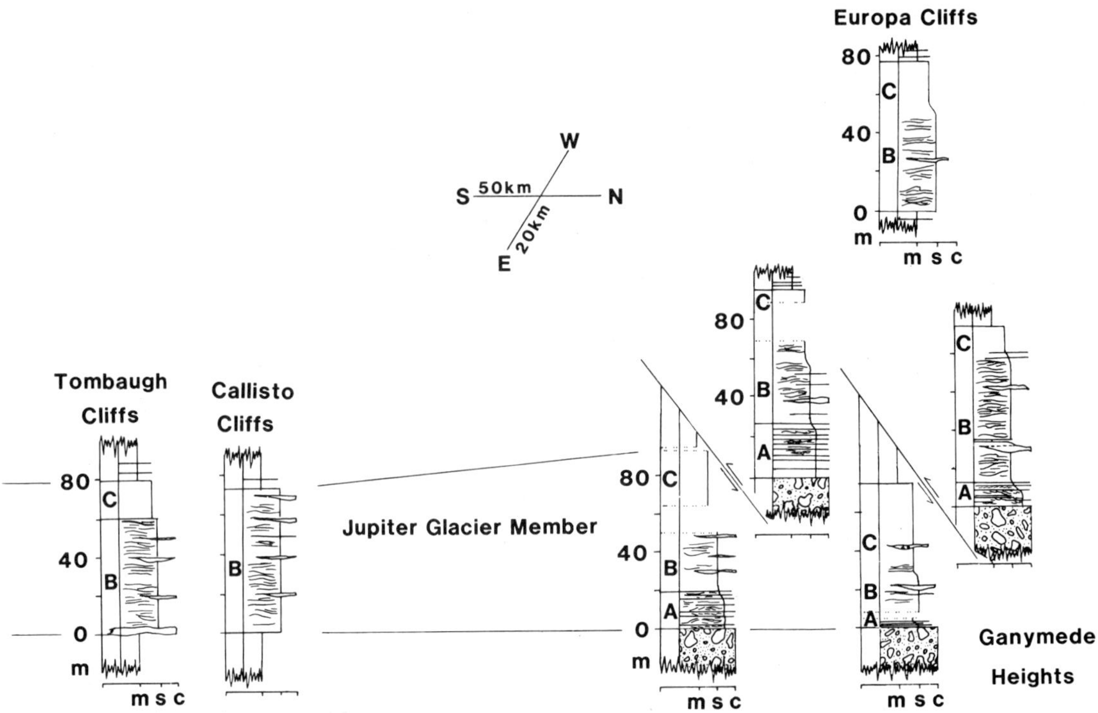


**Fig. 8.** Generalized three-dimensional sketch based on measured sections through the Jupiter Glacier Member shallow-marine sandstone packet, highlighting the spatial distribution of associations A–C. Conglomerate stipple indicates the underlying fan-channel deposits. Note the repetition of sections at Ganymede Heights by low-angle thrusts. Log localities shown in Fig. 2.

mented from modern shelf environments, and include a seaward decrease in grain size (Howard & Reineck, 1979), a seaward increase in faunal diversity (Carey, 1972), and a seaward increase in the amount of bioturbation (Kulm *et al.*, 1975; Howard & Reineck, 1979).

The upper boundary of the Jupiter Glacier Member with the overlying slope-mudstone sequence (Spartan Glacier Formation) can either be sharp (Fig. 6A) or gradational over 5–15 m, with thinly interbedded turbidite sandstones and mudstones. This event reflects a significant change in sediment supply to the basin, as the whole area was blanketed by fine-grained muds.

## Age

The timing of this basin-wide shallowing event is well constrained by the appearance of a relatively prolific fauna present within the outer shelf sediments (association C) of the Jupiter Glacier Member and the overyling slope mudstones of the Spartan Glacier Formation. This fauna is recognized throughout the basin, and comprises the ammonite genera *Olcostephanus*, *Raimondiceras*, *Spiticeras*, *Bochianites*, and *Haplophylloceras*, several species of the belemnite *Hibolithes*, and the belemnite *Belemnopsis gladiatoris* (Howlett, 1986,1989; Crame & Howlett, 1988). This relatively abundant macrofauna allows erection of a detailed biozonation scheme which can be compared to those of the Chilean, Andean, and Magallanes basins of South America, and also to Madagascar (Howlett, 1989). The Berriasian–Valanginian boundary is placed at approximately 25 m above the top of the Jupiter Glacier Member. Consequently, the base of the Jupiter Glacier Member, which forms a synchronous unit across the basin, is dated as latest Berriasian. Elsewhere within the succession, the biostratigraphic framework of the FBG is much less well constrained (Crame & Howlett, 1988; Howlett, 1989). No microfossils have, as yet, been isolated from the FBG.

## DISCUSSION

It is apparent that there was a strong tectonic control on sedimentation during much of the evolution of the FBG basin. Tectonic uplift resulted in at least three pulses of sedimentation during Tithonian–Berriasian times (Himalia Ridge Formation), which were the major control on submarine fan sedimentation (Butterworth, 1988). The subsequent development of a blanket of slope mudstone (Spartan Glacier Formation) is indicative of an increasingly tectonically quiescent setting, although the presence of large-scale slope-collapse deposits (Fig. 3B) and local angular unconformities (Fig. 3C) suggest that sedimentation was controlled, at least in part, by intra-basinal tectonic processes (Butterworth & Macdonald, in press). The final stage in the evolution of the basin in this northern area was the development of an extensive shelf sandstone/deltaic unit (Pluto Glacier Formation), which has a probable diachronous relationship with deltaic and terrestrial sequences farther south. The evolution of the FBG is similar to many other fore-arc regions, such as the Great Valley fore-arc basin of California, which shows a similarly well-developed, basin shallowing, regressive megasequence (Ingersoll, 1978b), and petrographic arc unroofing trend (Ingersoll, 1978a,1983).

The dramatic, sharp facies shift from a proximal, gravelly, inner fan channel-levee complex to a storm-dominated, transgressive shelf setting (Jupiter Glacier Member) during the late Berriasian reflects a rapid fall in relative sea-level at this time, followed by a marine transgression (slower, relative sea-level rise). This is the only synchronous event to affect sediment dispersal patterns that can be traced right across the northern area of the FBG basin, and it is this event which is under discussion here.

### Palaeobathymetry

An important consideration to quantify is the *amount* of sea-level fall sufficient to account for the facies change from turbidite to shelf sandstone. However, water depth is difficult to determine. The Himalia Ridge Formation has a limited fauna (Crame & Howlett, 1988), which includes a number of reworked shallow marine bivalves (e.g. *Pinna*, *Myophorella*), and locally abundant plant debris, which suggests a nearby shallow marine shelf. Rare conglomerate beds composed of very well-rounded cobbles contain a small number of relatively unabraded *Pinna*, which suggests a small transportation distance from the shelf edge. The trace fossil assemblage is diverse (Butterworth, 1988), and includes forms classically assigned to depths ranging from abyssal to shelf environments (Seilacher, 1967), although the dominant *Chondrites–Zoophycos–Planolites* assemblage is similar to that commonly reported from recent deep sea settings (e.g. Wetzel, 1984) and ancient deep sea settings (e.g. Buck & Bottjer, 1985). However, these do not provide any absolute bathymetric limits, as trace fossil communities are also the result of many other environmental factors, such as nutrient availability, grain size, redox conditions, and sediment supply. However, the association with the thick (minimum 440 m) laterally extensive slope-collapse deposit at the base of the sequence (Ablation Point Formation; Fig. 2) implies depths at the onset of sedimentation greater than 900 m.

The coarse-grained nature of the Himalia Ridge Formation conglomerates implies an appreciable relief at the basin margin, and the petrofacies indicate at least one major phase of arc uplift and dissection during Tithonian–Berriasian times (Fig. 5). Variations in sediment thickness across the basin between coarse-grained, channelled submarine fan and mud-rich interchannel environments are thought to reflect differential rates of sedimentation and subsidence within the FBG basin. This is very marked in Tithonian-aged strata (Fig. 4), where the rate of subsidence was greater than the rate of sedimentation in the deeper parts of the basin at Ganymede Heights. During the Berriasian, variations in sediment thickness are less apparent (Fig. 4), which suggests that the rate of subsidence was very much reduced, assuming a constant submarine-fan sedimentation rate of 250 m/Ma.

The trace fossil assemblage from the Jupiter Glacier Member sandstones is diverse (Table 1), and includes *Thalassinoides* and *Ophiomorpha*, which are characteristic of shallow subtidal shelves between fair-weather and storm wave-base (Frey *et al.*, 1978; Ekdale *et al.*, 1984), although these have been recorded in deeper water environments (Kern & Warme, 1974). However, the combination of wave ripples and swaley cross-stratification in association B suggests sedimentation within a storm dominated shoreface or shelf environment, in water depths restricted to those between storm and fair-weather wave-base (e.g. McCrory & Walker, 1986), although this precise environmental bracketing is becoming increasingly open to question (Allen &

Underhill, 1989). The diverse faunal community in association C includes both infaunal bivalves (*Pinna*, *Myophorella*, *Pleuromya*; Fursich, 1984) and epifaunal bivalves (*Grammatodon*, *Entolium*, pectinids; Kaufmann, 1969), which are often *in situ*. This assemblage is indicative of shelf depths (Wilson, 1982). It is important to note that this faunal assemblage does not exist within the submarine fan sediments of the Himalia Ridge Formation.

### Relationship to the Exxon sea-level curve

Ever since the concept of global sea-level cycles was first introduced for pre-Quaternary times (Vail *et al.*, 1977), there has been much debate on the validity, accuracy, and applicability of such a model to the ancient geological record (e.g. Hallam *et al.*, 1986; Plint, 1988). One of the major requirements for applying the concept of sequence stratigraphy to ancient onland basins is that there is good biostratigraphic control (Vail *et al.*, 1977). Although much of the FBG is relatively poorly dated, the timing of the basin-wide shallowing event represented by the Jupiter Glacier Member is marked by a diverse faunal assemblage, and correlates almost exactly with a relatively large, third-order, short-term eustatically controlled sea-level fall of approximately 115 m during the late Berriasian (Haq *et al.*, 1987; Fig. 9). It is envisaged that this dramatic basin-wide, synchronous facies shift from submarine fan to shelf sedimentation developed, at least partly, in response to this short-term sea-level fall.

The model proposed here (Fig. 10) is that the Jupiter Glacier Member represents a sedimentary response to a eustatic fall in sea-level. If sedimentation rates were greater than the rate of basin subsidence during conglomeratic submarine fan sedimentation (Fig. 4), accommodation would decrease. As growth of the submarine fan system seems to have been aggradational, this would bring large areas of the sediment surface to shallow depths simultaneously. A eustatic fall in sea-level of approximately 115 m during the late Berriasian could therefore bring the tops of the channel-fill complexes within the influence of shallow marine processes, generating the juxtaposition of turbidites directly overlain by shelf sandstones.

Initial reworking of the tops of the conglomerate channel-fill complexes is marked by a transitional facies (Jupiter Glacier Member, association A; Figs 7,8) resulting from the episodic input of conglomerate and sandstone by turbidity currents, with intermixed shelf sandstones. Continued eustatic sea-level fall created an extensive storm-dominated inner shelf with swaley cross-stratified sandstones and wave rippled beds (association B), which are indicative of shoreface sequences (McCrory & Walker, 1986). Mudflake intraclasts indicate some sediment reworking.

This was followed by sedimentation within a setting equivalent to an outer shelf, and the development of a diverse faunal assemblage (association C), as a result of the relatively slow sea-level rise (transgression). Eventually the narrow marginal shelf system was submerged, leading to the development of a starved basin dominated by the input of fine-grained sediments within an outer shelf-slope setting, which persisted throughout much of the Lower Cretaceous (Spartan Glacier Formation).

### Interpretative problems

Although it is quite plausible that the overriding control on sea level (base level) during the late Berriasian within the FBG was eustasy, there is a dilemma as to how this dramatic facies shift can be accommodated solely by this mechanism taking into account the evidence of active tectonism as the major control on sedimentation throughout basin evolution. However, one of the major problems with the southern Antarctic Peninsula region is the lack of data on the style and rates of crustal evolution, in particular the timing and magnitude of subduction processes and arc uplift, as these will dramatically influence sedimentation patterns in fore-arc regions (cf. Ingersoll, 1983). For example, the chapter by Fortuin & DeSmet provides an excellent example from the Banda Arc of Indonesia, where they show that pulses of tectonic uplift have a duration of between 0.1–1 Ma and rates exceeding 0.5 m/Ka, which are of a similar frequency as eustatic third-order sea-level changes, but which are orders of magnitude larger in vertical extent. Consequently, disentangling the tectonic and eustatic signals along

**Fig. 9.** *(opposite)* Summary chart of the main strato-tectonic components of the southern Antarctic Peninsula, with emphasis on fore-arc basin stratigraphy in comparison to arc evolution (Piercy & Harrison, 1990) and the eustatic sea-level curve of Haq *et al.* (1987). JGM, Jupiter Glacier Member.

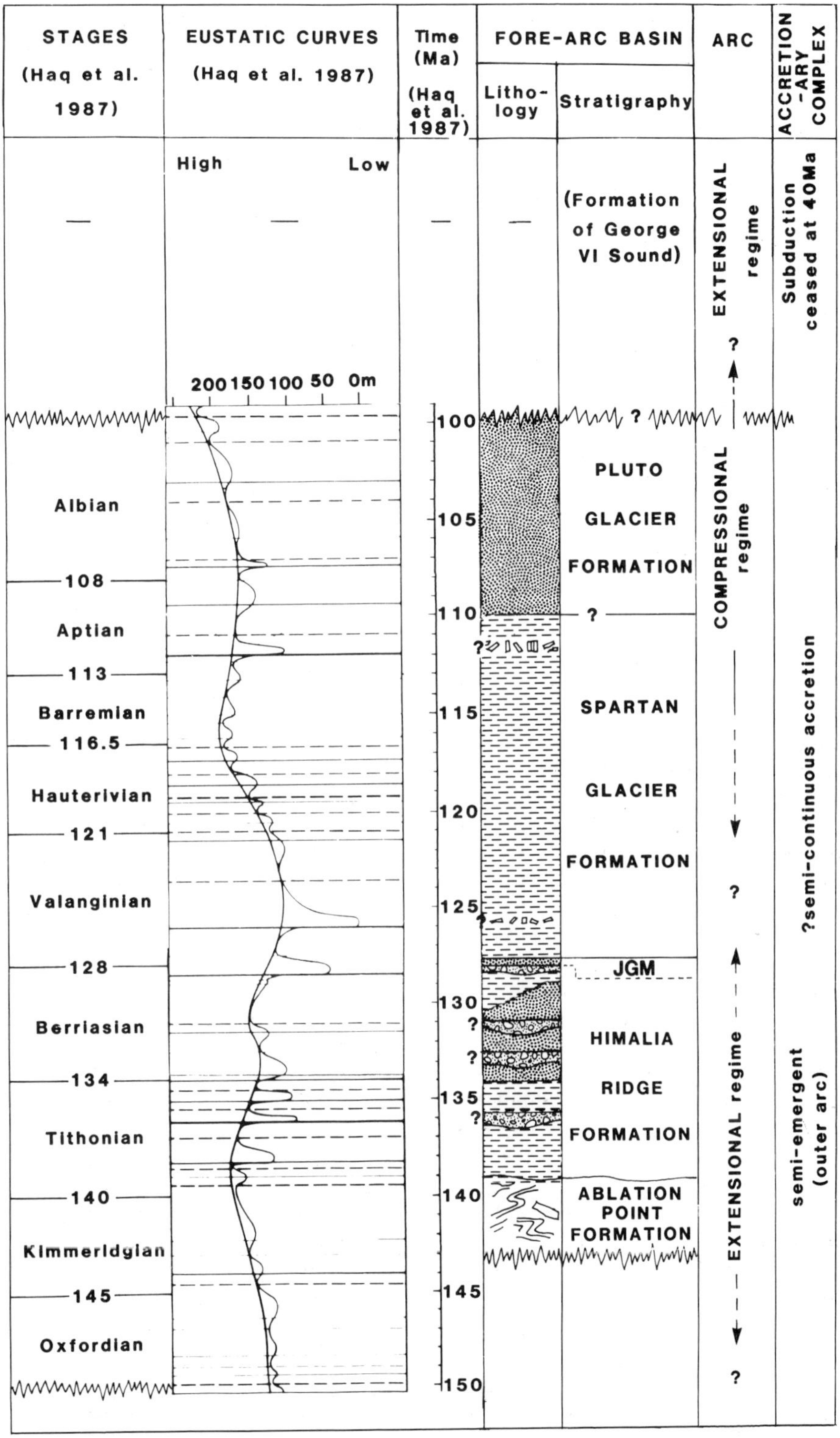
STAGES
(Haq et al. 1987)
EUSTATIC CURVES
(Haq et al. 1987)
Time (Ma)
(Haq et al. 1987)
FORE-ARC BASIN
Litho-logy
Stratigraphy
ARC
ACCRETION-ARY COMPLEX
High
Low
(Formation of George VI Sound)
EXTENSIONAL regime
Subduction ceased at 40Ma
200 150 100 50 0m
100
105
110
115
120
125
130
135
140
145
150
Albian
108
Aptian
113
Barremian
116.5
Hauterivian
121
Valanginian
128
Berriasian
134
Tithonian
140
Kimmeridgian
145
Oxfordian
PLUTO GLACIER FORMATION
COMPRESSIONAL regime
SPARTAN GLACIER FORMATION
JGM
HIMALIA RIDGE FORMATION
ABLATION POINT FORMATION
EXTENSIONAL regime
?semi-continuous accretion
semi-emergent (outer arc)
?

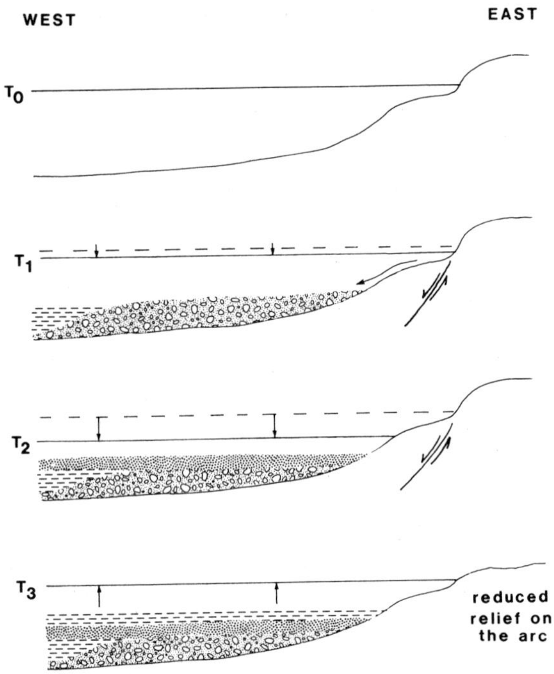


**Fig. 10.** Conceptual model to illustrate basin development through Himalia Ridge and Spartan Glacier Formation times. $T_0$, large initial basin subsidence; $T_1$, high sedimentation rate leads to reduced accommodation and raised sediment surface; $T_2$, sea-level falls below marginal shelf edge and brings large areas of the basin into shelfal conditions — Jupiter Glacier Member deposited; $T_3$, sea-level rises, drowning marginal shelf — this in combination with reduced arc relief cuts off clastic supply.

active margins is very difficult. In the paper by Hallam it is suggested, from a review of the Mesozoic basins of the southern Andes, that the two most important widespread regressions throughout the region occurred as a result of an eastward shift of the subduction zone related to the opening of the South Atlantic. However, during the much longer intervals of tectonic quiescence, the pattern of marine sedimentation within these Mesozoic basins substantially reflects global events, and as such eustasy was a significant controlling factor.

Throughout the evolution of the FBG basin-fill sequence there are relatively few periods that can be demonstrated to be tectonically quiescent, which makes it very difficult to delineate accurately any potential eustatic control on sedimentation. In addition, the present-day exposure of the FBG forms a section parallel to the basin margin, and any marginal facies are buried beneath the Tertiary rift of George VI Sound. This does not allow distinction of coastal onlap cycles that may have been created by eustatic sea-level fluctuations. A further complication is the role of strike-slip tectonics in basin evolution (cf. Nell & Storey, 1991) as this mechanism has been invoked elsewhere to create rapid basin inversion (Crowell, 1974). To date, strike-slip deformation of the FBG has only been proven post-sedimentation, and although it is difficult to recognize basin-fill sequences that can be attributed directly to strike-slip processes (Nilsen & McLaughlin, 1985), it is thought that the very regular palaeocurrent distribution throughout the period of submarine fan sedimentation indicates a relatively stable feeder canyon system, and as such it is most probable that synsedimentary strike-slip tectonics were a negligible component of basin evolution during this time interval (Butterworth, 1988).

A further important point to consider is that the transition from the Himalia Ridge Formation to the Spartan Glacier Formation, of which the shelf sandstone of the Jupiter Glacier Member forms the boundary, marks a significant change in sediment supply to the basin from coarse clastic gravels to fine-grained muds. This persists, with no conglomerates in the Spartan Glacier Formation, even though the eustatic sea-level rises back to the same levels as the Himalia Ridge Formation. The petrofacies trend (Fig. 5) suggests that there was probably a geomorphic control on sediment supply, as it seems likely that the source area (volcanic arc) had been largely planed off by late Berriasian times, which would create a low-relief hinterland. Although it is possible that there might have been a tectonic control on sediment supply, such as reactivation of a major structure that would block supply to the northern part of the basin, it seems unlikely as there is no evidence to suggest that this occurs at any other time throughout basin evolution.

## CONCLUSIONS

The dramatic facies shift recorded by storm-dominated shelf sandstones (Jupiter Glacier Member) sharply overlying submarine fan conglomerates of late Berriasian age from the FBG illustrates an example where an allocyclic mechanism is required to account for this rapid fall in base level. Many workers would automatically accommodate this major change in base level within a fore-arc basin setting by tectonic uplift. However, the facies distribution and style of sedimentation associated with this regressive shelf sequence is compatible with a eustatic control, and the special circumstances of basin fill created during the late Berriasian, due primarily to high sedimentation rates and a reduced rate of basin subsidence, are thought to have brought the sediment–water interface to within the limits (120–150 m) whereby a third-order eustatic sea-level fall would be the primary control on sedimentation. However, it is still plausible that this eustatic sea-level fall acted in combination with local tectonic uplift during the late Berriasian, but there are no data to support this. It is unlikely that many eustatically controlled sea-level fluctuations will be recognized from fore-arc basins, unless it is possible to demonstrate that there were prolonged periods of tectonic quiescence.

## ACKNOWLEDGEMENTS

This paper has benefited from comments on earlier drafts by D.I.M. Macdonald, D. Pirrie, A.C.M. Moncrieff, and A.G. Whitham. Ian Mills and Bish are thanked for their patience and enthusiasm during two highly enjoyable field seasons on Alexander Island. I thank C.J. Busby-Spera and G.T. Nicholls for their helpful criticism of an earlier version of this paper.

## REFERENCES

Allen, P.A. & Underhill, J.R. (1989) Swaley cross-stratification produced by unidirectional flows, Bencliff Grit (Upper Jurassic), Dorset, UK. *J. Geol. Soc. London* **146**, 241–252.

Barker, P.F. (1982) The Cenozoic subduction history of the Pacific margin of the Antarctic Peninsula: ridge crest–trench interactions. *J. Geol. Soc. London* **139**, 787–801.

Bergman, K.M. & Walker, R.G. (1987) The importance of sea-level fluctuations in the formation of linear conglomerate bodies: Carrot Creek member of Cardium Formation, Cretaceous Western Interior Seaway, Alberta, Canada. *J. Sediment. Petrol.* **57**, 651–665.

Bourgeois, J. (1980) A transgressive shelf sequence exhibiting hummocky stratification: the Cape Sebastian sandstone (Upper Cretaceous), southwestern Oregon. *J. Sediment. Petrol.* **50**, 681–702.

Brenchley, P.J., Romano, M. & Gutierrez-Marco, J.C. (1986) Proximal and distal hummocky cross-stratified facies on a wide Ordovician shelf in Iberia. In: *Shelf Sands and Sandstones* (Ed. Knight, R.J. & McLean, J.R.). *Mem. Can. Soc. Petrol. Geol.* **11**, 241–255.

Buck, S.P. & Bottjer, D.J. (1985) Continental slope deposits from a Late Cretaceous, tectonically active margin, southern California. *J. Sediment. Petrol.* **55**, 843–855.

Butterworth, P.J. (1988) *Sedimentology and stratigraphy of part of the Mesozoic Fossil Bluff Group, Alexander Island, Antarctica*. Unpublished PhD thesis, Council for National Academic Awards, London, 351 pp.

Butterworth, P.J. & Macdonald, D.I.M. (1991) Basin shallowing from the Mesozoic Fossil Bluff Group of Alexander Island and its regional tectonic significance. In: *Geological Evolution of Antarctica* (Eds Thomson, M.R.A., Crame, J.A. & Thomson, J.W.). Cambridge University Press, Cambridge (in press).

Butterworth, P.J., Crame, J.A., Howlett, P.J. & Macdonald, D.I.M. (1988) Lithostratigraphy of Upper Jurassic–Lower Cretaceous strata of eastern Alexander Island, Antarctica. *Cretaceous Res.* **9**, 249–264.

Carey, A.G. (1972) Ecological observations on the benthic invertebrates from the central Oregon continental shelf. In: *The Columbia River Estuary and Adjacent Ocean Waters* (Eds Pruter, A.T. & Anderson D.L.) pp. 422–431. University of Washington Press, Seattle.

Cherven, V.B. (1983) A delta-slope–submarine fan model for Maastrictian part of the Great Valley sequence, Sacramento and San Joaquin Basins, California. *Bull., Am. Assoc. Petrol. Geol.* **67**, 772–816.

Crame, J.A. & Howlett, P.J. (1988) Late Jurassic and Early Cretaceous biostratigraphy of the Fossil Bluff Formation, Alexander Island. *Br. Antarct. Surv. Bull.* **78**, 1–35.

Crowell, J.C. (1974) Origin of late Cenozoic basins in southern California. In: *Tectonics and Sedimentation* (Ed. Dickinson, W.R.). Soc. Econ. Paleontol. Mineral., Spec. Publ., 22, 190–204.

Dalziel, I.W.D., Dott, R.H., Winn, R.D. & Bruhn, R.L. (1975) Tectonic relations of South Georgia Island to the southernmost Andes. *Geol. Soc. Am. Bull.* **86**, 1034–1040.

Dickinson, W.R., Helmold, K.P. & Stein, J.A. (1979) Mesozoic lithic sandstones in central Oregon. *J. Sedim. Petrol.* **49**, 501–516.

Dickinson, W.R., Beard, L.S., Brakenridge, G.R., Erjavec, J.L., Ferguson, R.C., Inman, K.F., Knepp, R.A., Lindberg, F.A. & Rydberg, P.T. (1983) Provenance of North American Phanerozoic sandstones in relation to tectonic setting. *Geol. Soc. Am. Bull.* **94**, 222–235.

Duke, W.L. (1985) Hummocky cross-stratification, tropical hurricanes and intense winter storms. *Sedimentology* **32**, 167–194.

Edwards, C.W. (1980) New evidence of major faulting on Alexander Island. *Br. Antarct. Surv. Bull.* **49**, 15–20.

Ekdale, A.A., Bromley, R.G. & Pemberton, S.G. (1984) Ichnology: the use of trace fossils in sedimentology and stratigraphy. *Soc. Econ. Paleontol. Mineral. Short Course* **15**, 317 pp.

Frey, R.W., Howard, J.D. & Pryor, W.A. (1978) *Ophiomorpha*: its morphologic, taxanomic and environmental significance. *Palaeogeogr. Palaeoclim. Palaeoecol.* **23**, 199–229.

Fursich, F.T. (1984) Benthic macroinvertebrate associations from the Boreal Upper Jurassic of Milne Land, central East Greenland. *Bull. Grnl. Geol. Unders.* **149**, 72 pp.

Hallam, A., Biro-Bagoczky, L. & Perez, E. (1986) Facies analysis of the Lo Valdés Formation (Tithonian–Hauterivian) of the High Cordillera of central Chile, and the palaeogeographic evolution of the Andean Basin. *Geol. Mag.* **123**, 425–435.

Hamblin, A.P. & Walker, R.G. (1979) Storm-dominated shallow marine deposits: the Fernie–Kootenay (Jurassic) transition, southern Rocky Mountains. *Can. J. Earth Sci.* **16**, 1673–1690.

Haq, B.V., Hardenbol, J. & Vail, P.R. (1987) Chronology of fluctuating sea-levels since the Triassic. *Science* **235**, 1156–1167.

Hiscott, R.H. & Middleton, G.V. (1979) Depositional mechanics of thick-bedded sandstone at the base of a submarine slope, Tourelle Formation (Lower Ordovician), Quebec, Canada. In: *Geology of Continental Slopes* (Eds Doyle, L.J. & Pilkey, O.H.). Soc. Econ. Paleontol. Mineral., Spec. Publ. 27, 307–326.

Horne, R.R. (1969) Sedimentology and palaeogeography of the Lower Cretaceous depositional trough of southeastern Alexander Island. *Br. Antarct. Surv. Bull.* **21**, 51–60.

Horne, R.R. & Thomson, M.R.A. (1972) Airborne and detrital volcanic material in the Lower Cretaceous sediments of southeastern Alexander Island. *Br. Antarct. Surv. Bull.* **29**, 103–111.

Howard, J.D. & Reineck, H.E. (1979) Sedimentary structures of a 'high-energy' beach-to-offshore sequence, Ventura–Port Hueneme area, California, USA. *Bull., Am. Assoc. Petrol. Geol.* **63**, 468–469.

Howlett, P.J. (1986) *Olcostephanus* (Ammonitina) from the Fossil Bluff Formation, Alexander Island, and its stratigraphical significance. *Br. Antarct. Surv. Bull.* **70**, 71–77.

Howlett, P.J. (1989) Late Jurassic–Early Cretaceous cephalopods of eastern Alexander Island, Antarctica. *Spec. Pap. Palaeontol.* **41**, 72 pp.

Ingersoll, R.V. (1978a) Petrofacies and petrologic evolution of the Late Cretaceous fore-arc basin, northern and central California. *J. Geol.* **86**, 335–353.

Ingersoll, R.V. (1978b) Paleogeography and paleotectonics of the late Mesozoic forearc basin of northern and central California. In *Mesozoic Paleogeography of the Western United States* (Eds Howell, D.G. & McDougall, K.), *Paleogeography Symposium*, Vol. 2, pp. 471–482. Pacific Coast Section, Soc. Econ. Paleontol. Mineral.

Ingersoll, R.V. (1979) Evolution of the Late Cretaceous fore-arc basin, northern and central California. *Geol. Soc. Am. Bull.* **90**, 813–826.

Ingersoll, R.V. (1983) Petrofacies and provenance of the late Mesozoic forearc basin, northern and central California. *Bull., Am. Ass. Petrol. Geol.* **67**, 1125–1142.

Jefferson, T.H. (1982) Fossil forests from the Lower Cretaceous of Alexander Island, Antarctica. *Palaeontology* **25**, 681–708.

Kauffman, E.G. (1969) Form, function and evolution. In: *Treatise on Invertebrate Paleontology, Volume 1. Mollusca 6. Bivalvia* (Ed. Moore, R.C.) pp. 129–205. University of Kansas and the Geological Society of America, Kansas.

Kern, J.P. & Warme, J.E. (1974) Trace fossils and bathymetry of the Upper Cretaceous Point Loma Formation, San Diego, California. *Geol. Soc. Am. Bull.* **85**, 893–900.

Kulm, L.P., Roush, R.C., Harlett, J.C., Neudeck, R.H., Chambers, D.M. & Runge, E.J. (1975) Oregon continental shelf sedimentation: interrelationships of facies distribution and sedimentary processes. *J. Geol.* **83**, 145–175.

Leithold, E.L. & Bourgeois, J. (1984) Characteristics of coarse-grained sequences deposited in nearshore, wave-dominated environments — examples from the Miocene of south-west Oregon. *Sedimentology* **31**, 749–775.

Lowe, D.R. (1982) Sediment gravity flows: II. Depositional models with special reference to the deposits of high-density turbidity currents. *J. Sediment. Petrol.* **52**, 279–297.

Macdonald, D.I.M. & Butterworth, P.J. (1986) Slope collapse deposits in a Mesozoic marine fore-arc basin, Antarctica. *Abstracts, 12th International Sedimentological Congress*, Canberra, pp. 193–194.

Maslanyj, M.P. (1988) Gravity and aeromagnetic evidence for the crustal structure of George VI Sound, Antarctic Peninsula. *Br. Antarct. Surv. Bull.* **79**, 1–16.

McCrory, V.L.C. & Walker, R.G. (1986) A storm and tidally-influenced prograding shoreline — Upper Cretaceous Milk River Formation of southern Alberta, Canada. *Sedimentology* **32**, 47–60.

Meneilly, A.W., Harrison, S.M., Piercy, B.A. & Storey, B.C. (1987) Structural evolution of the magmatic arc in northern Palmer Land, Antarctic Peninsula. In: *Gondwana Six: Structure, Tectonics and Geophysics* (Ed. McKenzie, G.D.). Am. Geophys. Union, Geophys. Monogr. Ser. 40, 209–219.

Miall, A.D. (1984) *Principles of Sedimentary Basin Analysis*. Springer-Verlag, New York, 490 pp.

Nell, P.A.R. & Storey, B.C. (1991) Strike-slip tectonics within the Antarctic Peninsula forearc. In: *Geological Evolution of Antarctica* (Eds Thomson, M.R.A., Crame, J.A. & Thomson, J.W.). Cambridge University Press, Cambridge (in press).

Nilsen, T.H. & McLaughlin, R.J. (1985) Comparison of tectonic framework and depositional patterns of the Hornelen strike-slip basin of Norway and the Ridge and Little Sulphur Creek strike-slip basins of California. In: *Strike-slip Deformation, Basin Formation and Sedimentation* (Eds Biddle, K.T. & Christie-Blick, N.). Soc. Econ. Paleontol. Mineral., Spec. Publ. 37, 79–103.

Piercy, B.A. & Harrison, S.M. (1991) Mesozoic metamorphism, deformation and plutonism in the southern Antarctic Peninsula: evidence from north-western Palmer Land. In: *Geological Evolution of Antarctica* (Eds Thomson, M.R.A., Crame., J.A. & Thomson, J.W.). Cambridge University Press, Cambridge (in press).

Plint, A.G. (1988) Global eustacy and the Eocene sequence in the Hampshire Basin, England. *Basin Res.* **1**, 11–22.

Rosenthal, L.R.P. & Walker, R.G. (1987) Lateral and vertical facies sequences in the Upper Cretaceous Chungo Member, Waipiabi Formation, southern Alberta. *Can. J. Earth Sci.* **24**, 771–783.

Seilacher, A. (1967) Bathymetry of trace fossils. *Mar. Geol.* **5**, 413–428.

Stamp, L.D. (1921) On cycles of sedimentation in the Eocene strata of the Anglo-Franco-Belgian Basin. *Geol. Mag.* **58**, 108–114, 146–157, 194–200.

Storey, B.C. & Garrett, S.W. (1985) Crustal growth of the Antarctic Peninsula by accretion, magmatism and extension. *Geol. Mag.* **122**, 5–14.

Storey, B.C. & Macdonald, D.I.M. (1984) Processes of formation and filling of a Mesozoic back-arc basin on the island of South Georgia. In: *Marginal Basin Geology* (Eds Kokelaar, B.P. & Howells, M.F.). Geol. Soc. London, Spec. Publ. 16, 207–218.

Storey, B.C. & Nell, P.A.R. (1988) Role of strike-slip faulting in the tectonic evolution of the Antarctic Peninsula. *J. Geol. Soc. London* **145**, 333–337.

Surlyk, F. (1978) Submarine fan sedimentation along fault scarps on tilted fault blocks (Jurassic–Cretaceous boundary, East Greenland). *Bull. Grnl. Geol. Unders.* **128**, 108 pp.

Surlyk, F. (1984) Fan-delta to submarine fan conglomerates of the Volgian–Valanginian Wollaston Foreland Group, East Greenland. In: *Sedimentology of Gravels and Conglomerates* (Eds Koster, E.H. & Steel, R.J.). Mem., Can. Soc. Petrol. Geol. 10, 359–382.

Swift, D.J.P., Hudelson, P.M., Brenner, R.L. & Thompson, P. (1987) Shelf construction in a foreland basin: storm beds, shelf sand bodies, and shelf-slope depositional sequences in the Upper Cretaceous Mesaverde Group, Book Cliffs, Utah. *Sedimentology* **34**, 423–458.

Taylor, B.J. (1982) Sedimentary dykes, pipes and related structures in the Mesozoic sediments of south-eastern Alexander Island. *Br. Antarct. Surv. Bull.* **51**, 1–42.

Thomson, M.R.A. (1982) Mesozoic paleogeography of western Antarctica. In: *Antarctic Geoscience* (Ed. Craddock, C.) pp. 331–337. University of Wisconsin Press, Madison, WI.

Vail, P.R., Mitchum, R.M., Todd, R.G., Widmier, J.M., Thompson, S., Sangree, J.B., Bubb, J.N. & Hatlelid, W.G. (1977) Seismic stratigraphy and global changes of sea-level, parts 1–6. In: *Seismic Stratigraphy – Applications to Hydrocarbon Exploration* (Ed. Payton, C.E.). Mem. Am. Assoc. Petrol. Geol. 26, 49–212.

Valentine, P.C., Cooper, R.A. & Uzmann, J.R. (1984) Submarine sand dunes and sedimentary environments in Oceanographer Canyon. *J. Sediment. Petrol.* **54**, 704–715.

Wetzel, A. (1984) Bioturbation in deep-sea fine-grained sediments: influence of sediment texture, turbidite frequency and rates of environmental change. In: *Fine-grained Sediments* (Eds Stow, D.A.V. & Piper, D.J.W.). Geol. Soc. London, Spec. Publ. 15, 595–608.

Whitham, A.G. & Marshall, J.E.A. (1988) Syn-depositional deformation in a Cretaceous succession, James Ross Island, Antarctica. Evidence from vitrinite reflectivity. *Geol. Mag.* **125**, 583–591.

Wilson, J.B. (1982) Shelly faunas associated with temperate offshore tidal deposits. In: *Offshore Tidal Sands, Processes and Deposits* (Ed. Stride, A.H.) pp. 126–171. Chapman & Hall, London.

Winn, R.D. & Dott, R.H. (1979) Deep-water fan-channel conglomerates of late Cretaceous age, southern Chile. *Sedimentology* **26**, 203–228.

*Spec. Publs. int. Ass. Sediment.* (1991) **12**, 331–369

# The role of local tectonics versus global sea-level change in the Neogene evolution of the Cyprus active margin

A.H.F. ROBERTSON†, S. EATON‡, E.J. FOLLOWS† *and* J.E. McCALLUM*

† *Department of Geology and Geophysics, Grant Institute, West Mains Road, Edinburgh EH9 3JW, UK;*
‡ *K.S.E.P.L., Volmerlaan 6, Rijswijk, The Netherlands; and * Alastair Beach Associates, 11 Royal Exchange Square, Glasgow G1, UK*

## ABSTRACT

The Miocene–Pliocene stratigraphical record of a N–S transect of central Cyprus shows that tectonics exerted the dominant influence on sedimentation, modified by the effects of global eustatic sea-level changes. The area lies in a complex collision zone in the eastern Mediterranean that separates the African and Eurasian plates. The Cyprus active margin has undergone northward subduction, probably at least since early Miocene time. Subduction was limited, however, with no well-defined volcanic arc. Pulsed uplift of the Troodos ophiolite resulted in deposition of an overall regressive-upwards mega-sequence. Following Early Tertiary deep-water carbonate deposition, in a relatively stable setting, Miocene convergence gave rise to three subparallel, compressional lineaments in southern Cyprus. The best-exposed, central, lineament (Yerasa) had stabilized and was subsiding by late Miocene time, allowing reef limestone growth during inferred sea-level rise. To the north, another lineament (Ayia Mavri) was active in late Miocene (Messinian) to earliest Pliocene time, and several other deformation zones are recognized. Crustal weaknesses were exploited by these lineaments. A wide range of pelagic carbonates and minor channelized, ophiolite-derived conglomerates, and reefs and their associated debris accumulated in adjacent, tectonically controlled sub-basins (Maroni and Khalassa). Simultaneously, 60 km to the north, extension and growth faulting took place during early–late Miocene time, with Tortonian reef development on uplifted fault blocks. During the Messinian salinity crisis, sea-level dropped and gypsum was precipitated within pre-existing, silled basins, at depths from a few metres to several hundred metres. During a time of inferred northward subduction in the Pliocene, southern Cyprus remained relatively stable, but shallowed. A major channel was cut and then filled in the (?)late Pliocene, apparently following eustatic sea-level fall. Further north, strong extension in the early Pliocene created a major half-graben (Mesaoria Basin), that shallowed upwards, accommodating c. 900 m of marine clastics. In response to pulsed uplift, fan-deltas prograded northwards, off the eroding Troodos ophiolite, first in the early Pliocene, then, on a larger scale, in the (?)late Pliocene–(?)early Pleistocene, eventually giving way to fluvial deposition. More northerly areas of the Mesaoria Basin were dominated by bioclastic sand waves, derived from the north, with possibly eustatically controlled drowning of offshore barrier bars and coastal areas. The whole of Cyprus was drastically uplifted in mid-Pleistocene time, followed by relative tectonic quiescence in the Holocene.

## INTRODUCTION

The influence of eustatic sea-level changes on sedimentation is now well established for well-dated successions in tectonically stable areas, for example the European continental interior in the Jurassic (Hallam, 1988). The recognition of eustatic change, largely based on seismic profiles, has allowed global sea-level curves to be proposed for Phanerozoic time (e.g. Vail *et al.*, 1977,1984; Vail & Todd, 1981; Haq *et al.*, 1987; see paper by Haq). Attempts are now being made to recognize sea-level changes in tectonically active areas, for example active continental margins. However, lateral facies variation, poor preservation, and limited age dating of many active margin successions make this a difficult task (e.g. see paper by Dobson *et al.*).

The Neogene history of the Cyprus active margin offers the chance to evaluate the effects of sea-level changes in a relatively well-dated stratigraphic framework of an active margin in the eastern Mediterranean (Fig. 1, inset). Deep-sea carbonates and clastic sediments are well exposed on the uplifted, overriding plate that forms the island of Cyprus. In this case, subduction appears to have been limited, thus arc-derived volcaniclastic input is absent, by contrast with many other active margins, for example Alexander Island in the Antarctic (see paper by Butterworth). We will summarize the Neogene sedimentation of Cyprus in relation to the deformation associated with inferred northward subduction, identifying periods of relative sea-level change. We test our results for sea-level change against global sea-level curves. The Pleistocene record of Cyprus will be treated elsewhere (Poole *et al.*, 1990). Except where specified, data on the Miocene of southern Cyprus are from Eaton (1987), on Miocene reef facies from Follows (1990a), and on the Pliocene from McCallum (1989). It should be noted that our results mainly apply to a N–S transect across the central Cyprus area. Descriptions of west Cyprus (e.g. Robertson, 1977a; Orszag-Sperber *et al.*, 1980,1989) and southeast Cyprus (Follows & Robertson, 1990) can be found elsewhere. Absolute ages are from the Harland *et al.* (1989) time-scale.

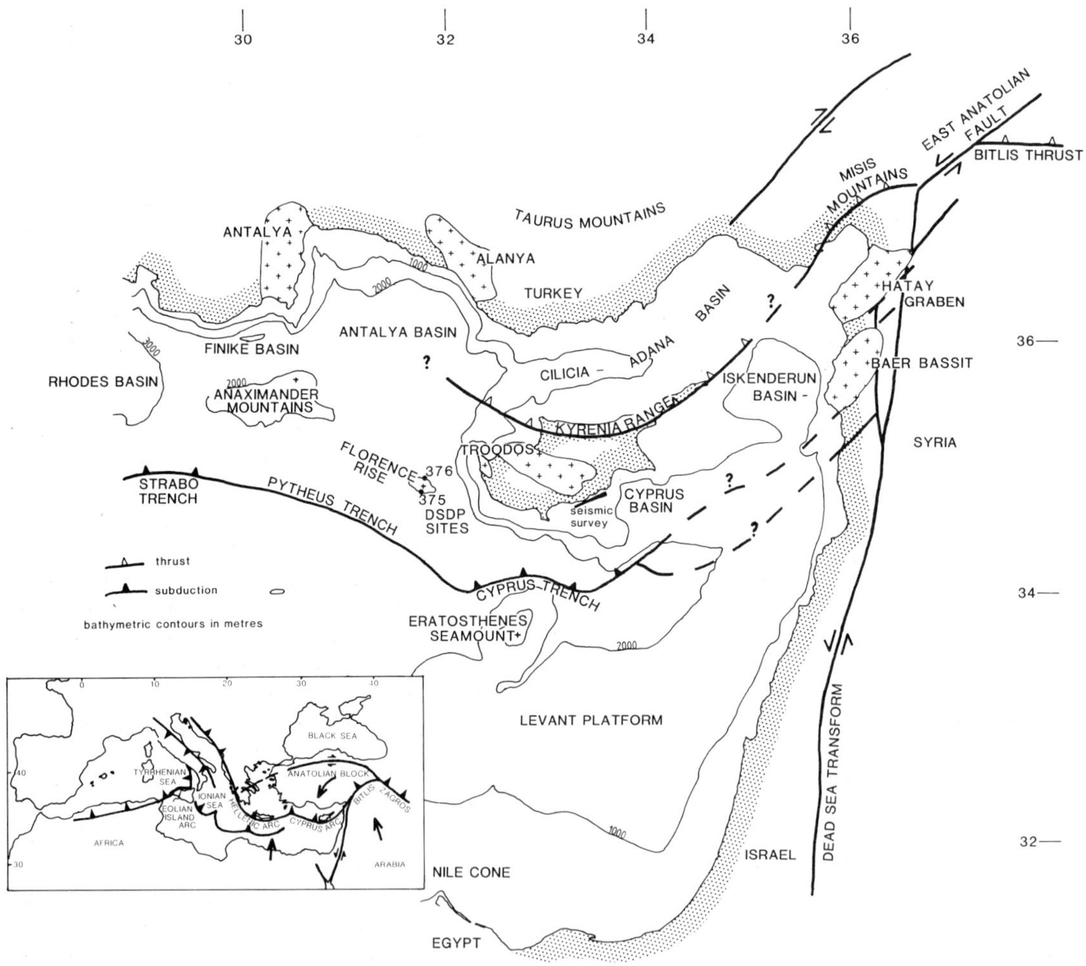


**Fig. 1.** Outline map of the eastern Mediterranean, showing the present-day Cyprus active margin, Deep Sea Drilling Project Sites 375 and 376, and locations of seismic data (location of structures from McCallum, 1989). Inset: the trends of active margins in the Mediterranean.

## SETTING FOR DEVELOPMENT OF THE CYPRUS ACTIVE MARGIN

During the Late Triassic to Early Cretaceous, the Cyprus area was located within the southerly Mesozoic Neotethys (Robertson & Dixon, 1984). Rocks of this age are found in the Mamonia Complex of southwestern Cyprus (Lapierre, 1975; Robertson & Woodcock, 1979; Swarbrick, 1980) and the Kyrenia Range of northern Cyprus (Ducloz, 1972; Baroz, 1979; Robertson & Woodcock, 1979; Fig. 2). The Troodos ophiolite formed at a spreading axis (Gass & Masson-Smith, 1963; Moores & Vine, 1971) in the Late Cretaceous (92–85 Ma; Mukasa & Ludden, 1987), probably above a subduction zone (Pearce, 1980; Clube & Robertson, 1986; Robertson, 1990). The southern outcrop of the Troodos ophiolitic massif was cut by a major E–W-trending fault zone, the Arakapas Fault Zone, considered to be a late Cretaceous oceanic fracture zone (Simonian & Gass, 1978; Murton & Gass, 1986; MacLeod, 1990), that was reactivated as an important compressional lineament in the Miocene. The Troodos ophiolite underwent 90° of anticlockwise rotation between late Cretaceous (Campanian) and early Eocene (Clube *et al.*, 1985; Clube & Robertson, 1986). The microplate boundaries are partly located in southwest and south Cyprus and formed zones of structural weakness that were also reactivated in the Neogene. In north Cyprus, the Kyrenia Range was first deformed in the late Cretaceous, then thrust southwards in the late Eocene (e.g. Ducloz, 1972; Baroz, 1979; Robertson & Woodcock, 1986), at the same time as suturing of Neotethyan ocean basins in southern Turkey (e.g. Şengör & Yilmaz, 1981; Aktaş & Robertson, 1990). Relatively thin pelagic carbonates (tens of metres) of early Tertiary age were deposited along the north margin of the Troodos ophiolite, while much thicker successions (> 1000 m) accumulated coevally in tectonically active (?transtensional) basins in south Cyprus, as the Maastrichtian–Oligocene Lefkara Formation (Mantis, 1970a; Robertson, 1976). Gradual shallowing upwards of the Troodos Massif took place in the Oligocene (Upper Lefkara Formation; Mantis, 1970a; Robertson, 1977a), during a time of relative tectonic quiescence. It is important to note that early formed structures were to exert an important control on tectonics and sedimentation in the Neogene of Cyprus.

## INITIATION OF THE NEOGENE CYPRUS ACTIVE MARGIN

To the east of Cyprus, the Neotethyan ocean finally sutured in middle to late Miocene time (Şengör & Yilmaz, 1981; Aktaş & Robertson, 1984; Şengör *et al.*, 1985), while the Red Sea began to open (Whitmarsh *et al.*, 1974; Purser *et al.*, 1987). The accompanying plate reorganization led to the initiation of northward subduction in the Mediterranean (e.g. south Aegean and Tyrrhenian Sea areas; McKenzie, 1972,1978; Le Pichon & Angelier, 1979; Jackson & McKenzie, 1984; Rotstein, 1984; Kempler & Ben-Avraham, 1987; Kastens *et al.*, 1988).

Stratigraphic evidence from Cyprus suggests an earliest Miocene age for onset of subduction, marked by a fundamental change in sedimentation, from relatively uniform pelagic carbonate deposition (Upper Lefkara Formation), to much more varied carbonate and clastic sedimentation in tectonically controlled small basins bordering the Troodos ophiolite (Pakhna Formation; Robertson, 1977a). This change is interpreted as the beginning of development of the present Cyprus active margin. Before going on to discuss this further, we will summarize the evidence that leads to Cyprus being interpreted as the overriding plate of an active plate margin.

## NATURE OF THE MODERN ACTIVE MARGIN

Cyprus today is located above a sluggishly subducting plate, a situation that has probably persisted throughout the Neogene. In consequence, no well-defined volcanic chain is present north of Cyprus, although possibly related Neogene volcanics exist in southern Turkey (Innocenti *et al.*, 1982). In general, Cyprus is now located between an area where Neotethyan basins were sutured in the mid to late Miocene in eastern Turkey and the Zagros (Dewey & Şengör, 1979; Şengör & Yilmaz, 1981; Aktaş & Robertson, 1984; Şengör *et al.*, 1985), and a region further west, where Neotethyan oceanic crust still remains (Robertson & Dixon, 1984; Dercourt *et al.*, 1986). The island now appears to be underlain by continental crust (Makris *et al.*, 1983), which may relate to the Eratosthenes Seamount, or some other microcontinental unit, underthrust at some earlier time (i.e. 'Olympus microcontinent'; Clube & Robertson, 1986).

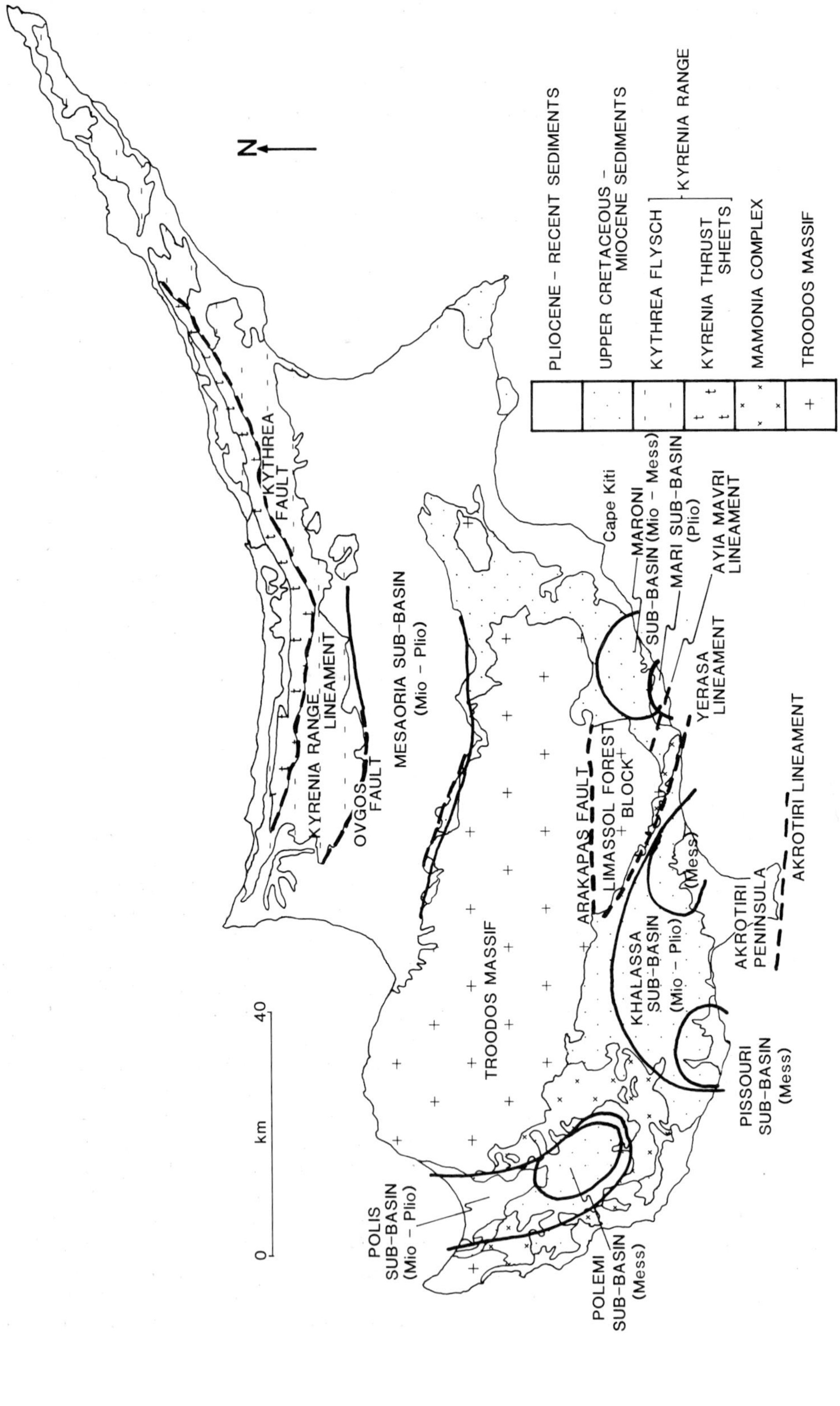


**Fig. 2.** Major structural elements and geological units of Cyprus. The main depositional sub-basins discussed in the text are outlined.

Based on bathymetric and seismic data, a subduction trench is clearly defined south of Cyprus. It curves around the north margin of the Eratosthenes Seamount (Fig. 1), broadens and shallows eastwards. The Eratosthenes Seamount is interpreted either as an oceanic plateau (Rotstein, 1984; Ben-Avraham & Nur, 1986), or a microcontinent (Kempler & Ben-Avraham, 1987), and could have considerable subsurface extent, based on magnetic data (Ben-Avraham *et al.*, 1988). The north wall of the subduction trench in this area is rather irregular, with perched sedimentary basins and intervening highs (McCallum, 1989). Basinal sediments on the overriding plate are slightly tilted. Deformation of underthrusting sediments is not observed on seismic profiles, except in the trench adjacent to the Eratosthenes Seamount where (?)Pliocene–Pleistocene sediments are markedly tilted. This could reflect northward subduction of the seamount (Kempler & Ben-Avraham, 1987; McCallum, 1989; Anastasakis & Kelling, in press). The seamount appears to be disintegrating into smaller, more easily subductable blocks (Kempler & Ben-Avraham, 1987), deforming the Cyprus active margin by collision, in a similar manner, perhaps, to the collapse and partial subduction of a seamount in the Japan Trench (Le Pichon *et al.*, 1987). Strong Pleistocene uplift of the Troodos ophiolite and its sediment cover could relate to partial subduction of the Eratosthenes Seamount already having occurred, or to earlier underthrusting of some other seamount unit (Robertson, 1990).

To the west of Cyprus, subduction, as indicated by earthquake data (Rotstein & Kafka, 1982; Kempler & Ben-Avraham, 1987) is revealed at the surface as a poorly defined chain of depressions and small highs (i.e. Pytheus Trench of Anastasakis & Kelling (in press)), which connect westwards with the Strabo Trench (Fig. 1). Seismic profiles are interpreted as indicating the existence of a NW–SE-trending lineament (Kempler & Ben-Avraham, 1987). This could reflect northeastwards (Dewey & Şengör, 1979; Jackson & McKenzie, 1984), or northwards (Rotstein & Kafka, 1982), subduction, or alternatively, strike-slip (Anastasakis & Kelling, in press).

To the east of Cyprus, dominantly strike-slip motion is inferred (except by Rotstein, 1984), but the exact location of the plate boundary and how it connects with the East Anatolian Fault and the Bitlis suture zone in eastern Turkey remains unclear. The sea floor shallows towards the Levantine coast where it is cut by NE–SW-trending escarpments. Onshore, these features connect with the Tethys suture zone in Baër-Bassit, and further east, with the Dead Sea transform fault. This, in turn, links up with the East Anatolian transform fault, via the Hatay graben (Şengör & Yilmaz, 1981; Şengör *et al.*, 1985; Fig. 1).

Another tectonic lineament, the Kyrenia Range, runs E–W, through the northern part of Cyprus (Fig. 1). Eastwards, this connects with the Misis Mountains in southern Turkey, via a submerged ridge (Kelling *et al.*, 1987). Pleistocene compression and uplift of the Kyrenia Range lineament was extensive (Baroz, 1979; Robertson & Woodcock, 1986), thrusting being visible on seismic sections across a submerged ridge to the east (Mulder, 1973). On either side of this lineament, the Cilicia, Adana and Iskenderun Basins (Fig. 1) contain thick Plio-Pleistocene sediments, and are inferred to be extensional, or strike-slip controlled basins (Mulder, 1973; Şengör *et al.*, 1985).

It should be noted that there is no evidence of any substantial plate convergence along the Kyrenia Range lineament after final southwards thrusting in the late Eocene (Robertson & Woodcock, 1986). The lineament did not act as convergent margin, or trench during Miocene–Pliocene time, contrary to some published tectonic maps of the area.

## MIOCENE STRATIGRAPHY OF SOUTHERN CYPRUS

The Neogene evolution of Cyprus began with deposition of the early to late Miocene Pakhna Formation, which crops out throughout central and southern Cyprus, south of the Ovgos Fault (Fig. 2), and probably also originally covered much of the Troodos ophiolite (Robertson, 1977a). The Pakhna Formation is, in turn, depositionally overlain by evaporitic lithofacies of the Messinian Kalavasos Formation (Eaton, 1987; Follows, 1990a; Table 1).

Eaton (1987) and Follows & Robertson (1990) demonstrated that reefal units, previously thought to both underlie ('Terra Formation') and overlie ('Koronia Formation') the Pakhna Formation, are in fact *interstratified* (Eaton, 1987; Follows, 1990a; Fig. 3). These units are now redefined as the Aquitanian–Burdigalian Terra Member and the Tortonian Koronia Member (see Bellamy & Jukes-Browne (1905), Henson *et al.* (1949), Bagnall, (1960), Pantazis (1967), and Eaton (1987) for earlier interpretations). The biostratigraphy of the Pakhna

**Table 1.** Stratigraphic nomenclature of the Neogene of Cyprus. Not to scale. Data sources are specified in the text. The time-scale is from Harland *et al.* (1989)

| Age (Ma) | | | Formation | Lithology |
|---|---|---|---|---|
| 2.0 | Pleistocene | | 'Fanglomerate' | Conglomerates and sandstones |
| | Pliocene | | Apalos Kakkaristra Athalassa | Calcarenite, sandstone, conglomerate |
| 5.2 | | | Nicosia | Marls, silts, muds, sandstone, conglomerate |
| | Miocene | Upper | Kalavasos | Evaporites |
| | | | Koronia Member | Reefal and bioclastic limestone |
| | | Middle | Pakhna | Pelagic chalks, marls, calcarenites, conglomerates |
| 23.3 | | Lower | Terra Member | Reefal and bioclastic limestone |
| 35.4 | Oligocene | | Upper Lefkara | Pelagic chalk and marls |

Formation is mainly based on data of planktonic and benthonic foraminifera, calcareous nannoplankton, and shelly fossils (Adams, 1959; Bagnall, 1960; Mantis, 1970a,b; Hadjistavrinou, 1974,1975; H. Stowe, personal communication, 1987; M. Simmons, personal communication, 1990), supplemented by unpublished studies (Herguera, 1982; Ortiz, 1982; Haskett, 1984). The age of the evaporites is largely by comparison with Messinian facies elsewhere in the Mediterranean (Hsü, 1973; Hsü *et al.*, 1978).

In the following sections, we will discuss, in turn, structures that controlled Neogene sedimentation in southern Cyprus, Miocene sedimentation in both southern and northern Cyprus (including the Messinian evaporites), and then summarize Pliocene facies. Finally, we will evaluate the importance of global sea level versus more local tectonic controls on sedimentation.

## COMPRESSIONAL LINEAMENTS OF SOUTHERN CYPRUS

Four WNW–ESE-trending structural lineaments (Figs 2, 5) of Miocene age are identified in southern Cyprus. These lineaments are: (1) the *Ayia Mavri lineament* (Eaton, 1987); (2) the *Yerasa lineament* (new name; Yerasa Fold and Thrust Belt of Morel, 1960); (3) the *Akrotiri lineament* (Akrotiri high of Eaton, 1987); and (4) the *Petounda lineament* (new name; McCallum, 1989).

### Yerasa lineament: reactivated ophiolite fracture zone

The most prominent of the four lineaments, Yerasa (Fig. 2) is a narrow, c. 30 km long by 2 km wide, WNW–ESE-trending zone of steep thrusts and folds (Fig. 4), that reactivated a late Cretaceous fracture zone in the Troodos ophiolite. The lineament underwent uplift in the Miocene, followed by deep erosion of the Limassol Forest Block to the south (Fig. 2). The Yerasa lineament extends east-southeast, beyond its zone of maximum expression, as a broad, gently eastwards plunging anticlinal structure, mainly developed within early Tertiary deep-water carbonates of the Lefkara Formation. It has also been traced offshore on shallow seismic data (McCallum, 1989).

Neogene structures in the ophiolitic basement, north of the Yerasa lineament, are not always easy to distinguish from those associated with late Cretaceous transform faulting. Deep-level mafic and ultramafic rocks there were already uplifted to high structural levels in the Late Cretaceous (Morel, 1960; Robertson, 1977a; Searle & Panayiotou, 1980; Murton & Gass, 1986). Some Miocene-aged structures, however, can be identified, and include ophiolitic pillow lavas thrust over Lefkara Formation

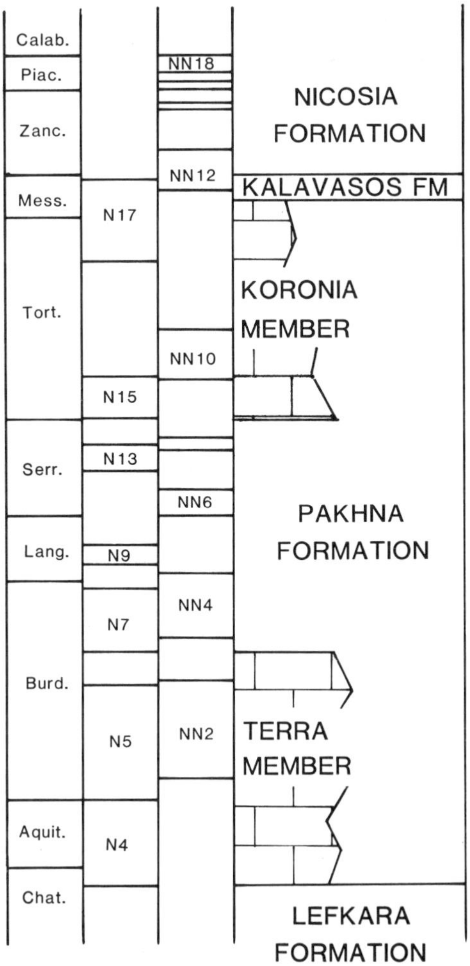


**Fig. 3.** Stratigraphy and age of the Miocene successions of Cyprus. The boundary between the Lefkara and the Pakhna Formations is apparently diachronous, at least in southern Cyprus.

chalks (e.g. at Akrounda, Fig. 5; Robertson, 1977a). Further west, microgabbros are thrust over lava breccias, and pillow lavas, umbers, and radiolarites (Upper Cretaceous Perapedhi Formation) are overturned (e.g. east of Apsiou, Fig. 5). Thrust planes dip north-northeast at 30° to >45°.

Immediately southeast of the Yerasa lineament, intense deformation is observed in the sedimentary succession. Bentonitic clays of the Upper Cretaceous Kannaviou Formation (Moni Member; Swarbrick & Robertson, 1980), which overlie the ophiolitic basement, are foliated and dip northwards at c. 60°. In the west, the base of the pelagic carbonate succession (Lefkara Formation) is a sharp tectonic contact that dips steeply northwards. Folding of the overlying succession is tight, with steep northwards dipping axial planes, and some overturned southern limbs (Fig. 4) cut by steep northwards dipping faults. Axial-plane cleavage is strongly developed (e.g. road section south of Armenokhori, Fig. 5). Further south, deformation intensity decreases rapidly, and within a kilometre or so, chalks of the Lefkara Formation exhibit only gentle southerly dips (Fig. 4), and then pass conformably upwards into the Pakhna Formation (e.g. near Amathus, Fig. 5).

Uplift of the Yerasa lineament resulted in deep erosion, estimated at up to 300 m. Detritus in the Pakhna Formation to the south was derived from all parts of the succession, down to and including lavas, chert, and diabase of the Troodos ophiolite (but not gabbro or ultramafics) (Eaton, 1987; see below).

Transgressive bioclastic limestones of the Koronia Member of the Pakhna Formation (Fig. 6) overlie the Yerasa lineament. These are not deformed, clearly indicating that transgression had taken place by late Miocene (Tortonian) time, possibly related to global sea-level rise (Vail *et al.*, 1984,1989; see later). Near the axis of the Yerasa lineament, the Koronia Member unconformably overlies 'basement', comprising Troodos ophiolite pillow lavas, near Parekklisha and late Cretaceous Moni Melange, near Armenokhori (Robertson, 1977b; Fig. 5). To the south, further from the lineament (e.g. Amathus, Fig. 5), a low-angle unconformity is observed between the Upper Lefkara Formation and the Koronia Member, with only the lower Pakhna Formation missing. Elsewhere, in adjacent basinal areas (e.g. Psematismenos, Fig. 5), the succession is complete and the Koronia Member is conformable on marls and calcarenites of the Pakhna Formation.

In summary, the Yerasa lineament was an important zone of structural uplift and deep erosion in southern Cyprus, and was active in early Miocene to pre-late Miocene (Tortonian) time (c. 16–10 Ma).

### Ayia Mavri lineament: inferred basement control

The second lineament, Ayia Mavri (Fig. 2), further northeast was active during both mid-Tertiary and Miocene time. It is recognized mainly on the basis of numerous faults that cut the Pakhna and Lefkara Formations. These faults generally throw down to the northeast in the area south of Kalavasos (Fig. 5).

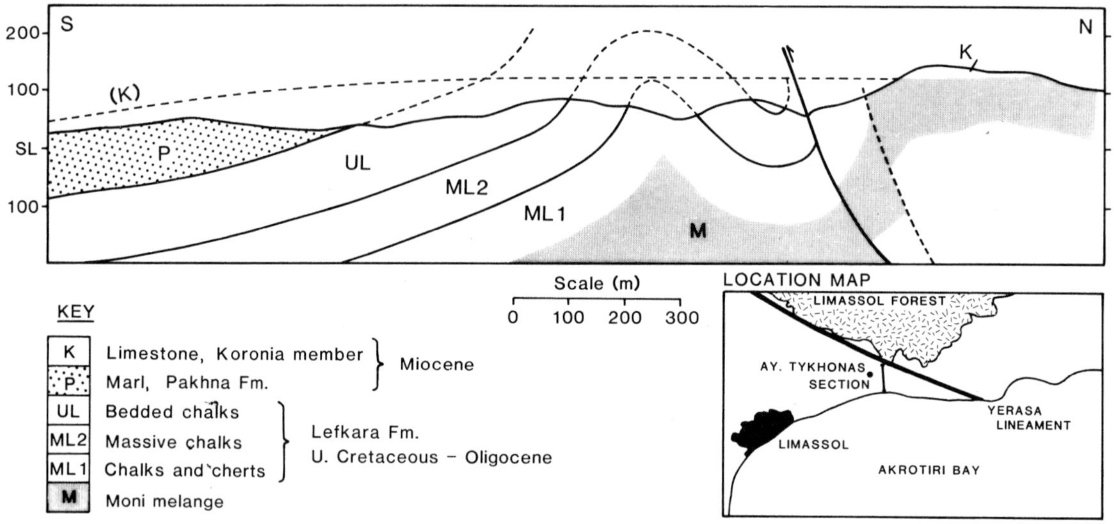


**Fig. 4.** N–S cross-section across the Yerasa lineament; inset gives the location. Modified from Eaton (1987).

In this area, Miocene strata of the Pakhna Formation are steep and define a monoclinal structure striking WNW–ESE. Near the faults that define the lineament, the Middle Lefkara Formation of Paleocene to early Eocene age contains conglomerates with clast-supported boulders (greater than 1 m in size), including Lefkara Formation replacement chert, within a chalky matrix. This facies is overlain by well-bedded calciturbidites. The conglomerates were presumably shed off an active submarine fault scarp. Nearby, a several-metre-deep, inferred depositional hollow in the sediment surface is underlain by numerous small faults, which do not extend into the overlying strata. This feature could represent a slump scar, underlain by gravity faults. Deep-water carbonates of the Lefkara Formation are ponded into this hollow.

The Ayia Mavri lineament was active again during the Miocene. The evidence for this later history is mainly the sediment fill of the adjacent Maroni sub-basin (Fig. 2), which includes reef debris shed from the inferred Ayia Mavri lineament (see below). The lineament continued to be active in late Miocene to (?)earliest Pliocene, as is documented by its eastward extension (near Mari, Fig. 5), where the Pakhna Formation *and* the overlying Messinian selenitic gypsum were disrupted, faulted, and folded into a mesoscopic open syncline that plunges southeastwards. This structure is overlain by a c. 60-m-thick conglomeratic interval of Messinian gypsum and Lefkara Formation chalk and chert. Deeper stratigraphical levels were exposed by erosion with time (i.e. Kalavasos Formation, then Lefkara Formation). Reactivation of zones of basement structural weakness zones is suspected to explain the location of the Ayia Mavri lineament.

## Akrotiri lineament: a concealed southerly structure

South of the major Yerasa lineament, a third WNW–ESE-trending lineament is inferred to run through the southern tip of the Akrotiri Peninsula (Bear & Morel, 1960) and adjacent offshore areas (Fig. 2). The main evidence for this concealed Akrotiri lineament comes from borehole and offshore shallow seismic data.

Several hydrogeological boreholes (Hadjistavrinou & Constantinou, 1977) in the Akrotiri Peninsula encountered successions that pass down from Pliocene and Pleistocene clastic sediments, 150 m thick, into carbonates of the Lefkara Formation, with no intervening Miocene Pakhna Formation. Some 3 km to the south another borehole revealed a Plio-Pleistocene succession c. 100 m thick, underlain by 230 m of Mesozoic shales and radiolarites, correlated with the Mamonia Complex. Along the south coast of the Akrotiri Peninsula, similar rocks of the Mamonia Complex, including folded radiolarian cherts, are unconformably overlain by Pliocene bioclastic carbonates and marls (Bear & Morel, 1960;

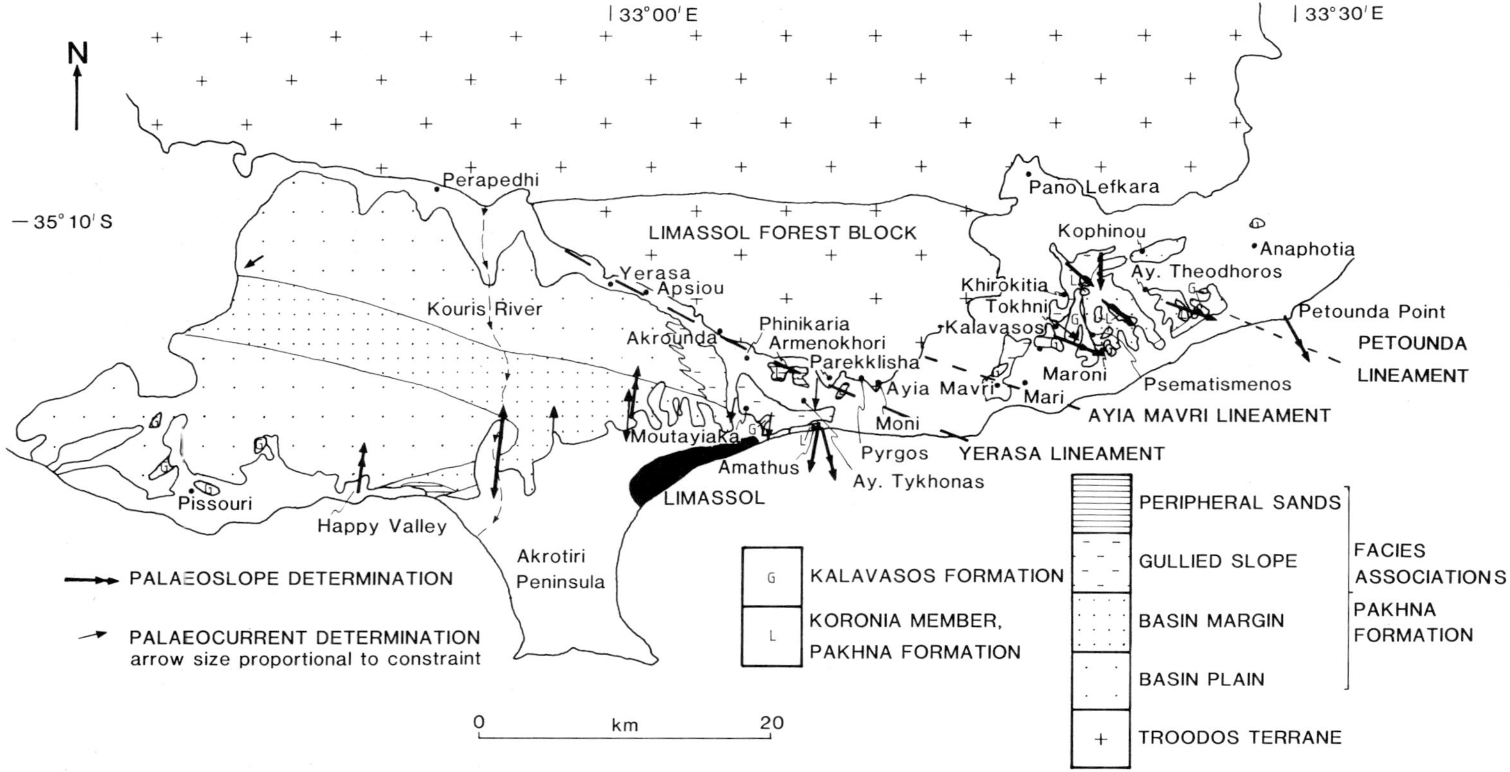


**Fig. 5.** Outline geological and tectonic map of southern Cyprus, showing the distribution of facies associations, palaeocurrent and palaeoslope data, and the locations of places mentioned in the text.

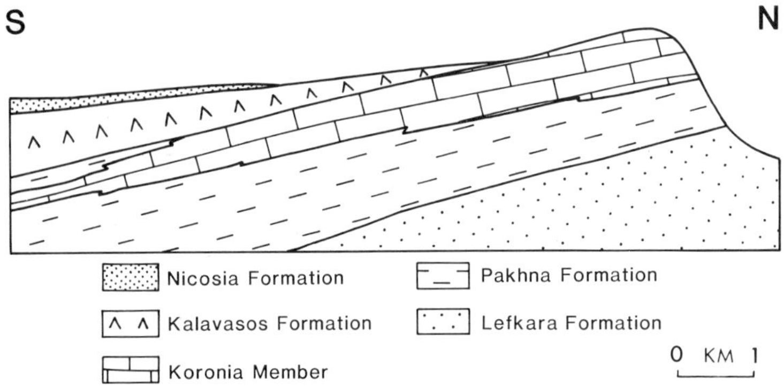


**Fig. 6.** Generalized stratigraphical relationships of the Pakhna Formation in the Khalassa sub-basin in the west. Note that the Late Miocene Koronia Member (Pakhna Formation) is conformably overlain by the Messinian evaporites of the Kalavasos Formation.

Robertson, 1977a; Eaton, 1987). Thus, in the area of the Akrotiri Peninsula, Miocene sediments of the Pakhna Formation were either completely eroded, or never deposited.

The Akrotiri lineament can be traced eastwards offshore on shallow seismic profiles, across an irregular slope. The inferred Miocene succession offshore is broadly folded, parallel to the lineament. Unconformably overlying sediments of assumed post-Messinian, Plio-Pleistocene age are not affected, however (McCallum, 1989). This suggests that the Akrotiri lineament was active during the late Miocene, but not subsequently.

Sedimentological evidence also supports the existence of the Akrotiri lineament. Palaeocurrent and slump data suggest sediment derivation from the south (see below). Also, numerous very small, well-rounded pebbles of grey and yellow chert, up to 1 cm sized, are concentrated in the pebbly bases of graded calcarenites exposed near the southern limit of the Pakhna Formation outcrop, close to the Akrotiri Peninsula (Figs 2,5). Petrographic studies (Eaton, 1987) reveal eight chert types, of which three are dissimilar to chert in the local Lefkara Formation and the underlying Perapedhi Formation. Similar cherts occur as detached blocks ('olistoliths') in Triassic–Early Cretaceous quartzitic sandstones (Swarbrick & Robertson, 1980) within the Moni Member in southern Cyprus (Moni Mélange of Robertson, 1977b). Derivation from the Mamonia Complex in southwest Cyprus is less likely.

### Petounda lineament: zone of folding and slumping

Finally, a fourth lineament is inferred further east, as revealed by the evidence of intense slumping of the Pakhna Formation beneath Pleistocene raised beach deposits in cliffs near Petounda Point (Fig. 5) and around the coast to Cape Kiti (Farrell & Eaton, 1987; Fig. 2). The slumped facies include pink pelagic chalks, marls, and sandstones with lignitic horizons, in units up to 12 m thick. In addition, marked folding is visible in adjacent offshore shallow seismic profiles, with a pinch-out of sediments, of inferred Pliocene age, to the northeast towards the lineament (McCallum, 1989).

### Summary

Four sub-parallel active structural lineaments, the Yerasa, Ayia Mavri, Akrotiri, and Petounda lineaments, are inferred in southern Cyprus in Miocene time. There was at least 300 m of uplift of the major Yerasa lineament prior to transgression by Tortonian lithofacies. Further northeast, the Ayia Mavri lineament remained active through late Messinian to earliest Pliocene time. The Akrotiri and Petounda lineaments were also active in Miocene to (?)earliest Pliocene time. The combined evidence of faulting, folding, uplift, erosion, and more localized thrusting and cleavage development is, in our view, explicable only in terms of significant N–S compressional tectonics affecting southern Cyprus during the Miocene.

## MIOCENE FACIES OF CYPRUS

In southern Cyprus the subparallel compressional lineaments discussed above exerted a dominant control on sedimentation during Miocene time and gave rise to a series of sub-basins with individual depocentres. The facies of the Aquitanian–

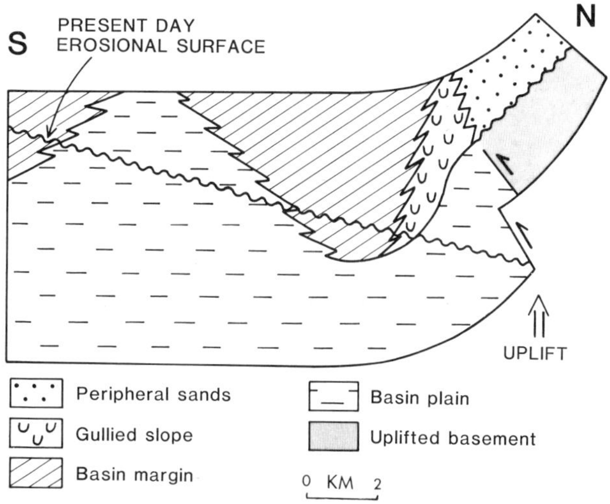


**Fig. 7.** Schematic basin profile showing the spatial relationships of the facies associations of the Pakhna Formation. Modified from Eaton (1987).

Tortonian Pakhna Formation (Henson *et al.*, 1949; Table 1; Fig. 3) delineate the following sub-basins, from the south clockwise (Fig. 2): *Maroni* (Eaton, 1987), *Khalassa* (Eaton, 1987), *Pissouri* (Elion, 1983), *Polemi* (Orszag-Sperber *et al.*, 1980), *Polis*, and *Mesaoria* (Gass & Cockbain, 1961; Orszag-Sperber *et al.*, 1989). Of these, we will discuss here only Miocene facies of the Maroni and Khalassa sub-basins of southern Cyprus and the Mesaoria sub-basin of northern Cyprus within our N–S transect (Fig. 2). The Yerasa and Ayia Mavri lineaments delineated the margins of the two southerly sub-basins. The Khalassa sub-basin in the west was located southeast of the strongly uplifed Yerasa lineament, and the Maroni sub-basin, further east was located northeast of the more subdued Ayia Mavri lineament (Eaton, 1987). Basinal sediments also accumulated more locally, between the Yerasa and Ayia Mavri lineaments (Fig. 2).

Within the Maroni and Khalassa sub-basins, the Pakhna Formation is dominated by chalky hemipelagic marls, argillites, and gravity deposited bioclastic carbonates, with subordinate channelized polymict, ophiolite-derived conglomerates. Four facies associations are recognized (Eaton, 1987; Fig. 7). In proximal to distal order, these are as follows: (1) *peripheral sands*, mainly thick-bedded, homogeneous calcarenites; (2) *gullied slope*, silty calcilutites, with minor channelized calcarenites and conglomerates; (3) *basin margin*, slumps, debris flows, and clast-supported conglomerates, interbedded with chalks and marls, that are lenticular, or sheet-like, with rare shallow channels; (4) *basin plain association*, mainly pelagic chalks and hemipelagic marls.

The details of these facies associations are outside the scope of this paper, but their deposition is summarized in Fig. 8. For example, the gullied slope association represents a bypass zone for clastic material transported from the more marginal depositional setting of the peripheral sands association, to the areas of accumulation of the basin margin and basin plain associations. Erosional unroofing of the Yerasa lineament in the west gave rise to deposition of channelized turbiditic sandstones and debris flow conglomerates further southeast (e.g. at Tokhni, Fig. 5). Another important feature, discussed by Farrell & Eaton (1987), is the common occurrence of major slumped intervals, registering gravitational instability of the marginal slopes of both the Maroni (e.g. at Ayios Theodhoros, Fig. 5) and Khalassa sub-basins (e.g. Happy Valley, Fig. 5).

## Reef facies

Associated reef facies are summarized here, since these are sensitive indicators of relative sea-level change. Reef limestones were originally widespread elsewhere in southern Cyprus, but were eroded. Their former existence is recorded by clastic interbeds and channel fills within the Pakhna Formation. The facies distribution indicates that reefs were located along the structural lineaments (Fig. 8) and were the main source of clastic sediment for axial

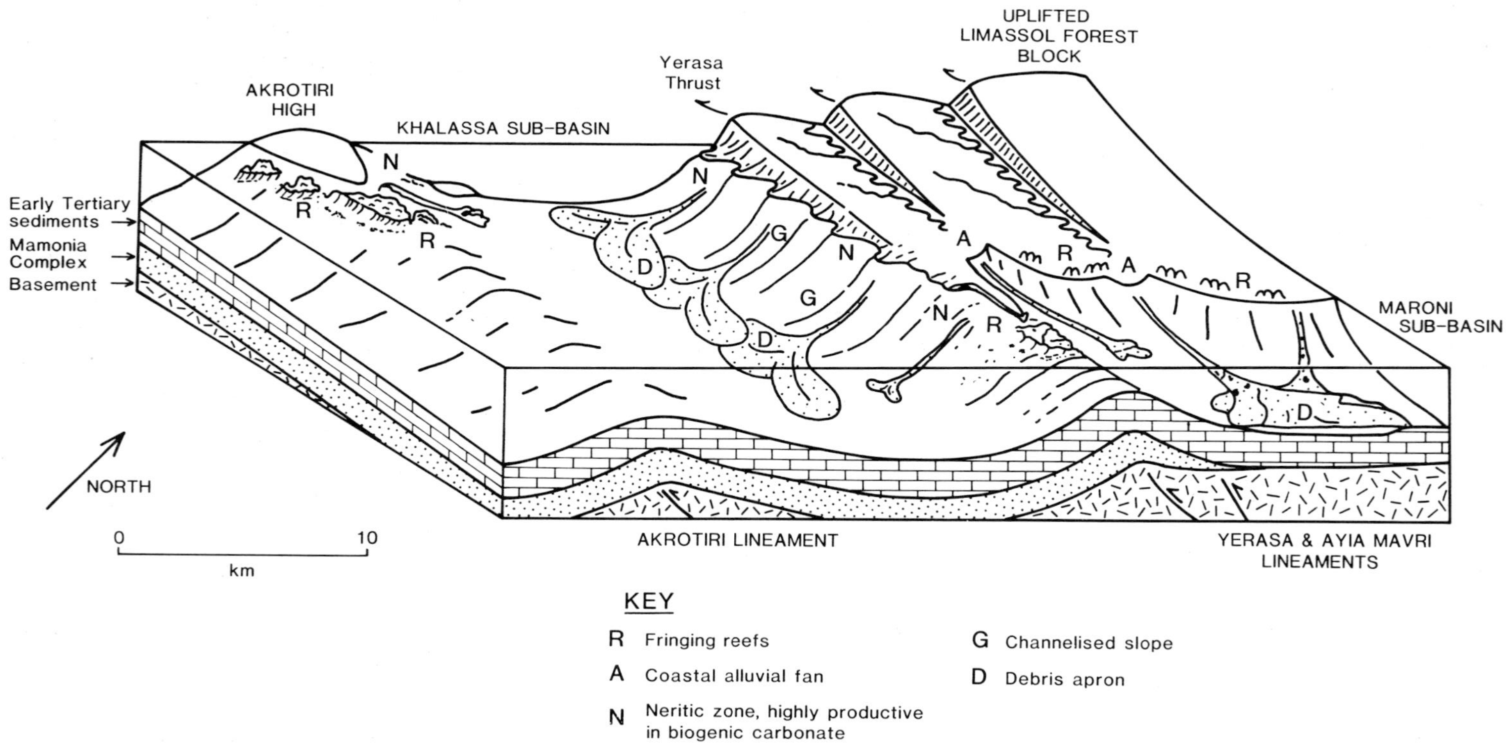


**Fig. 8.** Schematic reconstruction of depositional environments in mid-late Miocene time during deposition of the Pakhna Formation. See text for discussion. Modified from Eaton (1987).

sediments of the Khalassa and Maroni sub-basins. Reef detritus first appears near the base of the Pakhna Formation in both sub-basins and is interpreted as derived entirely from the Tortonian Koronia Member (Fig. 3).

Massive reef limestones are exposed at only one locality in southern Cyprus, Happy Valley (Fig. 5), near the inferred southeast margin of the Khalassa sub-basin. These are correlated with the Tortonian Koronia Member (Table 1). In Happy Valley, patch reefs up to tens of metres in size are interstratified with metre-thick beds of gravity deposited shallow-water derived carbonates, including calciturbidites and debris flows, forming irregular sheets. These units comprise bivalves, coralline algae, bryozoans, echinoids, benthic Foraminifera, serpulids, sponge spicules, and corals, together with granules of pelagic chalk, marl, calcarenite, and ophiolite-derived sediment, and are interpreted mainly as storm deposits. Thinner-bedded successions in this vicinity underwent considerable soft-sediment deformation and slumping, directed towards the south (Farrell & Eaton, 1987).

The coralline blocks were interpreted as fringing reefs by Robertson (1977a), but as detached blocks (olistoliths) by Eaton (1987). Recent work indicates that several of the larger limestone bodies are partly eroded patch reefs, essentially in growth position, while smaller blocks have foundered on a soft, unstable substratum (Follows, 1990a).

Transgression in the Tortonian is indicated by drowning of the, by then, inactive Yerasa lineament (see above). Similar transgression is documented in north (see below) and northwest Cyprus (e.g. Akamas Peninsula). In southern Cyprus, the Yerasa lineament became a submerged, broad, shallow-water ridge that was colonized by Poritid reefs of the Tortonian Koronia Member (Table 1; Fig. 3). Detritus was reworked southwards from these reefs into the higher levels of the Pakhna Formation, as talus wedges and channelized units. Reef debris is abundant close to the Yerasa lineament in the west (e.g. at Amathus), and along the northwest margin of the Maroni sub-basin (e.g. Khirokitia). By contrast, intervening basinal areas (e.g. Armenokhori and Pyrgos, Fig. 5) were dominated by fine-grained micrite with planktonic Foraminifera, and redeposited rhodoliths and bivalves.

In summary, Miocene (Aquitanian to Tortonian) lithofacies of the Maroni and Khalassa sub-basins in southern Cyprus record a marine basin-filling episode, with pelagic sedimentation, passing into facies representing shelf palaeoenvironments, with local reef development on bounding structural highs (Fig. 9). The tectonic lineaments appear to have exerted the dominant control on sedimentation. Global sea-level rise, however, may well have initiated submergence and reef colonization, particularly of the Yerasa lineament after uplift and erosion ended (see later).

**North Troodos extensional basin**

In strong contrast to the compressional tectonics that largely controlled Miocene sedimentation in southern Cyprus, c. 60 km further north, along the north margin of the Troodos ophiolite (Bear, 1960; Gass, 1960; Zomenis, 1972), a quite different regime of *extensional tectonics* was active during the same time period.

In this area, defined as the Mesaoria sub-basin (Fig. 2), the Late Cretaceous and Early Tertiary sedimentary cover (Perapedhi, Kannaviou, and Lower and Middle Lefkara Formations) is thin (c. 10 m) and ponded into hollows on the Late Cretaceous ocean floor (Robertson & Hudson, 1974; Robertson, 1976; Baroz & Bizon, 1977; Baroz, 1979). Above, the Upper Lefkara Formation forms a thicker (up to 75 m) and more laterally persistent blanket, that to the south overlaps directly on to the Troodos ophiolitic basement (Carr & Bear, 1960).

The generally disconformably overlying Pakhna Formation comprises laterally and lithologically variable facies, up to 150 m thick, and includes pelagic carbonate, bioclastic, and reefs limestones (Tortonian Koronia Member) and minor ophiolite-derived clastic sediments (Follows & Robertson, 1990; Fig. 3). Successions are considerably thinner and generally more muddy than the Pakhna Formation of southern Cyprus.

In the lower part of the Mesaoria sub-basin the Pakhna Formation comprises buff, argillaceous marls (e.g. on Kottaphi Hill, Fig. 10), with interbeds of calciturbidites and sheet-like sandstones derived from Troodos ophiolite extrusives to the south. Locally (e.g. north of Mitsero), channels pass up into debris flows, comprising reef-derived talus. The marls containing the channels are dated as mid-Miocene (O. Varol, personal communication, 1990). Upwards this talus becomes thicker-bedded and coarser-grained, with 2–20 cm sized clasts.

Marked thickness and facies variations across faults are indicative of syndepositional extensional faulting (i.e. growth faulting). The reef-derived car-

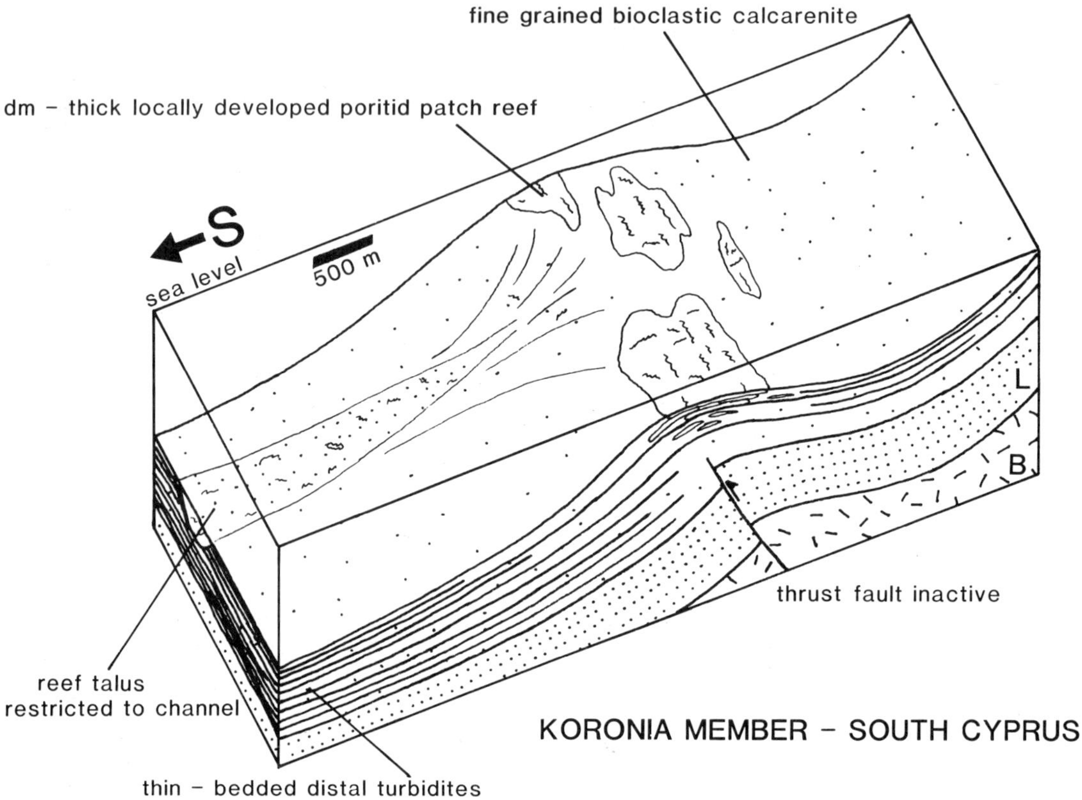


**Fig. 9.** Block diagram showing the reconstructed morphology of Miocene reefs along the south Troodos margin, capping the inactive Yerasa lineament. See text for further explanation (from Follows, 1990a).

bonates and marls of the Tortonian Koronia Member (Fig. 3) were redeposited on to topographically low sea-floor areas, while pelagic chalks accumulated on upfaulted blocks starved of clastic input (Follows & Robertson, 1990). *In situ* reefs are preserved on several hills (e.g. Kreatos; Fig. 10), where they, locally unconformably, overlie extrusives of the Troodos ophiolite. These limestones mainly comprise planar Poritid coral bindstones, alternating irregularly with layers of packstone and chalk-marl. Diverse shallow water fauna are developed in more distal off-reef facies, comprising well-bedded detrital packstones and grainstones (Follows & Robertson, 1990; Fig. 11).

In summary, in contrast to the compression recorded in southern Cyprus, deposition of the mid to late Miocene Pakhna Formation in the north was controlled by extensional faulting, accompanied by block rotation, with erosion of the underlying early Tertiary deep-sea cover, locally down to the Troodos extrusives. Tortonian-aged Poritid reefs then developed on structural highs and detritus was shed into extensional fault-controlled topographic lows (Fig. 11).

### Controls on Miocene reef development

The main controls on reef growth in Cyprus were local tectonic setting, sedimentary environment, global sea-level change, and climatic variation. Reefs flourished in many areas of the Mediterranean during the Miocene (e.g. Pedley, 1983,1987; Grasso & Pedley, 1988). Early Miocene patch reefs developed in southeast Cyprus (Cape Greco area) and around basin margins in western Cyprus, outside the region discussed here (Follows & Robertson, 1990; Follows, 1990a,b). These early reefs grew on an erosion surface, cut into underlying units. A phase of relative uplift, subaerial exposure, and erosion is implied, followed by sea-level rise in the

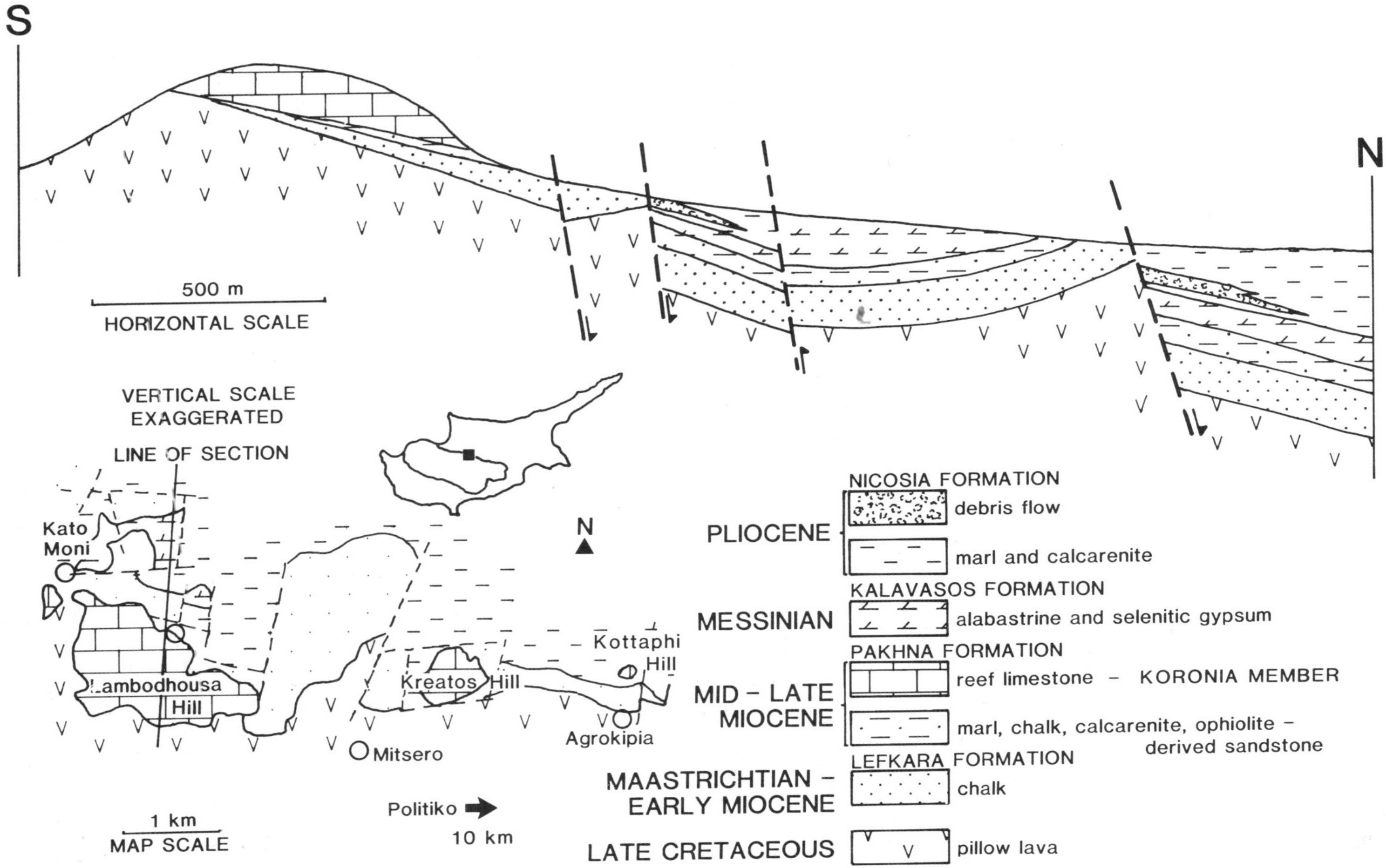


**Fig. 10.** N–S cross-section of the Miocene reefs along the north margin of the Troodos ophiolite, illustrating their relationship with other Miocene and Pliocene facies, including the Messinian gypsum. Below left, is an outline geological map of the Miocene reefs, also showing places mentioned in the text (from Follows, 1990a).

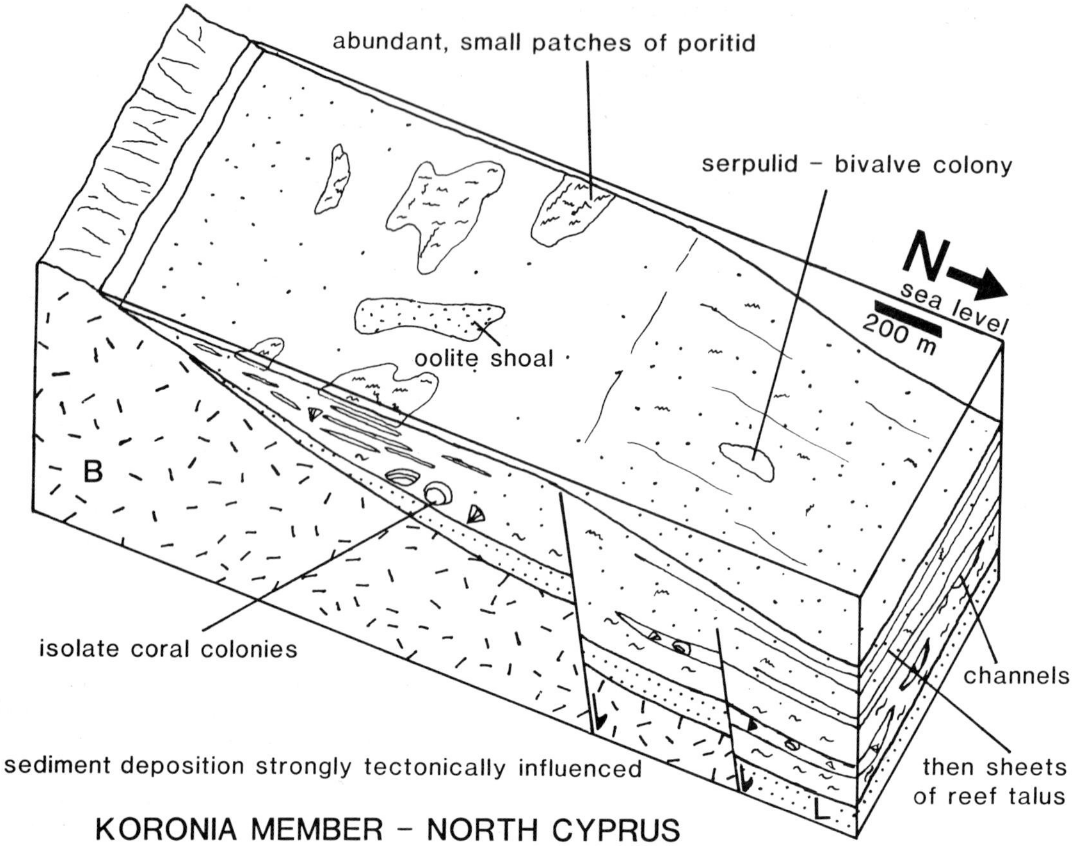


**Fig. 11.** Block diagram showing the reconstructed morphology of Miocene reefs along the north Troodos margin, dominated by extensional faulting. See text for further explanation (from Follows, 1990a).

Aquitanian and development of the reefs of the Terra Member. Mid-Miocene (Langhian) reefs are absent from Cyprus, as elsewhere in the Mediterranean (Chevalier, 1961). This could relate to drowning of the earlier reefs, in response to rapid sea-level rise (Vail *et al.*, 1984,1989). The later reefs of the Koronia Member again overlie a peneplaned surface (e.g. in north Cyprus), pointing to another cycle of emergence and erosion, followed by flooding and reef growth. These reefs, mainly recorded by talus, tolerated a considerable input of lithogenous sediment. Their location was again strongly structurally controlled, being located mainly along the uplifted periphery of the Troodos Massif in north and south Cyprus, and also in northwest Cyprus (Follows & Robertson, 1990; Follows, 1990b). Comparison with similar reefs in comparable, tectonically active areas in the western Mediterranean suggests a mainly Tortonian age for the Koronia Member, but the time range could be extended from Serrevalian to early Messinian in Cyprus.

## MESSINIAN SALINITY CRISIS IN CYPRUS

The Messinian salinity crisis involved isolation of the Mediterranean from the open ocean, followed by desiccation and precipitation of evaporites well below global sea level (Hsü, 1973; Ryan *et al.*, 1973; Hsü *et al.*, 1978; Cita, 1982). The presence of basinal evaporites was recently confirmed by drilling in the Tyrrhenian Sea (ODP Leg 107; Kastens *et al.*, 1988). The importance of local tectonics in controlling the location of evaporite depocentres, however, has only recently been appreciated. For example, the Tyrrhenian Sea was actively rifting during the Messinian (Kastens *et al.*, 1988) and thus the effects of sea-level fall and tectonics were superimposed.

It is now clear that tectonics also played a key role in evaporite deposition in Cyprus. The evaporitic Kalavasos Formation (Pantazis, 1967,1978) crops out in a number of sub-basins located around the Troodos periphery; moving clockwise, in the *Mesaoria* (Follows & Robertson, 1990), *Maroni, Khalassa* (Eaton, 1987), *Pissouri* (Elion, 1983), *Polemi* (Baroz *et al.*, 1978; Orszag-Sperber *et al.*, 1980,1989), and *Polis* (Gass & Cockbain, 1961) sub-basins (Fig. 2). These evaporites have recently been shown entirely to post-date the Pakhna Formation, including the Terra and Koronia Members (Eaton, 1987; Follows & Robertson, 1990; cf. the earlier work of Bagnall, 1960; Pantazis, 1967,1978; Cleintaur *et al.*, 1977; Fig. 6).

## Messinian evaporites in southern Cyprus

A varied Messinian history can be deciphered for southern Cyprus, including both pre- and post-evaporitic facies.

### *Transitional facies*

In southern Cyprus the gypsum of the Kalavasos Formation generally overlies the Pakhna Formation conformably, with no evidence of emergence, or a break in deposition across the boundary. In both the Khalassa and Maroni sub-basins, the transition to evaporite deposition is marked by reduced benthic activity and deposition of finely-laminated calcilutites, with plant debris (e.g. near Anaphotia, Fig. 5), and rare well-preserved small fossil fish (e.g. near Moutayiaka, Fig. 5). During this time, the bottom water perhaps became more stratified and anoxic, preserving organic matter and fish, without benthos.

Locally, in the Maroni sub-basin, gypsum is underlain by an unusual conglomeratic unit, composed of laminated platy clasts of pelsparite, up to 20 cm long, in a matrix containing pelagic Foraminifera, fragmented coralline algae, and peloids. A normal marine origin for these conglomerates is indicated by the presence of serpulids encrusting the clasts at one locality (near Kophinou, Fig. 5), of peloids coated by isopachous cement, and also by limited oxygen and carbon isotopic data (Eaton, 1987). Rouchy & Monty (1979), however, reported cryptalgal limestones with blue-green algae from one of these conglomeratic localities (Kalavasos, Fig. 5) and inferred a freshwater (marshy) palaeoenvironment prior to evaporite deposition. The cryptalgal carbonate presumably accumulated in a marginal freshwater setting and was then lithified, eroded, and redeposited on the margins of a marine (pre-evaporitic) basin. The presence of overlying clastic gypsum suggests the area was slope dominated. Overall, these early erosional events may relate to local tectonic events and/or sea-level fluctuations during the earlier stages of Mediterranean draw-down.

By comparison, in southwest Cyprus (e.g. near Pissouri; Fig. 5), the Pakhna Formation is conformably overlain by marine marls and diatomaceous laminites, with bioclastic carbonate lenticles and chaotic blocks of algal carbonate (up to 80 cm across). This facies is then overlain by evaporites of the Kalavasos Formation (Orszag-Sperber *et al.*, 1989).

### *Evaporitic facies*

The gypsum in southern Cyprus is interpreted as being deposited entirely subaqueously, on the basis of field relations and comparisons with other Mediterranean areas. The facies are similar to those described by Schreiber *et al.* (1976) and Vai & Ricci-Lucchi (1977), and include parallel laminated, fine-grained, alabastrine ('marmara' or 'balatino') gypsum and coarse, banded, stacked selenitic gypsum. These facies are interpreted as being essentially primary and *in situ*, while associated nodular gypsum is mainly diagenetic.

### *Basin morphology*

It is important to note that in southern Cyprus the topography of the Messinian evaporitic sub-basins was mainly inherited from the earlier depositional systems of the Pakhna Formation (Aquitanian–Tortonian). Thus, the various gypsiferous facies mirror the underlying facies associations (Fig. 8). For example, the relatively basinal fine-grained alabastrine gypsum is commonly seen above the basin plain facies association, while clastic, redeposited, and slumped gypsum tend to occur above the Pakhna Formation slope facies association. Thus gypsum deposition took place over a range of water depths, from very shallow, to perhaps tens to hundreds of metres. Algal filaments, reported locally from evaporites (e.g. in the Pissouri sub-basin;

Rouchy & Monty, 1979; Elion, 1983), indicate at least local deposition within the photic zone (i.e. 10–20 m; Schreiber *et al.*, 1976).

Further evidence of deposition on slopes and hence some relatively deep-water (i.e. tens to hundreds of metres) gypsum accumulation (but not necessarily initial precipitation, see below) comes from the Maroni sub-basin in the east (Fig. 2), where fold vergence in contorted gypsum (at Tokhni, Fig. 5) indicates a southeast palaeoslope. Thicker bedded, turbidity current-deposited successions occur down palaeoslope; i.e. towards the depocentre (e.g. at Maroni quarry, Fig. 5). By contrast, *in situ* banded selenite and clastic gypsum beds, with carbonate peloids and relict ooids, of possibly marginal, lagoonal origin were observed near the inferred western limit of the sub-basin (near Tokhni; Fig. 5). The southern margin of the Maroni gypsum sub-basin was presumably located offshore, since the evaporites thicken southwards in this direction (towards Maroni village, Fig. 5).

By contrast, during Messinian time the Khalassa sub-basin further west was broader and shallower than the Maroni sub-basin. Gypsum deposition there includes abundant *in situ* selenite, with evidence of local clastic sediment reworking, presumably by currents in shallow water. Interpretation of borehole data from the Akrotiri Peninsula (Hadjistavrinou & Constantinou, 1977) suggests that the evaporitic depocentre of this sub-basin lay within the Akrotiri Peninsula and was bounded by the Akrotiri lineament to the south.

### Messinian evaporites of the north Troodos margin

During the Messinian, the north Troodos margin continued to be affected by tensional faulting (Follows & Robertson, 1990). Gypsum of the Kalavasos Formation is exposed in small, fault-bounded units, generally at a much lower topographic level than the earlier reef limestones of the Koronia Member (Fig. 10). The gypsum appears abruptly, and unconformably overlies muddy marls of the Pakhna Formation. It comprises stubby selenite crystals, in successions up to tens of metres thick, with smaller volumes of laminated alabastrine gypsum (Follows & Robertson, 1990). The base of the evaporite succession locally contains large, tilted selenite botryoids, with crystals reaching 1.4 m long (e.g. near Kato Moni, Fig. 10, inset). Stratigraphically higher at this locality, terrigenous muds are interbedded with gravity flows, containing large (up to 50 cm long), broken and abraded selenite crystals lying parallel to bedding. Carbonate laminae, disrupted by diagenetic selenite also occur within laminated alabastrine gypsum, similar to that reported from the Western Mediterranean area (Hardie & Eugster, 1971).

### Modes of gypsum deposition

The finely laminated alabastrine gypsum settled as a fine precipitate from the upper, oxygenated layers of the water column (Schreiber & Decima, 1978). The salinity was increased by repeated influxes of sea water (i.e. refluxing) across the lips of small basins created by the pre-existing structural lineaments (Fig. 12). The largely monomineralic nature of the gypsum could relate to high rates of gypsum precipitation in a hypersaline setting of reduced carbonate production (Kinsman, 1976). By contrast, the nodular gypsum is seen as the result of subaqueous diagenesis (Dean *et al.*, 1975; Dean & Anderson, 1982), and is not considered further here.

Where slopes were steep near the basin margins, gypsum was redeposited by gravity. For example, laterally continuous, up to 30 cm thick, normally graded beds are rarely found in the Maroni and Khalassa sub-basins. Broken and abraded selenitic crystals, up to 2 cm long, show a marked preferred orientation parallel to bedding. Further west, in the Khalassa sub-basin (at Moutayiaka and north of Pissouri, Fig. 5), coarser, graded, gypsum turbidites, up to 2 m thick, comprise broken selenite crystals at the base, up to 30 cm long, grading up into acicular needles. Such acicular gypsum indicates derivation from primary gypsum precipitates (Schreiber *et al.*, 1976). Non-evaporite clasts are restricted to rare peloidal carbonate clasts, up to 5 mm long, with rare planktonic Foraminifera.

Further evidence of tectonic instability during evaporite deposition in southern Cyprus is observed in the form of chaotic selenite, characterized by poorly sorted clasts in a matrix of gypsum, or carbonate. For example, at Maroni quarry (Fig. 5) a bed of crudely graded selenite, more than 2 m thick, contains elliptical marl cobbles (containing sponge spicules) and a large boulder of banded selenite, c. 1 m in diameter. Elsewhere, at Tokhni and Kalavasos quarries (Fig. 5), coarse selenite laths, up to 15 cm long, are supported by an unfossiliferous micritic matrix. Comparable tectonically triggered chaotic units also occur in the Polemi and Pissouri sub-basins (Orszag-Sperber *et al.*, 1989).

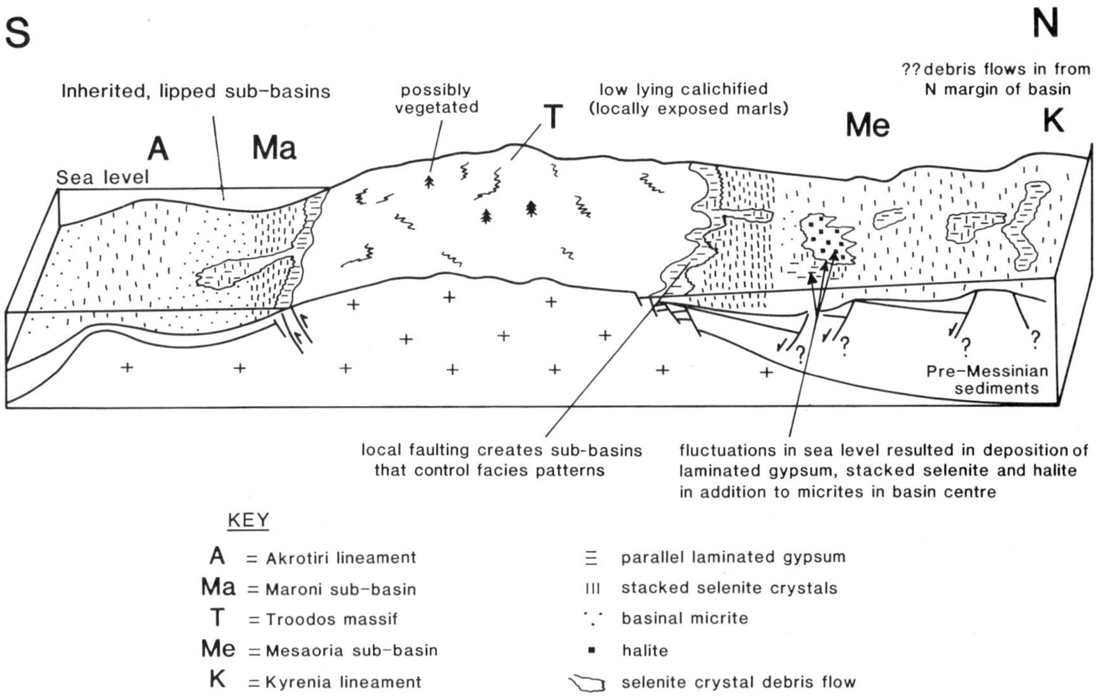


**Fig. 12.** Inferred depositional setting of evaporites along the southern and northern margins of the Troodos Massif. See text for further explanation.

In northern Cyprus the Messinian gypsum was deposited in small, extensional fault-controlled basins (Fig. 12). The sea floor was actively tilting *during* evaporite precipitation and selenite was reworked as debris flows. Thicker and more extensive evaporites, including halite, accumulated in the adjoining larger, and presumably deeper, Mesaoria basin further north, based on borehole data (Gass, 1960; Zomenis, 1972).

## Comparative evaporitic settings

In the Eastern Mediterranean, Messinian evaporitic deposits are present on the abyssal plain around Cyprus, as confirmed by seismic profiling and drilling (Mulder *et al.*, 1975; Evans *et al.*, 1978; Hsü *et al.*, 1978). However, the Cyprus gypsum differs somewhat from facies cored by the DSDP (Sites 375 and 376; Leg 42; Fig. 1; Hsü *et al.*, 1978), which include thick algal laminated sabkha-type deposits and higher evaporites (e.g. halite). Nodular andydrite, of sabkha origin, is also reported from wells off the coast of Israel (Gvirtzman & Buchbinder, 1978). Finely laminated alabastrine gypsum in the Tyrrhenian Sea, central Mediterranean (Site 654; Leg 107) is interpreted as having settled into moderately deep (tens to hundreds of metres) silled basins near the edge of a deeper extensional basin (Robertson *et al.*, 1990). Down dip, thicker evaporites, including halite, are inferred from seismic profiles (Kastens *et al.*, 1988). This setting is similar to the Cyprus Mesaoria sub-basin.

Finely laminated alabastrine gypsum similar to Cyprus also crops out widely elsewhere on land in the Mediterranean area, for example in Sicily (Hardie & Eugster, 1971; Decima & Wezel, 1973; Schreiber *et al.*, 1976), on the Ionian islands, western Greece (Braune *et al.*, 1973; Schilling *et al.*, 1978; Underhill, 1985), in the Italian Piedmont (Sturani, 1978), and on Crete (Meulenkamp & Zachariasse, 1973). Each of these settings is relatively marginal, similar to that inferred for Cyprus. Selenite banks in Sicily, as in Cyprus, are interpreted as having precipitated in shallow (i.e. several metres deep) subaqueous basins (Hardie & Eugster, 1971). On the other hand, gypsum turbidites like those of Cyprus are reported from the deep Mediterranean basin (Garrison *et al.*, 1978), from Sicily (Schreiber *et al.*, 1976), from the Appenines (Vai & Ricci-Lucchi, 1977), and, incidently, from the Zechstein of the North Sea area (Schlager & Boltz, 1977). Gypsum turbidites in Italy (Marche and Romagna areas) are interpreted as gravity flows into deep brines that were located in the foredeep of the rising Appenines (Ricci-Lucchi, 1973).

In summary, the Cyprus Messinian evaporites accumulated in very shallow to moderately deep basins (i.e. several metres to hundreds of metres deep) within silled basins around the periphery of the Troodos Massif. Prior to the Messinian, Cyprus was already uplifted far above the level of the surrounding Mediterranean abyssal plain (see earlier). In southern Cyprus, the evaporitic sub-basins were largely inherited from the preceding structural configuration, but were locally modified by continuing faulting (e.g. Ayia Mavri lineament). In northern Cyprus coeval extensional faulting controlled evaporite deposition in the Mesaoria sub-basin. In general, only a slight fall in sea level was needed to initiate gypsum precipitation by refluxing into marginal, silled sub-basins.

A curious feature, noted in passing, is that in southern Cyprus celestite (strontium carbonate) accumulated very locally (e.g. near Khirokitia), prior to, or contemporaneously with, the Messinian evaporites. The celestite was later locally reworked as detached blocks that immediately overlie selenitic gypsum (e.g. north of Maroni, Fig. 5).

### Late Messinian to early Pliocene transition

During the late Messinian, when surrounding deep basinal areas of the eastern Mediterranean were desiccating, the Troodos Massif and its sedimentary cover were emergent and eroding, resulting in development of caliche and karst, now preserved only locally (e.g. at Cape Greco, southeast Cyprus and south of Prodhromi in northwest Cyprus; Follows, 1990a). An erosional surface is also visible on shallow offshore seismic data covering the area of southern Cyprus, between the Akrotiri Peninsula and Petounda Point (Fig. 5; McCallum, 1989).

In southwest Cyprus, increased vegetation in the late Messinian is recorded by the development of five paleosol horizons in the Pissouri Basin (Rouchy *et al.*, 1980; Elion, 1983; Orszag-Sperber & Rouchy, 1983). During this time, the climate apparently changed from semi-arid to more humid and was less conducive to evaporite precipitation (Kinsman, 1976).

In southern and northern Cyprus, the basal Pliocene sediments transgress abruptly across Messinian facies. In southwest Cyprus, however, evaporite deposition was followed in the late Messinian by varied freshwater, brackish to marine palaeoenvironments (Lago Mare of Hsü, 1973). Fluctuating sea levels with frequent marine incursions are also implied by evidence from the Pissouri sub-basin. Facies there comprise lagoonal deposits with a restricted marine fauna, lacustrine marls, alluvial to fan deltaic ophiolite-derived clastics, reworked calcarenites, and palaeosols (Rouchy *et al.*, 1980; Elion, 1983; Orszag-Sperber *et al.*, 1989). In this area, pronounced rifting to form the Polis graben, with both axial subsidence and footwall uplift, was taking place during Messinian to earliest Pliocene time (Ward & Robertson, 1987) and the effects of Mediterranean-wide marine reincursion are thus not easily separable from equally short-lived local tectonics.

## PLIOCENE PALAEOENVIRONMENTS

By the earliest Pliocene, fully marine conditions were re-established throughout the eastern Mediterranean area. However, much of the Troodos Massif remained emergent as a low-lying land mass (Robertson, 1977a; Fig. 13A). Pliocene lithofacies are distinctly more muddy and ophiolite clast-rich (Ducloz, 1965) than their chalky Miocene counterparts, reflecting a combination of local input from the Troodos ophiolite, and the muddy nature of the Pliocene Mediterranean Sea as a whole. In Cyprus, Pliocene sediments in the south differ from those in the north (Fig. 13B; Table 2), mainly because of a greater input from the uplifting ophiolite in the north, and/or because much coarse material bypassed the southern margin to accumulate at greater depths further offshore, as suggested from a study of seismic profiles from southern offshore Cyprus (McCallum, 1989).

### South margin: relative tectonic quiescence

Where locally exposed (Mari area, Fig. 5), basal Pliocene silts of the Nicosia Formation unconformably overlie Miocene evaporites and marls, and record moderately deep basinal conditions (i.e. $>200$ m, based on ostracods ( Pavlakelli, 1987;

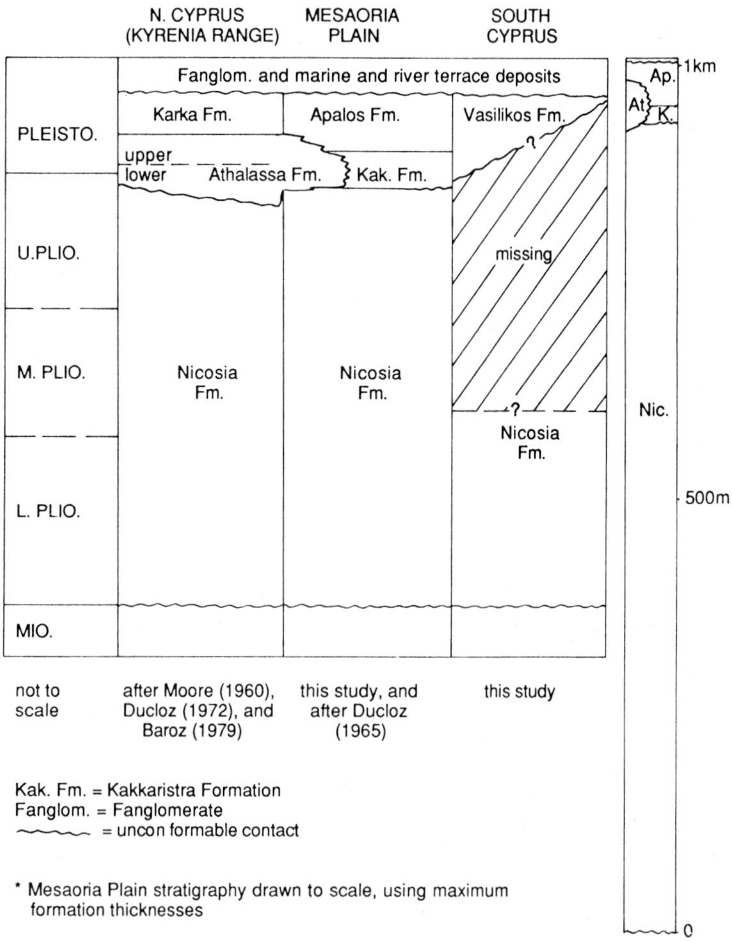


**Table 2.** Revised Pliocene and Pleistocene stratigraphy. Biostratigraphic data are referred to in the text; from McCallum (1989)

Fig. 14A). On seismic profiles of the shelf offshore to the south, Pliocene sediments are thickest adjacent to the coastline, near Mari, implying a local depocentre in this area, probably related to final movement along the Ayia Mavri lineament. Pliocene sediments are very thin or absent on the shelf further east (Fig. 14B), probably reflecting eastward shallowing of the Mari basin, followed by erosion (McCallum, 1989). On seismic profiles, Plio-Pleistocene sediments above the inferred Messinian unconformity are not deformed (McCallum, 1989) and thus the Akrotiri and Petounda lineaments were, by this time, inactive. The existence of the local early Pliocene depocentre near Mari was short-lived, since silts near the base of the Nicosia Formation show shallowing upward trends, and pass up into coarser grained, fan-deltaic sediments, still perhaps of early Pliocene age. The fan-delta facies prograded southeastwards into the basin and may record sharply accelerated uplift of the ophiolite to the north (Fig. 14A).

Locally in the Mari area (Fig. 5), the Nicosia Formation is unconformably overlain by a thin unit of transitional, unfossiliferous, braided-fluvial, ophiolite-derived conglomerates derived from the west. This unit records major uplift and unroofing of the Limassol Forest Block; apparently talus was shed southwards, through the now deeply incised Kouris River gorge (Greitzer & Constantinou, 1969; Fig. 5), possibly reflecting a sea-level lowstand. These clastics could correlate with the first appearance of thicker- and coarser-grained, ophiolite-derived conglomeratic sediments along the north Troodos margin, represented by the late Pliocene to early Pleistocene Kakkaristra Formation (Table 2; see below).

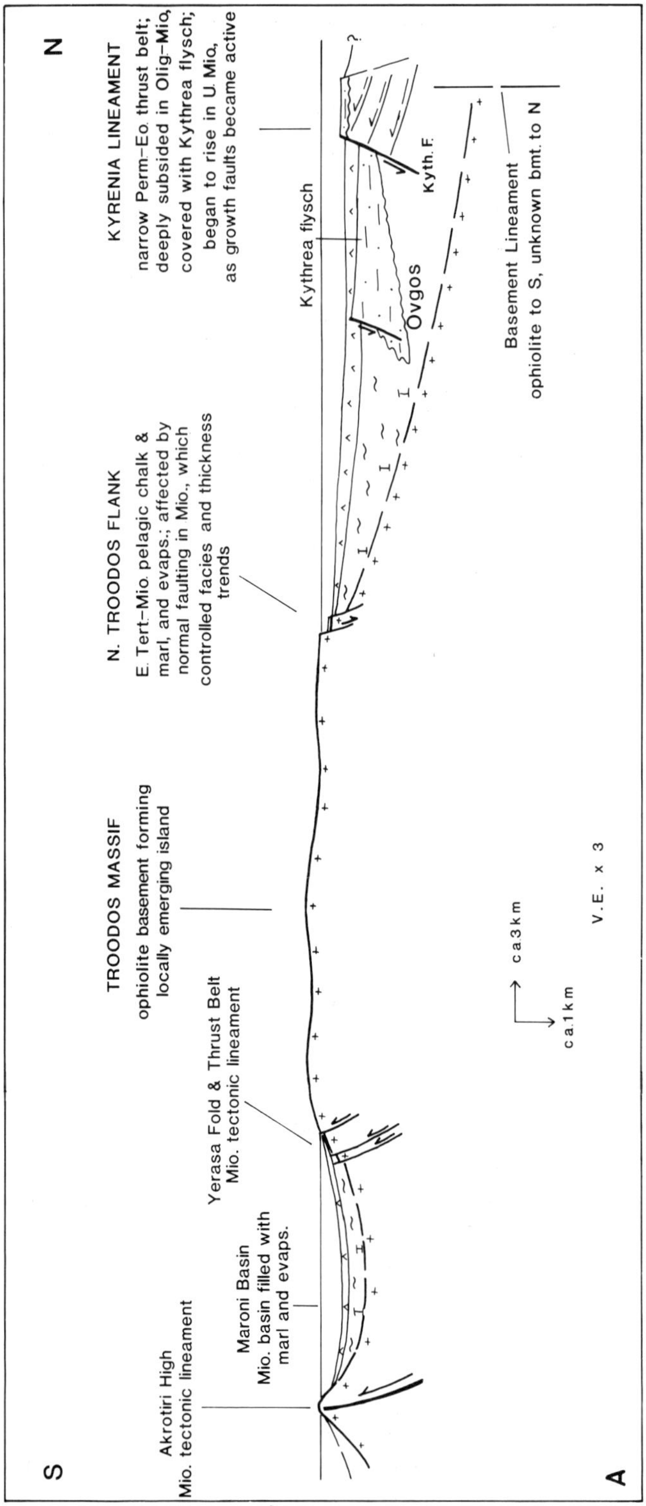
S
N
A
Akrotiri High
Mio. tectonic lineament
Maroni Basin
Mio. basin filled with
marl and evaps.
Yerasa Fold & Thrust Belt
Mio. tectonic lineament
TROODOS MASSIF
ophiolite basement forming
locally emerging island
N. TROODOS FLANK
E. Tert.-Mio. pelagic chalk &
marl, and evaps.; affected by
normal faulting in Mio., which
controlled facies and thickness
trends
KYRENIA LINEAMENT
narrow Perm-Eo. thrust belt;
deeply subsided in Olig-Mio,
covered with Kythrea flysch;
began to rise in U. Mio,
as growth faults became active
Kythrea flysch
Ovgos
Kyth. F.
Basement Lineament
ophiolite to S, unknown bmt. to N
ca 3 km
ca 1 km
V. E. x 3

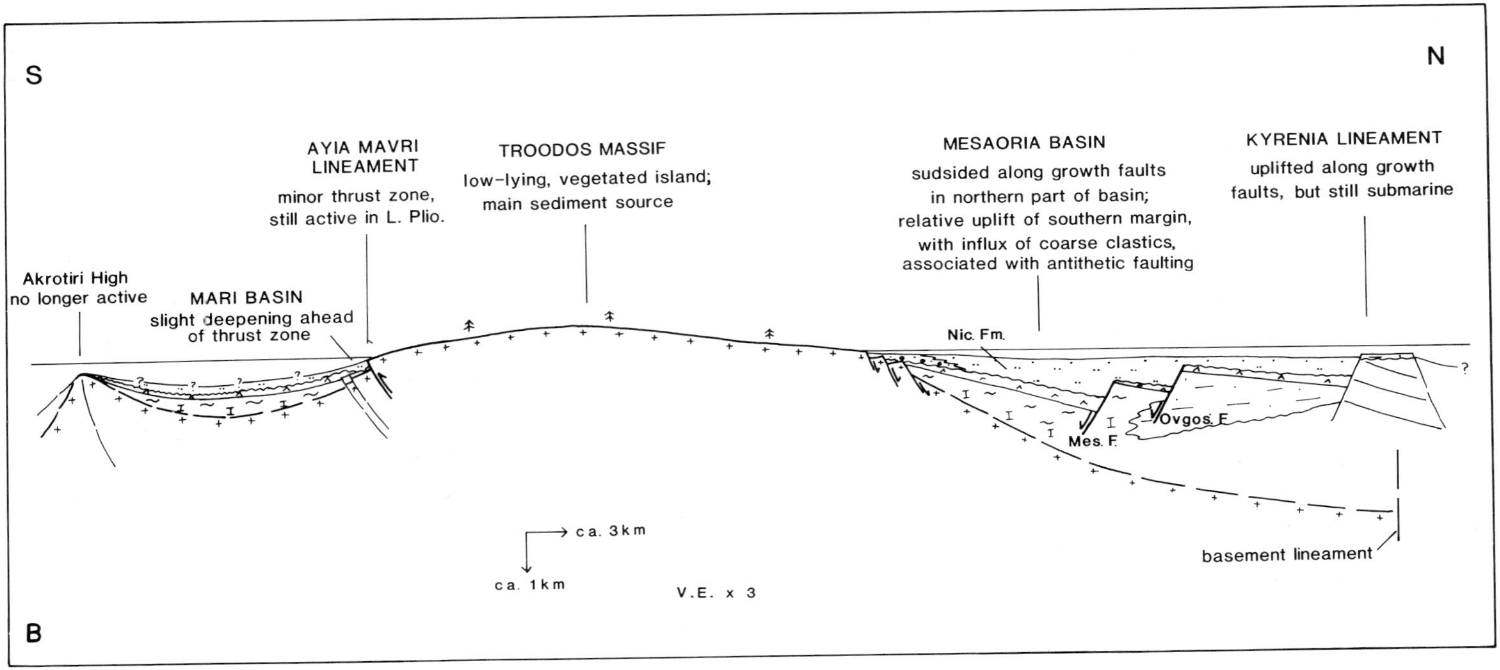


**Fig. 13.** (A) Pre-Pliocene setting of the N–S transect across central Cyprus discussed in this paper (data sources: Cleintaur *et al.*, 1977; Robertson & Woodcock, 1986; Eaton, 1987); from McCallum (1989). (B) Early to mid-Pliocene setting of the N–S transect across central Cyprus discussed in this paper (data sources: Cleintaur *et al.*, 1977; Robertson & Woodcock, 1986; Eaton, 1987); from McCallum (1989).

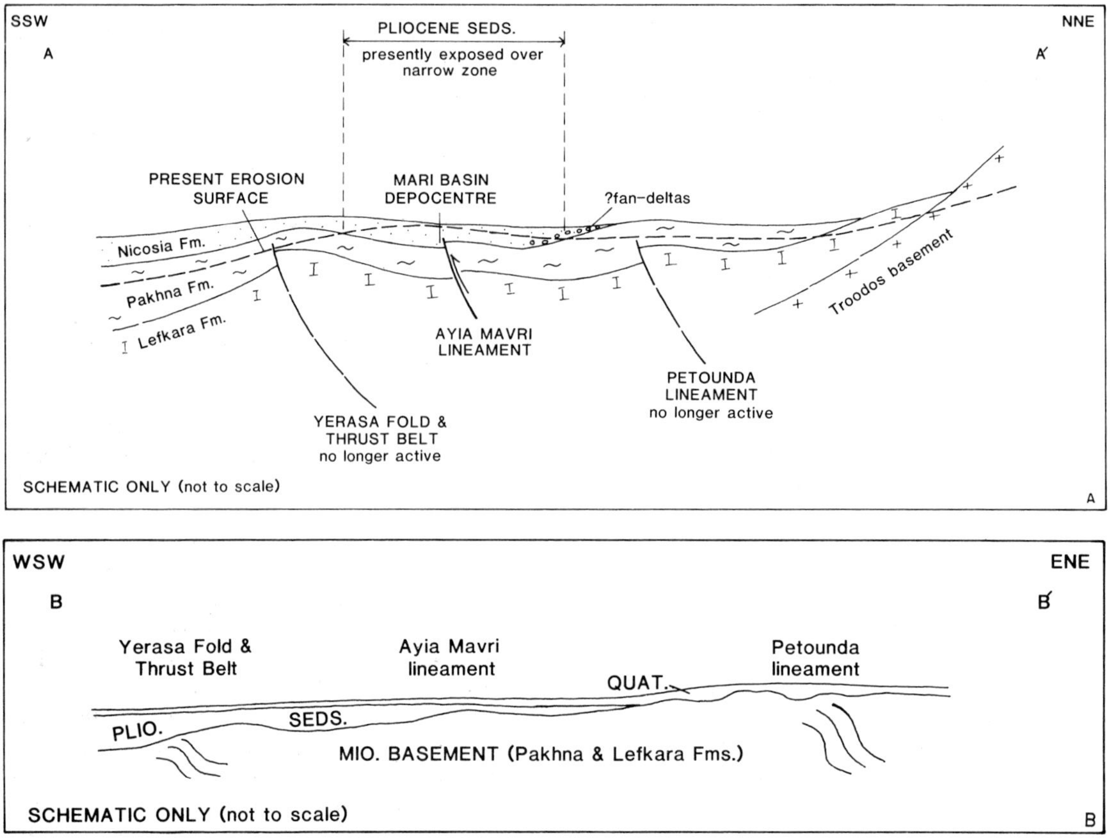


**Fig. 14.** Pliocene depositional setting in southern Cyprus. (A) Schematic cross-section of the south Troodos margin in latest Messinian to earliest Pliocene time. The Ayia Mavri lineament was active, and local subsidence to the south gave rise to the Mari Basin (see Fig. 2 for section location). (B) Schematic cross-section, based on shallow seismic data, showing the northeastward pinch-out of Pliocene sediments on the shelf adjacent to southern Cyprus near Petounda Point. Both figures from McCallum (1989).

At one locality, near Amathus (Fig. 5), the uppermost mudstones correlated with the Nicosia Formation, are dated as latest Pliocene, based on diagnostic calcareous nannoplankton (Houghton & Jenkins, personal communication). These sediments are disconformably overlain by bioclastic calcarenites that are thick-bedded and locally slumped. The carbonates presumably formed in productive coastal settings and were then reworked on to muds of the Nicosia Formation further offshore. There was thus an abrupt switch from relative deep-marine nannofossil muds with ophiolite-derived silt, to reworked shallow-water carbonate sedimentation with little ophiolite detritus. A possible explanation is that relative sea-level fall in the latest Pliocene caused rapid aggradation of bioclastic sediment.

### *(?)Late Pliocene channel complex*

An important recent discovery in southern Cyprus (near Khirokitia, Fig. 5) is the presence of a sediment-filled valley up to 30 m deep incised into Miocene sediments of the Pakhna Formation. The age of the channel incision is constrained by the fill, as ranging from Miocene, based on the abundance of derived Poritid coral clasts, to late Pliocene, based on calcareous nannofossils within interbedded hemipelagic muds (S.D. Houghton & D.G. Jenkins, personal communication, 1981).

Mapping allows the channel complex to be traced several kilometres in a N–S direction on both sides of the Limassol–Nicosia road (A.H.F. Robertson & A.J. Poole, unpublished data). The Pliocene channel fill is underlain by Pakhna Formation sediments

(Bagnall, 1960), which themselves include channelized conglomerates. These earlier conglomerates contain clasts of Lefkara Formation chalk and chert, Pakhna Formation sediments, including corals, and Troodos ophiolitic basic extrusives and red jasper. The clasts are up to 40 cm in size and show imbrication, indicating transport towards the south. The channelized conglomerates are interbedded with medium to thick-bedded, massive bioclastic sandstones of the Pakhna Formation gullied slope association (Eaton, 1987; Fig. 7).

Near Khirokitia, the channel walls are steep to subvertical, while the axis of the channel is paraconformable on the underlying Pakhna Formation. In general, the incised valley is infilled with a series of *en echelon*, or stacked channelized conglomerates, interbedded with marine sands and silts. The individual channel fills are up to 3 m thick and comprise matrix-supported conglomerates, including subangular to well-rounded clasts of Pakhna Formation lithologies, including abundant Poritid coral, Lefkara Formation chalk and chert, and Troodos ophiolitic pillow lava and diabase. Interbedded sands are rich in marine bioclastic material. Metre-sized tabular blocks of Pakhna Formation lithofacies also occur within some of the largest channelized units. The channel complex can be traced several kilometres south of Khirokhitia, where it broadens and the fill is generally finer-grained than further north.

The Khirokitia channel complex clearly acted as a conduit for southward transport of detritus eroded from marginal areas of the Troodos Massif to the north. The deep channel apparently resulted from fluvial incision. The probable age of downcutting is late Pliocene, more or less coeval with the sediment fill. A possible cause was global eustatic sea-level fall in the late Pliocene, resulting in emergence and erosion, followed by sea-level rise and infilling with marine sediments. The Kouris River channel was possibly incised at the same time. Evidence from the nearby Mari sub-Basin indicates that uplift of the Troodos Massif was already taking place by late Pliocene time (McCallum, 1989) and this may explain the presence of debris flows within the channel fill. An alternative explanation of the channel complex is that it was incised much earlier, in the Messinian. Messinian palaeochannels are reported from around the margins of the Mediterranean basin, for example beneath the Nile cone (Barber, 1981) and in offshore eastern Spain (Stampfli & Höcker, 1989). However, in this scenario an early, rather than late, Pliocene aged fill would be expected.

## North Troodos margin: extensional basin formation

### *South margin of the Mesaoria half-graben*

In northern Cyprus, extensional faulting persisted during the Messinian salinity crisis, and reached a climax in the early Pliocene. Along the southern margin of the Mesaoria sub-Basin (west of Agrokipia, Fig. 10), an irregular Messinian erosion surface is directly overlain, with angular discordance, first by pink, then white Pliocene foraminiferal chalks, up to several tens of metres thick. The chalk contains thin beds of bioclastic limestone (up to 10 cm thick), with a diverse shallow-marine biota (e.g. echinoderms and rhodoliths) and is, in turn, overlain by brown marls and mudstones, typical of the early Pliocene Nicosia Formation (Follows & Robertson, 1990; Fig. 15).

A spectacular feature of the base of the Pliocene succession is the occurrence of detached blocks, thick debris flows, and associated breccia units, intercalated with mudstone. These units dip at varying angles between a series of N–S-trending extensional faults. The detached blocks and the debris flows mainly comprise evaporitic lithologies (Kalavasos Formation), and more widespread detrital limestone, as well as chert-rich Early Tertiary chalks (Lefkara Formation). In the east, these facies dip northwards at up to 90°, while more westerly successions are back-rotated southwards at up to 35° (Follows & Robertson, 1990). The dominantly E–W faults are interpreted as important syndepositional extensional faults that were highly active in early Pliocene time. NNE–SSW-trending vertical faults are present and are interpreted as transverse (transfer) faults, that compartmentalized extension along the North Troodos margin. A small oblique (left-lateral) displacement component is possible, but there is no evidence of pure strike-slip faulting.

Extensional faulting greatly enlarged and deepened the existing sub-basin, and in the Pliocene this is termed the *Mesaoria Basin*. In the early Pliocene a major half-graben developed and this was bounded by the present north margin of the Troodos Massif in the south and by major faults in the north, including the Ovgos Fault (Figs 2, 13B, 16). Following pioneering work by Ducloz (1965), the sedimentology and tectonic evolution of the Mesaoria Basin was recently outlined by McCallum & Robertson (1990). Subsidence accommodated the deposition of more than 900 m of silts, rich in planktonic Foraminifera in the northern part of the half-graben, based

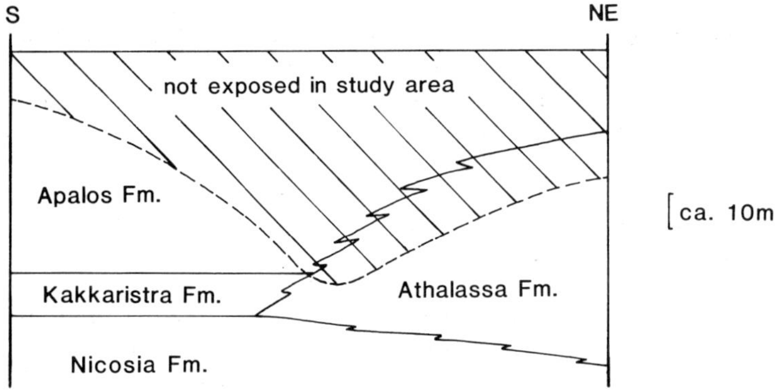


**Fig. 15.** Generalized facies relationships in the Mesaoria Basin, north Cyprus, between the mainly Pliocene Athalassa, Nicosia, Kakkaristra, and Apalos Formations; from McCallum (1989).

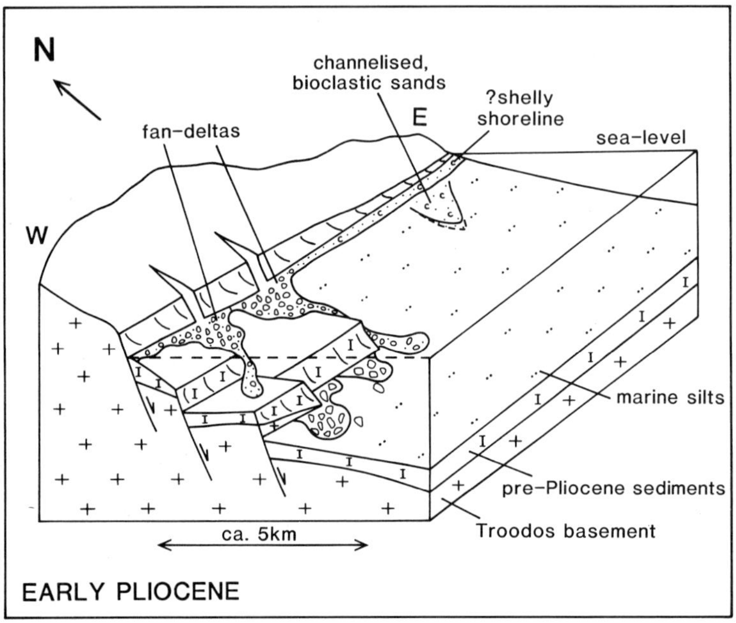


**Fig. 16.** Schematic block diagram illustrating deposition of the lower part of the early to mid-Pliocene Nicosia Formation along the south side of the Mesaoria Basin. This area was cut by antithetic faults. Modified from McCallum (1989).

on unpublished borehole data (Mantis, 1968), while much thinner successions were deposited in the south. There is little evidence of any significant clastic input from the north, either by reworking of the late Miocene Kythrea Flysch, or from the subdued Kyrenia lineament further north again (Weiler, 1965; Baroz, 1979; Robertson & Woodcock, 1986; Fig. 2).

In contrast, coarse-grained ophiolite-derived sediments are present in the lower part of the Nicosia Formation at intervals along the southern margin of the basin, near the present northerly limit of the Troodos ophiolite outcrop. Limestone breccias and conglomerates (e.g. slumped reef blocks at Politiko, Fig. 10) reflect syndepositional faulting of the half-graben system, antithetic to master footwall extensional faults (e.g. Ovgos Fault) further north. This south margin was steep and narrow, as confirmed by localized occurrences of marine fan-delta conglomerates, mainly deposited in channels by mass flow (Fig. 16). These clastics reflect erosion of the Troodos ophiolitic extrusives exposed to the south.

The channelized conglomerates of the lower Nicosia Formation give way upwards to sandier, though still channelized, facies, cut into marine silts. The presence of sandier facies suggests a decline in relief along the southern margin of the Mesaoria Basin, perhaps in response to declining fault activity.

However, the continued channelling may imply the persistence of relatively steep sea-floor gradients. An alternative explanation, however, is as follows: the channelized silty facies give way at the top of the Nicosia Formation to fine-grained, bioturbated sands representing shallow water conditions. The nature of these channelized sediments, followed by shallow-water deposition, would be consistent with incision during glacio-eustatic, lowered sea level (see below).

By the end of the early Pliocene, the Mesaoria Basin had largely filled, and had evolved into a relatively stable, sandy, shallow marine platform.

The Nicosia Formation is overlain by the coarser grained Kakkaristra Formation (Fig. 15). The two formations are separated along the south margin of the Mesaoria Basin by a low-angle unconformity. Tilting of the Nicosia Formation is, however, not evident towards the basin centre. *In situ* ostracods from mudstones near to top of the formation are provisionally dated as (?)late Pliocene to Pleistocene (A. Lord, personal communication, 1988). The Kakkaristra Formation comprises a series of coarse to fine-grained ophiolite-derived clastics, which document progradation of a large fan-delta system into the Mesaoria Basin (Fig. 17). Unlike the channelized fan-deltas of the Nicosia Formation, those of the Kakkaristra Formation reflect the progradation of a tabular, shelf-type fan-delta across the gently sloping southern margin of the, by then, largely filled Mesaoria Basin. This switch to coarser-grained deposition is explained in terms of a renewed pulse of uplift of the Troodos ophiolite to the south.

Deposition of the Kakkaristra Formation was soon followed by unfossiliferous fluvial facies of the Apalos Formation (Fig. 18), which comprise large volumes of overbank muds and silts, but with relatively little sand or conglomerate. This fining-upward trend is believed to reflect a decline in uplift of the Troodos hinterland to the south, peneplanation, and subsequent input of finer-grained sediment.

### *Carbonate-depositing north margin*

During the late Miocene the area to the north of the Mesaoria sub-basin was cut by the Ovgos Fault (Fig. 13A). This separated thick terrigenous tur-

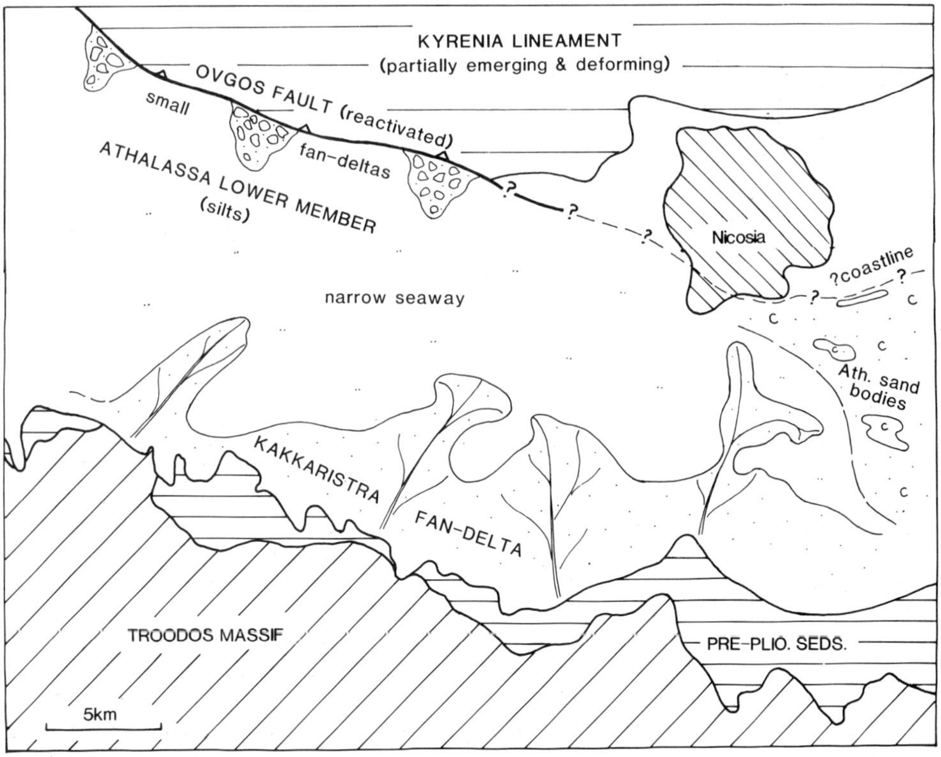


**Fig. 17.** Palaeogeographic reconstruction of the Mesaoria Basin, northern Cyprus, during lower Athalassa/Kakkaristra Formation times ((?)late Pliocene); from McCallum (1989).

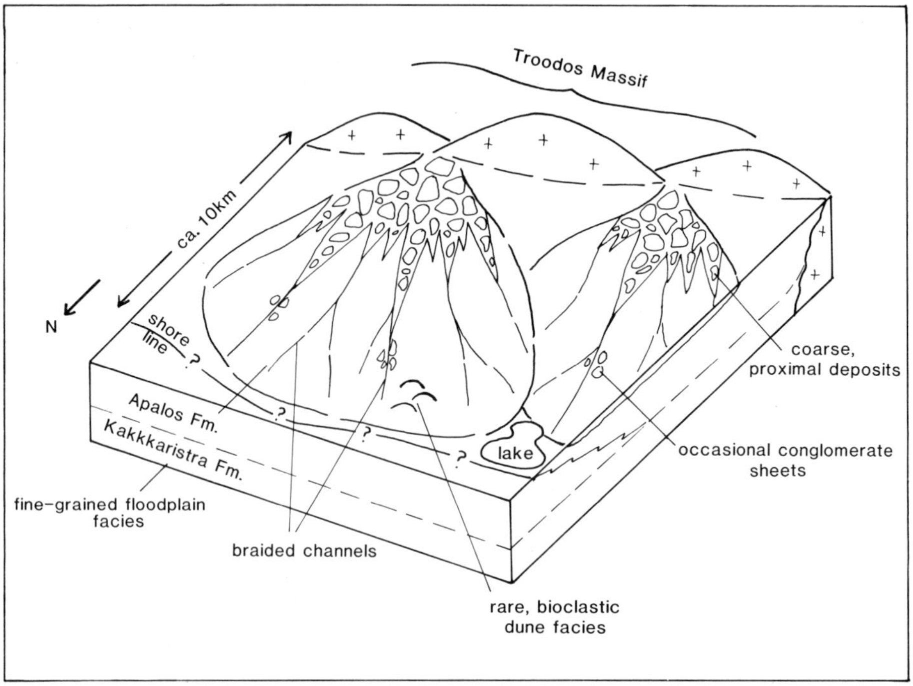


**Fig. 18.** Schematic block diagram illustrating deposition of the (?)late Pliocene to early Pleistocene Apalos Formation; from McCallum (1989). See text for further explanation.

bidites to the north (Kythrea Flysch) from much thinner pelagic carbonates to the south (Upper Lefkara Formation; Cleintaur *et al.*, 1977; Robertson & Woodcock, 1986). During the Pliocene (Fig. 13B) the Ovgos Fault continued to be active as an important extensional growth fault, with much larger downthrow than that recorded by faulting along the south margin of the Mesaoria Basin.

The northern coastal areas of the Mesaoria Basin, located between the Ovgos Fault and the Kyrenia lineament, were dominated by deposition of bioclastic sediments of the Athalassa Formation (Fig. 15). This unit is locally unconformable on the Nicosia Formation and is laterally equivalent to the Kakkaristra and Apalos Formations further south (McCallum, 1989; McCallum & Robertson, 1990). The bioclastic material mainly developed in shallow productive carbonate coastal areas, fringing a still subdued Kyrenia land mass (Fig. 19). However, small volumes of Kyrenia-derived material (e.g. dolomite) are also present in the Athalassa Formation, suggesting the Kyrenia Range lineament was beginning to emerge and be eroded.

In the northeastern part of the Mesaoria Basin, the Athalassa Formation was dominated by shallow-marine, bioclast-rich sand waves (and occasional barrier bars). These sand bodies, based on palaeocurrent evidence (McCallum, 1989; McCallum & Robertson, 1990), migrated southwards across the shallow Mesaoria Basin, under the dominant influence of southerly moving storms (Fig. 19).

In the northwestern area of the Mesaoria Basin, the Athalassa Formation comprises a thick, locally slumped, silt-dominated lower unit, and a thinner, cross-bedded, bioclastic upper unit (Moore, 1960). The thick lower unit of the Athalassa Formation was possibly deposited in a small flexural basin, formed in response to local reversed reactivation of the earlier extensional Ovgos Fault (McCallum & Robertson, 1990; Fig. 20). Thin conglomerates in these silts may represent small fan-delta lobes derived from the emerging western part of the Kyrenia lineament. This uplift was presumably coeval with renewed uplift of the Troodos ophiolite and northward progradation of the Kakkaristra fan-deltas. Fault reactivation, however, appears to have been short-lived, as the small inferred foredeep soon filled, and was transgressed (along with deformed

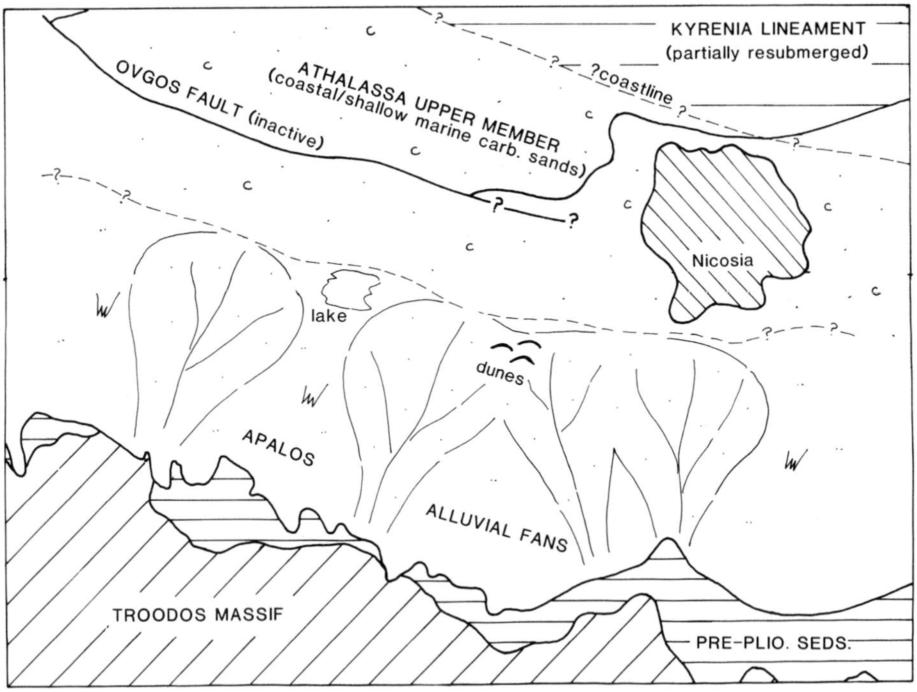


**Fig. 19.** Palaeogeographic sketch maps illustrating palaeoenvironments in the Mesaoria Basin in the late Pliocene to early Pleistocene; from McCallum (1989).

Kythrea Flysch) by shallow-marine to coastal carbonate sands forming the upper part of the Athalassa Formation in this area. During this relatively stable period (of (?)late Pliocene age), transgression may well have been in response to eustatic sea-level rise (see below).

## THE PLEISTOCENE FANGLOMERATES: ACCELERATED UPLIFT

The final stage of the evolution of the circum-Troodos sedimentary basins was marked by a widespread and sudden appearance of very coarse, immature clastic sediments, which overlie all older formations (Tables 1, 2). Along the south margin of the Troodos ophiolite, these sediments are coarse conglomerates (e.g. Bagnall, 1960; Pantazis, 1967). North of the Troodos ophiolite, a coarse-grained alluvial fan system prograded northwards far across the Mesaoria Basin (the Fanglomerate; Gass, 1960; Robertson, 1977a; McCallum, 1989). Thus, the Troodos ophiolite underwent dramatic tectonic uplift during (?)latest Pliocene to early Pleistocene time (Robertson, 1977a). Recent radiometric dating of corals in coastal terrace deposits indicates that rapid uplift was essentially over prior to Holocene time (Poole *et al.*, 1990).

## DISCUSSION: EUSTATIC SEA-LEVEL CHANGE VERSUS LOCAL TECTONICS

We now consider the possible influence of global sea-level changes on Neogene sedimentation in Cyprus. Vail *et al.* (1984,1989) argue that Phanerozoic stratigraphic successions can be packaged into 'globally correlative sequences and systems tracts', based on a combination of seismic profiles, well-logs and stage stratotype data from land outcrops. These authors contend that global sea-level changes exert the dominant control on the location of stratigraphic boundaries, even in areas of high tectonic activity, or where sediment rate has changed abruptly. They believe that when succession boundaries are traced from areas of such 'distortion' (e.g. with strong tectonic activity) to where they are *conformable*, the

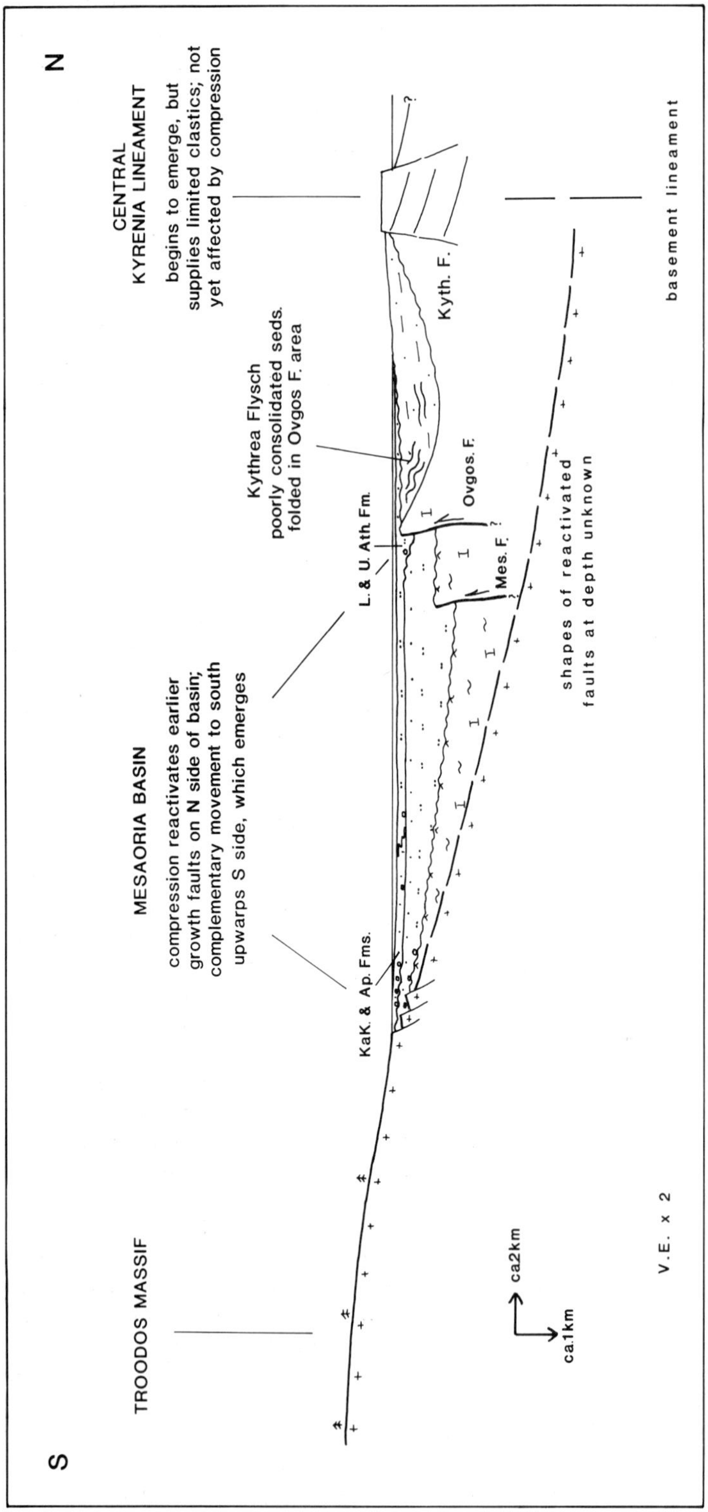


**Fig. 20.** Schematic cross-section illustrating the late Pliocene to early Pleistocene setting of the Mesaoria Basin (data source: Robertson & Woodcock, 1986); from McCallum (1989).

age of the discontinuities invariably corresponds to the global pattern.

### Possible role of global sea-level change

The Neogene global sea-level curve is compared with the stratigraphical record of Cyprus in Table 3. Our curve of relative sea-level change for the island is based on the most conformable stratigraphical contacts preserved between units for any given time interval. This is then compared with the global curve (Haq *et al.*, 1987) and with Vail *et al.*'s (1989) global stratigraphic signature of the Neogene. Difficulties in recognizing age-specific sequence boundaries include some inadequacies in the biostratigraphical data and the fact that marginal facies (e.g. reefs), sensitive environmental indicators, were largely eroded.

During the Oligocene the deep-sea carbonates of the Upper Lefkara Formation show evidence of shallowing and first emergence of the Troodos Massif, marked by input of ophiolite-derived silt, iron oxide, and fine plant detritus (e.g. east and southeast Troodos margins). The shallowing upwards trend previously was attributed to tectonic uplift of the Troodos Massif (Robertson, 1977a). Strong uplift apparently took place in earliest Miocene time, marking the boundary between the Lefkara and the Pakhna Formations. A tectonic rather than global eustatic control is inferred (see below).

The erosion surface below the Aquitanian–Burdigalian reefs of the Pakhna Formation (i.e. southeast and west Cyprus) could relate to tectonically controlled uplift at the start of deposition of the formation. However, the onset of pelagic deposition (e.g. west Cyprus) could relate to early Miocene flooding on the global curve. No break in reef deposition was indentified that could relate to Vail *et al.*'s early Miocene global lowstand deposition following aggradation. In the Burdigalian, reefs (Terra Member) flourished, consistent with gradual sea-level rise on the global curve (Table 3). The absence of mid-Miocene (Langhian) reefs in Cyprus could reflect coral growth failing to keep pace with rapid eustatic sea-level fluctuations on the global curve. Serrevalian sedimentation in south Cyprus was mainly pelagic chalk (Pakhna Formation), in keeping with a relatively high eustatic sea level, as on the global curve. Lowstand deposition in the late Serrevalian could correspond to emergence and erosion that predated colonization by the second phase of reefs in the Tortonian (Koronia Member).

The Messinian, Mediterranean-wide desiccation event is clearly at variance with the global curve. However, end-Messinian flooding is consistent with Vail *et al.*'s (1989) global eustatic sea-level rise at this time. On the other hand, tectonics were perhaps the dominant control on both creation and destruction of land barriers between the Mediterranean and the Atlantic (Hsü *et al.*, 1978).

The early Pliocene relatively deep marine deposition in north and south Cyprus is consistent with high global eustatic sea level at this time. Conglomerates in the lower part of the early Pliocene Nicosia Formation and then sands in the mid-part are both channelled and could reflect episodic eustatic sea-level fall, superimposed on a decline in tectonic uplift. The deep incision of the south Cyprus channels (at Khirokitia and (?)Kouris) could reflect global sea-level fall, followed by flooding, during a time of accelerating uplift. Bioclastic sediments accumulated subsequently in coastal areas in the latest Pliocene and were then reworked offshore, possibly in response to aggradation following continued global sea-level fall.

In north Cyprus the lower part of the late Pliocene to early Pleistocene Athalassa Formation documents filling of a small depocentre, with possible uplift and deformation of older rocks to the north. The depocentre filled very quickly, however, and both it and the deformed lithologies were transgressed by shallow marine facies. With no evidence of a tectonic cause for this transgression, it is perhaps attributable to sea-level rise. It is also probable that changes in climate associated with glacio-eustatic events triggered deposition of storm-influenced shallow-marine sand waves in the eastern part of the Mesaoria Basin.

Finally, the sudden emergence of the Troodos Massif and the Kyrenia Range in mid-Pleistocene time was clearly a response to tectonic uplift of Cyprus as a whole (Robertson, 1977a; Poole *et al.*, 1990).

In summary, the Neogene global sea-level changes postulated by Vail *et al.* (1989) are largely recognizable in the sedimentary record of Cyprus. This is in keeping with studies of other active margin areas. For example, in the paper by Hallam he recognizes a eustatic sea-level overprint on the Mesozoic Andean margin of South America and eustatic sea-level changes are also reported, for example from Antarctica (see paper by Butterworth) and Central America (see paper by Seyfried *et al.*).

| Ma (1) | | RELATIVE SEALEVEL CHANGES CYPRUS (2) | GLOBAL STRATIGRAPHIC (5) | ? GLOBAL VS LOCAL AGREE |
|---|---|---|---|---|
| 1 | PLEISTOCENE | Relative stability, local coastal subsidence<br>Intense pulsed uplift | High frequency sequences | NO |
| | PLIOCENE U | Gradual pulsed uplift<br>Channel incision due to sealevel fall, then flooding, climatic control on carbonate sand waves | Aggradation with multiple lowstand deposits | YES |
| 5 | PLIOCENE L | Marine, extension and subsidence in N, stability in S. Relative sealevel highstand | Flooding | YES |
| | MIOCENE Mess | Salinity crisis, gypsum deposition in silled sub-basins around margins of Troodos Massif | Aggradation commonly ending with lowstand deposits | NO |
| | MIOCENE Tort | Reefs develop on flooded tectonic lineaments in S and on tilted basement blocks in N,<br>Extension and subsidence in N Cyprus, reefs unconformably above, emersion and erosion then flooding. | | |
| 10 | MIOCENE Serr | | Major downward shift of onlap and lowstand deposits<br>Major progradatioin | ?YES |
| | | Reefs transgressed by pelagic carbonate in S E | | ? |
| 15 | MIOCENE Lang | Drowning of reefs by rapid sealevel rise | Flooding | YES |
| | MIOCENE Burd | Reefs develop on eroded surfaces in S E and W | Aggradation commonly ending with lowstand deposits | ? YES |
| 20 | MIOCENE Aquit | N-S compression produces en echelon fold and fault lineaments in S Cyprus, tectonic control<br>Sudden sealevel fall with exposure in N and W | Aggradation commonly ending with lowstand deposits<br>Flooding | ? YES |
| | OLIGOCENE | Gradual uplift and shallowing of the Troodos tectonically controlled. | Basinwards onlap wedge. | NO |

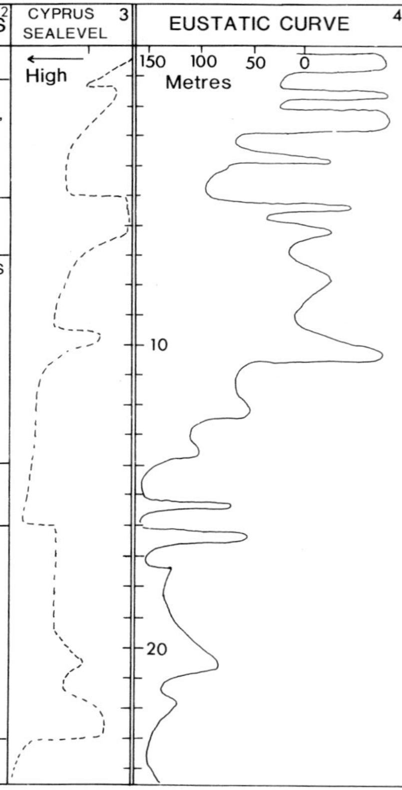


For explanation of 1–5 see caption

**Table 3.** Summary of the controls of Neogene sedimentation along the N–S transect of central Cyprus discussed in this paper. (1) Time-scale of Harland *et al.* (1989); (2) relative sea-level change curve is compiled from the most conformable successions in Cyprus for any time interval; (3) time-scale used by Haq *et al.* (1987); (4) global eustatic curve of Haq *et al.* (1987); (5) interpretation of Vail *et al.* (1989). A sea-level change signal can be recognized in Cyprus stratigraphy, but the dominant control on sedimentation was tectonic

**Dominant influence of tectonics**

Notwithstanding the identification of a signal of global sea-level change, the dominant control on Neogene sedimentation in Cyprus was tectonic. This was related to northward subduction and development of the Cyprus active margin in the eastern Mediterranean (Figs 21, 22).

The early compression of southern Cyprus, documented by switch from deposition of the Lefkara to the Pakhna Formation and by the presence of tectonically active lineaments, is attributed to onset of northward subduction along the Cyprus arc. Compression and uplift clearly persisted in only some areas until late Miocene. One explanation is that subduction was sluggish and/or pulsed during the Miocene. Another is that localized uplift resulted from episodic collision and underthrusting of seamounts beneath the overriding plate. In both scenarios, old basement weakness zones (e.g. transform faults) were reactivated, while the central core of the Troodos ophiolite remained as a rigid block, largely unaffected.

Following compression and uplift of southern

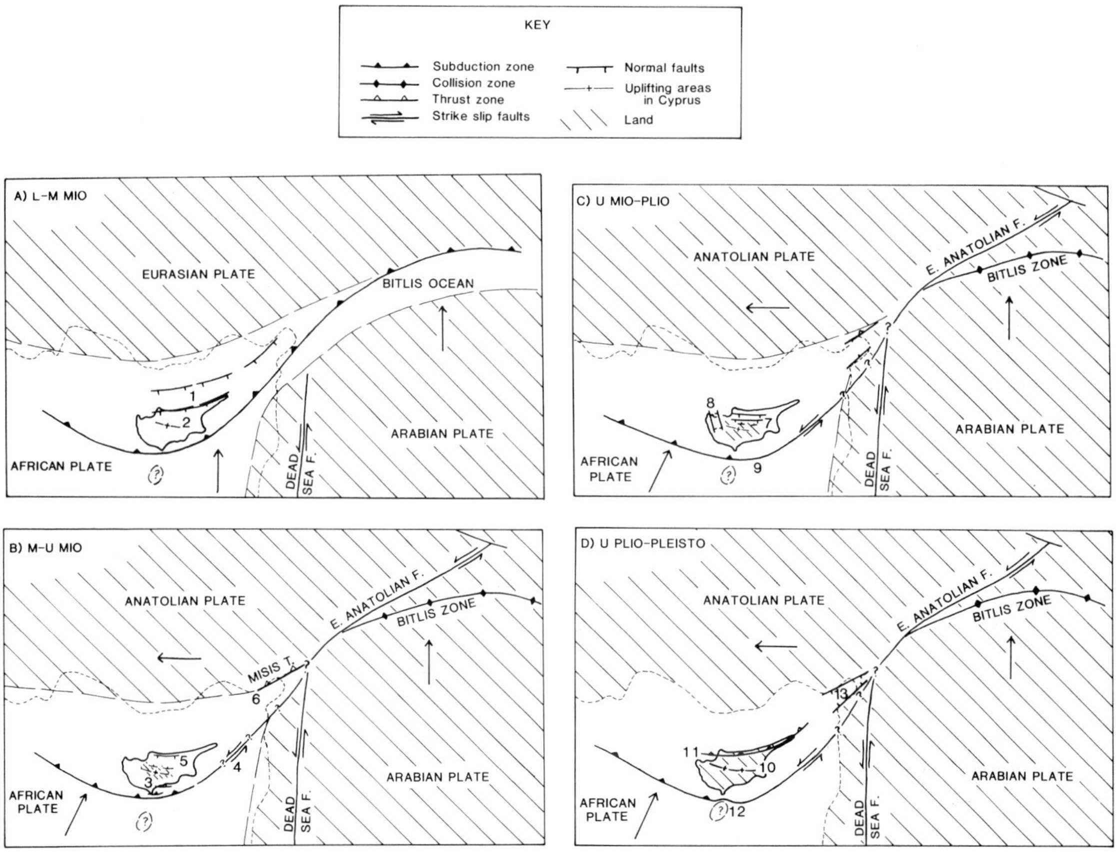


**Fig. 21.** Plate tectonic reconstructions for the Cyprus region. (A) Lower–middle Miocene. This period was marked by extension in the Kyrenia–Cilicia–Adana–Misis areas (1) and uplift of the Troodos ophiolite (2), which was still mainly submerged. (B) Middle–upper Miocene. This time was marked by tectonic shoaling of southern Cyprus (3), possible onset of strike-slip between Cyprus and Turkey (4), extension of the Mesaoria area (5) along the north Troodos margin, and by thrusting in the Misis Mountains (6). (C) Upper Miocene–Pliocene. This was a time of pronounced extension of the Mesaoria area (7), with significant emersion of the Troodos Massif, extension in the Paphos–Polis area (8) and, possibly, of southward migration ('roll-back') of the subduction zone (9). (D) Upper Pliocene–Pleistocene. This final phase saw severe uplift of the Troodos ophiolite (10), and the Kyrenia lineament, collision of the Eratosthenes Seamount with the subduction zone (11), and extension in the Iskenderun area (13). From McCallum (1989).

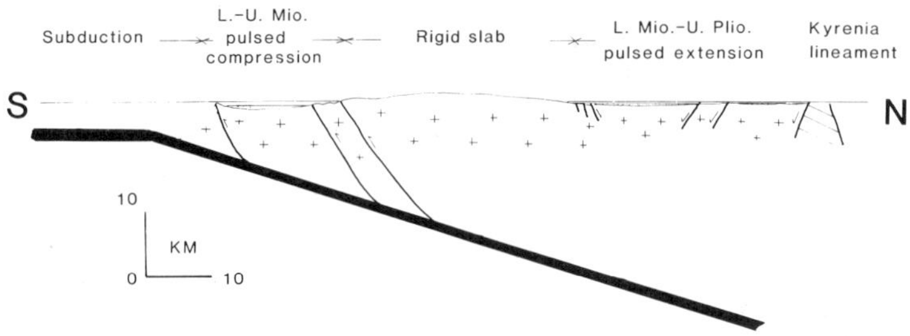


**Fig. 22.** Diagram illustrating the inferred role of northward subduction on the tectonic evolution of the overriding Cyprus plate. Compression probably relates to the early to mid-Miocene initiation of subduction, and extension to late Miocene to early Pliocene roll-back of the downgoing slab during continued subduction. Alternatives are discussed in the text.

Cyprus, extension dominated in the late Miocene and early Pliocene in northern Cyprus c. 60 km further north. A possible control was southward 'roll-back' of the downgoing slab. Study of shallow seismic data and deeper multichannel data from offshore southern Cyprus indicated no evidence of continued deformation of the continental slope in the Plio-Pleistocene (McCallum, 1989). By this time the subducting slab had apparently retreated southwards. Roll-back is a well-known potential control on fore-arc evolution (Dewey, 1980; Carlson & Melia, 1984). In Cyprus, the subducting slab is assumed to have been old, dense Mesozoic Neotethyan ocean crust, conducive to rapid roll-back. Theoretically, possible controls on roll-back include variations in the type of crust, the age and thus bouyancy of the subducting crust, the rate or angle of subduction, and changes in relative plate motion.

Pulsed extension and compression in fore-arc settings have been noted elsewhere, for example along the Aegean arc (Le Pichon & Angelier, 1981) and the Andean Cordillera (Mercier, 1981). On the north Chilean margin, phases of active faulting on land appear to correspond to times of rapid subduction. Uplift of the fore-arc in central Chile is explained in terms of collision of an aseismic ridge with the trench (see paper by Flint *et al.*), a process that could be applicable to Cyprus. Roll-back is also inferred as the dominant control on extension of the Cretan (Le Pichon & Angelier, 1979,1981) and Eolian arcs (Malinverno & Ryan, 1986; Kastens *et al.*, 1988). Finally, the drastic uplift of the whole of Cyprus in the mid-Pleistocene is explained in terms of collision of a large seamount or continental fragment with the overriding plate. This could be part of the Eratosthenes Seamount (Kempler & Ben-Avraham, 1987), presently south of Cyprus, or some other unit that collided earlier. Any interpretation must also take into account the complex local tectonic setting involving collision to the northeast (Bitlis suture zone), strike-slip to the east (Dead Sea Transform; e.g. Freund & Garfunkel, 1981; Quennell, 1984), rifting to the south (Whitmarsh *et al.*, 1974) and subduction to the west (Aegean arc).

## CONCLUSION

Sedimentation at active margins is typically very complicated and needs in-depth analysis of individual cases before the relative effects of global sea-level change on sedimentation can be evaluated. To this end, we have presented the first integrated account of Neogene sedimentation of the Cyprus active margin. A global eustatic sea-level change signal can, indeed, be discerned, above the 'noise' generated by local tectonic effects. However, tectonic processes dominated sedimentation overall. These tectonic events are interpreted in terms of the evolution of the Neogene Cyprus active margin, within the framework of progressive collision of the African and Eurasian plates. Early to mid-Miocene compression in southern Cyprus relates to onset of northward subduction, which was possibly initially sluggish. Roll-back of the downgoing slab then led to crustal extension in the late Miocene to mid-Pliocene in northern Cyprus. There was then pulsed uplift of the whole area, climaxing in early to mid-Pleistocene time, probably as a consequence of collision of a seamount, or microcontinent with the trench to the south.

## ACKNOWLEDGEMENTS

This summary and synthesis has benefited from discussions with I.G. Gass, H.M. Pedley, A.J. Poole, B.R. Rosen, T.P. Scoffin, R.A. Scrutton, and C. Xenophontos. Funding by N.E.R.C. is gratefully acknowledged, as is the continuing support of the Cyprus Geological Survey Department. The manuscript was improved considerably in the light of comments of G. Aktaş, J.R. Underhill, R.E. Swarbrick, and D.I.M. Macdonald.

## REFERENCES

ADAMS, C.G. (1959) Geological distribution of *Discospirina* (foraminifera) and occurrence of *D. italica* in the Miocene of Cyprus. *Palaeontology* **1**, 364–368.

AKTAŞ, G. & ROBERTSON, A.H.F. (1984) The Maden Complex, SE Turkey: evolution of a Neotethyan active continental margin. In: *The Geological Evolution of the Eastern Mediterranean* (Eds Dixon, J.E. & Robertson, A.H.F.). Geol. Soc. London, Spec. Publ. 17, 375–402.

AKTAŞ, G. & ROBERTSON, A.H.F. (1990) Tectonic evolution of the Tethys suture zone in SE Turkey: evidence from the petrology and geochemistry of Late Cretaceous and Middle Eocene extrusives. In: *Ophiolites and Oceanic Lithosphere* (Eds Moores, E.M. *et al.*) *Proceedings International Symposium, Nicosia Cyprus, October 1987* (in press).

ANASTASAKIS, G. & KELLING, G. (1990) Tectonic connection of the Hellenic and Cyprus Arcs and their probable landwards extension. *Mar. Geol.* (in press).

BAGNALL, P.S. (1960) The geology and mineral resources of the Pano Lefkara–Larnaca area. *Mem. Geol. Surv. Dept. Cyprus* **5**, 116 pp.

BARBER, P.M. (1981) Messinian subaerial erosion of the Proto-Nile Delta. *Mar. Geol.* **44**, 253–272.

BAROZ, F. (1979) *Étude géologique dans le Pentadactylos et de la Mésaoria (Chypre septentrionale)* unpublished doctoral thesis, Nancy University, France. 2 vols. 365 pp.

BAROZ, F. & BIZON, G. (1977) La couverture Tertiare du flanc nord du massif du Troodos et de la partie méridionale de la Mésaoria: étude stratigraphique et micropalaéontologique. *Rev. Inst. Fr. Pet.* **32**, 719–759.

BAROZ, F., BIZON, G., BIZON, J.-J., HEIMANN, K.O., MALLET, J.-P., MÜLLER, C., ORSZAG-SPERBER, F., ROUCHY, J.-M. & WEISGERBER, F. (1978) Le Miocène terminal et le Pliocène dans la région de Polemi (Chypre Ouest): milieux de dépôt. *C.R. Acad. Sci. Ser. D* **286**, 1771–1774.

BEAR, L.M (1960) The geology and mineral resources of the Akaki–Lythrodondha area. *Mem. Geol. Surv. Dept. Cyprus* **3**, 122 pp.

BEAR, L.M. & MOREL, S.W. (1960) The geology and mineral resources of the Agros–Akrotiri area. *Mem. Geol. Surv. Dept. Cyprus* **7**, 88 pp.

BELLAMY, C.V. & JUKES-BROWNE, A.J. (1905) *The Geology of Cyprus*. W. Brendon and Son, Plymouth, 72 pp.

BEN-AVRAHAM, Z. & NUR, A. (1986) Collisional processes in the East Mediterranean. *Geol. Rundsch.* **75**, 209– 217.

BEN-AVRAHAM, Z., KEMPLER, D. & GINZBURG, A. (1988) Plate convergence in the Cyprean Arc. *Tectonophysics* **146**, 231–240.

BRAUNE, K.F., FABRICIUS, F. & HEIMANN, K.O. (1973) Sedimentation and facies of Late Miocene strata on Cephalonia (Ionian Islands, Greece). In: *Messinian Events in the Mediterranean* (Ed. Drooger, C.W.) pp. 192–201. North Holland, Amsterdam.

CARLSON, R.L. & MELIA, P.J.A. (1984) Subduction hinge migration. *Tectonophysics* **102**, 399–411.

CARR, J.M. & BEAR, L.M. (1960) The geology and mineral resources of the Peristerona–Lagoudhera area *Mem. Geol. Surv. Dept. Cyprus* **2**, 79 pp.

CHEVALIER, J.P. (1961) Recherches sur les Madréporaires et les formations récifales Miocènes de la Méditerranée occidentale. *Mém. Soc. Géol. Fr.* **40/93**, 562 pp.

CITA, M.B. (1982) The Messinian salinity crisis in the Mediterranean: a review. In: *Alpine–Mediterranean Geodynamics* (Eds Berckhemer, H. & Hsü, K.). Am. Geophys. Union, Geodyn. Ser. 7, 113–140.

CLEINTAUR, M.R., KNOX, G.J. & EALEY, P.J. (1977) The geology of Cyprus and its place in the East Mediterranean framework. *Geol. Mijnbouw* **56**, 66–82.

CLUBE, T.M.M. & ROBERTSON, A.H.F. (1986) The palaeorotation of the Troodos microplate, Cyprus, in the Late Mesozoic–Early Cenozoic plate tectonic framework of the Eastern Mediterranean. *Surv. Geophys.* **8**, 375–437.

CLUBE, T.M.M., CREER, K.M. & ROBERTSON, A.H.F. (1985) The palaeorotation of the Troodos microplate, Cyprus. *Nature* **317**, 522–525.

DEAN, W.E. & ANDERSON, R.Y. (1982) Continuous subaqueous deposition of the Permian Castile evaporites, Delaware Basin, Texas and New Mexico. In: *Deposition and Diagenetic Spectra of Evaporites – a Core Workshop* (Eds Handford, C.R., Loucks, R.G. & Davies, G.R.). Can. Soc. Petrol. Geol., Calgary, 324–355.

DEAN, W.E., DAVIES, G.R. & ANDERSON, R.Y. (1975) Sedimentological significance of nodular and laminated anhydrite. *Geology* **3**, 367–372.

DECIMA, A. & WEZEL, F.C. (1973) Late Miocene evaporites of the central Sicilian basin, Italy. In: *Initial Reports of the Deep Sea Drilling Project*, Vol. 13 (Eds Ryan, W.B.F., Hsü, K.J. *et al.*) pp. 1234–1239. US Government Printing Office, Washington, DC.

DERCOURT, J., ZONENSHAIN, L.P., RICOU, L.-E., KAZMIN, V.G., LE PICHON, X., KNIPPER, A.L., GRANDJACQUET, C., SBORTSHIKOV, I.M., GEYSSANT, J., LEPVRIER, C., PECHERSKY, D.H., BOULIN, J., SIBUET, J.-C., SAVOSTIN, L.A., SOROKHTIN, O., WESTHAL, M., BAZHENOV, M.L., LAUER, J.P. & BIJU-DUVAL, B. (1986) Geological evolution of the Tethys Belt from the Atlantic to the Pamirs since the Lias. *Tectonophysics* **123**, 241–315.

DEWEY, J.F. (1980) Episodicity, sequence and style at convergent plate boundaries. In: *The Continental Crust and its Mineral Deposits* (Ed. Strangeway, D.F.). Spec. Publ., Geol. Ass. Canada 20, 553–573.

DEWEY, J.F. & ŞENGÖR, A.C.M. (1979) Aegean and surrounding areas: complex multi-plate and continuum tectonics in a convergent zone. *Geol. Soc. Am., Bull.* **90**, 84–92.

DUCLOZ, C. (1965) Revision of the Pliocene and Quaternary stratigraphy of the Central Mesaoria. *Cyprus Geol. Surv. Dept. Ann. Rep. for 1964*, pp. 31–42.

DUCLOZ, C. (1972) The Geology of the Bellapais-Kythrea area of the Central Kyrenia Range. *Bull. Cyprus Geol. Surv. Dept.* **6**, 75 pp.

EATON, S. (1987) *The sedimentology and mid to Late Miocene carbonates and evaporites in Southern Cyprus.*

Unpublished PhD thesis, University of Edinburgh, 240 pp.

Elion, P. (1983) *Étude structurale et sédimentologique du bassin Neogène de Pissouri (Chypre)*. Unpublished thesis Docteur 3em. Cycle, Université de Paris-Sud, 235 pp.

Evans, G., Morgan, P., Evans, W.E., Evans, T.R. & Woodside, J.M. (1978) Faulting and halokinesis in the northeastern Mediterranean between Cyprus and Turkey. *Geology* **6**, 392–396.

Farrell, S.G. & Eaton, S. (1987) Slump strain in the Tertiary of Cyprus and the Spanish Pyrenees. Definition of palaeoslopes and models of soft sediment deformation. In: *Deformation of Sediments and Sedimentary Rocks* (Eds Jones, M.E. & Preston, R.M.F.). Geol. Soc. London, Spec. Publ. 29, 181–196.

Follows, E.J. (1990a) *Sedimentology and tectonic setting of Miocene reefs and related sediments in Cyprus*. Unpublished PhD thesis, University of Edinburgh.

Follows, E.J. (1990b) Patterns of reef sedimentation and diagenesis in the Miocene of Cyprus. In: *Sedimentary Geology special issue* (Eds Gutteridge, P. & Harwood, G.), (in press).

Follows, E.J. & Robertson, A.H.F. (1990) Sedimentology and structural setting of Miocene reefal limestones in Cyprus. In: *Ophiolites and Oceanic Lithosphere* (Eds Moores, E. *et al.*) *Proceedings International Symposium, Nicosia, Cyprus, October 1987* (in press).

Freund, R. & Garfunkel, Z. (Eds) (1981) *The Dead Sea Rift. I.U.C.G. Sci. Rept. 60*. Tectonophysics 80, 310 pp.

Garrison, R.E., Schreiber, B.C., Bernoulli, D., Fabricius, F.H., Kidd, R.B. & Mélières, F. (1978) Sedimentary petrology and structures of Messinian evaporitic sediments in the Mediterranean Sea, Leg 42A, Deep Sea Drilling Project. In: *Initial Reports of the Deep Sea Drilling Project*, Vol. 42A (Ed. Hsü, K.J.) pp. 571–611. US Government Printing Office, Washington, DC.

Gass, I.G. (1960) The geology and mineral resources of the Dhali area. *Mem. Geol. Surv. Dept. Cyprus* **4**, 116 pp.

Gass, I.G. & Cockbain, A.E. (1961) Notes on the occurrence of gypsum in Cyprus. *Overseas Geol. Min. Resour.* **8**, 279–287.

Gass, I.G. & Masson-Smith, E.M. (1963) The geology and gravity anomalies of the Troodos Massif, Cyprus. *Philos. Trans. R. Soc. London, Ser. A* **255**, 417–467.

Grasso, M. & Pedley, H.M. (1988) The sedimentology and development of Terravecchia Formation carbonates (Upper Miocene) of North Central Sicily: possible eustatic influences on facies development. *Sediment. Geol.* **57**, 131–149.

Greitzer, Y. & Constantinou, C. (1969) *Report on the Geology of the Kouris River Area*. Internal Report, Republic of Cyprus and Tahal Consulting Engineers Ltd., Tel Ariv, Israel.

Gvirtzman, G. & Buchbinder, B. (1978) The Late Tertiary of the coastal plain and continental shelf of Israel and its bearing on the history of the Eastern Mediterranean. In: *Initial Reports of the Deep Sea Drilling Project*, Vol. 42 (Eds Ross, D.A. & Neprochnov, Y.P.) pp. 1195–1222. US Government Printing Office, Washington, DC.

Hadjistavrinou, Y. (1974) *Ann. Rep. Geol. Surv. Dept. Cyprus 1973*.

Hadjistavrinou, Y. (1975) *Ann. Rep. Geol. Surv. Dept. Cyprus 1974*.

Hadjistavrinou, Y. & Constantinou, G. (1977) Hydrogeology of the Akrotiri Peninsula. *Bull. Cyprus Geol. Surv. Dept.* **7**.

Hallam, A. (1988) A reevaluation of Jurassic eustasy in the light of new data and the revised Exxon Curve. In: *Sea-level Changes – an Integrated Approach* (Eds Wilgus, C.K., Hastings, B.S., Kendall, C.G.St.C., Posamentier, H., Ross, C.A. & van Wagoner, J.). Soc. Econ. Paleontol, Mineral. Spec. Publ. 42, 261–273.

Haq, B.U., Hardenbol, J. & Vail, P.R. (1987) Chronology of fluctuating sea levels since the Triassic. *Science* **235**, 1156–1167.

Hardie, L.A. & Eugster, H.P. (1971) The depositional environment of marine evaporites: a case for shallow, clastic accumulation. *Sedimentology* **16**, 187–220.

Harland, W.B., Armstrong, R.L., Cox, A.V., Craig, L.E., Smith, A.G. & Smith, D.G. (1989) *A Geological Time Scale*. Cambridge University Press, Cambridge.

Haskett, M.J. (1984) *Middle Miocene Foraminifera of the Kissousa to Pakhna Road Section, Southern Cyprus*. Unpublished Internal Report, University College London.

Henson, F.R.S., Browne, R.V. & McGinty, J. (1949) A synopsis of the stratigraphy and geological history of Cyprus. *Q. J. Geol. Soc. London* **105**, 1–41.

Herguera, J.C. (1982) *Brief notes on the biostratigraphy and palaeobathymetry of some sections in the Kouris River valley near Khalsasa (Cyprus)*. Unpublished Internal Report, University College London.

Hsü, K.J. (1973) The desiccated deep basin model for the Messinian events. In: *Messinian Events in the Mediterranean* (Ed. Drooger, C.W.) pp. 60–67. North Holland, Amsterdam.

Hsü, K.J., Montadert, L., Bernoulli, D., Cita, M.B., Erickson, A., Garrison, R.E., Kidd, R.B., Mélières, F., Müller, C. & Wright, R. (1978) History of the Mediterranean salinity crisis. In: *Initial Reports of the Deep Sea Drilling Project*, Vol. 42, pp. 1053–1078. US Government Printing Office, Washington, DC.

Innocenti, F., Manetti, P., Mazzuoli, R., Pasquare, G. & Villari, L. (1982) Anatolia and north-west Iran. In: *Andesites* (Ed. Thorpe, R.S.) pp. 328–349. Wiley, Chichester.

Jackson, J. & McKenzie, D. (1984) Active tectonics of the Alpine–Himalayan belt between western Turkey and Pakistan. *Geophys. J. R. Astron. Soc.* **77**, 185–264.

Kastens, K., Mascle, J., Auroux, C., Bonatti, E., Broglia, C., Channell, J., Curzi, P., Emeis, K.C., Glaçon, G., Hasegawa, S., Hieke, W., Mascle, G., McCoy, F., McKenzie, J., Mendelson, J., Müller, C., Réhault, J.-P., Robertson, A., Sartori, R., Sprovieri, R. & Torii, M. (1988) O.D.P. Leg 107 in the Tyrrhenian Sea: insights into passive margin and back-arc basin evolution. *Geol. Soc. Am. Bull.* **100**, 1140–1156.

Kelling, G., Gökçen, S.L., Floyd, P.A. & Gökçen, N. (1987) Neogene tectonics and plate convergence in the Eastern Mediterranean: new data from southern Turkey. *Geology* **15**, 425–429.

Kempler, D. & Ben-Avraham, Z. (1987) The tectonic evolution of the Cyprean Arc. *Ann. Tectonicae* **1**, 58–71.

Kinsman, D.J.J. (1976) Evaporites: relative humidity control of primary mineral facies. *J. Sediment. Pet.* **46**, 273–279.

Lapierre, H. (1975) Les formations sédimentaires et éruptives des nappes de Mamonia et leurs rélations avec le Massif de Troodos (Chypre occidentale). *Mém. Soc. Géol. Fr.* **123**, 132 pp.

Le Pichon, X. & Angelier, J. (1979) Hellenic arc and trench system. A key to the neo-tectonic evolution of the East Mediterranean. *Tectonophysics* **60**, 1–42.

Le Pichon, X. & Angelier, J. (1981) The Aegean Sea. *Philos. Trans. R. Soc. London, Ser. A* **300**, 357–372.

Le Pichon, X., Kobayashi, K., Cadet, J.-P., Iiyama, T., Nakamura, K., Pautot, G., Renard, V. & The Kaiko Scientific Crew (1987) Project Kaiko – Introduction. *Earth Planet Science Lett.* **83**, 183–185.

Makris, J., Ben-Avraham, Z., Behle, A., Ginzburg, A., Giese, P., Steinmetz, L., Whitmarsh, R.B. & Eleftheriou, S. (1983) Deep seismic soundings between Cyprus and Israel and their interpretation. *Geophys. J. R. Astron. Soc.* **75**, 575–591.

Malinverno, A. & Ryan, W. (1986) Extension in the Tyrrhenian Sea and shortening in the Apennines as a result of arc migration driven by sinking of the lithosphere. *Tectonics* **5**, 227–245.

Mantis, M. (1968) *Subsurface biostratigraphy of the Mesaoria, Cyprus.* Unpublished report, Cyprus Geological Survey Department, 13 pp.

Mantis, M. (1970a) Upper Cretaceous–Tertiary foraminiferal zones in Cyprus. *Epithris, Cyprus Research Centre, Nicosia* **3**, 227–241.

Mantis, M. (1970b) Composite stratigraphic range chart of foraminifera in the exploration boreholes from Ypsonas–Episkopi–Akrotiri area. *Ann. Rep. Geol. Surv. Dept. Cyprus 1969*, **Appendix 6**.

McCallum, J.E. (1989) *Sedimentation and tectonics of the Plio-Pleistocene of Cyprus.* Unpublished PhD thesis, University of Edinburgh, 263 pp.

McCallum, J.E. & Robertson, A.H.F. (1990) Pulsed uplift of the Troodos Massif; evidence from the Plio-Pleistocene Mesaoria basin. In: *Ophiolites and Oceanic Lithosphere* (Eds Moores, E. *et al.*) *Proceedings International Symposium, Nicosia, Cyprus, October 1987* (in press).

MacLeod, C.J. (1990) Extensional reactivation of an oceanic fracture fault: evidence from the Eastern Limassol Forest. In: *Ophiolites and Oceanic Lithosphere* (Eds Moores, E.M. *et al.*) *Proceedings International Symposium, Nicosia, Cyprus, October 1987* (in press).

McKenzie, D.P. (1972) Active tectonics of the Mediterranean region. *J.R. Astron. Soc.* **30**, 109–185.

McKenzie, D.P. (1978) Active tectonics of the Alpine–Himalayan belt: the Aegean Sea and sorrounding regions. *Geophys. J. R. Astron. Soc.* **55**, 217–254.

Mercier, J.L. (1981) Extensional–compressional tectonics associated with the Aegean Arc: comparison with the Andean Cordillera of south Peru – north Bolivia. *Philos. Trans. R. Soc. London, Ser. A* **300**, 337–355.

Meulenkamp, J.E. & Zachariasse, W.J. (1973) Stratigraphic and structural framework of the Messinian deposits on Crete. In: *Messinian Events in the Mediterranean* (Ed. Drooger, C.W.) pp. 202–205. North Holland, Amsterdam.

Moore, T.A. (1960) The Geology and Mineral resources of the Astromeritis–Kormakiti area. *Mem. Geol. Surv. Dept. Cyprus* **6**, 96 pp.

Moores, E.M. & Vine, F.J. (1971) The Troodos Massif, Cyprus, and other ophiolites as oceanic crust: evaluation and implications. *Philos. Trans. R. Soc. London, Ser. A* **268**, 433–466.

Morel, S.W. (1960) The geology and mineral resources of the Apsiou–Akrotiri area. *Mem. Geol. Surv. Dept. Cyprus* **7**, 51–88.

Mukasa, S.B. & Ludden, J.N. (1987) Uranium–lead ages of plagiogranites from the Troodos ophiolite, Cyprus and their tectonic significance. *Geology* **15**, 825–828.

Mulder, C.J. (1973) Messinian events in the Mediterranean. In: *Geodynamics Scientific Report*, Vol. 7 (Eds Broekman, *et al.*) pp. 44–59. Colloquium, Utrecht, March 1973. North Holland, Amsterdam.

Mulder, C.J., Lehner, P. & Allen, D.C.K. (1975) Structural evolution of the Neogene Mediterranean salt basins in the Eastern Mediterranean and the Red Sea. *Geol. Mijnbouw* **54**, 208–221.

Murton, B.J. & Gass, I.G. (1986) Western Limassol Forest Complex, Cyprus: part of an Upper Cretaceous leaky transform fault. *Geology* **14**, 255–258.

Orszag-Sperber, F. & Rouchy, J.-M. (1983) Un jalon dans la connaissance de l'évolution paléo-climatique à la limite Miocène-Pliocène en Mediterranée orientale: l'example du bassin de Pissouri (Chypre). *Paléobiol. Continent. (Montpellier)* **14**, 377–383.

Orszag-Sperber, F., Rouchy, J.-M., Bizon, G., Bizon, J.-J., Cravatte, J. & Müller, C.(1980) La sédimentation messinienne dans le bassin de Polemi (Chypre). *Géol. Méditerranéenne* **7**, 91–102.

Orszag-Sperber, F., Rouchy, J.-M. & Elion, P. (1989) The sedimentary expression of regional tectonic events during the Miocene–Pliocene transition in the southern Cyprus basins. *Geol. Mag.* **136**, 291–299.

Ortiz, P.A. (1982) *Study of the Miocene planktonic foraminifera of the Pakhna Formation.* Unpublished Internal Report, University College London.

Pantazis, Th.M. (1967) The geology and mineral resources of the Pharmakas–Kalavasos area. *Mem. Geol. Surv. Dept. Cyprus* **8**, 190 pp.

Pantazis, Th.M. (1978) Cyprus evaporites. In: *Initial Reports of the Deep Sea Drilling Project*, Vol. 42B (Eds Ross, D.A. & Neprochnov, Y.P.) pp. 1185–1194. US Government Printing Office, Washington, DC.

Pavlakelli, N. (1987) *Early Pliocene ostracods from the Mari area, south-central Cyprus.* Unpublished MSc thesis, University College London, 110 pp.

Pearce, J.A. (1980) Geochemical evidence for the genesis and setting of lavas from Tethyan ophiolites. In: *Ophiolites* (Ed. Panayiotou, A.) *Proceedings International Ophiolite Symposium, Nicosia*, pp. 261–272. Cyprus Geol. Surv. Dept. Nicosia, Cyprus.

Pedley, H.M. (1983) The petrology and palaeoenvironment of the Sortino Group (Miocene) of S.E. Sicily: evidence for periodic emergence. *J. Geol. Soc. London* **140**, 335–350.

Pedley, H.M. (1987) The Ghar Lapsi limestones: sedimentology of a Miocene intra-shelf graben. *Centro, University of Malta Press* **1**, 1–14.

Poole, A.J., Shimmield, G.B. & Robertson, A.H.F.

(1991) Late Quaternary uplift of the Troodos ophiolite, Cyprus: Uranium-series dating of Pleistocene coral. *Geology* **18**, 894–7.

Purser, B.M., Orszag-Sperber, F. & Plaziat, J.-C. (1987) Sédimentation et rifting des séries néogènes de la marge nord occidentale de la Mer Rouge (Égypte). *Notes Mém. Total Co. Fr. Pétrole* **21**, 111–144.

Quennell, A.M. (1984) The Western Arabia Rift System. In: *The Geological Evolution of the Eastern Mediterranean* (Eds Dixon, J.E. & Robertson, A.H.F.). Geol. Soc. London, Spec. Publ. 17, 775–788.

Ricci-Lucchi, F. (1973) Resedimented evaporites: indicators of slope instability and deep basin conditions in Peri-Adriatic Messinian (Appenines foredeep, Italy). In: *Messinian Events in the Mediterranean* (Ed. Drooger, C.W.) pp. 142–149. North Holland, Amsterdam.

Robertson, A.H.F. (1976) Pelagic chalks and calciturbidites from the Lower Tertiary of the Troodos Massif, Cyprus. *J. Sediment. Petrol.* **46**, 1007–1016.

Robertson, A.H.F. (1977a) Tertiary uplift history of the Troodos Massif, Cyprus. *Geol. Soc. Am. Bull.* **88**, 1763–1772.

Robertson, A.H.F. (1977b) The Moni Mélange, Cyprus: an olistostrome formed at a destructive plate margin. *J. Geol. Soc. London* **133**, 447–466.

Robertson, A.H.F. (1990) Tectonic evolution of Cyprus: In: *Ophiolites and Oceanic Lithosphere* (Eds Moores, E.M. *et al.*) *Proceedings International Symposium, Nicosia, October 1987.*

Robertson, A.H.F. & Dixon, J.E. (1984) Introduction: aspects of the geological evolution of the Eastern Mediterranean. In: *The Geological Evolution of the Eastern Mediterranean* (Eds Dixon, J.E. & Robertson, A.H.F.). Geol. Soc. London, Spec. Publ. 17, 1–74.

Robertson, A.H.F. & Hudson, J.D. (1974) Pelagic sediments in the Cretaceous and Tertiary history of the Troodos Massif, Cyprus. In: *Pelagic Sediments: On Land and Under the Sea* (Eds Hsü, K.J. & Jenkyns, H.C.). Spec. Publ. Int. Ass. Sediment. 1, 403–436.

Robertson, A.H.F. & Woodcock, N.H. (1979) The Mamonia Complex, southwest Cyprus: the evolution and emplacement of a Mesozoic continental margin. *Geol. Soc. Am. Bull.* **90**, 651–665.

Robertson, A.H.F. & Woodcock, N.H. (1986) The role of the Kyrenia Range lineament, Cyprus, in the geological evolution of the Eastern Mediterranean area. *Philos. Trans. R. Soc. London, Ser. A* **317**, 141–177.

Robertson, A.H.F., Hieke, W., Mascle, G., McCoy, F., McKenzie, J., Réhault, J.-P. & Sartori, R. (1990) Summary and synthesis of the Late Miocene to Recent sedimentology and palaeoceanographic evolution of the Tyrrhenian Sea, Western Mediterranean: Leg 107 of the Ocean Drilling Programme. *Proc. Ocean Drilling Project* **107**, 639–668.

Rotstein, Y. (1984) Counter-clockwise rotation of the Anatolian Block. *Tectonophysics* **108**, 71–91.

Rotstein, Y. & Kafka, A.L. (1982) Seismotectonics of the Cyprean Arc, Eastern Mediterranean region: subduction, collision and arc jumping. *J. Geophys. Res.* **87**, 7694–7706.

Rouchy, J.-M. & Monty, C.L.V. (1979) Stromatolites and cryptalgal laminites associated with Messinian gypsum of Cyprus. In: *Phanerozoic Stromatolites* (Ed. Monty, C.L.V.) pp. 155–179. Springer-Verlag, Berlin.

Rouchy, J.-M., Orszag-Sperber, F., Bizon, G. & Bizon, J.-J. (1980) Mise en évidence d'une phase d'émersion fini-messinienne dans le bassin de Pissouri (Chypre): une modalité de passage Miocène-Pliocène en Méditerranée orientale. *C.R. Acad. Sci. Ser. D* **291**, 729–732.

Ryan, W.B.F., Hsü, K.J., Cita, M.B., Dumitrica, P., Lort, J.M., Maync, W., Nesteroff, W.D., Pautot, G., Stradner, H. & Wezel, F.C. (1973) *Initial Reports of the Deep Sea Drilling Project*, Vol 13. p. 1256. US Government Printing Office, Washington, DC.

Schilling, A.V., Stradner, H., Cita, M.B. & Gaetani, M. (1978) Stratigraphic investigations of the Late Neogene of Corfu, with special reference to the Miocene–Pliocene boundary, and the geodynamic significance. *Mem. Soc. Geol. Ital.* **16**, 279–317.

Schlager, W. & Boltz, H. (1977) Clastic accumulation of sulphate evaporites in deep water. *J. Sedim. Petrol.* **47**, 600–609.

Schreiber, B.C. & Decima, A. (1978) Sedimentary facies produced under evaporitic environments: a review. *Mem. Soc. Geol. Ital.* **16**, 111–126.

Schreiber, B.C., Freidmann, G.M., Decima, A. & Schreiber, E. (1976) Depositional environments of Upper Miocene (Messinian) evaporite deposits of the Sicilian basin. *Sedimentology* **23**, 729–760.

Searle, D.L. & Panayiotou, A. (1980) Structural implications in the evolution of the Troodos Massif, Cyprus. In: *Ophiolites* (Ed. Panayiotou, A.) *Proceedings International Ophiolite Symposium, Nicosia, 1979*, pp. 50–60.

Şengör, A.M.C. & Yilmaz, Y. (1981) Tethyan evolution of Turkey: a plate tectonic approach. *Tectonophysics* **75**, 181–241.

Şengör, A.M.C., Görür, N. & Saroglu, F. (1985) Strike-slip faulting and related basin formation in zones of tectonic escape: Turkey as a case study. In: *Strike-slip Deformation, Basin Formation and Sedimentation* (Eds Biddle, K.T. & Christie-Blick, N.). Soc. Econ. Palaeont. Mineral., Spec. Publ. 37, 227–264.

Simonian, K.A. & Gass, I.G. (1978) Arakapas Fault Belt, Cyprus: a fossil transform belt. *Bull. Geol. Soc. Am.* **89**, 1220–1230.

Stampfli, G.M. & Höcker, C.F.W. (1989) Messinian palaeorelief from a 3-D seismic survey in the Tarraco concession area (Spanish Mediterranean Sea). *Geol. Mijnbouw* **68**, 201–210.

Sturani, C. (1978) Messinian facies in the Piedmont Basin. *Mem. Soc. Geol. Ital.* **16**, 11–25.

Swarbrick, R.E. (1980) The Mamonia Complex of SW Cyprus: a Mesozoic continental margin and its relationship with the Troodos Complex. In: *Ophiolites* (Ed. Panayiotou, A.) *Proceedings International Ophiolite Symposium, Nicosia, 1979*, pp. 86–92.

Swarbrick, R.E. & Robertson, A.H.F. (1980) Revised stratigraphy of the Mesozoic rocks of southern Cyprus. *Geol. Mag.* **117**, 547–563.

Underhill, J.R. (1985) *Neogene and Quaternary tectonics and sedimentation in W. Greece*. Unpublished PhD thesis, University College Cardiff, 425 pp.

Vai, G.B. & Ricci-Lucchi, F. (1977) Algal crusts, autochthonous and clastic gypsum in a cannibalistic evaporite basin: a case history from the Messinian of north Appenines. *Sedimentology* **24**, 211–244.

Vail, P.R. & Todd, R.G. (1981) Northern North Sea Jurassic unconformities, chronostratigraphy and sea-level changes from seismic stratigraphy. In: *Petroleum Geology of the Continental Shelf of North-West Europe* (Eds Illing, L.V. & Hobson, G.) pp. 216–235. Heyden, London.

Vail, P.R., Mitchum, R.M., Jr., Todd, R.G., Widmier, J.M., Thompson, S. III, Sangree, J.B., Bubb, J.N. & Hatlelid, W.G. (1977) Seismic stratigraphy and global changes of sea level: application of seismic reflection configuration to stratigraphic interpretations. In: *Seismic Stratigraphy – Applications to Hydrocarbon Exploration* (Ed. Payton, C.E.). Mem., Am. Assoc. Petrol. Geol. 26, 49–212.

Vail, P.R., Hardenbol, J. & Todd, R.G. (1984) Jurassic unconformities, chronostratigraphy and sea-level changes from seismic stratigraphy and biostratigraphy. In: *Interregional Unconformities and Hydrocarbon Accumulation* (Ed. Schlee, J.S.). Mem., Am. Assoc. Petrol. Geol. 36, 129–144.

Vail, P.R., Audemarch, F., Bartek, L.R., Bowman, S., Coterill, K., Emmet, P.A., Liu, G., Perez, G., Ross, M.R. & Wu, S. (1989) The global stratigraphic signature of the Neogene. In: *Abstracts Sea Level Changes at Active Plate Margins* (Ed. Macdonald, D.I.M.). Geological Society, London.

Ward, L.C. & Robertson, A.H.F. (1987) Structural and sedimentological development of the Miocene-Recent Polis extensional graben, N.W. Cyprus. In: *Abstracts Troodos '87, Ophiolites and Oceanic Lithosphere, Symposium Nicosia*, p. 45.

Weiler, Y. (1965) *The folded Kythrea Flysch in Cyprus*. Unpublished PhD thesis, The Hebrew University, Jerusalem, 86 pp.

Whitmarsh, R.B. (1974) *Initial Reports of the Deep Sea Drilling Project*, Vol. 23. US Government Printing Office, Washington, DC.

Zomenis, S.L. (1972) Stratigraphy and hydrogeology of the Neogene rocks in the northern foothills of the Troodos massif. *Bull. Geol. Surv. Dept., Cyprus* **5**, 22–90.

# Foreland, Foredeep and Cratonic Basins

*Spec. Publs. int. Ass. Sediment.* (1991) **12**, 373–390

# Miocene depositional sequences within a tectonically controlled transgressive–regressive cycle

PHILIPPE CRUMEYROLLE[†], JEAN-LOUP RUBINO[†] *and* GEORGES CLAUZON[‡]

[†] *TOTAL-CFP Centre Scientifique et Technique, Domaine de Beauplan, Route de Versailles, 78470 St Rémy-les-chevreuses, France; and* [‡] *Institut de Géographie, Université d'Aix-Marseille, Avenue R. Schumann, 13100 Aix en Provence, France*

## ABSTRACT

The Digne foreland basin is an elongate depression covering 1500 km$^2$. This triangular basin is bounded to the east and northeast by the Digne thrust sheet and to the west by the Durance Fault. The basin contains an assymetrical infill of marine and continental deposits, which have been termed Molasse. This Molasse is mainly Miocene and shows maximum thickness at the front of a basement-involved thrust system. The Molasse succession forms a large-scale structurally controlled transgressive–regressive cycle, which is punctuated by numerous unconformities. The unconformities bounding the major units within this Molasse correspond to depositional sequence boundaries. Most of them are thought to be eustatic in origin.

A variety of syntectonic features, such as progressive onlaps, wedging, and draping stratal patterns, are recognized. These are easily recognized on aerial photographs, and in outcrops.

Two main marine molasse units have been distinguished: the Lower and Upper Marine Molasse.

The Lower Marine Molasse is Burdigalian in age; it has been subdivided in turn into two depositional sequences. Biostratigraphic data and comparison with the sequences located in the Rhone Valley further west allow us to correlate these sequences with the TB2-1–TB2-2 eustatic cycles. The correlations proposed are tentative for the other depositional sequences of the Digne Basin, owing to paucity of biostratigraphic data. Nevertheless, good correlations with the Rhodanian Basin are established for the Upper Marine Molasse and the base of Yellow Molasse. These units probably correspond to the TB2-3–TB2-4 cycles.

The uppermost part of the Miocene deposits are not dated and the correlation proposed may be subject to modification if new biostratigraphic data are found.

The application of sequence-stratigraphic concepts within the context of an active margin, as illustrated by the Digne Basin, has allowed a better understanding of the geometry and patterns of the Miocene sedimentary cycle.

Transgression at the base of the first Burdigalian sequence is attributed to the combined effects of load-induced flexural subsidence and eustatic sea-level rise. As transgression progressed, a wave-dominated transgressive systems tract accumulated thick deposits on the more subsiding margin. This systems tract shows a landward stacking pattern of parasequences. The transgressive maxima are marked by glauconitic levels. When sediment shed by the rising topographic relief filled the accommodation space created by subsidence, successive highstand successions, bounded by lowstand unconformities, were laid down. This overall prograding evolution leads to the complete infill of the basin. The Miocene Digne basin fill is thus considered to be a tectonically controlled megacycle punctuated by eustatic unconformities.

## INTRODUCTION

Recent developments of sequence stratigraphy focus on the interpretation of unconformities and sedimentary cycles in terms of eustatism (Vail *et al.*, 1987). Most of these models are derived from the study of passive continental margins where tectonic activity was weak. The majority of the transgressive–regressive cycles in this context are related to global sea-level fluctuations (Vail *et al.*, 1987). However, as noticed by Mutti *et al.* (1988), the applicability of the above concepts to the study of convergent continental margins, and more generally to the sedimentary fill within tectonically active basins, has remained largely unexplored. The episodic influx of clastics in foreland basins is generally considered to be the result of either hinterland uplift or motion in the thrust zone. In this paper, we describe and discuss the origin of depositional sequences within the Digne foreland basin. We wanted to examine whether or not eustatic effects were overwhelmed by sedimentation and tectonic subsidence rates in the context of strong structural activity. The mapping and dating of unconformities within the Digne Basin and their more regional correlation across the Burdigalian seaway suggest a close link with the global cycle chart of Haq *et al.* (1987), and suggest that eustatic events are also recorded in foreland basin settings.

## METHODS

The Miocene molasse succession of the Digne Basin was studied along a strip 2 km wide and 25 km long, located on the northern limb of the basin.

Outcrop continuity allowed detailed mapping of strata between the different structural subunits in which the Tertiary molasse is involved. This study is based mainly on field data, providing more than 4 km of detailed sections and supplemented with aerial photographs. In addition, published and unpublished data on biostratigraphy and structural geology were used. The stratigraphic interpretation is based on the recognition of depositional sequences. Depositional sequences are understood here in the sense of Vail (Vail *et al.*, 1987; Posamentier & Vail, 1988); they correspond to groups of genetically related strata bounded by physical surfaces of basin-wide extent and with chronostratigraphic significance. Aerial photographs were particularly useful in investigating the complex stratal patterns observed on the field. This approach has been emphasized recently by Mutti *et al.* (1988) in the study of the Eocene South Pyrenean basin.

## GEOLOGICAL SETTING

### General structural framework

The Digne Miocene basin is an elongate depression covering an area of 1500 km$^2$: it is part of the Burdigalian seaway of the North Alpine foreland basin (Haug, 1891; Elliott *et al.*, 1985; Fig. 1). The North Alpine foreland basin comprises a suite of sub-basins, such as the peri-Alpine Digne Basin, or the Swiss molasse basin (Allen & Homewood, 1985), which are bounded on their active margin by a thrust belt. In this case, the Penninic frontal thrust resulted from the creation of the Alpine orogenic suture. Externally, foreland basins are bounded by a more stable zone, the Mesozoic 'sub-Alpine basin', representing part of the uplifted north Tethyan margin. In the southwest Alps, foreland basins occur adjacent to the Alpine chain, as a result of plate collision (Apulia and Eurasia) and associated lithospheric loading. This category of flexural basins was first recognized by Price (1973) and Beaumont (1981) in the Rocky Mountains. In the southwest Alps, the zone of maximum subsidence migrated westward in advance of deformation (Stanley, 1962; Elliott *et al.*, 1985; Fig. 1). Three main basins occurred, progressively younger toward the west: the 'Grès d'Annot' Basin, with deep-water turbiditic deposits (Stanley, 1962), is late Eocene to early Oligocene; the Barreme Basin, with shelf depositional systems (Elliott *et al.*, 1985; Apps *et al.*, 1985), is mainly Oligocene; the Digne Basin is late Oligocene to latest Tortonian and comprises a wide range of continental to shallow-marine facies known as Molasse (Goguel, 1936; de Lapparent, 1938; Gigot *et al.*, 1974; Beaudoin *et al.*, 1975).

Shallower facies therefore dominate basin-fill toward the west. Eastwards, the turbiditic 'Grès d'Annot' Basin can be interpreted as a foredeep related to the first stage of the collision; all the others are more typical molasse basins.

### Structural framework of the study

The most extensive outcrops are located on the northern limb of the Digne Basin (Fig. 2). In this area, the Digne Basin is bounded to the east by the Digne thrust sheet. The sedimentary infill is disrupted by a series of basement-involved thrusts associated with strike-slip movement (Goguel, 1963; Gigot *et al.*, 1974; Gidon & Pairis, 1988). These thrusts allow us to differentiate three main structural

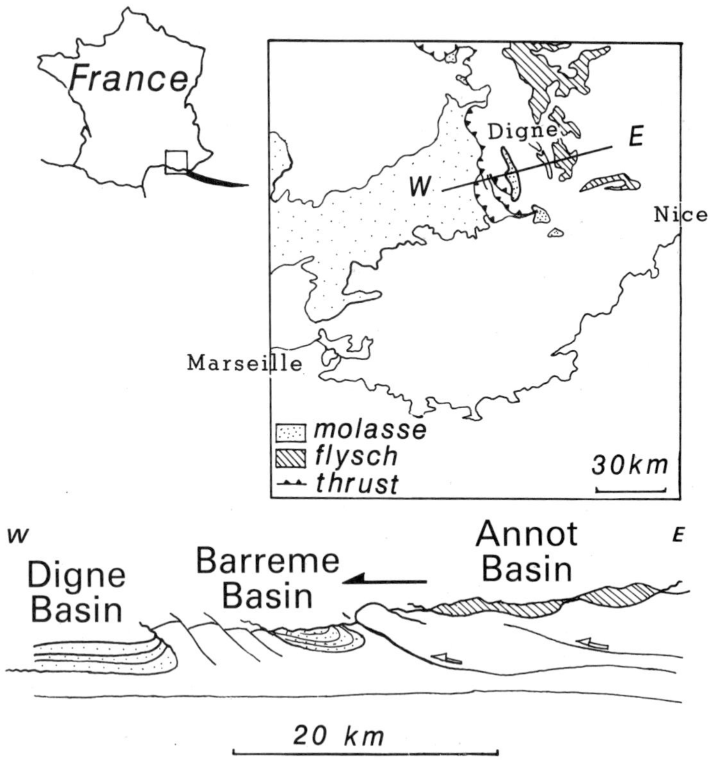


**Fig. 1.** Location map of the Digne Basin and schematic cross-section of the south Alpine foreland zone. Annot Basin, Barreme Basin, and Digne Basin occur successively in westward positions (modified from Elliott *et al.*, 1985).

subunits (Gigot *et al.*, 1974): the 'Nappe de Thuriers', the 'Melan-Clamensane', and the 'plateau de Chine' (Fig. 2). More to the south, in the centre of the basin, drape folds (the Mirabeau, Prefaisal, and Gaubert anticlines) with blind thrusts occur, having a NW–SE direction. These are interpreted as a result of the reactivation of E–W-trending 'Pyreneo-provençal' structures by later Alpine thrusts (Dubois & Curnelle, 1978; Siddans, 1979; Evans, 1985; Fig. 3). On the western margin, the Digne Basin is bounded by a reactivated lineament: the strike-slip Durance Fault. This fault is thought to be left lateral (Dubois & Curnelle, 1978; Rousset, 1981). West of this fault, the Forcalquier-Apt Miocene sub-basin is affected by E–W folds related to the Late Cretaceous to early Eocene 'Pyreneo-provençal' folding phase (Fig. 3). The structural geology and the gross sedimentary evolution of this part of the Digne Basin are given in Gigot & Haccard (1972) and Gigot *et al.* (1974). They provided a detailed three-dimensional reconstruction of the structure of the northern limb of the Digne Basin.

A well-known geological landmark in this area is the complex fold structure called the 'Velodrome' (bicycle racing track), because of its shape (Fig. 4). Within this structure a wide range of synsedimentary wedging stratal patterns and progressive marginal onlaps have been recognized; good outcrop allows us to reconstruct in detail the synsedimentary evolution of this complex northern limb of the Digne Basin in detail (Figs 5, 6). Similar syntectonic intraformational unconformities have been described by Anadon *et al.* (1986) in the Paleogene Ebro basin (Spain). Progressive and angular unconformities may have a regional tectonic significance in some cases but commonly record the occurrence of rapid localized uplift contemporaneously with sedimentation.

## THE CLASTIC INFILL OF THE DIGNE BASIN

### Regional lithostratigraphic subdivisions

The Digne Miocene succession is more than 2000 m thick and comprises a wide range of clastic sediments (Fig. 7). We have recognized five main molasse units, which are lower to upper Miocene (see later

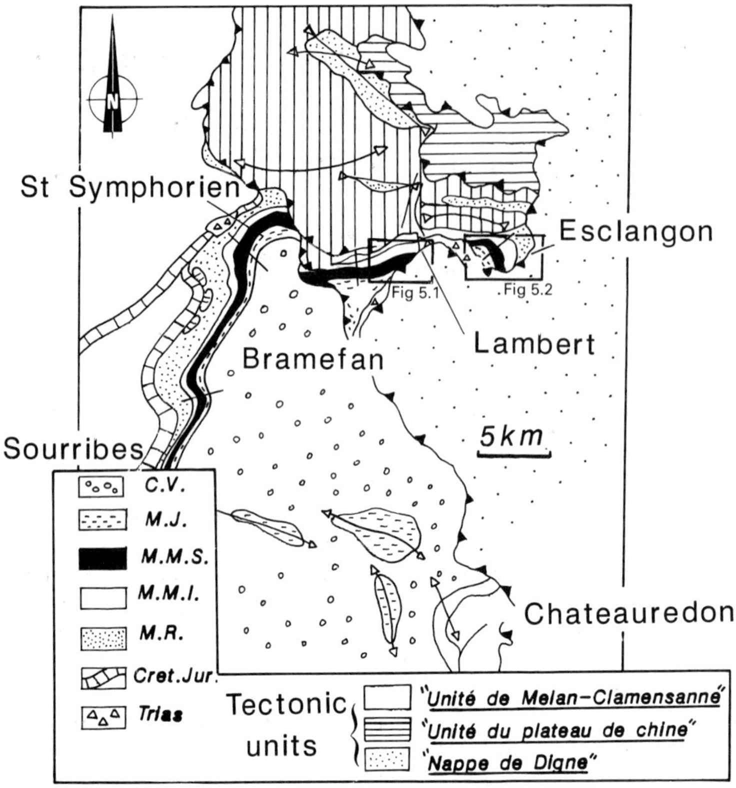


**Fig. 2.** Geological map of the Digne Basin. The Miocene deposits are involved in three main structural subunits in front of the Digne thrust sheet (modified from Gigot *et al.*, 1974).

biostratigraphic data). Each unit corresponds more or less to conventional stratigraphic units used in the literature (Beaudoin *et al.*, 1970,1975; Gigot *et al.*, 1974). These conventional units are bounded by abrupt facies changes, unconformities, subaerial or subaqueous erosional features, and the five units given here are depositional sequences in the sense of Vail *et al.* (1987).

In ascending order, the main Digne Basin units are as follows (Fig. 8).

**1** The Oligocene 'Red Molasse', which consists of an assemblage of fluvial channels and channelized alluvial fan conglomeratic bodies; chaotic facies also occur mainly at the base of the succession (Gigot

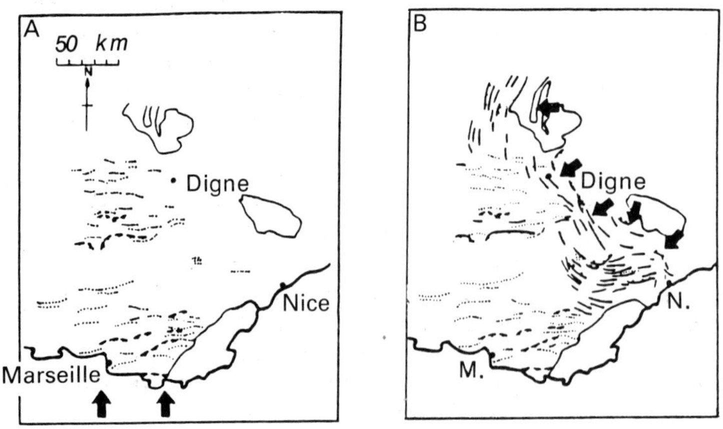


**Fig. 3.** Interpretation of the development of the arcuate structural geometries in southeast France (from Siddans, 1979, p. 123). (A) Pre-Nummulitique 'Pyreneo-provencal' phase. (B) End Miocene Alpine phase. Dotted lines, axial traces of Pyrenean provençal phase folds; broken lines, traces of Pyrenean provençal phase thrusts; solid lines, axial traces of main Alpine phase folds and thrusts.

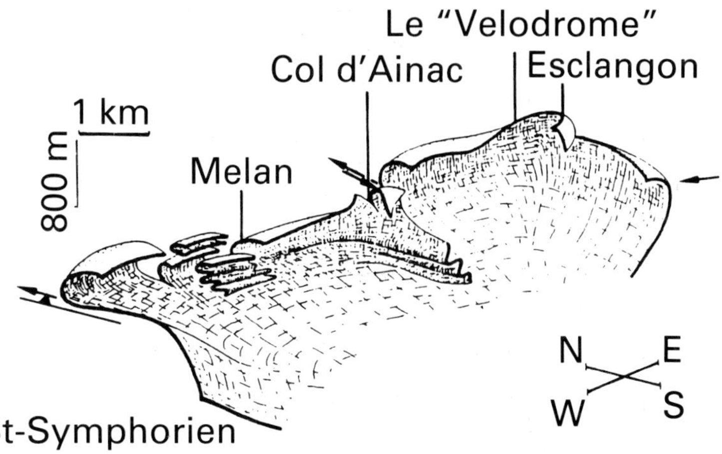


**Fig. 4.** Geometric reconstruction of the northern folded limb of the Digne Basin from Esclangon to St Symphorien. The large folded structure near Esclangon is called the 'Velodrome' (the bicycle racing track).

*et al.*, 1974). These thick deposits were not studied in detail.

**2** The Intermediate Molasse is represented by a relatively thin succession of fluvial to deltaic sandstones interbedded with brackish marls. The brackish marls have provided a rich assemblage of Miocene benthic Foraminifera, related to minor transgressive intervals. This is the most problematic unit; depending on its age, we can interpret it as either an individual depositional sequence, or part of the lower Burdigalian transgressive systems tracts.

This unit is present only in the thicker, eastern part of the basin (i.e. to the east)

**3** The Lower Marine Molasse is Burdigalian in age and attains a thickness of 800 m near Esclangon (Fig. 8). It conformably overlies the Intermediate Molasse strata. It comprises a succession of transgressive wave-dominated deltaic sequences that pass vertically to glauconitic levels. A sharp-based bioclastic bar also occurs in this deltaic succession and allows us to distinguish two depositional sequences within this unit (see biostratigraphic data below).

**4** The Upper Marine Molasse, although not dated directly, is supposed to be Langhian because it is capped by Serravallian deposits. It consists of a succession of wave and tide-dominated sand bodies interbedded with marls. These sandstone bodies show a wide range of spectacularly exposed sedimentary structures (Beaudoin *et al.*, 1970; Tessier & Gigot, 1989). These intertidal deposits grade into tidal inlet and littoral bar to shoreface sandstones.

**5** The Yellow Molasse reaches a thickness of 250 m and is dated as Serravallian. It forms a prograding succession of channelized conglomerates that grade westward into alluvial plain and shallow-marine deposits. Upward these deposits are unconformably overlain by the Valensole conglomerates, comprising more than 1500 m of massive channel-fill conglomerates. The Yellow Molasse deposits were previously included within the Valensole Formation (Beaudoin *et al.*, 1966; Clauzon *et al.*, 1987,1989). The Valensole conglomerates are interpreted as an alluvial fan system prograding southward (Mercier, 1979; Dubar, 1984), progressively infilling the entire Digne Basin area.

Recent biostratigraphic data have shown that the Valensole conglomerates are mainly Tortonian at the base (Clauzon *et al.*, 1987,1989) and reach the upper Pliocene (Dubar, 1984). A spectacular internal erosional unconformity, related to a base-level fall (Clauzon, 1979; Dubar, 1984) occurred during the Messinian salinity crisis.

A very large-scale infilled erosional feature occurs near Esclangon (see cross-section), filled by giant olistoliths dispersed in floodplain marls (Fig. 8). The Yellow Molasse is deeply incised by this fluvial palaeovalley. This feature could be equivalent to the base of Valensole conglomerates but this cannot be proved in the field. Moreover the scarcity of fauna does not allow this unconformity to be precisely dated.

In the south-west Alps, as sea level rose during Burdigalian time, the sea occupied an elongated gulf (Fig. 9; de Lapparent, 1938; Demarcq, 1970,1984). This seaway, the 'Rhodano-provençal' gulf (Demarco, 1970), was dissected by emergent areas and adjacent sub-basins. The Digne Basin forms a large embayment connected with the 'Rhodano-provençal' gulf (Fig. 9). A complex folded and faulted framework resulted from the reactivation of

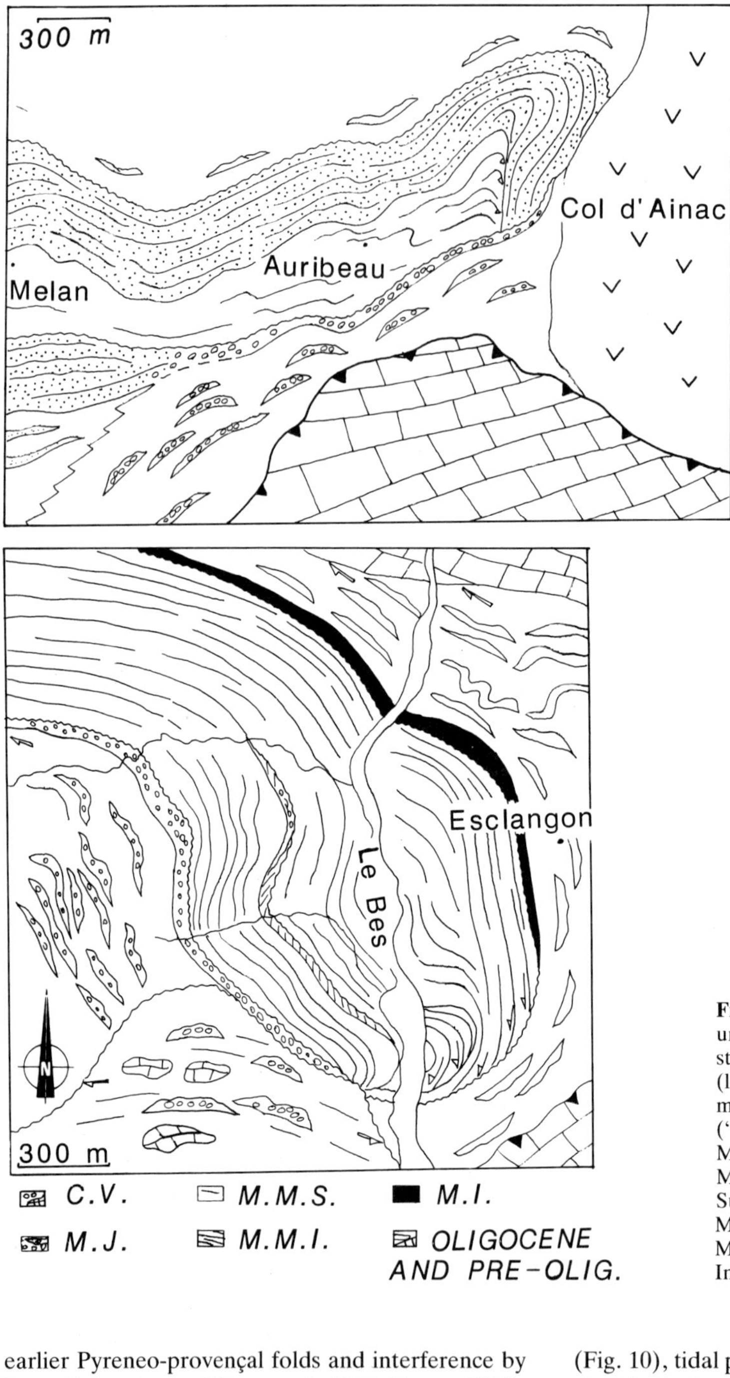


**Fig. 5.** Detailed geological map showing unconformities and associated, wedging stratal patterns within the 'Velodrome' (lower map) and near 'col d'Ainac' (upper map). C.V., Valensole conglomerate ('conglomérat de Valensole'); M.J., Yellow Molasse ('Molasse Jaune'); M.M.S., Upper Marine Molasse ('Molasse Marine Supérieure'); M.M.I., Lower Marine Molasse ('Molasse Marine Inférieure'); M.I., Intermediate Molasse ('Molasse Intermédiaire').

earlier Pyreneo-provençal folds and interference by later Alpine thrusts (Gigot *et al.*, 1974; Evans, 1985).

During the Burdigalian transgression, wave-dominated deltaic systems were built in the Digne embayment. During Langhian and Serravallian time, as shown on the palaeogeographic schemes (Fig. 10), tidal processes became predominant. Sand shed from the rising relief was reworked into shoal and tidal inlet systems. During Serravallian to Tortonian time, alluvial fan systems prograded (Clauzon *et al.*, 1987), leading to the complete infill of the basin. As these alluvial fans prograded,

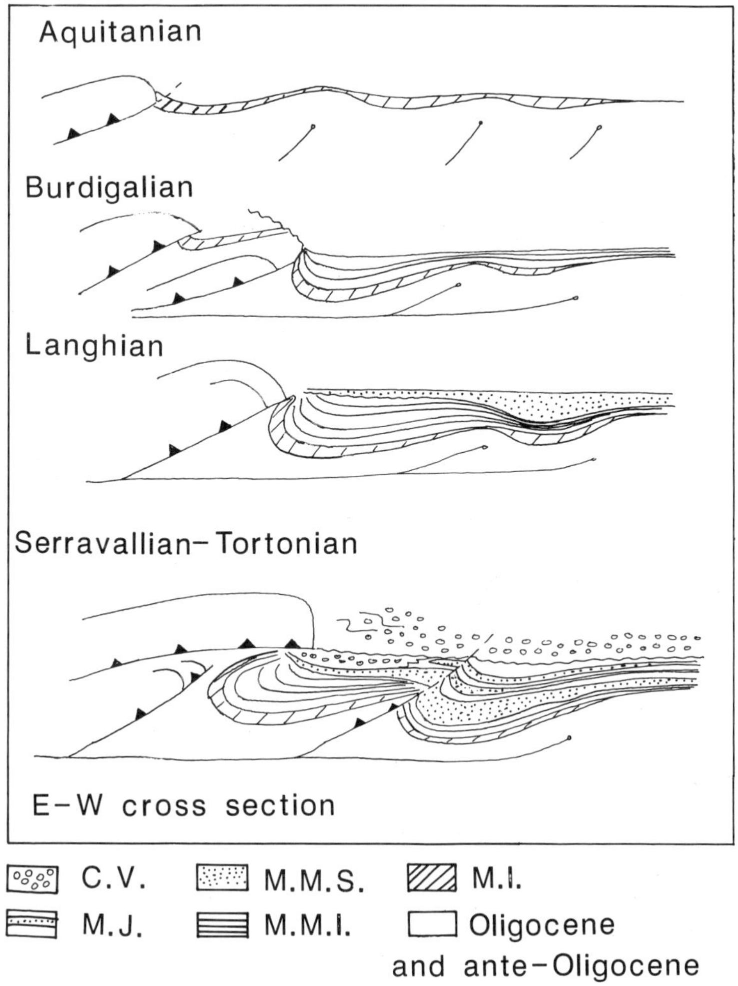


**Fig. 6.** Highly simplified structural evolution of the northern limb of the Digne Basin along an E–W cross-section during the Miocene.

the strike-slip Durance Fault played a key role in narrowing of the basin.

### Biostratigraphic data

The biostratigraphic framework of the Tertiary Molasse of the Digne Basin is based on numerous faunal groups, including Foraminifera, calcareous nanoplankton, micromammals, and Pelecypods.

The first level firmly dated indicates a general Miocene age for the Intermediate Molasse on the basis of *Ammonia beccarii* (Linné) in the marls. Another, more precise, age is given by *Riteneria manca* (Stehlin, Schaub) in the Lambert section of the Intermediate Molasse. This micromammal species was described by Gigot & Mein (1973) and belongs to the Laugnac zone. The same fauna has been found at the base of the Lower Marine Molasse in the Esclangon section. The Laugnac zone, previously attributed to the Aquitanian (Mein, 1975; Aguilar, 1981), has recently been considered as early Burdigalian (Wallez *et al.*, 1986; Gourinard *et al.*, 1987). Two other levels located within the first depositional sequence have yielded *Helicosphera ampliaperta* (Bramlette and Wilcoxson), a nanoplankton fauna that clearly indicates a Burdigalian age. A Burdigalian fauna is present throughout the second depositional sequence above the bioclastic bar deposit. A late Burdigalian age is indicated by *Chlamys praescabriuscula* (Fontannes) found in different sections located on the western margin of the basin (de Lapparent, 1938) and within the second

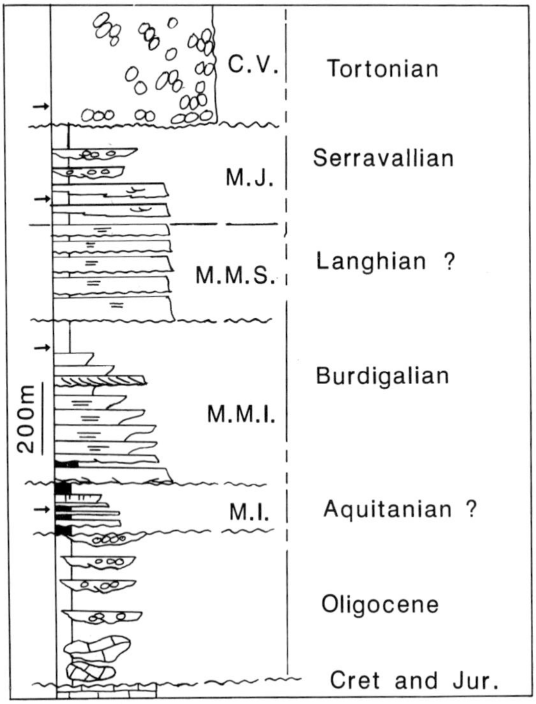


**Fig. 7.** Composite section of the Digne Molasse succession: arrows indicate point of dating. (See Fig. 5 for abbreviations.)

depositional sequence of Lower Marine Molasse. Higher up, at the base and top of the Yellow Molasse, dates are based on micromammal fossils (Clauzon *et al.*, 1987,1989). A Langhian to earliest Serravallian age is given by *Megacricetodon* aff. *bavaricus* (Fallbusch). *Megacricetodon* aff. *germanicus* (Aguilar) is present higher up and indicates a Serravallian age.

## DEPOSITIONAL SEQUENCES

### The Lower Marine Molasse depositional sequences

In the eastern and central part of the study area, the Lower Marine Molasse constitutes an easily recognizable stratigraphic unit, characterized by wave-dominated deltaic sandstone bodies, which pass laterally and vertically to offshore marine marls, together with less fossiliferous blue marls. Detailed stratigraphic and sedimentological data show that the Lower Marine Molasse can be subdivided into two depositional sequences (Fig. 12). From east to west, these two depositional sequences show a decrease in thickness associated with an increase of marls content.

*The Esclangon type section* (Figs 11,12)

Along the Bes Valley road, the Lower Marine Molasse is continously exposed. It reaches a maximum thickness of 800 m.

In this section, the Lower Marine Molasse overlies the Intermediate Molasse, which lies conformably on the fluvial Oligocene Red Molasse. No Aquitanian marine deposits are known between the present Mediterranean coastline and the Alpine front, although Lalaï (1986) has suggested that the Aquitanian Sea reached the Alpine front. Consequently, the Intermediate Molasse is more probably the base of the Burdigalian transgressive systems tract. The significant facies change that occurs at the base of the Lower Marine Molasse can be interpreted as a 'ravinement surface' (Vail & Sangree, 1988) related to beach migration above lagoonal deposits.

The base of the Lower Marine Molasse consists of stacked wave-dominated parasequences (Fig. 13). The observed landward-stepping retrograding stacking pattern (van Wagoner *et al.*, 1988) is characteristic of a transgressive systems tracts. Parasequences are 20–30 m thick; each parasequence forms an assymetrical shallowing – deepening upwards cycle of marine facies (Fig. 13).

From base to top, each parasequence consists of marine bioturbated marls that grade upward into hummocky cross-stratified and bioturbated fine sands. These are capped by massive fine to coarse, wave rippled and planar parallel-laminated sands with oyster accumulations. This facies is abruptly overlain by offshore marls. Intertidal deposits, characterized by fine-grained, laminated or rippled sands with bird tracks (Beaudoin *et al.*, 1970), sometimes occur at the top of parasequences and grade progressively into offshore marls. Each parasequence thus forms an assymetrical pattern due to small-scale relative sea-level fluctuations superimposed on a large scale. This is expressed as landward back-stepping of parasequences associated with a decrease in coarse sand and conglomerate content. Laterally in the St Symphorien area, a thinner shelfal marly succession is equivalent to this wave-dominated deltaic system. This transgressive succession is capped by a series of thinner strata that consist of fine to medium glauconitic burrowed sand

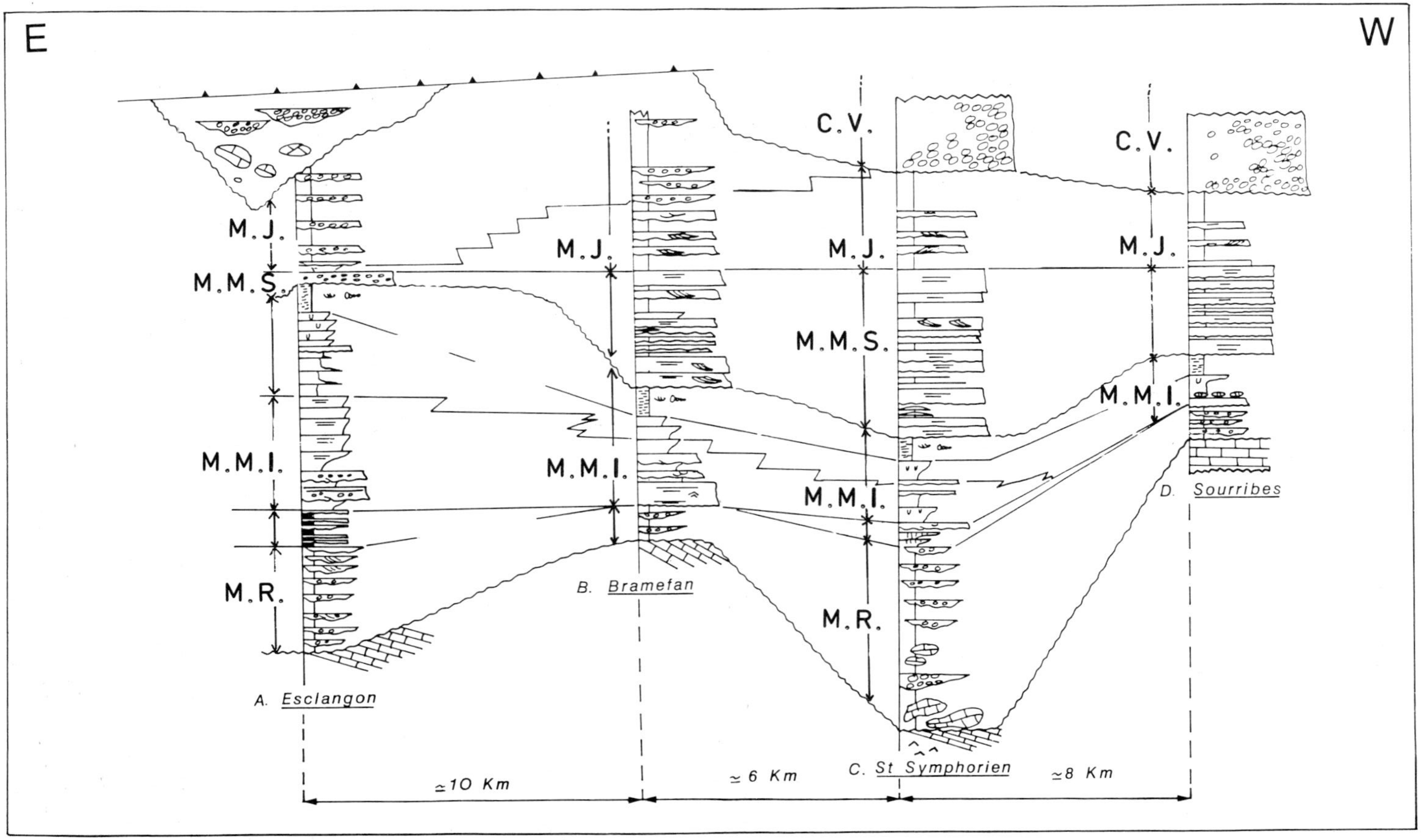


**Fig. 8.** Stratigraphic correlations and type sections in the Miocene Digne Basin. C.V., Valensole conglomerate; M.J., Yellow Molasse; M.M.S., Upper Marine Molasse; M.M.I., Lower Marine Molasse; M.I., Intermediate Molasse; M.R., Red Oligocene Molasse. The erosive feature in the left upper part of the scheme is the incised valley filled with olistoliths and red marls capped by the Digne thrust sheet.

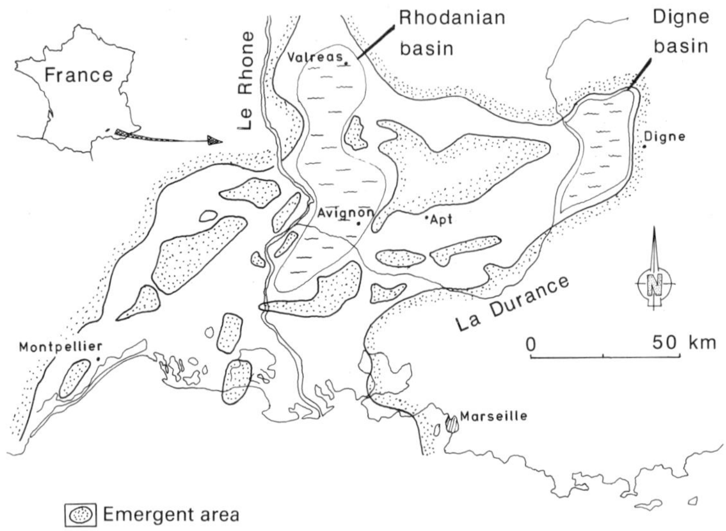


**Fig. 9.** Palaeogeographic scheme for the Burdigalian in the Digne and Rhodanian areas (from Demarcq, 1970).

and which are considered to represent a condensed section.

Condensed sections occur at the maximum stage of transgression when shelf starvation enhances mineral authigenesis (Loutit *et al.*, 1988). The glauconitic strata progressively onlap on to the top of the transgressive sandy systems tract within the 'Velodrome structure' (Fig. 5). More to the west, near Sourribes village, their equivalent is a glauconitized boulder level overlying the fluvial Oligocene Red Molasse with a sharp contact. Boulders are bored by phollads and covered with oysters. Glauconite occurs as a crust on the boulders or dispersed in silty sediment. Upward in the same succession these glauconitic levels pass to a 50 m thick, shallowing upward succession of deltaic parasequences (prodeltaic setting). They show highly burrowed sand and silt without sedimentary structures. These successive progradational events represent the highstand systems tract of the first Burdigalian depositional sequence.

### *The second Lower Marine Molasse depositional sequence*

In the Esclangon section, a bioclastic tidal bar deposit lies just above the highstand succession of the first depositional sequence (Fig. 12). The base of this bioclastic bar is a sharp erosion surface associated with a conglomeratic lag deposit. These deposits are interpreted as the transgressive systems tract of the second depositional sequence. They overlie the subaerial lowstand unconformity. The bioclastic transgressive deposits are abruptly overlain by shelfal mudstones through a distinct surface of maximum flooding expressed by a wave-reworked conglomeratic pavement. Upward, these deposits are capped by a prograding shallowing upwards succession more than 200 m thick. It consists of numerous stacked deltaic parasequences; these represent the early highstand succession of the second depositional sequence. The late highstand tract consists of brackish lagoonal marls with oysters and pelecypods. This unit is also present on the St Symphorien section. It also occurs on the Bramefan and Sourribes sections.

### *Discussion*

The two depositional sequences recognized within the Lower Marine Molasse are thickest in front of the thrust-loaded margin of the basin. Transgression at the base of the first sequence is attributed to the combined effects of the early Burdigalian sea-level rise together with load-induced flexural subsidence. The relative importance of each of these is not known.

On the less-subsiding western margin, the thick transgressive systems tract of the first sequence is entirely condensed (Fig. 14; see also Sourribes section, Fig. 8). It appears that local thrust movement together with regional load-induced flexural

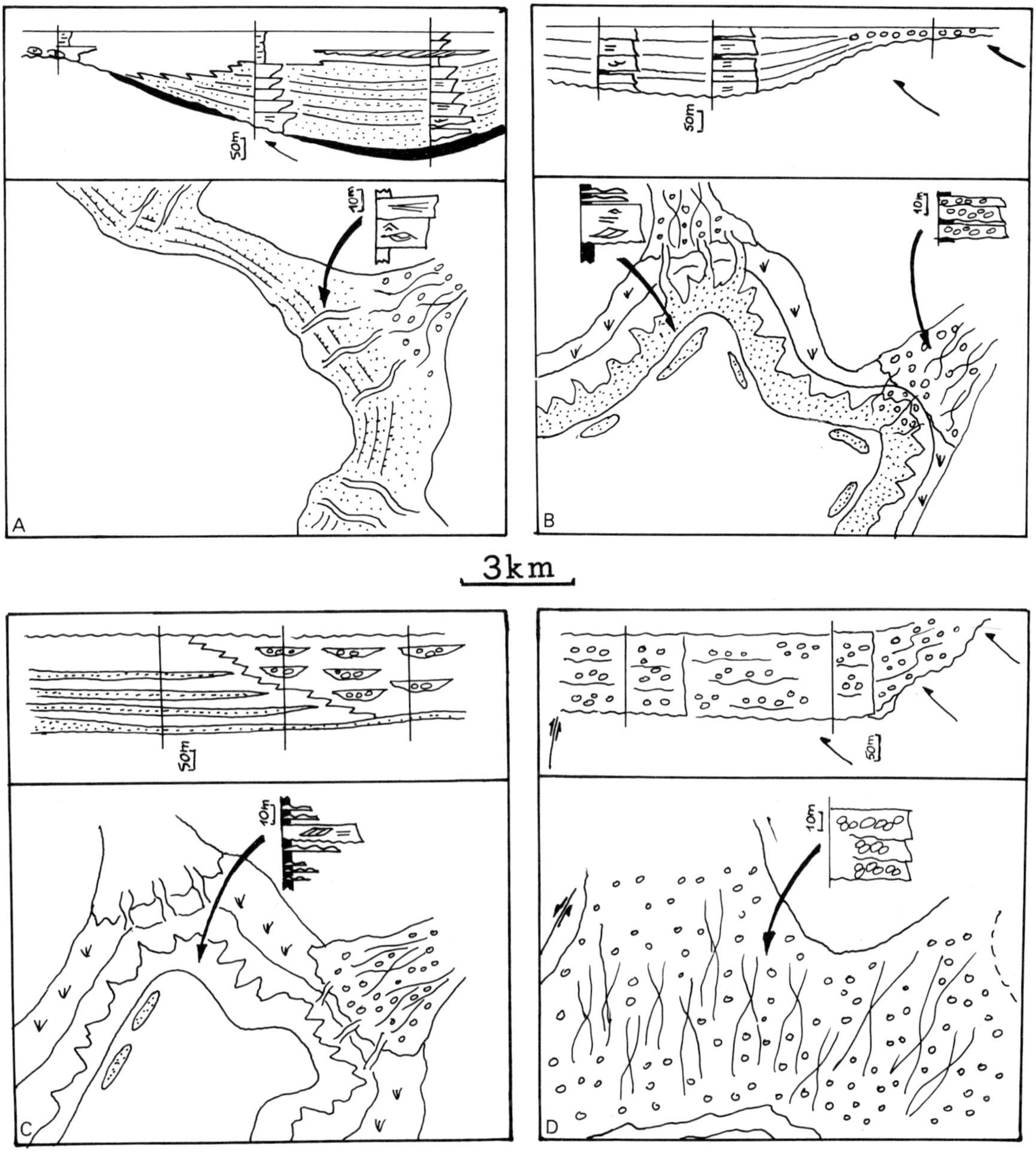


**Fig. 10.** Simplified palaeogeographic schemes and E–W cross-sections of the main molasse successions. (A) Burdigalian: Lower Marine Molasse; (B) Langhian: Upper Marine Molasse; (C) Serravallian: Yellow Molasse; (D) Tortonian Valensole conglomerates.

subsidence play a key role in the control of the gross geometry and thickness of the main systems tracts (Figs 11,14).

Available biostratigraphic data allow a tentative correlation between these two depositional sequences in the Lower Marine Molasse recognized here and to eustatic cycles TB2-1 and TB2-2 of Haq *et al.* (1987).

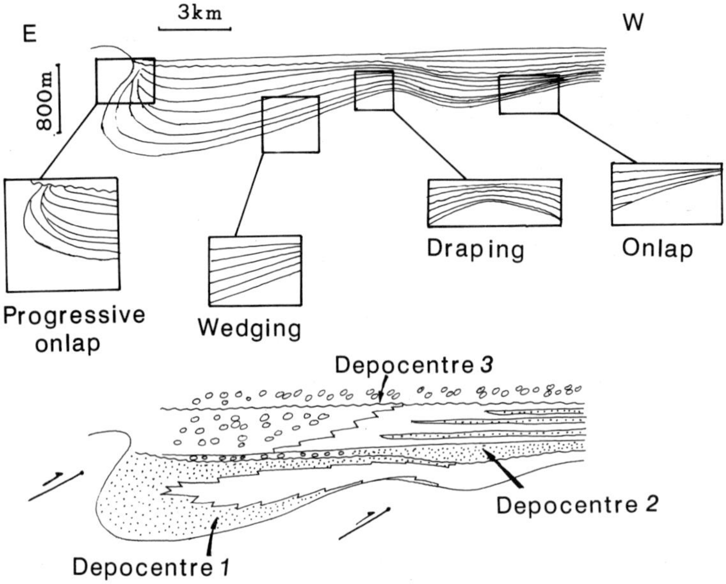


**Fig. 11.** Simplified model of stratal patterns and depocentre migration along an E–W cross-section within the Digne Basin.

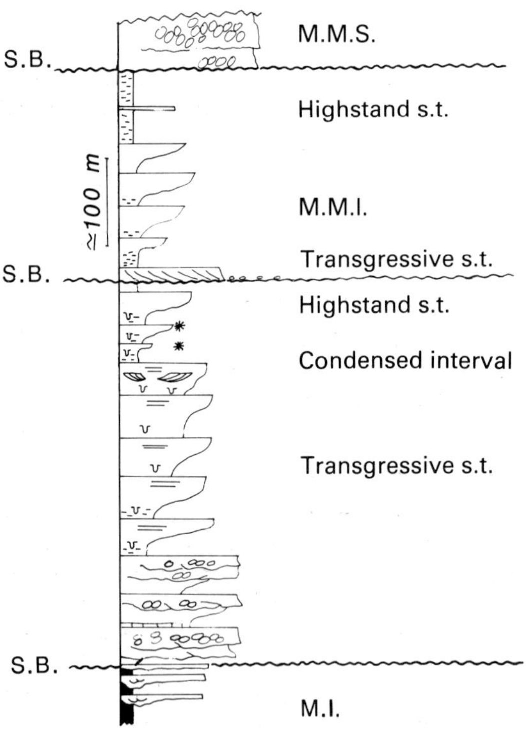


**Fig. 12.** Lower Marine Molasse type section at Esclangon. S.B., sequence boundary; M.M.S., Upper Marine Molasse; M.I., Intermediate Molasse; M.M.I., Lower Marine Molasse; s.t., systems tract.

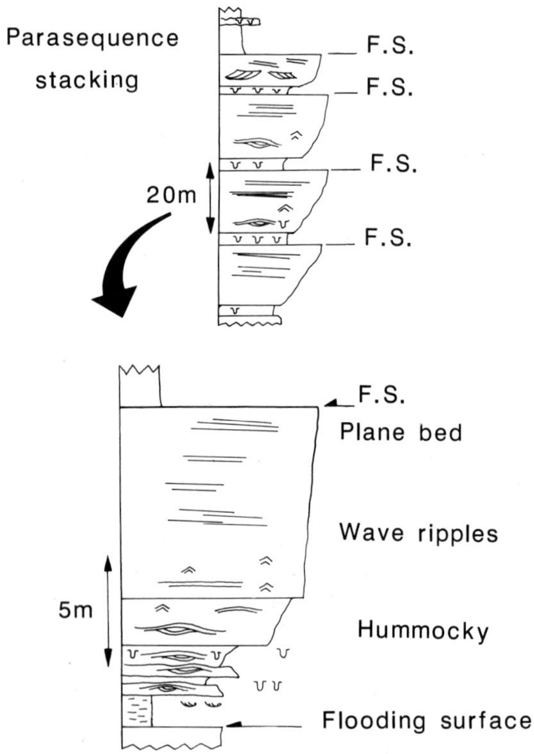


**Fig. 13.** Example of a wave-dominated deltaic parasequence within a transgressive systems tract of the first Lower Marine Molasse depositional sequence. F.S., flooding surface.

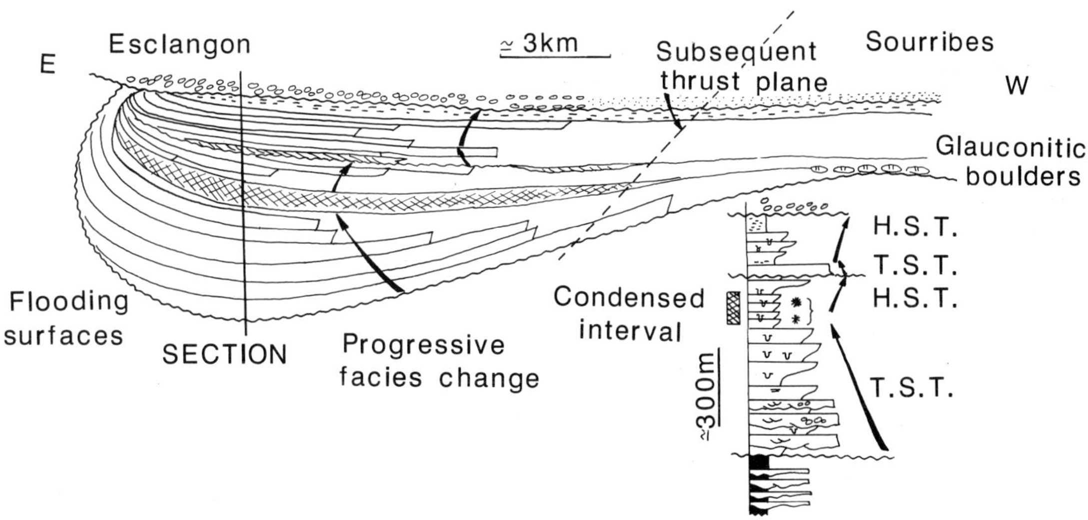


**Fig. 14.** Type section and geometric model of the Lower Marine Molasse depositional sequences. Notice the asymmetrical shape of the sequences in front of the eastern loaded margin of the basin. H.S.T., Highstand Systems Tract; T.S.T., Transgressive Systems Tract.

**The Upper Marine Molasse depositional sequence**

In the western part of the study area (Fig. 2) the Upper Marine Molasse forms a distinctive stratigraphic unit, bounded at the base by an erosive surface associated with an abrupt facies change. This Molasse unit consists of a stack of tabular sand bodies, each interpreted as subtidal, followed by intertidal inlet sequences.

*The St Symphorien type section* (Fig. 8)

Near St Symphorien village, the Upper Marine Molasse is continously exposed over a thickness of about 800 m. The base of the succession is conformable, sharp, and slightly erosive. Within the succession a wide range of well-preserved tide and wave-formed sedimentary structures are found, which have been described by Elliott *et al.* (1985). There are neap-spring bundled cross-beds, flaser to wavy ripple-beeding, plane beds, and wave ripples. More marine intervals are observed and comprise bioturbated silts and sands rich in macrofauna. Unfortunately this unit has no biostratigraphic markers.

Farther to the east, in the 'Velodrome' area, the Upper Marine Molasse is only represented by a 30 m thick bar composed of sandstones and conglomerates. These are partly subaerial deposits, interpreted as an alluvial fan to fan-delta depositional system. The entire Upper Marine Molasse is rather poorly defined as a sedimentary cycle, but is considered as a highstand succession in view of sedimentological data and regional correlations.

**The Yellow Molasse depositional sequence**

The Yellow Molasse is an overall prograding succession of conglomerates and sandstones interbedded with marls.

*The Esclangon section* (Fig. 8)

In the eastern part of the study area, the succession is characterized by channelized conglomerates and coarse sand; these are interbedded with red oxidized marls attesting subaerial depositional conditions. These deposits were laid down in a distal alluvial fan setting. Transgressive bioturbated silts and marls with oysters and gastropods occur at the base of the succession just above the conglomerate, which is stratigraphically equivalent to the Upper Marine Molasse. Micromammal teeth found just above these transgressive deposits indicate a late Langhian to earliest Serravallian age.

*The St Symphorien section*

More to the west, around Bramefan and St Symphorien (Fig. 8), the Yellow Molasse is characterized by lagoonal marls and tidal inlet sand bodies. These sand bodies are organized similarly to the tidal inlet deposits described from the Upper Marine Molasse.

*High-frequency cycles within the Yellow Molasse depositional sequence* (Fig. 15)

Continental facies intercalated with lagoonal and tidal inlet facies reveal transgressive–regressive cycles 10–40 m thick. Fluvial deposits consist of coarse-grained tabular decimetre-scale beds interbedded with red oxidized marls showing root marks. These deposits are interpreted as alluvial fan overbank deposits. There is an overall thickening upwards and thinning upwards pattern of strata. These continental deposits are interbedded with tidal inlet deposits bounded both at the top and at the base by lagoonal marls. Together these fluvial to shallow-marine deposits form symmetrical facies cycles. These cycles are significantly different from the asymmetrical shallowing upwards parasequences described above from the Upper Marine Molasse.

Deltaic asymmetrical cycles, which are similar to the parasequence of van Wagoner (1985; van Wagoner *et al.*, 1988), are generally interpreted as a response to a rapid stepwise sea-level rise. However, Cross (1988) has recently shown that irregular variations in sea-level or sudden tectonic movements are not necessary to produce asymmetrical sequences in terms of thickness. The symmetrical cycles recognized here may plausibly be related to regular (periodic?) small-scale sea-level changes superimposed on a high subsidence rate.

Following the convention of Vail (Vail *et al.*, 1987) these cycles correspond to 'fourth-order cycles' (or higher) occurring within third-order cycles, which are the main depositional sequences of this Molasse. These fourth-order cycles are similar in scale to the genetic increment of strata describe by Busch (1959, 1971), the depositional event of Frazier (1974), and the parasequences of van Wagoner (1985). The punctuated aggradational cycles of Goodwin & Anderson (1985) are thought to be shorter in

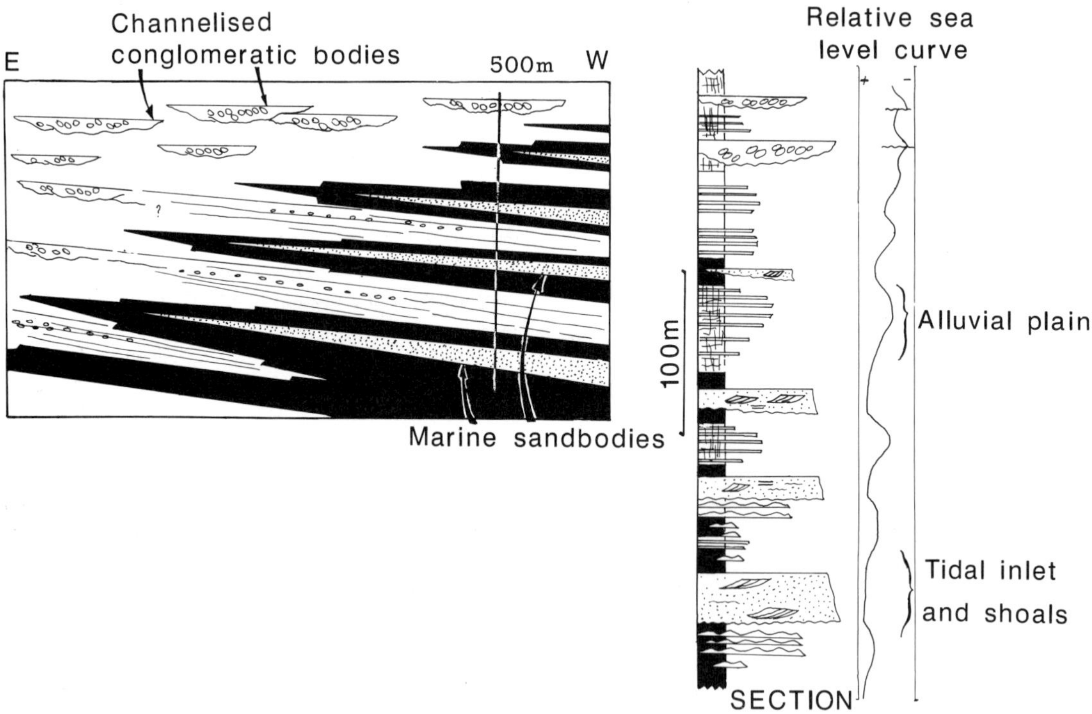


**Fig. 15.** Small-scale relative sea-level changes within the Yellow Molasse. The Yellow Molasse is an overall prograding succession punctuated by small-scale transgressive–regressive cycles.

duration. Small-scale sequences forming the Pennsylvanian cyclothems are interpreted by Klein & Willard (1989) in terms of the subsidence rate being the control on the character of each cycle (i.e. symmetrical or asymmetrical).

## CONCLUSIONS: COMPARISON BETWEEN DIGNE DEPOSITIONAL SEQUENCES AND EUSTATIC CYCLES

The Digne Miocene basin fill comprises groups of genetically related strata, bounded by a series of unconformities. These are depositional sequences in the sense of Vail (Vail *et al.*, 1987; Posamentier & Vail, 1988). In the Digne area, five main depositional sequences have been recognized, including the Valensole conglomerate which is not discussed here. Four of these sequences were deposited during Burdigalian to Serravallian times.

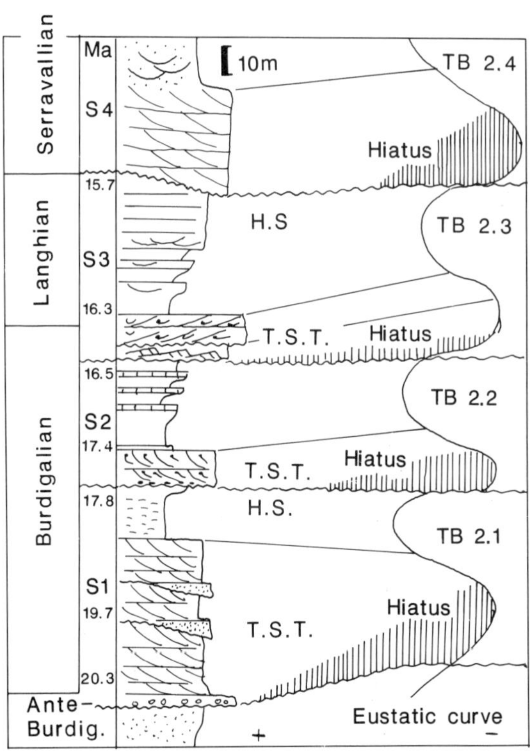


**Fig. 16.** Depositional sequences in the Miocene of the Rhodanian provençal basin and correlation with the cycles of Haq *et al.* (1987) (modified from Lesueur *et al.*, 1989).

In the Rhone Valley, which constitutes the westward part of the Miocene gulf, four depositional sequences have been recognized and precisely dated (Fig. 16; Anglada *et al.*, 1988; Lesueur *et al.*, 1989). In this area, which was marine throughout, planktonic Foraminifera allow precise dating (Gourinard *et al.*, 1985). Most of the unconformities bounding depositional sequences can be correlated with the global cycle of Haq *et al.* (1987).

In the Digne area, several stratal patterns, such as progressive onlaps and wedging, indicate that tectonic control was important during deposition. However, widely recognized depositional sequences are still recognized. In the Digne area, owing to the scarcity of fauna and to their poor preservation, it is more difficult to correlate precisely the observed depositional sequences and the global sea-level chart. The correlation scheme (proposed in Fig. 16) is therefore subject to future modifications. It must be emphasized that comparisons with the Rhodanian sequences are more important than the absolute palaeontological data in this scheme.

Apart from the Langhian depositional sequence, which remains speculative, all the depositional sequences observed in the Rhone and Digne areas can thus be correlated with the early and middle Miocene global eustatic events.

Two depositional sequences of the Lower Marine Molasse, also observed in the Rhone area, coincide respectively with the TB2-1 and TB2-2 cycles. The Upper Marine Molasse sequence correlates with the TB2-3 cycle, which is also observed in the Rhone area. Sequences identified in the Digne area are mappable regional stratigraphic units.

In the southwest Alps, orogenic processes caused the development of areas of topographic relief, which are source areas for sediment which fill the successive flexural basins (Elliott *et al.*, 1985; Evans, 1985). In these basins, episodic influx of coarse clastics is generally thought to reflect episodic motion and uplift in the thrust belt. However, as Evans (1985) noted, the situation is more complex: eustatic events are not entirely overshadowed, neither by the high sediment production rate nor by the amount of subsidence.

In the Digne Basin, load-induced flexural subsidence together with local thrust upwarping occur and are superimposed upon eustatic sea-level changes. When flexural and thrust-associated subsidence increases abruptly, the basin fill is characterized by an overall transgressive succession punctuated by eustatically controlled cycles. At this stage

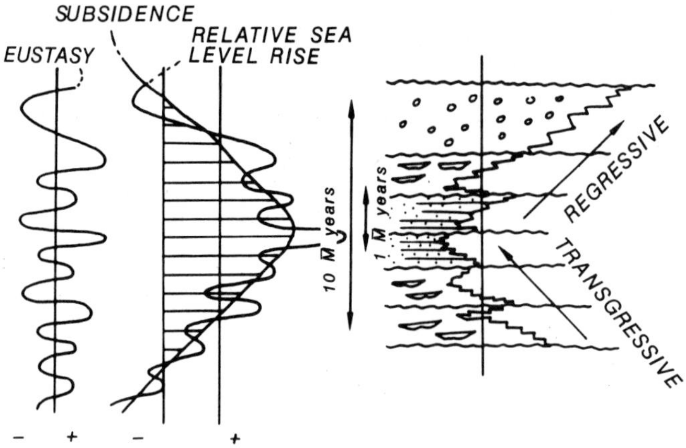


**Fig. 17.** Conceptual model of a megatectonic cycle punctuated by eustatic sea-level changes inspired from the Digne Miocene succession.

the basin evolves to underfilled conditions, and thick transgressive systems tract deposits accumulate on the subsiding margin. These form a retrogradational or landward-stepping stacking pattern of parasequences, as recorded by the first Lower Marine Molasse depositional sequence.

As basin margins are uplifted, and with increased deformation, clastic sediments are shed from the rising source area faster than the basin subsides. The basin evolves to overfilled conditions with rapidly prograding highstand successions. These are bounded by lowstand unconformities with a more pronounced tectonic imprint (e.g. the Valensole Conglomerate). The thickness of the different systems tract deposits within each area is controlled by the interaction of subsidence and eustasy.

The Digne Miocene succession is shown to be a tectonically induced transgressive–regressive cycle, more than 2000 m thick. It is punctuated by unconformities, which mostly can be correlated with global eustatic events. This study shows that eustatic events are also recorded in active tectonic settings (Fig. 17).

Tectonic subsidence and local thrust movements play a key role in the geometry and thickness of the main systems tracts. They also control the depocentre migration (Fig. 11), but unconformity genesis is mainly related to eustatic events.

## ACKNOWLEDGEMENTS

The authors express their thanks to TOTAL-Compagnie Française des Pétroles for permission to publish this work, which was carried out with the technical and financial support of the exploration-production laboratory. They also express their thanks to M.J. Altuna for typing the manuscript.

## REFERENCES

Aguilar, J.P. (1981) *Evolution des rongeurs miocènes et paléogéographie de la Méditerranée occidentale*. Thèse d'Etat, Université Montpellier, 184 pp.

Allen, P.A. & Homewood, P. (Eds) (1985) *Foreland Basins*. Spec. Publs. Int. Ass. Sediment. 8, 453 pp.

Anadon, P., Cabrera, L., Colombo, F., Marzo, M. & Riba, O. (1986) Syntectonic intraformational unconformities in alluvial fan deposits, eastern Ebro basin margins (N.E. Spain). In: *Foreland Basins* (Eds Allen, P.A. & Homewood, P.). Spec. Publs. Int. Ass. Sediment. 8, 259–271.

Anglada, R., Gourinard, Y., Lesueur, J.L. & Rubino, J.L. (1988) Les séquences de dépôt du Burdigalien et du Langhien de Basse-Provence. *Livret guide excursion A.S.F. 1988. Colloque: Paléobathymétrie, eustatisme et séquences de dépôts*, ASF, Paris.

Apps, G., Evans, M. & Davies, H. (1985) The south-west alpine foreland basin. *Abstracts, International Symposium on Foreland Basins, Fribourg*. International Association of Sedimentologists, p. 31.

Beaudoin, B., Cavelier, C., Haccard, D., Lanteaume, M. & Perrin, M. (1966) A propos de l'âge des molasses de Tanaron (Basses-Alpes). La redécouverte du gisement de Cérithes et ses conséquences structurales. *Bull. Soc. Géol. Fr.* **8**(7), 458–460.

Beaudoin, B., Gigot, P. & Haccard, D. (1970) Flysch et Molasse, approche sédimentologique. *Bull. Soc. Geol. Fr.* **XII**(7), 664–672.

Beaudoin, B., Campredon, R., Cotillon, P. & Gigot, P. (1975) Alpes méridionales françaises — reconstitution du bassin de sédimentation. *Proceedings IX International Congress Sedimentology — Nice Excursion*, 7, pp. 65–90.

Beaumont, C. (1981) Foreland basins. *Geophys. J.R.*

*Astron. Soc.* **65**, 291–329.

BUSCH, D.A. (1959) Prospecting for stratigraphic traps. *Bull., Am. Assoc. Petrol. Geol.* **43**, 2829–2843.

BUSCH, D.A. (1971) Genetic units in delta prospecting. *Bull., Am. Assoc. Petrol. Geol.* **55**, 1137–1154.

CROSS, T. (1988) Controls on coal distribution in transgressive–regressive cycles, Upper Cretaceous, western interior U.S.A. In: *Sea-level Changes: an Integrated approach* (Eds Wilgus, C.K., Hastings, B.S., Kendall, C.G.St.C., Posamentier, H., Ross, C.A. & van Wagoner, J.). Soc. Econ. Paleontol. Mineral., Spec. Publ. 42, 371–380.

CLAUZON, G. (1979) Le canyon messinien de la Durance (Provence, France): une preuve paléogéographique du bassin profond de dessication. *Palaeogeogr. Palaeoclimatol. Palaeoecol.* **29**(1–2), 15–40.

CLAUZON, G., AGUILAR, J.P. & MICHAUX, J. (1987) Mise en évidence du diachronisme du mur de la formation conglomératique de Valensole (Alpes de Haute-Provence, France): implications géodynamiques. *C.R. Acad. Sci., Ser. II*, **305**, 133–135.

CLAUZON, G., AGUILAR, J.P. & MICHAUX, J. (1989) Relations temps sédimentation dans le néogène méditerranéen français. *Bull. Soc. Geol. Fr. Ser.* **8** Fasc. 2, 361–372.

DE LAPPARENT, A.F. (1938) Etudes géologiques des régions provençales et alpines entre le Var et la Durance. *Bull. Serv. Carte Géol. Fr.*, **XL**, 198, 1–302.

DEMARCQ, G. (1970) Etude stratigraphique du Miocène rhodanien. *Mem. BRGM* **61**, 257 pp.

DEMARCQ, G. (1984) Paléogéographie. In: *Synthèse géologique du Sud-Est de la France* (Eds Debrant-Passart *et al.*). Mem. BRGM 125, 503–507.

DICKINSON, W.R. (1974) Plate tectonics and sedimentation. In: *Tectonics and Sedimentation* (Ed. Dickinson, W.R.). Soc. Econ. Paleontol. Mineral., Spec. Publ. 22, 1–27.

DUBAR, M. (1984) Chronologie et signification des dépôts continentaux du Néogène supérieur du bassin de Riez-Valensole (Alpes de Haute-Provence, France). *Bull. Soc. Géol. Fr.* **XXVI**, (7)5, 971–978.

DUBOIS, P. & CURNELLE, R. (1978) Résultats apportés par le forage les Mées 1 sur le plateau de Valensole. *C.R. Somm. Soc. Géol. Fr.* **4**, 181–184.

ELLIOTT, T., APPS, G., DAVIES, H., EVANS, M., GHIBAUDO, G. & GRAHAM, R.H. (1985) Field excursion: a structural and sedimentological traverse through the tertiary foreland basin of the external Alps of south-east France. *Excursion Guidebook International Symposium on Foreland Basins, Fribourg* (Eds Allen, Ph., Homewood, P. & Williams, G.) pp. 39–73. International Association of Sedimentologists.

EVANS, M. (1985) Syntectonic sedimentation in a thrust-sheet-top sedimentary basin. External S.W. Alps. *Abstracts, International Symposium on Foreland Basins, Fribourg.* International Association of Sedimentologists, p. 53.

FRAZIER, (1974) Depositional episodes: their relationship to the Quaternary stratigraphic framework in the north-western portion of the Gulf basin. *Univ. Tex. Bur. Econ. Geol. Handb.* **74–1**, 28 pp.

GIDON, M. & PAIRIS, J.L. (1988) La structure des environs de Digne (chaînes sub-alpines méridionales, Alpes de Haute-Provence): un exemple d'interférence entre l'avancée d'une nappe de charriage épiglyptique et la sédimentation sur son front. *C.R. Acad. Sci., Ser. II* **307**, 1283–1288.

GIGOT, P. & HACCARD, D. (1972) Nouvelles données sur l'origine des lambeaux exotiques de Courbons (Haute-Provence). Considérations morphotectoniques. *Bull. BRGM Ser. 2 Sect. I* **3**, 9–19.

GIGOT, P. & MEIN, P. (1973) Découvertes de mammifères aqurtaniens dans la molasse Burdigalienne du golfe de Digne. *C. R. Acad. Sci., Sér. D* **276**, 3293–3294.

GIGOT, P., GRANDJACQUET, C. & HACCARD, D. (1974) Evolution tectono-sédimentaire de la septentrionale du bassin tertiaire de Digne depuis l'Eocène. *Bull. Soc. Geol. Fr.* **16**, 126–139.

GOGUEL, J. (1936) Description tectonique de labordure des Alpes de la Bleone au Var. *Mem. Serv. Carte Geol. Fr.*, 360 pp.

GOGUEL, J. (1963) Les problèmes des chaînes sub-alpines. In: *Liv. à la Mém. du professeur P. Fallot*, Mém. H. S. Soc. Géol. Fr., pp. 301–308.

GOODWIN, P.W. & ANDERSON, E.J. (1985) Punctuated aggradational cycles: a general hypothesis of episodic stratigraphic accumulation. *J. Geology* **93**, 515–533.

GOURINARD, Y., MAGNE, J., RINGEADE, M. & WALLEZ, M.J. (1985) Chronologie de l'étage burdigalien. *C.R. Acad. Sci. Ser. II* **301**(10), 715–720.

GOURINARD, Y., MAGNE, J. & WALLEZ, M.J. (1987) Présence de la mer burdigalienne dans le centre de l'Aquitaine. *Bull. Soc. Hist. Nat. Toulouse* **123**.

HAQ, B.U., HARDENBOL, J. & VAIL, P.R. (1987) Chronology of fluctuating sea levels since the Triassic. *Science* **235**, 1156–1167.

HAUG, E. (1891) Les chaînes sub-alpines entre Gap et Digne. Contribution à l'histoire géologique des Alpes françaises. *Thèse. Bull. Serv. Carte Géol. Fr.* 197 pp.

JERVEY, M.T. (1988) Quantitative geological modeling of siliciclastic rock sequences and their seismic expression. In: *Sea-level Changes: an Integrated Approach* (Eds Wilgus, C.K., Hastings, B.S., Kendall, C.G.St.C., Posamentier, H., Ross, C.A. & van Wagoner, J.). Soc. Econ. Paleontol. Mineral., Spec. Publ. 42, 47–70.

KLEIN, G. DE V. & WILLARD, D.A. (1989) Origin of the Pennsylvanian coal-bearing cyclothems of North America. *Geology* **17**, 152–155.

LALAÏ, D. (1986) Nouvelles faunes de rongeurs de la fin du Miocène inférieur en Provence. Implications géologiques et paléogéographiques. *Paleovertebiota* **16**(2), 77–126.

LESUEUR, J.L., RUBINO, J.L. & GIRAUDMAILLET, M. (1989) Organisation et structures internes des dépôts tidaux du Miocène rhodanien. *Bull. Soc. Geol. Fr.* **VI**(1), 49–65.

LOUTIT, T.S., HARDENBOL, J., VAIL, P.R. & BAUM, G.R. (1988) Condensed sections: the key to age datings and correlation of continental margin sequences. In: *Sea-level Changes: an Integrated Approach* (Eds Wilgus, C.K., Hastings, B.S., Kendall, C.G.St.C., Posamentier, H., Ross, C.A. & van Wagoner, J.). Soc. Econ. Paleontol. Mineral., Spec. Publ. 42, 183–213.

MEIN, P. (1975) Résultats du groupe de travail des vertébrés. *Reports R.C.M.N.S. Working groups*, Bratislava, pp. 79–81.

MERCIER, H. (1979) Le Néogène et le Pleistocène inférieur duranciens. *Geol. Alpine* **55**, 111–132.

MUTTI, E., SGAVETTI, M. & SEGURET, M. (1988) Sedimentation and deformation in the tertiary sequences of the southern Pyrenees. *Mediterranean Basin Conference, Nice, Field Trip 7*, Am. Assoc. Petrol. Geol. 130 pp.

POSAMENTIER, H.W. & VAIL, P.R. (1988) Eustatic controls on clastic deposition, II: sequence and system tract models. In: *Sea-level Changes: an Integrated Approach* (Eds Wilgus, C.K., Hastings, B.S., Kendall, C.G.St.C., Posamentier, H., Ross, C.A. & van Wagoner, J.). Soc. Econ. Paleontol. Mineral., Spec. Publ. 42, 125–154.

PRICE, R.A. (1973) Large scale gravitational flow of supracrustal rocks. In: *Gravity and Tectonics* (Eds Dejong, K.A. & Scholten, R.) pp. 491–502. John Wiley, New York.

ROUSSET, C. (1981) Chronologie et géodynamique du bassin néogène de Digne. *C.R. Acad. Sci., Ser. II* **292**, 1213–1216.

SIDDANS, A.W.B. (1979) Arcuate fold and thrust patterns in the sub-alpine chains of south-east France. *J. Struct. Geol.* **7**, 117–126.

STANLEY, J.J. (1962) *Etudes sédimentologiques des Grès d'Annot et de leurs équivalents latéraux*. Thèse doctorat., Grenoble, 158 pp.

TESSIER, B. & GIGOT, P. (1989) A vertical record of different tidal cyclicities: an example from the Miocene Marine Molasse of Digne (Haute-Provence, France). *Sedimentology*.

VAIL, P.R. & SANGREE, J.B. (1988) Sequence stratigraphy interpretation of seismic, well and outcrop data, work book. *A.S.F. G.S.G.P. Meeting, Digne (NATO)*, 73 pp.

VAIL, P.R., COLIN, J.P., DUCHENE, R.J., KUCHLY, J., MEDIAVILLA, F. & TRIFILIER, V. (1987) La stratigraphie séquentielle et son application aux corrélations stratigraphiques dans le Jurassique du bassin de Paris. *Bull. Soc. Geol. Fr.*, **III**, (8)7, 1291–1299.

VAN WAGONER, J.C. (1985) Reservoir facies distribution as controlled by sea-level change. *Abstract Soc. Econ. Paleont. Mineral. Mid-Year Meeting, Golden, CO.*, pp. 91–92.

VAN WAGONER, J.C., POSAMENTIER, H.W., MITCHUM, R.M., VAIL, P.R., SARG, J.F., LOUTIT, T.S. & HARDENBOL, J. (1988) An overview of the fundamentals of sequence stratigraphy and key definition. In: *Sea-level Changes: an Integrated Approach* (Eds Wilgus, C.K., Hastings, B.S., Kendall, C.G.St.C., Posamentier, H., Ross, C.A. & van Wagoner, J.). Soc. Econ. Paleontol. Mineral., Spec. Publ. 42, 39–46.

WALLEZ, M.J., MAGNE, J., GOURINARD, Y. & DEMARCQ, G. (1986) Nouvelles grade-datations de sédiments burdigaliens des régions rhodaniennes. *Géologie Méditerranéenne* **XI–XIII** (1–2), 59–63.

*Spec. Publs. int. Ass. Sediment.* (1991) **12**, 391–407

# Carbonate–siliciclastic depositional systems in the Paleogene of the South Pyrenean foreland basin: a sequence-stratigraphic approach

H.P. LUTERBACHER†, H. EICHENSEER†, CH. BETZLER† *and* A.M. VAN DEN HURK‡

† *Institut und Museum für Geologie und Paläontologie, Sigwartstrasse 10, D-7400 Tübingen, Germany; and*
‡ *Dept. d'Estratigrafia, Facultat de Ciències, Universitat Autònoma de Barcelona, E-08193 Bellaterra, Spain*

## ABSTRACT

The Paleogene of the South Pyrenean foreland basin provides an outstanding example of eustatic and tectonic control on depositional sequences, systems, and systems tracts. The time interval studied in the South Pyrenean Paleogene coincides with the Tejas 2 Supercycle, which contains nine depositional sequences. During the Ilerdian and Cuisian, the South Pyrenean sequences correlate surprisingly well with the predicted third-order cycles of the 'Global Sea-level Chart'. The present data support the validity of this chart at the active margin of the South Pyrenean foreland basin, with some restrictions in the sequence boundary types. Tectonic enhancement transforms type 2 into type 1 unconformities.

Tectonism also controls the areal distribution of the depositional systems and influences the stacking pattern and thicknesses of parasequences. Sequence-internal onlap configurations result from tectonic tilting.

As in passive margin settings with a similar climate, the South Pyrenean lowstand systems tracts generally are composed of siliciclastics. The transgressive and highstand systems tracts are dominated by carbonates.

## INTRODUCTION

In recent decades, sequence stratigraphy has given rise to important conceptual changes in sedimentary basin analysis. The original concept of modern sequence stratigraphy, as introduced by Sloss (1963) and Wheeler (1958), allowed correlation of large-scale cratonic sequences by using interregional unconformities. Since this concept has been revitalized and refined by Vail *et al.* (1977), there has been great debate concerning the global significance of sea-level changes. Vail and his co-workers consider eustatic processes as responsible for the major breaks delimiting depositional sequences. Consequently, they emphasize the global synchroneity of sequence boundaries.

In contrast, Parkinson & Summerhayes (1985), Summerhayes (1986), and Hubbard (1988) argue that tectonic 'overprinting' of individual depositional areas prevents a global chronostratigraphic correlation of depositional sequences as shown in the 'Global Sea-level Chart' (Haq *et al.*, 1987). Furthermore, Cant & Stockmal (1989) conclude that unconformity bounded sequences in foreland basins are primarily tectonically controlled, with eustasy a minor factor.

The Paleogene strata in the southern Pyrenees were deposited in a foreland basin setting. The spectacular outcrops of the Paleogene sediments provide an excellent opportunity to examine depositional architecture of sedimentary bodies and entire depositional sequences over tens to hundreds of kilometres. In this context, the main emphasis is given to shallow-water carbonates, because they offer one of the most sensitive sea-level gauges in the geological record and commonly outpace sea-level rises (Kendall & Schlager, 1981).

The integration of large quantities of available biostratigraphic data (published and unpublished) into a physical stratigraphic approach provides an accurate test of eustatic and tectonic control on depositional sequences at an active plate margin. The present contribution synthesizes results of many years of biostratigraphic, sedimentological, and physical stratigraphic work in the southern Pyrenees. It is based on detailed facies, paleoecological, and biostratigraphic analysis of a large number of measured sections and field mapping of critical areas, especially where rapid lateral and vertical facies changes are observed. The observed facies are grouped into facies associations, which in turn are arranged in depositional systems. Subsequent sequence-stratigraphic interpretation includes the identification of system tracts, parasequences, and sequences, as well as the description of their respective boundaries and geometries. Whenever possible, biostratigraphic determinations have been carried to the zonal level. Extended documentations of these results are found, among others, in Ferrer *et al.* (1973), Fonnesu (1984), Eichenseer (1988), Betzler (1989a,b), and van den Hurk (1989).

**Paleogene sequence-stratigraphic background**

Outcrop studies using sequence stratigraphy have been carried out in the southern Pyrenees since the beginning of the 1980s (Mutti, 1984; Fonnesu, 1984). Mutti *et al.* (1985a,b), Remacha *et al.* (1986a,b), Costa (1985), Maestro (1985), Crumeyrolle & Mutti (1986), Crumeyrolle (1987), Eichenseer (1988), and Betzler (1989a) favour a eustatic controlling mechanism for the formation of the depositional sequences. On the other hand, Puigdefàbregas & Souquet (1986) and Puigdefàbregas *et al.* (1986) infer a mainly tectonic control for their sequences.

Most of the sequences defined in these papers lack precise biostratigraphic control of the major discontinuities and an exact definition of the depositional systems. Only Fonnesu (1984), Mutti *et al.* (1985a), Eichenseer (1988), van den Hurk (1988), and Betzler (1989a) provide a detailed definition of the various sequences and their boundaries.

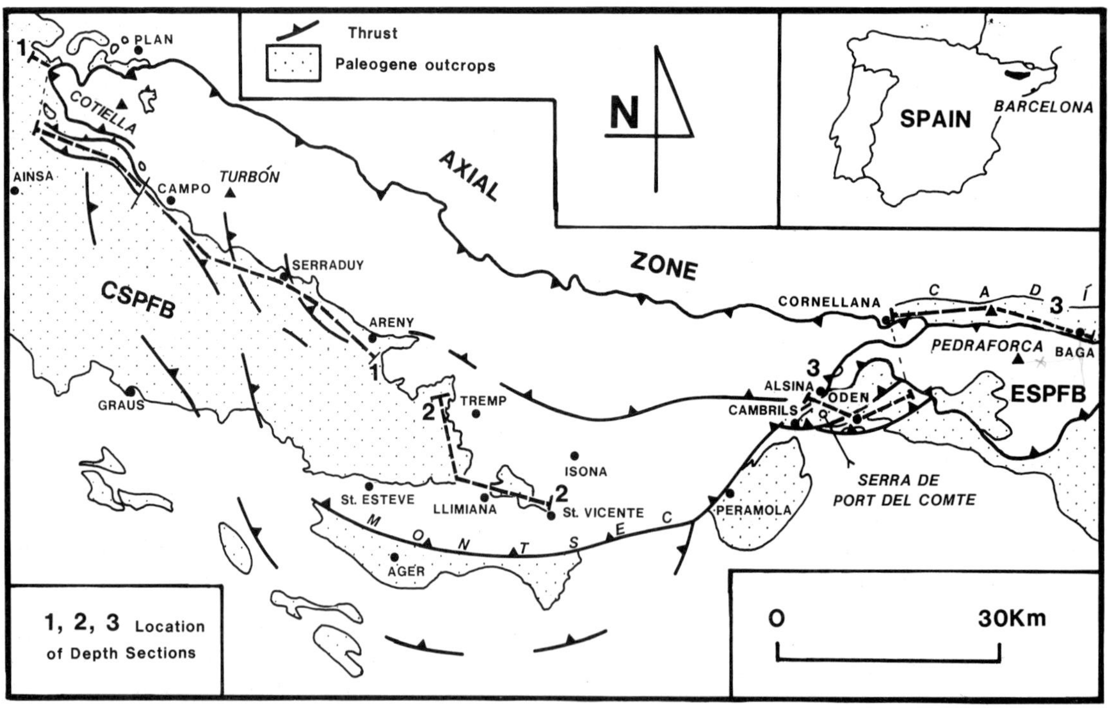


**Fig. 1.** Location of the study area. The dashed lines indicate the positions of the cross-sections (Figs 2–4) and the chronostratigraphic section (Fig. 5). CSPFB, Central South Pyrenean Foreland Basin; ESPFB, Eastern South Pyrenean Foreland Basin.

**Structural setting**

The Pyrenees extend from the Bay of Biscay to the Mediterranean Sea. This orogenic belt formed during Late Cretaceous to Miocene time owing to the rotation and oblique collision of Iberia with the European Plate (see Choukroune, 1976; Alvarado, 1980). The E–W-trending axial zone of the Pyrenean Mountain Range is bordered to the north and south by peripheral foreland basins (*sensu* Beaumont, 1981).

The Pamplona and Segre oblique wrench faults divide the south Pyrenean peripheral foreland basin (SPFB) into western, central (CSPFB), and eastern (ESPFB) units (Fig. 1). The tectonic evolution of these structural units took place at different times. Owing to the obliquity of plate encroachment, the early Paleogene basin configuration reflects a transition from a transpressional wrench basin to foreland basin conditions (Garrido-Megias & Ríos Aragües, 1972; Garrido-Megias, 1973,1988; Cuevas Gonzalo *et al.*, 1985; Puigdefàbregas & Souquet, 1986; Eichenseer, 1988).

The studied area in the CSPFB is located between Isona and Ainsa on the flanks of the present-day Tremp–Graus–Ainsa Syncline (Fig. 1). This is an allochtonous basin, which formed on the southwest-moving Montsec–Cotiella–Turbón thrusts (Seguret, 1972; Garrido-Megias & Ríos Aragües, 1972; Ori & Friend, 1984; Williams & Fischer, 1984; Camara & Klimowitz, 1985; Williams, 1985).

The studied area in the ESPFB is located between Cambrils (Serra de Port de Comte) and Bagà (Serra del Cadí, Fig. 1). According to Puigdefàbregas *et al.* (1986) and Muñoz *et al.* (1987), several structural subunits are differentiated in this area: lower thrust sheets (e.g. Cadí Nappe and Port del Comte unit) and upper thrust sheets (e.g. Pedraforca Nappe). The configuration of the main structural elements and the facies pattern in the Serra de Port del Comte suggest that this unit has undergone post-Cuisian transport directed towards south-southeast. As a consequence, the Serra del Cadí appears to be the eastward prolongation of the Serra de Port del Comte (Betzler, 1989a).

## DEPOSITIONAL SYSTEMS

The Paleogene strata of the southern Pyrenees comprise continental to relatively deep-water carbonates and siliciclastics (outer neritic to bathyal). These deposits are subdivided into different facies units based on sedimentological analysis and distribution of macro- and microfossils. Genetically related sedimentary facies are attributed to various depositional systems. The various carbonate and siliciclastic depositional systems are briefly outlined as follows.

**Carbonate depositional systems**

*(A1) Continental carbonates*

The continental carbonate rocks were deposited in coastal lakes or formed as soils.

*(A1.1) Carbonate coastal lakes.* Lake deposits are arranged in numerous lacustrine–palustrine cycles (Eichenseer & Betzler, 1987). Typical cycles start with charophyte limestones and end with pedogenic breccias (*sensu* Freytet & Plaziat, 1982).

*(A1.2) Calcimorphic and hypercalcimorphic soils* (Freytet & Plaziat, 1982). Calcimorphic soils are composed of vertically aligned caliche nodules (*sensu* Esteban & Klappa, 1983) embedded in reddish floodplain fines. The hypercalcimorphic soils are characterized by vertical successions of nodular and platy caliche. Mature soils are capped by a thick caliche hardpan (Esteban & Klappa, 1983). These soils and the coastal lake carbonates are excellent indicators for a warm semi-arid climate (Esteban & Klappa, 1983; Plaziat, 1984).

*(A2) Shallow marine carbonate depositional systems*

Tidal-flat systems with minor evaporites, carbonate beaches, carbonate banks, and coralgal reefs make up the shallow-marine carbonates.

*(A2.1) Tidal flats with evaporites.* These deposits consist of transgressive–regressive cycles of the type outlined in Shinn (1983), Wright (1984), and James (1984).

The subtidal facies comprise lagoonal carbonates (e.g. miliolid limestones) or bioclastic–siliciclastic tidal channel fills. Intertidal–supratidal intervals are characterized by algal marshes, bird's-eye limestones, and brackish tidal pond facies with charophyte limestones. Evaporitic supratidal ponds are represented by aphanitic dolomites with gypsum, celestite or silcretes, containing length-slow chalcedony (Khalaf, 1988). The lagoonal faunas and especially the brackish to evaporitic supratidal facies reflect semi-arid subtropical climatic conditions (e.g. Wright, 1984; Eichenseer, 1988).

*(A2.2) Carbonate beach systems.* The complex palaeogeography at the foreland-basin margins leads to the evolution of microtidal and mesotidal barrier beach systems and mainland beaches (Inden & Moore, 1983; Galloway & Cheng, 1985).

Regressive beach accretion wedges exhibit typical shoreface–foreshore facies sequences, and are composed of red-algal and foraminiferal packstones to grainstones. Complex diagenetic cementation is typically developed in the progradational beach wedges of mainland and barrier systems (Kerr, 1977; Inden & Moore, 1983; James & Choquette, 1984; van den Hurk, 1989). Brackish water to slightly hypersaline back-barrier lagoons are composed of foraminiferal limestones (e.g. sandy limestones with small rotaliids, miliolid limestones).

Transgressive mainland beach systems consist of thick wave-reworked sand sheets with strong basal lags (i.e. ravinement surfaces *sensu* Nummedal & Swift, 1987). These systems are composed of red-algal packstones rich in *Alveolina*. In transgressive mesotidal barrier systems only the basal inlet sequences are preserved (e.g. transgressive systems tract, Campo Sequence).

*(A2.3) Carbonate banks.* Wave–tide dominated, fringing and barrier banks (Read, 1985) cover structural highs. They are considerably larger in size and have more complex facies patterns than the carbonate beach systems. Red-algal sand bars, rich in *Alveolina*, compose the seaward edges of the fringing and barrier banks. Low-amplitude *Nummulites* bars occupy less inclined wave-exposed bank margins. Tide-influenced sand flats represent the interiors of both carbonate bank types. The barrier banks additionally contain back-barrier lagoonal limestones with miliolids. Coralgal patch reefs and reef mounds occur at the carbonate bank margins. These buildups are constructed by poritid and faviid corals, encrusting and branching solenomeris, and corallinacean algae. The reefs grade distally into grain flats (Kerr, 1977; Budd & Loucks, 1981) enriched in small *Nummulites* and *Assilina*.

*(A2.4) Carbonate homoclinal ramp systems.* Low-energy homoclinal ramps (Read, 1985) cover the entire basin margins during times of increased accommodation. These shallow ramps are comparable to shelf lagoonal ramp settings as outlined in Enos (1983). They contain packstones with *Alveolina* in their shallowest parts; with increasing water depth, this facies grades into muddy pack- to wackestones with *Operculina*, small *Nummulites*, *Assilina* and *Discocyclina*.

#### *(A3) Deeper carbonate depositional systems*

Slope and deeper (outer neritic to bathyal) intervals of distally steepened ramps reflect tectonic oversteepening. The corresponding facies consists of carbonate turbidites (i.e. thin-bedded couplets of argillaceous lime wacke- to mudstones and calcareous shales). In addition, megaturbidites linked to active tectonic elements occur. Megabreccias make up the basal units of the megaturbidites. These breccias contain intra-formational blocks of inner neritic limestones, or large components of Mesozoic carbonates and sandstones. Associated canyon-fill deposits are composed of intra-formational megabreccias and allodapic limestones.

### Siliciclastic depositional systems

The continental and shallow-marine siliciclastics belong to coarse and fine-grained deltaic systems. Both types of delta differ in their dimensions, source area types, transport paths, and subaerial components.

#### *(B1) Coarse-grained deltas*

The subaerial parts of these deltaic systems follow major tectonic elements (e.g. thrust ramps). Delta-front sands are deposited in structural lows (e.g. between blind thrust anticlines). The subaerial parts of the coarse-grained deltas comprise braided river deposits, embedded in red floodplain fines with calcimorphic soils. The dominating lithofacies of the braided channels consist of moderately sorted, subrounded–rounded sandy conglomerates and microconglomeratic sandstones.

The source areas of these braided rivers are located north of the foreland basin. Contemporaneous tectonic uplift causes the short, steep thalweg and coarse grain size of these braided rivers. The lower delta plain (Coleman & Prior, 1982) of the coarse-grained systems contains fluvial to tide-dominated distributary channels, marshes, and swamps. The delta-front sands and silts were dominated to various degrees by waves and tides, and the prodelta shales contain small *Nummulites* and *Operculina*. Interdistributary bay to lagoonal facies are composed of sandy limestones with *Alveolina* and calcareous shales with *Operculina*.

These coarse-grained deltaic systems are interpreted as braid deltas (*sensu* McPherson *et al.*, 1987).

### *(B2) Fine-grained deltaic system*

This deltaic system enters the CSPFB along its main axis, parallel to the thrust belt. The southeastern position of the source area suggests that this siliciclastic complex had its origin at the peripheral bulge. The subaerial interval of the fine-grained delta comprises low-sinuosity river channels associated with red to brownish floodplain fines. The channel-fill deposits consist of upward-fining well-sorted sands. The lower delta plain facies comprise fluvial to tide-dominated distributary channels (Fonnesu, 1984) in association with clastic tidal flats and marshes. The delta front is represented by estuarine sands (i.e. shoals, channels, and tidal bars). The distal zone of the delta front contains wave-reworked sands. Associated prodelta shales with small *Nummulites* and *Assilina* document relatively shallow water depths (middle neritic zone). This deltaic system is considerably larger in extent than the coarse-grained braid-deltas.

### *(B3) Deeper marine siliciclastics (outer neritic to bathyal)*

Sandy turbidites and canyon-fill sands occur in the western zone of the study areas. The channelized turbidites were derived most probably from the southeast. These deposits can be classified either as type 2 (*sensu* Mutti, 1985) or as delta-toe turbidites (*sensu* Erskine & Vail, 1988).

The canyon-fill is represented by massive sandy turbidites, derived from the north. These sediments are interpreted as B2 and C1 turbiditic beds *sensu* Mutti & Ricci-Lucchi (1972).

## STRATIGRAPHY

### Bio- and chronostratigraphy

The neritic and outer neritic deposits of the South Pyrenean Paleogene are subdivided by larger Foraminifera, planktonic Foraminifera, and calcareous nannoplankton. Larger Foraminifera (*Nummulites*, *Assilina*, *Operculina*, *Alveolina*, *Orbitolites*) are restricted generally to the inner neritic environment, whereas planktonic microfaunas and microfloras are mainly confined to outer neritic deposits. In a few cases, the larger Foraminifera have been reworked into deeper facies (beds of redeposited carbonates). Within the studied time-interval of approximately 10 Ma, the average duration of a larger foraminiferal zone is about 1 Ma.

In the Tremp Section (i.e. Ilerdian stratotype; Luterbacher, 1973) and in the Campo Section (i.e. Ilerdian parastratotype; Schaub, 1973), inner neritic facies alternate with outer neritic facies. These sections allow the correlation of benthonic and planktonic zonations. The correlation of the Eocene larger foraminiferal zones with the planktonic zones relies on data from Mutti *et al.* (1972), Rosell *et al.* (1973), Kapellos & Schaub (1973), Caro *et al.* (1975), and Cavelier & Pomerol (1986). However, an accurate calibration of Thanetian macroforaminiferal zones with planktonic foraminiferal zones remains difficult (see Hillebrandt, 1975).

The biostratigraphic subdivision used in both studied areas corresponds to the zonation of the South Pyrenean Paleogene by Schaub (1981). This zonation provides a definition, with larger Foraminifera, for the Thanetian, Ilerdian, Cuisian, and Lutetian. Magnetostratigraphic investigations of the Paleogene record are restricted mainly to deep-sea deposits (Berggren & Miller, 1988). Radiometric data for the Paleogene stratotypes are based largely on glauconitic sands from northwest Europe (see e.g. Hardenbol & Berggren, 1978; Odin & Curry, 1985; Aubry *et al.*, 1988). The validity of some of these radiometric ages is seriously questioned (see e.g. Odin & Curry, 1985; Cavelier & Pomerol, 1986; Obradorovich, 1988). In many cases, the exact positions of the glauconitic samples within the frame of Paleogene biozonations have not been established. These uncertainties, together with the somewhat fluctuating correlations of the biozonations, lead to additional errors.

Absolute ages attributed to biozones are merely 'best guesses'. Without doubt, they will undergo considerable changes as more data, especially on correlations with the magnetostratigraphic time-scale, become available.

### Physical stratigraphy

The physical stratigraphic approach is based on

sequence-stratigraphic methodology (e.g. Vail, 1987). This includes detailed facies analysis of depositional systems, tracing of parasequences, and estimation of the geometrical patterns of the parasequence sets (Eichenseer, 1988; Betzler, 1989a,b; van den Hurk, 1989).

Definition of systems tracts relies on identification of major discontinuity planes, including surfaces of maximum regression, transgressive surfaces, flooding surfaces, and sequence boundaries. Depth diagrams of the depositional sequences are given in Figs 2–4.

The various sequence boundaries and major discontinuity planes are calibrated with foraminiferal biozones. The ages of nine depositional sequences are shown in Fig. 5. The Campo, Navarri, and Serraduy sequences belong to the middle–late Thanetian. The Ager, Llimiana, Alinyà, Odén, and Cadí sequences are of Ilerdian to Cuisian age. The Bagà–Josa sequence corresponds to the oldest Lutetian.

In the following, the bounding unconformities, onlap patterns, and the systems tracts are described for each sequence.

*Campo sequence*

The lower boundary of the Campo sequence is marked by a basin-wide soil catena, the Colmenar–Tremp horizon (Eichenseer & Krauss, 1985; Eichenseer, 1988; van den Hurk & Eichenseer, 1988). This soil horizon is slightly older than the base of the *Glomalveolina primaeva* zone and indicates a considerable hiatus which overlaps the lower boundary of the Tejas 2 Supercycle (see Fig. 6). The Col de Vent High (Fig. 2) serves as an eastern hinge point of the marine onlap. Eastward, the sequence is entirely continental with low-sinuosity fluvial channels and floodplain deposits.

The lowstand wedge deposits consist of continental red beds merging to the west into sandy lagoonal limestones.

The transgressive systems tract starts with tidally and storm-reworked deltaic sands and grades upward into barrier beaches, carbonate banks, and reefs. The highstand systems tract includes prograding carbonate barrier beaches and regressive tidal flats. Tectonic uplifts at the Cerro Lumar and Collado del Santo highs cause local onlaps of parasequences

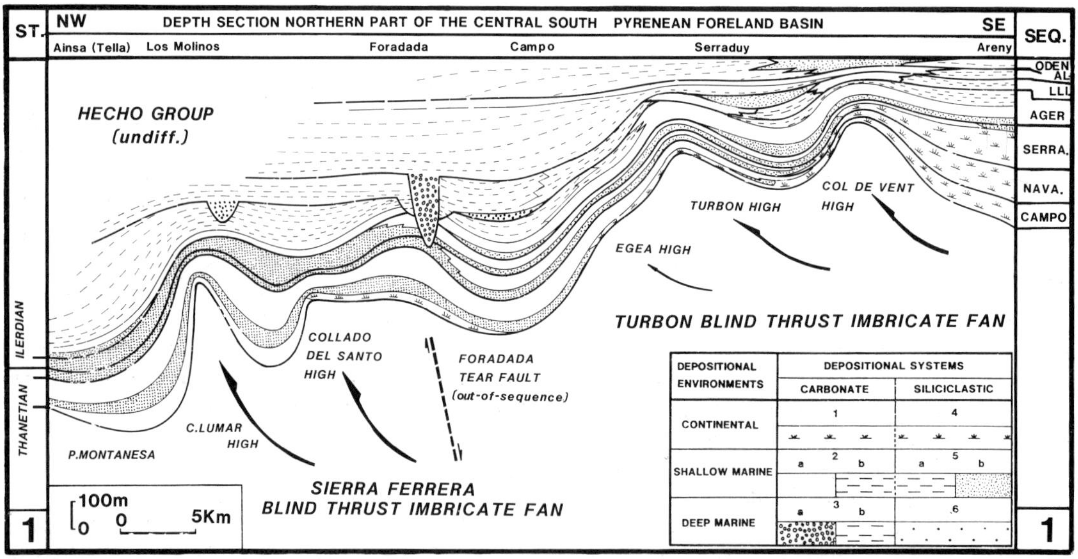


**Fig. 2.** Cross-section of the Paleogene in the central southern Pyrenees between Ainsa and Areny. In this section, the Serra Ferrera (Foradada–Ainsa sector) has been placed c. 5 km to the northeast (according to its original position). Canyon incisions (Foradada, Los Molinos) are bound to underlying structural elements (Foradada tear fault, Cerro Lumar blind thrust). Legend (Figs 2–5): 1, lacustrine–palustrine carbonates, calcimorphic and hypercalcimorphic soils; 2a, carbonate banks, reefs, tidal flats with evaporites, mainland and barrier beaches, back-barrier lagoons, mixed siliclastic–carbonate lagoons, homoclinal ramps; 2b, calcareous shales; 3a, megaturbidites, megabreccias, debris flows; 3b, outer ramp nummulitic marls, allodapic limestones; 4, gravelly braided streams, floodplains, sandy low-sinuosity rivers; 5a, prodelta shales; 5b, lower delta plain, tide and wave-dominated delta-front sands, estuarine sands; 6, turbidites.

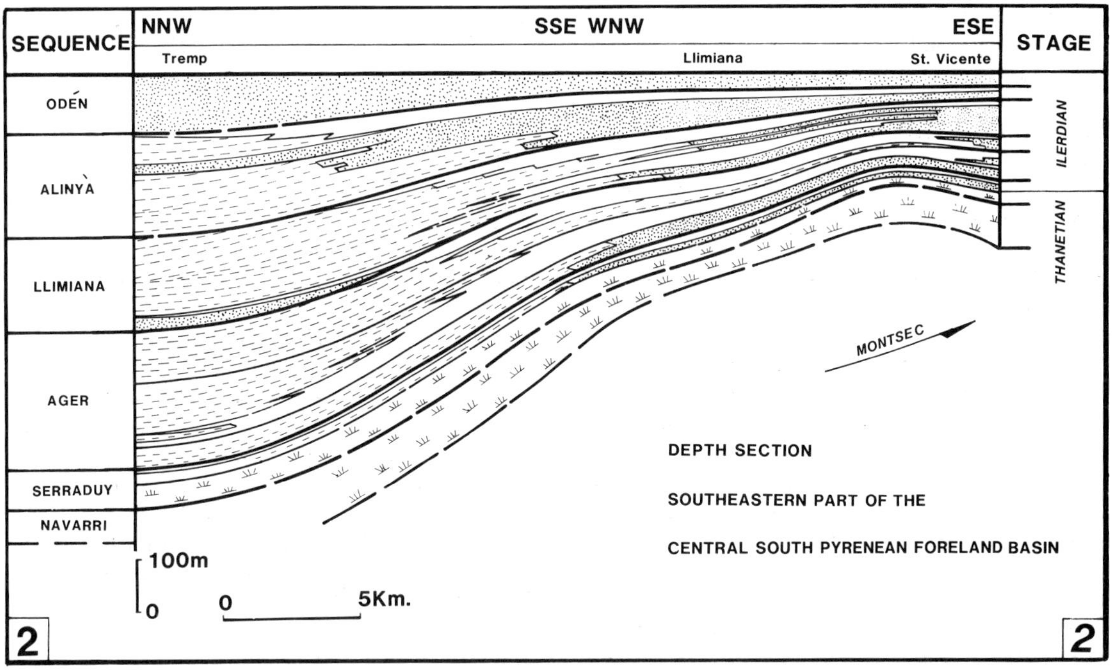


**Fig. 3.** Cross-section of the Paleogene in the eastern part of the central southern Pyrenees between Tremp and St Vicente (northern flank of the Montsec Mountain Range). For legend see Fig. 2.

(Fig. 2).

In the Eastern Pyrenees, between Alsina and Bagà, the transgressive systems tract (TST) and the highstand systems tract (HST) are represented by carbonate lake deposits.

*Navarri sequence*

In the Serraduy area, the sequence boundary is marked by calcimorphic soils overlying the older highstand deposits. In the Campo–Ainsa area, this boundary is indicated by a sudden change from carbonate to siliciclastic depositional systems, accompanied by local erosion at the intertidal level. The maximum regression of this deltaic depositional system delineates the upper limit of the lowstand systems tract (LST). These lowstand deltaic deposits were tide and wave-dominated.

The TST also comprises tide and wave-influenced deltaic siliciclastics. However, storm-induced reworking was of considerable importance in the distal part of the delta front. In the youngest part of the TST, a carbonate depositional system started to replace the siliciclastic one (Fig. 4).

The maximum flooding surface reaches the area of the Col de Vent high (Fig. 2), which again represents the eastern hinge point of the marine onlap. The HST consists of two highstand braid-deltas, separated by an extensive carbonate tidal bar belt (between Turbón and Foradada, west of Campo, Fig. 2).

The continental portion of the sequence comprises braided stream and floodplain deposits.

In the eastern Pyrenees (Serra de Port del Comte and Serra del Cadí), lacustrine and palustrine limestones make up the entire sequence.

*Serraduy sequence*

The lower sequence boundary is characterized by valley incision (near Serraduy) and submarine erosion (near the Cinca Valley in the easternmost part of the studied area).

The LST comprises prograding tide and storm-influenced braid-deltas. These deposits are considerably coarser than the HST deltaic sediments of the Navarri sequence and demonstrate an important change in palaeocurrent direction. The LST infills

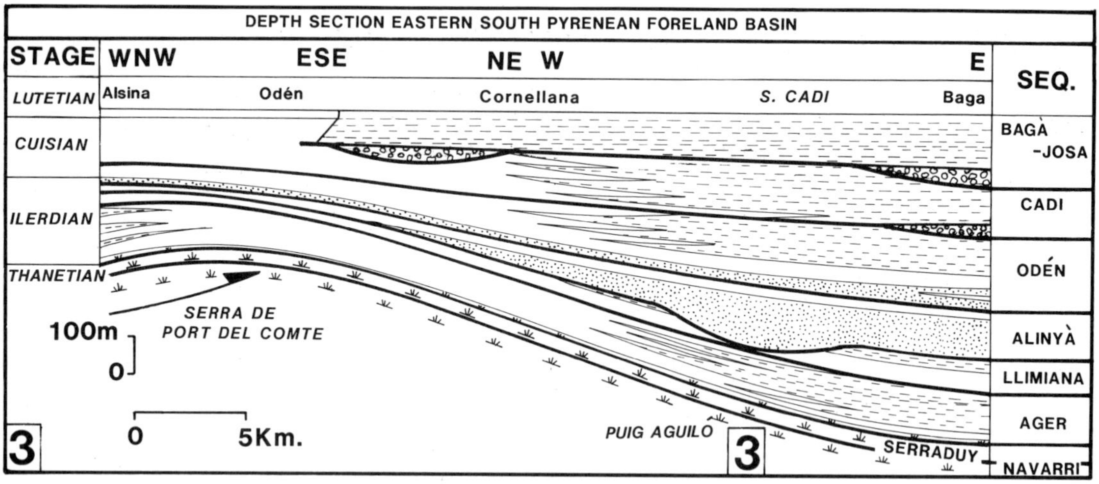


**Fig. 4.** Cross-section of the eastern southern Pyrenees between Alsina and Bagà. In this section, the Serra de Port del Comte has been placed c. 7 km to the northwest (according to its original position). For legend see Fig. 2.

structural lows with east and west-directed onlaps. A southwest-directed downlap is visible between Collado del Santo and Cerro Lumar (van den Hurk, 1989). The continental deposits are dominated by large, gravelly, braided stream deposits. The maximum regression is characterized by several soil horizons.

The TST consists of carbonate onlap cycles encroaching on the lowstand deltaic sands in an eastern direction. In the easternmost extension of the central southern Pyrenees, estuarine sands make up the main part of the TST (Fig. 4).

A relatively thin carbonate highstand complex exists north of Serraduy. (This HST is not included in Figs 2, 3, 5 because of its palaeogeographic position.)

In the ESPFB, this depositional sequence cannot be identified. In this area, the early Ilerdian deposits are arranged in two onlapping cycles of transgressive–regressive tidal-flat deposits.

### Ager sequence

The lower sequence boundary is best characterized near Serraduy and in the Peña Montañesa areas by submarine erosion of the TST of the Serraduy Sequence (see Fig. 2).

The LST is dominated east of the Turbón high by wave and tide-influenced deltaic sands with local intercalations of sandy lagoonal limestones.

The TST contains homoclinal carbonate ramps, which were retreating towards the basin margins. With continuous deepening, deltaic systems retreat shoreward and coralgal reef growth increases along the palaeohighs (e.g. Turbón). The depocentres near Tremp and Ainsa are infilled with transgressive shales. The youngest part of the TST consists of condensed bioclastic limestones rich in glauconite.

The HST is composed of prograding carbonate banks with reefs. The basinal areas are infilled by calcareous shales and marls with several condensed shell layers (i.e. grain flats). These horizons mark minor flooding surfaces.

### Llimiana sequence

In the CSPFB, the lower sequence boundary is characterized in the shallow marine areas by a sudden change from carbonate banks to prograding deltaic sands. Near Tremp, lowstand deltaic sands slightly erode the previous highstand calcareous shales (Fonnesu, 1984). Canyon incision occurs in the area between Foradada and Ainsa (Fig. 2); they are bound to underlying structural elements (Cerro Lumar blind thrust, Foradada tear fault). Sandy lowstand turbidites infill the structural low near Campo.

In the ESPFB, the LST consists of mixed siliclastic–carbonate lagoonal deposits (Fig. 4).

Aggrading to retrograding shallow-marine carbonates with patch reefs in the easternmost part of the CSPFB are interpreted as proximal TST (Fig. 3).

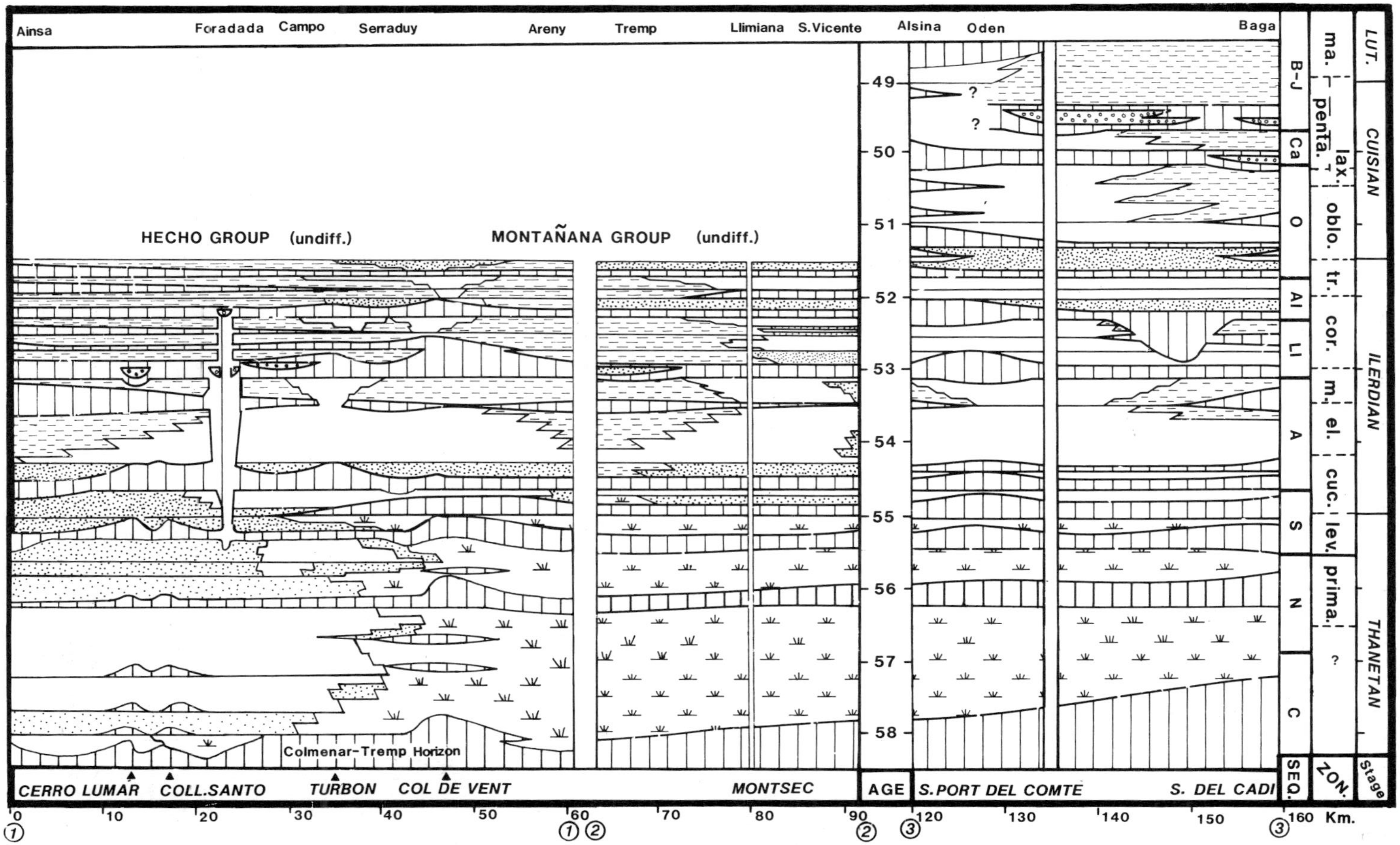


**Fig. 5.** Chronostratigraphic section of the central and eastern southern Pyrenees. The equivalent cross-sections are shown in Figs 2–4. For the location of the cross-sections see Fig. 1. For legend see Fig. 2. Vertical lines represent time intervals of non-deposition. ma., *Alveolina major* zone; penta., *Acarinina pentacamerata* zone; lax., *Assilina laxispira* zone; oblo., *Alveolina oblonga* zone; tr., *Alveolina trempina* zone; cor., *Alveolina corbarica* zone; el., *Alveolina ellipsoidalis* zone; lev., *Glomalveolina levis* zone; prima., *Glomalveolina primaeva* zone. Sequences: C, Campo; N, Navarri; S, Serraduy; A, Ager; Ll, Llimiana; Al, Alinyà; O, Odén; Ca, Cadí; B-J, Bagà–Josa. The *Assilina laxispira* zone is equivalent to the *Alveolina dainellii* zone; the *Assilina major* zone corresponds to the *Alveolina violae* and the lower part of the *Alveolina stipes* zones. For legend see Fig. 2.

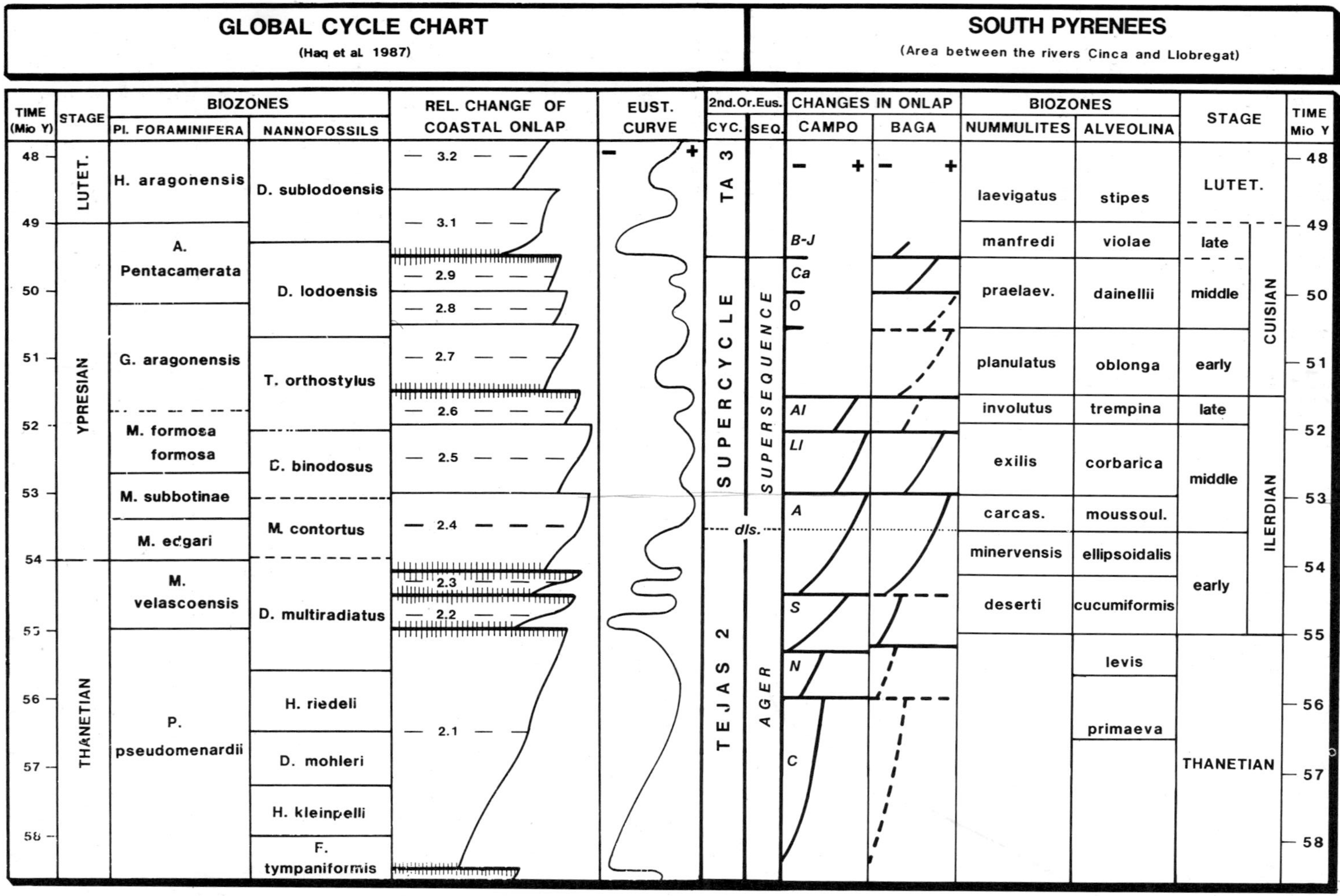


**Fig. 6.** This diagram shows the local changes in onlap (Campo section, Bagà section) with the corresponding sequence boundaries in the southern Pyrenees, compared with the global changes in onlap and the cycle limits given on the 'Global Sea-level Chart' (Haq *et al.*, 1987). The ages of the biozones correspond to those indicated on the 'Global Sea-level Chart'. The ages of zonal boundaries result from the correlation of the South Pyrenean zonation with planktonic foraminiferal and calcareous nanoplankton zones. Sequences: C, Campo; N. Navarri; S, Serraduy; A, Ager; Ll, Llimiana; Al, Alinyà, O, Odén, Ca, Cadí, B-J, Bagà–Josa. dls, downlap surface (Figure modified from Haq *et al.*, 1987).

Elsewhere, calcareous shales make up the transgressive interval, and are finally covered by a condensed calcareous horizon.

On the palaeohighs, the HST is dominated by prograding carbonate banks. The basinal areas consist of calcareous shales with slightly condensed intercalations of muddy nummulitic limestones. In the western half of the CSPFB, deeper marine facies of a distally steepened ramp occur (i.e. calcareous turbidites).

*Alinyà sequence*

The lower boundary of this sequence is characterized by strong submarine erosion of the HST of the Llimiana sequence (e.g. at Foradada and in the Serra del Cadí). These incisions are linked to synsedimentary fault systems (see Figs 2, 4).

The LST of the Alinyà sequence is represented by canyon-fill deposits (e.g. debris flows and megaturbidites). Deltaic sands and shales infill the shallower marine zone of the CSPFB (i.e. between Tremp and St Vicente, Fig. 3) and the eastern part of the ESPFB (Fig. 4). Reduced lowstand carbonates with minor reef mounds occur along the Col de Vent high (see Fig. 2).

The TST is dominated by calcareous shales. Minor coral reefs form in the vicinity of palaeohighs (e.g. Col de Vent high). Near Tremp and at the Serra de Port del Comte, the TST is composed of wave-reworked deltaic sands with abundant bioclasts.

The HST consists of strongly prograding carbonate bank units (tide-dominated in its proximal, wave-dominated in its distal parts). The sedimentation in the basinal areas is characterized by calcareous shales. In the westernmost areas, redeposited carbonates prevail.

*Odén sequence*

The lower sequence boundary coincides with an abrupt change from calcareous lagoonal deposits to prograding deltaic sands. The upper bounding unconformity is situated below the first carbonate debris-flow deposits appearing in the basinal Greixa–Bagà section. In the inner neritic zone, this boundary separates regressive tidal-flat deposits from overlying transgressive carbonate tidal shoals.

The LST of this sequence consists of red-coloured delta plain deposits. The surface of maximum regression is marked by a bleached soil horizon. Carbonate tidal flats onlapping on this soil horizon form the TST.

The upper part of the Odén sequence is dominated by westward back-stepping carbonate ramp systems. Outer ramp zones are characterized by marls with *Assilina* and *Discocyclina*. Inner ramp areas consist of bioclastic bars with abundant *Alveolina* and *Orbitolites*.

As shown in Fig. 6, this sequence could be further subdivided at the transition from the *Alveolina oblonga* to the *A. dainellii* zone. However, the corresponding sequence boundary cannot be reliably determined at the present state of investigations (see discussion).

*Cadí sequence*

The lower bounding unconformity is located at the base of the first carbonate debris flow (LST) in the Greixa–Bagà section. The upper bounding unconformity of this sequence is placed at the transition from the middle to the late Cuisian.

This sequence is composed of continuously back-stepping carbonate ramps. The retrogradational character of the parasequence sets is linked to a considerable increase in basinal subsidence. This distribution of parasequence sets excludes a subdivision into TST and HST.

Deeper ramp zones consist of marls with *Nummulites*, *Assilina*, *Discocyclina*, and sparse planktonic Foraminifera. Inner ramp areas are formed by wave-reworked bioclastic bars with abundant large *Assilina*.

*Bagà–Josa sequence*

In the western Serra del Cadí, the lower sequence boundary shows a sudden change from back-barrier lagoonal limestones to glauconitic sandstones with intraclasts, reworked *Nummulites*, and *Assilina*. This boundary coincides with the upper limit of the Tejas 2 Supercycle of the 'Global Sea-level Chart' of Haq *et al.* (1987) (see Discussion).

In the eastern Serra de Port del Comte, this unconformity is marked by the top of *Nummulites* packstones, which are overlain by carbonate debris-flow deposits. A glauconitic horizon separates the carbonate debris-flow bed from the *Nummulites* packstones. The Bagà–Josa sequence differs from underlying sequences by its clear differentiation into basinal facies and shallow carbonate platform deposits. The platform of the late Cuisian is characterized by relatively steep margins. The succession reflects progradational phases with bioclastic bars, which alternate with phases dominated by enhanced off-bank transport of bioclastic sands. The inner platform consists of lagoonal and peritidal limestones.

The basinal facies comprises marls with *Nummulites*, *Assilina*, *Discocyclina*, and planktonic Foraminifera. Numerous beds of redeposited carbonates and siliclastic turbidites are interbedded with the marls.

## CONTROLLING FACTORS

### Climate

The continental carbonate systems and the tidal flats containing evaporites are excellent indicators of a warm semi-arid climate (Plaziat, 1984; Eichenseer & Betzler, 1987). The subaerial facies (e.g. floodplains) of the deltaic systems confirm this interpretation.

The fossil contents of the shallow-marine carbonates are typical of subtropical latitudes (e.g. Hottinger, 1973).

### Sea level

Carbonate depositional systems dominated during highstand and transgressive intervals (Vail, 1987; e.g. lakes). Retreating homoclinal ramps and transgressive beaches were characteristic of transgressive systems tracts (Sarg, 1988). Carbonate banks were best developed during highstand conditions. Active tidal flats (*sensu* Tucker, 1985) occurred during transgressions, whereas regressive tidal flats are restricted to highstand systems tracts. Lowstand periods were characterized by increased activities of siliciclastic systems (Vail, 1987). Finer grained highstand braid-deltas evolved in areas of high sediment supply. Mature caliche soils commonly mark surfaces of maximum regression and sequence boundaries.

We think that the control of the global sea level on sedimentation in the South Pyrenean Basin is reflected in age equivalent sequence boundaries, systems tracts limits, and depositional systems in the two different tectonic settings to the east and the west of the Segre Fault (Fig. 5; Betzler *et al.*, 1987).

### Tectonic subsidence

As shown in Figs 2–4 structural elements in the different tectonic settings control the sediment thickness, the repetition of siliciclastics, and the type of sequence boundaries.

The evolution of mainland and barrier–beach systems is influenced by the tectonically controlled palaeogeography. Carbonate banks and coralgal reefs are located predominantly on blind thrust anticlines (e.g. Serra de Port del Comte, Fig. 4). Distally steepened ramps result from tectonic oversteepening.

The deltaic complexes are located in structural lows (e.g. deltaic deposits in Serraduy, Fig. 2). Deltaic systems with high sediment supply, caused by tectonic uplift, may spread over palaeohighs (e.g. Cerro Lumar high and Collado del Santo high, Fig. 2). Differential tectonic subsidence regulates the areal distribution, thicknesses, and hydrodynamic conditions of the depositional systems, sequences, and systems tracts.

## DISCUSSION

Figure 6 shows the close coincidence between the local South Pyrenean sequences and the 'Global Sea-level Chart' (Haq *et al.*, 1987) for the Paleogene. Good correspondence of sequence boundaries exists during the Ilerdian and the transition from the Cuisian to the Lutetian. Thanetian and early–middle Cuisian boundaries are difficult to correlate owing to problems of biostratigraphic correlations.

The middle and late Thanetian of the studied area is subdivided into two depositional sequences (Campo and Navarri sequences). However, only one sequence is predicted in the 'Global Sea-level Chart' for this part of the Thanetian (i.e. 2.1 cycle). The 2.2 and the 2.3 cycles cannot be distinguished in the studied area. For this time interval, only one depositional sequence can be defined (Serraduy sequence). The possibility that the 2.2. cycle could also coincide with the late Thanetian Navarri sequence in the southern Pyrenees (*Glomalveolina primaeva* and part of the *G.levis* zone) cannot be excluded.

The Ilerdian Ager, Llimiana, and Alinyà sequences, respectively, coincide with the global third-order cycles 2.4, 2.5, and 2.6. The downlap surface of the Ager sequence (i.e. limit of the *Alveolina ellipsoidalis–A.moussoulensis* zones) is well defined by planktonic Foraminifera and correlates with the downlap surface of the 2.4 cycle (*sensu* Haq *et al.*, 1987). This horizon represents the major continental flooding surface of the Tejas 2 Supercycle (P.R. Vail, personal communication, 1989). For the

Cuisian, the sequence-stratigraphic interpretation is restricted to the area east of the Segre Fault.

The Odén sequence of latest Ilerdian to middle Cuisian age could be further subdivided in its upper part. The corresponding sequence boundary, predicted in the 'Global Sea-level Chart' for this interval would be a type 2 unconformity. However, problems arise in defining this sequence boundary in the outcrop section. Subaerial and/or submarine erosions and a basinward shift in facies are lacking. Most likely, this sequence boundary is obscured owing to the very high rate of basinal subsidence in this structural unit.

The best-defined sequence boundary, between cycles 2.9. and 3.1, in this region is located within the *Acarinina pentacamerata* zone. This unconformity also represents the upper limit of the Tejas 2 Supercycle (Haq *et al.*, 1987).

All bounding unconformities of the South Pyrenean Paleogene sequences, with the exception of the obscured sequence boundary in the upper part of the Odén sequence, are associated with subaerial and/or submarine erosions. These boundaries are interpreted as type 1 unconformities in striking contrast to the 'Global Sea-level Chart', where six type 1 and four type 2 sequence boundaries are predicted for the studied time interval. This fact suggests that type 2 sequence boundaries are transformed into type 1 boundaries by tectonic uplift. In contrast, a very high rate of tectonic subsidence can obscure a type 2 sequence boundary. As demonstrated by Sleep (1976), even small sea-level variations can produce significant unconformities in basins with increased tectonic subsidence.

The Thanetian to Cuisian sequences (i.e. Campo to Cadí sequence) are grouped together in the Ager supersequence. The lower bounding unconformity of this supersequence is slightly older than the *Glomalveolina primaeva* zone. This unconformity (i.e. Colmenar–Tremp horizon) marks a considerable hiatus that overlaps with the lower boundary of the Tejas 2 Supercycle (i.e. 58.5 Ma boundary). The upper supersequence boundary correlates well with the upper limit of the Tejas 2 Supercycle (i.e. basal unconformity of the Bagà–Josa sequence, 49.5 Ma boundary). The associated major downlap surface has a well-defined age of 53.5 Ma (i.e. downlap surface of the Ager sequence).

The Tejas 2 supercycle differs from the second-order tectonic cycle of the early evolution of the South Pyrenean foreland basin. This tectonic cycle lasts from the Santonian (compare Intra-Santonian unconformity of Garrido-Megias (1973)) to the early Oligocene (see Garrido-Megias, 1973). The associated bounding unconformities are characterized by strong truncations of tilted strata (*sensu* Vail *et al.*, 1984, Fig. 8, section B). The megasequences of Hubbard *et al.* (1985a,b) and Hubbard (1988) reflect tectonic stages in basin evolution and seem to correspond to Vail's second-order tectonic cycles.

## CONCLUSIONS

The studied successions in the central and eastern South Pyrenean foreland basin belong to different tectonic units that were incorporated into the overthrust at different times: the central South Pyrenean foreland basin at the transition from the Ilerdian to the Cuisian, and the eastern South Pyrenean foreland basin at the transition from the Cuisian to the Lutetian.

Nevertheless, different major discontinuity planes are age-equivalent in both structural units. The intervals with good biostratigraphic data (Ilerdian, late Cuisian) contain major unconformities that are correlative to cycle limits of the 'Global Sea-level Chart'.

At least six local sequences coincide well with the corresponding third-order cycles. Additionally, most boundaries of larger foraminiferal zones correlate with transgressive surfaces and downlap surfaces.

These observations, together with the coincidence of the Ager supersequence with the Tejas 2 Supercycle, suggest that eustasy is the driving mechanism for the evolution of the Paleogene sequences in the South Pyrenean foreland basin.

Four third-order cycle boundaries of the 'Global Sea-level Chart' shown for the Tejas 2 Supercycle are type 2 unconformities. In the studied area, tectonic uplift enhances three of these unconformities and transforms them into type 1 sequence boundaries. One type 2 unconformity is probably obscured owing to strong basinal subsidence.

Many of the erosional features are controlled by structural elements and thus are indirectly related to tectonics. However, the major erosional features occur during relative sea-level falls.

Most siliciclastic depositional systems of the South Pyrenean sequences are related to relative sea-level lowstands. Carbonate depositional systems dominate during transgressive and highstand conditions. This overall distribution of carbonate and siliciclastic lithofacies is comparable to those of passive margin sequences in a similar climatic setting (e.g. Foz do Amazonas Basin, Brouwer & Schwander, 1988). We assume that this trend in lithofacies distribution is caused mainly by eustasy.

Tectonism also controls the geographical distribution of the depositional systems and influences the stacking patterns and thicknesses of parasequences and systems tracts. Highstand carbonate ramps may retrograde in response to strong basinal subsidence. Tectonic upwarping of a source area close by the depositional site can cause extremely high sediment supply. In this exceptional case, a siliciclastic (usually wave-reworked) highstand delta evolves, and sequence-internal onlap configurations may occur owing to tectonic tilting. These onlap patterns are best developed along the boundaries between lowstand and transgressive systems tracts.

## ACKNOWLEDGEMENTS

Financial support by the Deutsche Forschungsgemeinschaft is gratefully acknowledged. We thank Peter Vail for stimulating discussions and new ideas. We are also indebted to Jordi Ferrer, Joan Rosell, and Emiliano Mutti for their collaboration. We thank Lukas Hottinger and Hans Schaub for biostratigraphic data. The manuscript benefited from critical comments by J. Ferrer, R. Mitchum Jr., E. Mutti, J. Rosell, and P. Vail. We are grateful to P. Friend and D. Macdonald for their constructive reviews.

## REFERENCES

Alvarado, M. (1980) Introducción a la Geología General de España. *Bol. Geol. Min.* **XCI-I**, 1–65.

Aubry, M.P., Berggren, W.A., Kent, D.V., Flynn, J.J., Klitgordk, K.D., Obradovich, J.D. & Prothero, R. (1988) Paleogene geochronology: an integrated approach. *Paleoceanography* **3**, 707–742.

Beaumont, C. (1981) Foreland basins. *Geophys. J. R. Astron. Soc.* **65**, 219–329.

Berggren, W.A. & Miller, K.G. (1988) Paleogene tropical planktonic foraminiferal biostratigraphy and magnetobiochronology. *Micropaleontology* **345**, 362–380.

Betzler, Ch. (1989a) The upper Paleocene to middle Eocene between the Rio Segre and Rio Llobregat (Eastern South Pyrenees). Facies, stratigraphy and structural evolution. *Tübinger Geowiss. Arb.* **2**, 113 pp.

Betzler, Ch. (1989b) A carbonate complex in an active foreland basin. The Paleogene of the Sierra de Port del Comte and the Sierra del Cadi (southern Pyrenees). *Geodin. Acta* **3**, 207–220.

Betzler, Ch., Eichenseer, H. & Luterbacher, H.-P. (1987) The correlation of Early Eocene deposits across the Segre Fault (southern Pyrenees, Spain). *Neues Jahrb. Geol. Palaeontol. Monatsh.* **1**, 1–4.

Brouwer, M.J.N. & Schwander, M.M. (1988) Cenozoic carbonate banks, Foz do Amazonas Basin, Northeastern Brazil. In: *Atlas of Seismic Stratigraphy*, Vol. 2 (Ed. Bally, A.W.). Am. Assoc. Petrol. Geol., Stud. Geol. 27, 174–178.

Budd, D.A. & Loucks, G. (1981) Smackover and lower Buckner formations, South Texas: depositional systems on a Jurassic carbonate ramp. *Univ. Tex. Bur. Econ. Geol., Rep. Invest.* **112**. 38 pp.

Camara, P. & Klimowitz, J. (1985) Interpretación geodinámica de la vertiente centro-occidental surpirenaica (cuencas de Jaca-Tremp). *Estud. Geol.* **41**, 391–404.

Cant, D.J. & Stockmal, G.S. (1989) Tectonic controls of depositional environments and sedimentary sequences in foreland basins. *Bull., Am. Assoc. Petrol. Geol.* **73**, 340.

Caro, Y., Luterbacher, H.-P., Perch-Nielsen, K. & Premoli Silva, I. (1975) Zonations à l'aide de microfossiles du Paléocène supérieur et de l'Eocène infèrieur. *Bull. Soc. Géol. Fr.* **17**(7), 125–147.

Cavelier, C. & Pomerol, C. (1986) Stratigraphy of the Paleogene. *Bull. Soc. Géol. Fr.* **2**(8), 255–265.

Choukroune, Ph. (1976) Strain patterns in the Pyrenean chain. *Philos. Trans. R. Soc. London, Ser. A* **283**, 271–280.

Coleman, J.M. & Prior, D.B. (1982) Deltaic environments of deposition. In: *Sandstone Depositional Environments* (Eds Scholle, P.A. & Spearing, D.). Mem., Am. Assoc. Petrol. Geol. 31, 139–178.

Costa, J.M. (1985) Estratigrafía i facies del Paleogè de la unitat Cadí-Ripoll entre els Rius Greixer i Arija. *Publ. Fac. Ciències, Universitat Autònoma de Barcelona, Bellaterra*, 123 pp.

Crumeyrolle, Ph. (1987) *Stratigraphie physique et sédimentologie des systèmes dépôt de la séquence de Santa Lliestra (Eocène sudpyrénéen, Pyrénées aragonaises).* Unpublished PhD thesis, Université Bordeaux III, 216 pp.

Crumeyrolle, Ph. & Mutti, E. (1986) Stratigraphie et sédimentologie des systèmes de dépôt de plateforme de la séquence de Santa Lliestra (bassin éocène sudpyrénéen, province de Huesca, Espagne). *C.R. Acad. Sci., Ser. II* **303**, 581–584.

Cuevas Gonzalo, M., Donselaar, M.E. & Nio, S.D. (1985) Eocene clastic tidal deposits in the Tremp–Graus Basin (prov. of Lerida and Huesca). In: *International Association of Sedimentologists 6th European Regional Meeting. Excursion Guidebook* (Eds Milá, M. & Rosell, J.). Spec. Publs. Int. Ass. Sediment. 6, 217–305.

Eichenseer, H. (1988) Facies geology of late Maestrichtian to early Eocene coastal and shallow marine sediments, Tremp–Graus Basin, northeastern Spain. *Arb. Inst. Mus. Geol. Pal. Univ. Tübingen* **1**, 273 pp.

Eichenseer, H. & Betzler, Ch. (1987) Semiaride Gezeiten- und Strandseeablagerungen (Paläogen, Südpyrenäen). *Heidelberger Geowiss Abh.* **8**, 58–59.

Eichenseer, H. & Krauss, S. (1985) The Tremp Formation (Maestrichtian, Paleogene) and the lower Ager Group (Paleogene) of the northern flank of the Tremp–Graus embayment (NE Spain). In: *International Association of Sedimentologists 6th European Regional Meeting, Abstracts and Posters* (Eds Milà, M. & Rosell, J.) pp. 149–151.

Enos, P. (1983) Shelf environments. In: *Carbonate Depositional Environments* (Eds Scholle, P.A., Bebout, D.G. & Moore, C.H.). Mem., Am. Assoc. Petrol. Geol. 33, 267–295.

Erskine, R.D. & Vail, P.R. (1988) Seismic stratigraphy of the Exmouth Plateau. In: *Atlas of Seismic Stratigraphy*, Vol. 2 (Ed. Bally, A.W.). Am. Assoc. Petrol. Geol., Stud. Geol. 27, 163–173.

Esteban, M. & Klappa, C.F. (1983) Subaerial exposure environment. In: *Carbonate Depositional Environments* (Eds Scholle, P.A., Bebout, D.G. & Moore, C.H.). Mem., Am. Assoc. Petrol. Geol. 33, 1–54.

Ferrer, J., Le Calvez, Y., Luterbacher, H.-P. & Premoli Silva, I. (1973) Contribution à l'étude des foraminifères ilerdiens de la région de Tremp (Catalogne). *Mem. Mus. Hist. Nat. Paris, Ser. C* **24**, 78 pp.

Fonnesu, F. (1984) Estratigrafía física y análisis de facies de la secuencia de Figols entre el Rio Noguera Pallaresa e Iscles (Prov. de Lerida y Huesca). *Publ. Fac. Ciències, Universitat Autònoma de Barcelona, Bellaterra*, 317 pp.

Freytet, P. & Plaziat, J.C. (1982) Continental carbonate sedimentation and pedogenesis — Late Cretaceous and Early Tertiary of southern France. *Contrib. Sediment.* **12**, 213 pp.

Galloway, W.E. & Cheng, E.S. (1985) Reservoir facies architecture in a microtidal barrier system — Frio Formation, Texas Gulf Coast. *Univ. Tex. Bur. Econ. Geol., Rep. Invest.* **144**, 36 pp.

Garrido-Megias, A. (1973) Estudio geológico y relación entre tectónica y sedimentación del secundario y terciario de la vertiente meridional pirenaica en su zona central. Unpublished PhD thesis, Universidad de Granada, 395 pp.

Garrido-Megias, A. (1988) La tectónica pirenáica en relación con la evolución alpina del Margen Noribérico. *Rev. Soc. Geol. España* **1**, 365–372.

Garrido-Megias, A. & Rios Aragües, L.M. (1972) Síntesis geológica entre los ríos Cinca y Segre. *Bol. Geol. Min.* **LXXXIII**, 1–47.

Haq, B.U., Hardenbol, J. & Vail, P.R. (1987) Chronology of fluctuating sea levels since the Triassic. *Science* **235**, 1156–1167.

Hardenbol, J. & Berggren, W.A. (1978) A new Paleogene numerical time scale. In: *Contributions to the Geologic Time Scale* (Eds Cohee, G.V., Glaessner, M.F. & Hedberg, H.D.). Am. Assoc. Petrol. Geol., Stud. Geol. 6, 213–235.

Hillebrandt, A., von (1975) Corrélation entre les biozones de grands foraminifères et de foraminifères planctoniques de l'Ilerdien. *Bull. Soc. Géol. Fr.* **17**(7), 162–167.

Hottinger, L. (1973) Selected Paleogene larger Foraminifera. In: *Atlas of Palaeobiography* (Ed. Hallam, A.) pp. 443–452. Elsevier, Amsterdam.

Hubbard, R.J. (1988) Age and significance of sequence boundaries on Jurassic and Early Cretaceous rifted continental margins. *Bull., Am. Assoc. Petrol. Geol.* **72**, 49–72.

Hubbard, R.J., Pape, J. & Roberts, D.G. (1985a) Depositional sequence mapping as a technique to establish tectonic and stratigraphic framework and evaluate hydrocarbon potential on a passive continental margin. In: *Seismic Stratigraphy*, II: *an Integrated Approach to Hydrocarbon Exploration* (Eds Berg, O.R. & Woolverton, D.G.). Mem., Am. Assoc. Petrol. Geol. 39, 79–91.

Hubbard, R.J., Pape, J. & Roberts, D.G. (1985b) Depositional sequence mapping to illustrate the evolution of a passive continental margin. In: *Seismic Stratigraphy*, II: *an Integrated Approach to Hydrocarbon Exploration* (Eds Berg, O.R. & Woolverton, D.G.). Mem., Am. Assoc. Petrol. Geol. 39, 94–115.

Inden, R.F. & Moore, C.H. (1983) Beach environment. In: *Carbonate Depositional Environments* (Eds Scholle, P.A., Bebout, D.G. & Moore, C.H.). Mem., Am. Assoc. Petrol. Geol. 33, 211–265.

James, N.P. (1984) Shallowing upward sequences in carbonates. In: *Facies Models* (Ed. Walker, R.G.). Geosci. Can. Reprint Ser. 1, 213–228.

James, N.P. & Choquette, P.W. (1984) Diagenesis 9–limestones–the meteoric diagenetic environment. *Geosci. Can.* **11**, 161–194.

Kapellos, C. & Schaub, H. (1973) Zur Korrelation von Biozonierungen mit Grossforaminiferen und Nannoplankton im Paläogen der Pyrenäen. *Eclogae Geol. Helv.* **66**, 687–737.

Kendall, C.G.St.C. & Schlager, W. (1981) Carbonates and relative changes in sea level. *Marine Geol.* **44**, 181–212.

Kerr, R.S. (1977) Facies, diagenesis, and porosity development in a lower Cretaceous bank complex. Edwards Limestone, North-Central Texas. In: *Cretaceous Carbonates of Texas and New Mexico* (Eds Bebout, P.G. & Loucks, R.G.). Univ. Tex. Bur. Econ. Geol., Rep. Invest. 89, 216–233.

Khalaf, F.I. (1988) Petrography and diagenesis of silcretes from Kuwait, Arabian Gulf. *J. Sediment. Petrol.* **58**, 1014–1022.

Luterbacher, H.-P. (1973) La sección tipo del piso Ilerdiense. *XIII Coloquio Europeo de Micropaleontología, España* (Ed. Perconig, E.) pp. 113–140. Empresa Nacional 'Adaro' de Investigaciones Mineras S.A., Madrid.

McPherson, J.G., Shanmugam, G. & Moiola, R. (1987) Fan-deltas and braid deltas: varieties of coarse-grained deltas. *Geol. Soc. Am. Bull.* **99**, 331–340. (See also discussion and reply (1988) *Geol. Soc. Am. Bull.* **100**, 1308–1310.)

Maestro, E. (1985) Estratigrafía física i facies del Paleogé de la unitat Cadí-Ripoll entre els rius Segre i Greixer. *Publ. Fac. Ciències, Universitat Autònoma de Barcelona, Bellaterra*, 136 pp.

Muñoz, J.A., Puigdefabregas, C. & Fontbote, J.M. (1987) Orógenos alpínos. a) El Pirineo. b) El ciclo alpíno y la estructura tectónica del Pirineo. In: *Geología de España* (Ed. Comba, J.A.). Libro Jubilar J.M. Rios, 2, 161–205.

Mutti, E. (1984) The Hecho submarine fan systems, South-Central Pyrenees, Spain. *Geomar. Lett.* **3**, 199–202.

Mutti, E. (1985) Turbidite systems and their relations to depositional sequences. In: *Provenance of Arenites* (Ed. Zuffa, G.G.). NATO Advanced Science Institute Ser.

148, 65–93.

MUTTI, E., LUTERBACHER, H.-P., FERRER, J. & ROSELL, J. (1972) Scheme stratigrafico e lineamenti di facies del paleogene marino della zona centrale sudpirenica tra Tremp (Catalogna) e Pamplona (Navarra). *Mem. Soc. Geol. Ital.* **11**, 319–416.

MUTTI, E. & RICCI-LUCCHI, F.R. (1972) Le torbiditi dell Appenino settentrionale: introduzione all'analisi di facies. *Mem. Soc. Geol. Ital.* **11**, 161–199.

MUTTI, E., REMACHA, E., SGAVETTI, M., ROSELL, J., VALLONI, R. & ZAMORANO, M. (1985a) Stratigraphy and facies characteristics of the Eocene Hecho Group turbidite systems, South-Central Pyrenees. In: *International Association of Sedimentologists 6th European Regional Meeting. Excursion Guidebook* (Eds Milà, M. & Rosell, J.). Spec. Publs. Int. Ass. Sediment, 6, 521–578.

MUTTI, E., ROSELL, J., ALLEN, G., FONNESU, F. & SGAVETTI, M. (1985b) The Eocene Baronía tide dominated delta-shelf system in the Ager Basin. In: *International Association of Sedimentologists 6th European Regional Meeting. Excursion Guidebook* (Eds Milà, M. & Rosell, J.). Spec. Publs. Int. Ass. Sediment. 6, 579–600.

NUMMEDAL, D. & SWIFT, D.J.P. (1987) Transgressive stratigraphy at sequence-bounding unconformities: some principles derived from Holocene and Cretaceous examples. In: *Sea-level Fluctuation and Coastal Evolution* (Eds Nummedal, D., Pilkey, O.H. & Howard, J.D.). Soc. Econ. Paleontol. Mineral., Spec. Publ. 41, 241–260.

OBRADOROVICH, J.D. (1988) A different perspective perspective on glauconite as a chronometer for geologic time scale studies. *Paleoceanography* **3**, 757–770.

ODIN, G.S. & CURRY, D. (1985) The Paleogene time scale: radiometric dating versus magnetostratigraphic approach. *J. Geol. Soc. London* **142**, 1179–1188.

ORI, G. & FRIEND, P. (1984) Sedimentary basins formed and carried piggyback on active thrust sheets. *Geology* **12**, 475–478.

PARKINSON, N. & SUMMERHAYES, C. (1985) Synchronous global sequence boundaries. *Bull., Am. Assoc. Petrol. Geol.* **69**, 685–687.

PLAZIAT, J.C. (1984) *Le domaine Pyrénéen (de la fin du Crétacé à la fin de l'Eocène). Stratigraphie, paléoenvironements et évolution paléographique*. Unpublished thèse Doct., Sci. Nat. Université, Paris, 1362 pp.

PUIGDEFÀBREGAS, C. & SOUQUET, P. (1986) Tecto-sedimentary cycles and depositional sequences of the Mesozoic and Tertiary from the Pyrenees. *Tectonophysics* **129**, 173–203.

PUIGDEFÀBREGAS, C., MUÑOZ, J.A. & MARZO, M. (1986) Thrust belt development in the eastern Pyrenees and related depositional sequences in the southern foreland basin. In: *Foreland Basins* (Eds Allen, P.A. & Homewood, P.). Spec. Publs. Int. Ass. Sediment. 8, 319–336.

READ, J.F. (1985) Carbonate platform facies models. *Bull., Am. Assoc. Petrol. Geol.* **69**, 1–21.

REMACHA, E., ARBUES, P. & CARRERAS (1986a) Precisiones sobre los límites de la secuencia deposicional de Jaca. Evolución de las facies desde la base de la secuencia hasta el techo de la Arenisca de Sabiñanigo. *Bol. Geol. Min.* **XCVIII-I**, 40–48.

REMACHA, E., ROSELL, J. & ZAMORANO, M. (1986b) Aspectos estratigráficos y paleogeográficos del Eoceno Inferior y Medio en el Prepirineo Central. *Abstracts XI Congr. Español de Sedimentología, Barcelona*, p. 144.

ROSELL, J., FERRER, J. & LUTERBACHER, H.-P. (1973) El Paleogeno marino del noreste de España. *XIII Coloquio Europeo de Micropaleontología España* (Ed. Perconig, E.) pp. 29–61. Empresa Nacional 'Adaro' de Investigaciones mineras S.A. Madrid.

SARG, J.F. (1988) Carbonate sequence stratigraphy. In: *Sea-level Changes: an Integrated Approach* (Eds Wilgus, C.K., Hastings, B.S., Kendall, C.G.St.C., Posamentier, H., Ross, C.A. & van Wagoner, J.). Soc. Econ. Paleontol. Mineral., Spec. Publ. 42, 155–181.

SCHAUB, H. (1973) La sección de Campo. *XIII Coloquio Europeo de Micropaleontología, España* (Ed. Perconig, E.) pp. 151–170. Empresa Nacional 'Adaro' de Investigaciones Mineras S.A. Madrid.

SCHAUB, H. (1981) Nummulites et assilines de la Téthys paléogène. Taxonomie, phylogénèse et biostratigraphie. *Mem. Suiss. Paleontol.* **104–106**, 1–238.

SEGURET, M. (1972) Etude tectonique des nappes et series décollées de la partie centrale du versant sud des Pyrénées. *Publ. Université Sci. Techn. Languedoc, Sér. Géol. Struct.* **2**, 155 pp.

SHINN, E.A. (1983) Tidal flat environment. In: *Carbonate Depositional Environments* (Eds Scholle, P.A., Bebout, D.G. & Moore, C.H.). Mem., Am. Assoc. Petrol. Geol. 33, 171–210.

SLEEP, N.A. (1976) Platform subsidence mechanisms and 'eustatic' sea level changes. *Tectonophysics* **36**, 45–56.

SLOSS, L.L. (1963) Sequences in the cratonic interior of North America. *Geol. Soc. Am. Bull.* **74**, 93–114.

SUMMERHAYES, C.P. (1986) Sea level curves based on seismic stratigraphy: their chronostratigraphic significance. *Palaeogeogr. Palaeoclimatol. Palaeoecol.* **57**, 27–42.

TUCKER, M. (1985) Shallow marine carbonate facies and facies models. In: *Sedimentology. Recent Developments and Applied Aspects.* (Eds Brenchley, P.J. & Williams, B.). *Geol. Soc. London*, Spec. Publ. 18, 147–169.

VAIL, P.R. (1987) Seismic stratigraphy interpretation using sequence stratigraphy. In: *Atlas of Seismic Stratigraphy*, Vol. 1 (Ed. Bally, A.W.). Am. Assoc. Petrol. Geol., Stud. Geol. 27, 1–10.

VAIL, P.R., MITCHUM, R.M., JR., TODD, R.G., WIDMIER, J.M., THOMPSON, S., III, SANGREE, J.B., BUBB, J.N. & HATLELID, W.G. (1977) Seismic stratigraphy and global changes of sea level. In: *Seismic Stratigraphy – Applications to Hydrocarbon Exploration* (Ed. Payton, C.E.). Mem., Am. Assoc. Petrol. Geol. 26, 49–212.

VAIL, P.R., HARDENBOL, J. & TODD, R.G. (1984) Jurassic unconformities, chronostratigraphy and sea level changes from seismic stratigraphy and biostratigraphy. In: *Interregional Unconformities and Hydrocarbon Exploration* (Ed. Schlee, J.S.). Mem., Am. Assoc. Petrol. Geol. 36, 129–144.

VAN DEN HURK, A.M. (1989) Control eustatico y tectónico de los sistemas deposicionales Paleógenos siliciclasticos y carbonatados de la cuenca de antepais surpirenaica (rios Esera, Cinca y Cinqueta; Prov. de Huesca, España). *Publ. Fac. Cienciès, Universitat Autònoma de Barcelona, Bellaterra*, 237 pp.

van den Hurk, A.M. & Eichenseer, H. (1988) Los sedimentos del Paleoceno-Eoceno inferior entre los rios Esera, Cinca y Zinqueta. Consecuencias tectoestratigráficas. *Proc. II Congr. Nac. de Geol. de España*, Instituto de Geologia, Universidad de Bilbao, 95–98.

Wheeler, H.E. (1958) Time–stratigraphy. *Bull., Am. Assoc. Petrol. Geol.* **42**, 1047–1063.

Williams, G.D. (1985) Thrust tectonics in the South Central Pyrenees. *J. Struct. Geol.* **7**, 11–17.

Williams, G.D. & Fischer, M.W. (1984) A balanced section across the Pyrenean orogenic belt. *Tectonics* **3**, 773–780.

Wright, V.P. (1984) Peritidal carbonate facies models: a review. *Geol. J.* **19**, 309–325.

*Spec. Publs. int. Ass. Sediment.* (1991) **12**, 409–428

# High-frequency relative sea-level oscillations in Upper Cretaceous shelf clastics of the Alberta foreland basin: possible evidence for a glacio-eustatic control?

A. GUY PLINT

*Department of Geology, University of Western Ontario, London, Ontario, N6A 5B7, Canada*

## ABSTRACT

The Muskiki (mid–late Coniacian) and Marshybank (late Coniacian to early Santonian) Formations vary from about 50 to 100 m thick and are present throughout much of the Alberta foreland basin south of the Peace River. Together they form a sequence comprising transgressive and highstand systems tracts (TST and HST) that span about 1.5 Ma. Sequence-bounding unconformities developed during times of sea-level fall that appear to be regionally, if not globally, recognizable. The Muskiki–Marshybank sequence consists of at least 15, *regionally traceable* parasequences, each of which apparently lasted about 100 000 years. In much of the TST, parasequences generally shoal upward with little evidence of intervening relative sea-level fall. Towards the top of the TST, however, chert pebble beds mantle two parasequences and chert pebbles and/or erosive-based shoreface sand bodies cap most parasequences in the HST. These features suggest deposition as a result of relative sea-level fall, perhaps of only a few metres. Various lines of evidence suggest that the parasequences are due to eustatic fluctuations rather than to changes in subsidence or sediment-supply rates. The approximate 100 000 year frequency of the eustatic cycles is suggestive of an orbitally forced glacio-eustatic mechanism. Parasequences fall into five quasi-periodic groups that may reflect either changes in subsidence rate or eustatic fluctuations over several hundred thousand years.

## INTRODUCTION

The Cretaceous rocks of the Alberta Basin (see Fig. 1 for location of study area) provide one of the world's best natural laboratories in which to observe the relationships between sedimentation and relative sea-level fluctuations. Extensive outcrop along the deformed western margin of the basin facilitates detailed facies analysis, and public access to many thousands of well logs (and cores where available) permits regional mapping of facies and stratal geometry.

The Coniacian–Santonian Muskiki and Marshybank Formations appear to have been deposited in shallow, shoreface and inner to mid-shelf environments, which were very sensitive to even minor relative sea-level changes. The purpose of this paper is to illustrate the sedimentary response to such sea-level changes, and, in particular, to highlight the evidence for significant relative sea-level *falls* with a recurrence interval of approximately 100 000 years.

### Previous work

Plint *et al.* (1990) reviewed the history of work on the Muskiki, Marshybank, and Bad Heart Formations, which form a part of the Smoky (Alberta) Group (Fig. 2). Other than regional outcrop mapping (Stott, 1963,1967; Irish, 1965), little detailed sedimentological work on these rocks has been published. Plint & Walker (1987b) showed that the top of the Marshybank (then termed the 'Bad Heart') Formation was defined by a regional erosion surface with a relief of about 38 m. Subsequently, Norris (1989) and Plint *et al.* (1990), through detailed subsurface correlation, demonstrated that not only the Marshybank but also the underlying Muskiki Formation had suffered deep, post-Marshybank erosion that in places had completely removed both units, generating an erosional relief of at least 50 m over a lateral distance of about 300 km. In a regional allo-

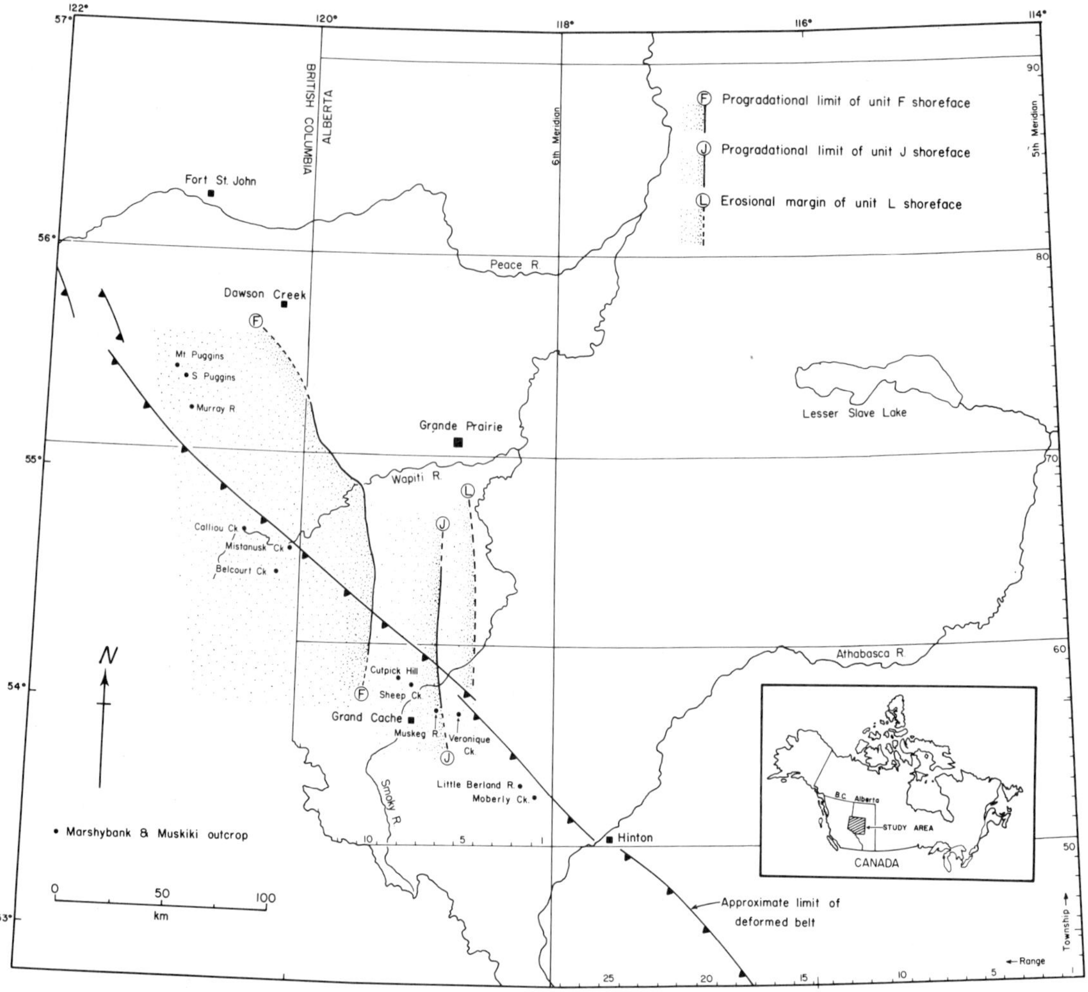


**Fig. 1.** Inset map shows the location of the study area. Main map shows study area and locations of principal outcrop sections in the Foothills. The progradational limits of shoreface sandstones of Marshybank units F, J, and L, determined from subsurface mapping (Plint, 1990), are also shown.

stratigraphic subdivision of the Muskiki and Marshybank Formations (Plint, 1990) documented, in subsurface, the stratal geometry of individual depositional units and showed how each unit could be correlated from subsurface into outcrop. Sedimentary facies successions and palaeocurrents in outcrop were related (Plint & Norris, 1991) to the geometry and distribution of individual depositional units in order to reconstruct depositional environments, palaeogeography, and sediment dispersal patterns.

## Methodology

Over 1500 gamma-ray and resistivity well-logs, 20 more or less complete outcrop sections, and eight cores form the data base for this study (see Plint (1990) and Plint & Norris (1991) for details).

The first stage of the analysis involved correlation of closely spaced lines of well-logs to determine the extent and stratal geometry of each discrete, mappable, upward-coarsening succession of rocks. Once a three-dimensional stratigraphic framework

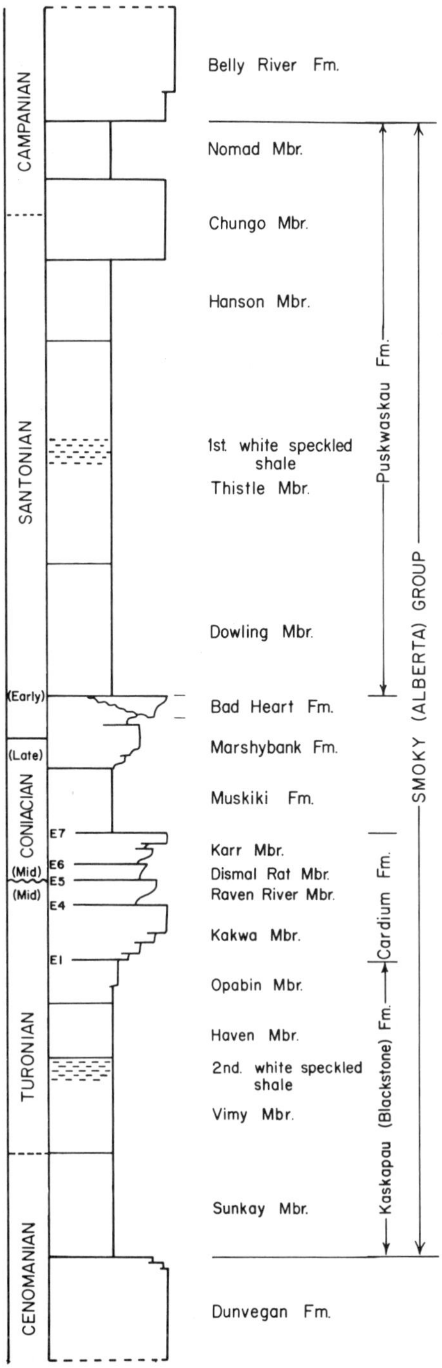


**Fig. 2.** Generalized stratigraphy of the Upper Cretaceous Smoky Group (= Alberta Group) of the Alberta Basin (modified after Stott, 1963,1967; Plint *et al.*, 1988,1990). The Smoky Group varies in thickness from about 380 to 1120 m.

had been established, it was possible to move to the second stage, that of outcrop to subsurface correlation, which enabled the well-logs to be 'calibrated' against known lithologies. Each outcrop section was correlated with the nearest (usually less than 10 km distant) available well-log. In every case, upward-coarsening successions recognizable in logs could be confidently matched with those in outcrop (Plint, 1990). With this stratigraphic and lithologic framework in place, it was then possible to determine the lateral variation in facies and environments, from which the palaeogeography could be inferred.

A comparable example of regional outcrop to subsurface correlation was presented by Plint *et al.* (1988) for the Turonian–Coniacian Cardium Formation in Alberta.

## MUSKIKI AND MARSHYBANK STRATAL GEOMETRY

Figure 3 summarizes the stratigraphic relationships of the major, regionally mappable upward-coarsening facies successions within the Muskiki and Marshybank Formations. The diagram is based on 19 detailed outcrop sections, eight cores, and over 1500 well-logs (data in Plint, 1990).

The Muskiki and Marshybank Formations are characterized by distinctly different stratal patterns. Stratification surfaces in the Muskiki are essentially parallel over 300 km from west to east, and subtly converge from south to north in response to regional differential subsidence. The base of the Marshybank Formation was defined, on lithostratigraphic grounds (Plint *et al.*, 1990), at a prominent gamma-ray and resistivity log deflection that corresponded to the base of the first prominent upward-coarsening facies succession that forms unit A of the Marshybank. The top of unit A (Fig. 3) is marked by a layer of chert pebbles overlying an extensively sideritized surface. Towards the east, the top of unit A converges very subtly with the underlying M2 marker horizon, which is also mantled with chert pebbles.

In contrast, the tops of Marshybank units B–F, which comprise a broadly progradational package of marine rocks, downlap on to the top of unit A from west to east (Fig. 3). Units H–L are also entirely marine; unfortunately, post-Marshybank–pre-Puskwaskau erosion has removed these units from the eastern part of the study area (Plint, 1990; Plint *et al.*, 1990), so downlap, if it existed, cannot be demonstrated. Unit F is overlain by non-marine

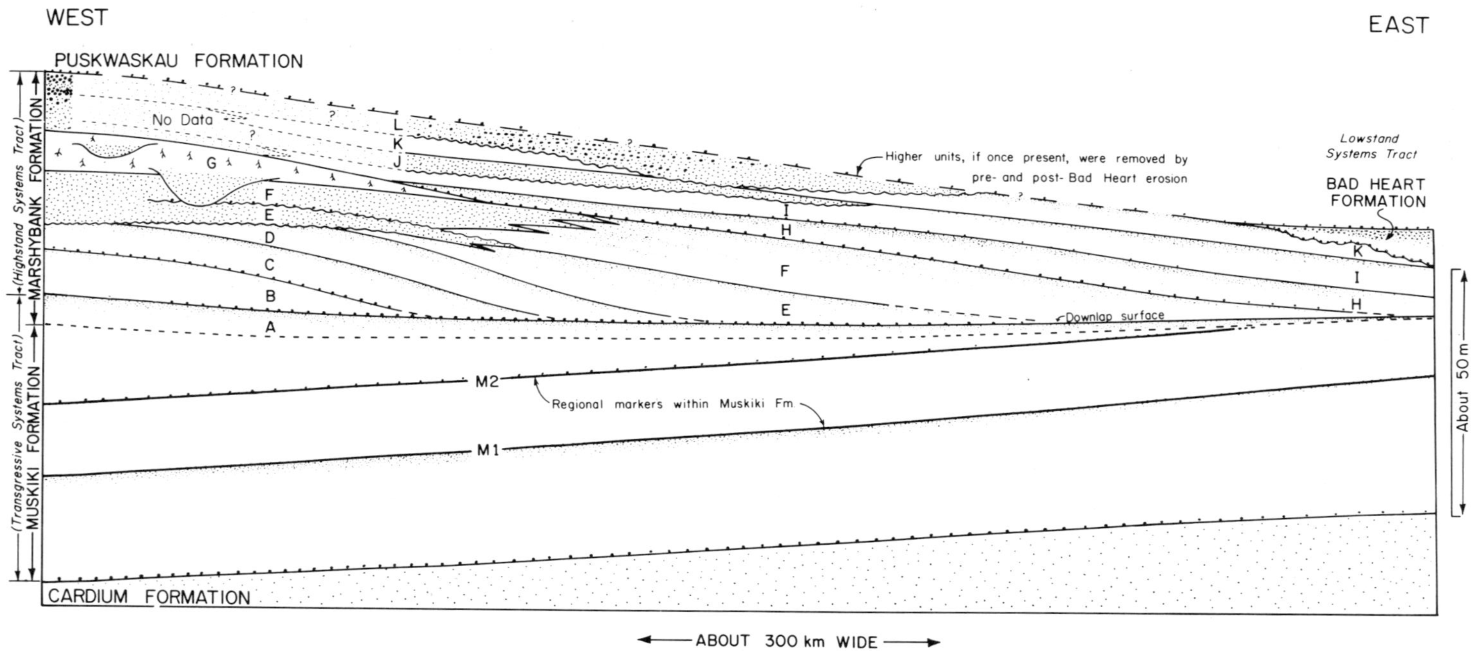


**Fig. 3.** Summary stratal geometry of the Muskiki and Marshybank Formations. The top of the Cardium Formation is a regional unconformity with about 40 m of erosional relief. M1 and M2 are regionally extensive marker horizons, the latter has a chert pebble veneer. The top of Marshybank unit A represents a 'maximum flooding (downlap) surface' separating the transgressive systems tract below from the highstand systems tract above. Light stipple indicates upward-coarsening; heavy stipple denotes swaley cross-stratified marine shoreface sandstones; roots indicate non-marine deposits. The extreme western part of the diagram represents the section at Mount Puggins, British Columbia (see Fig. 1 for location).

strata (unit G) representing lagoonal, lacustrine, and alluvial environments (Plint & Norris, 1991). Unit H comprises sandy siltstones and represents a marine transgression because it onlaps shoreface and non-marine strata of units F and G. Units I and K comprise silty shelf deposits, whereas units J and L consist of erosively based bodies of shoreface sandstone. Units H, I, J, K, and L form a second distinct progradational package.

In sequence-stratigraphic terms, the Muskiki plus Marshybank unit A comprise a 'transgressive systems tract'. The top of unit A represents the 'maximum flooding surface' on to which Marshybank units B–F (and once possibly units H–L) downlap from west to east (Fig. 3). Units B–L form a 'highstand systems tract', bounded at the top by a regional unconformity ('sequence boundary': van Wagoner *et al.*, 1988).

In the east, beneath the Plains, the Marshybank is unconformably overlain by siltstones, sandstones, and oolitic ironstones of the Bad Heart Formation (Plint *et al.*, 1990; Fig. 3). This formation, which appears to represent a shallow, clastic-starved environment, may constitute a lowstand systems tract. It is currently under investigation (W.S. Donaldson, PhD in progress) and will not be further discussed here. The Bad Heart and Marshybank are unconformably overlain by dark marine shales and siltstones of the mid-Santonian Puskwaskau Formation. Marker horizons in this transgressive systems tract progressively onlap the Marshybank from northeast to southwest.

## FACIES SUCCESSIONS

Two basic types of facies succession can be recognized in the Muskiki and Marshybank Formations: upward-coarsening heterolithic units, which may be erosively capped by a pebble bed of local or regional extent, and upward-coarsening heterolithic units erosively overlain by a shoreface sandstone.

### Upward-coarsening facies successions

Upward-coarsening facies successions, typically 5–10 m thick, characterize both the Muskiki Formation and Marshybank units A, B, C, D, H, I, and K, plus the more seaward portions of units E and F. Some of these units in turn consist of two or more, minor upward-coarsening successions. The base of each facies succession consists typically of dark laminated shale or bioturbated siltstone. This grades up into coarse-grained, bioturbated siltstone or heterolithic sediment comprising centimetre to decimetre thick interbeds of siltstone and ripple cross-laminated (wave and combined-flow types) or hummocky cross-stratified (HCS) very fine-grained sandstone; bioturbation has mixed these lithologies to a greater or lesser degree. The top of each succession is invariably sharp and may be overlain by a pebble bed consisting of a few centimetres of well-rounded chert pebbles and granules, sometimes including sandstone and sideritized siltstone intraclasts (Fig. 4). Where the top of the succession is sandy, pebble-filled *Thalassinoides* burrows may penetrate several decimetres below the erosion surface.

Chert pebbles are present at the base and on the M2 marker surface in the Muskiki (Fig. 3) and at the top of units A, B, E, F, and H, although *locally* they may be absent from any of these surfaces. Pebbles may also be locally present at the top of minor upward-coarsening successions within units B, H, and L. Chert pebbles are also present throughout the basin at the unconformable contact between the Marshybank and overlying Bad Heart and Puskwaskau Formations.

#### *Interpretation*

In the shelf setting represented by these rocks, each upward-coarsening facies succession may be interpreted in terms of shelf aggradation in response to shoreface progradation. This relationship is well illustrated by units E and F in which shoreface sandstones in the west (heavy stipple pattern, Fig. 3) pass laterally eastwards (offshore) into upward-coarsening successions. Shelf aggradation appears to have taken place when space (accommodation) was created by a relative rise in sea level. At such times, silt and clay deposition represents the temporally dominant 'fair weather' regime of the basin. The coarsest sediment reaching the shelf was fine-grained sand, transported during storms and forming centimetre to decimetre-scale rippled or HCS beds.

In contrast, the presence of pebbles mantling an eroded surface at the top of many of the upward-coarsening successions suggests much higher energy levels. The *regional* extent of erosion surfaces, plus the presence of pebbles, suggests formation as a result of relative sea-level *fall*. At such times, the shelf was sufficiently shallow that waves (?storm or fair weather) were able to erode early lithified sandstone beds and siderite nodules to form an intraclast

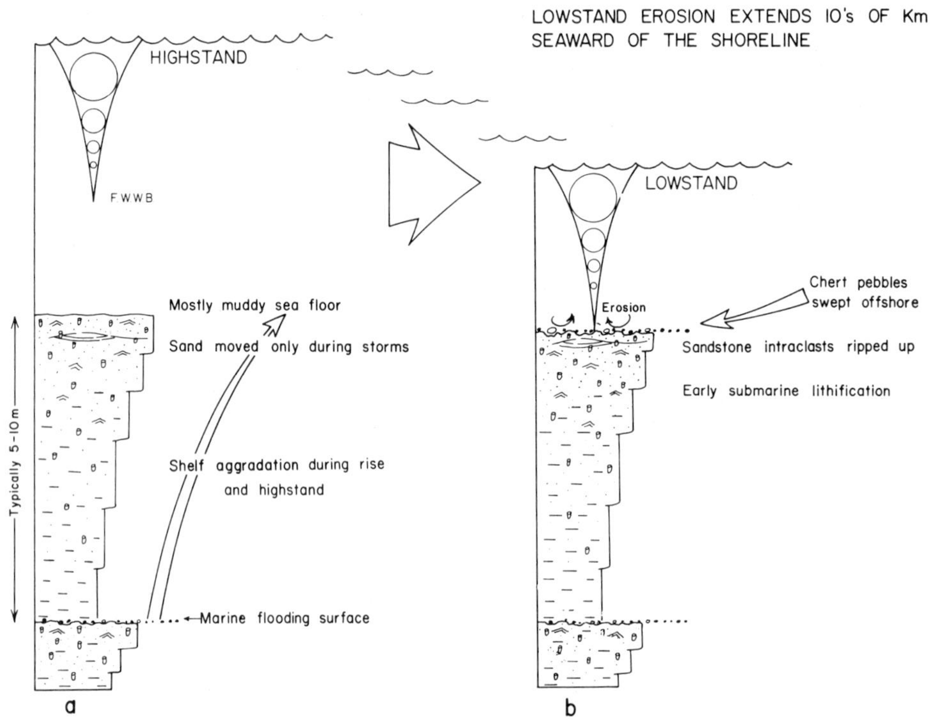


**Fig. 4.** Generalized features of upward-coarsening parasequences in the Muskiki and Marshybank. Relative sea-level rise generates space, allowing the shelf to aggrade, forming an upward-coarsening, upward-shallowing facies succession. Relative sea-level fall results in erosion of the top of the parasequence, generating an intraclast lag. In some instances, particularly in the highstand systems tract, chert pebbles were also transported offshore. It seems likely that, during lowstand, the inner shelf constituted a sediment-bypassing surface, which retained only the coarsest material (pebbles) in transit: finer sediments continued into deeper, offshore areas.

lag (Fig. 4). The presence of *chert* pebbles is more problematical. To some extent, the chert pebble beds in the Muskiki and Marshybank resemble those in the Cardium Formation in forming local to regional veneers that erosively overlie upward-coarsening facies successions (Plint *et al.*, 1986, 1988). The Cardium, however, also contains thick, erosively based linear conglomerate bodies, which are interpreted as lowstand shoreface deposits. Gravel was supplied to the lowstand shoreface by rivers that extended seawards across the exposed shelf (Bergman & Walker, 1987; Plint, 1988; Walker & Eyles, 1991). During transgression, a gravelly beach was driven landward from the lowstand shoreline position, leaving a veneer of pebbles on an erosion surface cut during passage of the surf zone. In contrast, gravelly lowstand shoreface deposits have not been recognized in either the Muskiki or Marshybank. Therefore, it is not easy to appeal to transgressive reworking of lowstand deposits to explain the chert pebble veneers capping upward-coarsening successions. Instead, the emplacement of gravel may have taken place when relative sea-level fall lowered fair-weather wave-base close to the seabed. Waves, particularly during storms, may then have been capable of moving pebbles (together with finer grained sediment) out from the shoreline for several tens of kilometres across an extremely low-gradient shelf comprising an eroded, 'firm ground' surface too shallow and energetic to retain anything but pebbles: silt and sand were 'bypassed' into deeper water offshore (cf. Nummedal & Swift, 1987).

### Erosive-based shoreface sand bodies overlying upward-coarsening successions

The more landward (western) portions of Marshybank units E and F, plus units J and L, comprise distinctive, sharp-based, fine-grained sandstone bodies (Fig. 3). Swaley cross-stratification (SCS) is the dominant sedimentary structure, although the sandstone may be structureless. The upper metre of unit F is parallel-laminated and capped by roots and non-marine deposits of unit G.

Each sand body rests on an erosion surface. Erosion on the scale of metres can be demonstrated through detailed correlation of closely spaced well-logs. In outcrop, flute casts and gutter casts up to about 70 cm deep are typically present on the basal surface of each sand body (Figs 5–7). Gutter casts are essentially parallel to each other and are orientated perpendicular to regional shoreline trends (determined from subsurface facies mapping; Plint & Norris, 1991; Fig. 1). Sand bodies vary from about 2 to 8 m thick. Of the four sand bodies, however, the tops of E, J, and L are defined by erosion surfaces. Only unit F, which averages 5.5 m thick, has a non-erosive (rooted) upper surface and apparently preserves the original thickness of the shoreface.

The upper parts of the heterolithic strata that underlie each SCS sandstone, comprising units D, I, and K (the latter being very poorly represented in outcrop), plus the non-SCS (offshore) portions of units E and F, usually contain large gutter casts, typically 10–30 cm deep and 20–80 cm wide (Fig. 5). The gutter casts commonly have steep walls and are usually isolated within thinly interbedded siltstones and sandstones (Fig. 8), or they may be attached to the base of tabular sandstone beds that may locally thicken from 1 to 2 cm into an isolated 'starved' hummock, 10–20 cm thick. Commonly the 'starved hummocks' directly overlie isolated gutter casts (Fig. 9). In every example, the isolated gutter casts were essentially parallel to those on the base of overlying or laterally equivalent sharp-based SCS sand bodies. Outcrop data indicate that in units E and F isolated gutter casts and hummocks are present for at least 50 km beyond the seaward limit of the SCS shoreface sand body.

### *Interpretation*

Although some workers (e.g. Allen & Underhill, 1989) have recently questioned the environmental significance of SCS, there is little doubt that, in the Cretaceous of the Alberta Basin, SCS is a reliable indicator of a shoreface environment, particularly as it is commonly overlain by beach and rooted non-marine deposits, and underlain by a thickening upwards succession of HCS sandstone and siltstone interbeds, representing the lower shoreface to offshore transition (e.g. Leckie & Walker, 1982; Plint & Walker, 1987a; McCrory & Walker, 1986; Rosenthal & Walker, 1987). Although only unit F of the Marshybank preserves unequivocal evidence of emergence, it seems reasonable to interpret each sharp-based SCS sandstone in the Marshybank as representative of a shoreface environment, despite the fact that the SCS sand body is never gradationally underlain by a succession of HCS beds (i.e. it is a 'base-absent' shoreface succession).

It seems unlikely that these tabular sand bodies originated as a vertically aggrading offshore sand sheet, or were deposited during transgression, when the shelf is usually starved of coarse clastics.

The relationship between sharp-based SCS sandstone bodies and relative sea-level fall was discussed by Plint (1988). Using subsurfaces data from the Cardium Formation, Plint (1988) showed that, as a result of relative sea-level fall, the space available to be filled by a 'normal' gradually upward-coarsening shelf to shoreline sequence was greatly reduced. Relative sea-level fall resulted in the erosion of formerly aggrading inner shelf areas as fair-weather wave-base (FWWB) approached the sea-bed. The bathymetric space below FWWB, formerly occupied by the lower shoreface and inner shelf environments (typically accumulating HCS), was eliminated by falling sea level, enabling SCS sandstones of the middle/upper shoreface to prograde directly over an erosion surface cut in offshore mudstones. Under these circumstances, it is important to recognize that the basinward movement of the shoreline is not so much the result of sediment 'passively' filling the space between sea level and sea-bed but is instead a 'forced' regression, brought about by a relative fall in sea level.

The explanation developed for sharp-based SCS sandstones in the Cardium is equally applicable to those in the Marshybank. In each of units E, F, J, and L, a shoreface sand body can be shown to have prograded many tens of kilometres into the basin (Fig. 1). The sharp base to each unit is evidence that relative sea-level fall caused the shoreface to prograde over a wave-scoured muddy shelf into water little deeper than the thickness of the sand body (< about 8 m). A Pleistocene analogue of the Marshybank sand bodies is provided by the sharp-based Ingleside sand, which prograded several tens of kilometres on to the Texas shelf during a mid-Wisconsinan sea-level fall (Wilkinson *et al.*, 1975).

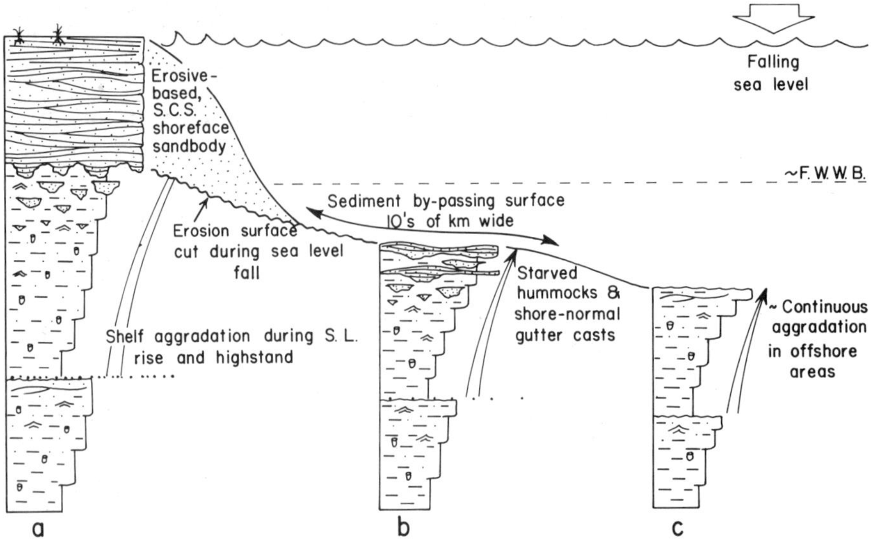


**Fig. 5.** Schematic illustration of the lateral variation in facies succession that existed during progradation of an erosive-based shoreface as a result of relative sea-level fall. (a) An upward-coarsening facies succession, typically 5–10 m thick, deposited during relative sea-level rise and highstand, is abruptly truncated by an erosive-based, SCS shoreface sand body, deposited during relative sea-level fall. Prominent, shore-normal gutter casts are present on the base of the sand body, and also commonly occur as isolated bodies in the underlying, heterolithic sediment (see Figs 6, 7). (b) The upward-coarsening succession terminates in 1–2 m of lenticular 'starved' hummocks and large, shore-normal gutter casts (see Figs 8, 9) that extend at least 50 km seaward of the shoreface. These structures indicate frequent erosion of the shelf and overall sand starvation, suggesting an environment dominated by sediment bypassing rather than aggradation. (c) An offshore area sufficiently deep so that it was unaffected by erosion at sea-level lowstand: here aggradation appears to have been continuous, generating a 'typical' parasequence.

The lateral passage, in units E and F, of SCS sandstones into heterolithic strata containing 'starved' HCS and isolated gutter casts suggests that, of the sand 'escaping' from the shoreface, little was retained in the inner shelf area, where it formed 'starved' hummocks and gutter casts; most was apparently bypassed into deeper water offshore (Figs 5, 10). A comparable assemblage of isolated

**Fig. 6.** Muskiki and Marshybank Formations exposed on Mistanusk Creek, British Columbia. Units A–G of the Marshybank are indicated. Note the large gutter casts (arrowed) on the base of shoreface sandstones of unit E.

**Fig. 7.** Marshybank Formation exposed on Muskeg River, Alberta. Unit J rests sharply on unit I; large gutter casts are present at the base of J and in the upper 3 m of I (arrows).

**Fig. 8.** Example of an isolated gutter cast near the top of Marshybank unit E, Muskeg River. Scale bar = 20 cm.

**Fig. 9.** An example of a 'starved' hummock immediately overlying a gutter cast at the top of unit E, Muskeg River. At this locality, unit E consists of heterolithic siltstones and sandstones, having been deposited seaward of the shoreface. Scale bar = 20 cm.

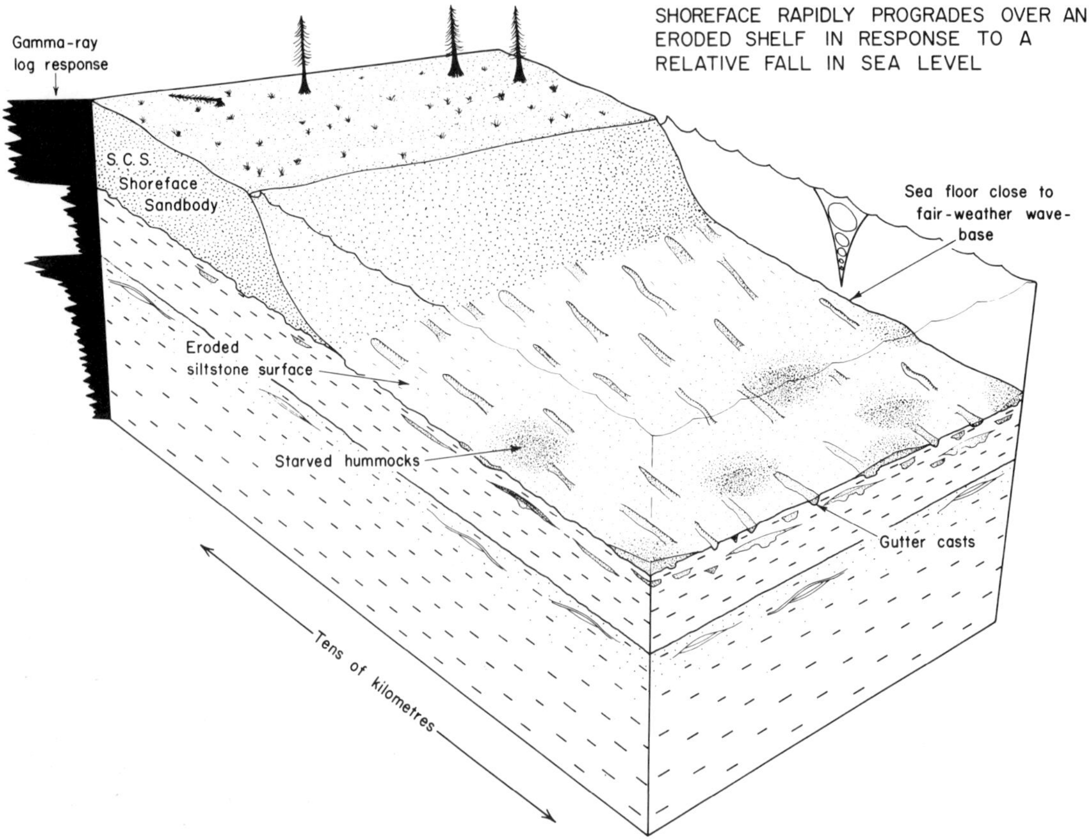


**Fig. 10.** Cartoon (not to scale) summarizing the features associated with sharp-based shoreface sandbodies. Relative sea-level fall exposes the inner shelf to ?fair weather wave action, which halts aggradation of muddy sediment and promotes rapid progradation of a sharp-based SCS shoreface sand body. During storms, oscillatory wave and wind forced currents scour the shelf, cut deep gutter casts (the planform of which is entirely speculative), and mould a patchy veneer of fine sand into 'starved' hummocks.

gutter casts and 'starved' HCS in Lower Cambrian shallow shelf sediments has been interpreted in similar terms (Myrow *et al.*, 1988). The origin of the large gutter casts is problematical. Although a detailed discussion of their genesis is beyond the scope of this paper, they are commonly interpreted in terms of high-energy events in which wave-action and/or storm-generated currents seem to have been important (e.g. Goldring, 1971; Aigner, 1985; Myrow *et al.*, 1988; Galli, 1989; Leckie & Krystinik, 1989).

## BIOSTRATIGRAPHIC CONTROL

Scaphitid ammonites and Inoceramid bivalves provide good biostratigraphic control in the Coniacian and Santonian rocks of the Alberta Basin. The stratigraphic distribution of zone fossils reported by Stott (1963,1967), has been integrated with the new allostratigraphic framework in Fig. 11.

The upper part of the Cardium Formation, comprising the Dismal Rat and Karr members (Fig. 2; Plint *et al.*, 1988; encompassing the Leyland and Sturrock members of Stott (1963)), contain the early Coniacian zone fossils *Scaphites preventricosus* and *Inoceramus deformis*. This fauna persists in the Muskiki Formation up to the M1 marker horizon (Fig. 11). The late Coniacian ammonite *Scaphites ventricosus* appears above marker M1 and is present through the remainder of the Muskiki. On Sheep Creek, Stott (1967) collected the basal Santonian ammonite *Scaphites depressus* from nodular siltstones (Marshybank unit E) containing abundant

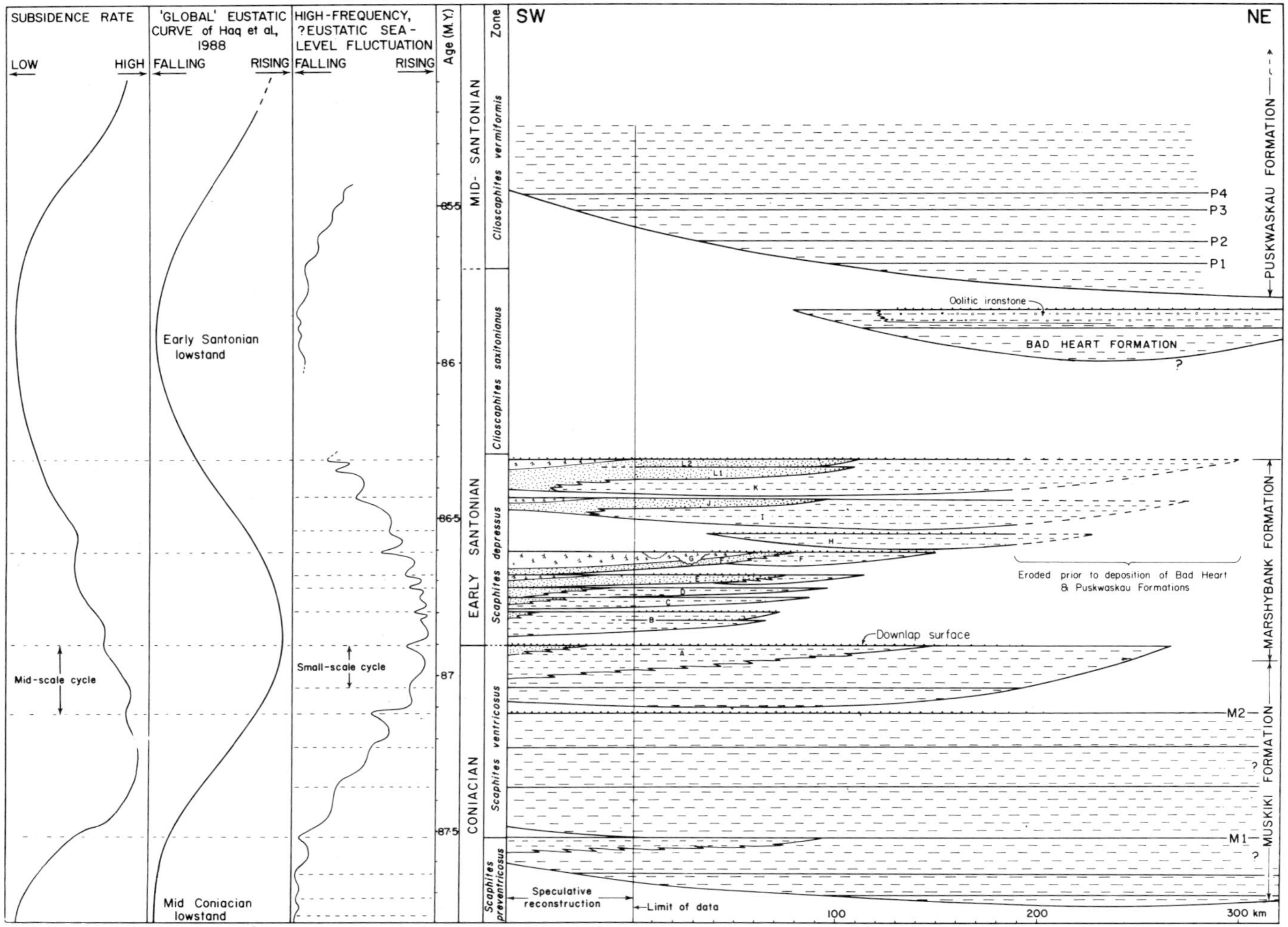


**Fig. 11.** Wheeler diagram illustrating the space–time distribution of rocks of the Muskiki, Marshybank, Bad Heart and lower Puskwaskau Formations. Time-scale averaged from four separate chronologies (see text); ammonite distribution data from Stott (1967). Third-order mid-Coniacian and early Santonian lowstands of Haq *et al.* (1988) correspond very closely with the parasequence stacking pattern and major regressive hiati observed in the rock record. The third-order cycle is modulated by numerous smaller scale sea-level oscillations with a frequency of approximately 100 000 years. A hypothetical subsidence rate curve incorporates a period of relatively rapid subsidence that may be expected to accompany accelerated thrust loading in response to accelerated spreading and subduction rates associated with a third-order eustatic cycle. Subtle fluctuations in subsidence rate *may* have generated mid-scale transgressive–regressive pulses over several hundred thousand years.

ammonites, immediately above unit A. At this locality, units B–D, which downlap further to the west, cannot be differentiated, but might be represented by a condensed interval, rich in ammonites, at the base of unit E. Thus, although the age of Marshybank units B–D has not been defined with certainty, it seems likely that the Coniacian–Santonian boundary lies at the top of Marshybank unit A.

The Bad Heart Formation contains the early Santonian zone fossil *Scaphites saxitonianus*, whereas overlying shales of the Dowling Member of the Puskwaskau Formation contain *Scaphites vermiformis*, typical of the mid-Coniacian (Fig. 11).

## CHRONOSTRATIGRAPHIC CONTROL

Figure 11 represents an attempt to integrate, in a Wheeler diagram (Wheeler, 1958), the spatial and temporal distribution of the rocks comprising the Muskiki, Marshybank, Bad Heart, and Lower Puskwaskau Formations. Four separate chronostratigraphic schemes have been averaged in a simplistic attempt to define the boundaries and durations (in Ma) of the Coniacian and Santonian stages, as shown in Table 1.

The duration of each stage was then divided by the number of faunal zones, on the further *assumption* that each zone represents approximately the same length of time. Thus the two ammonite zones of the Coniacian each represent about 625 000 years, and the six zones of the Santonian each represent about 603 000 years. Furthermore, it was assumed that only about one-quarter of the Coniacian *S. preventricosus* zone was represented by the lower part of the Muskiki Formation, the rest of the zone being represented by the Dismal Rat and Karr members of the underlying Cardium Formation (Fig. 2). *Within* any one faunal zone, it is impossible to accurately assess the amount of time missing between each depositional cycle: the duration of hiati have been assigned on a subjective basis.

## DEPOSITIONAL CYCLES

### Nomenclature

In their definitions of sequence-stratigraphic terms, van Wagoner *et al.* (1988) defined a *sequence* as comprising 'a relatively conformable succession of genetically related strata bounded by unconformities and their correlative conformities'. A *parasequence* was defined as 'a relatively conformable succession of genetically related beds or bedsets bounded by marine flooding surfaces. Siliciclastic parasequences are progradational and therefore shoal upward'. The marine flooding surface may constitute a minor hiatus, but no basinward shift in facies is seen.

In the following discussion, the stacked, upward-coarsening facies successions that comprise the Muskiki and Marshybank will be referred to as parasequences because in many cases, particularly in the Muskiki, they satisfy the definition of van Wagoner *et al.* (1988) in shoaling upward, and show no basinward shift in facies. However, many 'parasequences' in the Marshybank are bounded by chert pebble-veneered erosion surfaces, or include sharp-based shoreface sandstones, both of which indicate a significant relative *fall* in sea level, and a basinward shift in facies. They therefore more closely resemble true *sequences*. These examples highlight potential ambiguity in the definition of sequence and parasequence. In *nearshore* areas, a 'parasequence' may include sediments deposited during transgression (shales/siltstone), highstand (heterolithic facies), and falling (sharp-based shoreface, pebble bed) sea

**Table 1.** Boundaries and durations of the Coniacian and Santonian stages

| | Obradovitch & Cobban, 1975 | Decade of North American Geology, 1983 | Haq *et al.*, 1988 | International Union of Geological Sciences, 1989 | Average |
|---|---|---|---|---|---|
| Campanian | | | | | |
| | 82 | 84 | 84 | 83 | 83.25 |
| Santonian | (4) | (3.5) | (4) | (3) | (3.62) |
| | 86 | 87.5 | 88 | 86 | 86.87 |
| Coniacian | (1) | (1) | (1) | (2) | (1.25) |
| | 87 | 88.5 | 89 | 88 | 88.12 |
| Turonian | | | | | |

level, and therefore satisfy the definition of a sequence. Further offshore, the same 'parasequence' may simply shoal upwards to a flooding surface, and lack all evidence of relative sea-level fall.

## Scales of cycle

Within the Muskiki and Marshybank Formations, at least three scales of depositional cyclicity can be recognized.

### *Large-scale cycle*

The Muskiki and Marshybank Formations together comprise a third-order depositional sequence in the sense of van Wagoner *et al.* (1988). Parasequences in the Muskiki, plus Marshybank unit A, show a broadly retrogradational stacking pattern and form a transgressive systems tract. In contrast, Marshybank units B through to L are broadly progradational and downlap towards the northeast (basinwards, Figs 3, 11).

The mid-Coniacian and early Santonian unconformities that bound this third order sequence are recognized by Haq *et al.* (1988) as the product of global eustatic falls. Although a review of regional stratigraphy and sea-level changes is beyond the scope of this paper, it is pertinent to mention a few recent studies that document the presence of regional (if not global) hiatal surfaces related to relative sea-level changes in the Coniacian and Santonian. For example, the Coniacian–Santonian Kevin Member of the Marias River Formation is located over 800 km to the south-southeast in Montana. In the Kevin Member, Cobban *et al.* (1976) described a concretionary, pebble-strewn hiatal surface (the MacGowan Concretionary Bed) capping siltstones with a late Coniacian *S. ventricosus* fauna. The succeeding basal Santonian zone of *S. depressus*, equivalent to units B–L of the Marshybank, is represented by the MacGowan Concretionary Bed itself and a few metres of interbedded limestones and shales capped by a phosphatic pebble bed, overlain by shales with *S. vermiformis*. The *S. depressus* zone is apparently condensed and may be the basinal expression of the early Santonian lowstand that resulted in the dramatic regression in the upper part of the Marshybank.

In Colorado, Barlow & Kauffman (1985) recognize mid-Coniacian and late early Santonian regressive peaks (lowstands) in limestone and shale facies of the Niobrara cyclothem. Even farther to the south in New Mexico, Molenaar (1983) and Nummedal & Swift (1987) document a major mid-Coniacian sequence boundary between the Gallup Sandstone (time-equivalent to the Dismal Rat and Karr members of the Cardium Formation in Alberta) and the base of the Mulatto Tongue of the Mancos Shale (time-equivalent to the Muskiki Formation). Molenaar (1983) showed that regressive sandstones of the late Coniacian and early Santonian Dalton Sandstone Member (time-equivalent to the Marshybank) prograded up to 150 km before being flooded by the Satan Tongue of the Mancos Shale (time-equivalent to the Dowling member of the Puskwaskau Formation). In coastal basins of California, mid-Coniacian and early Santonian sequence boundaries have also been recognized in both shelf and deep water settings (May *et al.*, 1984).

### *Small-scale cycles*

All the formations discussed in this paper are characterized by *regionally traceable*, upward-coarsening facies successions (parasequences), typically 5–10 m thick. These parasequences vary in thickness by less than one order of magnitude. At least 15 parasequences can be recognized in the Muskiki and Marshybank (Fig. 11), although in some cases, two or more, apparently minor, upward-coarsening successions have been subjectively grouped as components of a more prominent parasequence. For example, Marshybank unit B appears to contain three minor cycles; units C and D are not separated by an erosion surface and may be grouped together, and unit L contains two minor cycles. Thus, from the base of the Muskiki to the top of unit A, eight *regionally traceable* parasequences are recognized, and in the Marshybank, seven major parasequences (comprising units B, C + D, E, F + G, H, I + J, and K + L) can be recognized. These small-scale cycles are indicated in Fig. 11.

### *Mid-scale cycles*

The parasequences defined above seem to fall into five distinct groups (indicated in Fig. 11), each marked at the top by evidence of more marked shallowing than in the intervening parasequences. Thus in the Muskiki, surface M1 marks the top of a very extensive and unusually sandy parasequence. Surface M2 caps a subtle, siltier upward parasequence and is unusual in being veneered with chert pebbles (which are present for over 400 km along strike). The top of unit A combines the characteristics of M1 and M2 in defining a major upward-coarsening parasequence, *and* having a pebble veneer. Units B, C + D, E, and F + G represent a set of parasequences that become progressively sandier and more progradational upwards: all but C + D are capped by a chert pebble bed. Finally, units H and I represent a marked flooding of the coastal plain represented by F + G, which was followed by overall regression through units J, K, and L.

**Mechanistic speculation**

A detailed discussion of the driving mechanisms of relative sea-level change is beyond the scope of this paper. Interested readers are referred to Devoy (1987) for a recent review.

*Large-scale cycle*

The origin of 1–10 Ma cycles is problematic. Haq *et al.* (1988) attribute such 'third order' cycles to 'glacio-eustacy': the growth and decay of continental ice sheets. In contrast, Harrison (1985) and Kauffman (1985) explained third-order cycles in terms of 'tectono-eustacy', involving changes in the volumes of ocean basins related to variations in rates of sea-floor spreading and attendant vulcanism and tectonism (including accelerated thrust-loading of foreland basins) at convergent margins. Nevertheless, Kauffman (1985) did acknowledge that the relative sea-level falls at the end of some third-order cycles appeared to be too rapid to be explained as a result of spreading ridge collapse. Fluctuations in intra-plate stress fields, related to plate interactions over periods of a few million years, have also been suggested (Cloetingh *et al.*, 1985) as a mechanism to explain the third-order sea-level cycles of Haq *et al.* (1988).

*Small-scale cycles*

It must be remembered that the assumptions and interpolations implicit in the construction of the chronometric time-scale in Fig. 11 render temporal subdivisions of < 1 Ma far from precise. This caveat aside, the 15 small-scale cycles defined in Fig. 11 appear to span approximately 1.5 Ma, suggesting an average periodicity of about 100 000 years. In considering the driving mechanism behind these cycles, it is necessary to consider the interplay between subsidence of the foreland basin and possible eustatic effects. As many of the parasequences can be traced along strike for over 400 km and basinwards for 100 to over 300 km, they are not considered to be the product of local autocyclic changes in sedimentation rate, and therefore this effect will not be considered further. In the following discussion, it is recognized that the rate of sediment supply to the basin may have fluctuated in harmony with Milankovitch-scale climatic cycles of ≤ 100 000 years (Barron *et al.*, 1985; Kauffman, 1988). Indeed, this mechanism has been advanced by Wright *et al.* (1989) to explain c. 100 000-year, fifth-order transgressive–regressive cycles in Cretaceous nearshore clastics of the San Juan Basin. In contrast, the evidence from the Muskiki and Marshybank Formations, discussed below, suggests a *eustatic* mechanism as the dominant control on 100 000-year parasequences.

The Muskiki–Marshybank sequence, corresponding to a third-order cycle of Haq *et al.* (1988), can be divided into two fundamentally different systems tracts (transgressive and highstand), separated by a downlap surface (Figs 3, 11). During global eustatic rise, the shoreline progressively back-stepped towards the southeast. The simple, upward-coarsening parasequences between the base of the Muskiki and marker M2 each reflect deepening followed by shallowing, during which coarser sediment was transported progressively further from the shore. Each shallowing phase could have been the result of *either* a *regional* decrease in subsidence rate leading to sea-floor aggradation, *or* a short-term eustatic fall superimposed on the overall rising limb of the third-order cycle.

The chert pebbles that rest on marker surface M2 cap a succession of relatively deep-water siltstones and imply a dramatic regional increase in energy at the sea floor. As this pebble bed lies *below* the maximum flooding (downlap) surface, it was probably formed on the *rising* limb of the third-order eustatic cycle. In order to produce the implied shallowing by tectonic means, it would be necessary to effect *regional uplift* of the basin floor at a rate *in excess* of the rate of eustatic rise. This seems mechanistically unlikely. Alternatively, and perhaps more plausibly, abrupt shallowing could be explained in terms of a short-term *eustatic fall*, superimposed on the longer term eustatic rise. The widespread chert pebbles capping unit A raise similar questions. Was eustatic sea level constant and subsidence negligible in order for the basin floor to aggrade to the point at which pebbles could be moved, or did a short-term eustatic fall briefly subject the shelf to a period of high wave and current energy, sufficient to move pebbles on the bed? Again, the latter explanation seems mechanistically more attractive.

The parasequences comprising Marshybank units B–L *could* be explained (ignoring changes in rate of sediment supply) simply in terms of an interaction between high-frequency ($\leq$ 100 000 year) fluctuations in subsidence rate, superimposed upon the third-order eustatic fall. If the subsidence rate were to slightly exceed the rate of eustatic fall, a transgression would result. Conversely, if the subsidence rate was slightly less than the rate of eustatic fall, a relative sea-level fall would result, giving rise to a rapidly prograding, sharp-based shoreface, and/or a pebble bed. Although this argument provides a plausible explanation of the parasequences, and in particular the pebble beds, formed on the *falling* limb of the third-order cycle, it is unlikely to explain the pebbles on top of unit A, and cannot explain those deposited during the *rising* limb (M2 marker). Is it therefore reasonable to invoke a high-frequency *eustatic* mechanism to explain parasequences and pebble beds on the rising limb of a third-order eustatic cycle, yet appeal to fluctuations in *subsidence rate* to explain similar parasequences and pebble beds on the falling limb? A parsimonious approach would suggest that *all* the small-scale parasequences could be explained by apparently regular, short-term eustatic fluctuations, superimposed on the longer, third-order, 'global' eustatic cycle.

The apparent 100 000-year periodicity of the eustatic cycles inferred above is reminiscent of one of the Earth's orbital rhythms (Milankovitch cycles) that have been identified at about 400 000 and 100 000 years (orbital eccentricity), 41 000 years (axial obliquity), and 21 000 years (axial precession; Imbrie & Imbrie, 1980). Detailed studies on a variety of rhythmically bedded deposits indicate that these insolation cycles have persisted at essentially modern frequencies for at least the last 300 Ma (Fischer, 1986; Park & Herbert, 1987). Depositional cycles involving a combination of the 21 000 and 100 000-year cycles are common in both shallow-marine clastics (e.g. Clifton, 1981), in shallow-marine carbonates (e.g. Goldhammer *et al.*, 1987), and in pelagic limestone–marl rhythms (e.g. Fischer *et al.*, 1985; Fischer, 1986). Although less commonly reported, the 41 000-year obliquity cycle has been recognized in several Jurassic and Cretaceous stages, notably the Cenomanian (Fischer & Schwarzacher, 1984; Fischer, 1986).

In the rocks described here, the 100 000-year cycle appears to be the dominant rhythm; a similar periodicity was found to dominate a sequence of rhythmically bedded pelagic limestones and shales of Albian age (Park & Herbert, 1987). Interpretation of the minor parasequences (e.g. within units B and L) in terms of higher frequency Milankovitch-style rhythms would be speculative, at least at present.

If the 100 000-year parasequences are indeed of eustatic origin, then their periodicity makes a glacio-eustatic driving mechanism attractive. Although unequivocal evidence of Cretaceous glaciation is lacking (Hambrey & Harland, 1981, p. 953), some recent studies (e.g. Matthews, 1984; Kemper, 1987; Spicer, 1987; Frakes & Francis, 1988; Pirrie & Marshall, 1990) have provided tantalizing geological evidence of subfreezing temperatures in the Cretaceous polar areas. Antarctica lay athwart the Cretaceous south pole (Barron, 1987), admitting the possibility of significant terrestrial ice accumulations. Recent palaeoclimatic simulations by Oglesby & Park (1989), incorporating a mid-Cretaceous palaeogeography and a 'global grassland' albedo, indicate subfreezing temperatures (about −15 to −25°C) for much of Antarctica in July (mid-winter). Unfortunately, these simulations were run in 'perpetual season' mode, from which, unfortunately, no conclusions can be drawn regarding the possible development of an ice sheet (Horrell, personal communication, 1990). However, Oglesby (1989) conducted another series of simulations using present-day geography and 'seasonal cycle' mode. Model results indicated that, under all but the most extreme conditions (unrealistic sea-surface temperatures of > 14°C and a topography of < 200 m), much of the interior of Antarctica would have accumulated a permanent snow cover. An ice-free topographic reconstruction of Antarctica (in Robin, 1988) places a large part of the continent above 1000 m, and indicates separation of east and west Antarctica by a shelf sea. Although Oglesby (1989) did acknowledge that this seaway could have brought relatively warm water close to the pole, realistic estimates of the elevation of the continent still appeared to favour the accumulation of a significant interior upland ice cap. It is possible that these model results are compatible with palaeontological evidence (e.g. Axelrod, 1984) for the existence of a temperate flora and fauna in low-lying *coastal* areas. More detailed modelling may improve our understanding of these seemingly rather steep climatic gradients.

Thus, the few metres (?< 10) of eustatic fluctuation inferred from the Muskiki and Marshybank Formations may conceivably be a reflection of the waxing and waning of continental ice masses in harmony with an orbital eccentricity cycle of about 100 000 years. A deliberate search for comparable scales of sea-level fluctuation, with particular emphasis on evidence for sea-level *fall* in widely spaced sections of coeval, shallow-marine strata, might strengthen this conclusion. The recent interpretation (King, 1990) of Campanian marl–limestone sequences in terms of high-frequency sea-level changes (interpreted to be of tectonic rather than glacio-eustatic processes) is nevertheless encouraging in this regard.

### *Mid-scale cycles*

Parasequences in the Muskiki and Marshybank appear to be grouped into five distinct 'packages', each of which terminates in an unusually prominent regression and contains between two and four parasequences (Fig. 11). It would appear that the 100 000-year cycles were modulated, over several hundred thousand years, by a ?quasi-cyclic control. This effect periodically resulted in an unusually marked transgressive–regressive pulse. The underlying mechanism is not clear. In Fig. 11, subtle, quasi-periodic fluctuations in subsidence rate are postulated as one possible cause of mid-scale cycles. Equally, they could be the result of another frequency of eustatic fluctuation.

## PALAEOGEOGRAPHY

Because the 1500 well-logs used in this study can be 'calibrated' with lithologic data from outcrop and core, it is possible to map, with some confidence, the regional subsurface distribution of major lithofacies (Plint, 1990). Of particular interest are the SCS sandstones of units E, F, J, and L as these allow reconstruction of the former extent of the strandplain. Unit E can be traced as a continuous sheet from the western extremity of the data (in the Mount Puggins area of northeastern British Columbia, Fig. 1) eastwards for a distance of 70 km. Similarly, unit F can be traced for 80 km before it shales out basinwards. Although post-Marshybank–pre-Puskwaskau erosion has removed much of units H–L, there exists in the Mount Puggins area (extreme western part of Fig. 3) an anomalously thick sequence of fine-grained SCS and medium-grained cross-bedded pebbly shoreface sandstones. It seems likely that this rather homogeneous sequence of rocks is the landward equivalent of units H–L, recognized about 120 km to the southeast (Plint, 1990). This interpretation is supported by the fact that in the east, only unit L contains significant volumes of chert pebbles, and that chert pebble conglomerates are present only in the uppermost portion of the undifferentiated Mount Puggins section, in the extreme northwest. These inferred relationships are shown by dashed lines in Fig. 3. If this interpretation is correct, it suggests that unit J once formed a sheet of sandstone extending from the Mount Puggins area to its (?slightly eroded) eastern margin about 100 km into the basin. Similarly, unit L appears to have extended about 115 km basinwards, although its depositional margin is never observed, owing to post-Marshybank erosion. The inferred palaeogeographic distribution of units F, J, and L is illustrated in Fig. 1.

### Rates of shoreline movement

Each shoreface sand body seems to have been deposited in response to small-scale (100 000 year) eustatic fluctuations that may not have exceeded a few metres in amplitude. In order to calculate rates of shoreface progradation, it is necessary to apportion time between the 'transgressive' and 'regressive' phases of each cycle Leckie (1986) suggested that, within an 'average' transgressive–regressive cycle, about one-third of the time is occupied by transgression and two-thirds by regression. On this basis, about 67 000 years may be assigned to each of units E, F, J, and L.

The shoreface sandstones of units E and F prograded a *minimum* of 70 and 80 km, which in the allotted 67 000 years indicates a progradation rate of about 1 and 1.2 m/yr, respectively. No muddy strata are present beneath the SCS sandstones within units E and F (Figs 3, 11). This suggests that the inner shelf was too shallow to accommodate or preserve any of the 'transgressive' and 'highstand' portions of these parasequences. If units J and L *are* equivalent to the thick sandstones seen at Mount Puggins, then a *minimum* of 100 and 115 km of progradation can be inferred. This represents progradation rates of about 1.5 and 1.7 m/yr, respectively.

Progradation rates in the Marshybank are comparable with the average rate of 3.0 m/yr calculated for the strandplain of Nayarit, Mexico (Curray *et al.*, 1969), which is probably the closest modern analogue for the Marshybank coastal system (despite the fact that the Nayarit system was deposited during the post-glacial eustatic *rise*). However, these rates are relatively high when compared with those calculated from other units in the Western Interior Seaway. For example, 0.21–0.43 m/yr for the Lower Cretaceous Moosebar–Gates Formation (Leckie, 1986); 0.2 m/yr for the Turner Sandy Member (Merewether *et al.*, 1979); 0.15–0.3 m/yr for the Mesaverde (Weimer, 1970); and 0.8 m/yr for the Fox Hills Sandstone (Gill & Cobban, 1973). The rapid progradation of the Marshybank strandplain is, however, entirely consistent with deposition on a very low-gradient shelf during relative sea-level fall: when the space available for *aggradation* is reduced, *progradation* must accelerate to accommodate the sediment, assuming a constant rate of supply.

## CONCLUSIONS

**1** The mid-Coniacian to early Santonian Muskiki and Marshybank Formations contain at least three distinct scales of depositional cycle. These two formations together comprise transgressive and highstand systems tracts, separated by a downlap surface. The overlying Bad Heart Formation (early Santonian) probably constitutes a lowstand systems tract. Marine shales of the mid-Santonian Puskwaskau Formation constitute another transgressive systems tract.

**2** Regional unconformities at the base of the Muskiki and top of the Marshybank correspond to third-order 'global' sequence boundaries recognized by Haq *et al.* (1988). This third-order cycle may be explicable in terms of 'tectono-eustacy' involving changes in ridge spreading rates and attendant convergent-margin tectonism and vulcanism.

**3** Within the third-order cycle, at least 15, regionally extensive, upward-shoaling parasequences are recognized. These span about 1.5 Ma, suggesting an average duration of about 100 000 years. Some parasequences contain two or more smaller depositional cycles, suggesting durations of the order of a few tens of thousands of years.

**4** The presence of chert pebble beds or erosive-based shoreface sandstones at the top of many parasequences is indicative of short-term relative sea-level fall. Various lines of reasoning suggest that eustatic fluctuations might provide the most reasonable explanation of these relative sea-level oscillations. The apparent 100 000-year periodicity may be evidence for glacio-eustacy, governed by a Milankovitch-style orbital forcing mechanism.

**5** The fifteen 100 000-year parasequences fall into five, quasi-periodic groups. The top of each group is defined by an unusually prominent regression. These enigmatic 'mid-size' cycles may reflect subtle changes in regional subsidence rate, or may be the product of eustatic fluctuations acting over a few hundred thousand years.

**6** In the upper part of the Marshybank, four shoreface sand bodies prograded at least 70, 80, 100, and 115 km into the basin at estimated rates varying from about 1 to 1.7 m/yr. The erosive base to each sand body suggests that relative sea-level fall forced the shoreface to rapidly prograde over a very low-gradient shelf into water little more than 5–8 m deep.

## ACKNOWLEDGEMENTS

Generous financial support for this work has been provided through grants from Imperial Oil Ltd, Texaco Canada Ltd, the Department of Energy, Mines and Resources, the Natural Sciences and Engineering Research Council of Canada, and the University of Western Ontario. Canadian Hunter Ltd, Esso Resources Ltd, and Unocal Canada Ltd contributed substantial funds to offset the cost of helicopters; subsurface data and basemaps were provided by Home Oil Ltd. I am very grateful to all these companies and institutions, without whose support this research would not have been possible.

In the field, Lorne Hunter provided logistical assistance, and pilots Elmer Aalgaard, Mike Bellamy, Spring Harrison, Bob Southworth, and Bryan Toso cheerfully ferried us to many of the more remote Foothills outcrops. Sections were measured with the help of Jennifer McKay who rendered many weeks of cheerful and competent assistance despite sometimes trying conditions.

My ideas have been sharpened through discussions with Bruce Hart, David James, Henry Posamentier, George Postma, and Rod Tillman. I am indebted to Mark Horrell for drawing my attention to the work of Oglesby, and for some results of his (as yet unpublished) Cretaceous climatic simulations. Jeremy Bartlett and Dale Leckie provided numerous constructive comments on an early draft of the paper, and the final version was further improved by the comments of Bruce Hart and of referees Janok Bhattacharya and Roger Walker: to all I express my thanks. Responsibility for facts and interpretations rests, nevertheless, with the author.

## REFERENCES


AIGNER, T. (1985) Storm depositional systems. *Lecture notes in Earth Sciences*, Springer-Verlag, Berlin, 174 pp.

ALLEN, P.A. & UNDERHILL, J.R. (1989) Swaley cross-stratification produced by unidirectional flows, Bencliff Grit (Upper Jurassic), Dorset, UK. *J. Geol. Soc. London* **146**, 241–252.

AXELROD, D.I. (1984) An interpretation of Cretaceous and Tertiary biota in polar regions. *Palaeogeogr. Palaeoclimatol. Palaeoecol.* **45**, 105–147.

BARLOW, L.K. & KAUFFMAN, E.G. (1985) Depositional cycles in the Niobrara Formation, Colorado Front Range. In: *Fine-grained Deposits and Biofacies of the Cretaceous Western Interior Seaway: Evidence of Cyclic Sedimentary Processes* (Eds Pratt, L.M., Kauffman, E.G. & Zelt, F.B.). Soc. Econ. Paleontol. Mineral., Field Trip Guidebook 4, 199–208.

BARRON, E.J. (1987) Global Cretaceous paleogeography — I.G.C.P. Project 191. *Palaeogeogr. Palaeoclimatol. Palaeoecol.* **59**, 207–214.

BARRON, E.J., ARTHUR, M.A. & KAUFFMAN, E.G. (1985) Cretaceous rhythmic bedding sequences: a plausible link between orbital variations and climate. *Earth Planet. Sci. Lett.* **72**, 327–340.

BERGMAN, K.M. & WALKER, R.G. (1987) The importance of sea level fluctuations in the formation of linear conglomerate bodies; Carrot Creek Member of Cardium Formation, Cretaceous Western Interior Seaway, Alberta, Canada. *J. Sediment. Petrol.* **57**, 651–665.

CLIFTON, H.E. (1981) Progradational sequences in Miocene shoreline deposits, southeastern Caliente Range, California. *J. Sediment. Petrol.* **51**, 165–184.

CLOETINGH, S., MCQUEEN, H. & LAMBECK, K. (1985) On a tectonic mechanism for regional sea level variations. *Earth Planet. Sci. Lett.* **75**, 157–166.

COBBAN, W.A., ERDMANN, C.E., LEMKE, R.W. & MAUGHAN, E.K. (1976) Type sections and stratigraphy of the members of the Blackleaf and Marias River formations (Cretaceous) of the Sweetgrass Arch, Montana. *U.S. Geol. Surv. Prof. Pap.* **974**, 63 pp.

CURRAY, J.R., EMMEL, F.J. & CRAMPTON, P.J.S. (1969) Holocene history of a strand plain, lagoonal coast, Nayarit, Mexico. In: *Coastal Lagoons, a Symposium* (Eds Castanares, A.A. & Phleger, F.B.) pp. 63–100. Univ. Nac. Autonoma de Mexico.

*Decade of North American Geology* (DNAG) (1983) Geologic timescale *Geology* **11**, 503–504.

DEVOY, R.J.N. (Ed.) (1987) *Sea Surface Studies, a Global View*. Croom Helm, New York, 649 pp.

FISCHER, A.G. (1986) Climatic rhythms recorded in strata. *Ann. Rev. Earth Planet. Sci.* **14**, 351–376.

FISCHER, A.G. & SCHWARZACHER, W. (1984) Cretaceous bedding rhythms under orbital control? In: *Milankovitch and Climate, Part 1* (Eds Berger, A.L., Imbrie, J., Hays, J., Kukla) pp. 163–175. D. Riedel, Hingham, MA.

FISCHER, A.G., HERBERT, T. & PREMOLI-SILVA, I. (1985) Carbonate bedding cycles in Cretaceous pelagic and hemipelagic sequences. In: *Fine-grained Deposits and Biofacies of the Cretaceous Western Interior Seaway: Evidence of Cyclic Sedimentary Processes* (Eds Pratt, L.M., Kauffman, E.G. & Zelt, F.B.). Soc. Econ. Paleontol. Mineral., Field Trip Guidebook 4, 1–10.

FRAKES, L.A. & FRANCIS, J.E. (1988) A guide to Phanerozoic cold polar climates from high-latitude ice-rafting in the Cretaceous. *Nature* **333**, 547–549.

GALLI, G. (1989) Depositional mechanisms of storm sedimentation in the Triassic Durrenstein Formation, Dolomites, Italy. *Sediment. Geol.* **61**, 81–93.

GILL, J.R. & COBBAN, W.A. (1973) Stratigraphy and geologic history of the Montana Group and equivalent rocks, Montana, Wyoming, and North and South Dakota. *U.S. Geol. Surv. Prof. Pap.* **776**, 37 pp.

GOLDHAMMER, R.K., DUNN, P.A. & HARDIE, L.A. (1987) High-frequency glacio-eustatic sea level oscillations with Milankovitch characteristics recorded in Middle Triassic platform carbonates in northern Italy. *Am. J. Sci.* **287**, 853–892.

GOLDRING, R. (1971) Shallow water sedimentation as illustrated by the Upper Devonian Baggy beds. *Mem. Geol. Soc. London* **5**, 80 pp.

HAQ, B.U., HARDENBOL, J. & VAIL, P.R. (1988) Mesozoic and Cenozoic chronostratigraphy and cycles of sea level change. In: *Sea-level Changes: an Integrated Approach* (Eds Wilgus, C.K., Hastings, B.S., Kendall, C.G.St.C., Posamentier, H., Ross, C.A. & van Wagoner, J.). Soc. Econ. Paleontol. Mineral. Spec. Publ. 42, 71–108.

HAMBREY, M.J. & HARLAND, W.B. (Eds) (1981) *Earth's pre-Pleistocene Glacial Record*. Cambridge University Press, Cambridge, 1004 pp.

HARRISON, C.G.A. (1985) Modelling fluctuations in water depth during the Cretaceous. In: *Fine-grained Deposits and Biofacies of the Cretaceous Western Interior Seaway: Evidence of Cyclic Sedimentary Processes* (Eds Pratt, L.M., Kauffman, E.G. & Zelt, F.B.). Soc. Econ. Paleontol. Mineral., Field Trip Guidebook 4, 11–15.

IMBRIE, J. & IMBRIE, J.Z. (1980) Modelling the climatic response to orbital variations. *Science* **207**, 943–953.

International Union of Geological Sciences (IUGS) (1989) Global Stratigraphic Chart, *Episodes*, 12, part 2 (supplement).

IRISH, E.J.W. (1965) Geology of the Rocky Mountain Foothills. *Geol. Surv. Can. Mem.* **334**, 241 pp.

KAUFFMAN, E.G. (1985) Cretaceous evolution of the Western Interior Basin of the United States. In: *Fine-grained Deposits and Biofacies of the Cretaceous Western Interior Seaway: Evidence of Cyclic Sedimentary Processes* (Eds Pratt, L.M., Kauffman, E.G. & Zelt, F.B.). Soc. Econ. Paleontol. Mineral., Field Trip Guidebook 4, iv–xiii.

KAUFFMAN, E.G. (1988) Concepts and methods of high-resolution event stratigraphy. *Ann. Rev. Earth Planet. Sci.* **16**, 605–654.

KEMPER, E. (1987) Das klima der Kreide-Zeit (The climate of the Cretaceous Period). *Geol. Jahrb., Reihe A*, **96**, 5–185.

KING, D.T. (1990) Upper Cretaceous marl–limestone sequences of Alabama: possible products of sea level change, not climate forcing. *Geology* **18**, 19–22.

LECKIE, D.A. (1986) Rates, controls, and sand-body geometries of transgressive–regressive cycles: Cretaceous Moosebar and Gates formations, British Columbia. *Bull., Am. Assoc. Petrol. Geol.* **70**, 516–535.

LECKIE, D.A. & WALKER, R.G. (1982) Storm and tide-dominated shorelines in Cretaceous Moosebar–Gates interval — outcrop equivalents of deep basin gas trap in western Canada. *Bull., Am. Assoc. Petrol. Geol.* **66**, 138–157.

LECKIE, D.A. & KRYSTINIK, L.F. (1989) Is there evidence for geostrophic currents preserved in the sedimentary record of inner to middle-shelf deposits? *J. Sediment. Petrol.* **59**, 862–870.

MATTHEWS, R.K. (1984) Oxygen isotope record of ice-volume history: 100 million years of glacio-eustatic sea-level fluctuation. In: *Interregional Unconformities and Hydrocarbon Accumulation* (Ed. Schlee, J.S.). Mem., Am. Assoc. Petrol. Geol. 36, 97–107.

MAY, J.A., YEO, R.K. & WARME, J.E. (1984) Eustatic control on synchronous stratigraphic development: Cretaceous and Eocene coastal basins along an active margin. *Sediment. Geol.* **40**, 131–149.

McCRORY, V.L.C. & WALKER, R.G. (1986) A storm and tidally-influenced prograding shoreline — Upper Cretaceous Milk River Formation of southern Alberta, Canada. *Sedimentology* **32**, 47–60.

MEREWETHER, E.A., COBBAN, W.A. & CAVANAUGH, E.T. (1979) Frontier Formation and equivalent rocks in eastern Wyoming. *Mountain Geol.* **16**, 67–101.

MOLENAAR, C.M. (1983) Major depositional cycles and regional correlations of Upper Cretaceous rocks, southern Colorado Plateau and adjacent areas. In: *Mesozoic Paleogeography of West-central United States* (Eds Reynolds, M.W. & Dolly, E.D.) pp. 201–223. Rocky Mountain Section, Denver, Soc. Econ. Paleontol. Mineral.

MYROW, P.M., NARBONNE, G.M. & HISCOTT, R.N. (1988) Storm-shelf and tidal deposits of the Chapel Island and Random formations, Burin Peninsula: facies and trace fossils. *Geol. Assoc. Can. Field Trip Guidebook* **B6**, 108 pp.

NORRIS, B. (1989) *Upper Cretaceous Muskiki, Marshybank and Bad Heart formations (Smoky Group), Western Interior Seaway, northwestern Alberta and northeastern British Columbia: relative sea level control on stratigraphy and sedimentology*. Unpublished MSc thesis, University of Western Ontario, London, Canada, 341 pp.

NUMMEDAL, D. & SWIFT, D.J.P. (1987) Transgressive stratigraphy at sequence-bounding unconformities: some principles derived from Holocene and Cretaceous examples. In: *Sea-level Fluctuations and Coastal Evolution* (Eds Nummedal, D., Pilkey, O.H. & Howard, J.D.). Soc. Econ. Paleontol. Mineral., Spec. Publ. 41, 241–260.

OBRADOVITCH, J.D. & COBBAN, W.A. (1975) A time-scale for the Late Cretaceous of the Western Interior of North America. *Geol. Assoc. Can., Spec. Pap.* **13**, 31–54.

OGLESBY, R.J. (1989) A GCM study of Antarctic glaciation. *Climate Dyn.* **3**, 135–156.

OGLESBY, R.J. & PARK, J. (1989) The effect of precessional insolation changes on Cretaceous climate and cyclic sedimentation. *J. Geophys. Res.* **94D**, 14793–14816.

PARK, J. & HERBERT, T.D. (1987) Hunting for paleoclimatic periodicities in a geologic time series with an uncertain time scale. *J. Geophys. Res.* **92B**, 14027–14040.

PIRRIE, D. & MARSHALL, J.D. (1990) High-paleolatitude Late Cretaceous paleotemperatures: new data from James Ross Island, Antarctica. *Geology* **18**, 31–34.

PLINT, A.G. (1988) Sharp-based shoreface sequences and 'offshore bars' in the Cardium Formation of Alberta: their relationship to relative changes in sea level. In: *Sea-level Changes: an Integrated Approach* (Eds Wilgus, C.K., Hastings, B.S., Kendall, C.G.St.C., Posamentier, H., Ross, C.A. & van Wagoner, J.). Soc. Econ. Paleontol. Mineral., Spec. Publ. 42, 357–370.

PLINT, A.G. (1990) An allostratigraphic correlation of the Muskiki and Marshybank formations (Coniacian–Santonian) in the Foothills and subsurface of the Alberta Basin. *Bull. Can. Petrol. Geol.* **38**, 288–306.

PLINT, A.G. & NORRIS, B. (1991) Anatomy of a ramp-margin sequence: facies, paleogeography and sediment dispersal patterns in the Muskiki and Marshybank formations, Alberta Foreland Basin. *Bull. Can. Petrol. Geol.* **39**, (in press).

PLINT, A.G. & WALKER, R.G. (1987a) Cardium Formation 8. Facies and environments of the Cardium shoreline and coastal plain in the Kakwa field and adjacent areas, northwestern Alberta. *Bull. Can. Petrol. Geol.* **35**, 48–64.

PLINT, A.G. & WALKER, R.G. (1987b) Morphology and origin of an erosion surface cut into the Bad Heart Formation during major sea-level change, Santonian of west-central Alberta, Canada. *J. Sediment. Petrol.* **57**, 639–650.

PLINT, A.G., WALKER, R.G. & BERGMAN, K.M. (1986) Cardium Formation 6. Stratigraphic framework of the Cardium in subsurface. *Bull. Can. Petrol. Geol.* **34**, 213–225.

PLINT, A.G., WALKER, R.G. & DUKE, W.L. (1988) An outcrop to subsurface correlation of the Cardium Formation in Alberta. In: *Sequences, Stratigraphy, Sedimentology: Surface and Subsurface* (Eds James, D.P. & Leckie, D.A.). Mem. Can. Soc. Petrol. Geol. 15, 167–184.

PLINT, A.G., NORRIS, B. & DONALDSON, W.S. (1990) Revised definitions for the Upper Cretaceous Bad Heart Formation and associated units in the Foothills and Plains of Alberta and British Columbia. *Bull. Can. Petrol. Geol.* **38**, 78–88.

ROBIN, G. DE Q. (1988) The Antarctic ice sheet, its history and response to sea level and climate changes over the past 100 million years. *Palaeogeogr. Palaeoclimatol. Palaeoecol.* **67**, 31–50.

ROSENTHAL, L.R.P. & WALKER, R.G. (1987) Lateral and vertical facies sequences in the Upper Cretaceous Chungo Member, Wapiabi Formation, southern Alberta. *Can. J. Earth. Sci.* **24**, 771–783.

SPICER, R.A. (1987) The significance of the Cretaceous flora of northern Alaska for the reconstruction of the climate of the Cretaceous. *Geol. Jahrb., Reihe A* **96**, 265–291.

STOTT, D.F. (1963) The Cretaceous Alberta Group and equivalent rocks, Rocky Mountain Foothills, Alberta. *Geol. Surv. Can. Mem.* **317**, 306 pp.

STOTT, D.F. (1967) The Cretaceous Smoky Group, Rocky Mountain Foothills, Alberta and British Columbia. *Geol. Surv. Can. Bull.* **132**, 133 pp.

VAN WAGONER, J.C., POSAMENTIER, H.W., MITCHUM,

R.M., Vail, P.R., Sarg, J.F., Loutit, T.S. & Hardenbol, J. (1988) An overview of the fundamentals of sequence stratigraphy. In: *Sea-level Changes: an Integrated Approach* (Eds Wilgus, C.K., Hastings, B.S., Kendall, C.G.St.C., Posamentier, H., Ross, C.A. & van Wagoner, J.). Soc. Econ. Paleontol. Mineral., Spec. Publ. Tulsa, 42, 39–45.

Walker, R.G. & Eyles, C.H. (1991) Topography and significance of a basin-wide sequence-bounding unconformity in Cretaceous Cardium Formation, Alberta, Canada. *J. Sediment. Petrol.* **61**, (in press).

Weimer, R.J. (1970) Rates of deltaic sedimentation and intrabasin deformation, Upper Cretaceous of Rocky Mountain Region. In: *Deltaic Sedimentation, Modern and Ancient* (Eds Morgan, J.P. & Shaver, R.H.). Soc. Econ. Paleontol. Mineral., Spec. Publ. 15, 270–292.

Wheeler, H.E. (1958) Time stratigraphy. *Bull., Am. Assoc. Petrol. Geol.* **42**, 1047–1063.

Wilkinson, B.H., McGowen, J.H. & Lewis, C.R. (1975) Ingleside strandplain sand of central Texas coast. *Bull., Am. Assoc. Petrol. Geol.* **59**, 347–352.

Wright, R., Katzman, D., Montz, M.J. & Zech. R.S. (1989) Coastal and shallow marine cyclicity. In: *Cretaceous Shelf Sandstones and Shelf Depositional Sequences, Western Interior Basin, Utah, Colorado and New Mexico*, pp. 25–33. 28th International Geologists Congress, Field Trip Guidebook T119, 87 pp.

*Spec. Publs. int. Ass. Sediment.* (1991) **12**, 429–450

# Jurassic–Cretaceous marine oscillations and geometry of back-arc basin fill, central Argentine Andes

LEONARDO LEGARRETA *and* MIGUEL A. ULIANA

*Astra C.A.P.S.A., Tucumán 744–16º Floor, 1049 Buenos Aires, Argentina*

## ABSTRACT

The Neuquén Basin of west-central Argentina is a northwest-trending back-arc depocentre implanted on Palaeozoic and possibly Precambrian continental crust. The Tithonian–Albian record of the region provides a good opportunity to gauge the incidence of eustasy on the dynamics of clastic, carbonate, and evaporite depositional systems in a convergent margin dominated by a tensional stress regime. Geohistory modelling indicates a relatively simple thermo-mechanical evolution, characterized by two broadly exponential subsidence cycles (Tithonian–Barremian and Aptian–Senonian).

The history of basin fill was reconstructed using a sequence-stratigraphic framework based on subsurface (seismic and wells) and surface data calibrated by ammonite biochronology, which allowed identification and regional mapping of 21 depositional units bounded by stratigraphic discontinuities. Several types of stratigraphic patterns, such a coastal onlap, regional offlap, gaps in paraconformable shelfal successions, lowstand wedges, morphology of the shelf to basin transitions, and carbonate or evaporite responses, developed according to a magnitude and frequency that compares favourably with those predicted by the global cycle chart, and thus point out that the local capacity to accommodate sediments was considerably influenced by eustatic changes. Conversely, the relative magnitude of the coastal encroachment was less than expected during the Aptian and early Albian. Because a coeval increase of magmatic activity was observed in the arc and a change in the rate of subsidence was detected in the basin, the deviation of the onlap pattern is considered an effect of the local tectonic activity. Smaller departures from the global trend, at the level of individual sequences and associated with evidence of total desiccation of the basin, suggest a local effect of enhanced marine fall related to evaporitic drawdown. Many of the sequences deposited under normal marine regime (Tithonian–Barremian) consistently show a less than complete development. Typically they lack accumulation comparable to basin-floor fans or slope fans, and display a relatively subdued transgressive deposit. These characteristics are explained as a local response to a depositional setting with a modest clastic supply and lacking tectonic hinges capable of generating steep slopes.

From a broader stratigraphic perspective, regional analysis allowed the recognition of sequence-sets or lithogenetic units that represent 10–30 Ma of sedimentary accumulation. The internal organization of these depositional entities reveals an ordered progression of stratigraphic properties that closely follows the trend of the secular (long-term or 'second order') eustatic dictator.

## INTRODUCTION

The present paper reports an attempt to unravel the influence of Late Jurassic to Early Cretaceous eustasy on the sedimentation of the Neuquén Basin, located near the western margin of southern South America (Fig. 1). The basin experienced relatively low rates of subsidence and modest clastic influx, and as a consequence provides an excellent opportunity to isolate the importance of eustasy on the dynamics of clastic, carbonate, and evaporite depositional systems active in the back-arc portion of a convergent plate boundary complex.

Earlier sequence-stratigraphic work in the region documented the significance of eustasy on the development of regional stratigraphic discontinuities (Legarreta & Kozlowski, 1984) and the effect of global eustatic trends on the depositional history of

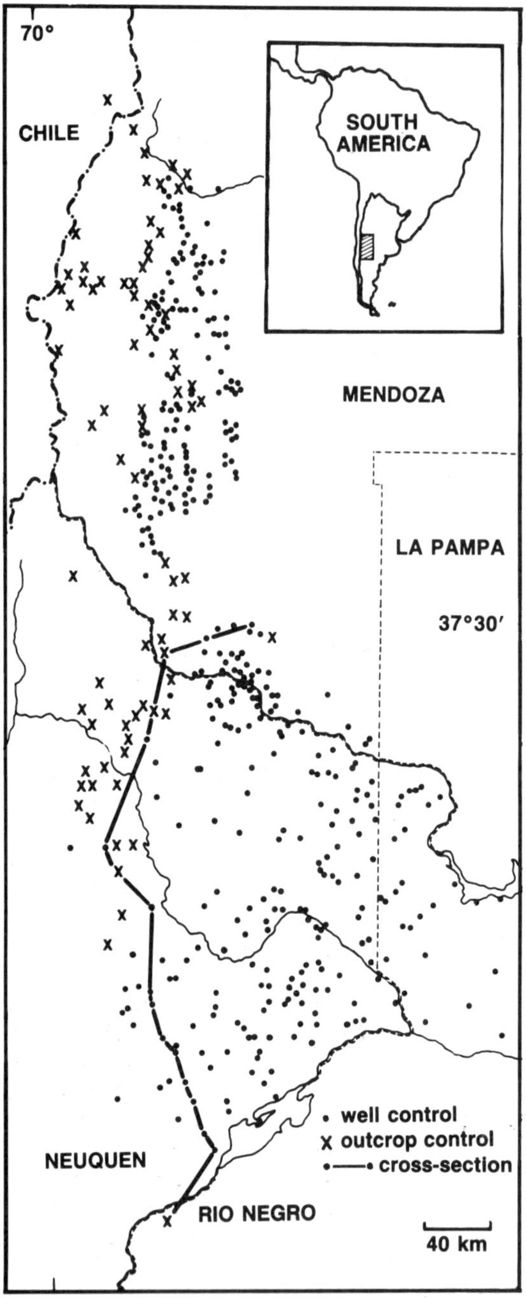


**Fig. 1.** Regional location of the Neuquén Basin in west-central Argentina showing the area contributing surface and well-log data for this study.

the Upper Jurassic to Neocomian sequences (Mitchum & Uliana, 1982,1985). In this paper an effort is made to show the changes in areal extent, palaeogeography, and internal stratification patterns of the successive sequences and key depositional systems tracts, as a response largely related to eustatic events.

The level of stratigraphic knowledge in the study area is considerable, as a result of academic and industrial research undertaken since the turn of the century (Burckhardt, 1903; Gerth, 1925a; Groeber, 1929,1953; Weaver, 1931; Herrero Ducloux, 1946; Marchese, 1971; Digregorio, 1972; Uliana *et al.*, 1977; Gulisano *et al.*, 1984). Parallel palaeontological investigations have led to the development of ammonite zonation that allows consistent dating of the Malm–Neocomian successions (Gerth, 1925b; Weaver, 1931; A. Leanza, 1945; H. Leanza, 1973,1981).

This study was organized as an integrated stratigraphic project based on the systematic analysis of 4000 km of reflection seismic profiles, data from 400 petroleum boreholes (cuttings, cores, and wireline logs) and detailed stratigraphic, sedimentological, and biostratigraphic data gathered from some 150 outcrop sections measured in the course of the last 15 years.

## GEOLOGICAL SETTING

The Neuquén Basin lies within what is conventionally known as the southern part of the central Andes (Gansser, 1973) and its foreland counterpart. During most of the Jurassic and the Cretaceous periods the western side of this portion of South America evolved as a continental margin under active convergence (Fig. 2). The edge of the continent was marked by the following zones from west to east: an active trench with small accretionary prism and fore-arc basin; a relatively narrow magmatic arc that operated mostly under an extensional crustal regime; and a relatively depressed segment of the foreland that formed a back-arc depocentre. As in other segments of the Andean belt, Mesozoic crustal stretching never led to the development of oceanic crust, and thus the back-arc remained as an entirely ensialic subsidence domain.

Deposition within the arc belt produced narrow but thick pods of marine and non-marine sediments, intricately interfingered with or locally replaced by large piles of flows and volcanic ejecta. Deposition in the back-arc was essentially devoid of volcanics and produced a 2500-m-thick package of largely marine clastic sedimentary rocks alternating with

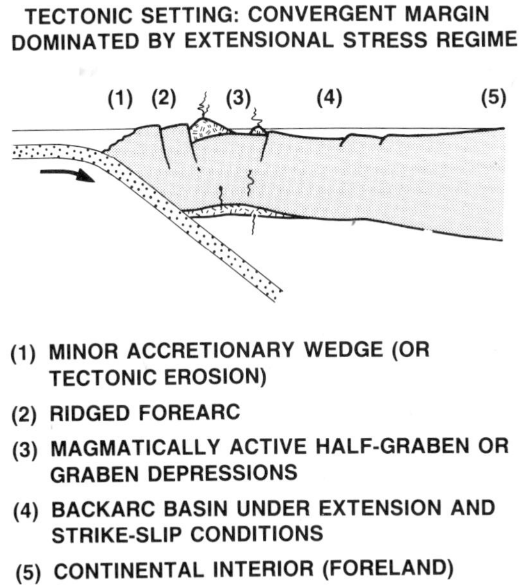


**Fig. 2.** Schematic diagram across the Pacific margin of Gondwanaland, to illustrate the location of the depositional setting described in the text. Not to scale.

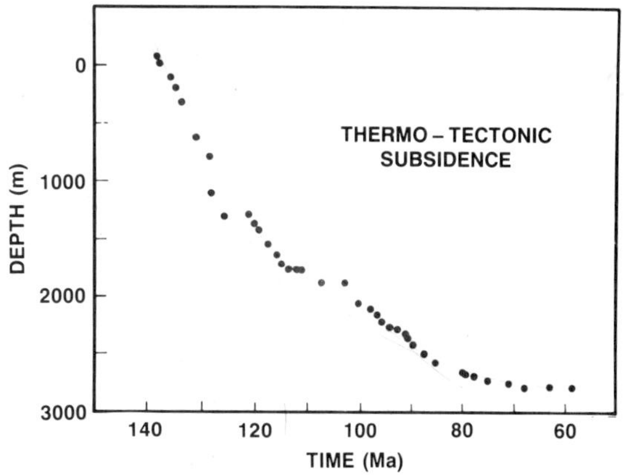


**Fig. 3.** Tithonian–Paleocene thermo-mechanical (back-stripped) subsidence history derived from surface and subsurface data gathered in west-central Neuquén. See text for discussion and implication.

carbonate and evaporite units, that constitute the subject of the present study.

## SUBSIDENCE REGIME AND GEOLOGICAL HISTORY

Subsidence of the back-arc region was focused around a basinal axis oriented NW–SE, oblique to the edge of the South American plate. Regional thickness analysis of chronostratigraphically constrained intervals indicates a rather simple subsidence history, suggestive of thermal drive related to a pre-Gondwanaland break-up tectonothermal event (Uliana & Biddle, 1988). Persistence of that pattern through time (Fig. 3) was presumably enhanced by preferential sediment-loading along the axis of an old rift trend, and caused a definite expansion of the back-arc shelfal domain known as the 'Neuquén Embayment'.

Precise subsidence analysis of the Cordilleran portion of the back-arc depocentre is hampered by lack of palinspastically restored base maps. Available data suggest the occurrence of at least two distinct episodes of regional downwarping along a N–S belt adjacent to the magmatic arc in Chile. These events occurred during the Albian–Cenomanian and Campanian–Paleocene. Regional knowledge on the timing and evolution of the arc system suggests that the increased subsidence events may be the result of dynamic coupling of the arc and near back-arc. Potential mechanisms to explain ensialic intra-arc and arc-margin subsidence include: development of a flexural moat as a response to the load exerted by the supracrustal emplacement of magmatic derivates, gravitational spreading, or regional subsidence due to net loss of crustal material erupted and removed through atmospheric transport (Hildebrand & Bowring, 1984).

Figure 3 shows an attempt to isolate the tectonic component of subsidence, by means of standard back-stripping techniques (e.g. Steckler & Watts, 1978). Data were gathered in west-central Neuquén and consisted of a mixture of stratigraphic information derived from surface sections and well-logs. Modelling produced a gradual, evenly varying curve that outlines a negative exponential thermomechanic subsidence pathway. The gently changing observed path is interpreted as a regional subsidence phenomenon, mostly derived from thermal recovery of the early Mesozoic extensional event (Uliana & Biddle, 1987,1988; Table 1). The change at around 105 Ma is thought to derive from the superimposition of mechanical (load) subsidence, related to mid-Cretaceous changes in the subduction–arc system (Uliana & Biddle, 1988; Table 1). The combination of those two sources of tectonic drive resulted in two main subsidence cycles that evolved along negative exponential trends.

**Table 1.** Regional evolution and subsidence style of the Neuquén Basin

| Time (Ma) | | Subsidence style: (W) Cordillera | Subsidence style: Embayment (E) | Regional evolution |
|---|---|---|---|---|
| 10 | | Shortening and uplift | | *Miocene–Pliocene*: Contractional deformation of the arc and proximal back-arc, and development of a fold and thrust belt. Depositional system became discontinuous and included mudflats or hypersaline to shallow-marine bodies on top of structural depressions, and volcanic aprons or alluvial fan wedges in front of structural uplifts. Morphogenesis of the cordilleran belt |
| 30 | | Flexural downwarp (?) | | *Oligocene–Miocene*: Pronounced eastward expansion of the magmatic belt, of relatively isolated to coalescent stratovolcanoes and dyke swarm complexes. Sedimentation resumed and produced a moderately thick cover of non-marine and pyroclastic deposits. Back-arc deformation was still limited |
| 50 | | Uplift (flexural rebound) | | *Eocene–Oligocene*: Dominated by flexural rebound after the episode of load-induced downwarp. Erosional regime was pervasive and produced regional truncation. Deposition was limited to thin gravel accumulation on top of a pediment surface |
| 80 | | Flexural downwarp | Regional (thermal sag) subsidence | *Campanian–Paleocene*: Changes in the arc belt induced an incipient fore-deep and reversal of the depositional polarity (west to east transport of coarse clastics). Depositional regime with carbonates and evaporites resumed owing to palaeogeographic connection with the Atlantic Ocean to the east |
| 95 | Studied interval | Regional sag | Regional (thermal sag) subsidence | *Turonian–Santonian*: The marine influx was cut off and deposition became entirely terrestrial, in a broad featureless back-arc depocentre dominated by fluvial depositional systems. Prevalent dispersal pattern was still to the west and northwest |
| 110 | Studied interval | Flexural moat | Regional (thermal sag) subsidence | *Late Barremian–early Cenomanian*: The marine connection became tenuous and favoured a dominantly negative water balance prone to evaporite deposition, presumably owing to growth of the magmatic arc region that fringed the South American Plate. The distal side of the back-arc embayment was regionally flexed and developed a subtle fore-bulge |
| 175 | Studied interval | Regional (thermal sag) subsidence | | *Pliensbachian–Neocomian*: Deposition proceeded under the influence of a well-developed arc and back-arc. Basin-wide subsidence in an increasingly large sag depocentre allowed the appearance of a wide marine embayment. Palaeogeographic connection was toward the Pacific Ocean (west), and hydrographic regime favoured a dominantly positive water balance, prone to preservation of organic-rich accumulations. The bulk of the clastic supply was from the southeast. Occasional presence of evaporites and sporadic influence of a western clastic source (usually volcanogenic) were associated with episodes of basin shallowing and downward shift of the depositional systems |
| 230 | | Transition<br>Local (fault induced) subsidence | | *Mid–late Triassic–Sinemurian*: Non-marine (alluvial to local lacustrine) clastic sedimentation and silicic to bimodal volcanics within a series of partially connected fault-bounded troughs. Graben and half-graben successor basins developed on top of collapsed portions of the late Palaeozoic arc and fore-arc complexes |

Regularity of subsidence across the foreland side of the embayment was only disrupted by differential displacement of blocks located in the vicinity of the east-trending 'Dorsal Huincul' wrench zone (Plozkiewicz *et al.*, 1984). Transpressional inversion of basement slices caught by the restraining bends of the dextral strike-slip zone was responsible for a local tendency toward stratigraphic condensation and structural enhancement of the unconformities seen in the rest of the area.

The left side of the chart in Table 1 shows the dominant pattern of subsidence through the evolution of the basin. The right side summarizes the main episodes of sedimentary accumulation as currently understood, and provides a perspective on the regional geological evolution.

## STRATIGRAPHIC FRAMEWORK

Considerable effort was directed toward the establishment of a consistent stratigraphic reference frame, suitable for mapping basin-wide palaeogeographic changes at 1–2 Ma intervals. To reach the required level of chronostratigraphic resolution, sequence stratigraphy was used, focusing on the identification and correlation of stratigraphic discontinuities that segregate successive genetic depositional units (sequences).

For the time interval considered in this paper there are several recently published time-scales. Although not universally accepted, the scheme by Haq *et al.* (1987) is followed by most sequence stratigraphers and has been used as the time reference frame for the present study. We remain aware that the absolute value of the ages quoted from Haq *et al.* (1987) may need adjustment; however, for the time being we find that reference to age numbers in the text and figures is useful to show time as the centrepiece of the themes analysed in the present paper.

Sequence mapping demanded the development of a regional correlation network, built upon previous stratigraphic studies (Mitchum & Uliana, 1982,1985; Legarreta *et al.*, 1981; Legarreta & Kozlowski, 1984; Gulisano *et al.*, 1984; Legarreta, 1985). Both areal coverage and level of consistency were expanded through the integration of subsurface (seismic and borehole) and outcrop information.

Timing of the depositional sequences was established by means of age-diagnostic fossil content (Fig. 4). To date ammonoids allow the most consistent local zonation, and correlation with the European biochronologic standard (Leanza, 1981; Riccardi, 1984). In contrast, the Aptian–Albian succession carries no ammonite faunas because of the provincial palaeogeographic restriction that affected the hydrological regime of the entire Andean back-arc (Riccardi, 1988; Uliana & Biddle, 1988). Dating of the Aptian–Albian succession is based on less diagnostic pollen and ostracod finds (Volkheimer & Salas, 1975; Musacchio & Palamarczuk, 1975).

The key biostratigraphic data shown in Fig. 4 provide support for intercontinental correlation and dating at the level of sequence-sets and sequence-set boundaries. With the exception of specific intervals in the Tithonian and Valanginian–Hauterivian, the Andean zonation is still not refined enough to date each one of the sequences already identified in the Neuquén Basin through physical stratigraphy. Loosely dated sequences, away from the biochronologically constrained intervals, were assigned ages by interpolation on the basis of sequence counting and comparison of onlap patterns (see also Mitchum & Uliana, 1985).

## FACIES AND DEPOSITIONAL SCENARIO

The Tithonian–Albian succession includes a heterogeneous suite of facies deposited behind the magmatic arc complex, over the broad Neuquén Embayment. During times of base-level highstand distal settings remained under moderately deep, quiet basinal conditions. These distal areas were confined by low to moderate relief ramps and depositional terraces, surrounded in turn by fringing coastal plain to terrestrial lowlands. During several significant but transient episodes of base-level lowstand the geographic configuration was much changed. Depositional sites were reduced by emergence of large portions or the entire length of the shelves, while the former basinal settings became shallow or entirely subaerial.

Figure 5 summarizes the distinctive properties of the facies subdivisions used to reconstruct the palaeogeography. For the purpose of regional mapping, two distinct facies assemblages were recognized. The 'dry shelf' assemblage is mostly represented by clastic subdivisions ranging from fluvial–aeolian to playa lake (associated with evaporite members). The 'wet shelf' periods during highstands are represented by a more conventional

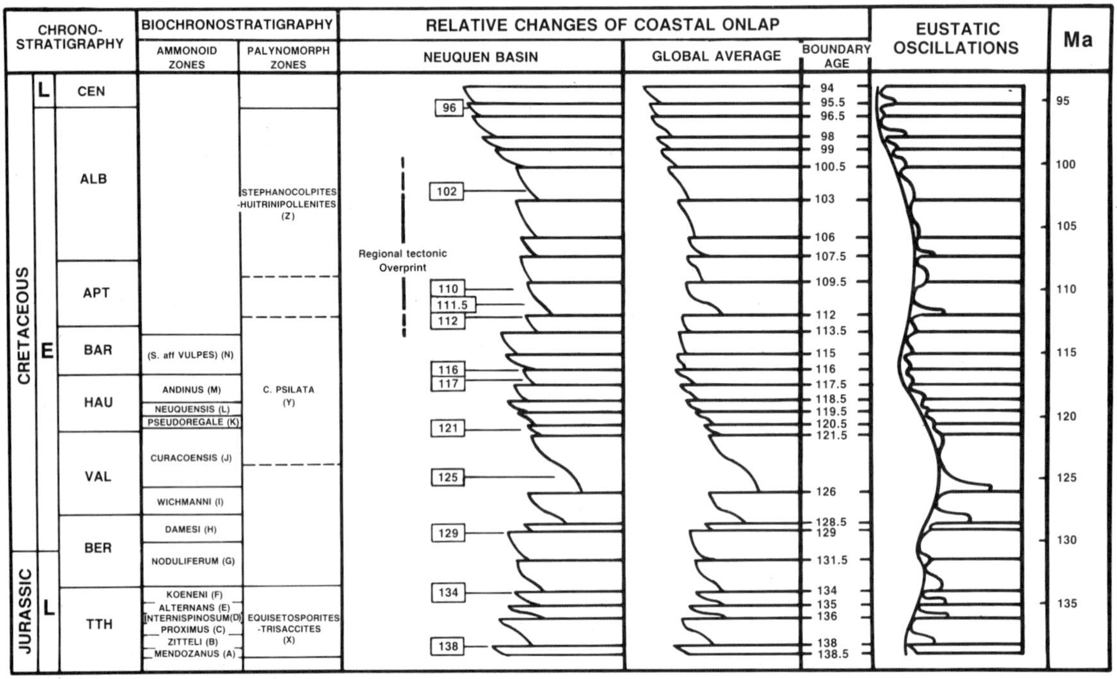


**Fig. 4.** Comparison of the Neuquén Basin coastal onlap and the global trend. Columns labelled 'Eustatic Oscillations' and 'Global Average' are after Haq *et al.* (1987). The 'Neuquén Basin' column shows the local stratigraphic signature, the extent of the mid-Cretaceous tectonic event, and the relative position of the slices illustrated by the palaeogeographic reconstructions in Figs 7–10. Fossil zonation used to date the depositional sequences is mainly after Leanza (1981), Riccardi (1984), and Volkheimer & Quattrocchio (1981). See Fig. 6 for additional information.

suite of nearshore to basinal facies assemblages, that were in addition differentiated into dominantly clastic and mostly carbonate facies tracts. The complex pattern of facies stacking and variation recorded within the Tithonian–Albian interval is demonstrated by the regional chronostratigraphic chart in Fig. 6, opposite, and is discussed further in the next section.

Throughout the Tithonian–Albian the depositional site maintained a basic asymmetry in terms of the subsidence rate and supply of siliciclastics. Most detritus entered the area across the faster subsiding southeast flank of the basin. As a consequence, considerable differentiation was maintained between a highly progradational southeast basin flank and a more stable northeast basin flank, prone to thin carbonate aggradation. In spite of its proximity to the west, the arc region played a limited role as a source terrain. Only during low base-level stands, basinward shift of the depositional systems allowed eastward transport from the arc down to the central portion of the basin.

## PALAEOGEOGRAPHIC EVOLUTION AND HISTORY OF BASIN FILL

### Tithonian (139–134 Ma)

After a period dominated by accumulation of evaporites and non-marine clastics (Oxfordian–Kimmeridgian; Riccardi, 1983, fig. 8), a major inundation episode occurred in the Tithonian. During the early stages of this transgression, water remained shallow and slightly hypersaline, and an area in the order of tens of thousands of square kilometres was covered by a thin stromatolite zone. As the transgression advanced conditions allowed the appearance of a well-diversified fauna, the depositional site became strongly differentiated and a wide basinal area was developed (Fig. 7A). A hydrological regime prone to induce euxinic conditions (Brongersma-Sanders, 1971), combined with flooding, promoted the development of widespread condensed sedimentation and preservation of organic matter (Veiga & Orchuela, 1988). Early in the

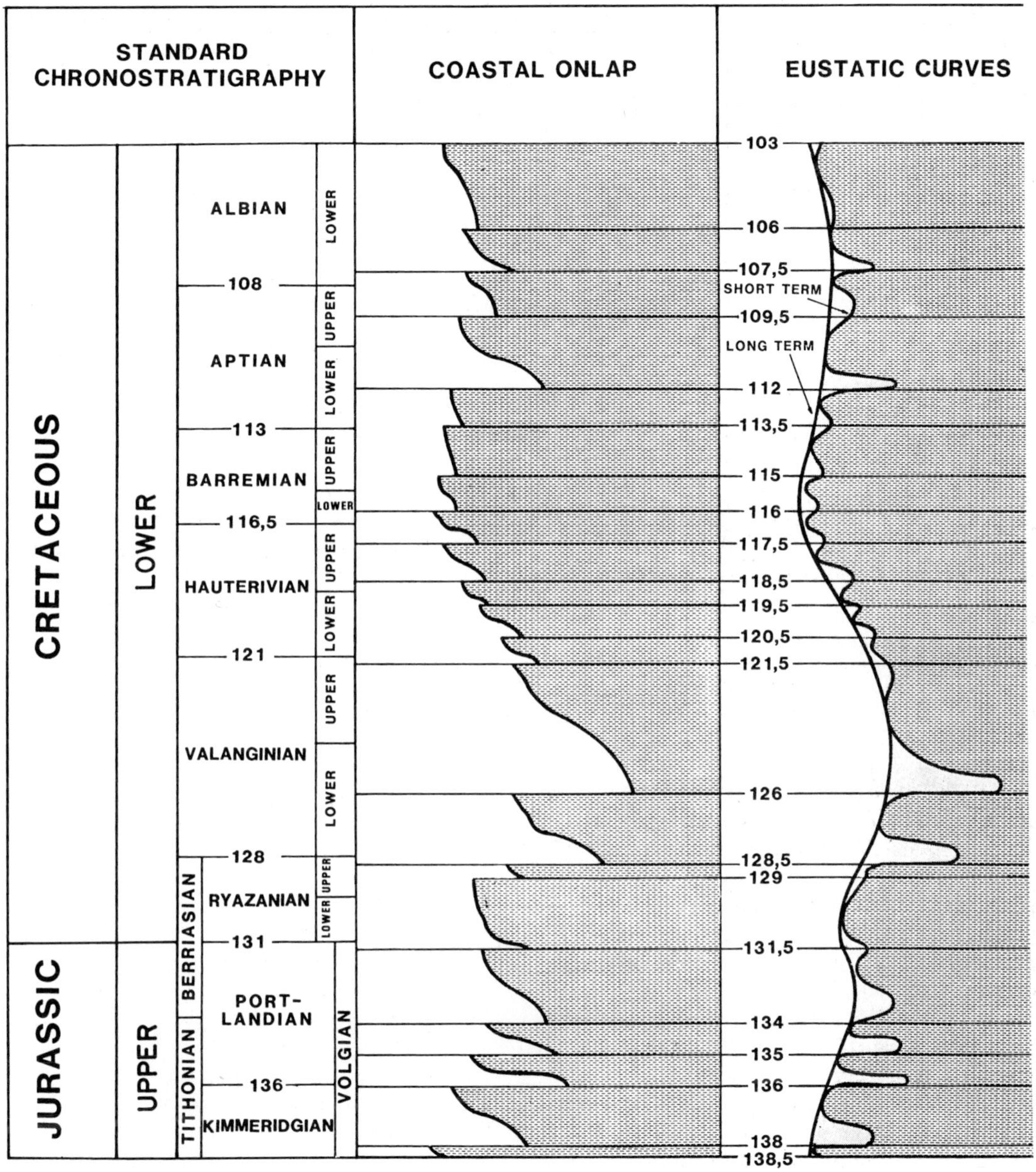


deep ⟷ shallow

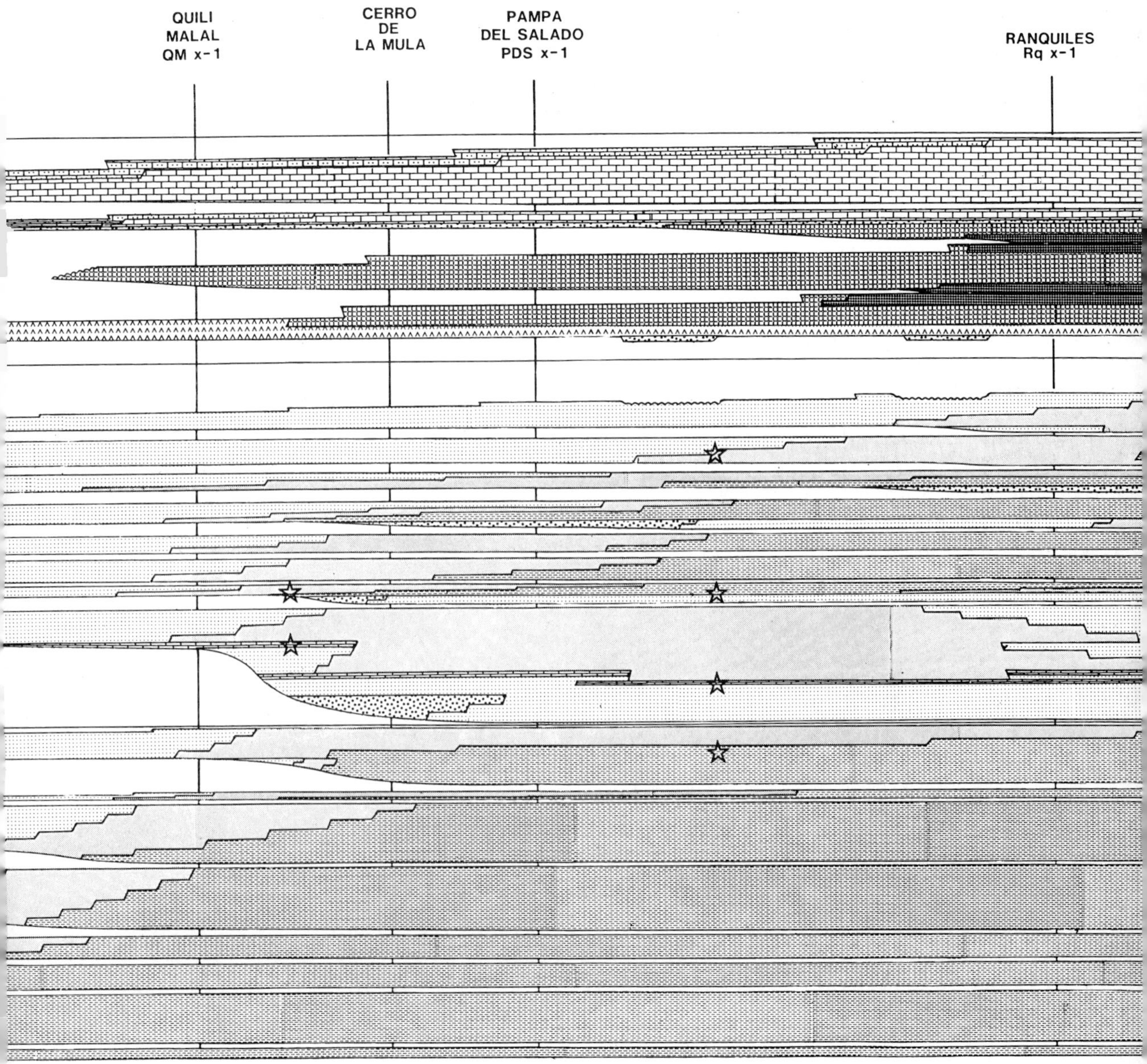
QUILI MALAL QM x-1
CERRO DE LA MULA
PAMPA DEL SALADO PDS x-1
RANQUILES Rq x-1

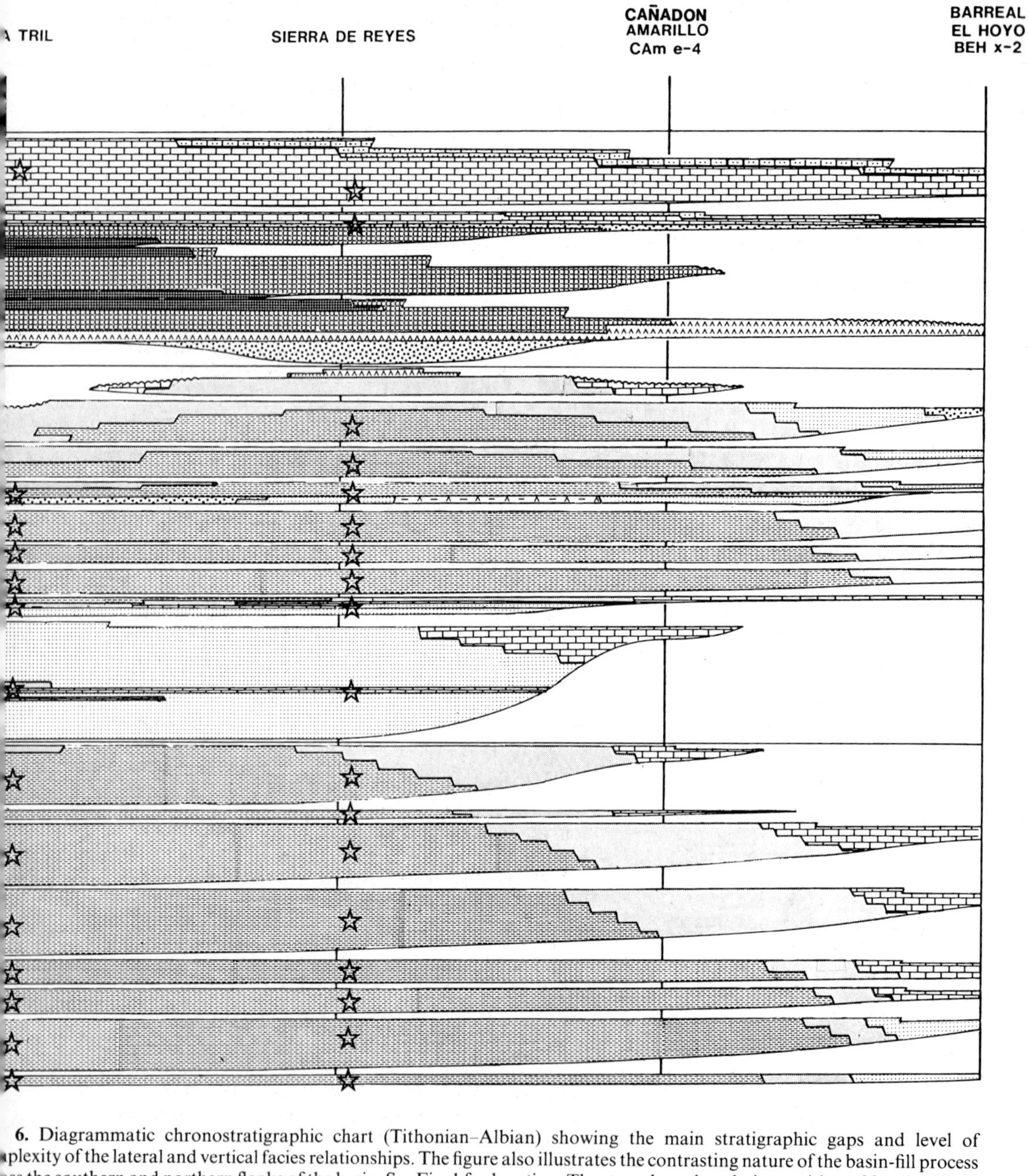


6. Diagrammatic chronostratigraphic chart (Tithonian–Albian) showing the main stratigraphic gaps and level of plexity of the lateral and vertical facies relationships. The figure also illustrates the contrasting nature of the basin-fill process ss the southern and northern flanks of the basin. See Fig. 1 for location. The stars show the relative position of fossil remains ing biochronological significance (see zonation in Fig. 4 and specific data in the papers by Archangelsky *et al.* (1981), Leanza 1), Musacchio (1981), Musacchio & Palamarczuk (1975), Ramos (1981), Riccardi (1984), Volkheimer & Quattrocchio 1), Volkheimer & Salas (1976) and Volkheimer *et al.* (1976).

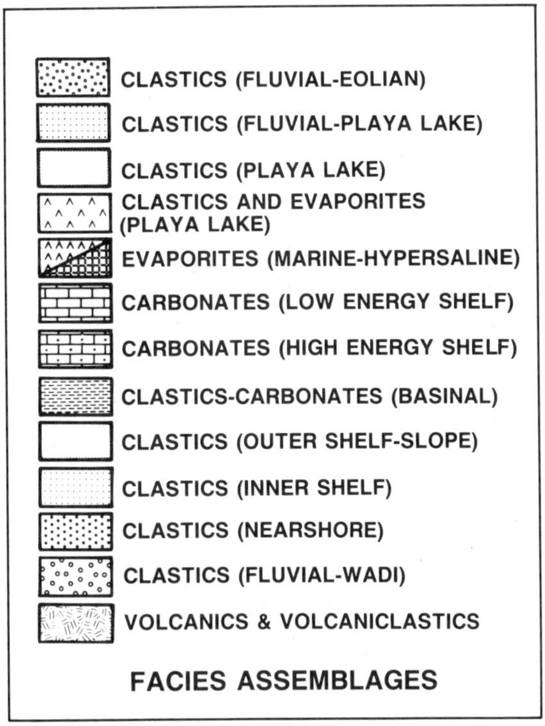


**Fig. 5.** Legend to facies associations shown in the chronostratigraphic chart (Fig. 6) and the palaeogeographic maps (Figs 7–10) and stratigraphic diagrams (Figs 11–14). *Clastics (fluvial–aeolian)*: medium sandstone partially cemented by calcite. 2–6 m thick, irregular to lenticular bodies bounded by sharp or erosional surfaces, with festoon to tabular cross-bedding. Upward thinning–fining internal organization. Laterally discontinuous bodies with large sweeping tangential foresets, large-scale planar cross-bedding, and climbing translatent stratification. Deposited by multichannel (braided) fluvial streams associated with dune fields. *Clastics (fluvial–playa lake)*: interbedded sandstone (wacke), claystone, and siltstone. Variegated (yellow, green, red). Claystones are thick and tabular, with mudcracks and halite hoppers. Sandstone bodies are lensoid, with erosional base, and feature cross-bedding and climbing-ripples. Deposition on mudflats and playa mudflats with ephemeral streams. *Clastics (playa lake)*: claystone, with fine-grained sandstone streaks and occasional evaporites. Green coloured and finely laminated, local evaporite 'varves', and halite hoppers. Rare presence of micro-invertebrates (forams, ostracods, and calcareous nannoplankton). *Clastics and evaporites (playa lake)*: mudstone and shale, green–grey–red coloured, with thin layers of cryptalgal boundstone and nodular or laminated anhydrite and halite layers or pseudomorphs. Mudstones are thick and massive and include wavy or lenticular sand streaks with flaser bedding. Saline mudflats and playas under episodic flash-flood or tidal regime. *Evaporites (marine–hypersaline)*: anhydrite, lime-boundstone (cryptalgal), halite, and rare sylvite. Thin to medium tabular strata with nodular to bedded fabric. Restricted basinal deposition under anternate subaqueous and playa conditions. *Carbonates (low-energy shelf)*: oyster lime-packstone and mudstone or shale. Grey to olive coloured. Thick to medium tabular stratofabric and rare domal to lensoid pelecypod-micrite build-ups. Highly bioturbated and capped by hardgrounds with glauconite and phosphate concentrations. Open quiet-water regime on the outer portions of depositional ramps or starved (drowned) shelves. *Carbonates (high-energy shelf)*: skeletal lime-packstone, oolitic–skeletal lime-grainstone, and minor coral boundstone. Light to dark grey coloured. Medium to very thick tabular strata, locally large imbricated lenses and occasional mounds and build-ups. Lateral replacement by sandy limestones with well diversified fauna (molluscs, corals, annelids, ammonoids, nautiloids). Thick and tabular strata, massive or cross-bedded. Open marine on top of ramps or gently sloped depositional terraces. *Clastics–carbonates (basinal)*: shales, lime-mudstones to skeletal wackestones. Dark coloured (black and shades of grey). Very thin tabular bedding, internally laminated, alternate nodular bedding zones. Scarce epifaunal and local infaunal organisms, low taxonomic diversity. Quiet anaerobic to dysaerobic bottoms. *Clastics (outer-shelf–slope)*: shale and mudstone, fine-grained silty sands, matrix-supported limestone, and marl. Dark green to dark grey coloured, even parallel bedding organized as upward-thickening–coarsening parasequences. Diversified (locally displaced) nectonic and benthonic epifauna and infauna. Large sigmoid and oblique clinoforms indicate progradational sedimentation under alternate quiet and storm dominated conditions. *Clastics (inner shelf)*: fine to coarse sandstone and siltstone, carbonate cement. Grey-greenish coloured. Reworked molluscs, corals, and ammonoids, lenses of coquinoid limestone. Thick-tabular to lenticular stratofabrics. Frequent bars and channels and some larger scours. Fining upward or coarsening upward packages. Very shallow marine under the influence of tidal currents, waves, and storms. *Clastics (nearshore)*: dirty sandstone, fine conglomerate, and siltstone. Pale green to olive and yellowish coloured. Thick to thin strata, lenticular to flaser bedding, some large-scale cross-stratification. Devoid of body fossils. Tidal flat and estuary to supratidal. *Clastics (fluvial–wadi)*: sandstones, fine to coarse conglomerate, and mudstone. Red coloured (purplish, brownish, brick). Tabular to lenticular stratofabric, common cross-bedding. Internally organized as 1–3 m thick, fining upward packages. Vertebrate remains (reptiles, mammals). Low–moderate sinuosity river systems. Locally linked to braided systems and alluvial fans.

Tithonian the deepening trend produced a transgressive stratigraphic pattern with retrogradational stacking of parasequences (Gulisano *et al.*, 1984) and a shallow to deep water transition with negligible relief (Mitchum & Uliana, 1985). At some time late in the early Tithonian, the process of marine flooding and drowning of the foreland must have been reduced to a rate low enough to allow the inception of

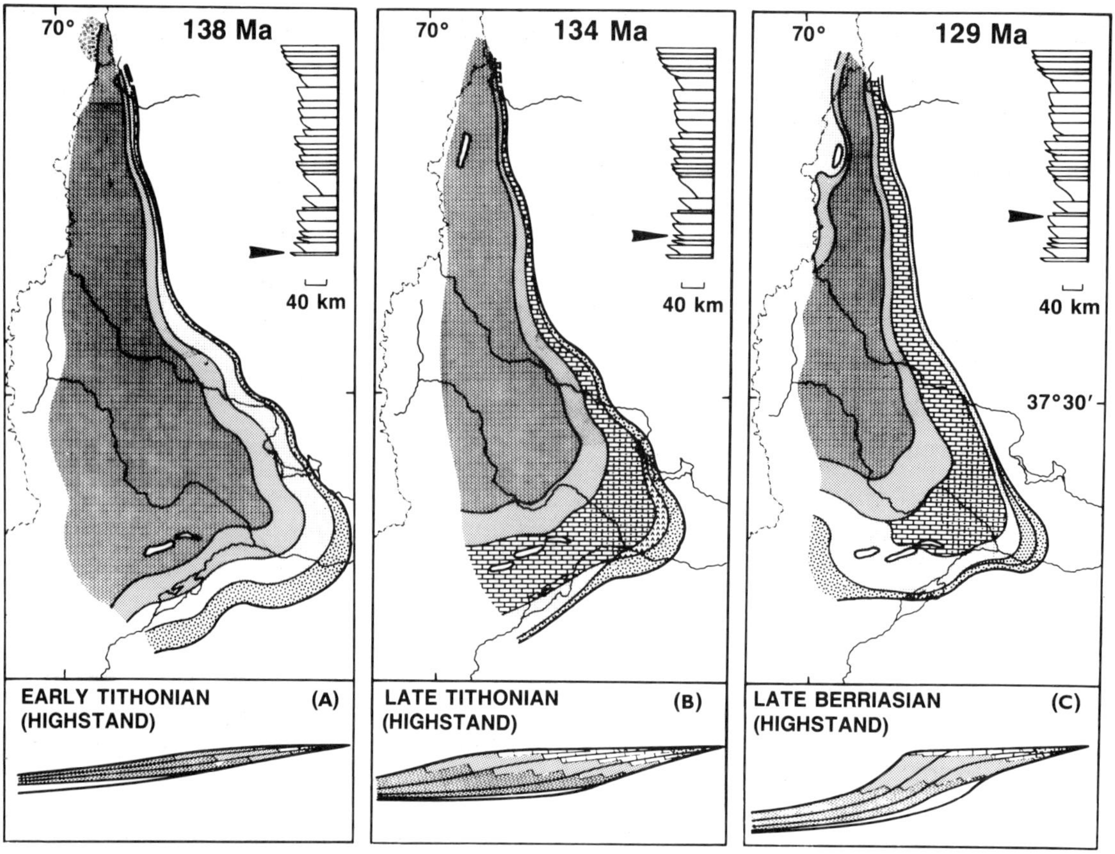


**Fig. 7.** Late Jurassic and early Neocomian palaeogeography of the Neuquén Basin. (A) Early Tithonian, about 138 Ma; (B) late Tithonian, about 134 Ma; and (C) late Berriasian, about 129 Ma. Key to facies belts in Fig. 5 and Table 1.

a shallow-water carbonate depositional system. From middle to late Tithonian (Fig. 7B) the carbonate shelf had low relief and extended down to a very subtle slope built on a basinal realm 200 m deep (Mitchum & Uliana, 1985). A depositional profile without shelf break, but featuring sigmoidal prograding geometry and occasional toplap, and the presence of a topset rich in grain-supported textural types, reveals a carbonate system that was able to catch-up and then slightly overcome the space available for shallow-water deposition. Whenever developed, the lowstand deposits of the Tithonian sequences are thin (5–10 m) and areally restricted. On the proximal side they appear as estuarine sandstones, filling subtle depressions incised in outer ramp facies, and on the distal side they consist of thin-bedded tempestites with occasional graded layers. The updip end of the carbonate systems was episodically affected by brief events of evaporite development (de Ferrariis & Montero, 1970; Zilli *et al.*, 1979). These anhydrites were formed within a sabkha setting, triggered by hydrographic restriction (Busson, 1978,1980) behind the wide and levelled mid–late Tithonian carbonate depositional terraces. Control on the core of the Chilean Andes (Biró, 1980; Hallam *et al.*, 1986) documents the proximity of the volcanically active region to the west. Details on the type of magmatic activity and its tectonic controls can be found in Levi *et al.* (1988) and papers cited therein.

### Berriasian (134–129 Ma)

In the course of the Berriasian there was an increased influx of clastic detritus to the shelf. As the inner shelf belt was invaded by siliciclastic sedimentation, carbonate deposition was inhibited over large portions of the southern embayment (Fig. 7C). The

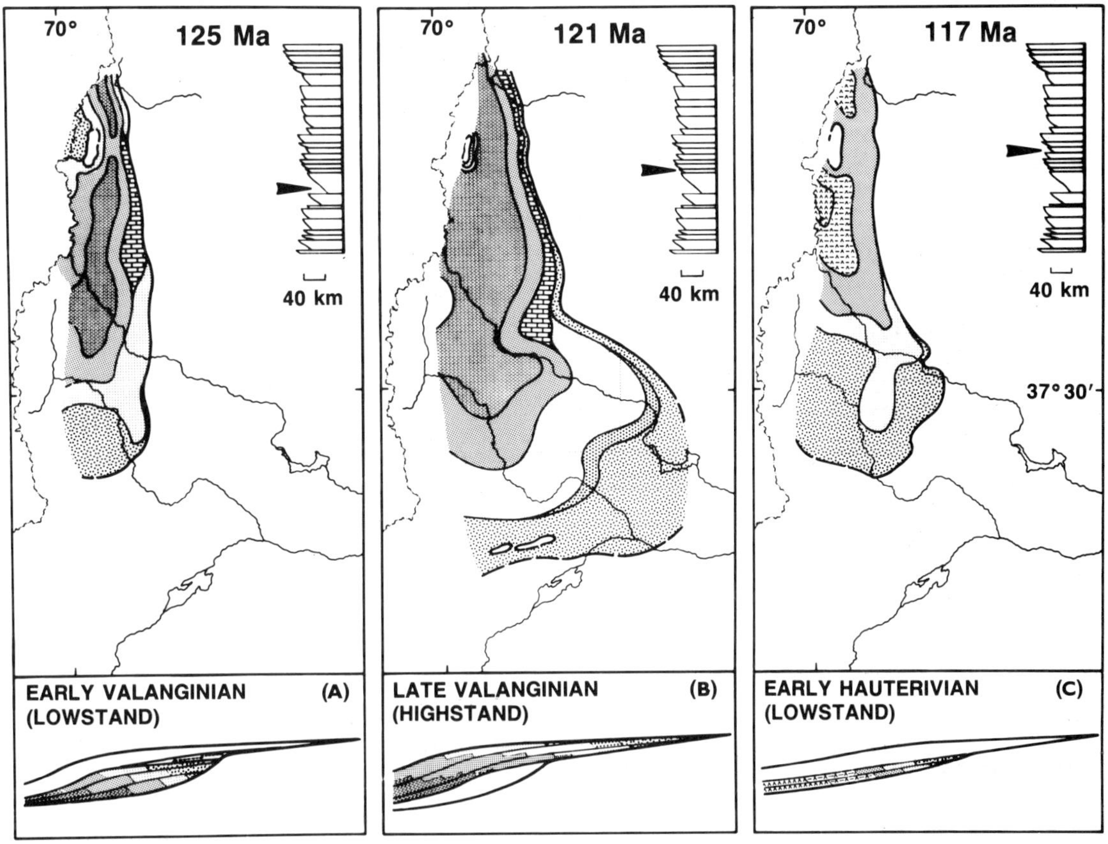


**Fig. 8.** Mid-late Neocomian palaeogeography of the Neuquén Basin. (A) Early Valanginian, about 125 Ma; (B) late Valanginian, about 121 Ma; (C) early Hauterivian, about 117 Ma. Key to facies belts in Fig. 5 and Table 1.

basinal area was much reduced by progradational sedimentation. This process, begun in the Tithonian, was accentuated in the Berriasian. While successive depocentres migrated basinward (Gulisano *et al.*, 1984; Mitchum & Uliana, 1982,1985), their shelfal counterparts became thin and areally restricted. During this period the prograding depositional tract developed well-defined shelf breaks, steep complex–oblique clinoforms, and common toplap. The increased trend toward bypass of the shelfal tract and expanded rate of lateral accretion has been interpreted as a result of progressive decrease in shelfal accommodation owing to the global eustatic fall during the early Neocomian (Mitchum & Uliana, 1985). Base-level lowering during the eustatic drop was enough to allow local development of tide and wave-dominated lowstand wedges on top of the previous lower slope deposits. A comparable trend is recorded on the arc side of the basin. Berriasian time also saw the development of an eastward prograding shelf, presumably perched on the toes of the volcanic complex. A different situation can be seen at the northeast end of the embayment, where subsidence was not matched by sedimentary influx. As a consequence Berriasian deposits consist of a starved-shelf assemblage that oversteps the basement beyond the Tithonian depositional edge (Legarreta & Kozlowski, 1984).

### Valanginian (129–126 Ma)

Late in the early Valanginian the tendency toward restriction in the areal extent of the depositional site reached a climax. Figure 8A depicts the sedimentation scenario at this time (125 Ma). The whole extent of the embayment updip of the late Berriasian (128.5 Ma) shelf break was exposed and remained above base level. A base-level drop is also recorded by pronounced downward shift of the depositional systems, locally in the order of 60 km (see also Hallam *et al.*, 1986). This change was enough to produce the abrupt appearance of coarse fluvial deposits on top of the outer slope facies assemblage (Gulisano *et al.*, 1984). Shallow water conditions, even at basin centre locations, are documented by the sporadic occurrence of thin volcaniclastic tempestites. Extreme shallowing during shelf exposures is perhaps the reason for the lack of well-developed submarine-fan complexes at this time. Exposure of the shelf must also have produced temporary increase of the clastic influx from the west, south, and north. The depositional sites reacted by drastic reduction in the areal extent of the basin centre and carbonate shelf domains. Again there is surprisingly little evidence of erosional scouring of the Berriasian shelf-edge deposits. The areas affected by erosional bevelling were restricted to the vicinity of blocks that showed persistent structural activity throughout the studied interval (Legarreta & Kozlowski, 1984; Ploszkiewicz *et al.*, 1984).

**Late Valanginian–early Hauterivian (126–118.5 Ma)**

Close to the end of the Valanginian a basin-wide reversal in the depositional regime came about, and the entire area of the embayment shifted to a submergent mode (Fig. 8B). Massive landward encroachment was followed by flooding and drowning of large portions of the shelf (Legarreta *et al.*, 1981). At the beginning of this phase the influx of clastic sediments was greatly reduced but the shelf remained under shallow water, as testified by a widespread glauconitic–phosphatic hardground cap (Mastandrea *et al.*, 1975; Gulisano *et al.*, 1984). During a short period the carbonate system developed on the northeast flank of the embayment was able to keep up by the development of a coralline shelf-margin shoal complex (Legarreta & Kozlowski, 1981). On the proximal side of the shelf and because of the restricted nature of the Valanginian wedge, the Hauterivian sequence stepped over the Berriasian accumulations (Legarreta & Kozlowski, 1984; Gulisano *et al.*, 1984; Fig. 6). Widening of the shelf environment belt at a time of base-level rise produced a situation where the depositional system developed an extremely wide ramp morphology (devoid of seismically recognizable clinoforms). Over this broad tract, accumulation occurred in a semi-starved setting that fluctuated between periods of 'catch-up' carbonate sedimentation and episodes of fine-grained clastic aggradation.

**Late Hauterivian–Barremian (118.8–115 Ma)**

Conditions favouring widespread shelfal accumulation in a mixed (clastic–carbonate) ramp depositional setting continued through the late Hauterivian and early Barremian. By middle–late Hauterivian time, however, these conditions were punctuated by at least two short episodes of shallowing that caused exposure of the shelf and abrupt basinward shift of the clastic-fringe depositional system. The first of these episodes occurred at around 118 Ma and the second one, leading to the situation depicted in Fig. 8C, preseumably took place at about 117 Ma. The early Hauterivian shallow-water assemblage was exposed down to the outer reaches of a ramp system, and clastics were transported across the emergent embayment down to a main depocentre located at the ramp to basin transition zone. The existence of a low-relief (ramp) depositional topography lacking a well-defined shelf break induced the deposition of a lowstand wedge consisting of a thin (average 10–30 m) but widespread aeolian–fluvial sheet sand ('Avilé Ss.', interpreted as a lowstand systems tract deposit in Legarreta & Gulisano, 1989). The facies characteristics of coeval deposits to the north show that the early Hauterivian euxinic basin had been temporarily transformed into a shallow evaporite pan fringed by extensive mudflats. The presence of halite hoppers and bedded rock-salt in the distal portions of the basin indicates that the marine connection was tenuous at best. In spite of the drastic changes implied by the lowstand facies distribution, the area was rapidly flooded and by 116 Ma (Fig. 9A) the basin-wide style of ramp deposition had resumed, including the presence of a quiet euxinic basin that accumulated organic-rich ammonite shales and micrites.

**Barremian–early Aptian (115–112 Ma)**

Late in the Hauterivian (Fig. 9A) a gradual change in the dominant depositional style began. Large segments of the shelf accumulated a monotonous alternation of thin unfossiliferous shales and skeletal–ooidal packstones–grainstones with a sharp base and hardground top. The facies assemblage is interpreted as representing a shallow water regime, fluctuating from a storm-dominated inner shelf to a muddy tidal flat. The widespread

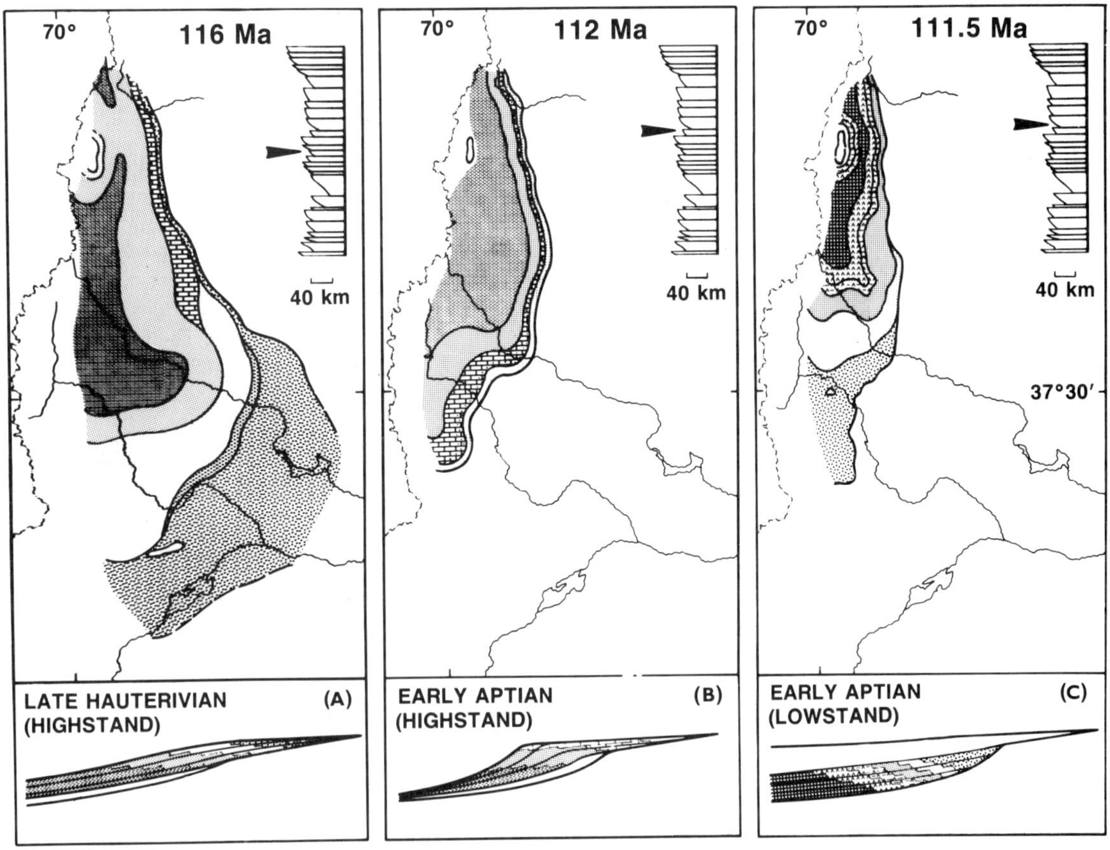


**Fig. 9.** Late Neocomian to Aptian palaeogeography of the Neuquén Basin. (A) Late Hauterivian, about 116 Ma; (B) early Aptian, about 112 Ma; (C) early Aptian, about 111.5 Ma. Key to facies belts in Fig. 5 and Table 1.

appearance of shallow-water accumulations implies a reduction in the shelfal accommodation capacity and suggests a general base-level lowering. The trend is further emphasized by the retraction seen in the shelfal extent of the 113.5 Ma sequence (Fig. 6), the appearance of toplap configurations, and the presence of sigmoidal to oblique prograding geometries, a depositional style not developed since the early Valanginian (125 Ma). By early Aptian the system had evolved to a situation as shown in Fig. 9B. The embayment was once again entirely exposed, and the depositional site was enormously restricted down to a location controlled by the morphology of the preceding slope. The proximal side of the depositional system included a narrow fringe of siliciclastic beaches and a well-developed carbonate belt, represented by pelletal packstone and oolitic grainstones with tidal-bar features. Low faunal diversity suggests a hydrological regime that induced considerable environmental stress. However, carbonate production was maintained, keeping up and forcing active progradation into a shale–wackestone basinal setting (Legarreta, 1985,1986). Development of steep clinoforms is clearly indicated by the presence of large-scale soft-sediment slumping of slope and hemipelagic lime-mudstone facies (Legarreta, 1985).

### Late Aptian (112–107.5 Ma)

The limited flooding during early Aptian time was followed by a base-level drop that allowed sands to pass over the carbonate belt. The depositional scenario developed after this event (Fig. 9C) much resembled the situation reached during the early–middle Hauterivian (Fig. 8C). The proximal tract encompassed a mixture of aeolian dunes and multi-

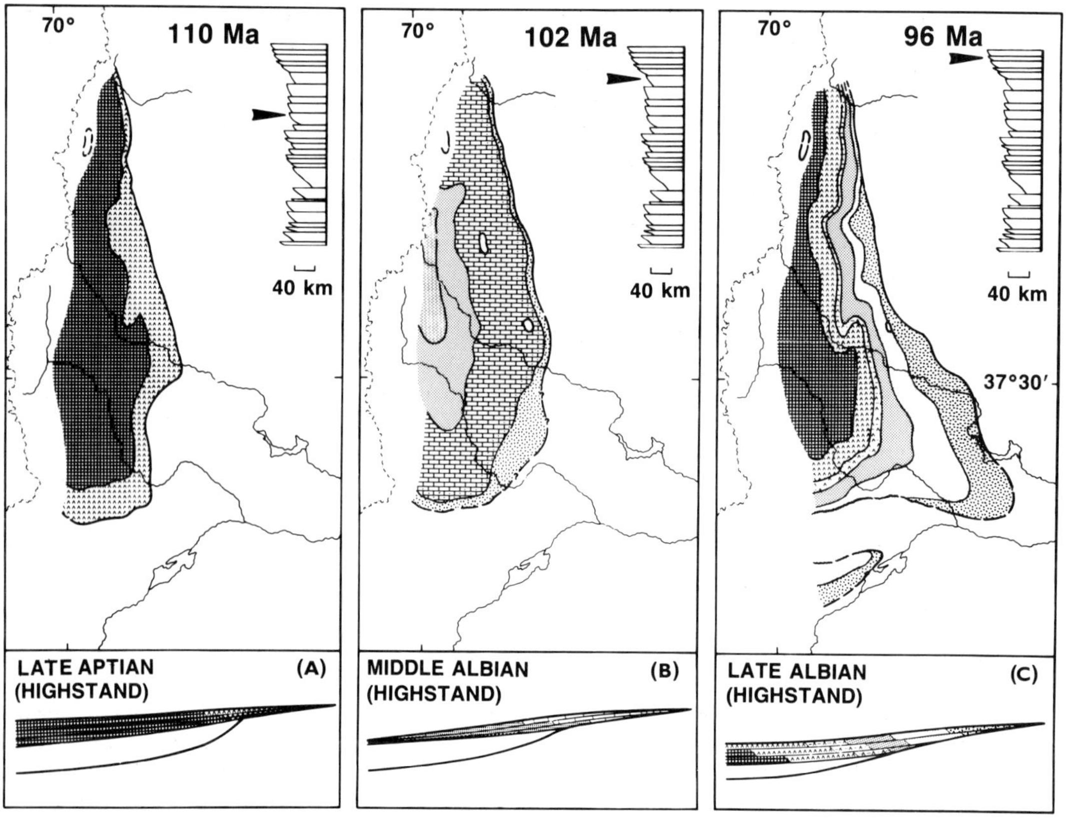


**Fig. 10.** Aptian–Albian palaeogeography of the Neuquén Basin. (A) Late Aptian, about 110 Ma; (B) middle Albian, about 102 Ma; (C) late Albian, about 96 Ma. Key to facies belts in Fig. 5 and Table 1.

storey cross-bedded fluvial sands. Facies tracts located basinward included ephemeral channels cut in mudstone, a saline mudflat assemblage, and finally a saline basin centre where bedded anhydrite and rock-salt were deposited (Legarreta, 1985). The topographic gradient at maximum base-level fall was enough to produce channel incision at outer shelf and upper slope locations, but further back on the shelf the stratigraphic discontinuity appears devoid of erosional relief. The southerly expansion of the sand-dune depositional tract is a striking palaeogeographic feature, interpreted as a response to the inception of a new subsidence pattern. From this perspective the dune field would have been preferentially accommodated above base level but within a subtle successor depression produced by downward flexion of the back-arc crust (see next section).

Stratigraphic patterns demonstrate that the base-level drop late in the early Aptian was followed by a renewed period of regional flooding during late Aptian and Albian time. Near the close of the Aptian only the Andean side of the basin was submerged (Fig. 10A). The hydrological regime was kept under negative balance as indicated by a marginal belt of nodular anhydrite in algal matrix and a basin-centre assemblage of 'varved' anhydrite and bedded halite. However, the eastward expansion of the depositional site and the enlargement of the area under subaqueous accumulation clearly indicate a trend of increased coastal onlap and the onset of a regional transgression under an evaporitic regime.

### Early–middle Albian (107.5–100.5 Ma)

A negative hydrological balance was sustained for some time, but by the beginning of the middle Albian, improvement in the circulation led to the disappearance of the conditions controlling evaporitic sedimentation. Still, a depauperate euryhaline invertebrate association suggests prevalence of fluctuating salinity conditions (Legarreta, 1985). The depositional tracts were arranged with a regular concentric pattern (Fig. 10B) and most of the accumulation occurred by stacking of small shoaling-up carbonate units deposited around a starved basin centre. The high-energy belt was near the shore and consisted of carbonate shoals fringing a tide-dominated shoreline and a band of coastal lagoons. The depositional morphology was one typical of a carbonate ramp system entirely devoid of shelf break ('distally steepened ramp'; Legarreta, 1986). Sequence analysis (Legarreta, 1985) revealed two series of shoaling-up packages stacked with a retrogradational pattern. This transgressive internal organization is in good agreement with evidence showing an increase in the magnitude of the coastal encroachment, with respect to the older Albian sequences (Fig. 6). The reduced size of the marine expansion along the axis of the embayment, so well developed in the course of previous flooding events, is remarkable. A possible mechanism to account for this palaeogeographic constriction could be found in the development of a peripheral bulge (as suggested by regional thickness control in Uliana *et al.* (1975a,b; Fig. 3), related to flexural yoking of the arc and the adjacent back-arc. Regional considerations suggest that this peculiar subsidence pattern may have developed by readjustment of the subducting geometry, triggered by the phase of global fast spreading during middle Cretaceous times (Larson & Pitman, 1972; Frutos, 1980; Uliana & Biddle, 1988).

### Late Albian (100.5–95.5 Ma)

The map in Fig. 10C shows the situation developed by late Albian. The depositional site had enlarged encroaching over the axis of the embayment. This type of palaeogeographic evolution reveals a trend toward increased accommodation on the shelf. In spite of the high base-level, the hydrological balance was again negative and evaporite deposition began again. The eastern fringe of the depositional site was dominated by craton-derived clastic sediments, deposited in an area with the characteristics of broad alluvial aprons dominated by ephemeral streams and flash-flood events (Legarreta, 1985). The position of the evaporitic tract across the central part of the embayment again suggests an effect of local constriction, presumably associated with the persistence of differential subsidence and the development of an upwarp at the edge of the successor deep.

## EUSTATIC TREND AND STRATIGRAPHIC PATTERNS

The Neuquén Embayment, with a Tithonian–Albian record of moderate, regular subsidence and sedimentary influx and its complex and well-dated stratigraphy, documenting the coeval existence of clastic, evaporite, and carbonate depositional systems, provides a unique opportunity to confront different types of local sedimentary response with the inferred record of global eustasy. Several kinds of stratigraphic pattern and the timing of their development were thus considered in an attempt to compare the local history of basin fill and relative sea-level fluctuations against the history of eustatic change as modelled by Haq *et al.* (1987) (Fig. 4).

### Coastal onlap

The proximal edge of the 12 key intervals shown in the palaeogeographic reconstructions delineate a gross time–space pattern that tends to mimic the onlap trend recognized by Haq *et al.* (1987) for the Tithonian–Albian interval. The recorded stratigraphic pattern, as illustrated in the local chronostratigraphic chart (Fig. 6), also shows that the relative position of the individual depositional sequences defines a series of cycles of progressively expanded and retracted coastal encroachment that follow the pattern of the global cycle chart. As discussed in the text, observed departures from the pattern of global onlap (Fig. 6) that developed during some periods of time and over specific parts of the basin can be explained reasonably well in terms of the local subsidence history.

### Regional offlap

Patterns of regional stratigraphic offlap indicated by stacking of sequences with progressively smaller (thinner and/or less widespread) shelfal component reveal a trend toward reduction in the potential of the shelf tract to accommodate the available sediments. The age control points out the occurrence of events of this kind during the Berriasian to early Valanginian and the Barremian to early Aptian, both periods coinciding with episodes of globally falling marine stand.

### Large gaps in paraconformable shelfal successions

At basin centre locations the tops of the offlapping wedges are sharply terminated at surfaces abruptly overlain by units that reflect a major shift toward progressively increased landward encroachment (stratigraphic 'turning points', cf. Schlarger (1981)). The main stratigraphic 'turning points' are dated as late Valanginian and late Aptian and when they are traced back toward the shelf they coincide with large stratigraphic breaks (missing sections in the order of 3–10 Ma) that separate the offlapping and the overlying onlapping sets of depositional sequences. Only very locally do these discontinuity surfaces separate truncated or tectonically disturbed deposits. Tilting and truncation are areally restricted to zones showing persistent synsedimentary tectonism, both on the foreland and near the arc (Fig. 6). The dominant parallelism and the general lack of erosional truncation along these large stratigraphic breaks, and also the circumstance that they are preceded by regional offlap, strongly suggest a genesis largely dictated by major eustatic drops.

### Lowstand wedges

Lowstand wedges, that is discrete wedges of relatively shallow-water sediments encased in deeper water accumulations, are deposited when the shelf is for the most part exposed and the depositional systems are forced to migrate basinward (Vail *et al.*, 1987). The Neuquén record shows that these wedges are best developed within the stratigraphic proximity of the 'turning points' (Fig. 6). They tend to appear as well-defined wedges at the base of the sequences deposited near the top of the offlapping successions and within the lowermost sequences of the succeeding onlapping set of sequences. Lowstand wedges are non-existent or minor at intermediate times, correlative with global (long-term) marine highstands. The appearance of two well-developed lowstand wedges during the Hauterivian represents a departure from this pattern and points out the possible occurrence of mid-Hauterivian eustatic drops with a relative magnitude larger than those estimated by Haq *et al.* (1987). Periods of sustained but gradual long-term fall, such as the late Tithonian to early Berriasian, induced the development of thin and areally restricted lowstand wedges, not shown in Fig. 6 for reasons of scale.

### Shelfal drowning and starvation

Major events of shelfal drowning and starvation occurred shortly after the stratigraphic 'turning points' dated as early Valanginian and early Aptian, and here correlated with episodes of sustained eustatic rise. During the latest Valanginian when flooding began, the shelf remained within the euphotic zone but conditions allowed only subdued carbonate development. By early Hauterivian time the drowning was complete and close to half of the Berriasian–Valanginian shelfal tract was under stagnant, starved conditions.

The second episode of massive flooding occurred after the early Aptian 'turning point', at a time when regional palaeogeographic isolation and evaporative drawdown turned the back-arc depocentre into a hypersaline regime. The flooding is thus recorded by laminated evaporites on top of the shallow subtidal Barremian shelfal deposits. The palaeogeographic reconstructions and the record of coastal onlap that developed in spite of the changes in the subsidence pattern strongly suggest that progressive marine encroachment tied to eustatic rise was once again the main reason behind the starvation of the basin and large portion of the shelf. High rates of evaporite deposition (Kendall & Schlager, 1982) may have contributed to masking of the clastic input.

### Morphology of the shelf to basin transition

Clinoform morphology is a very sensitive parameter that reflects closely how much of the space generated on top of the shelf (accommodation, Vail *et al.*, 1987) was occupied by the available sedimentary supply. In simple terms, accommodation greater than supply should produce depositional back-stepping and flat slope morphologies, whereas accommodation less than supply should produce sedimentary bypass, high rate of lateral accretion, and steep slope morphologies. The Neuquén record demonstrates that those depositional sequences accumulated during periods of sustained relative rise, such as the Hauterivian and the Albian, are characterized by low relief slope configurations (ramp or distally steepened ramp) coupled with aggradation or landward back-stepping of the successive depositional units. Sequences deposited during periods of gradual relative fall, such as the Tithonian, feature low to moderate slope relief and record sigmoidally shaped ramps and shelves with limited and late development of toplap terminations. Finally, sequences deposited under conditions of dominant relative fall, such as those representative of the late Berriasian to early Valanginian and late Barremian to early Aptian, show steep slope relief related to oblique-complex clinoforms and frequent toplap configurations indicative of high outbuilding rates.

### Carbonate response

Because carbonate sediments originate within the depositional environment, limestones can be diagnostic indicators of changes in relative sea level. Coexistence of clastic and carbonate depositional systems on the shelves of the Neuquén Embayment can be used as a tool to gauge variations in the arrival and lateral dispersal of siliciclastic supply across shelfal tracts. Shallow-water carbonate belts were broadest during and just after periods of major relative rise, such as the middle Tithonian, late early Hauterivian, and middle Albian. During these periods clastic sediment was presumably trapped within the proximal reaches of the shelf. Conversely, carbonate systems were at a minimum after periods of sustained relative fall, such as late in the early Valanginian, when reduced accommodation potential pushed siliciclastic deposits across the shelf terminating the growth of the carbonate 'factories'. Most of the time, the Neuquén carbonate systems operated under conditions of subdued growth and only episodically were able to 'catch up' and aggrade to sea level. The mid-late Tithonian sequences, however, provide a good example of departure from this style. Relative stillstand or slight fall after the early Tithonian flooding was enough to contain clastic encroachment and at the same time allowed the stacking and modest outbuilding of carbonate sand shoals on top of a wide and flat 'keep-up' shelfal tract aggraded to sea-level. The basic architecture of the successions dominated by carbonates does not depart from the stratal configurations found in areas under entirely clastic sedimentation. To a large extent the similarity seems related to the fact that the carbonate units were dominated by hydrodynamic accumulations and lacked frame-builders capable of resisting wave action and keeping up with rapid relative rises.

### Evaporites

Discrete evaporite units encased in non-evaporitic rocks, such as those deposited during the Hauterivian and early Aptian, are in stratigraphic association with lowstand deposits in areally restricted basin-centre locations. These associations strongly suggest evaporite accumulation under conditions of morphologic isolation (Busson, 1978,1980) caused by eustatic drops. The thin mid-late Tithonian anhydrites (Fig. 8B) are a good example of the shelfal evaporite assemblage (Sloss, 1959). Development of these anhydrites, as in the case of coeval evaporites in the Middle East, North Africa, and Europe (Murris, 1980; van Houten, 1980), seems related to a eustatic regime favouring the development of a flat-topped shelf that was wide enough to generate hydrographic restriction (Busson, 1978,1980) sufficient to trigger evaporite deposition at its inner tracts.

### Geometry of depositional sequences

Most studied sequences show a stratal fabric and a lithofacies organization different from those of the fully developed depositional sequences depicted by Vail (1987) and van Wagoner *et al.* (1987). As shown in Fig. 11, departures from the ideal development include absence of the basin-floor fan (bf) and slope fan (sf) deposits, and a much reduced development of the transgressive systems tract (TST) deposits. These peculiarities were presumably related to a specific combination of weak regional subsidence without tectonic hinges, and a modest clastic supply unable to match the increase in accommodation triggered during phases of eustatic rise (widespread shelfal starvation). The bulk of the sequences is made up of a progradational wedge or highstand systems tract (HST). Where thick enough to be resolved seismically, the successive clinoforms can be seen to become progressively steeper through time. This modification is understood to reflect a change from a low profile ramp toward a depositional terrace fringed by a sharper sigmoid or oblique slope tract, which resulted from a gradual decrease in the shelfal accommodation generated after the peak of the eustatic cycles.

## EUSTATIC CYCLES AND ANATOMY OF SEQUENCE-SET ROCK PACKAGES

Supersequences (Vail *et al.*, 1977) or sequence-sets are high-rank packages of sedimentary rocks, with thicknesses of few hundreds to thousands of metres and spanning 5–10 Ma, limited by significant inter-regional unconformities. The existence of strati-

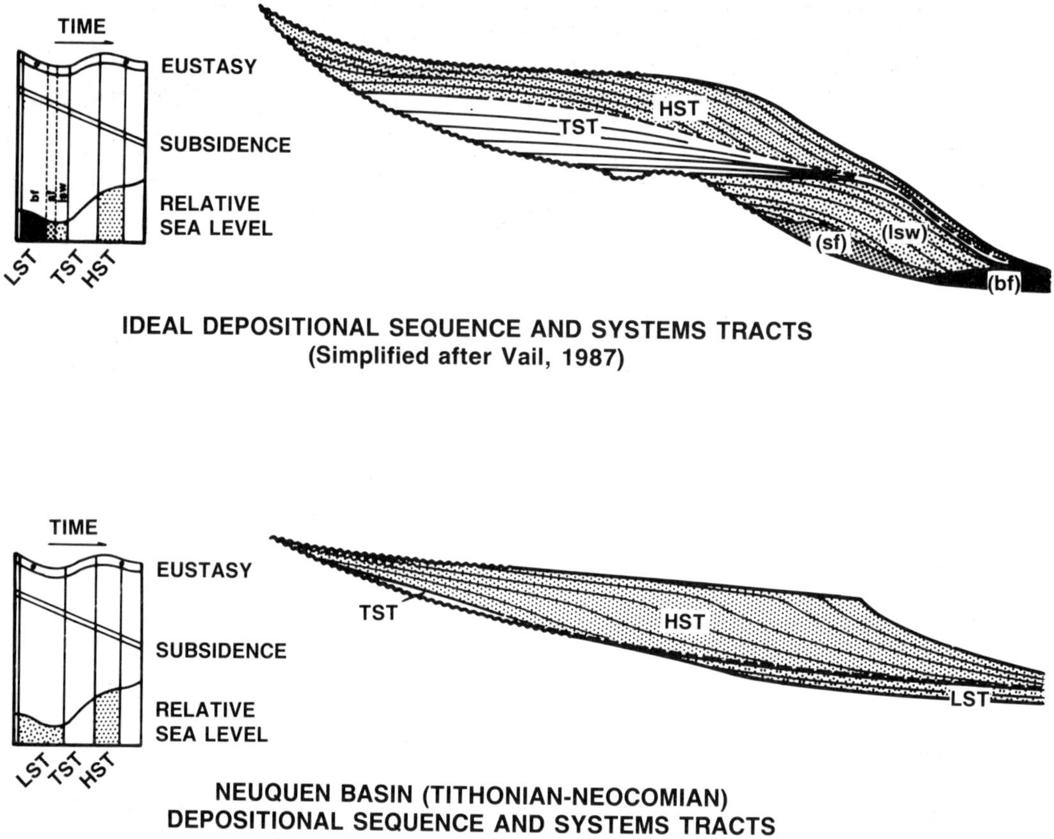


**Fig. 11.** The differences in facies and stratal patterns between the fully developed depositional sequences depicted by Vail (1987) and the average Tithonian–Neocomian sequence of the Neuquén Basin are seen as a result of the local rates of subsidence and sediment availability. See discussion in the text.

graphic units of this level of hierarchy, bounded by surfaces that do not co-ordinate with the standard chronostratigraphic divisions of European origin, has been recognized throughout the world (e.g. Grabau, 1936; Sloss *et al.*, 1949; Sloss, 1972; Chang, 1975; Soares *et al.*, 1978). Several authors have also pointed out that supersequence-rank units display some recurrent form of internal organization (e.g. Grabau, 1936; Sloss & Speed, 1974; Kendall & Schlager, 1982). Our own observations in the Neuquén Embayment and areas outside South America suggest that sequence-set packages feature a definite internal arrangement, comparable to what has been described in depositional units of lower hierarchical rank, such as sequences (Vail, 1987; Vail *et al.*, 1987; van Wagoner *et al.*, 1987) and parasequences (van Wagoner *et al.*, 1987; Strasser, 1988) or the punctuated aggradational cycles (Goodwin & Anderson, 1985).

Our perception of an ordered progression of stratigraphic properties and anatomy of rock packages of sequence-set scale hinges around four concepts.

**1** The idea that high-rank stratigraphic discontinuities separating paraconformable stratigraphic successions are largely related to major (second order) eustatic fluctuations (Suess, 1906; Grabau, 1936; Vail *et al.*, 1977).

**2** The realization that the large-scale stratigraphic architecture of many Phanerozoic successions, deposited over broad cratonic areas and continental margins, has developed as a response to an oscillatory alternation of submergent and emergent conditions (e.g. Grabau, 1936; Sloss & Speed, 1974).

**3** The observation that, at least in those areas affected by regular rates of thermo-tectonic subsidence, the rock packages lying between successive

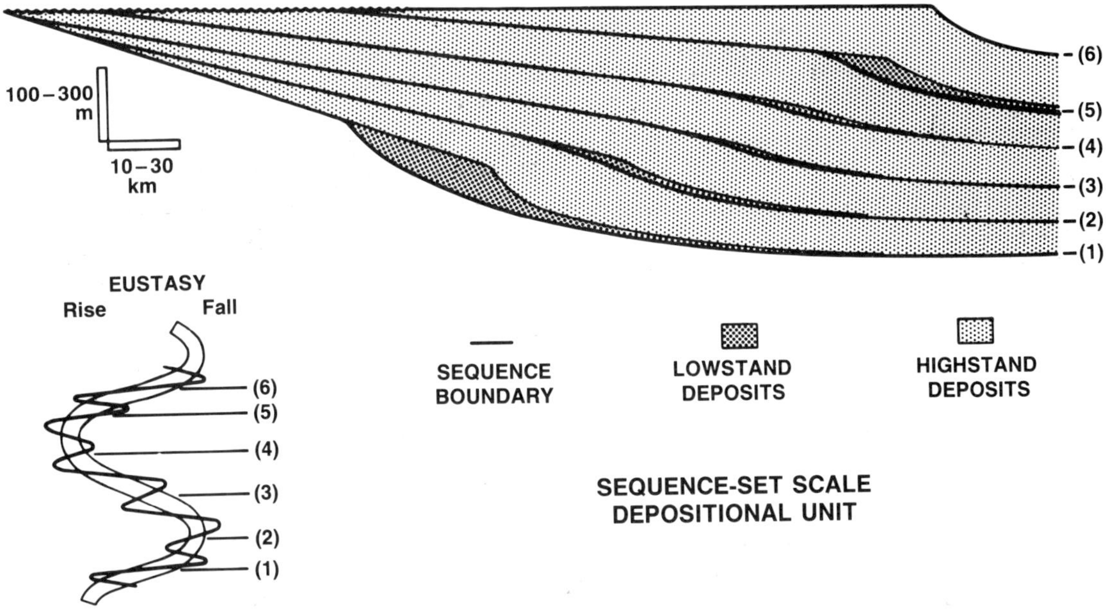


**Fig. 12.** Sequence-set scale of stratigraphic units: diagram illustrating relative changes in highstand and lowstand systems tracts deposits. Highly schematic.

interregional unconformities follow an orderly pattern of stacking that outlines globally synchronous cycles of coastal onlap (Grabau, 1936; Vail *et al.*, 1977; Haq *et al.*, 1987).

**4** The more recent idea of accommodation, or space available for sedimentary accumulation, as a factor modulated by the relative changes of sea level (Vail *et al.*, 1987) that represents the main controlling factor on stratigraphic anatomy (Vail, 1987).

Figure 12 is a conceptual diagram illustrating successive changes in the patterns of basin fill during the course of a second-order (5–10 Ma) eustatic cycle. The sketch was conceived to depict the sedimentary response in an area similar to the Neuquén Basin, which is a depocentre under moderately intense but relatively regular subsidence and sedimentary influx. The diagram shows the major stratal pattern of a depositional unit of sequence-set scale, bounded by high-rank stratigraphic discontinuities.

Figure 13 depicts in greater detail the changes in stratigraphic pattern of each one of the depositional sequences comprising a sequence-set unit. The changes shown at this level of resolution are assumed to reflect accommodation variations with a magnitude and frequency in the range of the third-order (0.5–2 Ma) eustatic oscillations. Figure 14 portrays the coeval response predicted for a carbonate depositional system. The internal organization of the sequence-set complex is thus envisioned as developing through systematic change in the properties of the successive sequences. Amongst others, those modifications should include variations in the areal extent of the shelfal member (coastal encroachment), in the relative size of the lowstand and highstand depositional system tracts, in the morphology of the shelf to basin transition (ramp versus depositional terrace), and in the associated geometry of the corresponding clinoforms (sigmoid–complex–oblique).

Our observations on depositional units of sequence-set scale indicate that, besides stratal configurations, some peculiar rock types and facies assemblages are more prone to develop around specific periods of time. As discussed in the first part of the paper, lowstand deltas (and occasionally turbidites), deposition of keep-up or catch-up carbonates, shelfal or basinal evaporites, episodes of shelfal drowning or exposure, and development of soils, coupled with aeolian deposits at basin centre locations, occur with a sequence that roughly parallels the inferred eustatic rhythm and suggest an important influence of global factors on local sedimentation.

Because sustained development of depositional

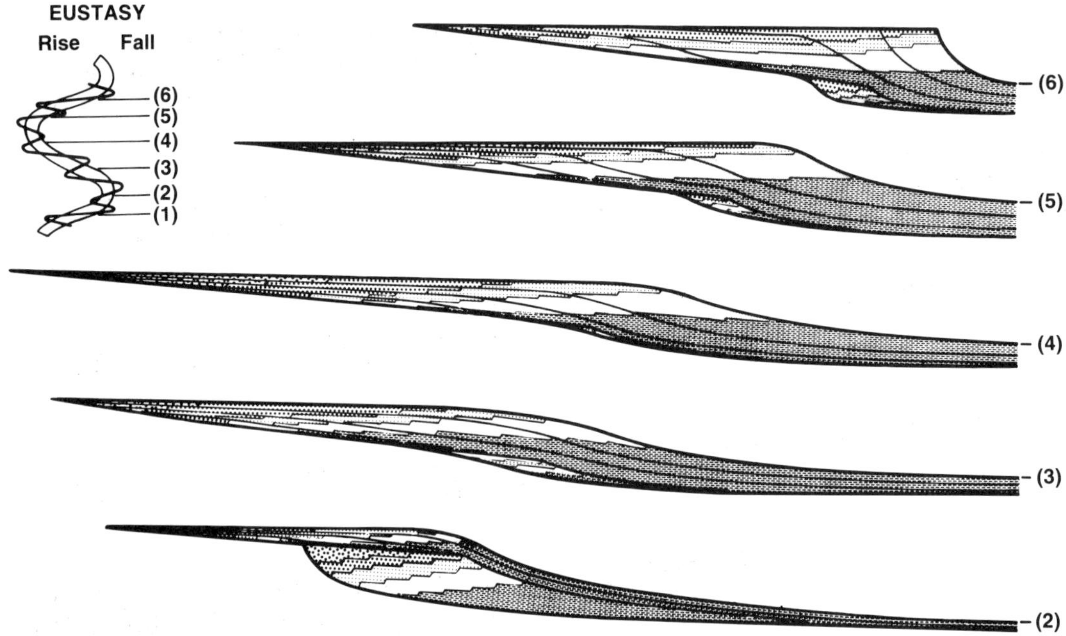


**Fig. 13.** Idealized succession of depositional events leading to the development of a sedimentary package of sequence-set scale in a clastic-dominated depositional scenario. Not to scale.

packages of sequence-set scale requires relative persistence of tectonic subsidence and sediment input over long periods of time, these units are less prone to stack-up along rapidly changing active plate margins. However, in areas featuring longstanding tectonic stability, such as the cratonic interior of North America (Sloss, 1963,1988) and the Arabian plate (Murris, 1980), several successive sequence-sets have remained preserved in the stratigraphic record and in our opinion they demonstrate the continued and decisive influence of eustasy on sedimentation over prolonged timespans. The long distance correlatability of the stratigraphic patterns typical of sequence-set packages, even between depocentres placed over different crustal blocks (Sloss, 1972; Soares *et al.*, 1978), further emphasizes the global nature of the control.

## CONCLUDING REMARKS

Regional analysis of the Tithonian–Albian succession accumulated within the back-arc Nauquén depocentre reveals that a number of properties, such as specific stratal patterns, facies assemblages, and peculiar rock types, follow a seemingly arranged trend of appearance and disappearance from the stratigraphic record. As far as it can be gauged by the available biochronologic control, the order of those occurrences closely matches the sequence expected in a depositional setting where changes in the space available to accommodate clastic, carbonate, and evaporite sedimentation were largely dictated by marine oscillations following a path that compares favourably with the third-order eustatic cycles proposed by Haq *et al.* (1987).

Beyond the general conformity between the global and the local onlap patterns, two types of departures have been detected. During the Aptian and the early Albian, over a 10–15 Ma interval, the magnitude of the coastal encroachment was significantly less than expected. The coincidence of this with a period of increased granitic emplacement in the arc domain and a change of subsidence rate detected in the basin strongly suggests that the discrepancy was a consequence of local tectonic overprint (see also the paper by Hallam). In spite of the increased rate of subsidence, the influence of eustasy on accommodation must have remained pervasive, because timing of the sequence boundaries and the relative range of coastal encroachment between successive sequences was apparently maintained.

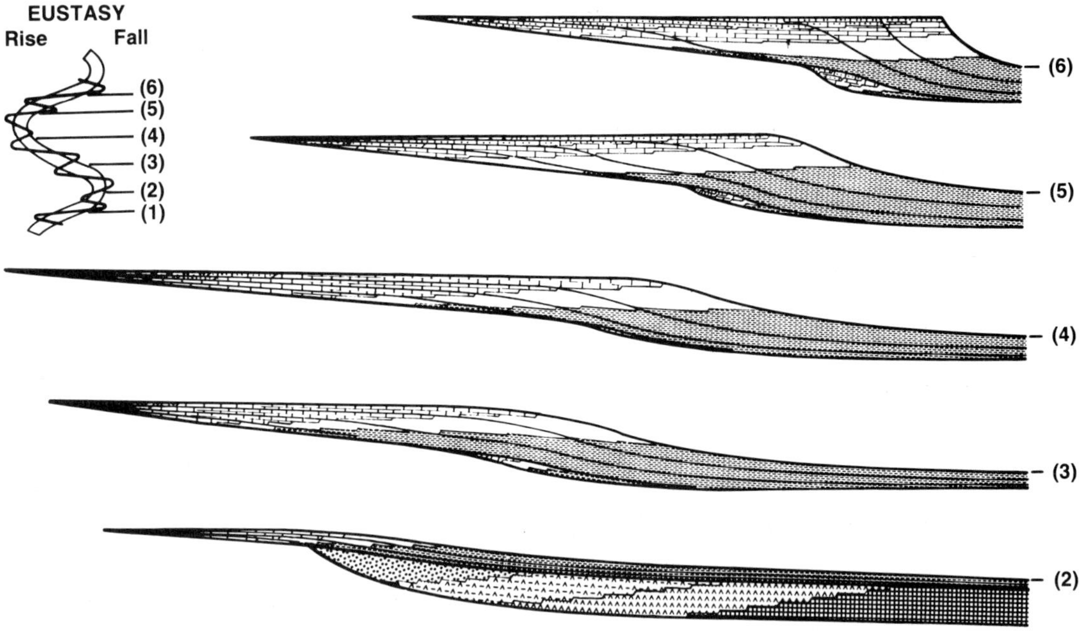


**Fig. 14.** Idealized succession of depositional events leading to the development of a sedimentary package of sequence-set scale in a carbonate-dominated depositional scenario. Not to scale.

At a much finer scale, and considering the level of stratigraphic resolution presently achievable in the study area, some additional deviations were noticed for specific depositional sequences (timespans in the order of 1–2 Ma). These discrepancies stand out because of the extreme areal restriction of a few depositional sequences, usually linked to the appearance of facies indicative of total or nearly complete desiccation of the basin. This evidence suggests the occurrence of local sea-level drops with a magnitude larger than the global falls, and they could be explained as a local effect owing to evaporitic drawdown, or alternatively they could indicate the need for some adjustment in the estimations by Haq *et al.* (1987).

When compared with the model of Vail (1987) the Neuquén sequences reveal some differences in their internal organization. Most noticeable are the volumetrically minor development or absence of basin-floor or slope fans and transgressive systems tracts. We view those peculiarities as the result of specific combinations of local subsidence rates and sediment supply, unable to generate steep slopes or keep pace with the increments of accommodation produced during periods of base-level rise.

When attention is focused on large-scale stratigraphic packages bounded by interregional unconformities, the local record can be divided rationally into slices representing 10–30 Ma of sedimentary accumulation (Sloss, 1988). These units (termed sequence-sets, megacycles, synthems, or allogroups) display a definite internal organization, defined by systematic changes in the stratigraphic properties of the successive smaller scale components or depositional sequences. Results of the present study indicate that the secular variations in features such as areal extension of the shelfal tracts or predominance of agradational versus progradational depositional modes, are tied to a large extent to the long-term eustatic drive (low-frequency or 'second order' eustatic component of accommodation). Thus, in areas where sedimentation proceeded under conditions of relatively steady subsidence rate, consideration of the eustatic signal within a sequence-set stratigraphic framework can be a helpful tool in forward modelling and predictive stratigraphy.

## ACKNOWLEDGEMENTS

The writers thank Astra C.A.P.S.A. management for permission to publish this paper and for providing an atmosphere suitable for geological investigation. Early versions of the manuscript benefited from critical reading and insightful comments by N. Christie-Blick, G.G. Bonorino, and M. Torres. Helpful reviews by D. Macdonald, A. Hallam, and B. Vining are gratefully acknowledged.

## REFERENCES


Archangelsky, S., Baldoni, A., Gamerro, J.C., Palamarczuk, S. & Seiler, J. (1981) Palinología estratigráfica del Cretácico de Argentina Austral. Diagramas de grupos polínicos del suroeste de Chubut y noroeste de Santa Cruz. *Actas VIII Congr. Geol. Argent.* **IV**, 719–742.

Biro, L. (1980) Algunos ammonites nuevos en la Formación Lo Valdés, Tithoniano–Neocomiano, Provincia de Santiago (33°50′ Lat. Sur), Chile. *Segundo Congr. Argent. Paleontol. Bioestratigraf. Actas* **1**, 223–235.

Brongersma-Sanders, M. (1971) Origin of major cyclicity of evaporites and bituminous rocks: an actualistic model. *Marine Geol.* **11**, 123–144.

Burckhardt, C. (1903) Beitrage zur kenntnis der Jura and Kreide formation der Cordillere. *Palaeontographica* **50**, 1–144.

Busson, G. (1978) L'unite des facies en milieu de plateforme carbonatée. In: *Les sédiments, leur histoire, leur environnement et leur devenir*. Doc. Lab. Geol. Fac. Sci. Lyon, Sér. 4, 87–112.

Busson, G. (1980) Les grandes cuvettes évaporitiques en milieu détritique: comment elles se creusent, comment elles se remplissent. *Bull. Cent. Rech. Expl. Prod. Elf-Aquitaine* **4**, 557–580.

Chang, K.H. (1975) Unconformity-bounded stratigraphic units. *Geol. Soc. Am. Bull.* **86**, 1544–1552.

de Ferrariis, C.I.C. & Montero, C.A. (1970) Los planos isoporosos obtenidos de perfiles en la evaluación y localización de trampas estratigráficas de hidrocarburos. *Petrotecnia (Inst. Argent. Petról., Buenos Aires)* **XX**, 30–43.

Digregorio, J.H. (1972) Neuquén. In: *Geología Regional Argentina* (Ed. Leanza, A.F.) pp. 439–505. Academia Nacional de Ciencias Córdoba.

Frutos, J. (1980) Andean tectonics as a consequence of sea-floor spreading. *Tectonophysics* **72**, T21–T32.

Gansser, A. (1973) Facts and theories on the Andes. *J. Geol. Soc. London* **129**, 93–131.

Gerth, H. (1925a) Estratigrafía y distribución de los sedimentos mesozoicos en los Andes Argentinos. *Actas Acad. Nac. Cienc. Córdoba* **19**, 11–55.

Gerth, H. (1925b) La fauna neocomiana de la Cordillera Argentina en la parte meridional de la Provincia de Mendoza. *Actas Acad. Nac. Cienc. Córdoba* **19**, 57–132.

Goodwin, P.W. & Anderson, E.J. (1985) Punctuated aggradational cycles: a general hypothesis of episodic stratigraphic accumulation. *J. Geol.* **93**, 515–533.

Grabau, A.W. (1936) Oscillation or pulsation. *16th International Geological Congress, Washington, DC, 1933* **1**, 539–552.

Groeber, P. (1929) *Líneas fundamentales de la geología del Neuquén, sur de Mendoza y regiones adyacentes*, pp. 1–109. Dirección de Minas. Geología e Hidrología, Buenos Aires, Publicación 58.

Groeber, P. (1953) Mesozoico. In: *Geografía de la República Argentina*, II, primera parte. Soc. Arg. Est. Geograf., 541 pp.

Gulisano, C.A., Gutierrez Pleimling, A.R. & Digregorio, R.E. (1984) Esquema estratigráfico de la secuencia jurásica del oeste de la Provincia del Neuquén. *Noveno Congr. Geol. Argent. Actas* **I**, 236–259.

Hallam, A., Biro-Bagoczky, L. & Perez, E. (1986) Facies analysis of the Lo Valdés Formation (Tithonian–Hauterivian) of the High Cordillera of central Chile, and the palaeogeographic evolution of the Andean Basin. *Geol. Mag.* **123**, 425–435.

Haq, B.U., Hardenbol, J. & Vail, P.R. (1987) Chronology of fluctuating sea levels since the Triassic. *Science* **235**, 1156–1166.

Herrero Ducloux, D.A. (1946) Contribución al conocimiento geológico del Neuquén Extraandino. *Bol. Inform. Petrol.* **266**, 245–281.

Hildebrand, R.S. & Bowring, S.A. (1984) Continental intra-arc depressions: a nonextensional model for the origin, with a Proterozoic example from Wopmay orogen. *Geology* **12**, 73–77.

Kendall, C.G.St.C. & Schlager, W. (1982) Relative changes in sea level and its effect upon carbonate deposition. *Shale Shaker, J. Geol. Soc. Okla.* **33**, 1–19.

Larson, R.L. & Pitman, W.C. (1972) World-wide correlation of Mesozoic magnetic anomalies and its implications. *Geol. Soc. Am. Bull.* **55**, 3645–3661.

Leanza, A.F. (1945) Ammonites del Jurásico Superior y Cretácico Inferior de la Sierra Azul de la Provincia de Mendoza. *An. Mus. La Plata (N.S.)* **1**, 1–99.

Leanza, H. (1973) Estudio sobre los cambios faciales de los estratos limítrofes jurásico-cretácicos entre Loncopué y Picún Leufú. Provincia de Neuquén, República Argentina. *Asoc. Geol. Argent. Rev.* **28**, 97–132.

Leanza, H. (1981) The Jurassic–Cretaceous boundary beds in west-central Argentina and their ammonite zones. *Neues Jahrb. Geol. Palaeontol., Abh.* **161**, 62–92.

Legarreta, L. (1985) *Análisis estratigráfico de la Formación Huitrin (Cretácico Inferior), Provincia de Neuquén.* Unpublished PhD thesis, Universidad de Buenos Aires, 247 pp.

Legarreta, L. (1986) Litogénesis de las secuencias deposicionales carbonáticas de la F. Huitrín (Cretácico Inferior), Provincia de Mendoza. *Primera Reunión Argentina de Sedimentología, Resúmenes expandidos*, pp. 173–176.

Legarreta, L. & Gulisano, C.A. (1989) Análisis estratigráfico secuencial de la Cuenca Neuquina (Triásico Superior–Terciario Inferior, Argentina). In: *Cuencas Sedimentarias Argentinas* (Eds Chebli, G.A. & Spalleti, L.A.) *Simposio de Cuencas Sedimentarias Argentinas, X Cong. Geol. Argent, Tucumán*, 221–243.

Legarreta, L. & Kozlowski, E. (1981) Estratigrafía y sedimentología de la Formación Chachao, Provincia de Mendoza. *Octavo Cong. Geol. Argent. Actas* **II**, 521–543.

Legarreta, L. & Kozlowski, E. (1984) Secciones condensadas del Jurásico Cretácico de los Andes del sur de

Mendoza: estratigrafía y significado tectosedimentario. *Noveno Cong. Geol. Argent. Actas* **I**, 286–297.

Legarreta, L., Kozlowski, E. & Boll, A. (1981) Esquema estratigráfico y distribución de facies del Grupo Mendoza en el ámbito surmendocino de la Cuenca Neuquina. *Octavo Cong. Geol. Argent. Actas* **III**, 389–409.

Levi, B., Nystrom, J.O., Thiele, R. & Aberg, G. (1988) Geochemical trends in Mesozoic–Tertiary volcanic rocks from the Andes in central Chile and tectonic implications. *J. S. Am. Earth Sci.* **1** (1), 63–74.

Marchese, H. (1971) Litoestratigrafía y variaciones faciales de las sedimentitas mesozoicas de la Cuenca Neuquina, Provincia del Neuquén, República Argentina. *Rev. Asoc. Geol. Argent.* **XXVI** (3), 343–410.

Mastandrea, O., Leanza, H., Hugo, C. & Spiegelman, A. (1975) Manifestaciones fosfáticas en la Sierra de Vaca Muerta y adyacencias, Provincia del Neuquén, República Argentina. *II Congr. Ibero-Americano Geol. Econ.* **II**, 461–486.

Mitchum, R.M. & Uliana, M.A. (1982) Estratigrafía sísmica de las formaciones Loma Montosa, Quintuco y Vaca Muerta, Jurásico Superior y Cretácico Inferior de la Cuenca Neuquina, República Argentina. *Primer Congreso Nacional de Hidrocarburos, Petróleo y Gas Conferencias*, pp. 439–484.

Mitchum, R.M. & Uliana, M.A. (1985) Seismic stratigraphy of carbonate depositional sequences, Upper Jurassic–Lower Cretaceous, Neuquén Basin, Argentina. In: *Seismic Stratigraphy, II: an Integrated Approach.* (Eds Berg, R.O. & Woolverton, D.G.). Mem., Am. Assoc. Petrol. Geol. 39, 255–274.

Murris, R.J. (1980) Hydrocarbon habitat of the Middle East. In: *Facts and Principles of World Petroleum Occurrence* (Ed. Miall, A.D.). Mem. Can. Soc. Petrol. Geol. 6, 765–800.

Musacchio, E. (1971) Hallazgo del Género Cypreda en Argentina y consdieraciones estratigráficas sobre la Formación La Amarga (Cretácico inferior) en la Provincia del Neuquén. *Ameghiniana* **VIII** (2), 105–125.

Musacchio, E. (1981) South American Jurassic and Cretaceous Foraminífera, Ostrácoda and Charophyta of Andean and Sub-Andean Regions. In: *Cuencas Sedimentarias del Jurásico y Cretácico de América del Sur*, (Eds Volkheimer, W. & Musacchio, E.A.) Comité Sudamericano del Jurásico y Cretácico. Vol. 2. Buenos Aires. pp. 461–498.

Musacchio, E. & Palamarczuk, S. (1975) Microfósiles calcáreos de la Formación Ranquiles (Cretácico inferior) en la Provincia del Neuquén, Argentina. *Ameghiniana* **XII** (4), 306–314.

Ploszkiewicz, J.V., Orchuela, I.A., Vaillard, J.C. & Viñes, R.F. (1984) Compresión y desplazamiento lateral en la zona de falla Huincul, estructuras asociadas, Provincia del Neuquén. *Noveno Cong. Geol. Argent. Actas* **II**, 163–169.

Ramos, V. (1981) Descripción geológica de la Hoja 33c. Los Chihuidos Norte, Provincia del Neuquén. *Serv. Geol. Nac. Bol.* **182**, 1–103.

Riccardi, A.C. (1983) The Jurassic of Argentina and Chile. In: *The Phanerozoic Geology of the World, II, Mesozoic B* (Eds Moullade M. & Nairn, A.E.M.). pp. 201–263. Elsevier, Amsterdam.

Riccardi, A.C. (1984) Las asociaciones de amonitas del Jurásico y Cretácico de la Argentina. *Noveno Cong. Geol. Argent. Actas* **IV**, 559–595.

Riccardi, A.C. (1988) The Cretaceous system of southern South America. *Geol. Soc. Am. Mem.* **168**, 161 pp.

Schlager, W. (1981) The paradox of drowned reef and carbonate platforms. *Geol. Soc. Am. Bull.* **92**, 197–211.

Sloss, L.L. (1959) Relationship of primary evaporites to oil accumulation. *Proc. Fifth World Petroleum Congress*, Section I, Paper 7, 123–138.

Sloss, L.L. (1963) Sequences in the cratonic interior of North America. *Geol. Soc. Am. Bull.* **74**, 93–114.

Sloss, L.L. (1972) Synchrony of Phanerozoic sedimentary–tectonic events of the North American craton and the Russian platform. *24th International Geological Congress, Montreal, 1972* **6**, 24–32.

Sloss, L.L. (1988) Forty years of sequence stratigraphy. *Geol. Soc. Am. Bull.* **100**, 1661–1665.

Sloss, L.L. & Speed, R.E. (1974) Relationships of cratonic and continental margin tectonics episodes. In: *Tectonics and Sedimentation* (Ed. Dickinson, W.R.). Soc. Econ. Paleontol. Mineral., Spec. Publ. 98–119.

Sloss, L.L., Krumbein, W.C. & Dapples, E.C. (1949) Integrated facies analysis. In: *Sedimentary Facies in Geologic History* (Ed. Longwell, C.R. (Chairman)). Geol. Soc. Am. Mem. 39, 91–124.

Soares, P.C., Landim, P.M.B. & Fulfaro, V.J. (1978) Tectonic cycles and sedimentary sequences in the Brasilian intracratonic basins. *Geol. Soc. Am. Bull.* **89**, 181–191.

Steckler, M.S. & Watts, A.B. (1978) Subsidence of the Atlantic-type continental margin off New York. *Earth Planet. Sci. Lett.* **41**, 1–13.

Strasser, A. (1988) Shallowing-upward sequences in Purbeckian peritidal carbonates (lowermost Cretaceous, Swiss and French Jura Mountains). *Sedimentology* **35**, 369–383.

Suess, E. (1906) *The Face of the Earth*, Vol. 2. Clarendon, Oxford, 535 pp.

Uliana, M.A. & Biddle, K.T. (1987) Permian to Late Cenozoic evolution of northern Patagonia: main tectonic events, magmatic activity and depositional trends. In: *Gondwana Six: Structure Tectonics, and Geophysics* (Ed McKenzie, G.D.) A.G.U. Geophys. Monogr. **40**, 271–286.

Uliana, M.A. & Biddle, K.T. (1988) Mesozoic–Cenozoic paleogeographic and geodynamic evolution of southern South America. *Rev. Brasileira Geocienc.* **18**, 172–190.

Uliana, M.A., Dellape, D.A. & Pando, G.A. (1975a) Distribución y génesis de las sedimentitas Rayosianas (Cretácico Inferior de las Provincias del Neuquén y Mendoza, República Argentina). *Actas Segundo Congr. Ibero-Americano Geol. Econ.* **1**, 151–176.

Uliana, M.A., Dellape, D.A. & Pando, G.A. (1975b) Estratigrafía de las sedimentitas Rayosianas (Cretácico Inferior de las Provincias del Neuquén y Mendoza). *Actas Segundo Congr. Ibero-Americano Geol. Econ.* **1**, 177–196.

Uliana, M.A., Dellape, D.A. & Pando, G.A. (1977) Análisis estratigráfico y evaluación del potencial petrolífero de las Formaciones Mulichinco, Chachao y Agrio (Provincias del Neuquén y Mendoza). *Petrotecnia (Inst. Argent. Petról., Buenos Aires)* **16** (1–2), 31–46; **16** (3), 25–33.

Vail, P.R. (1987) Seismic stratigraphy interpretation using

sequence stratigraphy, part I: seismic stratigraphy interpretation procedure. In: *Atlas of Seismic Stratigraphy* (Ed. Bally, A.W.). Am. Assoc. Petrol. Geol., Stud. Geol. 27, 1–70.

Vail, P.R., Mitchum, R.M. & Thompson, S. (1977) Seismic stratigraphy and global changes of sea level, part 4: global cycles of relative changes of sea level. In: *Seismic Stratigraphy – Applications to Hydrocarbon Exploration* (Ed. Payton, C.E.). Mem., Am. Assoc. Petrol. Geol. 26, 83–97.

Vail, P.R., Colin, J.P., DuChene, R.J., Kuchly, J., Mediavilla, J. & Trifílieff, V. (1987) La stratigraphie séquentielle et sons application aux corrélations chronostratigraphiques dans le Jurassique du bassin de Paris. *Bull. Soc. Geol. Fr.* **8** (3), 1301–1321.

van Houten, F.B. (1980) Latest Jurassic–Cretaceous regressive facies, Northeast Africa craton. *Bull., Am. Assoc. Petrol. Geol.* **64** (6), 857–857.

van Wagoner, J.C., Mitchum, R.M., Posamentier, H.W. & Vail, P.R. (1987) Seismic stratigraphy interpretation using sequence stratigraphy, part II: key definitions of sequence stratigraphy. In: *Atlas of Seismic Stratigraphy* (Ed. Bally, A.W.). Am. Assoc. Petrol. Geol., Stud. Geol. 27, 11–14.

Veiga, R.D. & Orchuela, I.A. (1988) Técnicas de perfilaje para identificar niveles generadores de hidrocarburos en la Formación Vaca Muerta. *Bol. Inform. Petrol. Tercera Epoca* **13**, 76–94.

Volkheimer, W. & Quattrocchio, M. (1981) Distribución estratigráfica de los palinomorfos jurásicos y cretácicos en la faja andina y áreas adyacentes de América del Sur Austral con especial consideración de la Cuenca Neuquina. In: *Cuencas Sedimentarias del Jurásico y Cretácico de América del Sur* (Eds Volkheimer, W. & Musacchio, E.) Comité Sudamericano del Jurásico y Cretácico. Vol. 2, Buenos Aires. pp. 407–444.

Volkheimer, W. & Salas, A. (1976) Estudio palinológico de la Formación Huitrin, Cretácico de la Cuenca Neuquina, en su localidad tipo. *Sexto Cong. Geol. Argent. Actas* **I**, 433–456.

Volkheimer, W., Quattrocchio, M., Salas, A. & Sepulveda, E. (1976) Consideración palinológica de las formaciones del Jurasico y Cretácico Inferior de la Cuenca Neuquina. República Argentina. *Actas VI Congr. Geol. Argent.* **I**, 593–608.

Weaver, C. (1931) Paleontology of the Jurassic and Cretaceous of West Central Argentina. *Mem. Univ. Washington* **1**, 496 pp.

Zilli, N., Orchuela, I., Dellape, D. & Otaño, R. (1979) Análisis de las formaciones Quintuco y Loma Montosa en el sector Centro oriental de la Cuenca Neuquina. *Séptimo Cong. Geol. Argent. Actas* **I**, 609–615.

*Spec. Publs. int. Ass. Sediment.* (1991) **12**, 451–464

# Sea-level change around the margins of the Catania–Gela Trough and Hyblean Plateau, southeast Sicily (African–European plate convergence zone): a problem of Plio-Quaternary plate buoyancy?

MARTYN PEDLEY[†] *and* MARIO GRASSO[‡]

[†] *Department of Geology (Leicester), School of Geography & Earth Resources, University of Hull, Hull, HU6 7RX, UK; and* [‡] *Istituto di Geologia e Geofisica, University of Catania, Corso Italia, 55, 95129 Catania, Italy*

## ABSTRACT

Neogene and Quaternary sea-level changes in the region were controlled by an interplay between neotectonism and Mediterranean eustacy.

After the Miocene Messinian dessication event, a broad, deep-water seaway was established in southeast Sicily between the developing Maghrebian thrust-fold belt and the emerging Hyblean Plateau to the south. After a regionally extensive pelagic chalk episode (early Pliocene, Trubi Formation), middle Pliocene to early Pleistocene coarse-grained bioclastic carbonates formed around the Hyblean coastal margins, whereas marls and clays accumulated in the Catania–Gela Trough (foredeep). Quartz arenites and silts became dominant, particularly in the west, during the closing stages of the early Pleistocene. Coastal conglomerates and sands typify the later Quaternary terrace deposits.

Neogene and Quaternary marine terraces associated with these sediments in the Hyblean Plateau margins provide the clearest record of onlap episodes. Additionally, they show rapid increases in elevation towards an uplift focus centred on the northwest Hyblean margin.

Plate convergence during the later stages of the Maghrebian orogeny was responsible for this southeast Sicilian regional neotectonism. Late Miocene Afro-European convergence was accompanied by syntectonic graben-associated sedimentation, particularly in the northern zone of the Scicli–Ragusa–Irminio fault line. This N–S to NE–SW orientated dextral wrench system extends from the northwest Hyblean Plateau south towards Malta. The syntectonic transtensional graben zone increased in effect as areas further west effected greater rates of crustal flexure than the Hyblean Plateau.

Major plate margin underthrusting ceased after the early Pleistocene and isostatic equilibrium was re-established rapidly in the north Hyblean Plateau. This was aided by extensive fault reactivation and uplift (c. 800 m) in northern marginal areas of the plateau.

## INTRODUCTION

That part of the central Mediterranean African plate lying north of the Saharan Platform and east of the N–S axis off eastern Tunisia is known as the Pelagian Block (Burollet *et al.*, 1978). To the east it falls away into the deep Ionian Basin, whereas to the north it terminates abruptly against the frontal part of the Sicilian segment of the Maghrebian thrust-fold belt. The block is cut by a complex shear zone to the south of the Medina Banks (Jongsma *et al.*, 1985; Ben-Avraham *et al.*, 1987). This shear zone could be the result of trench pull-apart of the Ionian lithosphere on the subduction zone beneath the Aegean Arc (Le Pichon, 1982).

The Sicily Channel Rift Zone (Finetti, 1984) comprising the Pantelleria, Malta, Linosa, and Medina grabens (Illies, 1980; Grasso *et al.*, 1985; see Fig. 1) lies to the north of the Medina Bank and Lampedusa Plateau. The volcanic islands of Pantelleria and Linosa occur within this neotectonic rift zone. The submerged grabens (Biju-Duval *et al.*, 1985; Finetti, 1985) contain a continuous marine carbonate and clay succession extending back into the Miocene (Colantoni & Borsetti, 1973; Colantoni, 1975).

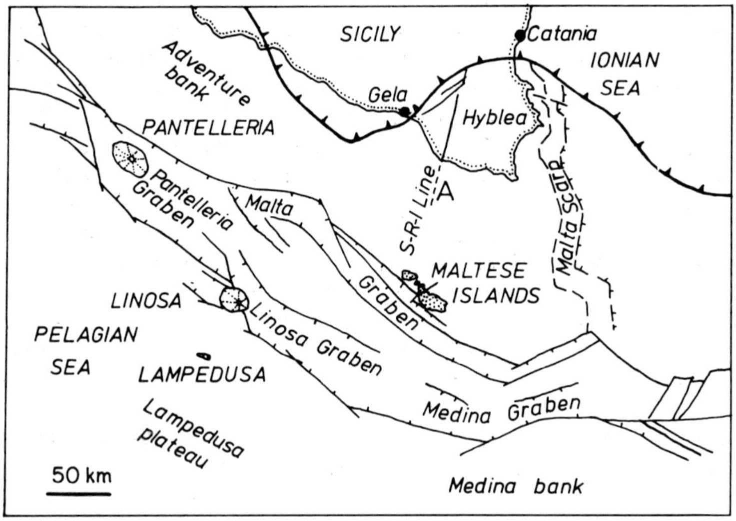


**Fig. 1.** Simplified structural model of the central Mediterranean region showing the various components of the Sicily Channel Rift Zone, and the allochthon margin (toothed heavy line separating the Hyblean Plateau from the rest of Sicily). A indicates the position of the Scicli–Ragusa–Irminio dextral wrench zone. The Pantelleria and Linosa volcanoes of the Sicily Channel Rift Zone are stylized for effect.

The Tertiary carbonate islands of Malta and Lampedusa (Pedley *et al.*, 1976; Grasso & Pedley, 1985,1988b) lie on either side of the rift zone, their respective northeast and south tectonic dips suggesting sea-floor bulging and shoulder upwarping in the vicinity of the rift zone (Illies, 1980; Grasso *et al.*, 1985).

The Hyblean Plateau of southeast Sicily (Figs 1, 2) provides one of the few examples of emergent Pelagian Block exposed close to its leading margin. Here, a thick pile of platform carbonates of Late Cretaceous to Quaternary age are exposed, with strata down to Late Triassic being known in the subsurface (Patacca *et al.*, 1979; Pedley, 1981; Grasso *et al.*, 1982; Bianchi *et al.*, 1987). From the highest areas in the northwest the land falls away rapidly westwards into the Vittoria Plain and northwards into the Catania Plain. These lowland areas

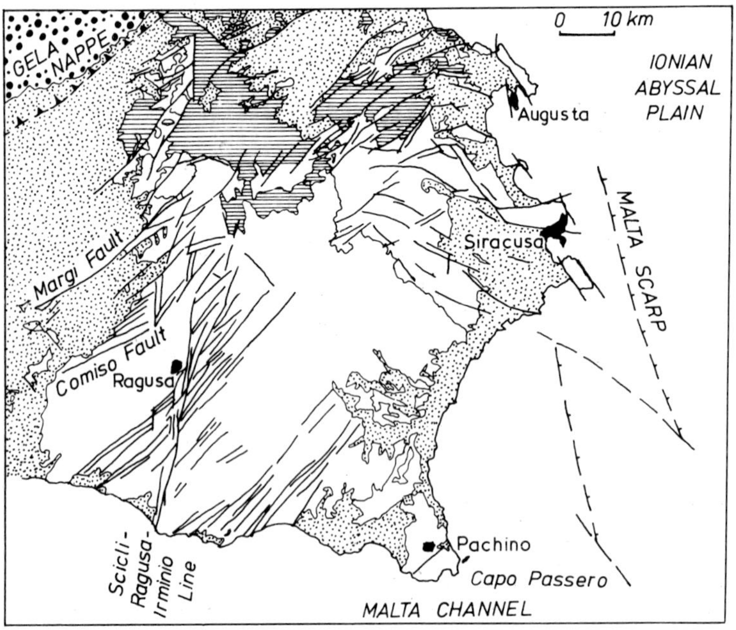


**Fig. 2.** General geology of southeast Sicily. Miocene sediments are without symbol. Plio-Pleistocene basic volcanics are shown by horizontal bars. Quaternary sediments are shown by a fine stipple, the allochthon (Gela nappe) by a bold stipple. Major faulting is indicated by solid lines.

are floored by thick Plio-Quaternary successions illustrated on the 1:100000 geological map of the region (Lentini *et al.*, 1984; see stippled areas in Fig. 2) containing rich molluscan faunas (Di Geronimo, 1975). The northern side of the Catania Plain is bordered by rugged hills, the subaerial expression of the Plio-Quaternary Gela Nappe, a melange accreted at the front of the foreland-verging Maghrebian thrust belt.

The structure of the Hyblean region is simple. Faulting appears predominantly to be post-depositional, with normal faults trending either northeast or northwest. An exceptional trending structure to this is the 'Scicli–Ragusa–Irminio (SRI) Line' (Ghisetti & Vezzani, 1980; Grasso *et al.*, 1986) a discrete N–S dextral wrench-fault complex splaying to the north into a branched complex. This, together with the Margi and Comiso Fault systems, marks the western margin of the Hyblean Plateau (Fig. 1).

The stability of this carbonate platform region contrasts markedly with the syntectonic and siliciclastic dominated Cenozoic sequences of the 'Allochthon' to the north, which are part of the Maghrebian thrust-fold belt (Lentini *et al.*, 1987).

This paper explores the relationships between sedimentation and tectonism associated with the leading margin of the Pelagian Block (African Plate foreland), particularly in the marginal areas of the Hyblean Plateau during the Plio-Quaternary interval (c. 5 Ma). In particular, the synsedimentary behaviour of the faulting around Palagonia, Mineo, Grammichele, and Licodia will be detailed, as this is the key area for elucidating the regional tectonic mechanism (see Fig. 3).

Exceptional rates of post-early Pleistocene uplift can be demonstrated for this area, which can be accounted for by crustal buoyancy subsequent to plate collision rather than solely to eustatic influences.

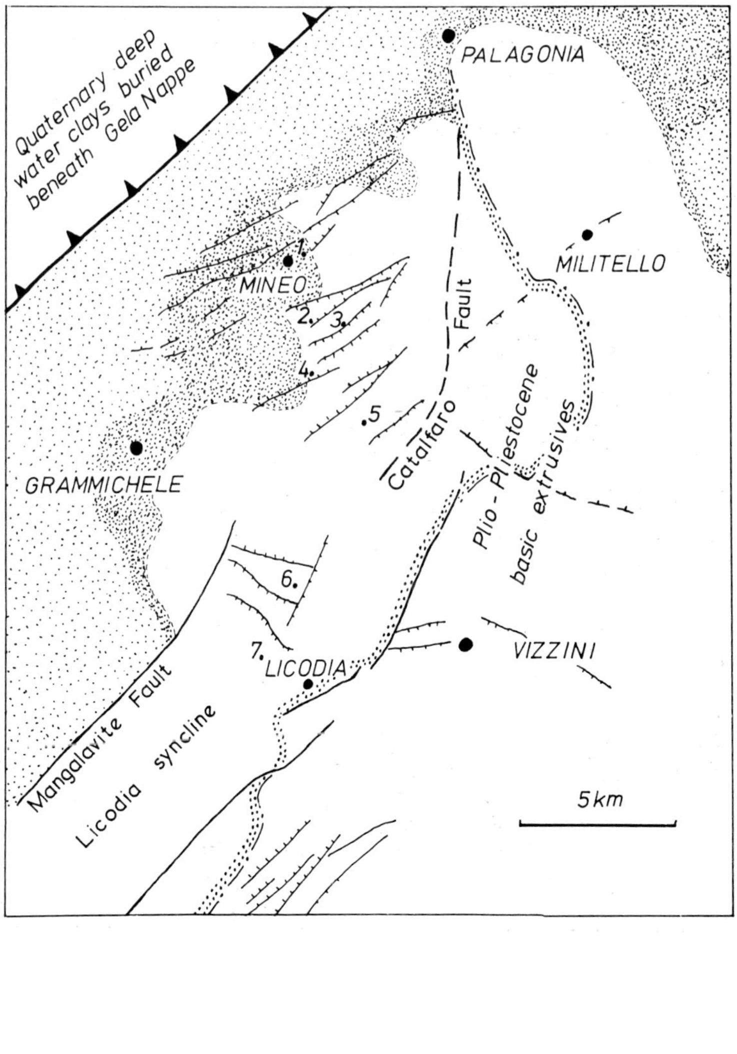


**Fig. 3.** Details of the northern segment of the Scicli–Ragusa–Irminio Line around Grammichele, Mineo, Palagonia, and Licodia. Numbers refer to the location of measured sections illustrated in Fig. 4. The Pliocene transgressive feather edge is indicated by a double line of stipples adjacent to a dot-dash line. The basal Pleistocene strata are shown in stipples, the denser stipple areas indicate lateral passage from fine siliciclastics to coastal carbonates.

## TERTIARY DEPOSITIONAL HISTORY

Many of the Plio-Quaternary tectonic events of the region have their origins in the late Miocene. Consequently, Miocene syntectonic sediment associations, where relevant, will also be considered. Structural details and stratigraphy are summarized in Fig. 4.

### Late Miocene

Throughout the Tertiary and for much of the late Mesozoic, eastern parts of the Hyblean region were sites of carbonate platform sedimentation (Pedley, 1981; Grasso & Lentini, 1982). Typically these successions built out southwestwards as carbonate ramp associations.

Isopachyte studies of the Miocene strata in the northwest Hyblean region (see Grasso & Pedley (1991) for details; summarized in Fig. 4) illustrate the following.

During the Serravallian–Tortonian (middle to late Miocene), up to 200 m of pelagic marl (Tellaro Formation of Rigo & Barbieri, 1959) was deposited on the Hyblean foreland over an undulose sea-floor topography. Thicknesses were modified by incipient northeast oriented 'highs' (Grasso & Pedley, 1991). Initially, these manifest themselves as attenuated developments (< 50 m) occurring on NE–SW-oriented sea-floor swells (e.g. Vallone Lamia High, east of Mineo town; Monte Caratabia High and Cameme High, in the area under special consideration — sea sites 1, 3, and 6 in Figs 3, 5).

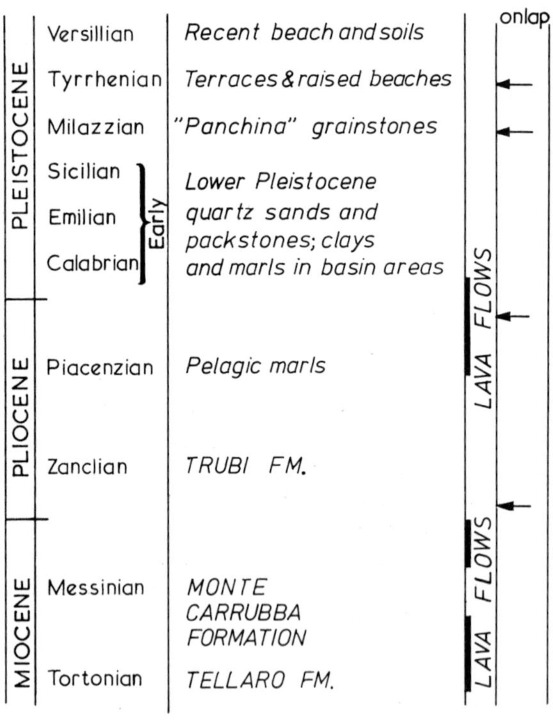


**Fig. 4.** Stratigraphy and range chart showing the local sedimentary and volcanic successions of the Hyblean Plateau and Catania–Gela foredeep.

Extrusive submarine basaltic lavas (0–25 m) occur locally at the top of the Tellaro Formation in western and northern marginal areas of the Hyblean Plateau, with subaerial flows locally on the sea-floor swells (see Fig. 5). These have been ascribed to magnetic interval 7 (6–7 Ma) by Grasso *et al.* (1983) on the basis of magnetostratigraphy.

Early Messinian Monte Carrubba Formation (Grasso *et al.*, 1982) succeeds the Tellaro marls. The Monte Carrubba mudstones contain a typical miliolid dominated benthic shelf fauna. Their constant thickness (c. 4 m) clearly indicates a temporary levelling of the sea-floor highs at this time.

The closing stage of the Miocene (Messinian) is represented by renewed tectonism associated with the three 'highs', which developed steep marginal slopes. Slumping of unconsolidated Monte Carrubba Formation, local debris flows, and calciturbidite development are common. Immediately above this occur localized gypsum deposits (up to 80 m) within narrow NE–SW-oriented hanging wall areas between developing horst blocks (Fiume Caldo and Contrada San Nicola grabens) and the Licodia syncline (a local syndepositional trough confined between the Mangalavite and Catalfaro dextral wrench faults).

No evaporite deposition is recorded from the footwall blocks (Fig. 5). Here, the latest Miocene is represented by continental deposits of fluvial and paludal clay and conglomerate containing a diagnostic *Congerie* fauna (Cafici, 1883). These thin deposits cover both horst and graben areas alike. Messinian deep karstic weathering affects the top of both the evaporites and the Monte Carrubba carbonates.

### Pliocene

Thick Pliocene pelagic chalks (Trubi Formation) were deposited on the African foreland after the late Miocene dessication event. Up to 100 m bury marginal areas of the Hyblean Plateau. In the area

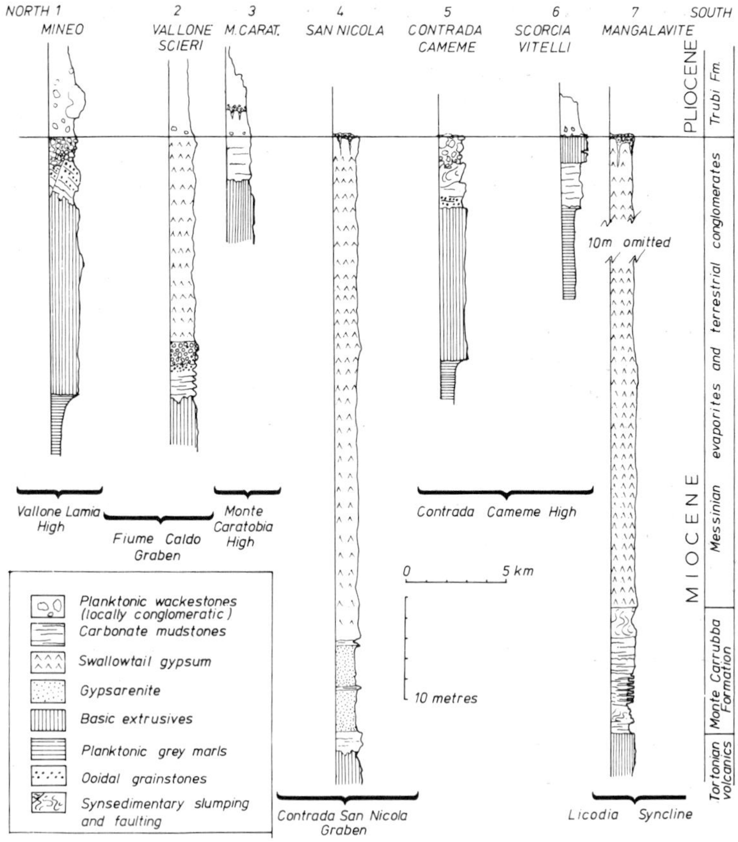


**Fig. 5.** Bar column profiles demonstrating the infill successions associated with the Miocene pull-apart structures connected with the Scicli–Ragusa–Irminio Line.

under consideration the basal Pliocene marine transgression terminated at Licodia and Militello (see Fig. 3 for position of the Pliocene feather edge). Only in the vicinity of Mineo are these early Pliocene carbonate mudstones and wackestones affected by synsedimentary movement, though the formation margins appear to coincide remarkably with major NNE–NE-swinging lineaments marked on Fig. 3.

The middle and upper marine Pliocene is represented locally by up to 350 m of pale to bluish grey pelagic marls. On the northern Hyblean flanks they are only developed in the Licodia syncline and the Fiume Caldo graben though they are much thicker in the Catania Plain to the northwest. To the south and east the post-Trubi Pliocene sequence consists of isolated patches of coarse bioclastic packstones.

Extensive Pliocene submarine pillow lavas passing up into subaerial flows (Monte Lauro flows) occur throughout the northern margins of the Hyblean region (horizontal bar symbol in Fig. 2). They belong to the 3 Ma 'Mammoth' magnetostratigraphic event of the early Pliocene according to Grasso *et al.* (1983).

## Quaternary

Several hundred metres of early Pleistocene sands and sandy clays are developed in the Catania–Gela foredeep of the Hyblean foreland (Di Geronimo, 1969; Di Geronimo & Costa, 1978). The succession towards Catania is characteristically 500–600 m of clay and marl. Quartz arenites and silts dominate the shallower depositional areas (Amore *et al.*, 1984).

Di Geronimo (1969,1975) locally divided the basinal successions into Calabrian and Sicilian stages separated by an erosional disconformity near Grammichele and Caltagirone. Early Pleistocene pillow lavas and submarine volcanics continued to develop along the northern Hyblean margins. Associated subaerial lavas, dated at 2 Ma (pre-Kaena event of Matuyama epoch) by Grasso *et al.* (1983), occur intermittently along the northern Hyblean margins and are included in the horizontal symbol areas of Fig. 2.

There is a clearly defined embayment-like distribution of shallow-marine early Pleistocene carbonate facies around the northwestern Hyblean margins (see Fig. 3). Plio-Pleistocene basic extrusives also created emergent areas marginal to these bioclastic sequences along the northern edge of the Hyblean Plateau and this also constrained the distribution of the carbonates.

'Panchina' is the only regionally significant younger Pleistocene deposit. These coarse bioclastic grainstones are principally found along the eastern and southern Hyblean flanks and attain a maximum elevation in the region of c. 190 m (Carbone *et al.*, 1982a).

## CORRELATION

### Biostratigraphy

Precise age determinations of the earliest marginal Hyblean Pleistocene strata are difficult. Many are too shallow water in aspect to yield more than occasional *Arctica* bivalves. In the Catania–Gela Trough the earliest Pleistocene cool-water zone fossils *Hyalinea balthica*, *Bulimina etnea*, and *Bolivina catanensis* are used for local correlation. These are unsatisfactory as *H. balthica* does not appear until the top of the earliest Pleistocene. *Arctica islandica* is restricted to the shallower, quartz arenitic strata, in particular to the areas west and south of Grammichele (Malatesta, 1960). Attempts to ascribe strata to the earliest Pleistocene stages on malacological grounds are applicable locally (Ruggieri, 1965; Di Geronimo, 1975,1979). A coherent regional stratigraphy (*sensu* Raffi, 1986) has yet to be applied to southeast Sicily.

The shallow-marine Panchina contains undiagnostic pectinid associations and is only generally correlative with the Milazzian stage. Finally, the presence of *Strombus bubonius* and associated warm water 'Senagalian' faunas in the youngest terraces along the eastern Hyblean coastline clearly place these deposits as Tyrrhenian in age (late Pleistocene).

### Eustatic event stratigraphy

The eustatic signal is of much greater value in correlating African foreland neotectonic events than is biostratigraphy. The Neogene eustatic record is punctuated by several rapid, basin-wide Mediterranean falls in sea level, each of which was of brief duration. It is essential, however, to identify clearly the signal before details on tectonic event timing and local rates of uplift can accurately be deduced.

This signal is not always clear in the Hyblean region. At times the Mediterranean generated its own isolated eustatic record during world glacial events. On other occasions tectonic uplift outstripped eustatic fluctuations in sea level. Nevertheless, the following account provides the essential information on the regional eustatic signature necessary for the interpretation of the tectonic mechanism.

During the late Miocene the Atlantic connection was lost and partial dessication of the Mediterranean resulted in several minor Tortonian sea-level lowstands (Pedley, 1983; Grasso & Pedley, 1988a) prior to early Messinian evaporite precipitation (early Messinian salinity crisis of Ruggieri & Sprovieri (1978), and Sonnenfeld (1985)).

Marine conditions in the basins were re-established in the early Pliocene during the later phases of the 5.0 Ma highstand event (Global Cycle TB3.4; Haq *et al.*, 1987). This is the oldest of the major regional late Neogene eustatic markers. Neither this nor later transgressive events onlapped beyond marginal areas of the relatively low subaerial Hyblean interior. Late Pliocene onlap cycles are impossible to recognize from present data.

The development of deeper water, earliest Pleistocene conditions (Di Geronimo, 1979) at c. 1.8 Ma in the Catania–Gela foredeep is connected with the later stages of onlap associated with cycle TB3.8. This is the second major eustatic marker of regional significance.

A disconformity between this and the terminal early Pleistocene has been interpreted by Ruggieri *et al.* (1984) as the Sicilian transgression of 0.7 Ma. This may, however, be a local tectonic event.

It is in marginal Hyblean areas where the Quaternary eustatic signal is frequently clearest. Here, sporadic but continuing tectonic uplift has ensured that the earliest Pleistocene onlap events are preserved at the highest subaerial elevations south of Palagonia and east of Mineo (marine planation surface without sediments at elevations up to 650 m above sea level (a.s.l.). This cambered terrace descends to c. 200 m a.s.l. around the southwest flanks of the Hyblean Plateau (Carbone *et al.*, 1982a; Carbone, 1984; Lentini *et al.*, 1984).

Successively younger events are developed on to the lower hillslopes thus ensuring minimal reworking of the earlier deposits (see Fig. 9). In the eastern borders of the Hyblean region a well-developed planation surface and paleosol associated with *Elephas mnaidriensis* (Accordi, 1963) was developed at the close of the early Pleistocene. Ruggieri & Unti (1974) consider this 'Grande Terrazzo Superiore' to be synchronous with the 'Monoclinale di Niscemi', a broad marine planation surface dipping gently south from c. 500 m a.s.l. in the Caltagirone area west of the Hyblean margin. The regional magnitude of this drawdown event is evident from the fact that the same insular elephant species occurs 200 km south in Malta, and is associated with an influx of pulmonate molluscs of Sicilian origin (Thake, 1985).

This major late early Pleistocene central Mediterranean regression is succeeded by Milazzian (middle Pleistocene) coastal onlap and deposition of the 'Panchina' terraces of the eastern Hyblean flanks (TB3.9 cycle; Haq *et al.*, 1987). This is the third major regional eustatic marker event. These terraces have been mapped in some detail (Ruggieri & Unti, 1974; Carbone *et al.*, 1982a). Around the southern and eastern Hyblean margins persistent terraces of up to 200 m a.s.l. are well developed (see Fig. 9).

Subsequent marine offlap is recorded by up to six regressive middle Pleistocene terraces, the lower ones carrying Milazzian 'Panchina' deposits. These terraces descend from c. 200 m a.s.l. in the northeastern Hyblean margins and around Gela in the south to c. 130 m a.s.l. east of the SRI Line where it cuts the south coast (see Fig. 8).

Tyrrhenian (late Pleistocene) marine terraces occur principally along the east Hyblean coastline. The highest descends from c. 30 m a.s.l. in the north to present sea level in the extreme south. A thin veneer of cobbles and sand covers part of those exposed in eastern coastal sections. The characteristic 'Senagalian' (*Strombus bubonius*) warm water fauna occurs principally in terrace II (DiGrande & Scamarda, 1973). Collectively these represent the fourth regionally important eustatic event.

## TECTONIC ASPECTS

The sedimentation patterns discussed earlier are subtly controlled by regional neotectonic events. By applying the eustatic event stratigraphy and magnetostratigraphy to the sedimentation patterns of the Hyblean region it is possible interpret in detail the timing and relative magnitude of the tectonic events.

### Miocene details

The Miocene horst and graben structures of the northwest Hyblean region appear to have developed progressively, but only penetrated contemporaneous cover during the late Miocene (Grasso & Pedley, 1991). The earliest Tortonian basic extrusives penetrate along these propagating planes of weakness.

Figure 6 shows a typical fault associated with the pull-apart tectonics of this area. Of particular note is the development of Messinian (late Miocene) evaporites (showing depositional thinning towards the fault) followed by thin fluviatile conglomerates in the hanging wall sequence. A thin sheet of latest Miocene fluvial conglomerates alone occurs in the footwall succession. Furthermore, these conglomerates lie unconformably on marine Monte Carrubba Formation and an attenuated Tortonian marl and volcanic sequence on the footwall. After Trubi (early Pliocene) times, fault propagation recommenced but sedimentation kept pace with renewed movement. A volcanic feeder dyke gained access to the sediment–water interface along the fault plane in mid-Pliocene times. The dyke fed hyaloclastic pillow lavas, which were extruded on to late Messinian conglomerate ramanié on the footwall but concordantly on to mid-Pliocene marls of the hanging wall succession.

When viewed on a regional scale (Fig. 2) these northeast orientated structures are seen to be an integral part of a much more extensive fault complex. Detailed work in the Ragusa, Comiso, and Scicli areas to the south (Grasso *et al.*, 1986; Grasso & Reuther, 1988) demonstrates that this neotectonic pattern is intimately associated with dextral wrench faulting along an approximately N–S trajectory (SRI Line). This fault zone, together with the associated Comiso and Margi Faults, is responsible for the western stepped scarp margin of the region, against which the Plio-Quaternary cover onlaps. However, only in the Palagonia, Mineo, Grammichele, and Licodia area are the Neogene synsedimentary relationships clearly seen (Grasso & Pedley, 1991). Figure 7 illustrates the intimate relationship between dextral wrench faulting along the SRI Line and synsedimentation at tensional sites within the northwestern Hyblean fault complex.

Synsedimentary extensional faulting associated

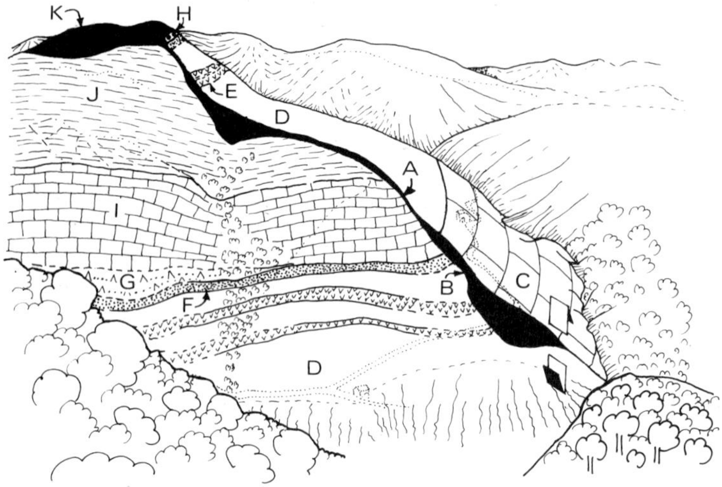


**Fig. 6.** Sketch view looking southeast (from field notes and photographs) of the Monte Palermo area south of Mineo (right bank of the Fiume Caldo, near location 4 in Figs 3, 4) to show the relationships between horst and graben tectonics and syntectonic deposition. The synsedimentary horst margin fault (A) is followed by a basic dyke (B). The successions comprise: C, Ragusa Formation; D, Tellaro Formation with volcanic levels (E) towards the top; F, Monte Carrubba Formation; G, Messinian evaporites; H, late Messinian conglomerates and breccias; I, Trubi Formation; J, unnamed middle Pliocene. Note particularly the pinchout of the graben evaporites towards the fault and the capping sequence of upper middle Pliocene extrusive volcanics (mainly hyaloclastics) on the hill top (K), which have been fed by the fault controlled dyke (B). Roads and houses are dotted in for scale.

with the northern end of the SRI Line continued into the Quaternary and is responsible for many of the local unconformities in the basal part of the marine succession. Early Pleistocene hanging wall successions in the area rest on middle and late Pliocene strata with only minor hiatus. Onlapping

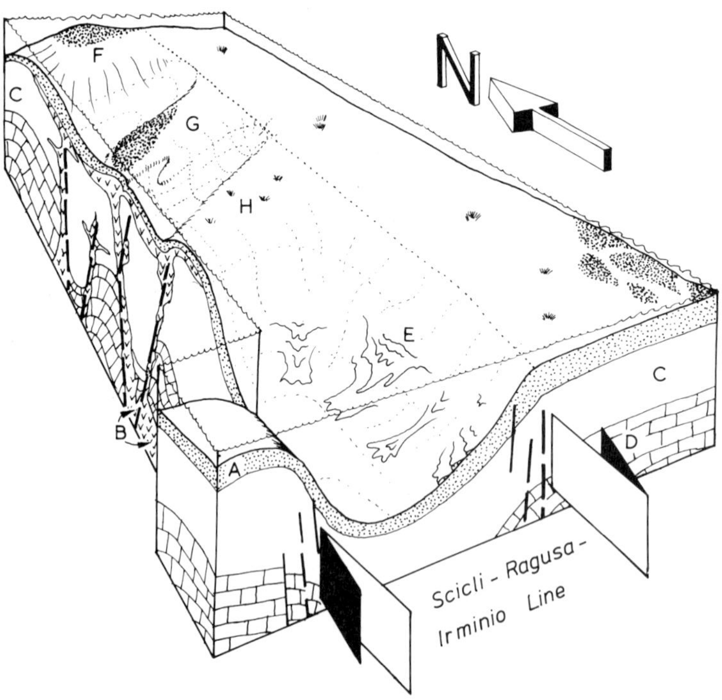


**Fig. 7.** Block model demonstrating the relationship between the northeast-oriented Miocene horsts and grabens, near Mineo and Licodia, and the dextral shear along the Scicli–Ragusa–Irminio Line. A, late Miocene Monte Carrubba Formation; B, basalt following incipient faults between horst and graben blocks; C, Tortonian marls; D, older Miocene carbonates; E, slump structures observed in the Licodia syncline margin; F, the position of the Vallone Lamia High; G, the Monte Caratabia High; H, the Contrada Camerne High.

Pleistocene deposits in the footwall sequences cover early Pliocene Trubi Formation or older strata.

### Plio-Quaternary

The early Pliocene pelagic Trubi Formation is found both above the Hyblean sequence (now partly underthrusted beneath the Gela Nappe, see Fig. 9) and also above the Messinian evaporites rafted on the Gela Nappe sequence (Bianchi *et al.*, 1987).

Fault propagation continued in the Hyblean region during the remainder of the Pleistocene. This is characterized by exceptionally high rates of regional uplift (see Fig. 8). Up to 1000 m of vertical offset (post-Pliocene in part) is apparent on a north-northeast fault near Licodia (Grasso & Reuther, 1988). Additionally, vertical movement of over 700 m occurs on the northern flank fault of the Vallone Lamia horst at Mineo. Here, on the ramparts of Mineo Hill, early Pleistocene marls with a deep-water fauna (Di Geronimo, 1979) are unconformably overlain by inshore early Pleistocene bioclastic carbonates. On a broader regional scale, tectonic uplift is demonstrated by the development of younger, middle Pleistocene bioclastic grainstones ('Panchina facies') around the eastern Hyblean margins. Their base elevation lies around 200 m in the northeastern Hyblean Plateau margins but decreases towards the south (Carbone *et al.*, 1982a).

Another manifestation of this syndepositional faulting is the development of widespread basic lavas and hyaloclastite deposits around the northern margins of the Hyblean Plateau (see Lentini *et al.*, 1984). In some cases these can clearly be related to feeder dykes (cf. Fig. 6) lying along northeast graben margin fault lines associated with northward splaying of the SRI Line. They reach over 600 m in thickness to the northeast of Fig. 4, around Palagonia, and in the Lentini graben further east (Grasso & Reuther, 1988). In other cases (e.g. immediately to the east of the Catalfaro Fault, a northern branch of the SRI Line) thick extrusive piles are developed where N–S dextral wrench faults intersect northeast extensional structures (Grasso *et al.*, 1986).

## THE REGIONAL TECTONIC MECHANISM

The complex local neotectonic features of the Hyblean Plateau provide key information linking regionally important African plate margin neotectonic structures. The significance of the local features becomes clear when the major structural

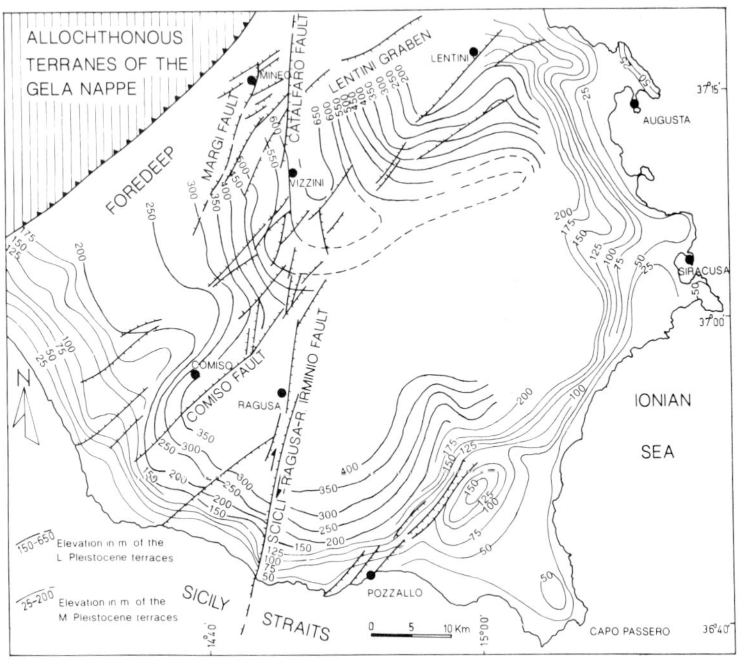


**Fig. 8.** Map of southeast Sicily illustrating elevations of the major marine Pleistocene terraces in the Hyblean region. Rapid post-early Pleistocene uplift is indicated by the high elevations and steep, cambered terrace surfaces around the Mineo–Palagonia and Vizzini area (east of the Catalfaro Fault, a branch of the SRI Line). Note also the swing of the lower Pleistocene terrace surface upon entering the syntectonic Lentini graben. Post-middle Pleistocene Hyblean uplift is regionally more uniform; nevertheless there are clear indications of significant uplift over this brief time interval.

elements of the Pelagian Block are considered collectively.

### African foreland (Pelagian Block)

During the early Neogene, foreland areas of the African plate lay at the southern margins of a wide palaeo-Mediterranean seaway. This progressively narrowed during the Pliocene, by continental collision and associated development of the Maghrebian thrust-fold belt of northern Sicily. A narrowing NE–SW-oriented foredeep developed above the African foreland margins. By early Quaternary times the depocentral axis of the basin had migrated southwards to lie along a line from Gela to Catania (the Catania–Gela foredeep, see Fig. 1).

The Gela nappe, a southeast-verging, thrust-fold wedge extending around the northern Hyblean margin and southwest to Gela (Argnani, 1987; Bianchi *et al.*, 1987) is partly responsible for narrowing of this African foreland basin. The nappe was emplaced mainly during late Pliocene to late early Pleistocene. Emplacement of the nappe within the northwest side of the Catania–Gela Trough is associated with a regionally important early Pleistocene tectonic phase recorded along the northern Hyblean margin (Carbone *et al.*, 1982a).

Gravity maps (AGIP, 1978; Finetti, 1984) show that the Hyblean Plateau and associated Malta Plateau are characterized by a positive gravity anomaly. In contrast, Pelagian Block areas to the west of the Hyblea show a marked negative gravity signature and are deeply buried beneath the Maghrebian allochthon. This Pelagian Block crustal inhomogeneity morphologically manifests itself in the development of the northern margins of the Hyblean Plateau into a cusp-like promontory. This part of the Pelagian Block corresponds to the area showing positive gravity anomaly values (see Fig. 1). Here, in contrast to Pelagian Block areas further west, the foreland has not been extensively overridden, by the allochthonous units of the Maghrebian arc (Ben-Avraham & Grasso, 1989a,b).

The intra-plate division between the plateau and the adjacent negative gravity anomaly terrain to the west is sharply defined by the SRI Line and other faults to the west.

### Sicily Channel Rift Zone

The Sicily Channel Rift Zone (comprised of the Pantelleria, Linosa, Malta, and Medina grabens, see Fig. 1) lies c. 200 km south of the exposed African plate leading margin, close to the southwest of the Maltese Islands. Syntectonic sedimentation patterns in the Maltese Islands show that rift-related movements became dominant in the late Miocene (Illies, 1980; Reuther, 1984; Grasso *et al.*, 1985,1986; Pedley, 1985,1987; Reuther & Eisbacher, 1985).

Bulging and early rift margin cambering is well documented in the Maltese early Miocene sediments ('Western Offshore Rise'; Pedley, 1987) and in the younger Miocene of Lampedusa, lying on the opposing rift shoulder (Grasso & Pedley, 1985). Finetti (1984) concludes that a total of 18 m of post-Miocene extension has developed in the Sicily Channel Rift, associated with mantle uplift.

A comparison with the Red Sea Rift (Finetti, 1984) concluded that the Sicily Channel Rift Zone is a neotectonic spreading centre within the African foreland. Regional crustal thinning from c. 35 to 20 km along the Sicily Channel Rift Zone lends further support to this conclusion (Colombi *et al.*, 1973; Nicolich, 1981; Boccaletti *et al.*, 1984).

### The Scicli–Ragusa–Irminio Line

The SRI Line not only separates the two contrasting areas of the Pelagian Block but also links the structural elements involved in the continental collision process. Development of the late Miocene synsedimentary horst and graben structures to the south of Palagonia and Mineo was controlled by dextral wrench faulting along the SRI fault complex. The local syntectonic sediment associations monitor specific episodes of shear progression along this N–S line (see interrelationships in Figs 1, 7).

The SRI Line and other closely associated dextral wrench structures along the western margin of the plateau as a whole (Grasso & Reuther, 1988) extend offshore towards the Maltese Islands. They comprise a complex dextral transfer zone linking the southeast margin of the Sicily Channel Rift Zone with the zone of Recent Pelagian Block underthrusting located to the northwest.

The buoyant Hyblean–Maltese Plateau, driven north against the European plate, resisted subduction. Pelagian Block areas to the west were potentially more readily subductable. During the closing stages of Afro-European continental collision this transfer zone accommodated continuing underthrusting of Pelagian Block areas west of the Hyblean Plateau (Grasso & Reuther, 1988).

Non-parallel opening of the Sicily Channel Rift

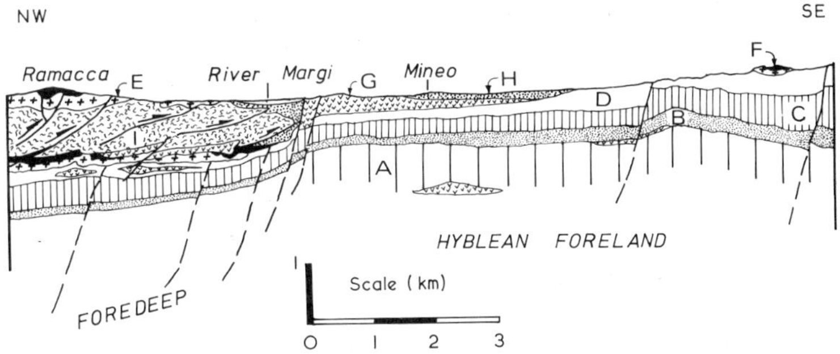


**Fig. 9** Simplified N–S section, from the northern margin of the Hyblean Plateau towards the allochthon, showing the interrelationships between the allochthon (Gela nappe, I), the Catania–Gela foredeep, and the Hyblean region after the Pleistocene uplift phase (modified after Carbone *et al.*, 1982c). Symbols: A (broad vertical bars), Triassic dolomites; B (fine stipple), Jurassic carbonates and marls; C (fine vertical bars), Cretaceous–Eocene carbonates; D (no symbol), Oligocene to upper Miocene carbonates and marls; E ('+' symbol), late Miocene (Messinian) evaporites; F (solid black), lower Pliocene Trubi chalks; G ('V' symbol), Pliocene–Pleistocene mafic volcanics; H (coarse stipple), Pleistocene clays, sands, and carbonates. Vermiform symbol at 'I' = Gela nappe. Alluvium covers the Margi River plain.

Zone may have provided an additional driving force during this intra-plate motion process.

### *Plio-Pleistocene crustal failure*

Late Miocene to early Pleistocene volcanicity in the north Hyblean margins can be related to the local development of an atypical 'hot spot' of rising low-density mantle material (confirmed by xenoliths; Scribano, 1987) within a crust weakened by faults. Subcrustal thinning and erosion perhaps facilitated this magma injection. However, progression of lithospheric flexure probably inhibited further magma upwelling after the early Pleistocene (Carbone *et al.* 1982b; Lentini, 1982; Grasso & Reuther, 1988).

The present study confirms the termination of this regime in the early Pleistocene in the north Hyblean margins. In part, this was induced by loading, and concomitant sinking, of peripheral parts of the African foreland promontory as the result of Gela nappe emplacement from the northwest (see Fig. 9).

Residual buoyancy, particularly in the unburied Hyblean region southeast of the Gela nappe, reactivated northeast faults along the weakened northern margin of the Hyblean Plateau ('braided' northern segment of the SRI Line; Grasso & Reuther, 1988). High rates of post-early Pleistocene uplift occurred along the northern Hyblean margins and rapid draining of the Catania–Gela Trough followed. Isostatic uplift was greatest around Mineo and Vizzini where reactivation of the SRI Line (e.g. the westerly throwing Catalfaro Fault, Figs 3, 8) facilitated maximum crustal rebound by relieving stress build-up along the western Hyblean margin.

The precise timing and local detail of this rapid crustal adjustment is recorded around the Hyblean Plateau margins by the elevated Quaternary onlap sequences now occupying terraces up to 650 m a.s.l. (Fig. 8). The focal points and relative uplift rates of each event can readily be calculated from the regional distribution of the terraces. The timing of each event can be estimated from the onlap events related to global sea-level curves and the local magnetostratigraphy.

The graben areas linked with the SRI Line in the northwest Hyblean Plateau (Palagonia to Licodia areas) continue to be affected by frequent minor earth tremors (Carbone *et al.*, 1982b), though no significant fault movement has been recorded in historical times.

## SUMMARY

The neotectonic history of the region can be summarized as follows.

**1** Inhomogeneities within the Pelagian Block continental crust caused differential resistance to African foreland underthrusting starting in the late Miocene.
**2** A complex transfer zone, the mechanism detailed in the northern end of the SRI Line, developed along the western margins of the Hyblean Plateau in order to accommodate more rapid subduction of crustal areas to the west.
**3** Loading of the bouyant leading margin of the Hyblean Plateau (African foreland) resulted from emplacement of the Gela nappe at the close of the early Pleistocene. Local synsedimentary tensional basins (grabens), which developed in response to earlier intermittent right lateral wrenching, became extinct at this time, concurrently with the cessation of African plate underthrusting.
**4** Residual buoyancy of the African plate margin caused local reactivation of older fault systems, including the SRI Line. Rapid uplift and erosion of the northern Hyblean margins followed, restoring crustal buoyancy equilibrium to the Hyblean Plateau.

## ACKNOWLEDGEMENTS

The authors were assisted in this study by M. P. I. (art 65–60%) grant number 20120146/88 and by (art 65–40%) grants 2012026/87 and 201202230/89 under the framework of the Italian national project: Tettonica fragile Recente e attiva nel Mediterraneo Occidentale (project co-ordinator, Professor M. Boccaletti, University of Florence).

## REFERENCES

Accordi, B. (1963) Some data on the Pleistocene stratigraphy and related pygmy mammalian faunas of Eastern Sicily. *Quaternaria* **6**, 415–430.

AGIP (1978) Carta gravimetrica della Sicilia, anomalie di Bouguer, scale 1:500000. *Boll. Serv. Geol. d'Ital.* **99**.

Amore, C., Casale, V., Costa, B., Di Geronimo, I., Giuffrida, E. & Zanini, A. (1984) Evoluzione sedimentologica e faunistica dei sedimenti pleistocenici di Poggio Spica (Mineo, Catania). *Boll. Soc. Geol. Ital.* **103**, 755–786.

Argnani, A. (1987) The Gela Nappe: evidence of accretionary melange in the Maghrebian foredeep of Sicily. *Mem. Soc. Geol. Ital.* **38**, 419–428.

Ben-Avraham, Z. & Grasso, M. (1989a) Neogene plate collisional processes in Sicily: different modes of consumption as the result of variable crustal structures. Union of Geosciences, *Strassbourg.*

Ben-Avraham, Z. & Grasso, M. (1989b) Plate collisional processes in Sicily. *28th International Geological Congress, Washington* **1**, 125.

Ben-Avraham, Z., Nur, A. & Cello, G. (1987) Active transcurrent fault systems along the north African passive margin. *Tectonophysics* **141**, 249–260.

Bianchi, F., Carbone, S., Grasso, M., Inverizzi, G., Lentini, F., Langaretti, G., Merlini, S. & Mostardini, F. (1987) Sicilia orientale: profilo geologico Nebrodi-Iblei. *Mem. Soc. Geol. Ital.* **38**, 429–458.

Biju-Duval, B., Borsetti, A.M., Colantoni, P. (1985) Geologie des fosses du detroit Siculo-Tunisien (Mer Pelagienne). *Reve de L'Inst. Francais du Petrol.* **40** (6), 691–722.

Boccaletti, M., Nicolich, R. & Tortorici, L. (1984) The Calabrian Arc and the Ionian Sea in the dynamic evolution of the Central Mediterranean. *Mar. Geol.* **55**, 219–245.

Burollet, P.F., Mugniot, J.M. & Sweeney, P. (1978) The geology of the Pelagian Block: the margins and basins of southern Tunisia and Tripolitania. In: *The Ocean Basins and Margins (4C) The Western Mediterranean* (Eds Nairn, A.E.M., Kanes, W.H. & Stehli, F.G.) pp. 331–359. Plenum Press, New York.

Cafici, I. (1883) La formazione miocenica nel territorio di Licodia Eubea (provincia di Catania). *Atti. R. Accad. Naz. Lincei. Mem. Cl. Sci. Fis. Mat. Nat.* **14**, 2–42.

Carbone, S. (1984) I depositi pleistocenici del settore nord-orientale ibleo tra Agnone e Melilli (Sicilia SE): relazione tra facies e lineamenti strutturali. *Boll. Soc. Geol. Ital.* **104**, 405–420.

Carbone, S., Di Geronimo, I., Grasso, M., Iozzia, S. & Lentini, F. (1982a) I terrazzi marini quaternari dell'area iblea (Sicilia sub-orientale). *Contributi conclusivi per la realizzazione della carta neotettonica d'Italia (publicazione 506 del Progetto Finalizzato Geodinamica).* F. Giannini, Naples, 35 pp.

Carbone, S., Cosentino, M., Grasso, M., Lentini, F., Lombardo, G. & Patane, G. (1982b) Elementi per una prima valutazione dei caratteri sismotettonici dell'Avampaese Ibleo (Sicilia sud-orientale). *Mem. Soc. Geol. Ital.* **24**, 507–520.

Carbone, S., Grasso, M. & Lentini, F. (1982c) Considerazioni sulla evoluzione geodinamica della Sicilia sud-orientale del Cretaceo al Quaternario. *Mem. Soc. Geol. Ital.* **24**, 367–386.

Colantoni, P. (1975) Note di geologia marina sul Canale di Sicilia. *G. Geol. Ser. 2* **40**, 181–207.

Colantoni, P. & Borsetti, A.M. (1973) Some notes on the geology and stratigraphy of the Straits of Sicily. *Rapp. Comm. Int. Mer. Medit. Monaco* **22**, 70–72.

Colombi, B., Giese, P., Luongo, G., Morelli, C., Riuscetti, M., Scarascia, S., Schutte, K.G., Strowald, J. & De Visentini, G. (1973) Preliminary report on the seismic refraction profile Gargano–Palermo–Pantelleria. (1971) *Boll. Geof. Teor. Appl.* **15**, 225–254.

Di Geronimo, I. (1969) La sezione stratigrafica plio-pleistocenica di Monte Navone (Piazza Armerina, Enna). *Atti Accad. Gioenia Sci. Nat. Catania, Ser. 6* **20**, 81–146.

Di Geronimo, I. (1975) La malacofauna Siciliana del Ciaramitaio (Grammichele, Catania). *Conchiglie* **11**

(5–6), 101–137.
Di Geronimo, I. (1979) Il Pleistocene in facies batiale di Valle Palione (Grammichele, Catania). *Boll. Malacologico* **15** (5–6), 85–156.
Di Geronimo, I. & Costa, B. (1978) Il Pleistocene di Monte dell'Apa (Gela). *Riv. Ital. Paleontol. Stratigr.* **84**, 1121–1158.
Digrande, A. & Scamarda, G. (1973) Segnalazione di livelli a *Strombus bubonius* LAMARCK nei dintorni di Augusta (Siracusa). *Atti Accad. Gioenia Sci. Nat. Catania, Ser. 4* **11**, 157–172.
Finetti, I. (1984) Geophysical study of the Sicily Channel Rift Zone. *Boll. Geof. Teor. Appl.* **26** (101–102), 3–27.
Finetti, I. (1985) Structure and geological evolution of the Central Mediterranean. In: *Geological Evolution of the Mediterranean Basins* (Ed. Stanley, D.J. & Wezel, F.C.) pp. 215–230. Springer-Verlag, Berlin.
Ghisetti, F. & Vezzani, L. (1980) The structural features of the Iblean Plateau and of the Mt. Judica area (South-Eastern Sicily): a microtectonic contribution to the deformational history of the Calabrian Arc. *Boll. Soc. Geol. Ital.* **99**, 57–102.
Grasso, M. & Lentini, F. (1982) Sedimentary and tectonic evolution of the Eastern Hyblean Plateau (southeast Sicily) during Late Cretaceous to Quaternary time. *Palaeogeogr. Palaeoclimatol. Palaeoecol.* **39**, 279–300.
Grasso, M. & Pedley, H.M. (1985) The Pelagian Islands: a new geological interpretation from sedimentological and tectonic studies and its bearing on the evolution of the Central Mediterranean Sea (Pelagian Block). *Geol. Romana* **24**, 13–34.
Grasso, M. & Pedley, H.M. (1988a) The sedimentology and development of the Terravechia Formation carbonates (Upper Miocene) of north central Sicily: possible eustatic influence on facies development. *Sediment. Geol.* **57**, 131–149.
Grasso, M. & Pedley, H.M. (1988b) *Geological map of the island of Lampedusa (Central Mediterranean), scale 1:10 000.* Ed. S. EL. CA., Florence.
Grasso, M. & Pedley, H.M. (1991) Neogene and Quaternary sedimentation patterns in the northwestern Hyblean Plateau (SE Sicily): the effects of a collisional tectonic process on a foreland margin. *Riv. Ital. Paleontol. Stratigr.*, (in press).
Grasso, M. & Reuther, C. (1988) The western margin of the Hyblean Plateau: a neotectonic transform system on the S E Sicilian foreland. *Ann. Tectonicae* **2**(2), 107–120.
Grasso, M., Lentini, F. & Pedley, H.M. (1982) Late Tortonian–Lower Messinian (Miocene) palaeogeography of S. E. Sicily: information from two new formations of the Sortino Group. *Sediment. Geol.* **32**, 279–300.
Grasso, M., Lentini, F., Nairn, A.E.M. & Vigliotti, L. (1983) A geological and paleomagnetic study of the Hyblean volcanic rocks, Sicily. *Tectonophysics* **98**, 271–295.
Grasso, M., Pedley, H.M. & Reuther, C. (1985) The geology of the Pelagian Islands and their structural setting related to the Pantelleria Rift (Central Mediterranean Sea). *Centro* **1**(2), 1–19.
Grasso, M., Reuther, C., Baumann, H. & Becker, A. (1986) Shallow crustal stress and neotectonic framework of the Malta Platform and the southeastern Pantelleria Rift (Central Mediterranean). *Geol. Romana* **25**, 191–212.
Haq, B.U., Hardenbol, J. & Vail, P.R. (1987) Chronology of fluctuating sea levels since the Triassic. *Science* **235**, 1156–1167.
Illies, J.H. (1980) Form and formation of graben structures: the Maltese Islands. In: *Mobile Earth.* (Cloos, H., Gehlen, K.V., Illies, J.H., Kuntz, E., Neumann, J. & Seibold, E.) pp. 161–184. Harold Boldt Verlag, Bonn.
Jongsma, D., van Hinte, J. & Woodside, J.M. (1985) Geologic structure and neotectonics of the North African continental margin south of Sicily. *Mar. Petrol. Geol.* **2**, 156–179.
Lentini, F. (1982) The geology of Mt. Etna basement. *Mem. Soc. Geol. Ital.* **23**, 7–25.
Lentini, F., Carbone, S., Grasso, M., Di Geronimo, I., Scamarda, G., Bommarito, S., Iozzia, S., La Rosa, N. & Sciuto, F. (1984) *Carta Geologica della Sicilia sud-orientale, scale 1:100 000.* S.EL.CA., Florence.
Lentini, F., Grasso, M. & Carbone, S. (1987) Introduzione alla geologia della Sicilia e guida all'escursione. *Convegno Soc. Geol. Italia, Naxos-Pergusa, 22–25 Aprile*, 51 pp.
Le Pichon, X. (1982) Land locked ocean basins and continental collision: the eastern Mediterranean, a case example. In: *Mountain Building Processes* (Ed. Hsü, K.J.) pp. 201–211. Academic Press, New York.
Malatesta, A. (1960) Malacofauna pleistocenica di Grammichele (Sicilia). *Mem. Descr. Carta Geol. Ital. Roma* **12**, 392 pp.
Nicolich, R. (1981) Crustal structure in the Italian peninsula and surrounding seas: a review of DSS data. In: *Sedimentary Basins of Mediterranean Margins* (Ed. Wezel, F.C.) pp. 3–17, Tecnoprint, Bologna.
Patacca, E., Scandone, P., Guinta, G. & Liguori, V. (1979) Mesozoic paleotectonic evolution of the Ragusa Zone (southeastern Sicily). *Geol. Romana* **18**, 331–69.
Pedley, H.M. (1981) The sedimentology and palaeoenvironment of the southeast sicilian Tertiary platform carbonates. *Sediment. Geol.* **28**, 273–291.
Pedley, H.M. (1983) Petrology and palaeoenvironment of the Sortino Group (Miocene) of SE Sicily: evidence for periodic emergence. *J. Geol. Soc. London* **140**(3), 335–350.
Pedley, H.M. (1985) The Ghar Lapsi Limestones: sedimentology of a Miocene intra-shelf graben, Malta. *Centro* **3**, 1–14.
Pedley, H.M. (1987) Controls on Cenozoic sedimentation in the Maltese Islands: review and reinterpretation. *Mem. Soc. Geol. Ital.* **38**, 81–94.
Pedley, H.M., House, M.R. & Waugh, B. (1976) The geology of Malta and Gozo. *Proc. Geol. Assoc. London* **20**, 209–234.
Raffi, S. (1986) The significance of marine boreal molluscs in the Early Pleistocene faunas of the Mediterranean area. *Palaeogeogr. Palaeoclimatol. Palaeoecol.* **52**, 267–289.
Reuther, C. (1984) Tectonics of the Maltese Islands. *Centro* **1**(1), 1–20.
Reuther, C. & Eisbacher, G.H. (1985) Pantelleria Rift: crustal extension in a convergent intraplate setting. *Geol. Rundsch.* **74**, 585–597.

RIGO, M. & BARBIERI, F. (1959) Stratigrafia pratica applicata in Sicilia. *Boll. Serv. Geol. Ital.* **80**, 351–442.

RUGGIERI, G. (1965) A contribution to the stratigraphy of the marine lower Quaternary sequences in Italy. *Geol. Soc. Am. Spec Pap.* **84**, 141–152.

RUGGIERI, G. & SPROVIERI, R. (1978) The 'dessication theory' and its evidence in Italy and Sicily. In: *Messinian Evaporites in the Mediterranean* (Eds Catalano, R., Ruggieri, G. & Sprovieri, R.). Mem. Soc. Geol. Ital. **16**, 165–169.

RUGGIERI, G. & UNTI, M. (1974) Pliocene e Pleistocene nell'entroterra di Marsala. *Boll. Soc. Geol. Ital.* **96**, 723–733.

RUGGIERI, G., RIO, D. & SPROVIERI, R. (1984) Remarks on the chronostratigraphic classification of the Lower Pleistocene. *Boll. Soc. Geol. Ital.* **103**, 251–259.

SCRIBANO, V. (1987) Ultramafic and mafic xenoliths in the alkaline volcanic rocks from the Hyblean Plateau (Sicily): a contribution to the knowledge of the upper mantle and deep crust. *Mem. Soc. Geol. Ital.* **38**, 475–482.

SONNENFELD, P. (1985) Models for upper Miocene evaporite genesis in the Mediterranean. In: *Geological Evolution of the Mediterranean Basins* (Eds Stanley, D.J. & Wezel, F.C.) pp. 323–346. Springer-Verlag, Berlin.

THAKE, M.A. (1985) The biogeography of the Maltese Islands, illustrated by the Clausiliidae. *J. Biogeogr.* **12**, 269–287.

VAIL, P.R., MITCHUM, R.M., JR., TODD, R.G., WIDMIER, J.M., THOMPSON, S. III, SANGREE, J.B., BUB, J.N. & HATLELID, W.G. (1977) Seismic stratigraphy and global changes of sea level. In: *Seismic Stratigraphy – Applications to Hydrocarbon Exploration* (Ed. Payton, C.E.). Mem. Am. Assoc. Petrol. Geol. **26**, 49–212, Tulsa.

*Spec. Publs. int. Ass. Sediment.* (1991) **12**, 465–484

# The Ordovician back-arc to foreland successor basin in the Argentinian–Chilean Puna: tectono-sedimentary trends and sea-level changes

HEINRICH BAHLBURG

*Institut für Geologie und Paläontologie, Technische Universität, Ernst-Reuter-Platz 1, 1000 Berlin 10, Germany*

## ABSTRACT

New sedimentological and stratigraphic data are used to evaluate the roles of tectonism and sea-level change in the Ordovician Puna Basin of northwestern Argentina and northern Chile. The basin was located to the west of a broad, shallow marine shelf in the Argentinian Cordillera Oriental. Subsidence (Tremadoc to late Arenig) of the shelf edge is documented by shallow-marine quartz sandstones giving way to turbidites.

A submarine volcaniclastic apron c. 3500 m thick formed during the Arenig in the Puna Basin, which then was in a back-arc position above an east-dipping subduction zone. Overthrusting of the arc complex over its back-arc basin began in the late Arenig. This was accompanied by the extinction of volcanism and increased subsidence at rates of up to 1100 m/Ma, which are typical of foreland basins. The emergence of the eastern shelf in the Cordillera Oriental during the early Llanvirn (Guandacol Diastrophic Phase) is best explained by the formation of a flexural bulge in response to the thrusting in the west. Contemporaneously, deposition of two successive turbidite systems of 3500 m combined thickness occurred in the Puna, lasting until the (?)early Caradoc.

The dominant detrital input to the turbidite basin was derived from the volcanic arc complex (line source) and was transported axially towards the north-northwest. Coarsening and fining upwards megacycles correlate well with local tectonic events related to thrusting. Immaturity and poor sorting of the turbidite arenites in coarsening and fining upwards cycles of various orders attest to short transport paths and little reworking during both cycles. This, together with differential cyclicity from section to section, indicates that global sea-level highstands did not lead to the formation of large shelf and sediment storage areas withholding sediment from the basin. Global sea-level changes in the Ordovician had only complementary influence on sedimentary patterns. These were controlled by local tectonics related to the transformation of the back-arc basin to a foreland successor basin.

## INTRODUCTION

In the Andes of northwestern Argentina and northern Chile, Ordovician clastic sedimentary rocks are well exposed in the Argentinian Cordillera Oriental and to the west in the adjacent Puna Highlands (Figs 1, 2). The highlands of northwestern Argentina are subdivided into the northern and southern Puna by the NW–SE-trending Calama–Olacapato–El Toro lineament (Fig. 1). This lineament has been active repeatedly since the Precambrian (Salfity, 1985) and is seen as an important factor in controlling the differing structural histories of the southern and northern Puna (Alonso *et al.*, 1984).

This study concentrates on the Ordovician basin in the northern Puna between 22 and 24°S. There are only very few papers that deal with the influences of relative or global sea-level changes on Ordovician sedimentation, either in the Puna or the Cordillera Oriental. As the wide distribution of Ordovician strata in the Puna only became known in the late 1960s, Harrington & Leanza's (1957) reconstruction of palaeogeography and of regressive and transgressive events is now applicable only to parts of the Cordillera Oriental. Salfity *et al.* (1984) showed that uplift and emergence of the Cordillera Oriental during the 'Guandacol Diastrophic Phase' at the

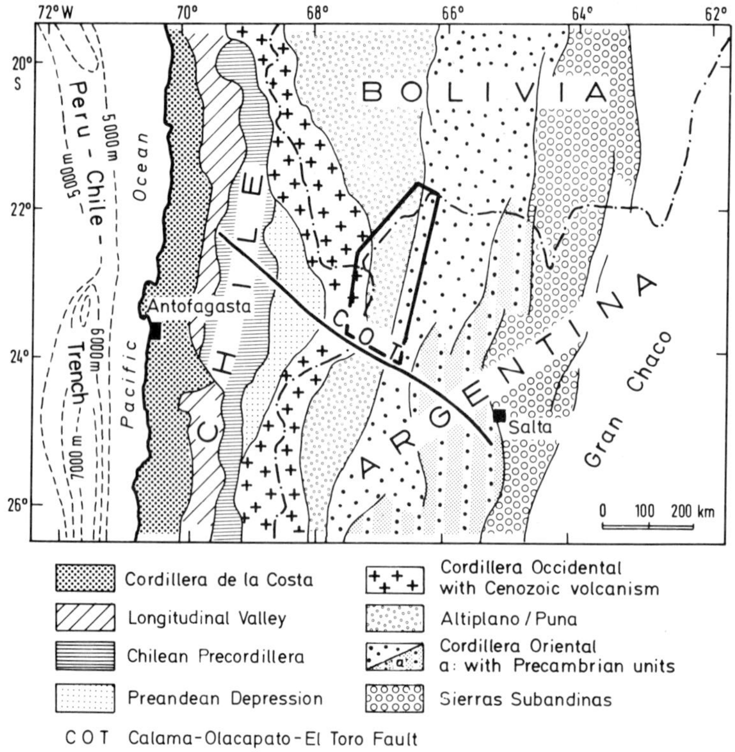


**Fig. 1.** Schematic map of the main morphostructural units of the southern Central Andes between 20 and 26°S. Frame indicates the study area.

Arenig–Llanvirn transition led to the formation of two distinct faunal provinces to the east and west of the positive area. Moya (1988) described the sequence of transgressions and regressions in the southern Cordillera Oriental from the very Late Cambrian to the final regression in the early Llanvirn. Schwab (1973) portrayed the evolutionary trends of the Ordovician and younger basins in the Puna. New sedimentological data presented here emphasize the dominance of 'local' tectonic factors over effects of global sea-level changes in an active margin basin, which evolved from back-arc to foreland basin during the Early and Middle Ordovician.

## Geological setting

The Ordovician Puna Basin evolved on sialic crust that was deformed and metamorphosed during the Brazilian (Pampean) Orogeny in the Middle Cambrian (Willner *et al.*, 1987; Fig. 3). In the Late Cambrian, basement stretching in the area of the Cordillera Oriental led to the formation of a north-dipping shallow marine platform on which a succession of quartz sandstones was deposited (Mesón Group: Turner, 1960; Kumpa & Sanchez, 1988). Above an erosional unconformity of latest Cambrian age (Irúyica Phase: Turner & Mendez, 1975), which predates the earliest Tremadocian eustatic regression (Fortey, 1984), the sediments of the 'Ordovician Cycle' (Moya, 1988) developed in the Cordillera Oriental and Puna. The Ordovician rocks of both regions are in turn overlain unconformably by Jurassic and Cretaceous continental clastic and volcanic rocks. Only in one small outcrop in the southern Puna has an angular unconformity between Ordovician and Lower Silurian strata been observed (Isaacson *et al.*, 1976). This is attributed to the Oclóyic Orogeny at the Ordovician–Silurian boundary (Turner & Mendez, 1979). Apart from a few marginal transgressions in the late Palaeozoic the whole region has been a positive area since the Early Silurian.

In the Cordillera Oriental, shallow-water pelites with intercalated arenites of Tremadoc to Llanvirn age (Harrington & Leanza, 1957; Turner, 1960) were deposited on a neritic shelf, and are exposed in tectonic slices trending north-northeast (Moya, 1988). The succession of the Eastern Cordillera was

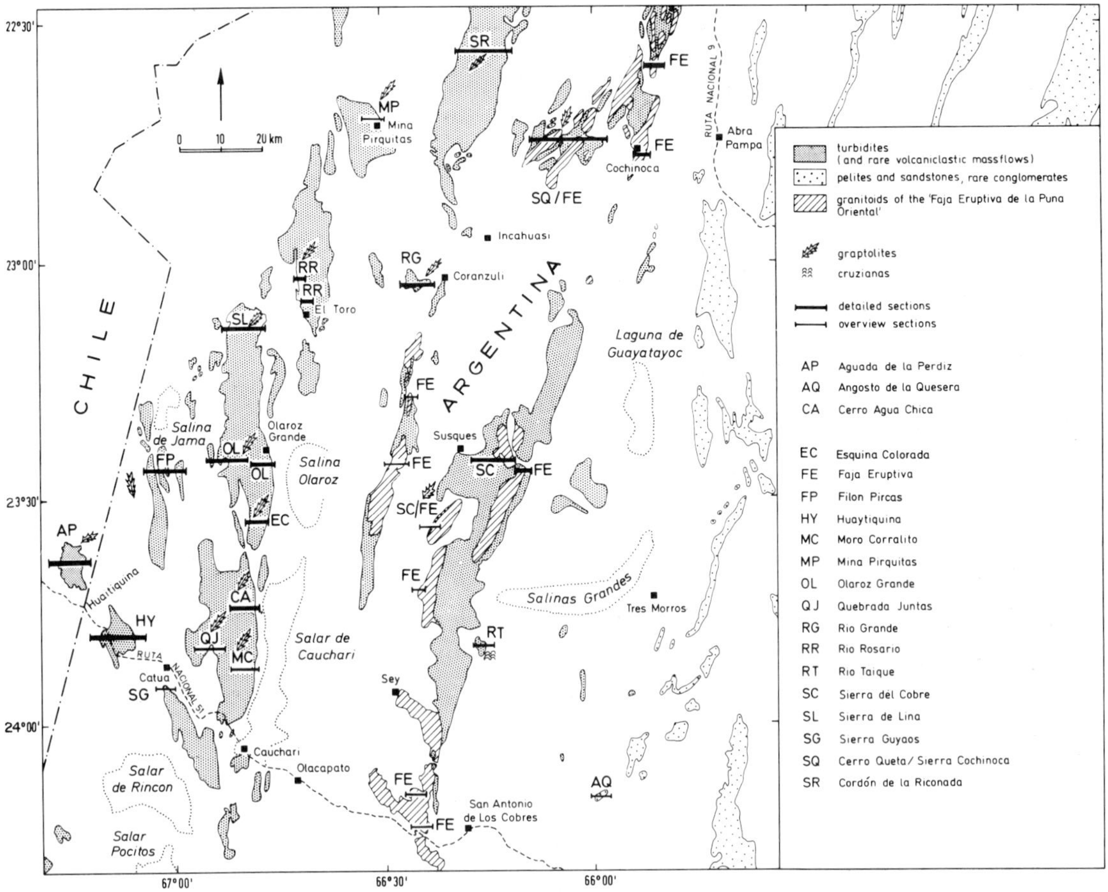


**Fig. 2.** Outcrop map of Ordovician rocks and lower Palaeozoic intrusives in the northern Puna of Argentina and Chile.

named the Santa Victoria Group and includes the Santa Rosita (Tremadoc) and Acoite Formations (Arenig–Llanvirn; Turner, 1960). Ordovician rocks younger than Llanvirn do not occur.

In the Puna, shallow-marine sandstones and deep-marine volcaniclastic turbidites are exposed in horsts trending north-northeast (Fig. 2). In the eastern regions of the northern Puna, Tremadoc to (?middle) Arenig rocks occur in the area between the salars of Cauchari and Olaroz in the west and the Salinas Grandes and Laguna de Guyatayoc in the east (sections SC and RT, Fig. 2). To the west of the salars of Cauchari and Olaroz, the AP, HY, and SG outcrops were previously assigned to the Arenig Aguada de la Perdiz Formation, whereas all other outcrops had been grouped in the Arenig–Llanvirn Coquena and the Llandeilo Falda Ciénaga Formations (Fig. 4; Schwab, 1973; Aceñolaza & Baldis, 1987). An exception to the latter represented the SL outcrop in the northern Sierra de Lina (Lina Formation, (?)lower Caradoc, Ramos, 1972; Figs 2, 4). Recent graptolite finds and sedimentological data have allowed a new, more precise, subdivision of the Ordovician successions into the 'Volcanosedimentary Succession' (middle to upper Arenig), the 'Lower Turbidite System' (uppermost Arenig–Llanvirn) and the 'Upper Turbidite System' (Llandeilo–(?)lower Caradoc; Fig. 4; Bahlburg *et al.*, 1990).

## Conflicting geodynamic interpretations

The turbidite units of the northern and southern Puna are usually described as fore-arc deposits of an

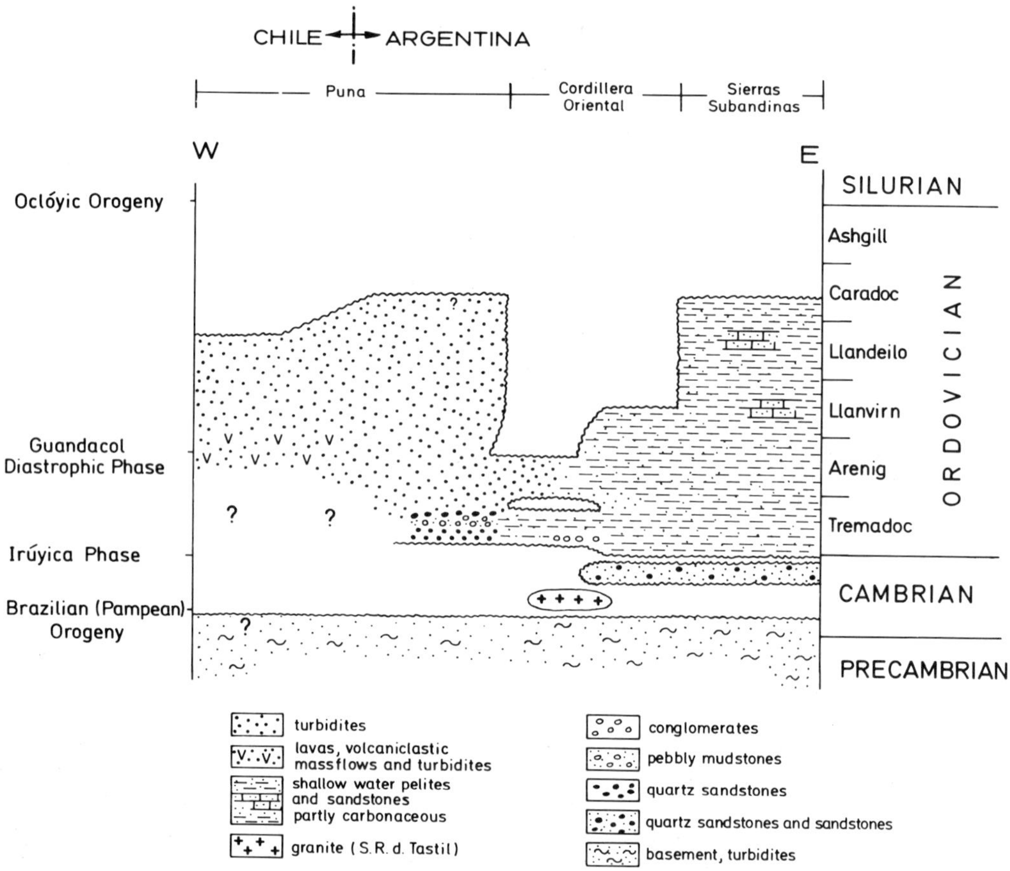


**Fig. 3.** Time–stratigraphic diagram for the northwestern Argentinian and northern Chilean morphostructural units. For further details see text.

east-dipping (relative to present-day coordinates) Ordovician subduction zone at the western margin of Gondwanaland (e.g. Coira *et al.*, 1982; Allmendinger *et al.*, 1983; Bell, 1987; Hervé *et al.*, 1987). In these models the magmatic belt of the 'Faja Eruptiva de la Puna Oriental' (FE; Mendez *et al.*, 1973; Fig. 2) is quoted as the magmatic arc of the subduction complex. Subduction ceased with collision of the allochtonous Arequipa Massif terrane (see below) during the Oclóyic Órogeny at the Ordovician–Silurian boundary (e.g. Ramos, 1988). The Arequipa Massif now forms an outcrop of Proterozoic basement in southern Peru near the city of Arequipa (Shackleton *et al.*, 1979). It is part of the post-Ordovician basement of northern Chile and northwestern Argentina (Baeza & Pichowiak, 1988).

The age and origin of the rocks of the FE have been the subject of debate. In the first paper on the FE, Mendez *et al.* (1973) described the rocks as post-Ordovician intrusives. According to Coira (1973), Koukharsky & Mirre (1974), and Coira *et al.* (1982), the rocks are submarine ignimbrites, rhyolites, and spilites intercalated with Ordovician sedimentary rocks. Results of radiometric dating vary between 471 and 300 Ma (Mendez *et al.*, 1979; Omarini *et al.*, 1984). However, in the course of this study in the northern Puna (FE sections, Fig. 1), porphyric and equigranular high-level granitoids with locally very well-developed intrusive contacts to the previously folded Ordovician sedimentary rocks were observed. Basic dykes and sills cut both intrusives and sedimentary rocks. With regard to the above-mentioned Late Ordovician folding phase, a post-Ordovician age has to be assigned to the intrusives of the FE (Turner, 1960; Schwab, 1973). This supports the observations of Mendez *et al.* (1973).

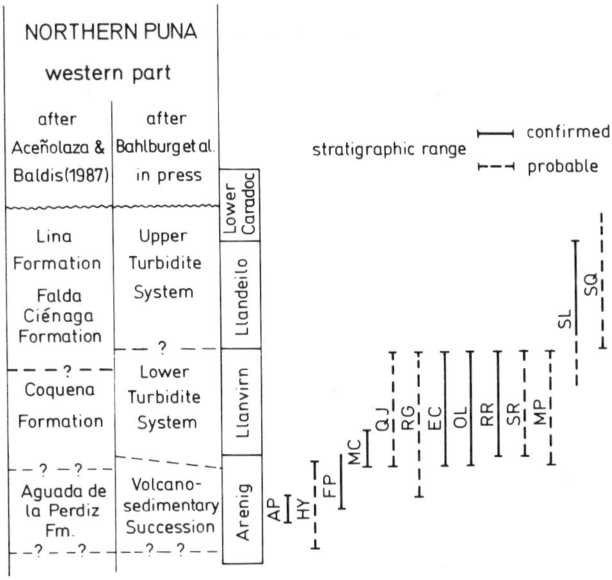


**Fig. 4.** Ordovician stratigraphic units and stratigraphic range of sections for the western part of the northern Puna, modified after Aceñolaza & Baldis (1987) and Bahlburg *et al.* (1990). Abbreviations indicate sections marked in Fig. 2.

Furthermore, our observations show that the granites are cut by north-trending, subvertical shear zones that accommodated left lateral slip of unknown extent. Thus, the post-Ordovician FE cannot be regarded as the arc of an Ordovician subduction zone. Consequently, the Ordovician clastic rocks are not fore-arc sediments and another geodynamic model has to be found to explain their genesis.

## SEDIMENTOLOGY OF SELECTED SECTIONS

Apart from general descriptions (e.g. Turner, 1960; Garcia *et al.*, 1962; Ramos, 1972; Schwab, 1973), the first sedimentological studies of the Ordovician units in the Puna were only carried out recently, and concentrated on the volcanosedimentary successions in the western part of the northern Puna (e.g. Mendez *et al.*, 1979; Breitkreuz, 1986; Coira & Barber, 1987; Breitkreuz *et al.*, 1989). The first results of our work on detailed sedimentological sections in the overlying turbidites are presented in Bahlburg *et al.* (1988).

The turbidite units were divided into five facies associations (Fig. 5).

*Facies association 1.* This group of sediments is composed of (very) coarse sandy or gravelly sandstones and conglomerates. The deposits are generally scoured and amalgamated to packages a few metres thick and display normal, inverse, and coarse-tail grading (facies A of Mutti & Ricci-Lucchi (1972) and facies group A2 of Pickering *et al.* (1986)) or are disorganized (facies group A1 of Pickering *et al.* (1986)). They were deposited by high-density turbidity currents and are interpreted to represent channel-fill sediments (Mutti & Normark, 1987).

*Facies association 2.* This subdivision contains successions of relatively even bedded (very) coarse and medium sand turbidites and pebbly sandstones. Bed thicknesses are in the range of 10–100 cm. The rocks are equivalent to the massive sandstones of Walker (1984) (also facies E of Mutti & Ricci-Lucchi (1972) and facies B2.1 of Pickering *et al.* (1986)). In rare cases, trough-bedded dune structures up to 1 m thick occur (facies B2.2 of Pickering *et al.* (1986)). They may be attributed to the hydraulic jump that occurs when sand-rich turbidity currents emerge from the confines of a channel in the channel-lobe transition zone (Mutti & Normark, 1987). Thick sandstones with very thin shale intercalations are included in this facies association.

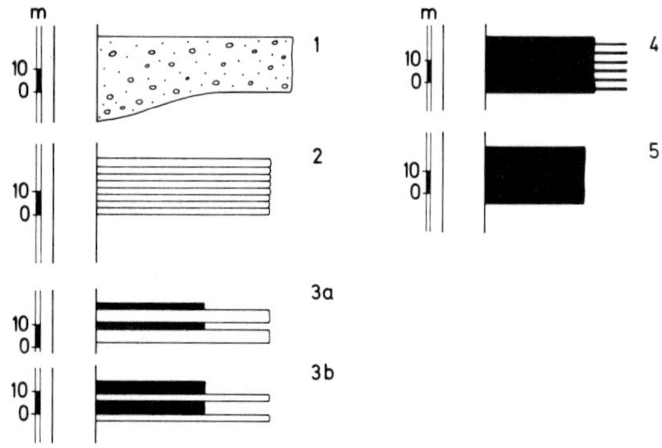


**Fig. 5.** Key to facies associations 1–5 (see text) and schematic logs of Figs 8–10.

*Facies association 3.* Association 3 comprises (1) turbidites that consist of successions of 15–60 cm thick $T_{a-c}$ of the Bouma sequence (facies association 3a; facies C1 of Mutti & Ricci-Lucchi (1975) and facies C2.1 of Pickering *et al.* (1986)), and (2) sequences of up to 70 cm thick, base-absent $T_{b-(d)e}$ and rare $T_{a-(d)e}$ or $T_{a-c(d)e}$. Bouma sequences $T_b$ and, if present, $T_a$ are of minor thickness (facies association 3b; facies C2 of Mutti & Ricci-Lucchi (1975) and facies C2.2 and C2.3 of Pickering *et al.* (1986)). Owing to tectonic and diagenetic effects it is commonly not possible to distinguish between $T_d$ and $T_e$; therefore the first is denominated $T_{(d)}$ or included in $T_e$ (Walker, 1978). Association 3a is attributed to depositional lobes but may also occur as the upper part of channel fill; association 3b is more 'distal' in character and also contains sediments of depositional lobes and levees (Mutti & Normark, 1987).

*Facies association 4.* Here, fine sand and mud turbidites, consisting of $T_{c(d)e}$ sequences, are grouped together (facies D of Mutti & Ricci-Lucchi (1972) and facies group D2 of Pickering *et al.* (1986)). Single turbidites are generally not thicker than 20 cm. They are assigned to lobe fringe and overbank/levee environments (Mutti & Normark, 1987).

*Facies association 5.* Structureless shales up to 80 m thick occur in some sections (facies E1.1 'structureless muds' or G2.1 'hemipelagites' of Pickering *et al.* (1986)). They are interpreted as shales of (hemi)pelagic origin representing background sedimentation.

The interpretation of the shales of facies association 5 as being of (hemi)pelagic origin is based on the following circumstantial evidence: (1) the lack of bioturbation may be due to the fact that Ordovician oceans were commonly anoxic at depths below 100 m (Wilde & Berry, 1984). The interbedding of the shales with turbidites points to a far greater deposition depth. (2) The shales are devoid of any sedimentary structures indicative of deposition from mud turbidites (Pickering *et al.*, 1986). (3) As discussed below, lobe switching as a cause of mud deposition seems unlikely as sediment was derived from a basin-parallel line source (volcanic arc: Breitkreuz *et al.*, 1989) and deposition was dominated by axial currents in a relatively confined basin. Consequently, if presumably coeval shales occur in the basin at different loci in a downcurrent direction, they most likely represent phases of tectonic quiescence or sea-level highstands.

## The eastern part of the northern Puna

The oldest Ordovician rocks of the Puna occur in the valley of the Rio Taique and rest unconformably on the Upper Proterozoic to Lower Cambrian Puncoviscana Formation (RT section, Fig. 2; Schwab, 1973; Aceñolaza *et al.*, 1988). Quartz sandstones with approximately symmetrical herringbone cross-stratification give way to a c. 200-m-thick succession of up to 2-m-thick pebbly mudstones with rare intercalations of slumped quartz sandstones. This succession, in turn, is overlain by massive, structureless quartz sandstones. Both quartz sandstone packages yielded Ordovician cruzianas of the *Rugosa* group (M. Kumpa, personal communication). This threefold unit is c. 1100 m thick. It has been assigned to the Tremadoc in view of the cruzianas and the conformable contact with overlying upper Tremadoc to Arenig turbidites (Schwab, 1973; RT and SC sections, Fig. 2). The thickness of the turbidite unit is estimated to be 2000 m (Schwab, 1973); its upper contact is faulted.

This succession forms a transgressive sequence starting with quartz sandstones deposited by approximately symmetrical tidal currents (Allen, 1982). The overlying pebbly mudstones and slumped sandstone beds, which give way to turbidites, record increased instability and presumably greater depth of the depositional site. It was located on the western flank of a subsiding platform/outer shelf high region (Cobres High: Moya, 1988) on the edge of the shelf in the Cordillera Oriental. Quartz sandstones intercalated in the lower part of the turbidite unit probably originated from the outer shelf high, which collapsed during subsidence.

## The western part of the northern Puna

### *Arenig*

*Aguada de la Perdiz (AP) and Huaytiquina (HY) sections* (Fig. 2). The oldest Ordovician rocks of this area crop out in a volcano-sedimentary succession along the Chilean–Argentine border. In the 'Aguada de la Perdiz' outcrop (Garcia *et al.*, 1962; AP section, Fig. 2), middle Arenig graptolites occur in various thin siltstone horizons (Breitkreuz, 1986); underlying strata are not known. These rocks also crop out along strike in the HY and SG sections (Fig. 2). The series is between 2700 m (AP section) and 3800 m (HY section) thick (Fig. 6). It is probable that the HY section contains lower as well as middle and upper Arenig strata.

The marine volcano-sedimentary succession fines upward. It begins with a basal sequence of c. 300 m of subaqueous vesicular basalt lavas and associated hydroclastic rocks, with mostly graded monomict basaltic debris flows, and breccia tuffs containing rafts of siliceous ash tuffs. These strata are overlain by up to 3500 m of fine and coarse-grained, graded volcaniclastic sandstones alternating with packages 20–50 cm thick of siliceous ash tuffs and occasional basal lavas. The submarine tuffs were most probably deposited by 'dilute slow-moving lutite flows' (White & Busby-Spera, 1987) whereas the coarse clastics were sedimented by debris flows and turbidity currents.

Occasional stromatolites (Coira & Barber, 1987) together with the low-diversity graptolite fauna (Breitkreuz, 1986) and the subaqueous vesicular basalts indicate a shallow marine depositional environment at least for the lower part of the section. Volcaniclastic turbidites become more abundant towards the top of the section and point to increasing subsidence of the depositional site. Farther east, slightly younger (late Arenig) deep-marine turbidites occur in the MC section (Fig. 2; see below). Therefore it can be inferred that the basin deepened in this direction.

The depositional environment can be characterized as a submarine volcaniclastic apron on the eastern flank of a middle Arenig zone of presumably subaerial volcanoes. Geochemical features characterize the basic lavas and the siliceous ash tuffs as products of a volcanic arc (Breitkreuz *et al.*, 1989; Bahlburg *et al.*, 1990).

*Filon Pircas (FP) and Morro Corralito (MC) sections* (Fig. 2). The FP and MC sections comprise (?)upper Arenig strata. The 660-m-thick FP section (Fig. 7) forms the stratigraphic and lithologic continuation of the AP and HY sections. Here, volcaniclastic turbidites alternate with thin ash tuffs. The turbidites are up to 50 cm thick, predominantly fine grained and $T_{a-c}$ dominated (facies association 3a). Coarse-grained deposits of high-density turbidity currents are rare. The MC section is intensely folded and also comprises apparently laterally continuous $T_{ab}$ dominated turbidites of facies association 3a. The deposits are assigned to a depositional lobe environment. Graptolites indicate roughly the same age for both sections. However, the MC section lacks evidence of active volcanism. It is presumably a downslope equivalent of the FP section to the west and the probable upper Arenig parts of the HY section.

### *Lower Turbidite System*

*Olaroz Grande (OL) and Cordón de la Rinconada (SR) sections* (Fig. 2). Most of the measured sections belong entirely or to a large extent to the Llanvirn (CA, QJ, EC, OL, RG, RR, MP, SR sections; Figs 2, 4). Of these the OL and SR sections are presented here, because they are relatively well-documented stratigraphically and, together with the EC section, they are the only sections that have (hemi)pelagic shales in their upper parts. Unfortunately, only the uppermost shale of the OL and EC sections are fossiliferous. Therefore the correlation of the shales is tentative. The determination of sedimentary cyclicity was checked by plots of cumulative thickness of sand against cumulative thickness of mud (Kimura, 1966).

At 2700 m, the OL section is the thickest measured. It can be divided into members A–D (Fig. 8). Members A (0–300 m) is poorly exposed. It comprises 20–60 cm thick, basc-absent, fine sand turbidites (facies association 3b) forming thinning upward cycles 2–3 m thick. Member B (c. 1500 m) follows with an erosive base; it consists of medium and fine sand turbidites of facies associations 3a, 3b, and 4. In submember $B_1$ they are generally arranged in thickening/coarsening upwards megacycles up to

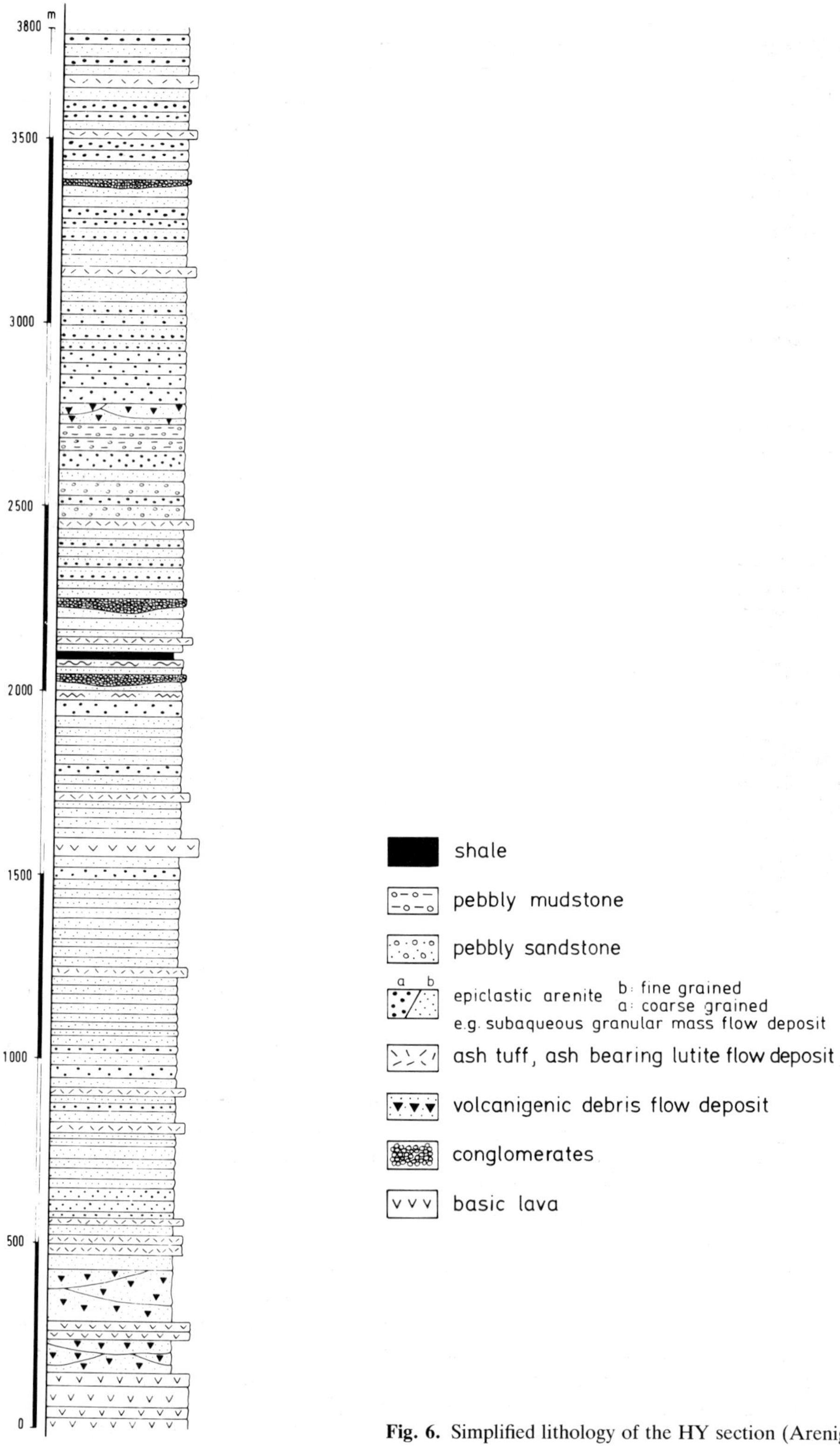


**Fig. 6.** Simplified lithology of the HY section (Arenig; Fig. 2).

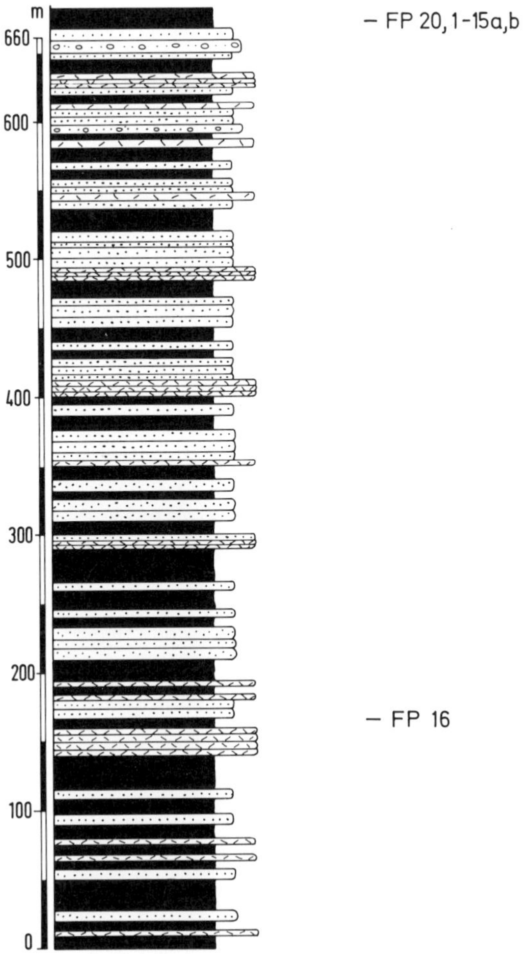


**Fig. 7.** Simplified lithology of the FP section (upper Arenig; Fig. 2); for key see Fig. 6. Sample numbers at the right side of the column indicate fossiliferous beds.

100 m thick. Individual cycles are c. 2–10 m. Submember $B_2$ comprises the same facies associations as $B_1$ but here the strata typically form thinning/fining upwards cycles. These tabular deposits are interpreted as depositional lobe sediments. The depositional lobes of both submembers are cut by channels (facies associations 1 and 2) up to 150 m thick.

Member C (425 m) comprises parallel-bedded, fine-grained turbidites of facies associations 3a and 3b. They are arranged in thickening upwards cycles 10 m thick and form part of depositional lobes. Member C grades into member D, which is 425 m thick and forms the top of the exposed section. This member is composed of $T_{c-e}$ turbidites of a lobe fringe and (hemi)pelagic shales (facies association 4) of up to 70 m in thickness. Repeated, 20 m thick, successions of massive sandstones as well as $T_{a-e}$ and base-absent turbidites are intercalated. They represent a waning in sedimentary activity in this turbidite system (Mutti & Normark, 1987).

The SR section (Fig. 9) is subdivided into the members A–E. Member A is 600 m thick and contains deposits of facies associations 3a, 3b, and 4. They are commonly arranged in thickening upward cycles of 2–8 m thickness formed by depositional lobes. In its upper part, channels 4–15 m thick (facies association 1) are intercalated. The following member B (160 m) consists predominantly of $T_{c-e}$ turbidites, 15–30 cm thick (facies association 4), which are cut by a channel up to 16 m thick (facies association 2). A few of the beds 4–5 m below and c. 3 m above the channel are slumped, indicating steeper slopes. They may be interpreted as the levees of the channel, which possibly shifted across the section (Walker, 1985).

A succession of predominantly parallel-bedded, fine to medium-grained, massive sandstones (facies association 2), alternations of $T_{a-c}$ and $T_{b-e}$ turbidites (facies association 3b) pertaining to depositional lobes, and small channels form member C (700 m). This member is overlain with erosive contact by member D, a channel complex 350 m thick. In a few cases, $T_c$ divisions 10–15 cm thick consist of steep climbing ripples typical of rapid deposition of large volumes of sediment.

Above a covered interval, member E (350 m) concludes the section with fine sand to mud turbidites of facies associations 3a, 3b, and 4, which form fining upwards cycles c. 50 m thick. One hundred metres above the base of member E the first of three (hemi)pelagic shales occurs. The turbidites belong to a lobe fringe areas whereas the shales represent tectonic lulls or sea-level highstands.

*Upper Turbidite System*

*Sierra de Lina (SL) section* (Fig. 2). The stratigraphically youngest units belong to the Llandeilo and possibly the very early Caradoc. They are exposed in the SL and SQ sections and form a turbidite system overlying the Lower Turbidite System. However, as the strata of the SQ section

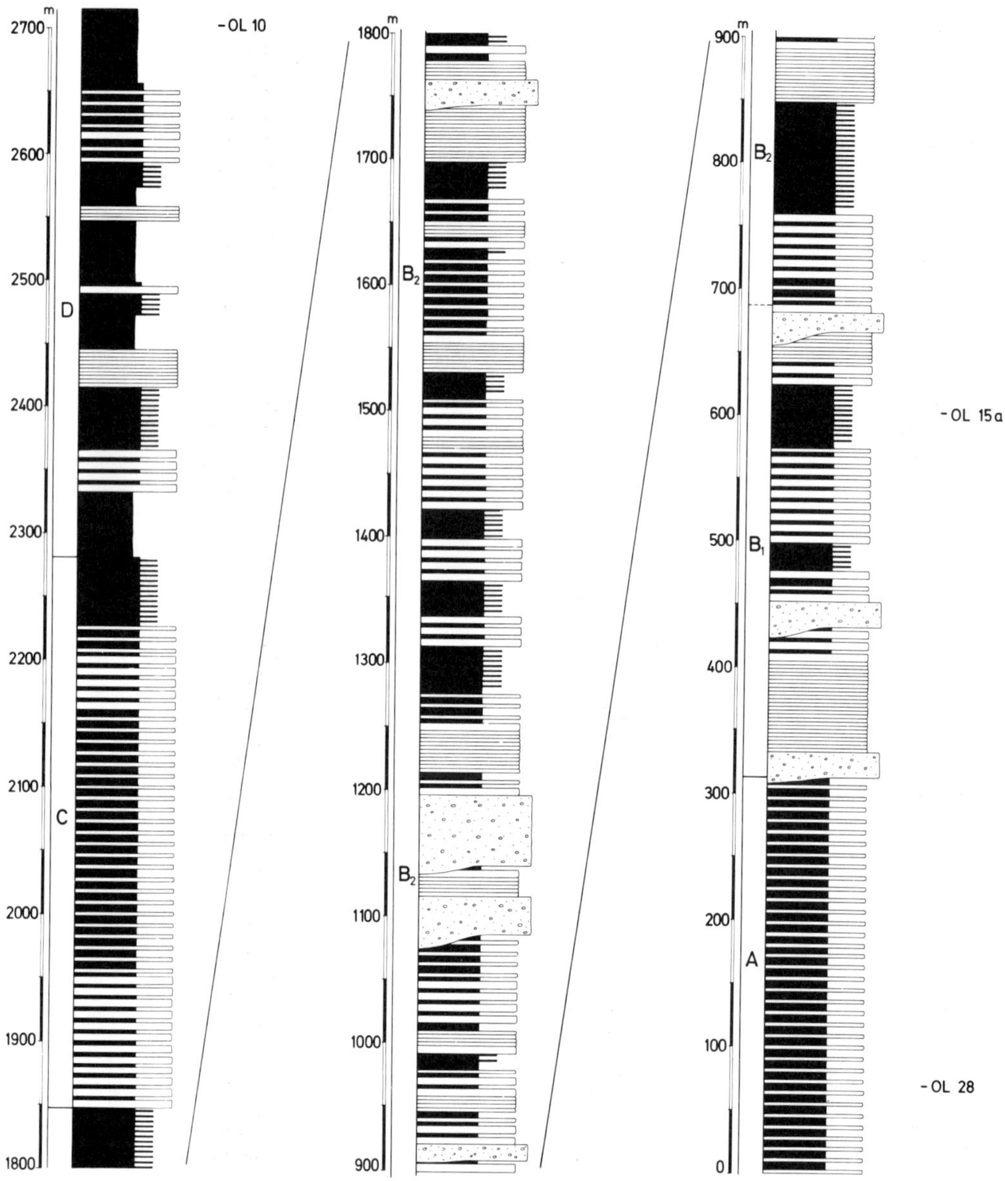


**Fig. 8.** Schematic sedimentological log of the OL section (Fig. 2) in the Lower Turbidite System (uppermost Arenig to Llanvirn). Sample numbers at the right side of the column indicate fossiliferous beds.

are intensely folded, the 920-m-thick SL section (Fig. 10) is the only continuous section available in the youngest Ordovician sedimentary rocks of the Puna.

The section comprises members A–C. Member A is 240 m thick and is made up of fine sandy and silty turbidites of facies associations 2, 3a, 3b, and 4, which form two equally thick coarsening upwards megacycles culminating in a lenticular bedded coarse sandy channel and massive sandstones.

Member B (590 m) is composed of successions, up to 100 m thick, of $T_c$ and $T_{c-e}$ turbidites (facies association 4), which alternate with (hemi)pelagic shale units up to 70 m thick. Turbidites of facies

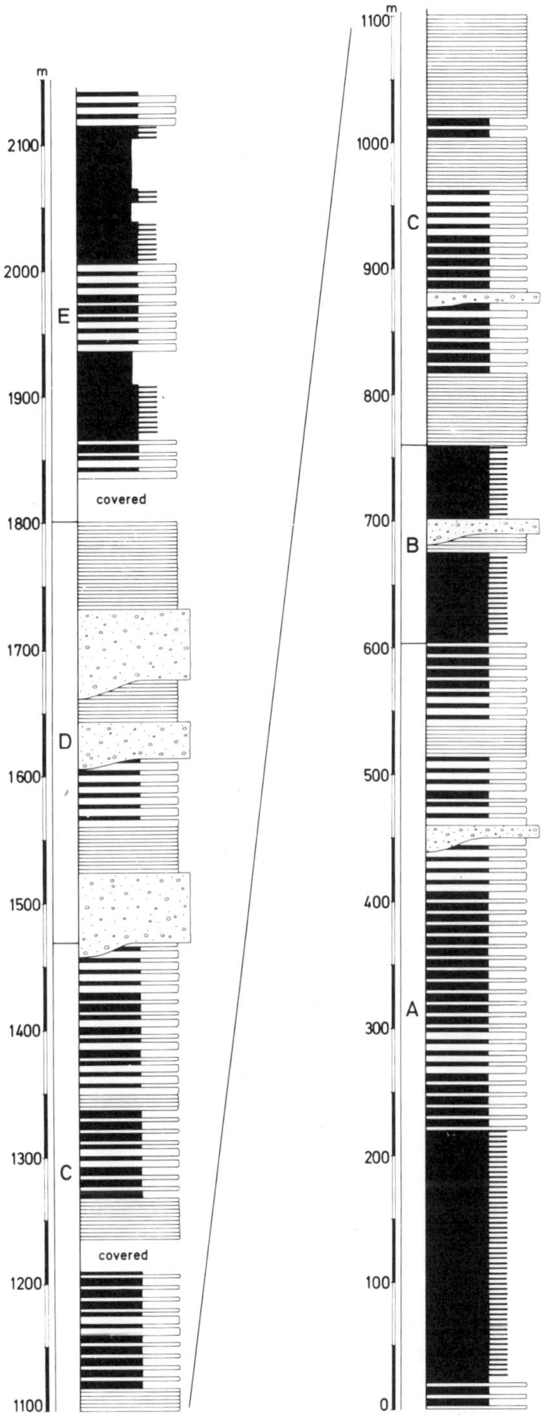


**Fig. 9.** Schematic sedimentological log of the SR section (Fig. 2) in the Lower Turbidite System (uppermost Arenig to Llanvirn).

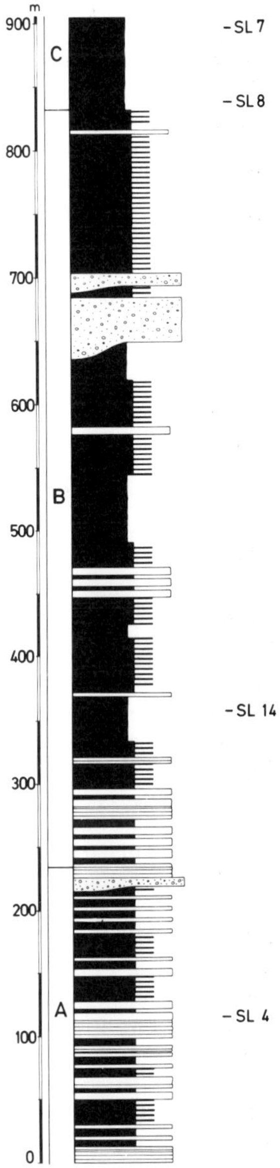


**Fig. 10.** Schematic sedimentological log of the SL section (Fig. 2) in the Upper Turbidite System (Llandeilo to (?)lower Caradoc). Sample numbers at the right side of the column indicate fossiliferous beds.

associations 3a and 3b are intercalated. A prominent feature of this member is an 80 m thick, coarse sandy to gravelly, amalgamated channel complex. The channel complex is overlain by 100 m of $T_c$ and $T_{c-e}$ turbidites. Supercritical climbing ripples are typical of their lower part. It may be interpreted as the levee of the channel that had migrated across the section, or as the levee of a laterally adjacent channel, whereas the upper part consists of depositional lobe or lobe fringe deposits. The channel cuts a fining upwards megacycle and is the base of a succeeding one. The upper cycle culminates in the 80 m thick, (hemi)pelagic shales of member C, which end the exposed section.

The sediments of the SL section were deposited in distal areas of the depositional system. They belong predominantly to depositional lobes that were periodically attached to advancing or shifting channels. The channel in the upper part of member B in particular is supposed to be the result of instantaneous sediment instability in up-channel or shelf regions.

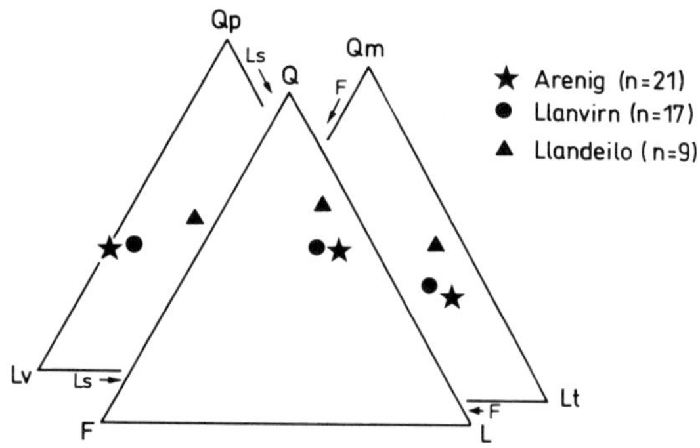


**Fig. 11.** Averages of framework modes of turbidite arenites pertaining to the Arenig, Llanvirn, and Llandeilo time slices, plotted after Dickinson (1985). Q, total quartz; F, total feldspar; L, total rock fragments; Qm, monocrystalline quartz; Lt, total rock fragments plus polycrystalline quartz; Qp, polycrystalline quartz; Lv, (meta)volcanic rock fragments; Ls, (meta)sedimentary rock fragments.

## Petrofacies

In the eastern Puna, the Tremadoc to Arenig sedimentary rocks lack evidence of contemporary volcanism. They comprise quartz sandstones, pebbly mudstones containing clasts of older sedimentary rocks and quartz-rich turbidites (*sensu* Crook, 1974). Sparse palaeocurrent data indicate that the sediments were derived from the east (Bahlburg *et al.*, 1990).

In the western Puna, non-volcanic arc detritus is almost completely missing in the volcanigenic Arenig rocks (AP, HY, FP, MC sections). Their framework minerals consist of fragments of basalts, andesites, rhyolites, and associated volcaniclastic rocks (Fig. 11). From the Arenig–Llanvirn transition to the Llandeilo, the abundance of (meta)sedimentary and plutonic components increases (QpLvLs diagram, Fig. 11). However, erosional debris of the Arenig volcanic arc (Breitkreuz *et al.*, 1989) continues to dominate the component spectrum. Detritus of carbonates and of high-grade metamorphic rocks was not observed. A characteristic petrofacial feature of the rocks exposed in the western Puna is their overall consistent composition through time and along the basin axis. Furthermore, it does not differ in sediments deposited from palaeocurrents of opposing directions (see next section and Fig. 12), thus giving further evidence of the presence of a line source (volcanic arc). The distribution of framework minerals (Fig. 11) in the turbidite sandstones reflects the following sequence of events for the Arenig, Llanvirn, and Llandeilo intervals: (1) activity of the volcanic arc on the western margin of the basin, (2) and (3) erosion of the arc and an increased but still minor input of older sedimentary and basement rocks.

## Sedimentological model

The Ordovician basin in the Argentinian–Chilean Puna developed as a north-trending elongate and relatively narrow structure between the Arenig arc to the west and the shelf of the Cordillera Oriental to the east (Figs 1, 2). During the period of active volcanism in the N–S striking arc, lasting at least until the middle Arenig (Niemeyer, 1989; Bahlburg *et al.*, 1990), the shelf and adjacent basin were in a back-arc position. The very immature detritus was probably derived by tectonically triggered processes from the line sources close to or on the basin margins, in particular from the arc region to the west. The line sources inhibited the development of regional proximal to distal trends (Macdonald, 1986), whereas the elongate shape of the basin and its apparently generally north-dipping floor led to deviation of palaeocurrents at the basin axis (Ricci-Lucchi, 1985).

The turbidite systems formed in the basin are characterized by the dominance of depositional lobe deposits (Figs 8–10), which may have been temporarily channel-detached or -attached in a similar manner to the stage I and stage II deposits, respectively, of Mutti (1985). In the description of the turbidite systems, however, the term 'fan' has been consciously omitted for the following reasons: (1)

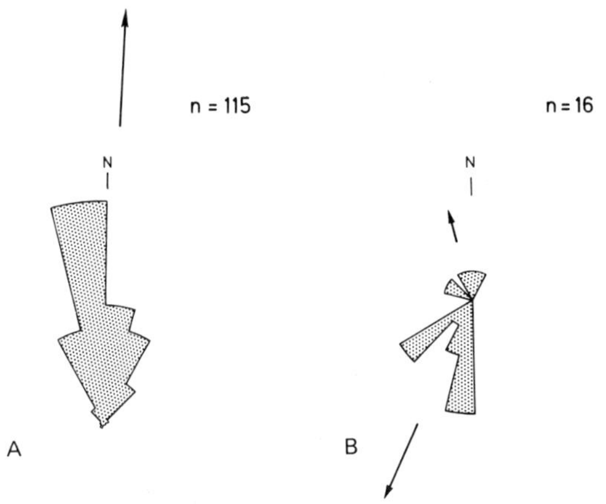


**Fig. 12.** Palaeocurrent directions as indicated by flutes: (A) Lower Turbidite System (uppermost Arenig to Llanvirn; Fig. 4); (B) Upper Turbidite System (Llandeilo to (?)lower Caradoc; Fig. 4).

the shape of the depositional system is unknown, (2) it is not known whether there were one or several feeder channels, and (3) the palaeocurrents documented by flutes (Fig. 12a,b) display a uniform axial sediment transport towards the north-northwest in all the differentiated time slices. Only in the Upper Turbidite System (Llandeilo to (?)early Caradoc; Fig. 4) was there also transport towards the south-southwest (Fig. 12b). In both cases the palaeocurrents developed along the basin axis and parallel to the north-trending tectonic axes of the Oclóyic orogeny (Turner & Mendez, 1979; Bahlburg *et al.*, 1988). In view of the marginal line sources, the presence of a point-sourced fan seems improbable. However, according to a fan model (Mutti & Ricci-Lucchi, 1972; Mutti, 1985) this basin would have been oversupplied and the shape of the fan would have been determined by basin geometry and not by depositional processes (e.g. Hiscott *et al.*, 1986).

## SEDIMENT CYCLICITY AND SEA-LEVEL CHANGES

Sedimentary cycles within the Puna Basin are arranged in alternating fining and coarsening upwards megacycles (Fig. 13). Stratigraphic dating of megacycles is relatively reliable in the Arenig and Llanvirn, but inaccurate in the Llandeilo to (?)lower Caradoc (Bahlburg *et al.*, 1988, 1990). However, apart from evidence that the AP section belongs to the *bifidus* zone (Breitkreuz, 1986), biostratigraphic resolution is never more precise than approximately 5 Ma, which is poor in relation to sequence stratigraphic 'showcases' (e.g. Wilgus *et al.*, 1988).

The waning of volcanic and associated tectonic activity is expressed in the middle and late Arenig strata as a fining upwards megacycle c. 3000 m thick (Figs 6, 7, 13). This is followed in the very late Arenig by a coarsening upwards trend at least 700 m thick (OL section, Fig. 8), which reaches even greater thickness in the SR section (Fig. 9). This megacycle at the Arenig–Llanvirn transition coincides with the tectonic event of the 'Guandacol Diastrophic Phase' (Fig. 13), which led to the development of a pronounced unconformity in the

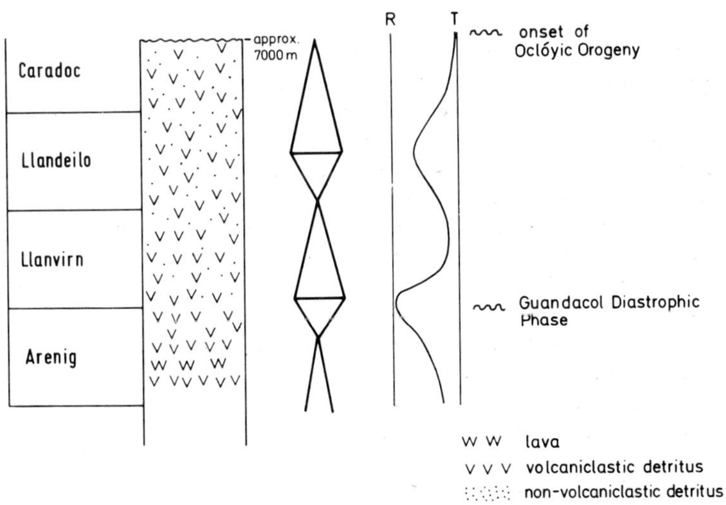


**Fig. 13.** Synopsis of petrographic trend, and eustatic and tectonic events as recorded in the Ordovician units of the western Puna Basin, and the global sea-level curve of Fortey (1984) for the Ordovician. From left to right: petrographic trend; coarsening and fining upwards megasequences as determined from the sedimentological sections (see text) and stratigraphic data (Bahlburg *et al.*, 1990); sea-level curve of Fortey (1984; R, regression; T, transgression); tectonic events (for references see text).

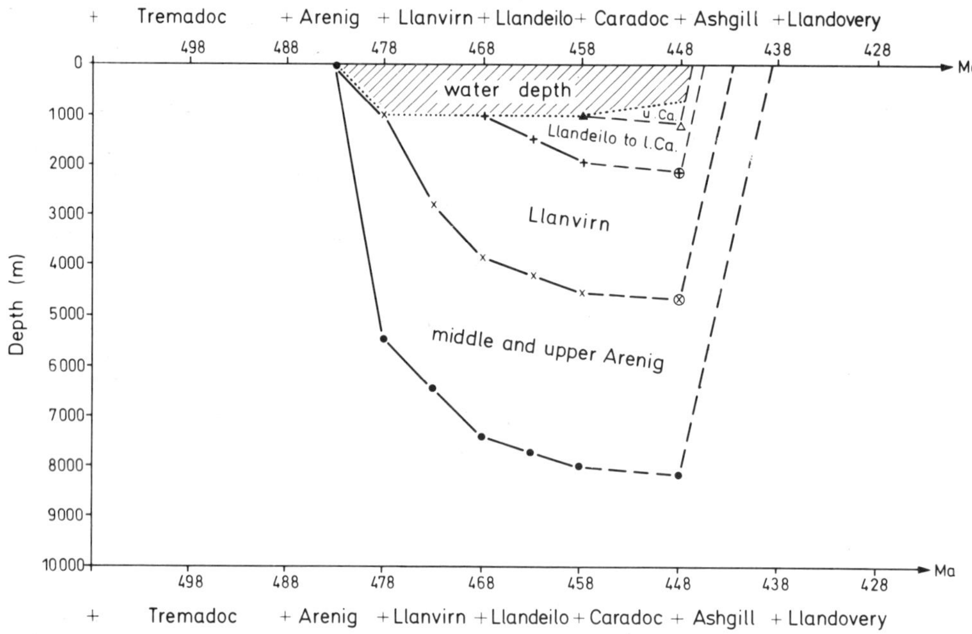


**Fig. 14.** Geohistory plot for the western Puna Basin, based on incremental decompaction according to Sclater & Christie (1980). Bathymetry estimated from facies analysis, potential basement compaction ignored. For further details see Bahlburg (1990).

Argentine Precordillera further south (Furque & Cuerda, 1982). In the area under consideration it is seen as an uplift event leading to the emergence of the Cordillera Oriental area in the course of the Llanvirn (Salfity *et al.*, 1984). Coeval with these events, a global regression took place (Fortey, 1984; Fig. 13). However, shallow-marine sedimentation apparently continued uninterrupted further east in the Sierras Subandinas (Cuerda, 1973; Figs 1, 3).

This coarsening upwards trend is succeeded by a fining upwards megacycle that is thickest in the OL section (2000 m), and which culminates in both the OL and SR sections in (hemi)pelagic shales. A global transgression occurred contemporaneously (Fig. 13). The Upper Turbidite System represented by the SL section begins with a coarsening upwards megacycle c. 250 m thick, which can be interpreted tentatively as being coeval with a global regression in the early Llandeilo. The remaining 650 m are arranged in a fining upwards megacycle that is topped by (hemi)pelagic shales of presumed late Llandeilo to very early Caradoc age (Bahlburg *et al.*, 1990).

The youngest fining upwards megacycle is paralleled by a global transgression that began in the late Llandeilo. The transgression was succeeded by a pronounced global regression in the Ashgill, which was caused by the Late Ordovician glaciation centred in North Africa (Barnes, 1986). Roughly at the same time the Oclóyic orogeny in northwestern Argentina and southern Bolivia occurred.

Immaturity and poor sorting of the turbidites in coarsening upwards as well as in fining upward cycles and megacycles attest to short transport paths, high relief, and minor reworking in shelf areas during both cycles. Shelf areas and coastal lowlands were probably rather small and inefficient in reworking and weathering sediment owing to continued uplift of the Cordillera Oriental and the extinct volcanic arc. However, the marginal repositories must still have been sufficiently large to store the necessary quantities of sediment to be periodically supplied as gravity flows to the turbidite systems. Sediment might have been delivered directly to the basin via fan-delta systems (Mutti, 1985; McPherson *et al.*, 1987).

At first glance there is a good correlation of eustatic sea-level changes and sedimentary cyclicity. The succession of coarsening and fining upwards megacycles can be grouped to megasequences. The coarsening upwards megacycles would represent a lowstand wedge and can be interpreted as belonging to a lowstand systems tract (Posamentier & Vail, 1988). This was succeeded in the fining upwards megacycles by the transgressive systems tract leading to (hemi)pelagic shales of a highstand systems tract. Each of the two megasequences had an approximate duration of 10 Ma, which would class them as second-order cycles.

In this basin, however, formation of thick turbidite systems was not restricted to or solely determined by phases of eustatic or relative sea-level lowstand, as is generally suggested by Posamentier & Vail (1988). In contrast, turbidite deposition continued during periods of sea-level rise and parts of highstands and strongly points to closely linked tectonic and sedimentary processes (Kolla & Macurda, 1988). Increased tectonism (uplift) on the basin margins would have led to more frequent sediment discharge from the small marginal repositories, and would have equally lowered their recharge time. Consequently, the amount of available sediment became progressively smaller, thus producing tectonically controlled thinning upwards and, probably, fining upwards cycles, as are present in the upper parts of the two described turbidite systems, at a time of sea-level rise or highstand (Mutti, 1985). Furthermore, phases of enhanced subsidence (see below) are characterized by the deposition of thick successions of fine-grained turbidites (Beaumont *et al.*, 1982).

In conclusion, sedimentary trends and sea-level changes documented in the stratigraphic record of the basin are primarily the result of the tectonic instability of source regions and of the basin itself (see below). The global regressions at the Arenig–Llanvirn transition and in the middle Llandeilo (Fortey, 1984) most likely merely complemented the influence of tectonics on depositional patterns.

In the Sierras Subandinas, to the east of the actual Cordillera Oriental, shallow-marine sedimentation occurred in the Ordovician until the early Caradoc (Fig. 3; Cuerda, 1973; Aceñolaza, 1978). No transgressive or regressive cycles of any scale are described in the literature. However, as the shallow marine conditions prevailed here throughout the Ordovician it can be inferred that the alleged 'global' sea-level changes, and regressions in particular, had very little effect in this area. It is probable that the sea-level changes were either not truly global or were outweighed by the interplay of local tectonics and sedimentation rate.

## EVOLUTION FROM BACK-ARC TO FORELAND BASIN

### The eastern part of the northern Puna

The shallow marine shelf in the Cordillera Oriental formed in the Late Cambrian (Moya, 1988), presumably as a response to extension of Brazilian basement. The shelf edge was marked by an outer shelf high (Cobres High, Moya 1988; RT section, Fig. 2). On the western flank of this high, the c. 3000-m-thick transgressive sequence formed. Western equivalents of this succession are unknown, as the stratigraphic record in the western Puna only begins in the middle Arenig. Furthermore, the exact stratigraphic position of probable Ordovician units west of the Cordillera Occidental (Bahlburg *et al.*, 1987,1988; Fig. 1) in northern Chile is also unclear, thus prohibiting their correlation with the successions of the northern Puna. However, from the middle Arenig onwards, the volcanic-arc domain formed the western margin of the basin, which at that time was a back-arc basin.

On the Cordillera Oriental shelf, sedimentation ended in the early Llanvirn after a marked regression (Moya, 1988). An explanation for both the deposition of a sedimentary column at least 2200 m thick on the shallow shelf, and the marked regression is offered by the 'intra-plate stress model' of Cloetingh (1988). According to this model, basin areas close to basin margins can accommodate thick sedimentary piles owing to relatively rapid subsidence in the course of back-arc extension. After a change to compression (see below), the margins are uplifted rapidly while the basin interior begins to subside preferentially.

### The western part of the northern Puna

The cumulative stratigraphic thickness of the units in the western Puna Basin is in the order of 8000 m. However, especially in the late Arenig, there is a facies equivalency between the sections influenced by active volcanism (Figs 6, 7) and those of the MC section further downslope lacking in this evidence. In order to account for this facies equivalency, an estimated 1000 m (500 m each of the volcano sedimentary succession (AP and HY sections) and the Lower Turbidite System) were subtracted from the cumulative thickness column, leaving an overall stratigraphic thickness of 7000 m.

In the course of deposition of this succession, from the middle Arenig to at least the late Llandeilo (c. 25 Ma), the bathymetry and facies of the depositional site changed from a nearshore environment to a deep marine environment. Figure 14 depicts this evolution in a geohistory plot based on incremental decompaction using the method of Sclater & Christie (1980). In this plot a constant level of post-Arenig turbidite deposition at 1000 m depth is assumed. A greater depth estimate would result in even higher rates of subsidence. Global sea-level changes were not considered as their influence on subsidence can be neglected in view of the high sedimentation rate averaging 280 m/Ma (Arenig, 890 m/Ma; uppermost Arenig to Llanvirn, 540 m/Ma; Llandeilo to (?)lower Caradoc, 180 m/Ma (cf. Fig. 14); Lundberg & Dorsey, 1988). Furthermore, no compaction was calculated for the underlying basement. The stratigraphic record ends with deep-water sediments in the early Caradoc at the latest. At the Ordovician–Silurian boundary the basin and its fill were folded. A pronounced regression leading most probably to non-marine sedimentation must have occurred prior to this (Covey, 1986). The sedimentary products of the regression were presumably cannibalized during orogeny. The hypothetical regression and uplift curves depicted in Fig. 14 take account of these inferences.

Thus the middle Arenig basin floor of the western Puna subsided within c. 25 Ma from a depth not exceeding 500 m to a depth of approximately 8000 m at the end of the Llandeilo to (?)lower Caradoc, and c. 7000 m of sediment were deposited on top of the middle Arenig reference level. Water depth did not exceed an estimated 1000 m (Fig. 14). This corresponds to an average subsidence rate of 320 m/Ma (Arenig, 1100 m/Ma; uppermost Arenig to Llanvirn, 190 m/Ma; Llandeilo to (?)lower Caradoc, 60 m/Ma (cf. Fig. 14)). In comparison with subsidence patterns of rifted basins, this is a very rapid development, typical of foreland basins (Allen *et al.*, 1986). The modern Timor–Tanimbar foreland basin off north Australia subsided in the last 2.3 Ma at rates of 1000 m/Ma (Audley-Charles, 1986).

Rapid subsidence starting in the late Arenig was probably connected to the beginning of collision of the allochtonous Arequipa Massif (Fig. 15) with Gondwanaland. This led to eastward overthrusting of the arc complex on to its back-arc basin resulting in foreland-type subsidence further to the east. In response to thrusting, a flexural bulge formed in the region of the Cordillera Oriental. The formation of the flexural bulge offers another explanation for uplift and emergence of the Cordillera Oriental shelf during the 'Guandacol Diastrophic Phase' (Figs 3, 15). Recent results of palaeomagnetic investigations of presumably Ordovician basalts and Silurian granitoids in northern Chile suggest that the Arequipa Massif reached its present position in the Late Ordovician Oclóyic orogeny (Mpoclozis, personal communication). In broad terms, this collision, with subsequent formation of a turbiditic foreland basin, may be comparable to the situation of the Middle Ordovician Appalachian margin in Quebec (Hiscott *et al.*, 1986; Pickering, 1987) or the modern north Australian margin, which is overthrust by the Banda Arc (Audley-Charles, 1986).

As the back-arc basin most probably originated by extension, the arc complex was thrust over attenuated continental crust. The lack of outcrops of ophiolitic rocks in the northern Puna suggests that no oceanic crust was generated in the back-arc basin. The resulting marine foreland basin formed as a narrow elongate structure as predicted by the 'visco-elastic lithospheric flexural model' of Beaumont *et al.* (1982). It is therefore likely that sedimentary patterns in the turbidite systems were controlled by the narrow basin geometry. Rapid sedimentation and cyclicity of the turbidite complexes as well as sea-level changes were a consequence of episodic thrusting, lithospheric relaxation, and migration of the flexural bulge (Fig. 15). The end of the volcanic activity was either caused by compression during thrusting or by termination of the subduction process. Increased input of non-volcanic debris in the Upper Turbidite System, in particular, indicates that erosion reached pre-volcanic levels in the arc complex to the west and/or was active on the flexural bulge in the east. The formation of a 'back-arc to foreland successor basin' couplet at the Arenig–Llanvirn transition constitutes a pronounced change in the geodynamic regime.

The lack of pronounced fold vergences in the Oclóyic fold belt of the northern Puna (Mon & Hongn, 1987; Bahlburg *et al.*, 1988) could be an indication that no fully fledged fold and thrust belt developed during overthrusting of the arc complex. According to Beaumont *et al.* (1982), the joint action of visco-elastically induced migration of the flexural bulge towards the thrust load, and uplift of the basin owing to erosion of the load, could lead to the closure of a basin. In addition, large quantities of erosional debris derived from the thrust arc complex would then be trapped in the basin and would

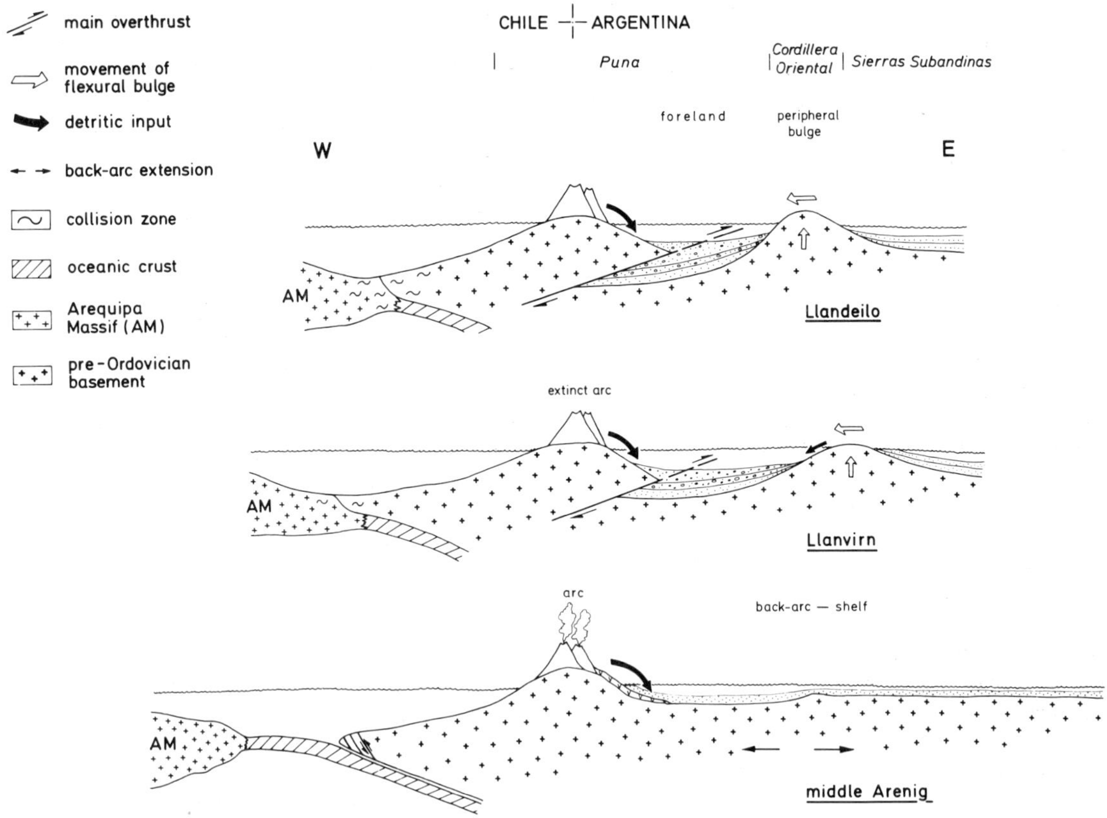


**Fig. 15.** Hypothetical cross-sections depicting tectonic evolution of the Gondwanaland margin in northwestern Argentina and northern Chile during the Ordovician.

hasten its evolution to non-marine conditions. Thus the folding of the strata was probably caused by eastward thrusting of the arc complex. However, eastward thrusting of the Cenozoic thrust belt of the Central Andes (e.g. Mingramm *et al.*, 1979) could have led to secondary anticlockwise rotation (seen from the south) of previously east-verging Palaeozoic folds in the Puna along early tectonic west-dipping listric thrusts, thus obliterating at least parts of the characteristics of previous fold systems.

## ACKNOWLEDGEMENTS

I thank C. Breitkreuz, Berlin, and K. Pickering, Leicester, for their comments on an early version of the manuscript, and B. Dunker and B. Kleeberg, Berlin, for technical assistance. The paper benefited from the reviews of H.G. Reading, Oxford, and R.D.A. Smith, Liverpool. The work was funded by the Deutsche Forschungsgemeinschaft (ref. nos. Gi 31/51–3+5) and was carried out in cooperation with M.C. Moya and J.A. Salfity, Universidad Nacional de Salta, Argentina. F. Jurtan, Berlin, checked the English text. This paper is a contribution to IGCP programme no. 279 'Terranes of Latin America'.

## REFERENCES

Aceñolaza, F.G. (1978) El Paleozoico inferior de Argentina segun sus trazas fosiles. *Ameghiniana* **15**, 15–64.

Aceñolaza, F.G. & Baldis, B. (1987) *The Ordovician System of South America. Correlation Chart and Explanatory Notes*. International Union of Geological Sciences, Publication No. 22, 68 pp, Ottowa.

Aceñolaza, F.G., Miller, H. & Toselli, A.J. (1988) The

Puncoviscana Formation (Late Precambrian–Early Cambrian). Sedimentology, tectonometamorphic history and age of the oldest rocks of NW Argentina. In: *The Southern Central Andes: Contributions to Structure and Evolution of an Active Continental Margin* (Eds Bahlburg, H., Breitkreuz, C. & Giese, P.). Lect. Notes Earth Sci. **17**, 25–38.

Allen, J.R.L. (1982) Sedimentary structures. *Developments in Sedimentology* **30A**. 593 pp.

Allen, P.A., Homewood, P. & Williams, G.D. (1986) Foreland basins: an introduction. In: *Foreland Basins* (Eds Allen, P.A. & Homewood, P.). Spec. Publs. Int. Ass. Sediment. **8**, 3–12.

Allmendinger, R.W., Ramos, V.A., Jordan, T.E., Palma, M. & Isacks, B.L. (1983) Paleogeography and Andean structural geometry, northwest Argentina. *Tectonics* **2**, 1–16.

Alonso, R., Viramonte, J. & Gutierrez, R. (1984) Puna Austral — Bases para el subprovincialismo geológico de la Puna Argentina. *IX Congr. Geol. Argent., Actas* **1**, 43–63.

Audley-Charles, M.G. (1986) Timor–Tanimbar Trough: the foreland basin of the evolving Banda orogen. In: *Foreland Basins* (Eds Allen, P.A. & Homewood, P.). Spec. Publs. Int. Ass. Sediment. 8, 91–102.

Baeza, L. & Pichowiak, S. (1988) Ancient crystalline basement provinces in the North Chilean Central Andes — relics of continental crust development since the Mid-Proterozoic. In: *The Southern Central Andes: Contributions to Structure and Evolution of an Active Continental Margin* (Eds Bahlburg, H., Breitkreuz, C. & Giese, P.). Lect. Notes Earth Sci. 17, 3–24.

Bahlburg, H. (1990) The Ordovician basin in the northern Puna of Argentina and Chile: geodynamic evolution from back-arc to foreland basin. *Geotektonische Forsch.*

Bahlburg, H., Breitkreuz, C. & Zeil, W. (1987) Paleozoic basin development in northern Chile (21°–27°S). *Geol. Rundsch.* **76**(2), 633–646.

Bahlburg, H., Breitkreuz, C. & Zeil, W. (1988) Geology of the Coquena Formation (Arenigian–Llanvirnian) in the NW Argentina Puna: constraints on geodynamic interpretation. In: *The Southern Central Andes: Contributions to Structure and Evolution of an Active Continental Margin* (Eds Bahlburg, H., Breitkreuz, C. & Giese, P.). Lect. Notes Earth Sci. 17, 71–86.

Bahlburg, H., Breitkreuz, C., Maletz, J., Moya, M.C. & Salfity, J.A. (1990) The Ordovician sedimentary rocks in the northern Puna of Argentina and Chile: new stratigraphical data based on graptolites. *Newsl. Stratigr.*

Barnes, C.R. (1986) The faunal extinction event near the Ordovician–Silurian boundary: a climatically induced crisis. In: *Global Bioevents* (Ed. Walliser, O.H.). Lect. Notes Earth Sci. 8, 121–126.

Beaumont, C., Keen, C.E. & Boutilier, R. (1982) A comparison of foreland and rift margin sedimentary basins. *Philos. Trans. R. Soc. London, Ser. A* **305**, 295–317.

Bell, C.M. (1987) The Late Paleozoic evolution of the Gondwanaland continental margin in northern Chile. *Geophys. Monogr., Am. Geophys. Union* **40**, 261–270.

Breitkreuz, C. (1986) Das Paläozoikum in den Kordilleren Nordchiles (21°–25°S). *Geotektonische Forsch.* **70**, 88 pp.

Breitkreuz, C., Bahlburg, H., Pichowiak, S. & Delakowitz, B. (1989) Volcanic events in the Paleozoic central Andes. *J. S. Am. Earth Sci.* **2**, 171–189.

Cloetingh, S. (1988) Intraplate stresses: a new element in basin analysis. In: *New Perspectives in Basin Analysis* (Eds Kleinspehn, K.L. & Paola, C.) pp. 205–230. Springer, New York.

Coira, B. (1973) Resultados preliminares sobre la petrología del ciclo eruptivo concomitante con la sedimentación de la Formación Acoite en la zona de Abra Pampa, provincia de Jujuy. *Rev. Asoc. Geol. Argent.* **23**, 85–88.

Coira, B. & Barber, E. (1987) Vulcanismo submarino Ordovícico (Arenigiano–Llanvirniano) del Rio Huaytiquina, Provincia de Salta, Argentina. *X Congr. Geol. Argent., Actas* **4**, 305–307.

Coira, B., Davidson, J., Mpodozis, C. & Ramos, V. (1982) Tectonic and magmatic evolution of the Andes of northern Argentina and Chile. *Earth Sci. Rev.* **18**, 303–332.

Covey, M. (1986) The evolution of foreland basins to steady state: evidence from the western Taiwan foreland basin. In: *Foreland Basins* (Eds Allen, P.A. & Homewood, P.). Spec. Publ. Int. Ass. Sediment. 8, 77–90.

Crook, K.A.W. (1974) Lithogenesis and geotectonics: the significance of compositional variation in flysch arenites (greywackes). In: *Modern and Ancient Geosynclinal Sedimentation* (Eds Dott, R.H. & Shaver, R.H.). Soc. Econ. Paleontol. Mineral. Spec. Publ. 19, 304–310.

Cuerda, A.J. (1973) Reseña del Ordovícico Argentino. *Ameghiniana* **10**, 272–312.

Dickinson, W.R. (1985) Interpreting provenance relations from detrital modes of sandstones. In: *Provenance of Arenites* (Ed. Zuffa, G.G.) pp. 333–361. Nato ASI Series C 148, Reidel, Dordrecht.

Fortey, R.A. (1984) Global earlier Ordovician transgressions and regressions and their biological implications. In: *Aspects of the Ordovician System* (Ed. Bruton, D.L.). Palaeontol. Contrib. Univ. Oslo, 295, 37–50.

Furque, G. & Cuerda, A.J. (1982) Extensión y edad de los movimientos de la Fase Guandacol en la Precordillera del oeste argentino. *V Congr. Latinoam. Geol., Actas* **1**, 191–200.

Garcia, A.F., Perez D'Angelo, E. & Ceballos, S.E. (1962) El Ordovícico de Aguada de la Perdiz, Puna de Atacama, provincia de Antofagasta. *Rev. Mineral.* **77**, 52–61.

Harrington, H. & Leanza, A.F. (1957) Ordovician trilobites of Argentina. *Spec. Publ. Univ. Kansas Press* **1**, 276 pp.

Hervé, F., Godoy, E., Parada, M.A., Ramos, V. Rapela, C., Mpodozis, C. & Davidson, J. (1987) A general view on the Chilean–Argentine Andes, with emphasis on their early history. *Geodyn. Ser.*, **18**, 97–113.

Hiscott, R.N., Pickering, K.T. & Beeden, D.R. (1986) Progressive filling of a confined Middle-Ordovician foreland basin associated with the Taconic orogeny, Quebec, Canada. In: *Foreland Basins* (Ed. Allen, P.A. & Homewood, P.). Spec. Publs. Int. Ass. Sediment. 8, 165–181.

Isaacson, P.E., Antelo, B. & Boucot, A.J. (1976) Implications of a Llandovery (Early Silurian) brachiopod fauna from Salta province, Argentina. *J. Palaeontol.* **50**, 1103–1112.

Kimura, T. (1966) Thickness distribution of sandstone beds and cyclic sedimentations in the turbidite sequences at two localities in Japan. *Bull. Earthquake Res. Inst. Tokyo* **44**, 561–607.

Kolla, V. & Macurda, D.B. (1988) Sea-level changes and timing of turbidity-current events in deep-sea fan systems. In: *Sea-level Changes: an Integrated Approach* (Eds Wilgus, C.K., Hastings, B.S., Kendall, C.G., Posamentier, H.W., Ross, C.A. & van Wagoner, J). Soc. Econ. Paleontol. Mineral., Spec. Publ. 42, 381–392.

Koukharsky, M. & Mirre, J.C. (1974) Nuevas evidencias del vulcanismo Ordovicico en la Puna. *Rev. Asoc. Geol. Argent.* **26**, 128–134.

Kumpa, M. & Sanchez, M.C. (1988) Geology and sedimentology of the Cambrian Grupo Mesón (NW Argentina). In: *The Southern Central Andes: Contributions to Structure and Evolution of an Active Continental Margin* (Eds Bahlburg, H., Breitkreuz, C. & Giese, P.). Lect. Notes Earth Sci 17, 39–54.

Lundberg, N. & Dorsey, R.J. (1988) Synorogenic sedimentation and subsidence in a Plio-Pleistocene collisional basin, Eastern Taiwan. In: *New Perspectives in Basin Analysis* (Eds Kleinspehn, K.L. & Paola, C.) pp. 265–280. Springer, New York.

Macdonald, D.I.M. (1986) Proximal to distal sedimentological variation in a linear turbidite trough: implications for the fan model. *Sedimentology* **33**, 243–259.

McPherson, J.G., Shanmugam, G. & Moiola, R.J. (1987) Fan-deltas and braid deltas: varieties of coarse-grained deltas. *Geol. Soc. Am. Bull.* **99**, 331–340.

Mendez, V., Navarini, A., Plaza, D. & Viera, V. (1973) Faja Eruptiva de la Puna Oriental. *V Congr. Geol. Argent., Actas* **4**. 89–100.

Mendez, V., Turner, J.C.M., Navarini, A., Amengual, R. & Viera, V. (1979) *Geología de la región noroeste, Provincias Salta y Jujuy, Republica Argentina*. Dir. Gral. Fab. Milit., 118 pp.

Mingramm, A., Russo, A., Pozzo, A. & Cazau, L. (1979) Sierras Subandinas. *II Symp. Geol. Regional Argent.* **1**, 95–138. Acad. Nac. Cieuc. Cordoba.

Mon, R. & Hongn, F. (1987) Estructura del Ordovícico de la Puna. *Rev. Asoc. Geol. Argent.* **42**, 31–38.

Moya, C. (1988) Lower Ordovician in the southern part of the Argentine Eastern Cordillera. In: *The Southern Central Andes: Contributions to Structure and Evolution of an Active Continental Margin* (Eds Bahlburg, H., Breitkreuz, C. & Giese, P.). Lect. Notes Earth Sci. **17**, 55–70.

Mutti, E. (1985) Turbidite systems and their relation to depositional sequences. In: *Provenance of Arenites* (Ed. Zuffa, G.G.) pp. 65–93. Nato ASI Series C 148, Reidel, Dordrecht.

Mutti, E. & Normark, W.R. (1987) Comparing examples of modern and ancient turbidite systems: problems and concepts. In: *Marine Clastic Sedimentology* (Eds Leggett, J.K. & Zuffa, G.G.) pp. 1–38. Graham & Trotman, London.

Mutti, E. & Ricci-Lucchi, F. (1972) Le torbiditi del l'Appennino settentrionale: introduzione all'analisi di facies. *Mem. Soc. Geol. Ital.* **11**, 161–199.

Mutti, E. & Ricci-Lucchi, F. (1975) Turbidite facies and facies associations. In: *Examples of Turbidite Facies and Associations from Selected Formations of the Northern Apennines* (Eds Mutti, E. *et al.*). Field Trip Guidebook A-11, 9th Int. Ass. Sediment. Congr., 21–36.

Niemeyer, H. (1989) El complejo ingneo-sedimentario del Cordón de Lila, region de Antofagasta: significado tectonico. *Rev. Geol. Chile* **16**, 163–181.

Omarini, R., Viramonte, J., Cordani, U., Salfity, J. & Kawashita, K. (1984) Estudio geochronológico Rb–Sr de la Faja Eruptiva de la Puna en el sector de San Antonio de los Cobres, Provincia de Salta. *IX Congr. Geol. Argent., Actas* **3**, 146–158.

Pickering, K.T. (1987) Deep-marine foreland basin and forearc sedimentation: a comparative study from the Lower Palaeozoic Northern Appalachians, Quebec and Newfoundland. In: *Marine Clastic Sedimentology* (Eds Leggett, J.K. & Zuffa, G.G.) pp. 190–211. Graham & Trotman, London.

Pickering, K.T., Stow, D.A.V., Watson, M. & Hiscott, R.N. (1986) Deep-water facies, processes and models: a review and classification scheme for modern and ancient sediments. *Earth Sci. Rev.* **22**, 75–174.

Posamentier, H.W. & Vail, P.R. (1988) Eustatic controls on clastic deposition, II: sequence and systems tract models. In: *Sea-level Changes: an Integrated Approach* (Eds Wilgus, C.K., Hastings, B.S., Kendall, C.G., Posamentier, H.W., Ross, C.A. & van Wagoner, J.). Soc. Econ. Paleontol. Mineral., Spec. Publ. 42, 125–154.

Ramos, V.A. (1972) El Ordovícico fosilífero de la Sierra de Lina, departamento de Susques, provincia de Jujuy, Republica Argentina. *Rev. Asoc. Geol. Argent.* **27**, 84–94.

Ramos, V.A. (1988) Late Proterozoic–Early Paleozoic of South America — a collisional history. *Episodes* **11**, 168–174.

Ricci-Lucchi, F. (1985) Influence of transport processes and basin geometry on sand composition. In: *Provenance of Arenites* (Ed. Zuffa, G.G.) pp. 19–45. Nato ASI Series C 148, Reidel, Dordrecht.

Salfity, J.A. (1985) Lineamentos transversales al rumbo andino en, el noroeste argentino. *IV Congr. Geol. Chileno, Actas* **2**(2), 119–137.

Salfity, J.A., Malanca, S., Brandan, M.E., Monaldi, C.R. & Moya, C. (1984) La Fase Guanadacol en el norte de la Argentina. *IX Congr. Geol. Argent., Actas* **1**, 555–567.

Schwab, K. (1973) Die Stratigraphie in der Umgebung des Salar de Cauchari (NW Argentinien). Ein Beitrag zur erdgeschichtlichen Entwicklung der Puna. *Geotektnische Forsch.* **43**, 168 pp.

Sclater, J.G. & Christie, P.A.F. (1980) Continental stretching: an explanation of the post-Mid Cretaceous subsidence of the central North Sea Basin. *J. Geophys. Res.* **85**, 3711–3739.

Shackleton, R.M., Ries, A.C., Coward, M.P. & Cobbold, P.R. (1979) Structure, metamorphism and geochronology of the Arequipa Massif of coastal Peru. *J. Geol. Soc. London* **136**, 195–214.

Turner, J.C.M. (1960) Estratigrafía de la Sierra de Santa Victoria y adyaciencias. *Bol. Acad. Nac. Cienc.* **41**, 163–196.

Turner, J.C.M. & Mendez, V. (1975) Geología del sector oriental de los departamentos de Santa Victoria e Iruya, Provincia de Salta, Republica Argentina. *Acad. Nac. Cienc.* **51**, 11–24.

Turner, J.C.M. & Mendez, V. (1979) Puna. *II Symp. Geol. Regional Argent.* **1**, 13–56. Acad. Nac. Cieuc. Cordoba.

Walker, R.G. (1978) Deep-water sandstone facies and ancient submarine fans: models for exploration for stratigraphic traps. *Bull., Am. Assoc. Petrol. Geol.* **62**, 932–966.

Walker, R.G. (1984) Turbidites and associated coarse clastic deposits. In: *Facies Models* (Ed. Walker, R.G.). Geosci. Can. Reprint Ser. 1, 91–104.

Walker, R.G. (1985) Mudstones and thin-bedded turbidites associated with the Upper Cretaceous Wheeler Gorge conglomerates, California: a possible channel-levee complex. *J. Sediment. Petrol.* **55**, 279–290.

White, J.D.L. & Busby-Spera, C. (1987) Deep marine arc apron deposits and syndepositional magmatism in the Alisitos Group at Punta Cono, Baja California, Mexico. *Sedimentology* **34**, 911–927.

Wilde, P. & Berry, W.B.N. (1984) Destabilization of the oceanic density structure and its significance to marine 'extinction' events. *Palaeogeogr. Palaeoclimatol. Palaeoecol.* **48**, 143–162.

Wilgus, C.K., Hastings, B.S., Kendall, C.G., Posamentier, H.W., Ross, C.A. & van Wagoner, J. (1988) Sea-level changes: an integrated approach. *Soc. Econ. Paleontol. Mineral., Spec. Publ.* **42**, 407 pp.

Willner, A., Lottner, U. & Miller, H. (1987) Early Paleozoic structural development in the NW Argentine basement of the Andes and its implications for geodynamic reconstructions. *Geophys. Monogr. Am. Geophys. Union* **40**, 229–239.

*Spec. Publs. int. Ass. Sediment.* (1991) **12**, 485–511

# Sedimentation and tectonics in the Dinantian limestones of South Wales

A.T.S. RAMSAY

*Department of Geology, University of Wales College of Cardiff, P.O. Box 914, Cathays Park, Cardiff, CF1 3YE, UK*

## ABSTRACT

The Dinantian succession in South Wales was deposited on the southward-sloping hanging wall of a half-graben, which probably formed by northerly downthrow on a major fault system in the Bristol Channel. This basin formed and evolved in a N–S orientated extensional stress field, which pervaded Britain during the Dinantian. Deposition was controlled by movements on basement faults or fault complexes that formed active hinge lines with E–W, NE–SW, NNW–SSE, and N–S trends, and by a buoyant palaeohigh that developed along the Carreg Cennen Disturbance. The hinge-lines and palaeohigh influenced the thickness of the succession, and patterns of onlap during transgression. Erosion was enhanced on uplifted fault footwalls and over the active Caledonoid-trending palaeohigh. Variations in the geometry of the basin-margin floor (ramp or platform) are attributed to variations in the regional pattern of asymmetrical subsidence and to rifting phases associated with E–W orientated structures. The N–S stress field, half-graben basin, and tectonic activity on the south-sloping hanging wall developed in response to subduction and/or lithospheric loading associated with an orogeny to the south.

The pattern of Dinantian cyclicity is attributed to the interaction between uplift, episodic (pulsed subsidence) sedimentation rates, and changes in relative base level, which were probably tectonic and related to rifting and subsidence. The process of marine transgression involved rifting subsidence and shoreface and barrier retreat. Regressive phases are progradational and contain barrier and back barrier deposits when progradation coincided with high sedimentation rates and relative rise in sea level. Progradation and high sedimentation rates during relatively static base levels produced regressive sequences minus a back-barrier lagoonal–peritidal facies.

## INTRODUCTION

This account is based on the investigation of the Dinantian limestones that outcrop around the margin of the main South Wales Coalfield. The location of the study area and the regional names referred to in the text are shown in Fig. 1.

The Dinantian rocks of South Wales were deposited on a broad shelf situated between the Cornubian Basin to the south and a northern St George's land massif (George *et al.*, 1976). Marine conditions began in this area with the late Devonian marine transgression (Wilson *et al.*, 1988) and the contact between late Devonian continental clastics and lower Dinantian marine deposits is both transitional and unconformable (George, 1958; Gayer *et al.*, 1973; George *et al.*, 1976; Waters & Lawrence, 1987). The onset of conditions conclusive to the accumulation of shelf carbonates was marked by deposition of the mixed clastic–carbonate sequences of the Lower Limestone Shale. Asymmetrical subsidence of the basin floor during the Dinantian produced a southwards thickening sediment wedge that onlaps against the southern margin of St George's Land.

The succession is characterized by cyclic sequences in which shallow-water barrier and shoal grainstones and peritidal deposits alternate with deeper water bioclastic limestones. This cyclicity was first documented by Dixon (Dixon & Vaughan, 1912) in the Dinantian succession of Gower. Robertson & George (1929) and George (1933) provided further evidence from the Carboniferous Limestone of South

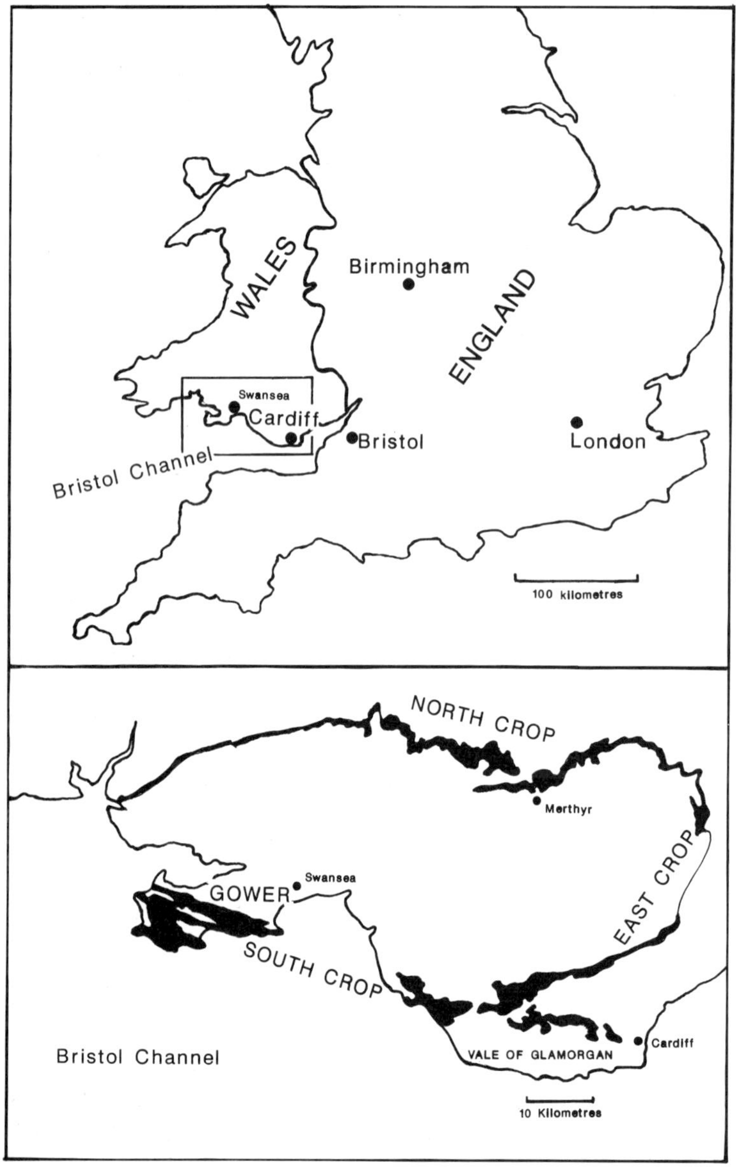


**Fig. 1.** Location of study area showing the position of regional names that are used in the text.

Wales. Alternative views concerning the origin of these cycles were proposed by Ramsbottom (1973, 1979) who favoured a eustatic mechanism, by George (1974,1978) who invoked the influence of local tectonic controls on sedimentation, and by Wilson *et al.* (1988) who proposed a combination of eustatic and tectonic processes. The above arguments also pertain to the origin of pedogenic horizons, which represent repeated episodes of emergence and subaerial erosion during the Dinantian. Some of these horizons coincide with stage boundaries, for example between the Chadian and Arundian, Holkerian and Asbian, and Asbian and Brigantian, and extend across the basin; others are restricted to its northern margin.

The E–W orientated Dinantian basins of the southwest Province of George (1958) probably formed by crustal extension behind a southern orogenic belt. Southwest Province basins were similar to the extensional graben and half-graben described by Miller & Grayson (1982), Grayson & Oldham (1987), and Leeder (1982,1987) for the region north of St George's Land. The history of formation of intrashelf extensional basins varied within the southwest Province. In southwest England a phase of extension during the early to mid-Devonian (late Caledonian movements) produced E–W orientated basins that contain a succession of marine middle Devonian to Visean deposits (Sellwood & Thomas, 1988). In the South Wales area, late Caledonian (early to mid-Devonian) uplift and erosion was followed by a late Devonian phase of extension and basin formation. The trend of the Dinantian depositional slope (at right angles to the trend of major Caledonoid structures) was probably established during uplift, and enhanced by extensional tectonics. The basin margin of the South Wales area probably represents the landward component of a hanging wall of a half-graben that formed by northerly downthrow on the Bristol Channel Fault Zone. Deposition on the hanging wall was accompanied by the reactivation of E–W, NE–SW, NNW–SSE, and N–S orientated basement structures.

The lithostratigraphy used in this account is expressed within the chronostratigraphical framework of stages proposed by George *et al.* (1976). It also incorporates recent refinements in the interpretation of the South Wales Dinantian succession that are contained in work reported by Riding & Wright (1981), Burchette (1987), Waters & Lawrence (1987), Barclay *et al.* (1988), and Wilson *et al.* (1988).

The construction of representative and accurate N–S profiles for the Carboniferous Limestone across the South Wales Coalfield is excluded by the Dinantian outcrop pattern (Fig. 2) and the absence of subsurface data from boreholes or continuous seismic reflection profiles. In this study a broad impression of the influence of active structures on thickness and erosion patterns for the Dinantian succession has been achieved by comparing sequences across individual structures and by combining thickness and facies data from localities on the North, East, and South Crops (Fig. 1) within single composite sections (Fig. 2). Sections for the Courceyan–Chadian, Arundian, and Holkerian represent a compilation of data from the North and East Crops and Vale of Glamorgan (north of the Cardiff–Cowbridge anticline), which are projected on to a common NW–SE line, and from the Gower

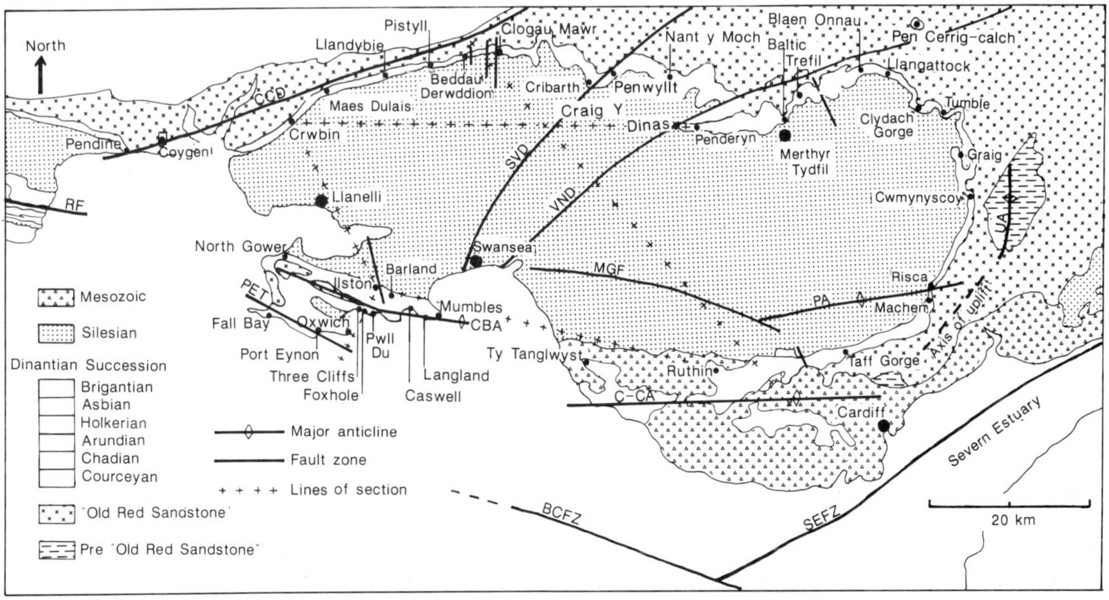


**Fig. 2.** The geology of the main South Wales Coalfield showing the locations of the major exposures and the lines of section on to which these are projected. Named structures are: BCFZ, Bristol Channel Fault Zone; CBA, Caswell Bay anticline; C–CA, Cardiff–Cowbridge anticline; CCD, Carreg Cennen Disturbance; MGF, Moel Gilau Fault; PA, Pontypridd anticline; PET, Port Eynon thrust; SEFZ, Severn Estuary Fault Zone; SVD, Swansea Valley Disturbance; UA, Usk anticline; VND, Vale of Neath Disturbance.

sections, which are projected along strike within a tectonically restored section. The line of section across the main coalfield is at right angles to the NE–SW depositional strike obtained from isopachyte data for the Arundian Llanelli Formation. Namurian erosion and overstep on the east crop has resulted in the erosion of the Asbian–Brigantian succession and much of the Arundian and Holkerian sequence. The Arundian and Holkerian successions were reconstructed by extrapolating the rates of thickening of these successions between the North Crop and the northern limb of the Cardiff–Cowbridge anticline. The Asbian–Brigantian sections combine data from the North Crop between Penderyn and Crwbin and the Gower sections and between the Vale of Glamorgan and Gower.

## TECTONICS AND SEDIMENTATION

In late Devonian, Dinantian, and Silesian times in South Wales there was a strong connection between the history of tectonically active structures, basin evolution, and sedimentation.

### Structural trends

Late Palaeozoic syndepositional activity can be demonstrated for present-day structures (Fig. 2) with the following trends.

**1** WNW–ESE-trending structures include the Ritec Fault (Sullivan, 1966), probably the Moel Gilau Fault Complex (Owen & Weaver, 1983), the Cefn Bryn anticline, and Port Eynon thrust.

**2** ENE–WSW-trending structures include the Cardiff–Cowbridge anticline (Waters & Lawrence, 1987; Wilson *et al.*, 1988) and the Pontypridd anticline. The latter runs into the Moel Gilau Fault via the Maesteg anticline (Owen & Weaver, 1983) and defines the trend of a major structural hinge line that has been proposed to explain the thickness variation in Dinantian and Silesian deposits across a structure at Risca.

**3** NE–SW-trending structures, traditionally described as Caledonoid, include the Carreg Cennen, Swansea Valley, and Vale of Neath Disturbances. All are defined by narrow zones of aligned fault-fold complexes (Owen & Weaver, 1983). The influence of the Vale of Neath Disturbance on late Dinantian and Namurian sedimentation has been discussed by Owen (1954), Jones (1974), and Ramsay (1989). Wilson *et al.* (1988) postulated that intra-Dinantian dextral strike-slip movements along a Caledonoid Severn Estuary Fault Zone were related to the tectonic history of the Cardiff–Cowbridge anticline.

**4** There is a pronounced swarm of cross faults that are orientated N–S or NNW–SSE in western and central areas of the coalfield and NW–SE in eastern areas. On the North Crop of the coalfield this fault system influenced late Dinantian sedimentation (Ramsay, 1989) and the pattern of Namurian overstep. In the Vale of Glamorgan, Wilson *et al.* (1988) demonstrated the influence of major NW–SE structures (e.g. the Miskin Fault) on Namurian overstep.

**5** N–S orientated Usk anticline, which according to George (1956b) and Owen (1970) passes into the Cardiff–Cowbridge anticline. Wilson *et al.* (1988), in a recent structural synthesis, suggested that these structures were not continuous. This view is supported by the differences between the synsedimentary histories of these structures. The Cardiff–Cowbridge anticline was active from the Devonian to Arundian (Wilson *et al.*, 1988) whereas the Usk anticline was probably not active before the Holkerian (Barclay & Jones, 1978; Wilson *et al.*, 1988).

### Evidence for synsedimentary tectonics

The evidence for tectonically controlled sedimentation is based on the following criteria.

**1** The substantial difference in thickness and facies patterns between the North Crop and South Crop successions, which reflects the N–S asymmetry in basin subsidence.

**2** The occurrence of lateral changes in the thickness of the succession on either side of a fault. (NB Changes in facies rarely accompany changes in thickness.)

**3** Variations in thickness between successions on the opposite limbs of the Cardiff–Cowbridge (Wilson *et al.*, 1988) and Cefn Bryn anticlines. In both cases, the length of the depositional slopes, calculated on the basis of estimates of tectonic shortening (Vale of Glamorgan 15–20%, Gower 25–30%), is too short to explain these variations.

**4** The occurrence of tectonic hiatuses, which involved increased erosion on an uplifted fault footwall or on the uplifted components of fault blocks characterized by scissors movement. In these situations post-rift deposits either rest on formations of different ages or on younger and older units of the same formation.

The evolution of basin floor topography over a large area of Dinantian South Wales was probably controlled by movements on concealed basement faults or fault complexes. These movements were expressed at the surface by active hinge lines that separated subdued palaeohighs on the areas of footwall uplift from shallow depressions created by hanging wall subsidence. The occurrence of the hinge lines and their tectonic origin is indicated by changes in the thickness of the succession across present structural trends. Deposition on the subdued, hinge-line generated, sea-floor topography produced lateral changes in facies on a scale of kilometres to tens of kilometres; facies changes across individual faults are rare. Evidence for the interaction between sedimentary and tectonic processes at active fault scarps, such as slumped horizons and limestone and conglomerate breccias of mass flow origin (Gawthorpe, 1986; Gutteridge, 1987), is not recorded in the studied area.

During phases of enhanced tectonic activity and uplift, hinge-line influenced erosion patterns developed with increased erosion on the fault footwall. The rotation and tilting of concealed fault blocks also influenced erosion and thickness patterns within the basin.

## RELATIONSHIP BETWEEN E–W-TRENDING STRUCTURES AND SEDIMENTATION

Structures with WNW–ESE trend are grouped under this heading. These were active from the Late Devonian to Silesian and were associated with the rifting and asymmetrical subsidence of the basin floor. During the Dinantian, active hinge-lines associated with these major basement structures influenced thickness and erosion patterns and probably contributed to the development of a cyclic succession. From the Courceyan to the Holkerian the activity of these structures occurred in conjunction with the development of a tectonically buoyant NE–SW orientated structure that developed between the Carreg Cennen and Vale of Neath Disturbances. During the Asbian and Brigantian, activity along E–W-trending structures was accompanied by movements along the Caledonoid Vale of Neath Disturbance and a series of cross faults.

The widespread Late Devonian to early Dinantian phase of rifting is documented in Ireland (Diemer & Bridge, 1988; Naylor *et al.*, 1989; Nolan, 1989) and in basins north of St George's Land (Gawthorpe *et al.*, 1989).

### Early Courceyan

The early Courceyan of this study comprises two major shoaling cycles capped by oolitic grainstone (Fig. 3). The deposition of the lower Courceyan was initiated during a phase of extension accompanied by subsidence and marine transgression. In the Gower area the effects of this extensional phase are indicated by significant thickening of the Lower Limestone Shale across the Cefn Bryn anticline from approximately 130 m on its northern limb to approximately 180 m on its southern limb. Evidence for the activity of other E–W structures is unclear, although the increased thickness of the Brofiscin Oolite (=Pwll-y-Cwm Oolite) between Cwmynyscoy and Risca may be associated with an active hinge-line approximately 1 km to the north of Risca. Over most of the studied area the N–S asymmetrical subsidence influenced facies distribution, the pattern of transgression, and the thickness of the succession. Uplift along the North Crop associated with contemporaneous movements along NE–SW orientated Caledonoid structures resulted in localized emergence and pedogenesis.

On Gower the contact between the marine Cefn Bryn Shales and underlying fluvial quartz conglomerates (QRS) is transitional. The coarse conglomerates are overlain by unfossiliferous, fine sandstones and siltstones, which pass upward into 5 m of lagoonal, ostracod-rich, bioturbated, calcareous mudstone with plant fragments. The lagoonal sequence is separated from a succession of offshore mudstones by a 2 m-thick oolitic grainstone with a sharp planar erosive base. This succession was probably generated by shoreface and barrier retreat (Fischer, 1961; Swift, 1968), and represents a transgressive lagoonal fill overlain by a ravinement surface, lower shoreface oolitic sands, and offshore shelf mudstones. This type of sequence is indicative of a slow rise in relative sea level accompanied by enhanced shoreface erosion (Fischer, 1961; Swift, 1968; Heward, 1981). In the Cardiff area, the contact between the Lower Limestone Shale Group and fluvial sandstones is also transitional (Gayer *et al.*, 1973; Burchette, 1987; Waters & Lawrence, 1987). The transgression was complex and the deposition of the Lower Limestone Shale commenced with

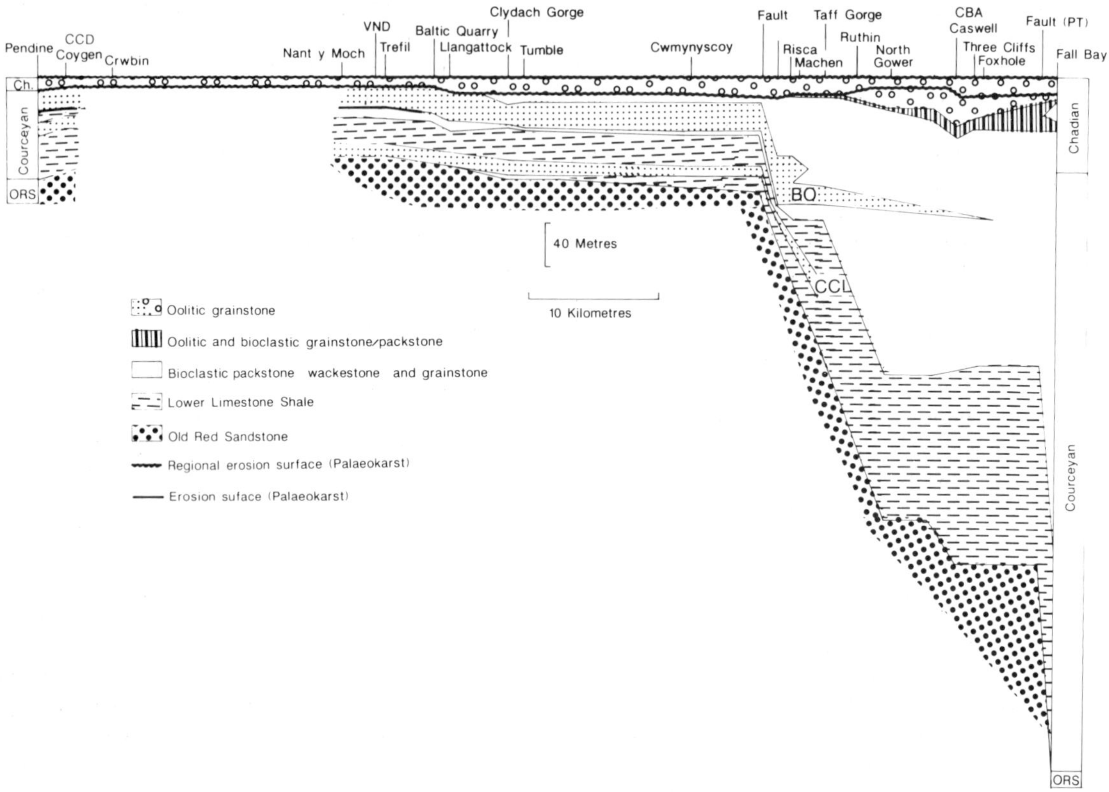


**Fig. 3.** The late Courceyan to Chadian succession (prior to Arundian and Holkerian erosion) shown in a composite section between the North Crop and south Gower. Named formations are: BO, Brofiscin Oolite; CCL, Castell Coch Limestone. Named structures are CBA, Caswell Bay anticline; CCD, Carreg Cennen Disturbance; PT, Port Eynon thrust.

lagoon development and two phases of shoaling followed by ravinement, and the establishment of a muddy shelf sequence with a rich marine fauna (Burchette, 1987; Waters & Lawrence, 1987).

On Gower the few exposures of the Cefn Byrn Shales above the transgressive oolite consist of offshore mudstones. Data from other parts of the studied area reveal a relationship between depositional slope, water depth, and the thickness and nature of the sedimentary facies. In the south (Cardiff area; southern Vale of Glamorgan), transgression and a relative rise in sea level generated an upward-deepening sequence 40 m thick of neritic shelf mudstones with clastic and carbonate storm sheets (Burchette, 1987; Waters & Lawrence, 1987). Towards the north, the equivalent succession is represented increasingly by oolitic grainstones. On the North Crop, the oolitic Castell Coch Formation transgressed, probably with onlap, the Quartz Conglomerate Group of the Old Red Sandstone (Burchette, 1987; Barclay *et al.*, 1988). At Coygen and Trefil, pedogenic horizons in the Castell Coch Oolite developed in response to movement on a Caledonoid structure.

The northward-transgressive oolite also spread southwards across the shelf mudstones and is approximately 20 m thick at Taff Gorge and 9 m thick at Barry (Waters & Lawrence, 1987). This regional shoaling event was attributed by Burchette (1987) to the southward progradation of oolite shoal complexes rather than to major marine regression. The change from a transgressive to a regressive mode probably reflects a reduction in the rate of relative sea-level rise. Renewed deposition of Lower Limestone Shale facies records transgression and the establishment of neritic shelf conditions (Burchette, 1987; Waters & Lawrence, 1987). These were characterized by deposition of an interbedded succession of dark mudstones and sheet-like beds of calcareous sandstone and bioclastic packstone that were probably emplaced during storm events. Evidence of shoaling at the top of the Lower Limestone Shale is provided by an increase in the number and thickness of storm-generated beds, a decrease in the interbedded mud component, and eventual amalgamation of the storm sheets in the beds of the thicker Barry Harbour Limestone in the south (Waters & Lawrence, 1987) and the equivalent thinner Sychnant Dolomite (Barclay *et al.*, 1988) on the East and North Crops. Between Baltic Quarry and the South Crop, the succession shoals from offshore muds, through near wave-base storm sheets to the above-wave-base oolitic deposits of a southward prograding oolitic shoal (Waters & Lawrence, 1987). The variation in the thickness of this unit between Cwmynyscoy (7 m) and Risca (22 m) may be associated with an active hinge line. The oolite does not extend into the southern Vale of Glamorgan (Waters & Lawrence, 1987). From the Baltic Quarry to Coygen the upward-shoaling, offshore deposits of the Sychnant Dolomite are truncated by an emergent surface and associated palaeosol. The contact between this horizon and the overlying Pwll-y-Cwm Oolite is clearly transgressive. The Pwll-y-Cwm Oolie also terminates at an emergent surface and palaeosol. The nature of the sequence between the Baltic Quarry and Coygen is attributed to local uplift associated with a NE–SW-trending palaeohigh.

**Late Courceyan to Chadian**

Variations in the thickness of this succession (Fig. 4) are associated with synsedimentary activity along an E–W structure at Risca, the Cardiff–Cowbridge anticline (Waters & Lawrence, 1987; Wilson *et al.*, 1988), the Cefn Bryn anticline, and the Port Eynon thrust.

The regional thickness, facies distribution, number of cycles, and pattern of transgression were strongly influenced by asymmetrical subsidence. On the North Crop, continued localized movement associated with a Caledonoid structure determined the distribution of emergent surfaces and palaeosols. The vertical and lateral facies distribution patterns for this interval indicate deposition on a northward-shoaling ramp (Wright, 1986,1987; Ramsay, 1987). Deposition in offshore areas was influenced predominantly by the interplay between storm and fair weather processes (Ramsay, 1987). Oolitic shoal facies are particularly evident north of the Cardiff–Cowbridge anticline and are concentrated along the North Crop (Wright, 1986).

From Machen to the Clydach Gorge this interval contains two upward-coarsening cycles (Fig. 4). Each commences with transgressive offshore bioclastic packstones with storm-generated bedding that pass upwards via thicker-bedded packstone/grainstones into progradational oolitic grainstones. The thickness of these cycles is unaffected by the fault at Risca. In the sequence between Llangattock and Trefil the oolitic component of the first cycle (Blaen Onnen Oolite) is transgressive and overlies an erosion surface and palaeosol. This is transgressed by the oolite of the second cycle. The variation between the East Crop and North Crop sequence (i.e. the distribution of palaeosols and the transgressive or progradational emplacement of the oolites) is attributed to asymmetrical regional subsidence, coupled with contemporaneous movements of the North Crop Caledonoid-trending palaeohigh.

Between Llangattock and Trefil (Fig. 4), an erosion surface and palaeosol (regional unconformity) separate the Courceyan oolites of the second cycle from the peritidal mudstones and/or overlying Gilwern Oolite, which were assigned a Chadian age (George *et al.*, 1976). South of Llangattock the Gilwern Oolite rests unconformably on Courceyan offshore deposits. This relationship can be traced to the Taff Gorge.

The deposition of continuous Courceyan to Chadian sequences without prominent shoaling events in the Vale of Glamorgan and Gower (Fig. 4) is possibly the result of extension-enhanced subsidence, associated with hinge lines developed along the trends of the Cardiff–Cowbridge anticline, Cefn Bryn anticline, and Port Eynon thrust. In Gower (Ramsay, 1987) and the Vale of Glamorgan (Waters & Lawrence, 1987) the offshore deposits of the Langland Dolomite and Friars Point Limestone shoal upwards. In both areas the Chadian oolites are progradational and have transitional and conformable contacts with the underlying offshore facies. The oolites contain upward-shoaling cycles consisting of lower shoreface cross-bedded packstones and upper planar laminated oolite shoal deposits. On Gower the lateral variation in the number of cycles may be autocyclic in origin (Ramsay, 1987). Within the Gower sequence the second cycle is truncated by an erosion surface with a marked topography and an associated palaeosol (Fig. 4). This surface probably correlates with the erosion surface and palaeosol recorded at the top of the third shoaling cycle of the

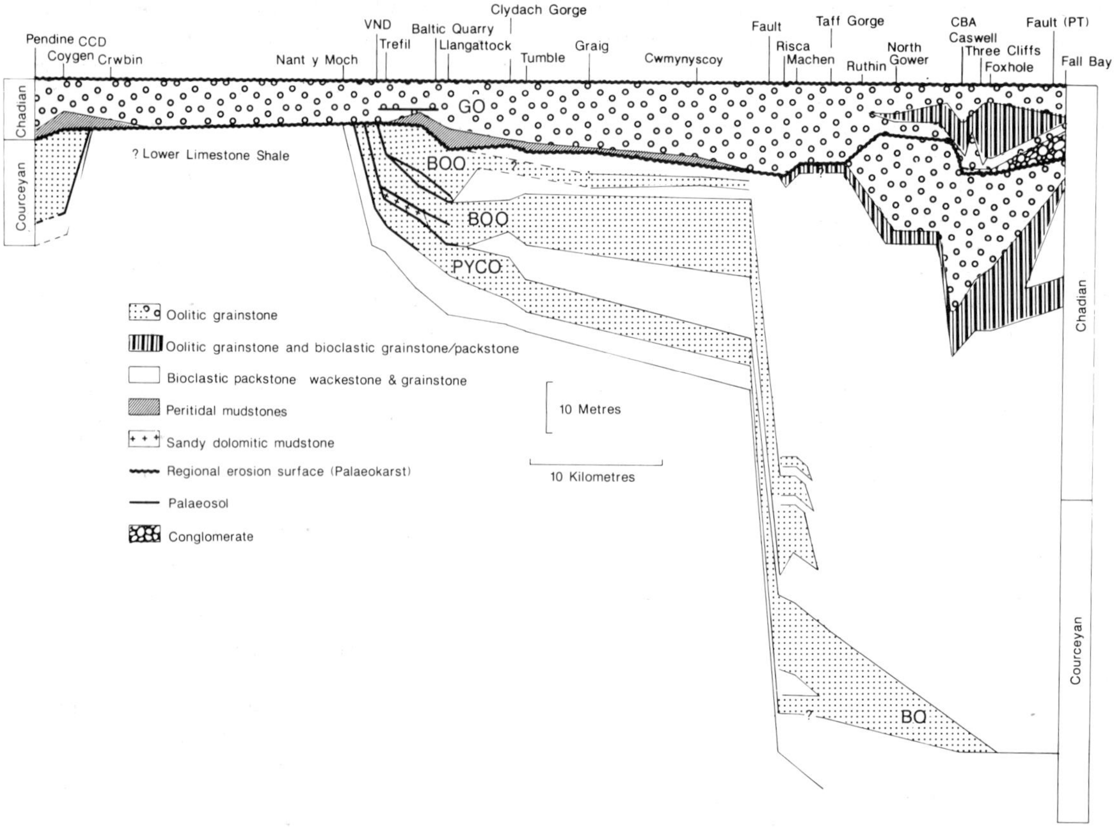


**Fig. 4.** The late Courceyan to Chadian succession (prior to Arundian and Holkerian erosion) shown in a composite succession between the North Crop and south Gower. Named formations are: BO, Brofiscin Oolite; BOO, Blaen Onnen Oolite; GO, Gilwern Oolite; PYCO, Pwll-y-Cwm Oolite. Named structures are: CBA, Caswell Bay anticline; CCD, Carreg Cennen Disturbance; PT, Port Eynon thrust; VND, Vale of Neath Disturbance. The variation in ornament is used to emphasize the relationship between the Courceyan (fine stipple) and Chadian (open circle) oolites.

St Lythan's Borehole (Waters & Lawrence, 1987) and with the major non-sequence that can be traced between the Taff Gorge and Coygen.

This unconformity is tectonic in origin. There is evidence of footwall uplift and enhanced erosion associated with the fault at Risca (Fig. 4). On the North Crop the Gilwern Oolite oversteps older units of the underlying Courceyan oolites between Baltic Quarry and Nany-y-Moch. This relationship is the product of uplift associated with the Caledonoid-trending palaeohigh. Late Chadian uplift and erosion in South Wales may be part of a regional tectonic event. Evidence of late Chadian tectonism is recorded from the north of England (Gawthorpe, 1986) and Ireland (Nolan, 1989).

In South Wales, uplift and erosion were followed by marine transgression and deposition of a sequence in which the facies pattern is probably the product of coastal onlap followed by progradation (Fig. 4). In Gower the transgressive–regressive nature of the succession is distinct and a transgressive sheet of limestone conglomerate deposited in the wake of the retreating barrier is overlain by a coarsening upward sequence of cross-bedded shoreface packstones and prograding oolitic grainstones. In the north, back-barrier lagoonal mudstones preserved beneath the northward-migrating oolites are overstepped between the Baltic Quarry and Trefil, and the Gilwern Oolite rests directly on the Curveyan Blaen Onnen Oolite. The transgressive facies pattern on the North Crop probably reflects the imprint of the active Caledonoid trending palaeohigh on the depositional slope. At Coygen and Pendine the transgressive sequence also contains back-barrier deposits.

The late Chadian oolites are truncated by a widespread erosion surface (Fig. 4), identified as a paleokarst (Riding & Wright, 1981; Wright, 1982,1984a; Ramsay, 1987). This horizon is associated with a regolith of rounded and angular fragments of oolite and laminar calcrete set in a clay matrix (Ramsay, 1987; Waters & Lawrence, 1987). The regolith profile thickens northward, and is a distinctive feature of North Crop sequences. This surface is absent in south Gower and the southern Vale of Glamorgan (Waters & Lawrence 1987; Wilson *et al.*, 1988). In South Wales this widespread palaeokarst platform probably formed in response to uplift and N–S tilting, which was possibly accompanied by faulting.

In Ireland this unconformity is associated with movements on northeast-trending faults (Philcox, 1989; Philcox *et al.*, 1989) and the graben geometry of the Dublin Basin (Nolan, 1989). The late Chadian to Arundian break also has a tectonic origin in the southern Lake District and parts of the Bowland Basin (Gawthorpe *et al.*, 1989).

**Arundian**

The proposed model for Arundian tectonics and sedimentation on the East Crop is based on an extrapolation of the rate of thickening of the Arundian sequence between the North Crop and the fault at Risca. In this model (Fig. 5), variation in thickness, facies patterns, and depositional history are attributed to extension and footwall uplift associated with the fault at Risca and a postulated active E–W hinge axis between the Taff Gorge and Ruthin.

Arundian deposition was probably initiated during a phase of extension, subsidence, and tilting accompanied by a relative rise in sea level. The combined effects of tilting and development of active hinge lines produced a ramp, where significant thicknesses of lagoonal–peritidal carbonates accumulated on landward areas of footwall uplift. The active E–W orientated hinge line at Risca probably restricted the lateral extent of the two lower cycles of the Arundian succession. The three Arundian cycles (Fig. 5) are as follows.

*Cycle 1.* This is an upward shoaling cycle initiated during extension and associated marine transgression. In the southern Vale of Glamorgan and south Gower this event resulted in a marked change in facies from Chadian oolites to open shelf bioclastic limestones of the High Tor Limestone (Ramsay, 1987; Waters & Lawrence, 1987). In the northern Gower sections and to the north of the Cardiff–Cowbridge anticline, flooding of the palaeokarst at the top of the Chadian oolites was followed by accumulation of transgressive back-barrier peritidal and lagoonal deposits of the Caswell Bay Mudstone. The strongly erosional contact between these deposits and the open marine High Tor Limestone facies is attributed to shoreface erosion and barrier retreat (Riding & Wright, 1981; Ramsay, 1987). Continued transgression resulted in establishment of a typical ramp succession in which nearshore lagoonal–peritidal facies pass laterally downramp into deeper water deposits (Ahr, 1973; Read, 1985; Wright, 1986). Maximum transgression resulted in the landward migration of shoal facies (oolitic and bioclastic grainstones) across the back barrier flats. This was followed by emergence, development of a palaeokarst between Risca and the Taff Gorge, and shoaling across the rest of the basin, either in response to progradation or aggradation.

*Cycle 2.* This typical ramp succession also represents deposition during a transgressive to 'regressive' phase. In the landward back barrier area the top of this cycle is defined by a palaeokarst horizon. This is possibly continuous with the palaeokarst erosion surface that terminates the upward-shoaling sequence in the offshore basin deposits between Ruthin and the Three Cliffs Bay.

*Cycle 3.* A third phase of extension and subsidence followed by continued northward transgression resulted in the overstep of cycle 2 deposits and establishment of carbonate deposition in a ramp setting that extended over the whole of the studied area. In the 'coastal' succession of the Llanelly Formation, drowning was accompanied by the deposition of a wedge-like succession of calcareous sandstones and siltstones with a thin calcareous quartz conglomerate at its base. At Llangattock the unit is approximately 4 m thick and consists of sheet-like sandstones 30–40 cm thick with interbedded calcareous siltstones. In the Trefil section this unit is 3 m thick and consists predominantly of lagoonal–peritidal carbonates with pedogenic features and thin, interbedded lenticular and sheet-like sands. In the Clydach area the unit is approximately 1 m thick and lagoonal peritidal carbonates predominate. The sandy unit may represent a minor bay-fill sequence associated with the

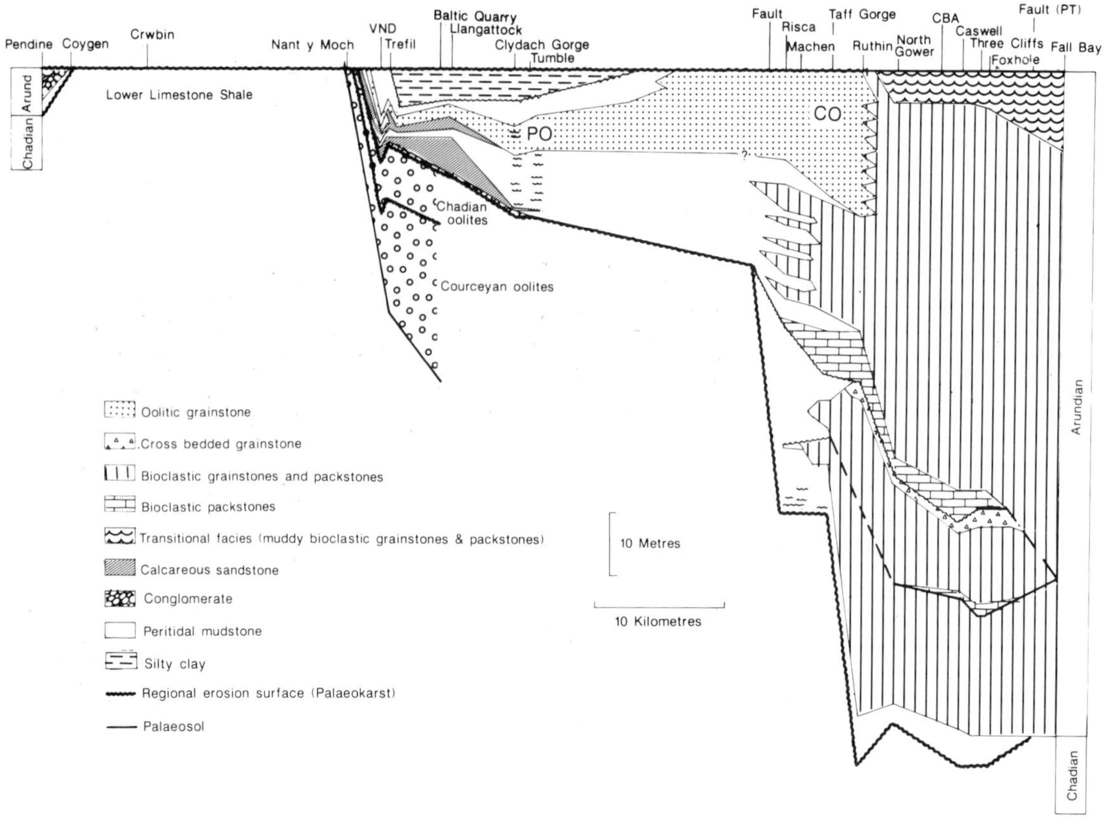


**Fig. 5.** The Arundian succession shown in a composite section between the North Crop and south Gower. Named formations are: CO, Cefnyrhendy Oolite; PO, Penllwyn Oolite. Named structures are: CBA, Caswell Bay anticline; PT, Port Eynon thrust; VND, Vale of Neath Disturbance.

restricted development of a fluvial delta. The sand sheets are possibly crevasse-splay lobes that formed in a depositional system analogous to the bay-fill sequences of the Yoredale succession (Elliott, 1974). These deposits were overstepped by lagoonal–peritidal carbonate deposits during transgression and coastal onlap and the northward retreat of the clastic shoreline. Wright (1986) interprets the siliciclastic unit as a product of fluvial erosion that accompanied a fall in base level.

The coastal carbonates consist predominantly of a succession characterized by 'poorly ordered facies transitions' (Wright, 1986) in which sheets of marine-derived carbonates (oolitic grainstones, peloidal grainstone/packstones, and granular conglomerates) and siliciclastic limestones are interbedded with lagoonal/peritidal mudstones and pedogenic clays. The marine carbonates probably represent storm-emplaced washovers of barrier and shallow subtidal sediments. The siliciclastic component indicates continued fluviodeltaic activity. Typical upward-shoaling lagoon–peritidal sequences of James (1979) are rare (Wright, 1984b,1986). The deposition of the Penllwyn Oolite (Wright, 1986), here equated with the Cefnyrhendy Oolite of the Vale of Glamorgan, may record the maximum northward migration of an oolitic barrier. Thereafter on the North Crop, siliclastic floodplain deposition (Wright, 1986) followed in the wake of the southward prograding barrier–back barrier sequences. The floodplain sequence is truncated by an erosion surface that defines the Arundian–Holkerian boundary on the North Crop (Figs 5, 6).

In the south, the Arundian succession varies between the Vale of Glamorgan (Taff Gorge, Ruthin) and Gower (Fig. 5). In the Vale, transgression was

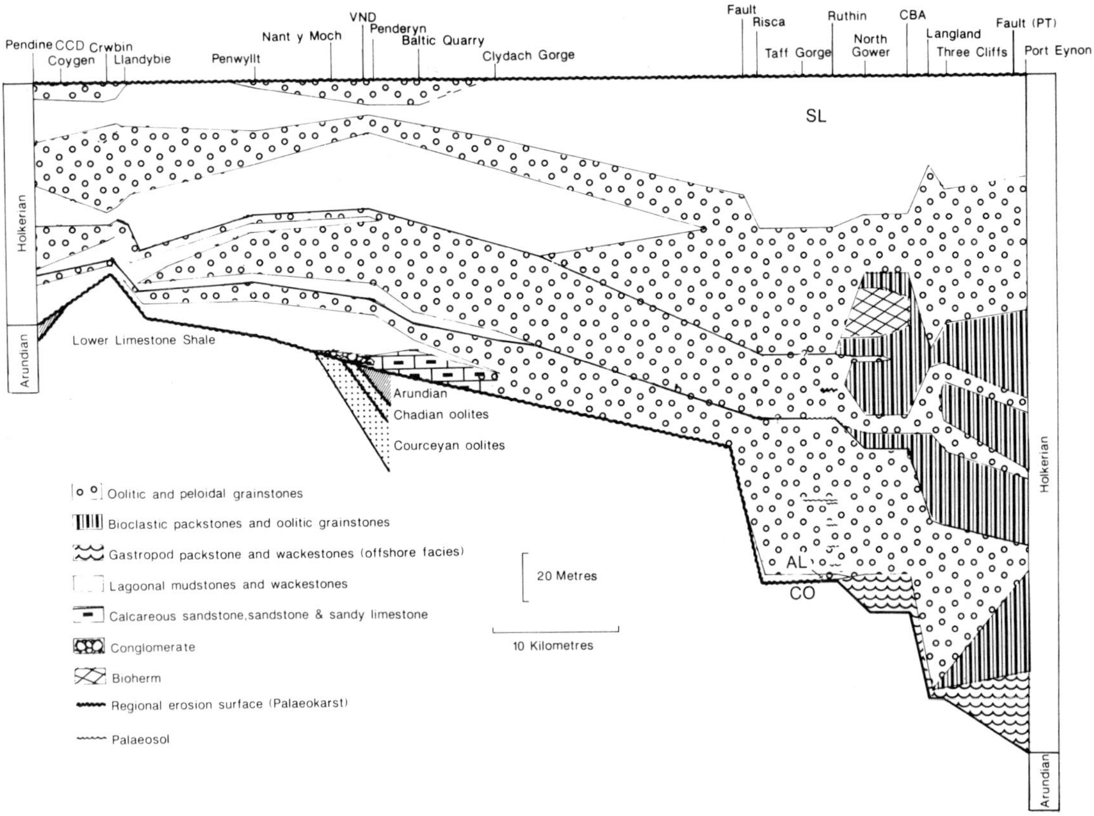


**Fig. 6.** The Holkerian succession shown in a composite section between the North Crop and south Gower. Named formations are: AL, Argoed Limestone; CO, Cefnyrhendy Oolite; SL, Stormy Limestone. Named structures are: CBA, Caswell Bay anticline; CCD, Carreg Cennen Disturbance; PT, Port Eynon thrust; VND, Vale of Neath Disturbance.

followed by deposition of a sequence that coarsens upwards from offshore packstone with storm-generated bedding and hummocky cross-bedding to an oolitic unit (Cefnyrhendy Oolite) with an irregular palaeokarstic top and overlying dark clay (?palaeosol). In Gower, transgression, probably with shoreface erosion and barrier retreat, was followed by deposition of a sequence in which upward-shoaling deposits pass via a transitional succession of muddier limestones into an offshore ramp sequence (Holkerian Overton Cliff Formation of Junghans *et al.* (1984)).

The nature of the Arundian–Holkerian boundary varies across the area (Figs 5, 6). On the North Crop it is strongly erosive. In the Vale of Glamorgan there is evidence of emergence and pedogenesis without major erosion. In Gower, the boundary lies within a deepening sequence with no evidence of emergence (Fig. 5). This variation is attributed to the influence of the tectonic regime on the development of the depositional slope. The North Crop was characterized by localized uplift, the Vale of Glamorgan by subsidence and seaward extension of a vegetated strandplain. In Gower, enhanced subsidence and tilting associated with movements on tectonic hinge lines produced a northward-shoaling slope or ramp (Ramsay, 1987).

## Holkerian

The part of the reconstructed section between Clydach Gorge and Taff Gorge shown in Fig. 6 is based on extrapolating the observed rate of thickening of the Holkerian sequence between Penwyllt and Baltic Quarry. In this restored section, thickness variations are attributed to synsedimentary movements on an E–W hinge line at Risca and on hinge lines developed along the present trends of the Cefn Bryn anticline and Port Eynon thrust in Gower. The thickness of the succession in the Vale of Glamorgan (Ruthin) is less than the thickness of Holkerian strata shown in sections on British Geological Survey maps. Detailed field logging in the Vale of Glamorgan indicates that the thickness of this sequence has been exaggerated by fault repetition.

Holkerian sedimentation was initiated during a phase of extension, subsidence, and tilting, and a concomitant relative rise in sea level. Combined asymmetrical subsidence and active hinge lines established a ramp across the studied area.

On Gower the cyclic nature (Fig. 6) of the succession is interpreted as the product of barrier migration and progradation in response to relative changes in sea level, which were tectonic in origin (Ramsay, 1987). The three major cycles in Gower are correlated with three cycles on the North Crop (Fig. 6). The depositional setting consisted of a northern, open, 'deep', coastal lagoon in which peritidal deposits were rare. The lagoonal succession contains laterally juxtaposed complex facies mosaics; in vertical sections, facies are ordered into upward-coarsening and shoaling units. The deep lagoon developed behind an oolitic barrier. The locus of oolite deposition was not static but shifted in response to changes in relative sea level. This resulted in barrier retreat followed by progradation and the development of the three upward-shoaling cycles in Gower. Barrier retreat probably occurred in response to rapid subsidence, which resulted in drowning and maintained high base-level stands.

*Cycle 1.* In Gower this cycle is represented by a sequence that shoals upward from storm-dominated offshore ramp deposits, through shoreface bioclastic packstone/grainstones, to oolitic intraclast grainstones. The variation between the thickness of the storm ramp sequence at Three Cliffs and Port Eynon (Fig. 6) may be due to movements of the hinge line along the present Port Eynon thrust. The discrepancy between the thicknesses of the Langland and North Gower sequences was produced by footwall uplift associated with a hinge line along the present Cefn Bryn anticline. In the Vale of Glamorgan the sequence shoals upward from the lagoonal mudstones and bioclastic packstone of the Argoed Limestone (Waters & Lawrence, 1987). In this area, however, the succession above the Argoed Limestone consists of a series of upward-shoaling cycles. Each cycle coarsens up from bioclastic mudstones and/or packstones to oolitic grainstones, which in places terminate at palaeokarst surfaces with associated clay palaeosols. The boundary between the deposits of the first and second cycles in the Vale of Glamorgan is drawn arbitrarily at one of these palaeosol horizons. Cycles of this type are absent in Gower and on the North Crop. This facies difference between Gower and the Vale of Glamorgan may reflect a developing E–W asymmetry in the basin, possibly in response to uplift along a N–S-trending axis of uplift that represents a continuation of the Usk anticline lineament. This uplift would have introduced a component of E–W tilt within existing fault blocks. Alternatively the cyclic deposits and palaeosols developed within a broad tectonically defined linear zone.

On the North Crop (Fig. 6) the beds of the first cycle onlap on to the northern, Caledonoid-trending palaeohigh with increasing unconformity towards its crest. Drowning was accompanied by the deposition of a lenticular unit of calcareous sandstone between the Clydach Gorge and Nant-y-Moch. This unit consists of sheet-like sandstones 30–50 cm thick, and calcareous sandstones with interbedded dark shales, yellow clay, sandy packstones, and ostracod mudstones. The sandstone sheets contain fragments of brachiopod shells, *Chondrites* burrows, and commonly have plant fragments at their tops. The lenticular body is approximately 10 m thick at Trefil. This unit probably represents Wright's (1986) upper fluvial unit, which he interpreted as the product of fluvial incision in response to a fall in sea level. These sands do not have the characteristics of fluvial sand bodies, i.e. erosive-based, thick, upward-fining units. The sandstone sheets are interpreted in this paper as crevasse-splay lobes that formed part of a minor bay-fill sequence. This developed in response to the restricted development of a fluvial delta during transgression. Evidence of a crevasse-splay channel was observed at Blaen Onnen. The fluvial delta was drowned during continued transgression and development of a muddy carbonate lagoon. At Nant-y-Moch (Fig. 6) the base of the cycle contains a carbonate conglomerate with clasts of Courceyan oolite and calcrete. Over the rest of the area lagoonal carbonate mudstone/packstones rest directly on the underlying rocks. The first cycle coarsens upwards and shoaling terminates with the deposition of oolite grainstones.

*Cycle 2.* These deposits also show evidence of tectonic activity along the Cefn Bryn anticline. This cycle was also initiated by barrier retreat associated with extension, subsidence, and relative sea-level rise followed by progradation. On Gower, this resulted in the deposition of an upward-shoaling sequence. In the Vale of Glamorgan, cycle 2 deposits are predominantly oolitic, strongly cyclic with some palaeokarst development. The difference between this and the Gower sequence may reflect different subsidence and/or sedimentation rate histories in response to E–W tectonic asymmetry. The boundary between cycle 2 and cycle 3 deposits in the Vale of Glamorgan is drawn arbitrarily at the contact between the cyclic unit and a thick sequence of massive oolite limestone. The equivalent deposits on the North Crop consist of a sequence of lagoonal deposits.

*Cycle 3.* The third cycle also accumulated during a period of tectonic activity and extension, with evidence of movements associated with the Cefn Bryn anticline faults. Again the Gower sequence is initiated by a phase of barrier retreat and incursion of interbedded/shoreface oolitic grainstones and bioclastic packstone grainstones. During this transgressive phase, *Lithostrotion* bioherms developed in North Gower. These are not recorded in the sequence to the south of the Cefn Bryn axis. The upward-shoaling Gower succession (lower shoreface via oolite shoal to lagoonal deposition) probably formed as a result of progradation and a seaward shift in the position of the barrier. In the Vale of Glamorgan the equivalent sequence is represented by massive oolites overlain by lagoonal (Stormy Limestone) deposits. On the North Crop a lagoon sequence is capped by isolated oolite shoal deposits.

The third cycle sequence probably formed in response to a continued rise in relative sea level, combined with sedimentation rates that enabled the barrier to prograde and aggrade and thus develop an extensive lagoon. This type of sequence could only form in a dynamic regime. A relative sea-level rise, coupled with a static aggrading barrier, would not extend the lagoon. A static base level would probably result in lagoonal infilling (Heward, 1981; Burchette, 1987) and a normally saline marine lagoon would not be maintained with a fall in base level.

The Holkerian succession terminates at a 'basin-wide' erosion surface that formed on a palaeokarst platform. In the studied area there is no evidence of tectonic movement associated with the development of this platform. This may not be of great significance, however, since so much of the sequence is hidden beneath the coalfield and obliterated by Namurian erosion and overstep on the East Crop.

## Asbian and Brigantian

The deposition of the late Dinantian succession was initiated during a phase of subsidence and relative rise in sea-level. Thickness variations within the area (Figs 7, 8) were produced by the following factors:

**1** N–S and E–W orientated asymmetrical subsidence, which established a depocentre in the Gower area;

**2** late Asbian and Brigantian movements along the E–W, NE–SW (Caledonoid) and N–S or NNW–SSE-trending cross faults;

**3** late Dinantian and Namurian uplift particularly towards the northern and eastern margin of the main South Wales Coalfield.

Asbian and Brigantian sedimentation is characterized by its minor cyclicity in which individual, commonly upward-shoaling, cycles are truncated by palaeokarst surfaces and associated clay palaeosols. This cyclicity is generally attributed to drowning and exposure of a mature carbonate shelf in response to minor fluctuations in sea level (Ramsbottom, 1973; Walkden, 1987). In South Wales, minor cycles form the components of three major cyclic phases.

*Phase 1.* This cyclic succession was deposited on an asymmetrically subsiding surface that shoaled towards the northern and eastern margins of the basin (Figs 7, 8). Subsidence was probably accompanied by rifting associated with an E–W orientated fault along the trend of the present Port Eynon thrust. The asymmetrical nature of the basin subsidence is reflected by the landward onlap of facies within cycle 1 and by the differentiation of facies patterns within the two upward-shoaling cycles that constitute phase 1. To the east of the Vale of Neath Disturbance, on the North Crop, and in the Vale of Glamorgan the initial transgressive deposits are represented by calcareous sandstones (Figs 7, 8). The oolitic grainstone/calcareous sandstone-dominated North Crop successions differ from the South Crop and particularly south Gower succession, in which dark offshore packstones form a major component. The calcareous sandstones of cycle 2 (Fig. 7) represent a minor bay-fill sequence associated with fluvial dominated deltas (Ramsay, 1989) that developed along the southern margin of the St George's Land massif.

Phase 1 cycles are truncated by erosion surfaces

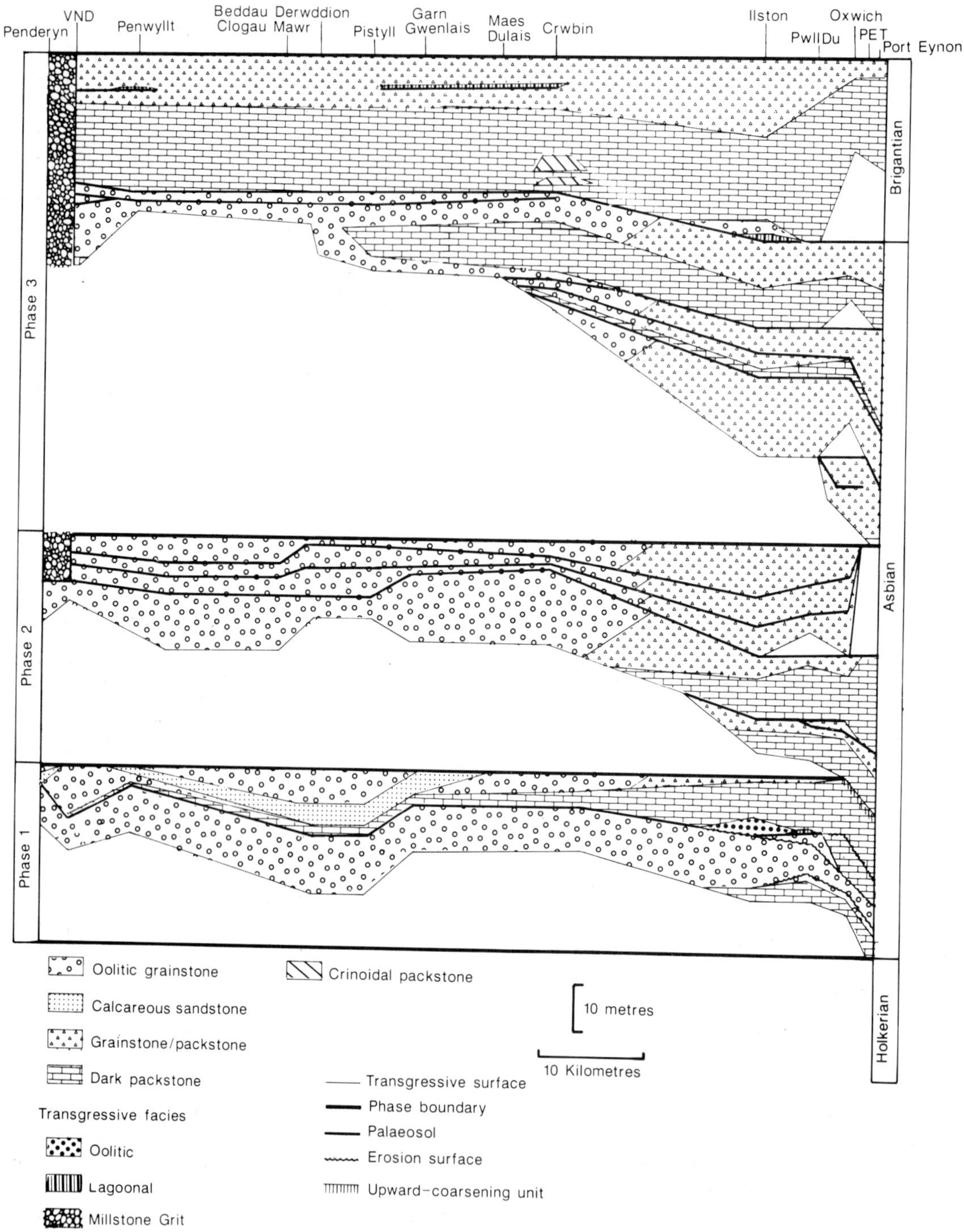


**Fig. 7.** Late Dinantian depositional phases based on a proposed correlation of palaeosol horizons between the North Crop and south Gower. The section at Garn Gwenlais is situated 1 km to the west of Llandybie. Named structures are: PET, Port Eynon thrust; VND, Vale of Neath Disturbance.

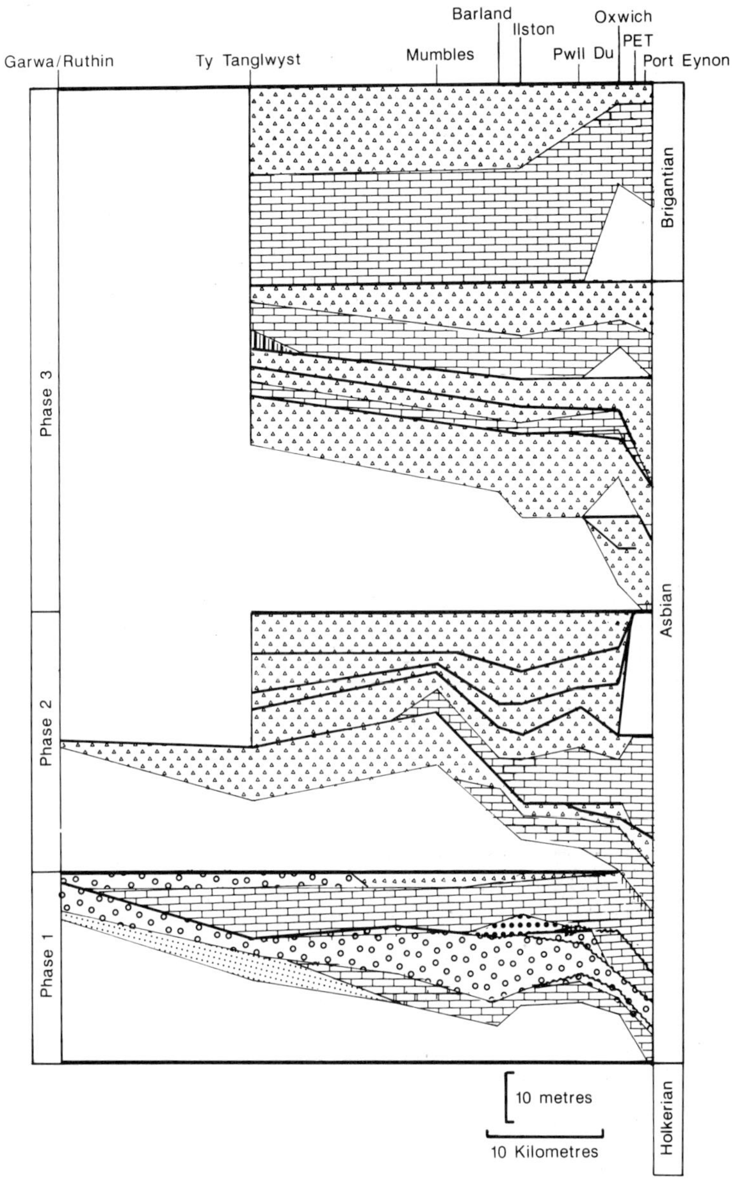


**Fig. 8.** Late Dinantian depositional phases based on proposed correlations of palaeosol horizons between the Vale of Glamorgan and Gower. The lithological key is shown in Fig. 7.

and associated palaeosols. The exposure surfaces do not extend into south Gower (Figs 7, 8) but probably correlate with an erosion surface (first palaeokarst) and facies change (second palaeokarst) within the dark packstones at Oxwich and Port Eynon.

Phase 1 deposition was terminated by uplift, tilting, and erosion. The effects of this event are evident on the North Crop and in the Vale of Glamorgan and absent in south Gower where deposition of phase 1 is not continuous.

*Phase 2.* This cyclic depositional phase was initiated by landward onlap of cycle 1 deposits and their overstep by the deposits of cycle 2 (Figs 7, 8). Thus phase 2 consists of five or six cycles (the palaeosol that truncates cycle 1 splits between Ilston and Pwll Du) on the South Crop and four on the North Crop. The attenuation of the succession and absence of the three upper cycles between Oxwich and Port Eynon is attributed to inversion, exposure, and erosion of the southern hanging wall of the Port Eynon axis.

During phase 2, the depositional regime appears to have evolved from a northward and eastward shoaling gentle ramp to a platform. During the ramp state, dark offshore packstones formed a significant basal component in the upward-shoaling deposits of cycles 1 and 2 on the South Crop. The platform cycles were dominated by grainstone and grainstone/packstone cycles on the South Crop and by oolitic grainstones on the North Crop. The exposure surfaces that truncate cycles 2–5 extend across the whole area (Figs 7, 8).

Phase 2 deposition was terminated by a period of uplift, tilting, and erosion that affected the whole area. The influence of this tectonic event is more pronounced towards the northern and eastern basin margins but is also represented by the development of a palaeosol at Port Eynon.

This interval of tectonic activity possibly coincides with tectonism in the Dublin Basin (Nolan, 1989) and with tectonism recorded by Gawthorpe *et al.* (1989) from the Bowland Basin, North Wales, and Derbyshire.

*Phase 3.* The deposition of this phase was initiated during transgression and rifting associated with the E–W orientated Port Eynon hinge axis (Figs 7, 8). Evidence of short-lived syndepositional tectonic activity associated with NW–SE orientated cross-faults between Barland and Ilston on Gower and Beddau Derwyddion and Clogau Mawr on the North Crop is recorded from phase 3 deposits. Contemporaneous activity on the NE–SW orientated Vale of Neath Disturbance produced a roll-over embayment (Fig. 7) between Penwyllt and this structure (Ramsay, 1989). The Port Eynon hinge axis and the Vale of Neath Disturbance remained active during phase 3 deposition.

The phase 3 succession is characterized by progressive onlap and attenuation of upward-shoaling cycles against the tectonic depositional surface. This effect is particularly evident from the South to the North Crop, where the deposits of cycle 4 overstep the deposits of the first three cycles (see Fig. 7). There is some eastward attenuation in the thickness of phase 3 deposits between Gower and the Vale of Glamorgan (Fig. 8).

The phase 3 depositional regime evolved from a ramp with differentiated shoreward and basinward facies to a platform characterized by undifferentiated facies. Dark offshore packstones became an increasingly important component of the succession during the progression from ramp to platform.

The origin of minor cyclicity in the late Dinantian of South Wales is attributed to pulsed subsidence followed by marine transgression and the deposition of upward-shoaling prograding sequences. Palaeosols originated in lowland marshes that formed when plants began to colonize the intermittently exposed surfaces on seaward margins of prograding strandplains. Eventually the depositional surface was elevated above sea level as a result of pedogenic processes.

The Dinantian carbonate shelf terminated with deposition of Namurian conglomerates and sandstones on the North Crop and muds on the South Crop. The contact between the Namurian and Brigantian in the western Vale of Glamorgan and Gower is probably transitional. Pronounced uplift on the northern and eastern basin margins resulted in the erosion of much or all of the late Dinantian succession.

## RELATIONSHIP BETWEEN SEDIMENTATION AND NE–SW-TRENDING STRUCTURES

During the Dinantian a palaeohigh developed along the trend of the Carreg Cennen Disturbance (Fig. 9) and influenced the thickness of the succession. This feature, associated with a basement high, is defined (Fig. 10) in the magnetic anomaly pattern for South Wales (British Geological Survey, 1980; Williams, 1985). A shallow depth to basement is also indicated by the seismic refraction data of Mechie & Brooks (1984), which show a minimum depth to Precambrian basement of approximately 3 km for the Crwbin area. The NE–SW orientation of the palaeohigh is also evident in the geographical extent of the Holkerian–Courceyan unconformity and in isopach data for Holkerian (*Seminula* Zone) deposits (George, 1958). This palaeohigh probably contributed to the formation of the promontory on the southern flanks of the St George's Massif postulated by George (1974).

From the Courceyan to the Holkerian, activity of this feature and of E–W structures contributed to the topography of the Dinantian basin floor. Uplift and erosion of the Caledonoid-trending palaeohigh probably influenced the concentration of palaeokarst surfaces and disconformable contacts within North Crop Courceyan successions. Three phases of enhanced uplift resulted in the Courceyan to late Chadian, late Chadian to Arundian, and Arundian–

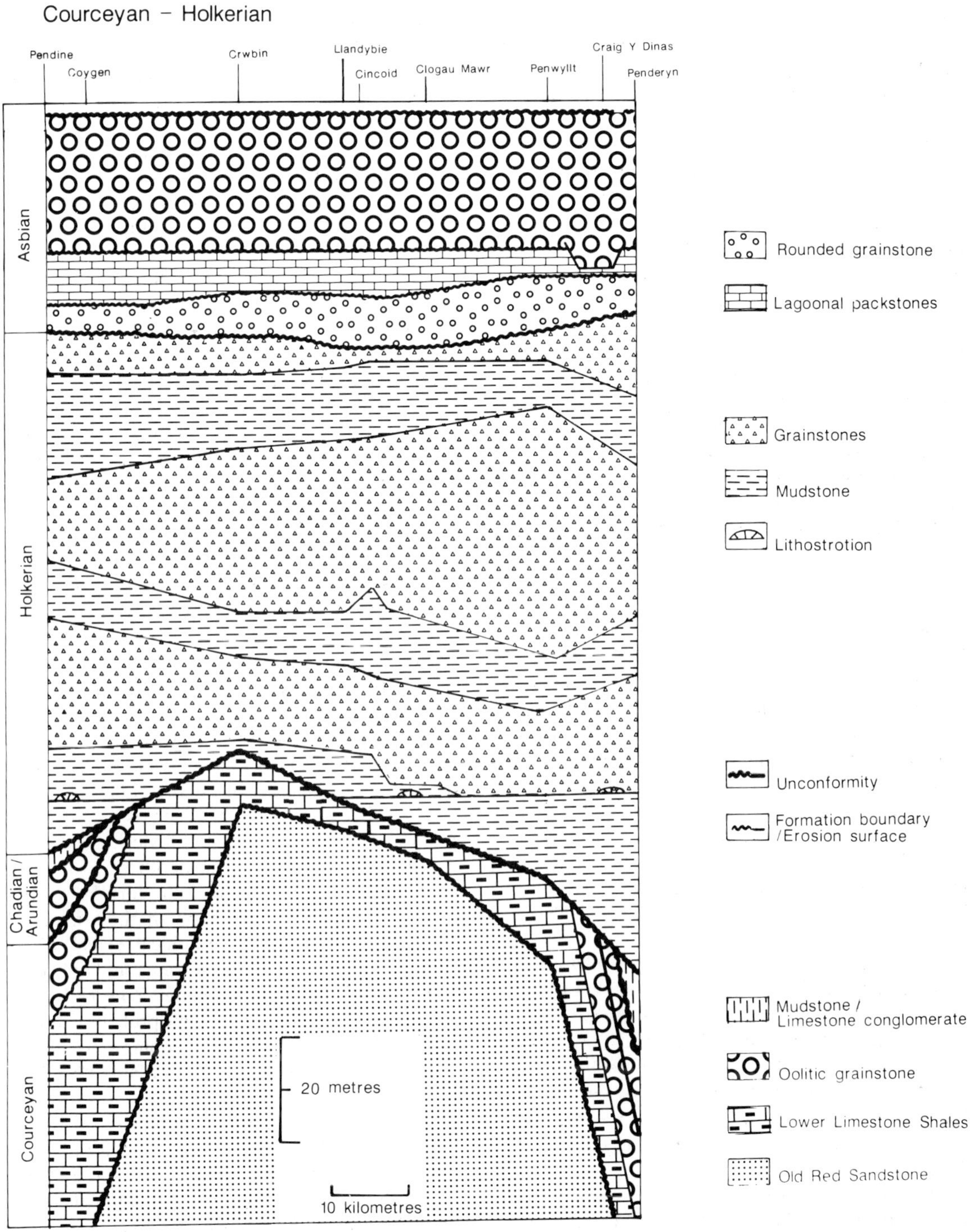


**Fig. 9.** The Courceyan to Asbian section between Pendine and Penderyn showing the influence of a Caledonian-trending palaeohigh on the development of the succession. The coral *Lithostrotion* is the main component of a biostrome that forms a marker horizon near the base of the Holkerian along the North Crop. The Cincoid section is a disused quarry 3 km to the east of Llandybie.

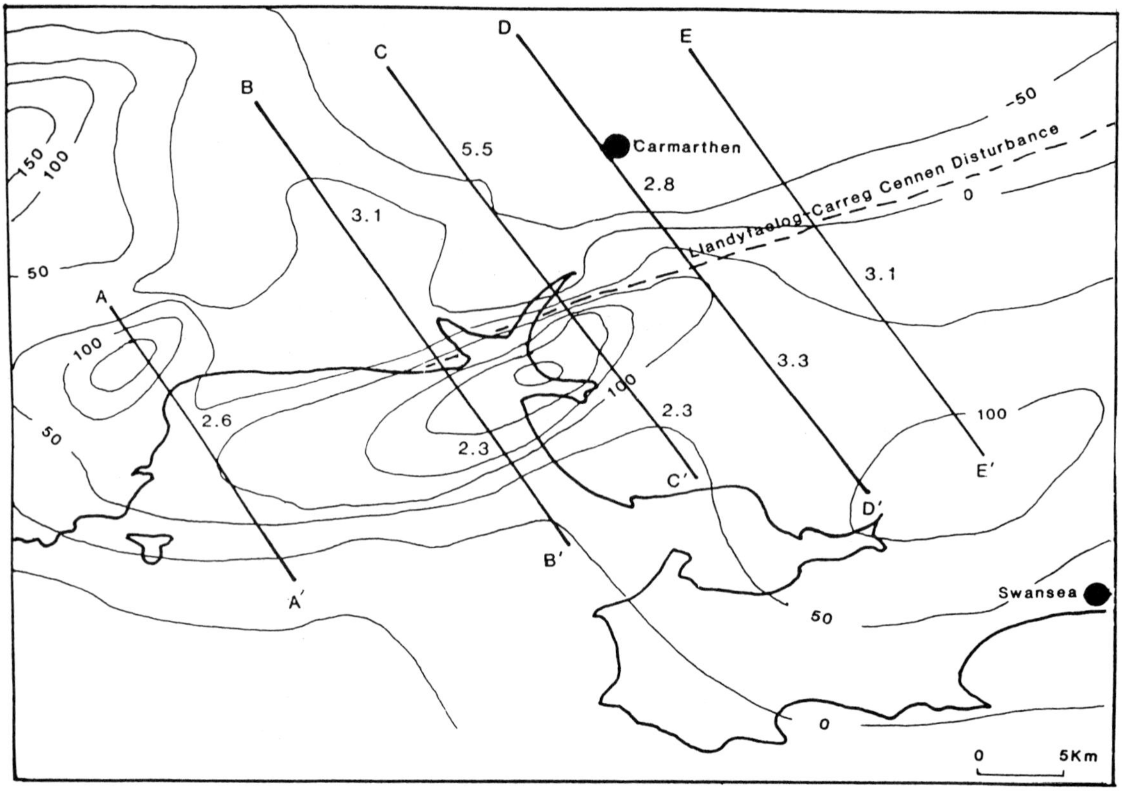


**Fig. 10.** Magnetic anomaly contour map showing profile lines (A–A′) and magnetic depth estimates in kilometres. Magnetic contour intervals 50 nT (Williams, 1985).

Holkerian unconformities. The late Chadian and Holkerian deposits transgressed on to the flanks of this high with progressive overstep (Figs 4, 6). Unfortunately, the continuity of exposure is insufficient to observe whether the Arundian transgression was associated with overstep, though evidence of onlap is recorded within the Arundian sequence. It is possible that the breaching of the Courceyan limestone succession and exposure of the Lower Limestone Shale occurred during the late Chadian tectonic episode (Fig. 4). The cessation of tectonic activity associated with this structure coincided with its drowning during the Holkerian transgression. Holkerian limestones rest on progressively older deposits from the flanks to the west of the palaeohigh (Figs 6, 9). Asbian deposits transgressed across a palaeokarst platform with low relief (Figs 7, 9).

The influence of an active Vale of Neath Disturbance on the late Dinantian succession is recorded by Owen (1954). There is no evidence from thickness data to suggest pre-Asbian activity of this structure. During the Asbian, tectonic activity associated with this fault produced a roll-over embayment that was flooded during phase 3 transgression (Fig. 7). Differential subsidence associated with roll-over also resulted in the splitting of post-transgressive palaeosols. The scissors-like rotation of the fault blocks along this structure had a profound influence on the sub-Namurian erosion pattern and Namurian overstep (Fig. 11). On the northern block, Namurian conglomerates and sandstones rest on the Pwll-y-Cwm Oolite (Courceyan) at Pen Cerrig Calch and on Brigantian limestones at Craig-y-Dinas. On the southern block the Namurian deposits overly the Holkerian at Blaen Onnau and Llangattock and Asbian Oolites at Penderyn.

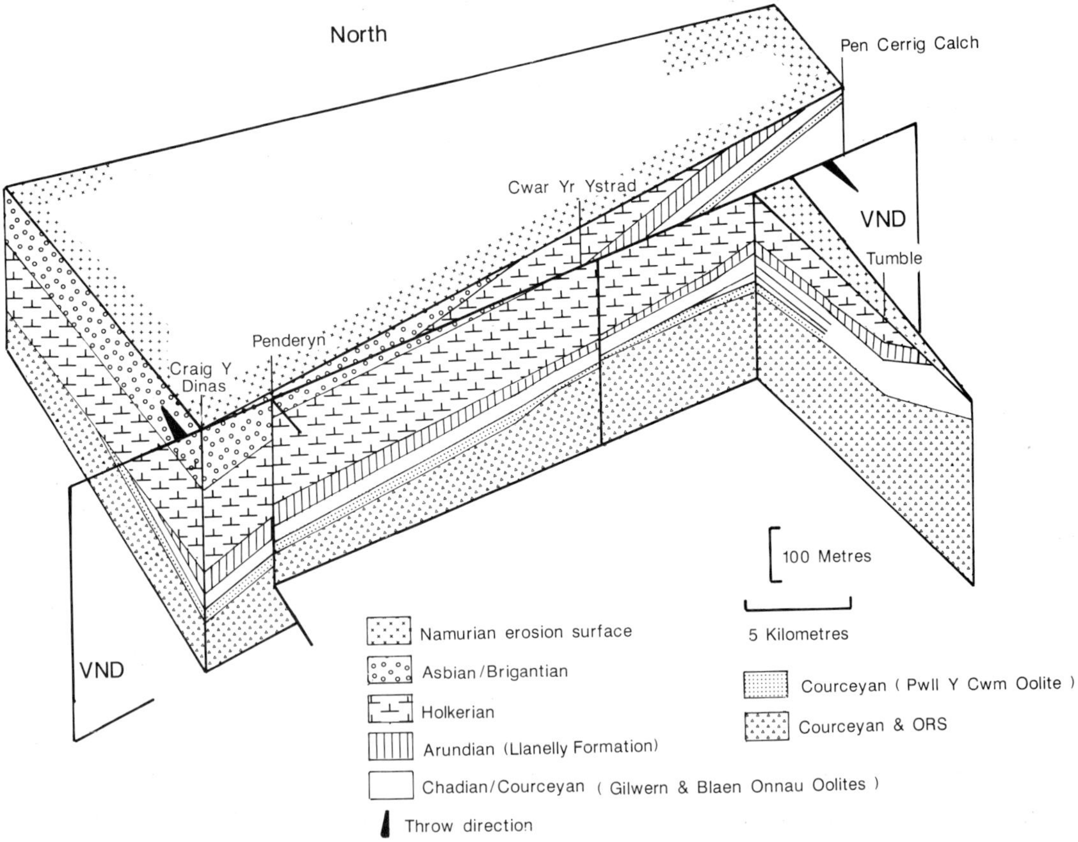


**Fig. 11.** The influence of the Vale of Neath Disturbance on the sub-Namurian erosion pattern. Cwar Yr Ystrad is a quarry section and comprises one of several exposures of Dinantian rocks in the Trefil area (Fig. 1).

## RELATIONSHIP BETWEEN SEDIMENTATION AND N–S AND NNW–SSE-TRENDING CROSS FAULTS

These faults are mainly normal with downthrows to the west. Evidence of strike-slip movement was rarely observed on the North Crop (Barclay *et al.*, 1988) but the Gower cross faults are primarily strike-slip (George, 1940; Owen & Weaver, 1983). Contemporaneous movements along these structures were initiated during a major period of uplift and tilting, which preceded and possibly accompanied the third phase of Asbian and Brigantian deposition. These faults had a minor impact on thickness variations in late Dinantian deposits but, in conjunction with active E–W and NE–SW orientated faults and a buoyant Usk axis, had a major influence on the patterns of sub-Namurian erosion and overstep. On the South Crop the degree of unconformity between the Namurian clastics and underlying Dinantian rocks is a consequence of increased erosion in response to variable footwall uplift (Fig. 12). The preservation of the Plastic Clay Formation on the North Crop is also a product of differential erosion across variously uplifted fault blocks (Fig. 13).

## RELATIONSHIP BETWEEN SEDIMENTATION AND THE N–S-TRENDING USK ANTICLINE

Previous workers (Moore, 1948; George, 1956a,b; Squirrel & Downing, 1964,1969; Barclay & Jones, 1978) have demonstrated the influence of synsedimentary movements of the Usk axis on the pattern of Namurian overstep and attenuation of the Namurian and Silurian successions of the East Crop.

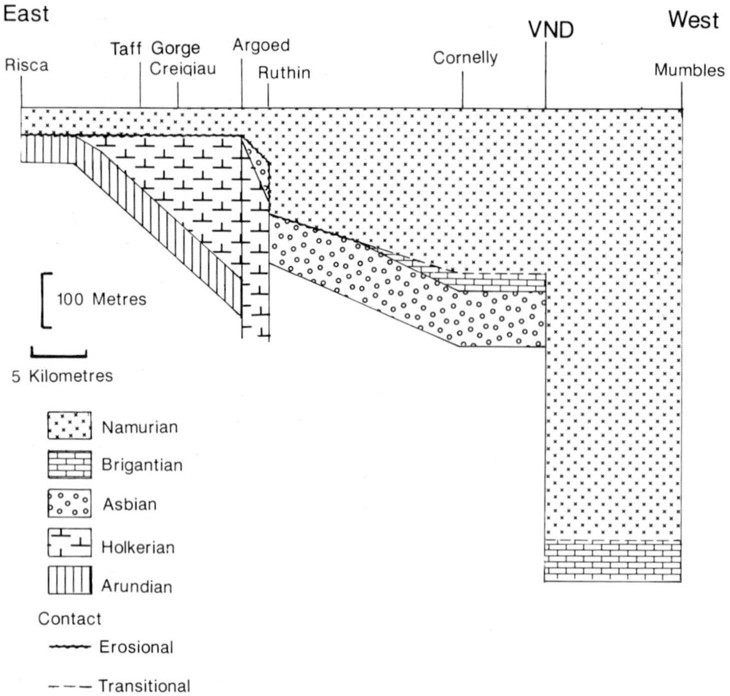


**Fig. 12.** The influence of the Vale of Neath Disturbance and NNW–SSE cross faults on the sub-Namurian erosion pattern between Gower and the East Crop. Creigiau and Argoed are quarry sections situated at 4 km and 14 km to the southwest of the Taff Gorge. Cornelly is synonymous with Ty Tanglwyst of Figs 1, 8. VND, Vale of Neath Disturbance.

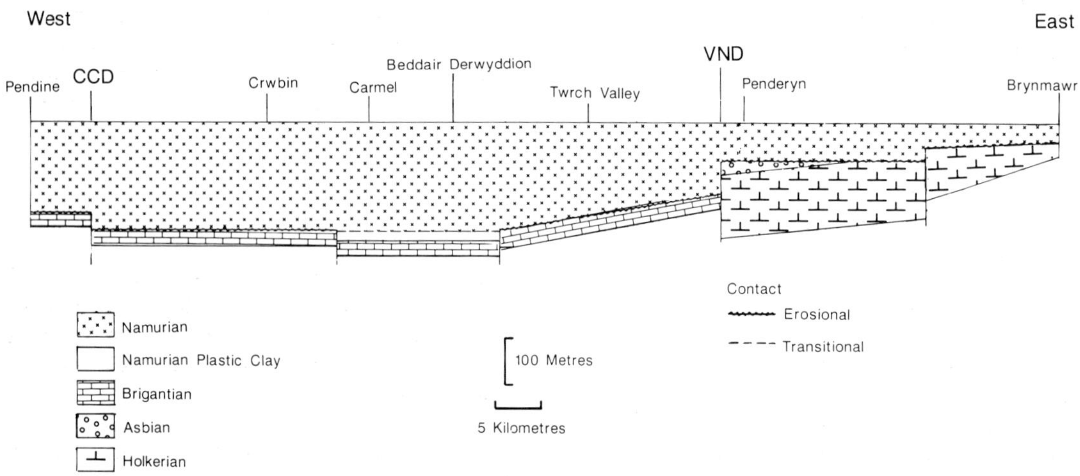


**Fig. 13.** The influence of the Carreg Cennen Disturbance (CCD), Vale of Neath Disturbance (VND) and N–S to NNW–SSE cross faults on the sub-Namurian erosion pattern along the North Crop.

There are conflicting views concerning the pre-Namurian activity along this N–S trend. George (1956a) noted a significant northward and eastward reduction in the thickness of the Carboniferous Limestone on the East Crop but considered evidence for Dinantian uplift and easterly control as equivocal. George (1956b) proposed that the first major significant emergence of the Usk anticline was immediately pre-Namurian. Wilson *et al.* (1988) invoked early emergence and uplift along the Usk anticline during the Holkerian to explain the subdivision of the Southwest Province into discrete sedimentary basins. In this investigation the variation in facies patterns between the Holkerian successions of north Gower and the Vale of Glamorgan and eastward attenuation of the late Dinantian sequence are attributed to uplift along a N–S orientated structure that is probably an extension of the Usk anticline.

## THE NATURE OF MARINE TRANSGRESSION, REGRESSION, AND FACIES SUCCESSIONS

The sequential development of facies successions in the major Dinantian cycles (equivalent to stratigraphical stages) in South Wales is a product of relative changes in sea-level. The deposition of each major cycle commenced with a phase of rifting and subsidence followed by marine transgression with onlap. A terminal phase of subaerial exposure and pedogenesis at the basin margin is associated with upward shoaling in places. Tectonic effects, i.e. rifting, subsidence, and uplift, are also evident in the development of minor cyclicity. Palaeokarsts that terminate minor upward-shoaling cycles within each stage are either confined to the northern, landward margin or extend across the whole region.

### Marine transgression

The process of marine transgression is attributed to shoreface retreat. Four types of transgressive sequence are recognized.

*Type 1.* Shoreface or offshore deposits that rest on a pre-transgressive palaeokarst. This sequence is attributed to enhanced shoreface erosion either in response to scour at the bases of laterally migrating tidal inlets (Kumar & Sanders, 1974), low rates of sedimentation, or slow subsidence and relative rise in sea-level (Fischer, 1961; Swift, 1968; Heward, 1981).

*Type 2.* Type 2 sequences comprise: basal, pre-transgressive erosion surface; peritidal–lagoonal facies, which include washover grainstones; shoreface grainstones or offshore bioclastic limestones.

The peritidal–lagoon deposits of the Caswell Bay Mudstone were first recognized as a transgressive unit by Riding & Wright (1981). Subsequently, Wright (1986) demonstrated that peritidal–lagoon deposits are characteristic of transgressive phases in the Dinantian of South Wales. The generation and preservation of this type of sequence are attributed to restricted rates of shoreface erosion (Fischer, 1961; Swift, 1968; Riding & Wright, 1981). Shoreface erosion rates were probably inhibited by rapid rates of subsidence and relative sea-level rise, and by rates of sedimentation which ensured that the elevation of back-barrier deposits maintained pace with the upward growth of the barrier.

*Type 3.* A sequence in which 'deep-water' open lagoonal mudstones overlie the pre-transgressive surface, for example the base of the Holkerian on the North Crop. This transgressive sequence is attributed to rapid rates of subsidence and relative sea-level rise during which the vertical component of barrier accretion exceeded the rate of back-barrier sedimentation to produce a deep broad lagoon.

*Type 4.* Transitional or sharp contacts between prograded shoreface grainstones and offshore packstones and mudstones. This type of sequence was produced by relative sea-level rise in an offshore setting. Sequences of this nature occur in the Courceyan and at the top of the Arundian on Gower.

On the North Crop, minor siliclastic deltas and bay-fill sequences developed locally in response to relative rises in sea level during the Arundian, Holkerian, and Asbian.

### Marine regression

The process of marine regression, which includes the deposition of upward-shoaling sequences, subaerial exposure, and the development of palaeokarsts, is complex. Four types of unconformity are distinguished.

*Type 1.* Extensive palaeokarst platforms with low relief cut across rocks and facies of different ages. These formed by subaerial erosion of a tectonically generated relief with enhanced erosion on areas of footwall uplift. The intra-Chadian unconformity is an example.

*Type 2.* Type 2 unconformities are extensive, flat palaeokarst platforms that truncate oolite shoal and lagoonal facies. This type of surface is represented by the Holkerian–Asbian boundary. The mechanism responsible for the development of this surface, i.e. tectonic or eustatic, is uncertain.

*Type 3.* These are surfaces where topographic relief is demonstrably a product of uplift, faulting, and tilting. This type of surface truncates shoaling-upward sequences and passes downslope, i.e. seawards, from an erosional palaeokarst to a transitional and conformable contact between shallow and deeper water facies. Type 3 surfaces are recorded in the Courceyan and Asbian and form the boundaries between the Chadian–Arundian and Arundian–Holkerian stages.

*Type 4.* Flat palaeokarst platforms that terminate the upward-shoaling cycles. These show no evidence of tectonic processes and characterize Asbian and Brigantian successions.

The type 3 and 4 surfaces that terminate regressive phases are not associated with peritidal–lagoonal deposits. Wright (1986) attributed this absence of peritidal limestones to the rapid fill of back-barrier areas in a progradational setting. The landward transport of coarse sediment is, however, a characteristic of transgressive rather than regressive systems. This process would also not necessarily preclude the development of peritidal deposits. Walkden (1987), in his discussion of late Dinantian cyclicity, invoked eustatic sea-level changes to promote drowning and sudden emergence. He equated the absence of peritidal facies with rapidly falling sea levels, which allowed insufficient time for the development of peritidal–lagoonal facies.

The generation of back-barrier peritidal–lagoon facies in prograding systems probably requires that the seaward migration of the barrier is accompanied by a rising sea-level and excess sediment supply. In this account upward coarsening sequences that terminate at pedogenic horizons are attributed to high rates of sedimentation and relatively static sea levels. In this setting the basin margins became choked with rapidly prograding and even aggrading deposits. Vegetated lowland marshes developed on broad platforms in the wake of the prograding and/or aggrading sequences and initiated pedogenesis.

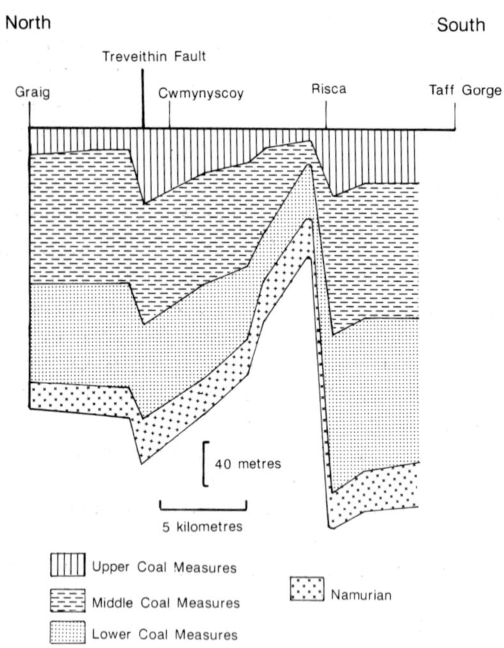


**Fig. 14.** The Silesian succession on the East Crop of the coalfield between the base of the Namurian and Rhondda No. 2 coal seam, based on the following BGS 1:10 000 geological maps: SO, 20 NE, 20 SW; ST, 29 NE, 29 SW, 28 NE, 28 SW, 18 NE, 18 SW.

## POSSIBLE CAUSAL MECHANISMS OF DINANTIAN CYCLICITY AND TECTONICS

### Tectonic and sedimentary mechanisms

Dinantian South Wales was probably situated on the hanging wall of a half-graben, which formed by the northerly downthrow on a major fault system to the south, possibly the Bristol Channel Fault Zone. The nature of the hanging wall succession, including deposits and unconformities, was influenced by the reactivation of E–W, NE–SW, NNW–SSE (cross faults), and N–S orientated structures in the basement. This reactivation occurred in response to a predominantly N–S orientated extensional stress field, thus E–W orientated structures formed dip-slip faults with a southerly throw.

The major pattern of Dinantian cyclicity in South Wales is a product of the evolution of sea-floor geometry on the northern hanging wall margin of the Cornubian Basin. Between the Courceyan transgression and intra-Chadian unconformity, deposition occurred on a broad ramp. This developed in conjunction with rifting and extension, which is particularly evident in the south of the area. The facies distribution pattern, in which shallow-water oolites are concentrated on the North Crop, is a consequence of the ramp setting (Wright, 1986). These, commonly transgressive, deposits were confined in their 'shoreline' setting by high sea-level stands associated with rapid subsidence. The basinward progradation of oolitic shoals probably occurred in response to high rates of sedimentation and relatively static sea level.

Courceyan minor cyclicity is attributed to the combined effects of pulsed regional subsidence and progradation on a tectonically differentiated depositional surface. Subsidence-generated transgressions were followed by high rates of sedimentation, relatively static sea level and the southward progradation of the oolite shoals. Palaeokarst surfaces were restricted to the North Crop on the crest and flanks of a NE–SW-trending palaeohigh. These surfaces pass laterally into sharp or transitional oolitic-grainstone–offshore-packstone contacts.

By late Chadian times oolitic deposition had extended into Gower and the Vale of Glamorgan. Intra-Chadian uplift, faulting, tilting, and erosion were followed by transgression and deposition of oolitic grainstones across a broad gentle ramp or platform. The causal mechanism of this transgressive event is unknown. It could be either eustatic or related to regional subsidence. Deposition was possibly short lived as the late Chandian oolites are truncated at the intra-Avonian unconformity. This type 3 surface forms a widespread palaeokarst to the north of the Cardiff–Cowbridge anticline and Port Eynon thrust and developed in response to uplift faulting and tilting.

Arundian sedimentation was initiated on a ramp that developed during phases of rifting, pulsed subsidence, and transgression. Pulsed subsidence associated with rifting resulted in the deposition of three shoaling cycles, the third of which extended on to the North Crop. In each ramp cycle, peritidal–lagoonal sequences were probably contained in a coastal setting by continued subsidence associated with relative rises in sea level and maintenance of highstand Arundian deposition. This situation culminated in uplift, exposure, and erosion along the Caledonoid-trending palaeohigh on the North Crop. In contrast, there was subsidence, accompanied by deepening and a facies change, in the Gower area. The Arundian–Holkerian boundary is a type 3 erosion surface.

The Holkerian ramp was also initiated during a phase of rifting, subsidence, and transgression. The three Holkerian cycles record landward and seaward shifts of an oolitic shoal, probably in response to rifting, subsidence, and relative sea-level changes. The rapid vertical accretion of the oolite barriers maintained pace with relative rises in sea level and were associated with broad deep-water open marine lagoons. Continued sea-level rise resulted in the seaward migration of the barrier and back barrier lagoon. The Holkerian terminates at a widespread exposure surface and palaeokarst that is either eustatic or tectonic in origin.

Asbian sedimentation was initiated by a marine transgression associated with rifting and subsidence. During the Asbian–Brigantian interval the three phases of deposition were associated with uplift and tilting towards the eastern margin of the basin where the successions of each phase are attenuated. The onlap of the minor cycles during phase 2 and phase 3 is attributed to pulsed subsidence and continued cyclic deposition, which eventually resulted in ramp–platform evolution.

## Structural processes and mechanisms

The causal mechanisms of tectonic processes are complex and still unresolved. The initiation of the Hercynian orogeny in the Middle Devonian (Haszeldine, 1989) suggests that the proposed extensional southern half-graben formed and developed in an overall compressional regime. Haszeldine (1989) highlights the absence of crustal thinning beneath or adjacent to Carboniferous basins as a major problem in applying McKenzie's (1978) model of lithospheric stretching and thermal subsidence to explain the origin of these basins, for example Leeder (1982,1987) and Leeder & McMahon (1988). The proposition that a Dinantian regime of crustal extension was superceded by a Silesian regime of cooling and thermal subsidence is belied by the occurrence of Namurian and Westphalian vulcanicity (Haszeldine, 1989). In South Wales, Silesian deposition continued to be influenced by the complex interaction between E–W orientated structures (see Fig. 14) and structures with other trends.

Two contrasting mechanisms have been proposed to explain the N–S extensional stress field. These are back-arc stretching associated with northward subduction and closure of the Rheic Ocean (Leeder, 1982,1987) and slab-pull forces in response to the southward subduction of the same ocean (Bott, 1987). Haszeldine (1989) argues that the Dinantian and Silesian were dominated by a major component of E–W orientated extensional stress with a minor N–S orientated stress field. This argument can be refuted on the grounds that the E–W-trending faults in South Wales have dip-slip displacements rather than the strike-slip displacements that would have occurred in an E–W orientated extensional stress field. Gawthorpe *et al.* (1989) have also used similar evidence from fault systems in the North Wales and northern England basins to dismiss the concept of dominant E–W extensional stress during the Dinantian.

Wilson *et al.* (1988) propose a model that involves interaction between dextral strike-slip movements on the Severn Estuary Fault Zone and synthetically developed E–W orientated structures, e.g. the Cardiff–Cowbridge axis. It is possible that NE–SW and NNE–SSW cross faults responded to the dominant N–S extensional stress regime by varying amounts of strike-slip deformation. The NE–SW-trending palaeohigh along the Carreg Cennen Disturbance was possibly generated and maintained in a strike-slip regime. The timing of dextral strike-slip movements along the Vale of Neath Disturbance and the NNW–SSE system of cross faults is debatable and may date from the Late Carboniferous or younger, i.e. Permo-Triassic or even Alpine. A model that attributes the origin of the stress field to large-scale dextral shear (Wilson *et al.*, 1988) is inapplicable in South Wales for the following reasons:

**1** the absence of major lateral offsets along faults that are unequivocally Dinantian in age;
**2** the absence of pull-apart basins at potential releasing bends;
**3** strike-slip shear was probably an intermittent rather than a continuous component of stress, thus the Vale of Neath Disturbance and cross faults functioned as synsedimentary dip-slip faults with westerly downthrows during parts of their history.

The Dinantian basin, which includes South Wales, probably formed and evolved in a N–S orientated extensional stress field. The origin of this field is uncertain. Possible causal mechanisms are either subduction or lithospheric loading in response to an orogeny to the south. In the author's opinion, the development of Dinantian fault activity occurred as basement fault systems were reactivated in response to developments in the evolution of a southern orogenic belt. The orogenic effect extended beyond South Wales and probably accounts for the apparent synchroneity between the following tectonic events in Britain and Ireland:

**1** the initiation of a N–S stress field and development of grabens and half-grabens in the Late Devonian and early Courceyan;
**2** widespread reactivation in the late Chadian and between the late Chadian and early Arundian;
**3** extensive reactivation during the middle and late Asbian;
**4** major inversion between the Brigantian and Namurian. In South Wales this event probably involved all fault systems.

Tectonic activity between the Arundian and Holkerian stages is recorded from Northumberland and the southern Lake District (Gawthorpe *et al.*, 1989). Intra-Asbian tectonism is recorded from the Northumberland basin and possibly Bowland Basin (early Asbian) (Gawthorpe *et al.*, 1989). Intra-Holkerian movements have not been recorded outside South Wales. Evidence of inter-Holkerian–Asbian tectonism is documented from the southern Lake District and Derbyshire (Gawthorpe *et al.*, 1989; Horbury, 1989).

### Eustatic tectonic mechanisms

The succession of Dinantian facies architectures generated by sedimentation within the tectonically dynamic South Wales basin margin is similar to the siliclastic facies geometries attributed by Haq *et al.* (1987) and van Wagoner *et al.* (1988) to a combination of eustatic fluctuations in sea level and constant rates of subsidence. Most of the cyclic facies successions recognized in this study correspond to sequences composed of transgressive and highstand systems tracts. During the early Holkerian and late Dinantian phases 2 and 3, uplift and restriction of sedimentation to the South Crop resulted in the deposition of carbonate successions that are probably tectonic analogues of the lowstand systems tract of van Wagoner *et al.* (1988).

## CONCLUSIONS

In this account, Dinantian cyclicity is attributed to the interaction between tectonic uplift, regional subsidence (which may have been episodic rather than constant), sedimentation rates, and a relatively constant base level.

This model does not require recourse to a eustatic mechanism. The existence of eustatic influence cannot be entirely excluded, but is not testable within the framework of a single basin. Even if eustasy were the dominant component in the production of Dinantian cyclicity, it is unlikely that this

process would generate sequences of globally or interbasinally synchronous higher and lower order cycles by overriding the following major factors:
**1** differential subsidence rates between basins that formed within the same or different tectonic regimes;
**2** differential sedimentation rates related to variations in organic production and/or supply of clastic detritus.

The British Dinantian basins contain successions in which the imprint of tectonism is evident in the similarities and variations between Dinantian sedimentary patterns both within and between basins. Thus transgression is related to rifting and subsidence. Regressive phases are progradational and contain barrier and back barrier deposits when progradation coincided with high sedimentation rates and relative rise in sea level related to tectonic processes. During intervals of relatively static sea level, for example as a result of tectonic quiessence or high sedimentation rates, progradation produced regressive sequences without back barrier peritidal–lagoonal components. The intrabasinal geography of major and minor emergent surfaces in South Wales is dependent on the interaction between tectonics and basin geometry, i.e. ramp or platform. Some major tectonic events and their associated emergent surfaces are recorded in widely separated basins, for example the Chadian–Arundian boundary. In the Asbian and Brigantian, emergent surfaces, the origin of which is considered to be eustatic by Walkden (1987) and Horbury (1989) and tectonic by Ramsay (1989), vary in number within the South Wales basin. In South Wales the Asbian succession contains nine basin-wide emergent surfaces (Fig. 7). In Derbyshire, Walkden (1987) documents 29 persistent horizons. Even if these surfaces were eustatic in origin their number distribution and preservation are related to an overriding tectonic influence.

## ACKNOWLEDGEMENTS

This research has benefited considerably from stimulating discussions with my colleagues Norman Fry, Richard Lisle, and Geraint Owen. Gareth Williams generously provided me with Fig. 10 from his unpublished MSc thesis. The manuscript was typed by Helen Thompson; Steve Hibbs and Alan Cutliffe photographed my diagrams. I also wish to thank Mr Clive Hopkins of Sinclair Garages, Gorseinon for the use of his office in which much of the first draft was written. The fieldwork was funded in part by a grant from the University College of Swansea.

## REFERENCES

Ahr, W.M. (1973) The carbonate ramp: an alternative to the shelf model. *Trans. Gulf Coast Assoc. Geol. Socs. 23rd Ann. Conv.* 221–225.

Barclay, W.J. & Jones, D.G. (1978) Recent boreholes in the attenuated Carboniferous strata of the Blaenavon–Pontypool area, Gwent. *Bull. Geol. Surv. G.B.* **67**, 1–17.

Barclay, W.J., Taylor, K. & Thomas, L.P. (1988) Geology of the South Wales Coalfield Part V, the country around Merthyr Tydfil (3rd Edn). *Mem. Br. Geol. Surv.* Sheet 231 (England and Wales).

Bott, M.H.P. (1987) Subsidence mechanisms of Carboniferous basins in northern England. In: *European Dinantian Environments* (Eds Miller, J., Adams, A.E. & Wright, V.P.) pp. 21–32. Wiley, Chichester.

British Geological Survey (1980) *Aeromagnetic Anomaly Map*, Sheet 51°N–04°W (Bristol Channel), 1:250 000 Series.

Burchette, J.P. (1987) Carbonate barrier shorelines during the basal Carboniferous transgression: the Lower Limestone Shale Group. South Wales and western England. In: *European Dinantian Environments* (Eds Miller, J., Adams, A.E. & Wright, V.R.) pp. 239–263. Wiley, Chichester.

Diemer, J.A. & Bridge, J.S. (1988) Transition from alluvial plain to tide dominated coastal deposits associated with the Tournaisian marine transgression in south-west Ireland. In: *Tide Influenced Sedimentary Environments and Facies* (Eds Boer, P.L., van Gelder, A. & Nio, S.D.) pp. 359–388. Reidel, Dordrecht.

Dixon, E.E.L. & Vaughan, A. (1912) The Carboniferous Limestone in Gower (Glamorganshire). *Q. J. Geol. Soc. London* **67**, 477–571.

Elliott, T. (1974) Interdistributary bay sequences and their genesis. *Sedimentology* **21**, 611–622.

Fischer, A.G. (1961) Stratigraphic record of transgressing seas in the light of sedimentation on Atlantic coast of New Jersey. *Bull., Am. Assoc. Petrol. Geol.* **45**, 1656–1666.

Gawthorpe, R.L. (1986) Sedimentation during carbonate ramp to slope evolution in a tectonically active area: Bowland Basin. *Sedimentology* **33**, 185–206.

Gawthorpe, R.L., Gutteridge, P. & Leeder, M.R. (1989) Late Devonian and Dinantian basin evolution in northern England and North Wales. In: *The Role of Tectonics in Devonian and Carboniferous Sedimentation in the British Isles* (Eds Arthurton, R.S., Gutteridge, P. & Nolan, S.C.) pp. 1–23. Occasional Publication 6, Yorks. Geol. Soc.

Gayer, R., Allen, K.C., Basset, M.G. & Edwards, D.E. (1973) The structure of the Taff Gorge area, Glamorgan, and the stratigraphy of the Old Red Sandstone–Carboniferous Limestone Transition. *Geol. J.* **8**, 345–374.

George, T.N. (1933) The Carboniferous Limestone Series in the west of the Vale of Glamorgan. *Q. J. Geol. Soc.*

*London* **89**, 221–272.

George, T.N. (1940) The geology of Gower. *Q.J. Geol. Soc. London* **96**, 131–198.

George, T.N. (1956a) Carboniferous main limestone of the East Crop in South Wales. *Q.J. Geol. Soc. London* **111**, 309–322.

George, T.N. (1956b) The Namurian Usk Anticline. *Proc. Geol. Assoc.* **66**, 297–316.

George, T.N. (1958) Lower Carboniferous palaeogeography of the British Isles. *Proc. Yorks. Geol. Soc.* **31**, 227–318.

George, T.N. (1974) Lower Carboniferous rocks in Wales. In: *The Upper Paleozoic Rocks of Wales* (Ed. Owen, T.R.) pp. 85–115. University of Wales Press, Cardiff.

George, T.N. (1978) Eustasy and tectonics: sedimentary rhythms and stratigraphical units in British Dinantian correlation. *Proc., Yorks. Geol. Soc.* **42**, 229–253.

George, T.N., Johnson, G.A.L., Mitchell, M., Prentice, J.E., Ramsbottom, W.H.C., Sevastopulo, G.D. & Wilson, R.B. (1976) A correlation of Dinantian rocks in the British Isles. *Spec. Rep. Geol. Soc., London* **7**, 1–87.

Grayson, R.F. & Oldham, L. (1987) A new structural framework for the northern British Dinantian as a basis for oil, gas and mineral exploration. In: *European Dinantian Environments* (Eds Miller, J., Adams, A.E. & Wright, V.P.) pp. 33–59. Wiley, Chichester.

Gutteridge, P. (1987) Dinantian sedimentation and the basement structure of the Derbyshire Dome. *Geol. J.* **22**, 25–41.

Haq, B.U., Hardenbol, J. & Vail, P.R. (1987) Chronology of fluctuating sea levels since the Triassic. *Science* **235**, 1156–1167.

Haszeldine, R.S. (1989) Evidence against crustal stretching, north–south tension and Hercynian collision, forming British Carboniferous basins. In: *The Role of Tectonics in Devonian and Carboniferous Sedimentation in the British Isles* (Eds Arthurton, R.S., Gutteridge, P. & Nolan, S.C.) pp. 25–33. Occasional Publication 6, Yorks. Geol. Soc.

Heward, A.P. (1981) A review of wave-dominated clastic shoreline deposits. *Earth Sci. Rev.* **17**, 223–276.

Horbury, A.D. (1989) The relative role of tectonism and eustasy in the deposition of the Urswick Limestone in South Cumbria and North Lancashire. In: *The Role of Tectonics in Devonian and Carboniferous Sedimentation in the British Isles* (Eds Arthurton, R.S., Gutteridge, P. & Nolan, S.C.) pp. 153–169. Occasional Publication 6, Yorks. Geol. Soc.

James, N.P. (1979) Shallowing upward sequences in carbonates. In: *Facies Models 1* (Ed. Walker, R.G.) pp. 109–119. Geosci. Can. Reprint Ser., Geological Association of Canada, Ontario.

Jones, D.G. (1974) The Namurian series in South Wales. In: *The Upper Palaeozoic and Post-Palaeozoic Rocks of Wales* (Ed. Owen, T.R.) pp. 117–132. University of Wales Press, Cardiff.

Junghans, T., Goldring, R. & Simpson, J. (1984) The Dinantian between Rhosilli and Port Eynon (Gower, South Wales). In: *Abstract, European Dinantian Environments 1st Meeting, 1984*, pp. 61–63. Department of Earth Science, Open University, Milton Keynes.

Kumar, N. & Sanders, J.E. (1974) Inlet sequence: a vertical succession of sedimentary structures and textures created by the Internal migration of tidal inlets. *Sedimentology* **21**, 491–532.

Leeder, M.R. (1982) Upper Palaeozoic basins of the British Isles — Caledonide inheritance versus Hercynian plate margin processes. *Q. J. Geol. Soc. London* **139**, 479–491.

Leeder, M.R. (1987) Tectonic and palaeogeographic models for Lower Carboniferous Europe. In: *European Dinantian Environments* (Eds Miller, J., Adams, A.E. & Wright, V.P.) pp. 1–20. Wiley, Chichester.

Leeder, M.R. & McMahon, A.H. (1988) Upper Carboniferous (Silesian) basin subsidence in northern Britain. In: *Sedimentation in a Synorogenic Complex. The Carboniferous of Northwest Europe* (Eds Besly, B.M. & Kelling, G.) pp. 43–52. Blackie, Glasgow.

McKenzie, D.P. (1978) Some remarks on the development of sedimentary basins. *Earth Planet. Sci. Lett.* **40**, 25–32.

Mechie, J. & Brooks, M. (1984) A seismic study of deep geological structure in the Bristol Channel area S.W. Britain. *Geophys. J.R. Astron. Soc.* **78**, 661–689.

Miller, J. & Grayson, R.F. (1982) The regional context of Waulsortian facies in northern England. In: *Symposium on the Palaeoenvironmental Setting and Distribution of the Waulsortian Facies* (Eds Bolton, K., Lane, H.R. & Le Mone, D.V.) pp. 17–33. University of El Paso, Texas and El Paso Geological Society.

Moore, L.R. (1948) The sequence and structure of the southern portion of the east crop of the south Wales coalfield. *Q. J. Geol. Soc. London* **103**, 261–300.

Naylor, J.D., Sevastopulo, G.P. & Sleeman, A.G. (1989) Subsidence history of the South Munster Basin, Ireland. In: *The Role of Tectonics in Devonian and Carboniferous Sedimentation in the British Isles*. (Eds Arthurton, R.S., Gutteridge, P. & Nolan, S.C.) pp. 99–121. Occasional Publication 6, Yorks. Geol. Soc.

Nolan, S.C. (1989) The style and timing of Dinantian synsedimentary tectonics in the eastern part of the Dublin Basin Ireland. In: *The Role of Tectonics in Devonian and Carboniferous Sedimentation in the British Isles* (Eds Arthurton, R.S., Gutteridge, P. & Nolan, S.C.) pp. 83–97. Occasional Publication 6, Yorks. Geol. Soc.

Owen, T.R. (1954) The structure of the Neath Disturbance between Bryniau Gleison and Glynneath, South Wales. *Q.J. Geol. Soc. London* **109**, 333–365.

Owen, T.R. (1970) The relationship of Carboniferous sedimentation to structure in South Wales. *Compte Rendu sixieme Congres International de Stratigraphie et de Geologie du Carbonifere*, Sheffield, 1967, Vol. 3, pp. 1305–1315.

Owen, T.R. & Weaver, J.D. (1983) The structure of the South Wales Coalfield. In: *The Variscan Fold Belt* (Ed. Hancock, P.) pp. 74–87. Adam Hilger, Bristol.

Philcox, M.E. (1989) The mid-Dinantian unconformity at Navan, Ireland. In: *The Role of Tectonics in Devonian and Carboniferous Sedimentation in the British Isles* (Eds Arthurton, R.S., Gutteridge, P. & Nolan, S.C.) pp. 67–87. Occasional Publication 6, Yorks. Geol. Soc.

Philcox, M.E., Sevastopulo, G.D. & MacDermot, C.V. (1989) Intra-Dinantian tectonic activity on the Curlew Fault, north-west Ireland. In: *The Role of Tectonics in Devonian and Carboniferous Sedimentation in the British*

*Isles* (Eds Arthurton, R.S., Gutteridge, P. & Nolan, S.C.) pp. 55–66. Occasional Publication 6, Yorks. Geol. Soc.

Ramsbottom, W.H.C. (1973) Transgression and regression in the Dinantian: a new synthesis of British Dinantian stratigraphy. *Proc. Yorks. Geol. Soc.* **39**, 567–607.

Ramsbottom, W.H.C. (1978) Rates of transgression and regression in the Carboniferous of NW Europe *Q. J. Geol. Soc. London* **136**, 147–153.

Ramsay, A.T.S. (1987) Depositional environments of Dinantian Limestones in Gower, South Wales. In: *European Dinantian Environments* (Eds Miller, J., Adams, A.E. & Wright, V.P.) pp. 265–308. Wiley, Chichester.

Ramsay, A.T.S. (1989) Tectonics and sedimentation of Late Dinantian limestones in South Wales. In: *The Role of Tectonics in Devonian and Carboniferous Sedimentation in the British Isles* (Eds Arthurton, R.S., Gutteridge, P. & Nolan, S.C.) pp. 225–241. Occasional Publication 6, Yorks. Geol. Soc.

Read, J.F. (1985) Carbonate platform facies models. *Bull., Am. Assoc. Petrol. Geol.* **66**, 860–878.

Riding, R. & Wright, V.P. (1981) Palaeosols and tidal flat/lagoon sequences on a carbonate shelf: sedimentary associations of triple disconformities. *J. Sediment. Petrol.* **51**, 1323–1339.

Robertson, T. & George, T.N. (1929) The Carboniferous Limestone of the North Crop of the South Wales Coalfield. *Proc. Geol. Assoc.* **40**, 18–40.

Sellwood, E.B. & Thomas, J.M. (1988) The tectonic framework of Upper Carboniferous sedimentation in central southwest England. In: *Sedimentation in a Synorogenic Complex. The Carboniferous of Northwest Europe*. (Eds Besley, B.M. & Kelling, G.) pp. 18–23. Blackie, Glasgow.

Squirrel, H.C. & Downing, R.A. (1964) The attenuation of the Coal Measures in the southeast part of the South Wales Coalfield. *Bull. Geol. Surv. G.B.* **21**, 119–132.

Squirrel, H.C. & Downing, R.A. (1969) The geology of the South Wales Coalfield. Part 1. The country around Newport (Monmouthshire) (3rd Edn). *Mem. Geol. Surv. G.B.*

Sullivan, R. (1966) The stratigraphical effects of the mid-Dinantian movements in south-west Wales. *Palaeogeogr. Palaeoclimatol. Palaeoecol.* **2**, 213–244.

Swift, D.J.P. (1968) Coastal erosion and transgressive stratigraphy. *J. Geol.* **76**, 444–456.

van Wagoner, J.C., Posamantier, H.W., Mitchum, R.M., Sarg, J.F., Loutit, T.S., Hardenbol, J. & Vail, P.R. (1988) An overview of the fundamentals of sequence stratigraphy and key definitions. In: *Sea-level Changes: an Integrated Approach* (Eds Wilgus, C.K., Hastings, B.S., Kendall, C.G.St.C., Posamentier, H., Ross, C.A. & van Wagoner, J.). Soc. Econ. Palaeontol. Mineral., Spec. Publ. 42, 39–45.

Walkden, G.M. (1987) Sedimentary and diagenetic styles in Late Dinantian carbonates of Britain. In: *European Dinantian Enviroments* (Eds Miller, J., Adams, A.E. & Wright, V.P.) pp. 131–155. Wiley, Chichester.

Waters, R.A. & Lawrence, D.J.D. (1987) Geology of the South Wales Coalfield, Part III, the country around Cardiff (3rd Edn). *Mem. Br. Geol. Surv.*, Sheet 263, 24–50.

Williams, D.G. (1985) *An investigation of the Precambrian basement in south-west Wales.* Unpublished MSc thesis, University of Wales.

Wilson, D., Davies, J.R., Smith, M. & Waters, R.A. (1988) Structural controls on Upper Palaeozoic sedimentation in south-east Wales. *Q. J. Geol. Soc. London* **145**, 901–914.

Wright, V.P. (1982) The recognition and interpretation of palaeokarsts: two examples from the Lower Carboniferous of South Wales. *J. Sediment. Petrol.* **52**, 83–94.

Wright, V.P. (1984a) The significance of needle-fibre calcite in a Lower Carboniferous palaeosol. *Geol. J.* **19**, 23–34.

Wright, V.P. (1984b) Peritidal carbonate facies models: a review. *Geol. J.* **19**, 309–325.

Wright, V.P. (1986) Facies sequences on a carbonate ramp: the Carboniferous Limestone of South Wales. *Sedimentology* **33**, 221–241.

Wright, V.P. (1987) The evolution of the early Carboniferous Limestone province in southwest Britain. *Geol. Mag.* **124**, 477–480.

# Index

Page numbers in *italic* refer to figures.